SCHAUM'S®
outlines

Schaum's Outline of Calculus for Business, Economics, and Finance

Fourth Edition

Luis Moisés Peña-Lévano, PhD

Schaum's Outline Series

Mc
Graw
Hill

New York   Chicago   San Francisco   Athens
London   Madrid   Mexico City   Milan   New Delhi
Singapore   Sydney   Toronto

Luis Peña-Lévano, PhD, (St. Paul, MN) is Assistant Professor of Agricultural Economics at the University of Wisconsin River-Falls. He received his M.Sc. in applied economics from the University of Georgia and his PhD is from Purdue University. As a teacher, he has taught mathematics for economics, microeconomics, international trade and policy, environmental and natural resource economics, finance, and contemporary issues in business. Dr. Peña-Lévano is an award-winning professional and has been recognized globally. In 2020, he received the EAERE Award for the Best Paper Published in Environmental and Resource Economics. He is also a member of the Global Trade Analysis Project, and currently serves as treasurer and secretary of the Teaching, Learning and Communication section of the Agricultural and Applied Economic Association, and has served as reviewer and guest editor for multiple peer-reviewed journals.

1 2 3 4 5 6 7 8 9   LOV   25 24 23 22 21

ISBN   978-1-264-26685-2
MHID   1-264-26685-5

e-ISBN   978-1-264-26686-9
e-MHID   1-264-26686-3

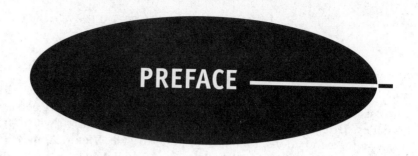

PREFACE

The mathematics needed for the study of economics and business continues to grow with each passing year, placing ever more demands on students and faculty alike. *Schaum's Outline of Calculus for Business, Economics, and Finance,* Fourth edition introduces one new chapter on sequences and series, and two sections on applications in Microsoft Excel.

The end of book also presents a new section of problems called Additional Practice Problems which provide a template that can be used to test students' knowledge on each chapter, with solutions in the subsequent pages (called 'Solution to Practice Problems').

The objectives of the book have not changed over the 31 years since the introduction of the first edition, originally called *Mathematics for Economists. Schaum's Outline of Calculus for Business, Economics, and Finance,* Fourth edition is designed to present a thorough, easily understood introduction to the wide array of mathematical topics that economists, social scientists, and business majors need to know today, such as linear algebra, differential and integral calculus, nonlinear programming, differential and difference equations, the calculus of variations, and optimal control theory. The book also offers a brief review of basic algebra for those who are rusty and provides direct, frequent, and practical applications to everyday economic problems and business situations.

The theory-and-solved-problem format of each chapter provides concise explanations illustrated by examples, plus numerous problems with fully worked-out solutions. The topics and related problems range in difficulty from simpler mathematical operations to sophisticated applications. No mathematical proficiency beyond the high school level is assumed at the start. The learning-by-doing pedagogy will enable students to progress at their own rates and adapt the book to their own needs.

Those in need of more time and help in getting started with some of the elementary topics may feel more comfortable beginning with or working in conjunction with my *Schaum's Outline of Mathematical Methods for Business, Economics, and Finance,* Second Edition, which offers a kinder, gentler approach to the discipline.

Schaum's Outline of Calculus for Business, Economics, and Finance, Fourth Edition can be used by itself or as a supplement to other texts for undergraduate and graduate students in economics, business, and the social sciences. It is largely self-contained. Starting with a basic review of high school algebra in Chapter 1, the book consistently explains all the concepts and techniques needed for the material in subsequent chapters.

Since there is no universal agreement on the order in which differential calculus and linear algebra should be presented, the book is designed so that Chapters 10 and 11 on linear algebra can be covered immediately after Chapter 2, if so desired, without loss of continuity.

This book contains over 1600 problems, all solved in considerable detail. To get the most from the book, students should strive as soon as possible to work independently of the solutions. This can be done by solving problems on individual sheets of paper with the book closed. If difficulties arise, the solution can then be checked in the book.

For best results, students should never be satisfied with passive knowledge—the capacity merely to follow or comprehend the various steps presented in the book. Mastery of the subject and doing well on exams requires active knowledge—the ability to solve any problem, in any order, without the aid of the book.

Experience has proven that students of very different backgrounds and abilities can be successful in handling the subject matter presented in this text if they apply themselves and work consistently through the problems and examples.

In closing, I would like to thank my mother Betty, my father Alberto, my brother William, and my sister Mirella, for raising me to be the man I am today. My nephew Fitzgerald, my goddaughters Ashley, Briggitte and Valentina, and my godson Adrian, who I envision to be the future of the upcoming generations. I am also grateful to José, Víctor, Ahmed, Edward, Irvin, Ernesto, Mario, Daniele, Shaheer, Jared and Andrés, my 11 friends who are part of my family and have been my support during all my career, and for their unfailing encouragement over these last 15 years.

LUIS MOISÉS PEÑA-LÉVANO

CONTENTS

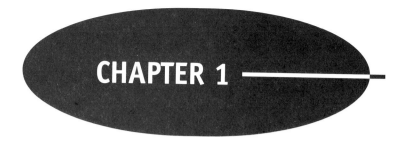

CHAPTER 1

Review

1.1 EXPONENTS

Given n a positive integer, x^n signifies that x is multiplied by itself n times. Here x is referred to as the *base* and n is termed an *exponent*. By convention an exponent of 1 is not expressed: $x^1 = x$, $8^1 = 8$. By definition, any nonzero number or variable raised to the zero power is equal to 1: $x^0 = 1$, $3^0 = 1$. And 0^0 is undefined. Assuming a and b are positive integers and x and y are real numbers for which the following exist, the *rules of exponents* are outlined below and illustrated in Examples 1 and 2 and Problem 1.1.

1. $x^a(x^b) = x^{a+b}$

2. $\dfrac{x^a}{x^b} = x^{a-b}$

3. $(x^a)^b = x^{ab}$

4. $(xy)^a = x^a y^a$

5. $\left(\dfrac{x}{y}\right)^a = \dfrac{x^a}{y^a}$

6. $\dfrac{1}{x^a} = x^{-a}$

7. $\sqrt{x} = x^{1/2}$

8. $\sqrt[a]{x} = x^{1/a}$

9. $\sqrt[b]{x^a} = x^{a/b} = (x^{1/b})^a$

10. $x^{-(a/b)} = \dfrac{1}{x^{a/b}}$

EXAMPLE 1. From Rule 2, it can easily be seen why any variable or nonzero number raised to the zero power equals 1. For example, $x^3/x^3 = x^{3-3} = x^0 = 1$; $8^5/8^5 = 8^{5-5} = 8^0 = 1$.

EXAMPLE 2. In multiplication, exponents of the same variable are added; in division, the exponents are subtracted; when raised to a power, the exponents are multiplied, as indicated by the rules above and shown in the examples below followed by illustrations in brackets.

a) $x^2(x^3) = x^{2+3} = x^5 \neq x^6$

Rule 1

$$[x^2(x^3) = (x \cdot x)(x \cdot x \cdot x) = x \cdot x \cdot x \cdot x \cdot x = x^5]$$

b) $\dfrac{x^6}{x^3} = x^{6-3} = x^3 \neq x^2$

Rule 2

$$\left[\dfrac{x^6}{x^3} = \dfrac{x \cdot x \cdot x \cdot x \cdot x \cdot x}{x \cdot x \cdot x} = x \cdot x \cdot x = x^3\right]$$

c) $(x^4)^2 = x^{4 \cdot 2} = x^8 \neq x^{16}$ or x^6

Rule 3

$$[(x^4)^2 = (x \cdot x \cdot x \cdot x)(x \cdot x \cdot x \cdot x) = x^8]$$

1

d) $(xy)^4 = x^4y^4 \neq xy^4$ Rule 4

$$[(xy)^4 = (xy)(xy)(xy)(xy) = (x \cdot x \cdot x \cdot x)(y \cdot y \cdot y \cdot y) = x^4y^4]$$

e) $\left(\dfrac{x}{y}\right)^5 = \dfrac{x^5}{y^5} \neq \dfrac{x^5}{y}$ or $\dfrac{x}{y^5}$ Rule 5

$$\left[\left(\frac{x}{y}\right)^5 = \frac{(x)}{(y)}\frac{(x)}{(y)}\frac{(x)}{(y)}\frac{(x)}{(y)}\frac{(x)}{(y)} = \frac{x^5}{y^5}\right]$$

f) $\dfrac{x^3}{x^4} = x^{3-4} = x^{-1} = \dfrac{1}{x} \neq x^{3/4}$ Rules 2 and 6

$$\left[\frac{x^3}{x^4} = \frac{x \cdot x \cdot x}{x \cdot x \cdot x \cdot x} = \frac{1}{x}\right]$$

g) $\sqrt{x} = x^{1/2}$ Rule 7

Since $\sqrt{x} \cdot \sqrt{x} = x$ and from Rule 1 exponents of a common base are added in multiplication, the exponent of $\sqrt{x}$, when added to itself, must equal 1. With $\frac{1}{2} + \frac{1}{2} = 1$, the exponent of $\sqrt{x}$ is $\frac{1}{2}$. Thus, $\sqrt{x} \cdot \sqrt{x} = x^{1/2} \cdot x^{1/2} = x^{1/2+1/2} = x^1 = x$.

h) $\sqrt[3]{x} = x^{1/3}$ Rule 8

Just as $\sqrt[3]{x} \cdot \sqrt[3]{x} \cdot \sqrt[3]{x} = x$, so $x^{1/3} \cdot x^{1/3} \cdot x^{1/3} = x^{1/3+1/3+1/3} = x^1 = x$.

i) $x^{3/2} = (x^{1/2})^3$ or $(x^3)^{1/2}$ Rule 9

$$[4^{3/2} = (4^{1/2})^3 = (\sqrt{4})^3 = (\pm 2)^3 = \pm 8, \text{ or equally valid, } 4^{3/2} = (4^3)^{1/2} = (64)^{1/2} = \sqrt{64} = \pm 8]$$

j) $x^{-2/3} = \dfrac{1}{x^{2/3}} = \dfrac{1}{(x^{1/3})^2}$ or $\dfrac{1}{(x^2)^{1/3}}$ Rule 10

$$\left[27^{-2/3} = \frac{1}{(27^{1/3})^2} = \frac{1}{(3)^2} = \frac{1}{9}, \text{ or equally valid, } 27^{-2/3} = \frac{1}{(27^2)^{1/3}} = \frac{1}{(729)^{1/3}} = \frac{1}{9}\right]$$

See Problem 1.1.

1.2 POLYNOMIALS

Given an expression such as $5x^3$, x is called a *variable* because it can assume any number of different values, and 5 is referred to as the *coefficient* of x. Expressions consisting simply of a real number or of a coefficient times one or more variables raised to the power of a positive integer are called *monomials*. Monomials can be added or subtracted to form *polynomials*. Each of the monomials comprising a polynomial is called a *term*. Terms that have the same variables and exponents are called *like terms*. Rules for adding, subtracting, multiplying, and dividing polynomials are explained in Examples 3 through 5 and treated in Problems 1.2 to 1.4.

EXAMPLE 3. Like terms in polynomials can be added or subtracted by adding their coefficients. Unlike terms cannot be so added or subtracted. See Problems 1.2 and 1.3.

   a) $4x^5 + 9x^5 = 13x^5$ *b)* $12xy - 3xy = 9xy$

   c) $(7x^3 + 5x^2 - 8x) + (11x^3 - 9x^2 - 2x) = 18x^3 - 4x^2 - 10x$

   d) $(24x - 17y) + (6x + 5z) = 30x - 17y + 5z$

EXAMPLE 4. Like and unlike terms can be multiplied or divided by multiplying or dividing both the coefficients and variables.

   a) $(5x)(13y^2) = 65xy^2$ *b)* $(7x^3y^5)(4x^2y^4) = 28x^5y^9$

   c) $(2x^3y)(17y^4z^2) = 34x^3y^5z^2$ *d)* $\dfrac{15x^4y^3z^6}{3x^2y^2z^3} = 5x^2yz^3$

$e)$ $\dfrac{4x^2 y^5 z^3}{8x^5 y^3 z^4} = \dfrac{y^2}{2x^3 z}$

EXAMPLE 5.   In multiplying two polynomials, each term in the first polynomial must be multiplied by each term in the second and their products added. See Problem 1.4.

$$(6x + 7y)(4x + 9y) = 24x^2 + 54xy + 28xy + 63y^2$$
$$= 24x^2 + 82xy + 63y^2$$
$$(2x + 3y)(8x - 5y - 7z) = 16x^2 - 10xy - 14xz + 24xy - 15y^2 - 21yz$$
$$= 16x^2 + 14xy - 14xz - 21yz - 15y^2$$

1.3 EQUATIONS: LINEAR AND QUADRATIC

A mathematical statement setting two algebraic expressions equal to each other is called an *equation*. An equation in which all variables are raised to the first power is known as a *linear equation*. A linear equation can be solved by moving the unknown variable to the left-hand side of the equal sign and all the other terms to the right-hand side, as is illustrated in Example 6. A *quadratic equation* of the form $ax^2 + bx + c = 0$, where a, b, and c are constants and $a \neq 0$, can be solved by factoring or using the *quadratic formula*:

$$x = \frac{-b \pm \sqrt{b^2 - 4ac}}{2a} \tag{1.1}$$

Solving quadratic equations by factoring is explained in Example 7 and by the quadratic formula in Example 8 and Problem 1.6.

EXAMPLE 6.   The linear equation given below is solved in three easy steps.

$$\frac{x}{4} - 3 = \frac{x}{5} + 1$$

1. Move all terms with the unknown variable x to the left, here by subtracting $x/5$ from both sides of the equation.

$$\frac{x}{4} - 3 - \frac{x}{5} = 1$$

2. Move any term without the unknown variable to the right, here by adding 3 to both sides of the equation.

$$\frac{x}{4} - \frac{x}{5} = 1 + 3 = 4$$

3. Simplify both sides of the equation until the unknown variable is by itself on the left and the solution is on the right, here by multiplying both sides of the equation by 20 and subtracting.

$$20 \cdot \left(\frac{x}{4} - \frac{x}{5} \right) = 4 \cdot 20$$
$$5x - 4x = 80$$
$$x = 80$$

EXAMPLE 7.   Factoring is the easiest way to solve a quadratic equation, provided the factors are easily recognized integers. Given

$$x^2 + 13x + 30 = 0$$

by factoring, we have

$$(x + 3)(x + 10) = 0$$

For $(x+3)(x+10)$ to equal 0, $x+3$ or $x+10$ must equal 0. Setting each in turn equal to 0 and solving for x, we have

$$x + 3 = 0 \qquad x + 10 = 0$$
$$x = -3 \qquad x = -10$$

Those wishing a thorough review of factoring and other basic mathematical techniques should consult another of the author's books, *Schaum's Outline of Mathematical Methods for Business and Economics*, for a gentler, more gradual approach to the discipline.

EXAMPLE 8.   The quadratic formula is used below to solve the quadratic equation

$$5x^2 - 55x + 140 = 0$$

Substituting $a = 5$, $b = -55$, $c = 140$ from the given equation in (*1.1*) gives

$$x = \frac{-(-55) \pm \sqrt{(-55)^2 - 4(5)(140)}}{2(5)}$$
$$= \frac{55 \pm \sqrt{3025 - 2800}}{10} = \frac{55 \pm \sqrt{225}}{10} = \frac{55 \pm 15}{10}$$

Adding $+15$ and then -15 to find each of the two solutions, we get

$$x = \frac{55 + 15}{10} = 7 \qquad x = \frac{55 - 15}{10} = 4$$

See Problem 1.6.

1.4   SIMULTANEOUS EQUATIONS

To solve a system of two or more equations simultaneously, (1) the equations must be *consistent* (noncontradictory), (2) they must be *independent* (not multiples of each other), and (3) there must be as many consistent and independent equations as variables. A system of simultaneous linear equations can be solved by either the *substitution* or *elimination* method, explained in Example 9 and Problems 2.11 to 2.16, as well as by methods developed later in linear algebra in Sections 11.8 and 11.9.

EXAMPLE 9.   The equilibrium conditions for two markets, butter and margarine, where P_b and P_m are the prices of butter and margarine, respectively, are given in (*1.2*) and (*1.3*):

$$8P_b - 3P_m = 7 \tag{1.2}$$
$$-P_b + 7P_m = 19 \tag{1.3}$$

The prices that will bring equilibrium to the model are found below by using the substitution and elimination methods.

Substitution Method

1.   Solve one of the equations for one variable in terms of the other. Solving (*1.3*) for P_b gives

$$P_b = 7P_m - 19$$

2.   Substitute the value of that term in the *other* equation, here (*1.2*), and solve for P_m.

$$8P_b - 3P_m = 7$$
$$8(7P_m - 19) - 3P_m = 7$$
$$56P_m - 152 - 3P_m = 7$$
$$53P_m = 159$$
$$P_m = 3$$

3. Then substitute $P_m = 3$ in either (1.2) or (1.3) to find P_b.

$$8P_b - 3(3) = 7$$
$$8P_b = 16$$
$$P_b = 2$$

Elimination Method

1. Multiply (1.2) by the coefficient of P_b (or P_m) in (1.3) and (1.3) by the coefficient of P_b (or P_m) in (1.2). Picking P_m, we get

$$7(8P_b - 3P_m = 7) \qquad 56P_b - 21P_m = 49 \qquad (1.4)$$
$$-3(-P_b + 7P_m = 19) \qquad 3P_b - 21P_m = -57 \qquad (1.5)$$

2. Subtract (1.5) from (1.4) to eliminate the selected variable.

$$53P_b = 106$$
$$P_b = 2$$

3. Substitute $P_b = 2$ in (1.4) or (1.5) to find P_m as in step 3 of the substitution method.

1.5 FUNCTIONS

A *function f* is a *rule* which assigns to each value of a variable (x), called the *argument* of the function, one and only one value $[f(x)]$, referred to as the *value of the function at x*. The *domain* of a function refers to the set of all possible values of x; the *range* is the set of all possible values for $f(x)$. Functions are generally defined by algebraic formulas, as illustrated in Example 10. Other letters, such as g, h, or the Greek letter ϕ, are also used to express functions. Functions encountered frequently in economics are listed below.

Linear function:

$$f(x) = mx + b$$

Quadratic function:

$$f(x) = ax^2 + bx + c \qquad (a \neq 0)$$

Polynomial function of degree n:

$$f(x) = a_n x^n + a_{n-1} x^{n-1} + \cdots + a_0 \qquad (n = \text{nonnegative integer}; a_n \neq 0)$$

Rational function:

$$f(x) = \frac{g(x)}{h(x)}$$

where $g(x)$ and $h(x)$ are both polynomials and $h(x) \neq 0$. (*Note*: Rational comes from *ratio*.)

Power function:

$$f(x) = ax^n \qquad (n = \text{any real number})$$

EXAMPLE 10. The function $f(x) = 8x - 5$ is the rule that takes a number, multiplies it by 8, and then subtracts 5 from the product. If a value is given for x, the value is substituted for x in the formula, and the equation solved for $f(x)$. For example, if $x = 3$,

$$f(x) = 8(3) - 5 = 19$$

If $x = 4$,

$$f(x) = 8(4) - 5 = 27$$

See Problems 1.7 to 1.9.

EXAMPLE 11.   Given below are examples of different functions:

Linear: $f(x) = 7x - 4$ $g(x) = -3x$ $h(x) = 9$

Quadratic: $f(x) = 5x^2 + 8x - 2$ $g(x) = x^2 - 6x$ $h(x) = 6x^2$

Polynomial: $f(x) = 4x^3 + 2x^2 - 9x + 5$ $g(x) = 2x^5 - x^3 + 7$

Rational: $f(x) = \dfrac{x^2 - 9}{x + 4}$ $(x \neq -4)$ $g(x) = \dfrac{5x}{x - 2}$ $(x \neq 2)$

Power: $f(x) = 2x^6$ $g(x) = x^{1/2}$ $h(x) = 4x^{-3}$

1.6   GRAPHS, SLOPES, AND INTERCEPTS

In graphing a function such as $y = f(x)$, x is placed on the horizontal axis and is known as the *independent variable*; y is placed on the vertical axis and is called the *dependent variable*. The graph of a linear function is a straight line. The *slope* of a line measures the *change in y* (Δy) divided by a *change in x* (Δx). The slope indicates the steepness and direction of a line. The greater the absolute value of the slope, the steeper the line. A positively sloped line moves up from left to right; a negatively sloped line moves down. The slope of a horizontal line, for which $\Delta y = 0$, is zero. The slope of a vertical line, for which $\Delta x = 0$, is undefined, i.e., does not exist because division by zero is impossible. The *y intercept* is the point where the graph crosses the y axis; it occurs when $x = 0$. The *x intercept* is the point where the line intersects the x axis; it occurs when $y = 0$. See Problem 1.10.

EXAMPLE 12.   To graph a linear equation such as

$$y = -\tfrac{1}{4}x + 3$$

one need only find two points which satisfy the equation and connect them by a straight line. Since the graph of a linear function is a straight line, all the points satisfying the equation must lie on the line.

To find the y intercept, set $x = 0$ and solve for y, getting $y = -\tfrac{1}{4}(0) + 3$, $y = 3$. The y intercept is the point $(x, y) = (0, 3)$. To find the x intercept, set $y = 0$ and solve for x. Thus, $0 = -\tfrac{1}{4}x + 3$, $\tfrac{1}{4}x = 3$, $x = 12$. The x intercept is the point $(x, y) = (12, 0)$. Then plot the points $(0, 3)$ and $(12, 0)$ and connect them by a straight line, as in Fig. 1-1, to complete the graph of $y = -\tfrac{1}{4}x + 3$. See Examples 13 and 14 and Problems 1.10 to 1.12.

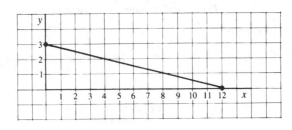

Fig. 1-1

EXAMPLE 13.   For a line passing through points (x_1, y_1) and (x_2, y_2), the slope m is calculated as follows:

$$m = \frac{\Delta y}{\Delta x} = \frac{y_2 - y_1}{x_2 - x_1} \qquad x_1 \neq x_2$$

For the line in Fig. 1-1 passing through $(0, 3)$ and $(12, 0)$,

$$m = \frac{\Delta y}{\Delta x} = \frac{0 - 3}{12 - 0} = -\frac{1}{4}$$

and the vertical intercept can be seen to be the point $(0, 3)$.

EXAMPLE 14. For a linear equation in the *slope-intercept form*

$$y = mx + b \qquad m, b = \text{constants}$$

the slope and intercepts of the line can be read directly from the equation. For such an equation, m is the slope of the line; $(0, b)$ is the y intercept; and, as seen in Problem 1.10, $(-b/m, 0)$ is the x intercept. One can tell immediately from the equation in Example 12, therefore, that the slope of the line is $-\frac{1}{4}$, the y intercept is $(0, 3)$, and the x intercept is $(12, 0)$.

Solved Problems

EXPONENTS

1.1. Simplify the following, using the rules of exponents:

a) $x^4 \cdot x^5$

$$x^4 \cdot x^5 = x^{4+5} = x^9$$

b) $x^7 \cdot x^{-3}$

$$x^7 \cdot x^{-3} = x^{7+(-3)} = x^4$$

$$\left[x^7 \cdot x^{-3} = x^7 \cdot \frac{1}{x^3} = x \cdot x \cdot x \cdot x \cdot x \cdot x \cdot x \cdot \frac{1}{x \cdot x \cdot x} = x^4 \right]$$

c) $x^{-2} \cdot x^{-4}$

$$x^{-2} \cdot x^{-4} = x^{-2+(-4)} = x^{-6} = \frac{1}{x^6}$$

$$\left[x^{-2} \cdot x^{-4} = \frac{1}{x \cdot x} \cdot \frac{1}{x \cdot x \cdot x \cdot x} = \frac{1}{x^6} \right]$$

d) $x^2 \cdot x^{1/2}$

$$x^2 \cdot x^{1/2} = x^{2+(1/2)} = x^{5/2} = \sqrt{x^5}$$

$$[x^2 \cdot x^{1/2} = (x \cdot x)(\sqrt{x})$$
$$= (\sqrt{x} \cdot \sqrt{x} \cdot \sqrt{x} \cdot \sqrt{x})(\sqrt{x}) = (x^{1/2})^5 = x^{5/2}]$$

e) $\dfrac{x^9}{x^3}$

$$\frac{x^9}{x^3} = x^{9-3} = x^6$$

f) $\dfrac{x^4}{x^7}$

$$\frac{x^4}{x^7} = x^{4-7} = x^{-3} = \frac{1}{x^3}$$

$$\left[\frac{x^4}{x^7} = \frac{x \cdot x \cdot x \cdot x}{x \cdot x \cdot x \cdot x \cdot x \cdot x \cdot x} = \frac{1}{x^3} \right]$$

g) $\dfrac{x^3}{x^{-4}}$

$$\dfrac{x^3}{x^{-4}} = x^{3-(-4)} = x^{3+4} = x^7$$

$$\left[\dfrac{x^3}{x^{-4}} = \dfrac{x^3}{1/x^4} = x^3 \cdot x^4 = x^7\right]$$

h) $\dfrac{x^3}{\sqrt{x}}$

$$\dfrac{x^3}{\sqrt{x}} = \dfrac{x^3}{x^{1/2}} = x^{3-(1/2)} = x^{5/2} = \sqrt{x^5}$$

i) $(x^2)^5$

$$(x^2)^5 = x^{2 \cdot 5} = x^{10}$$

j) $(x^4)^{-2}$

$$(x^4)^{-2} = x^{4 \cdot (-2)} = x^{-8} = \dfrac{1}{x^8}$$

k) $\dfrac{1}{x^5} \cdot \dfrac{1}{y^5}$

$$\dfrac{1}{x^5} \cdot \dfrac{1}{y^5} = x^{-5} \cdot y^{-5} = (xy)^{-5} = \dfrac{1}{(xy)^5}$$

l) $\dfrac{x^3}{y^3}$

$$\dfrac{x^3}{y^3} = \left(\dfrac{x}{y}\right)^3$$

POLYNOMIALS

1.2. Perform the indicated arithmetic operations on the following polynomials:

a) $3xy + 5xy$ b) $13yz^2 - 28yz^2$ c) $36x^2y^3 - 25x^2y^3$

d) $26x_1x_2 + 58x_1x_2$ e) $16x^2y^3z^5 - 37x^2y^3z^5$

a) $8xy$,   b) $-15yz^2$,   c) $11x^2y^3$,   d) $84x_1x_2$,   e) $-21x^2y^3z^5$

1.3. Add or subtract the following polynomials as indicated. Note that in subtraction the sign of every term within the parentheses must be changed before corresponding elements are added.

a) $(34x - 8y) + (13x + 12y)$

$$(34x - 8y) + (13x + 12y) = 47x + 4y$$

b) $(26x - 19y) - (17x - 50y)$

$$(26x - 19y) - (17x - 50y) = 9x + 31y$$

c) $(5x^2 - 8x - 23) - (2x^2 + 7x)$

$$(5x^2 - 8x - 23) - (2x^2 + 7x) = 3x^2 - 15x - 23$$

d) $(13x^2 + 35x) - (4x^2 + 17x - 49)$

$$(13x^2 + 35x) - (4x^2 + 17x - 49) = 9x^2 + 18x + 49$$

1.4. Perform the indicated operations, recalling that each term in the first polynomial must be multiplied by each term in the second and their products summed.

a) $(2x + 9)(3x - 8)$

$$(2x + 9)(3x - 8) = 6x^2 - 16x + 27x - 72 = 6x^2 + 11x - 72$$

b) $(6x - 4y)(3x - 5y)$

$$(6x - 4y)(3x - 5y) = 18x^2 - 30xy - 12xy + 20y^2 = 18x^2 - 42xy + 20y^2$$

c) $(3x - 7)^2$

$$(3x - 7)^2 = (3x - 7)(3x - 7) = 9x^2 - 21x - 21x + 49 = 9x^2 - 42x + 49$$

d) $(x + y)(x - y)$

$$(x + y)(x - y) = x^2 - xy + xy - y^2 = x^2 - y^2$$

SOLVING EQUATIONS

1.5. Solve each of the following linear equations by moving all terms with the unknown variable to the left, moving all other terms to the right, and then simplifying.

a) $5x + 6 = 9x - 10$

$5x + 6 = 9x - 10$

$5x - 9x = -10 - 6$

$-4x = -16$

$x = 4$

b) $26 - 2x = 8x - 44$

$26 - 2x = 8x - 44$

$-2x - 8x = -44 - 26$

$-10x = -70$

$x = 7$

c) $9(3x + 4) - 2x = 11 + 5(4x - 1)$

$9(3x + 4) - 2x = 11 + 5(4x - 1)$

$27x + 36 - 2x = 11 + 20x - 5$

$27x - 2x - 20x = 11 - 5 - 36$

$5x = -30$

$x = -6$

d) $\dfrac{x}{3} - 16 = \dfrac{x}{12} + 14$

$\dfrac{x}{3} - 16 = \dfrac{x}{12} + 14$

$\dfrac{x}{3} - \dfrac{x}{12} = 14 + 16$

Multiplying both sides of the equation by the least common denominator (LCD), here 12, gives

$$12 \cdot \left(\frac{x}{3} - \frac{x}{12}\right) = 30 \cdot 12$$

$$4x - x = 360$$

$$x = 120$$

e) $\dfrac{5}{x} + \dfrac{3}{x + 4} = \dfrac{7}{x}$ $[x \neq 0, -4]$

$$\frac{5}{x} + \frac{3}{x + 4} = \frac{7}{x}$$

Multiplying both sides by the LCD, we get

$$x(x + 4) \cdot \left(\frac{5}{x} + \frac{3}{x + 4}\right) = \frac{7}{x} \cdot x(x + 4)$$

$$5(x + 4) + 3x = 7(x + 4)$$

$$8x + 20 = 7x + 28$$

$$x = 8$$

1.6. Solve the following quadratic equations, using the quadratic formula:

a) $5x^2 + 23x + 12 = 0$

Using (1.1) and substituting $a = 5$, $b = 23$, and $c = 12$, we get

$$x = \frac{-b \pm \sqrt{b^2 - 4ac}}{2a}$$

$$= \frac{-23 \pm \sqrt{(23)^2 - 4(5)(12)}}{2(5)} = \frac{-23 \pm \sqrt{529 - 240}}{10}$$

$$= \frac{-23 \pm \sqrt{289}}{10} = \frac{-23 \pm 17}{10}$$

$$x = \frac{-23 + 17}{10} = -0.6 \qquad x = \frac{-23 - 17}{10} = -4$$

b) $3x^2 - 41x + 26 = 0$

$$x = \frac{-(-41) \pm \sqrt{(-41)^2 - 4(3)(26)}}{2(3)} = \frac{41 \pm \sqrt{1681 - 312}}{6}$$

$$= \frac{41 \pm \sqrt{1369}}{6} = \frac{41 \pm 37}{6}$$

$$x = \frac{41 + 37}{6} = 13 \qquad x = \frac{41 - 37}{6} = \frac{2}{3}$$

FUNCTIONS

1.7. a) Given $f(x) = x^2 + 4x - 5$, find $f(2)$ and $f(-3)$.

Substituting 2 for each occurrence of x in the function gives

$$f(2) = (2)^2 + 4(2) - 5 = 7$$

Now substituting -3 for each occurrence of x, we get

$$f(-3) = (-3)^2 + 4(-3) - 5 = -8$$

b) Given $f(x) = 2x^3 - 5x^2 + 8x - 20$, find $f(5)$ and $f(-4)$.

$$f(5) = 2(5)^3 - 5(5)^2 + 8(5) - 20 = 145$$
$$f(-4) = 2(-4)^3 - 5(-4)^2 + 8(-4) - 20 = -260$$

1.8. In the following graphs (Fig. 1-2), where y replaces $f(x)$ as the dependent variable in functions, indicate which graphs are graphs of functions and which are not.

For a graph to be the graph of a function, for each value of x, there can be one and only one value of y. If a vertical line can be drawn which intersects the graph at more than one point, then the graph is not the graph of a function. Applying this criterion, which is known as the *vertical-line test*, we see that (a), (b), and (d) are functions; (c), (e), and (f) are not.

1.9. Which of the following equations are functions and why?

a) $y = -2x + 7$

$y = -2x + 7$ is a function because for each value of the independent variable x there is one and only one value of the dependent variable y. For example, if $x = 1$, $y = -2(1) + 7 = 5$. The graph would be similar to (a) in Fig. 1-2.

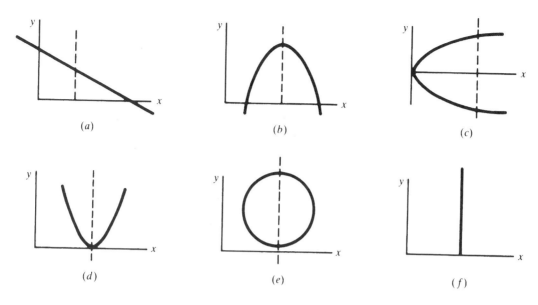

Fig. 1-2

b) $y^2 = x$

   $y^2 = x$, which is equivalent to $y = \pm\sqrt{x}$, is not a function because for each positive value of x, there are two values of y. For example, if $y^2 = 9$, $y = \pm 3$. The graph would be similar to that of (c) in Fig. 1-2, illustrating that a parabola whose axis is parallel to the x axis cannot be a function.

c) $y = x^2$

   $y = x^2$ is a function. For each value of x there is only one value of y. For instance, if $x = -5$, $y = 25$. While it is also true that $y = 25$ when $x = 5$, it is irrelevant. The definition of a function simply demands that for each value of x there be one value of y, *not* that for each value of y there be only one value of x. The graph would be like (d) in Fig. 1-2, demonstrating that a parabola with axis parallel to the y axis is a function.

d) $y = -x^2 + 6x + 15$

   $y = -x^2 + 6x + 15$ is a function. For each value of x there is a unique value of y. The graph would be like (b) in Fig. 1-2.

e) $x^2 + y^2 = 64$

   $x^2 + y^2 = 64$ is not a function. If $x = 0$, $y^2 = 64$, and $y = \pm 8$. The graph would be a circle, similar to (e) in Fig. 1-2. A circle does not pass the vertical-line test.

f) $x = 4$

   $x = 4$ is not a function. The graph of $x = 4$ is a vertical line. This means that at $x = 4$, y has many values. The graph would look like (f) in Fig. 1-2.

GRAPHS, SLOPES, AND INTERCEPTS

1.10.   Find the x intercept in terms of the parameters of the slope-intercept form of a linear equation $y = mx + b$.

   Setting $y = 0$,

$$0 = mx + b$$
$$mx = -b$$
$$x = -\frac{b}{m}$$

Thus, the x intercept of the slope-intercept form is $(-b/m, 0)$.

1.11. Graph the following equations and indicate their respective slopes and intercepts:

  a) $3y + 15x = 30$ *b)* $2y - 6x = 12$ *c)* $8y - 2x + 16 = 0$ *d)* $6y + 3x - 18 = 0$

  To graph an equation, first set it in slope-intercept form by solving it for y in terms of x. From Example 14, the slope and two intercepts can then be read directly from the equation, providing three pieces of information, whereas only two are needed to graph a straight line. See Fig. 1-3 and Problems 2.1 to 2.10.

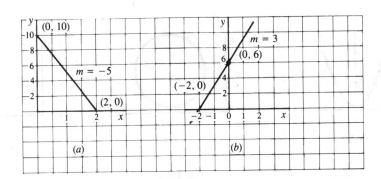

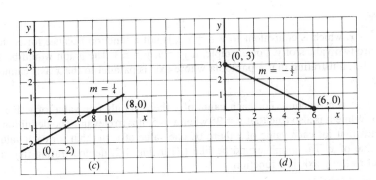

Fig. 1-3

a) $3y + 15x = 30$

$\qquad 3y = -15x + 30$

$\qquad y = -5x + 10$

Slope $m = -5$

y intercept: $(0, 10)$

x intercept: $(2, 0)$

b) $2y - 6x = 12$

$\qquad 2y = 6x + 12$

$\qquad y = 3x + 6$

Slope $m = 3$

y intercept: $(0, 6)$

x intercept: $(-2, 0)$

c) $8y - 2x + 16 = 0$

$\qquad 8y = 2x - 16$

$\qquad y = \frac{1}{4}x - 2$

Slope $m = \frac{1}{4}$

y intercept: $(0, -2)$

x intercept: $(8, 0)$

d) $6y + 3x - 18 = 0$

$\qquad 6y = -3x + 18$

$\qquad y = -\frac{1}{2}x + 3$

Slope $m = -\frac{1}{2}$

y intercept: $(0, 3)$

x intercept: $(6, 0)$

1.12. Find the slope m of the linear function passing through: a) $(4, 12)$, $(8, 2)$; b) $(-1, 15)$, $(3, 6)$; c) $(2, -3)$, $(5, 18)$.

a)   Substituting in the formula from Example 13, we get

$$m = \frac{y_2 - y_1}{x_2 - x_1} = \frac{2 - 12}{8 - 4} = \frac{-10}{4} = -2\frac{1}{2}$$

b)

$$m = \frac{6 - 15}{3 - (-1)} = \frac{-9}{4} = -2\frac{1}{4}$$

c)

$$m = \frac{18 - (-3)}{5 - 2} = \frac{21}{3} = 7$$

1.13. Graph (a) the quadratic function $y = 2x^2$ and (b) the rational function $y = 2/x$.

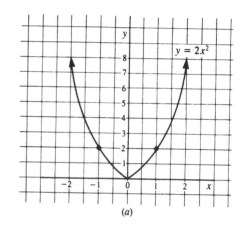

(a)	x	$f(x)$	$=$	$2x^2$	$=$	y	Points
	-2	$f(-2)$	$=$	$2(-2)^2$	$=$	8	$(-2, 8)$
	-1	$f(-1)$	$=$	$2(-1)^2$	$=$	2	$(-1, 2)$
	0	$f(0)$	$=$	$2(0)^2$	$=$	0	$(0, 0)$
	1	$f(1)$	$=$	$2(1)^2$	$=$	2	$(1, 2)$
	2	$f(2)$	$=$	$2(2)^2$	$=$	8	$(2, 8)$
	3	$f(3)$	$=$	$2(3)^2$	$=$	18	$(3, 18)$

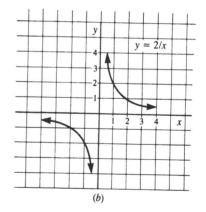

(b)	x	$f(x)$	$=$	$2/x$	$=$	y	Points
	-4	$f(-4)$	$=$	$2/(-4)$	$=$	$-\frac{1}{2}$	$(-4, -\frac{1}{2})$
	-2	$f(-2)$	$=$	$2/(-2)$	$=$	-1	$(-2, -1)$
	-1	$f(-1)$	$=$	$2/(-1)$	$=$	-2	$(-1, -2)$
	$-\frac{1}{2}$	$f(-\frac{1}{2})$	$=$	$2/(-\frac{1}{2})$	$=$	-4	$(-\frac{1}{2}, -4)$
	$\frac{1}{2}$	$f(\frac{1}{2})$	$=$	$2/\frac{1}{2}$	$=$	4	$(\frac{1}{2}, 4)$
	1	$f(1)$	$=$	$2/1$	$=$	2	$(1, 2)$
	2	$f(2)$	$=$	$2/2$	$=$	1	$(2, 1)$
	4	$f(4)$	$=$	$2/4$	$=$	$\frac{1}{2}$	$(4, \frac{1}{2})$

Fig. 1-3

CHAPTER 2

Economic Applications of Graphs and Equations

2.1 ISOCOST LINES

An *isocost line* represents the different combinations of two inputs or factors of production that can be purchased with a given sum of money. The general formula is $P_K K + P_L L = E$, where K and L are capital and labor, P_K and P_L their respective prices, and E the amount allotted to expenditures. In isocost analysis the individual prices and the expenditure are initially held constant; only the different combinations of inputs are allowed to change. The function can then be graphed by expressing one variable in terms of the other, as seen in Example 1 and Problems 2.5 and 2.6.

EXAMPLE 1. Given:

$$P_K K + P_L L = E$$
$$P_K K = E - P_L L$$
$$K = \frac{E - P_L L}{P_K}$$
$$K = \frac{E}{P_K} - \left(\frac{P_L}{P_K}\right) L$$

This is the familiar linear function of the form $y = mx + b$, where $b = E/P_K$ = the vertical intercept and $m = -P_L/P_K$ = the slope. The graph is given by the solid line in Fig. 2-1.

From the equation and graph, the effects of a change in any one of the parameters are easily discernible. An increase in the expenditure from E to E' will increase the vertical intercept and cause the isocost line to shift out to the right (dashed line) parallel to the old line. The slope is unaffected because the slope depends on the relative prices $(-P_L/P_K)$ and prices are not affected by expenditure changes. A change in P_L will alter the slope of the line but leave the vertical intercept unchanged. A change in P_K will alter the slope and the vertical intercept. See Problems 2.5 and 2.6.

14

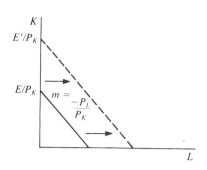

Fig. 2-1

2.2 SUPPLY AND DEMAND ANALYSIS

Equilibrium in supply and demand analysis occurs when $Q_s = Q_d$. By equating the supply and demand functions, the equilibrium price and quantity can be determined. See Example 2 and Problems 2.1 to 2.4 and 2.11 to 2.16.

EXAMPLE 2. Given:

$$Q_s = -5 + 3P \qquad Q_d = 10 - 2P$$

In equilibrium,

$$Q_s = Q_d$$

Solving for P,

$$-5 + 3P = 10 - 2P$$
$$5P = 15 \qquad P = 3$$

Substituting $P = 3$ in either of the equations,

$$Q_s = -5 + 3P = -5 + 3(3) = 4 = Q_d$$

2.3 INCOME DETERMINATION MODELS

Income determination models generally express the equilibrium level of income in a four-sector economy as

$$Y = C + I + G + (X - Z)$$

where Y = income, C = consumption, I = investment, G = government expenditures, X = exports, and Z = imports. By substituting the information supplied in the problem, it is an easy matter to solve for the equilibrium level of income. *Aggregating* (summing) the variables on the right allows the equation of be graphed in two-dimensional space. See Example 3 and Problems 2.7 to 2.10 and 2.17 to 2.22.

EXAMPLE 3. Assume a simple two-sector economy where $Y = C + I$, $C = C_0 + bY$, and $I = I_0$. Assume further that $C_0 = 85$, $b = 0.9$, and $I_0 = 55$. The equilibrium level of income can be calculated in terms of (1) the general parameters and (2) the specific values assigned to these parameters.

1. The *equilibrium equation* is

$$Y = C + I$$

Substituting for C and I,

$$Y = C_0 + bY + I_0$$

Solving for Y,

$$Y - bY = C_0 + I_0$$
$$(1 - b)Y = C_0 + I_0$$
$$Y = \frac{C_0 + I_0}{1 - b}$$

The solution in this form is called the reduced form. The *reduced form* (or *solution equation*) expresses the endogenous variable (here Y) as an explicit function of the exogenous variables (C_0, I_0) and the parameters (b).

2. The specific equilibrium level of income can be calculated by substituting the numerical values for the parameters in either the original equation (a) or the reduced form (b).

$$a) \quad Y = C_0 + bY + I_0 = 85 + 0.9Y + 55 \qquad b) \quad Y = \frac{C_0 + I_0}{1 - b} = \frac{85 + 55}{1 - 0.9}$$

$$Y - 0.9Y = 140$$

$$0.1Y = 140 \qquad\qquad\qquad\qquad = \frac{140}{0.1} = 1400$$

$$Y = 1400$$

The term $1/(1-b)$ is called the *autonomous expenditure multiplier* in economics. It measures the multiple effect each dollar of autonomous spending has on the equilibrium level of income. Since $b = $ MPC in the income determination model, the multiplier $= 1/(1 - $ MPC$)$.

Note: Decimals may be converted to fractions for ease in working with the income determination model. For example, $0.1 = \frac{1}{10}$, $0.9 = \frac{9}{10}$, $0.5 = \frac{1}{2}$, $0.2 = \frac{1}{5}$, etc.

2.4 *IS-LM* ANALYSIS

The *IS schedule* is a locus of points representing all the different combinations of interest rates and income levels consistent with equilibrium in the goods (commodity) market. The *LM schedule* is a locus of points representing all the different combinations of interest rates and income levels consistent with equilibrium in the money market. *IS-LM analysis* seeks to find the level of income and the rate of interest at which both the commodity market and the money market will be in equilibrium. This can be accomplished with the techniques used for solving simultaneous equations. Unlike the simple income determination model in Section 2.3, *IS-LM* analysis deals explicitly with the interest rate and incorporates its effect into the model. See Example 4 and Problems 2.23 and 2.24.

EXAMPLE 4.   The commodity market for a simple two-sector economy is in equilibrium when $Y = C + I$. The money market is in equilibrium when the supply of money (M_s) equals the demand for money (M_d), which in turn is composed of the transaction-precautionary demand for money (M_t) and the speculative demand for money (M_z). Assume a two-sector economy where $C = 48 + 0.8Y$, $I = 98 - 75i$, $M_s = 250$, $M_t = 0.3Y$, and $M_z = 52 - 150i$.

Commodity equilibrium (*IS*) exists when $Y = C + I$. Substituting into the equation,

$$Y = 48 + 0.8Y + 98 - 75i$$

$$Y - 0.8Y = 146 - 75i$$

$$0.2Y + 75i - 146 = 0 \qquad\qquad (2.1)$$

Monetary equilibrium (*LM*) exists when $M_s = M_t + M_z$. Substituting into the equation,

$$250 = 0.3Y + 52 - 150i$$

$$0.3Y - 150i - 198 = 0 \qquad\qquad (2.2)$$

A condition of simultaneous equilibrium in both markets can be found, then, by solving (*2.1*) and (*2.2*) simultaneously:

$$0.2Y + 75i - 146 = 0 \qquad\qquad (2.1)$$

$$0.3Y - 150i - 198 = 0 \qquad\qquad (2.2)$$

Multiply (2.1) by 2, add the result (2.3) to (2.2) to eliminate i, and solve for Y

$$0.4Y + 150i - 292 = 0$$

$$\underline{0.3Y - 150i - 198 = 0}$$ (2.3)

$$0.7Y \qquad\quad - 490 = 0$$

$$Y = 700$$

Substitute $Y = 700$ in (2.1) or (2.2) to find i.

$$0.2Y + 75i - 146 = 0$$

$$0.2(700) + 75i - 146 = 0$$

$$140 + 75i - 146 = 0$$

$$75i = 6$$

$$i = \tfrac{6}{75} = 0.08$$

The commodity and money markets will be in simultaneous equilibrium when $Y = 700$ and $i = 0.08$. At that point $C = 48 + 0.8(700) = 608$, $I = 98 - 75(0.08) = 92$, $M_t = 0.3(700) = 210$, and $M_z = 52 - 150(0.08) = 40$. $C + I = 608 + 92 = 700$ and $M_t + M_z = 210 + 40 = 250 = M_s$.

Solved Problems

GRAPHS

2.1. A complete demand function is given by the equation

$$Q_d = -30P + 0.05Y + 2P_r + 4T$$

where P is the price of the good, Y is income, P_r is the price of a related good (here a substitute), and T is taste. Can the function be graphed?

Since the complete function contains five different variables, it cannot be graphed as is. In ordinary demand analysis, however, it is assumed that all the independent variables except price are held constant so that the effect of a change in price on the quantity demanded can be measured independently of the influence of other factors, or *ceteris paribus*. If the other variables (Y, P_r, T) are held constant, the function can be graphed.

2.2 (*a*) Draw the graph for the demand function in Problem 2.1, assuming $Y = 5000$, $P_r = 25$, and $T = 30$. (*b*) What does the typical demand function drawn in part (*a*) show? (*c*) What happens to the graph if the price of the good changes from 5 to 6? (*d*) What happens if any of the other variables change? For example, if income increases to 7400?

a) By adding the new data to the equation in Problem 2.1, the function is easily graphable. See Fig. 2-2.

$$Q_d = -30P + 0.05Y + 2P_r + 4T = -30P + 0.05(5000) + 2(25) + 4(30) = -30P + 420$$

b) The demand function graphed in part (*a*) shows all the diferent quantities of the good that will be demanded at different prices, assuming a given level of income, taste, and prices of substitutes (here 5000, 30, 25) which are not allowed to change.

c) If nothing changes but the price of the good, the graph remains exactly the same since the graph indicates the different quantities that will be demanded at all the possible prices. A simple change in the price of the good occasions a movement along the curve which is called a *change in quantity demanded*. When the price goes from 5 to 6, the quantity demanded falls from 270 [420 − 30(5)] to 240 [420 − 30(6)], a movement from A to B on the curve.

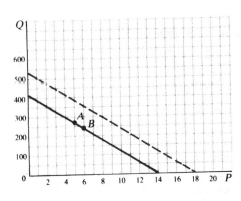

Fig. 2-2

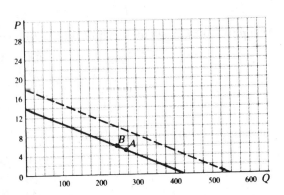

Fig. 2-3

d)   If any of the other variables change, there will be a shift in the curve. This is called a *change in demand* because it results in a totally new demand function (and curve) in response to the changed conditions. If income increases to 7400, the new demand function becomes

$$Q_d = -30P + 0.05(7400) + 2(25) + 4(30) = -30P + 540$$

This is graphed as a dashed line in Fig. 2-2.

2.3.   In economics the independent variable (price) has traditionally been graphed on the vertical axis in supply and demand analysis, and the dependent variable (quantity) has been graphed on the horizontal. (a) Graph the demand function in Problem 2.2 according to the traditional method. (b) Show what happens if the price goes from 5 to 6 and income increases to 7400.

a)   The function $Q_d = 420 - 30P$ is graphed according to traditional economic practice by means of the *inverse function*, which is obtained by solving the original function for the independent variable in terms of the dependent variable. Solving algebraically for P in terms of Q_d, therefore, the inverse function of $Q_d = 420 - 30P$ is $P = 14 - \frac{1}{30}Q_d$. The graph appears as a solid line in Fig. 2-3.

b)   If P goes from 5 to 6, Q_d falls from 270 to 240.

$$
\begin{array}{ll}
P = 14 - \frac{1}{30}Q_d & \qquad P = 14 - \frac{1}{30}Q_d \\
5 = 14 - \frac{1}{30}Q_d & \qquad 6 = 14 - \frac{1}{30}Q_d \\
\frac{1}{30}Q_d = 9 & \qquad \frac{1}{30}Q_d = 8 \\
Q_d = 270 & \qquad Q_d = 240
\end{array}
$$

The change is represented by a movement from A to B in Fig. 2-3.

   If $Y = 7400$, as in Problems 2.2(d), $Q_d = 540 - 30P$. Solving algebraically for P in terms of Q, the inverse function is $P = 18 - \frac{1}{30}Q_d$. It is graphed as a dashed line in Fig. 2-3.

2.4.   Graph the demand function

$$Q_d = -4P + 0.01Y - 5P_r + 10T$$

when $Y = 8000$, $P_r = 8$, and $T = 4$. (b) What type of good is the related good? (c) What happens if T increases to 8, indicating greater preference for the good? (d) Construct the graph along the traditional economic lines with P on the vertical axis and Q on the horizontal axis.

a) $$Q_d = -4P + 0.01(8000) - 5(8) + 10(4) = -4P + 80$$

This is graphed as a solid line in Fig. 2-4(a).

b)   The related good has a negative coefficient. This means that a rise in the price of the related good will lead to a decrease in demand for the original good. The related good is, by definition, a complementary good.

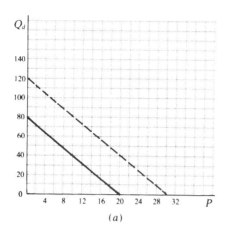 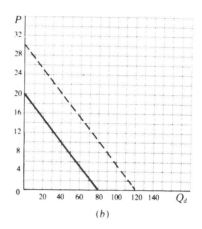

Fig. 2-4

c) If $T = 8$, indicating greater preference, there will be a totally new demand.

$$Q_d = -4P + 0.01(8000) - 5(8) + 10(8) = -4P + 120$$

See the dashed line in Fig. 2-4(a).

d) Graphing P on the vertical calls for the inverse function. Solving for P in terms of Q_d, the inverse of $Q_d = 80 - 4P$ is $P = 20 - \frac{1}{4}Q_d$ and is graphed as a solid line in Fig. 2-4(b). The inverse of $Q_d = 120 - 4P$ is $P = 30 - \frac{1}{4}Q_d$. It is the dashed line in Fig. 2-4(b).

2.5   A person has \$120 to spend on two goods (X, Y) whose respective prices are \$3 and \$5. (a) Draw a *budget line* showing all the different combinations of the two goods that can be bought with the given budget (B). What happens to the original budget line (b) if the budget falls by 25 percent, (c) if the price of X doubles, (d) if the price of Y falls to 4?

a)   The general function for a budget line is $P_x X + P_Y Y = B$

If $P_x = 3$, $P_Y = 5$, and $B = 120$, $3X + 5Y = 120$

Solving for Y in terms of X in order to graph the function, $Y = 24 - \frac{3}{5}X$

The graph is given as a solid line in Fig. 2-5(a).

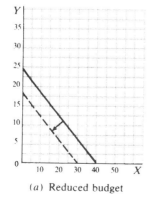 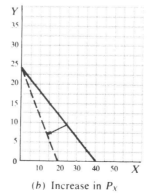 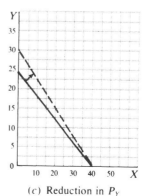

(a) Reduced budget   (b) Increase in P_X (c) Reduction in P_Y

Fig. 2-5

$b)$ If the budget falls by 25 percent, the new budget is 90 [$120 - \frac{1}{4}(120) = 90$]. The equation for the new budget line is

$$3X + 5Y = 90$$
$$Y = 18 - \tfrac{3}{5}X$$

The graph is a dashed line in Fig. 2-5(a). Lowering the budget causes the budget line to shift parallel to the left.

$c)$ If P_X doubles, the original equation becomes

$$6X + 5Y = 120$$
$$Y = 24 - \tfrac{6}{5}X$$

The vertical intercept remains the same, but the slope changes and becomes steeper. See the dashed line in Fig. 2-5(b). With a higher price for X, less X can be bought with the given budget.

$d)$ If P_Y now equals 4,

$$3X + 4Y = 120$$
$$Y = 30 - \tfrac{3}{4}X$$

With a change in P_Y, both the vertical intercept and the slope change. This is shown in Fig. 2-5(c) by the dashed line.

2.6. Either coal (C) or gas (G) can be used in the production of steel. The cost of coal is 100, the cost of gas 500. Draw an isocost curve showing the different combinations of gas and coal that can be purchased (a) with an initial expenditure (E) of 10,000, (b) if expenditures increase by 50 percent, (c) if the price of gas is reduced by 20 percent, (d) if the price of coal rises by 25 percent. Always start from the original equation.

$a)$
$$P_C C + P_G G = E$$
$$100C + 500G = 10{,}000$$
$$C = 100 - 5G$$

The graph is a solid line in Fig. 2-6(a).

$b)$ A 50 percent increase in expenditures makes the new outlay 15,000 [$10{,}000 + 0.5(10{,}000)$]. The new equation is

$$100C + 500G = 15{,}000$$
$$C = 150 - 5G$$

The graph is the dashed line in Fig. 2-6(a).

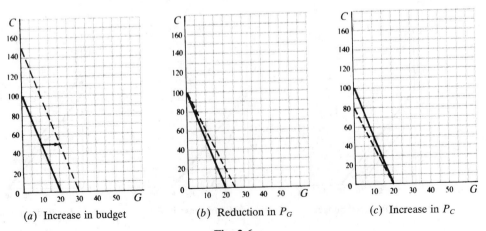

(a) Increase in budget   (b) Reduction in P_G (c) Increase in P_C

Fig. 2-6

c) If the price of gas is reduced by 20 percent, the new price is 400 $[500 - 0.2(500)]$, and the new equation is

$$100C + 400G = 10,000$$
$$C = 100 - 4G$$

The graph is the dashed line in Fig. 2-6(b).

d) A 25 percent rise in the price of coal makes the new price 125 $[100 + 0.25(100)]$.

$$125C + 500G = 10,000$$
$$C = 80 - 4G$$

The graph appears as a dashed line in Fig. 2-6(c).

GRAPHS IN THE INCOME DETERMINATION MODEL

2.7.   Given: $Y = C + I$, $C = 50 + 0.8Y$, and $I_0 = 50$. (a) Graph the consumption function. (b) Graph the aggregate demand function, $C + I_0$. (c) Find the equilibrium level of income from the graph.

a) Since consumption is a function of income, it is graphed on the vertical axis; income is graphed on the horizontal. See Fig. 2-7. When other components of aggregate demand such as I, G, and $X - Z$ are added to the model, they are also graphed on the vertical axis. It is easily determined from the linear form of the consumption function that the vertical intercept is 50 and the slope of the line (the MPC or $\Delta C/\Delta Y$) is 0.8.

b) Investment in the model is *autonomous investment*. This means investment is independent of income and does not change in response to changes in income. When considered by itself, the graph of a constant is a horizontal line; when added to a linear function, it causes a parallel shift in the original function by an amount equal to its value. In Fig. 2-7, autonomous investment causes the aggregate demand function to shift up by 50 parallel to the initial consumption function.

c) To obtain the equilibrium level of income from a graph, a 45° dashed line is drawn from the origin. If the same scale of measurement is used on both axes, a 45° line has a slope of 1, meaning that as the line moves away from the origin, it moves up vertically (ΔY) by one unit for every unit it moves across horizontally (ΔX). Every point on the 45° line, therefore, has a horizontal coordinate (*abscissa*)

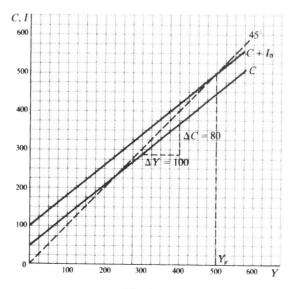

Fig. 2-7

exactly equal to its vertical coordinate (*ordinate*). Consequently, when the aggregate demand function intersects the 45° line, aggregate demand (as graphed on the vertical) will equal national income (as graphed on the horizontal). From Fig. 2-7 it is clear that the equilibrium level of income is 500, since the aggregate demand function $(C + I)$ intersects the 45° line at 500.

2.8. Given: $Y = C + I + G$, $C = 25 + 0.75Y$, $I = I_0 = 50$, and $G = G_0 = 25$. (*a*) Graph the aggregate demand function and show its individual components. (*b*) Find the equilibrium level of income. (*c*) How can the aggregate demand function be graphed directly, without having to graph each of the component parts?

a)   See Fig. 2-8.

b)   Equilibrium income = 400.

c)   To graph the aggregate demand function directly, sum up the individual components,

$$\text{Agg. } D = C + I + G = 25 + 0.75Y + 50 + 25 = 100 + 0.75Y$$

The direct graphing of the aggregate demand function coincides exactly with the graph of the summation of the individual graphs of C, I, and G above.

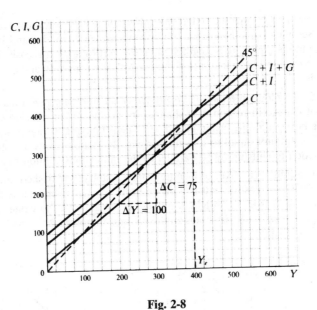

Fig. 2-8

2.9. Use a graph to show how the addition of a lump-sum tax (a tax independent of income) influences the parameters of the income determination model. Graph the two systems individually, using a solid line for (1) and a dashed line for (2).

1) $Y = C + I$ 2) $Y = C + I$ $Yd = Y - T$
 $C = 100 + 0.6Y$ $C = 100 + 0.6Yd$ $T = 50$
 $I_0 = 40$ $I_0 = 40$

The first system of equations presents no problems; the second requires that C first be converted from a function of Yd to a function of Y.

1)   Agg. $D = C + I$ 2)   Agg. $D = C + I$
 $= 100 + 0.6Y + 40$ $= 100 + 0.6Yd + 40 = 140 + 0.6(Y - T)$
 $= 140 + 0.6Y$ $= 140 + 0.6(Y - 50) = 110 + 0.6Y$

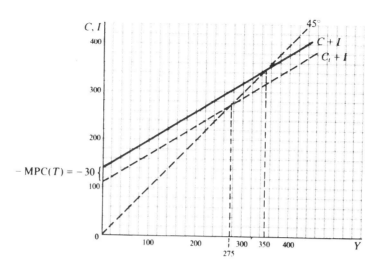

Fig. 2-9

A lump-sum tax has a negative effect on the vertical intercept of the aggregate demand function equal to $-MPC(T)$. Here $-0.6(50) = -30$. The slope is not affected (note the parallel lines for the two graphs in Fig. 2-9). Income falls from 350 to 275 as a result of the tax. See Fig. 2-9.

2.10. Explain with the aid of a graph how the incorporation of a *proportional* tax (a tax depending on income) influences the parameters of the income determination model. Graph the model without the tax as a solid line and the model with the tax as a dashed line.

1) $Y = C + I$ 2) $Y = C + I$ $Yd = Y - T$
     $C = 85 + 0.75Y$ $C = 85 + 0.75Yd$ $T = 20 + 0.2Y$
     $I_0 = 30$ $I_0 = 30$

1)   Agg. $D = C + I$ 2)   Agg. $D = C + I$
          $= 85 + 0.75Y + 30$ $= 85 + 0.75Yd + 30 = 115 + 0.75(Y - T)$
          $= 115 + 0.75Y$ $= 115 + 0.75(Y - 20 - 0.2Y)$
                                         $= 115 + 0.75Y - 15 - 0.15Y = 100 + 0.6Y$

Incorporation of a proportional income tax into the model affects the slope of the line, or the MPC. In this case it lowers it from 0.75 to 0.6. The vertical intercept is also lowered because the tax structure includes a lump-sum tax of 20. Because of the tax structure, the equilibrium level of income falls from 460 to 250. See Fig. 2-10.

EQUATIONS IN SUPPLY AND DEMAND ANALYSIS

2.11. Find the equilibrium price and quantity for the following markets:

    a) $Q_s = -20 + 3P$ *b)* $Q_s = -45 + 8P$
        $Q_d = 220 - 5P$ $Q_d = 125 - 2P$

    c) $Q_s + 32 - 7P = 0$ *d)* $13P - Q_s = 27$
        $Q_d - 128 + 9P = 0$ $Q_d + 4P - 24 = 0$

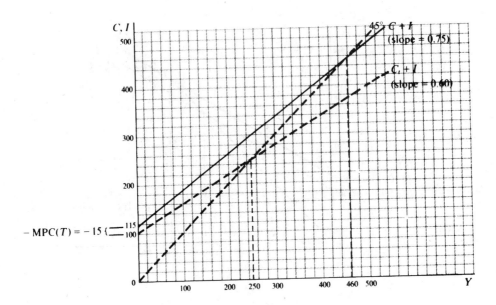

Fig. 2-10

Each of the markets will be in equilibrium when $Q_s = Q_d$.

a)
$$Q_s = Q_d$$
$$-20 + 3P = 220 - 5P$$
$$8P = 240$$
$$P = 30$$
$$Q_s = -20 + 3P = -20 + 3(30)$$
$$Q_s = 70 = Q_d$$

b)
$$Q_s = Q_d$$
$$-45 + 8P = 125 - 2P$$
$$10P = 170$$
$$P = 17$$
$$Q_d = 125 - 2P = 125 - 2(17)$$
$$Q_d = 91 = Q_s$$

c)
$$Q_s = 7P - 32$$
$$Q_d = 128 - 9P$$
$$7P - 32 = 128 - 9P$$
$$16P = 160$$
$$P = 10$$
$$Q_s = 7P - 32 = 7(10) - 32$$
$$Q_s = 38 = Q_d$$

d)
$$Q_s = -27 + 13P$$
$$Q_d = 24 - 4P$$
$$-27 + 13P = 24 - 4P$$
$$17P = 51$$
$$P = 3$$
$$Q_d = 24 - 4P = 24 - 4(3)$$
$$Q_d = 12 = Q_s$$

2.12. Given the following set of simultaneous equations for two related markets, beef (B) and pork (P), find the equilibrium conditions for each market, using the substitution method.

1) $Q_{dB} = 82 - 3P_B + P_P$ 2) $Q_{dP} = 92 + 2P_B - 4P_P$
    $Q_{sB} = -5 + 15P_B$ $Q_{sP} = -6 + 32P_P$

Equilibrium requires that $Q_s = Q_d$ in each market.

1)
$$Q_{sB} = Q_{dB}$$
$$-5 + 15P_B = 82 - 3P_B + P_P$$
$$18P_B - P_P = 87$$

2)
$$Q_{sP} = Q_{dP}$$
$$-6 + 32P_P = 92 + 2P_B - 4P_P$$
$$36P_P - 2P_B = 98$$

This reduces the problem to two equations and two unknowns:

$$18P_B - P_P = 87 \qquad\qquad (2.4)$$
$$-2P_B + 36P_P = 98 \qquad\qquad (2.5)$$

Solving for P_P in *(2.4)* gives

$$P_P = 18P_B - 87$$

Substituting the value of this term in *(2.5)* gives

$$-2P_B + 36(18P_B - 87) = 98 \qquad -2P_B + 648P_B - 3132 = 98$$
$$646P_B = 3230 \qquad\qquad P_B = 5$$

Substituting $P_B = 5$ in *(2.5)*, or *(2.4)*,

$$-2(5) + 36P_P = 98$$
$$36P_P = 108 \qquad P_P = 3$$

Finally, substituting the values for P_B and P_P in either the supply or the demand function for each market,

1) $\quad Q_{dB} = 82 - 3P_B + P_P = 82 - 3(5) + (3)$ 2) $\quad Q_{dP} = 92 + 2P_B - 4P_P = 92 + 2(5) - 4(3)$
   $\qquad Q_{dB} = 70 = Q_{sB}$ $Q_{dP} = 90 = Q_{sP}$

2.13. Find the equilibrium price and quantity for two complementary goods, slacks (S) and jackets (J), using the elimination method.

1) $\quad Q_{dS} = 410 - 5P_S - 2P_J$ 2) $\quad Q_{dJ} = 295 - P_S - 3P_J$
   $\qquad Q_{sS} = -60 + 3P_S$ $Q_{sJ} = -120 + 2P_J$

In equilibrium,

1) $\qquad\qquad Q_{dS} = Q_{sS}$ 2) $\qquad\qquad Q_{dJ} = Q_{sJ}$
$\quad 410 - 5P_S - 2P_J = -60 + 3P_S$ $\quad 295 - P_S - 3P_J = -120 + 2P_J$
$\quad 470 - 8P_S - 2P_J = 0$ $\quad 415 - P_S - 5P_J = 0$

This leaves two equations

$$470 - 8P_S - 2P_J = 0 \qquad\qquad (2.6)$$
$$415 - P_S - 5P_J = 0 \qquad\qquad (2.7)$$

Multiplying *(2.7)* by 8 gives *(2.8)*. Subtract *(2.6)* from *(2.8)* to eliminate P_S, and solve for P_J.

$$3320 - 8P_S - 40P_J = 0 \qquad\qquad (2.8)$$
$$\underline{-(+470 - 8P_S - \;\;2P_J = 0)}$$
$$2850 \qquad - 38P_J = 0$$
$$\qquad\qquad P_J = 75$$

Substituting $P_J = 75$ in *(2.6)*,

$$470 - 8P_S - 2(75) = 0$$
$$320 = 8P_S \qquad P_S = 40$$

Finally, substituting $P_J = 75$ and $P_S = 40$ into Q_d or Q_s for each market,

1) $\quad Q_{dS} = 410 - 5P_S - 2P_J = 410 - 5(40) - 2(75)$ 2) $\quad Q_{dJ} = 295 - P_S - 3P_J = 295 - 40 - 3(75)$
   $\qquad Q_{dS} = 60 = Q_{sS}$ $Q_{dJ} = 30 = Q_{sJ}$

2.14. Supply and demand conditions can also be expressed in quadratic form. Find the equilibrium price and quantity, given the demand function

$$P + Q^2 + 3Q - 20 = 0 \qquad\qquad (2.9)$$

and the supply function

$$P - 3Q^2 + 10Q = 5 \qquad\qquad (2.10)$$

Either the substitution method or the elimination method can be used, since this problem involves two equations and two unknowns. Using the substitution method, (2.10) is solved for P in terms of Q.

$$P - 3Q^2 + 10Q = 5$$
$$P = 3Q^2 - 10Q + 5$$

Substituting $P = 3Q^2 - 10Q + 5$ in (2.9),

$$(3Q^2 - 10Q + 5) + Q^2 + 3Q - 20 = 0$$
$$4Q^2 - 7Q - 15 = 0$$

Using the quadratic formula $Q_1, Q_2 = (-b \pm \sqrt{b^2 - 4ac})/(2a)$, where $a = 4$, $b = -7$, and $c = -15$, $Q_1 = 3$, and $Q_2 = -1.25$. Since neither price nor quantity can be negative, $Q = 3$. Substitute $Q = 3$ in (2.9) or (2.10) to find P.

$$P + (3)^2 + 3(3) - 20 = 0 \qquad P = 2$$

2.15. Use the elimination method to find the equilibrium price and quantity when the demand function is

$$3P + Q^2 + 5Q - 102 = 0 \qquad\qquad (2.11)$$

and the supply function is

$$P - 2Q^2 + 3Q + 71 = 0 \qquad\qquad (2.12)$$

Multiply (2.12) by 3 to get (2.13) and subtract it from (2.11) to eliminate P.

$$\begin{array}{r} 3P + Q^2 + 5Q - 102 = 0 \\ -(3P - 6Q^2 + 9Q + 213 = 0) \\ \hline 7Q^2 - 4Q - 315 = 0 \end{array} \qquad (2.13)$$

Use the quadratic formula (see Problem 2.14) to solve for Q, and substitute the result, $Q = 7$, in (2.12) or (2.11) to solve for P.

$$P - 2(7)^2 + 3(7) + 71 = 0 \qquad P = 6$$

2.16. Supply and demand analysis can also involve more than two markets. Find the equilibrium price and quantity for the three substitute goods below.

$$Q_{d1} = 23 - 5P_1 + P_2 + P_3 \qquad Q_{s1} = -8 + 6P_1$$
$$Q_{d2} = 15 + P_1 - 3P_2 + 2P_3 \qquad Q_{s2} = -11 + 3P_2$$
$$Q_{d3} = 19 + P_1 + 2P_2 - 4P_3 \qquad Q_{s3} = -5 + 3P_3$$

For equilibrium in each market,

$Q_{d1} = Q_{s1}$	$Q_{d2} = Q_{s2}$	$Q_{d3} = Q_{s3}$
$23 - 5P_1 + P_2 + P_3 = -8 + 6P_1$	$15 + P_1 - 3P_2 + 2P_3 = -11 + 3P_2$	$19 + P_1 + 2P_2 - 4P_3 = -5 + 3P_3$
$31 - 11P_1 + P_2 + P_3 = 0$	$26 + P_1 - 6P_2 + 2P_3 = 0$	$24 + P_1 + 2P_2 - 7P_3 = 0$

This leaves three equations with three unknowns:

$$31 - 11P_1 + P_2 + P_3 = 0 \qquad\qquad (2.14)$$
$$26 + P_1 - 6P_2 + 2P_3 = 0 \qquad\qquad (2.15)$$
$$24 + P_1 + 2P_2 - 7P_3 = 0 \qquad\qquad (2.16)$$

Start by eliminating one of the variables (here P_2). Multiply (2.14) by 2 to get

$$62 - 22P_1 + 2P_2 + 2P_3 = 0$$

From this subtract (2.16).

$$
\begin{array}{r}
62 - 22P_1 + 2P_2 + 2P_3 = 0 \\
-(24 + P_1 + 2P_2 - 7P_3) = 0 \\
\hline
38 - 23P_1 + 9P_3 = 0
\end{array}
$$

$$(2.17)$$

Multiply (2.16) by 3.

$$72 + 3P_1 + 6P_2 - 21P_3 = 0$$

Add the result to (2.15).

$$
\begin{array}{r}
26 + P_1 - 6P_2 + 2P_3 = 0 \\
72 + 3P_1 + 6P_2 - 21P_3 = 0 \\
\hline
98 + 4P_1 - 19P_3 = 0
\end{array}
$$

$$(2.18)$$

Now there are two equations, (2.17) and (2.18), and two unknowns. Multiply (2.17) by 19 and (2.18) by 9; then add to eliminate P_3.

$$
\begin{array}{r}
722 - 437P_1 + 171P_3 = 0 \\
882 + 36P_1 - 171P_3 = 0 \\
\hline
1604 - 401P_1 = 0 \\
P_1 = 4
\end{array}
$$

Substitute $P_1 = 4$ in (2.18) to solve for P_3.

$$98 + 4(4) - 19P_3 = 0$$
$$19P_3 = 114 \qquad P_3 = 6$$

Substitute $P_1 = 4$ and $P_3 = 6$ into (2.14), (2.15), or (2.16), to solve for P_2.

$$31 - 11(4) + P_2 + (6) = 0 \qquad P_2 = 7$$

EQUATIONS IN THE INCOME DETERMINATION MODEL

2.17. Given: $Y = C + I + G$, $C = C_0 + bY$, $I = I_0$, and $G = G_0$, where $C_0 = 135$, $b = 0.8$, $I_0 = 75$, and $G_0 = 30$. (a) Find the equation for the equilibrium level of income in the reduced form. (b) Solve for the equilibrium level of income (1) directly and (2) with the reduced form.

a)   From Section 2.3,

$$
\begin{aligned}
Y &= C + I + G \\
&= C_0 + bY + I_0 + G_0 \\
Y - bY &= C_0 + I_0 + G_0 \\
(1 - b)Y &= C_0 + I_0 + G_0 \\
Y &= \frac{C_0 + I_0 + G_0}{1 - b}
\end{aligned}
$$

b)   1) $Y = C + I + G = 135 + 0.8Y + 75 + 30$ 2) $Y = \dfrac{C_0 + I_0 + G_0}{1 - b}$

$$
\begin{aligned}
Y - 0.8Y &= 240 \\
0.2Y &= 240 \\
Y &= 1200
\end{aligned}
$$

$$
\begin{aligned}
&= \frac{135 + 75 + 30}{1 - 0.8} \\
&= 5(240) = 1200
\end{aligned}
$$

2.18. Find the equilibrium level of income $Y = C + I$, when $C = 89 + 0.8Y$ and $I_0 = 24$.

$$Y = \frac{C_0 + I_0}{1 - b} = 5(89 + 24) = 565$$

From Problem 2.17, the value of the multiplier $[1/(1-b)]$ is already known for cases when $b = 0.8$. Use of the reduced form to solve the equation in this instance is faster, therefore, although the other method is also correct.

2.19. (a) Find the reduced form of the following income determination model where investment is not autonomous but is a function of income. (b) Find the numerical value of the equilibrium level of income (Y_e). (c) Show what happens to the multiplier.

$$Y = C + I \qquad C = C_0 + bY \qquad I = I_0 + aY$$

where $C_0 = 65$, $I_0 = 70$, $b = 0.6$, and $a = 0.2$.

a)
$$Y = C + I$$
$$= C_0 + bY + I_0 + aY$$
$$Y - bY - aY = C_0 + I_0$$
$$(1 - b - a)Y = C_0 + I_0$$
$$Y = \frac{C_0 + I_0}{1 - b - a}$$

b)
$$Y = C + I$$
$$= 65 + 0.6Y + 70 + 0.2Y$$
$$Y - 0.6Y - 0.2Y = 65 + 70$$
$$0.2Y = 135$$
$$Y = 675$$

c) When investment is a function of income, and no longer autonomous, the multiplier changes from $1/(1-b)$ to $1/(1-b-a)$. This increases the value of the multiplier because it reduces the denominator of the fraction and makes the quotient larger, as substitution of the values of the parameters in the problem shows:

$$\frac{1}{1-b} = \frac{1}{1-0.6} = \frac{1}{0.4} = 2.5 \qquad \frac{1}{1-b-a} = \frac{1}{1-0.6-0.2} = \frac{1}{0.2} = 5$$

2.20. Find (a) the reduced form, (b) the numerical value of Y_e, and (c) the effect on the multiplier when a lump-sum tax is added to the model and consumption becomes a function of disposable income (Yd).

$$Y = C + I \qquad C = C_0 + bYd \qquad I = I_0 \qquad Yd = Y - T$$

where $C_0 = 100$, $b = 0.6$, $I_0 = 40$, and $T = 50$.

a)
$$Y = C + I = C_0 + bYd + I_0 = C_0 + b(Y - T) + I_0 = C_0 + bY - bT + I_0$$
$$Y - bY = C_0 + I_0 - bT$$
$$Y = \frac{C_0 + I_0 - bT}{1 - b}$$

b)
$$Y = 100 + 0.6Yd + 40 = 140 + 0.6(Y - T) \qquad \text{or} \qquad Y = \frac{100 + 40 - 0.6(50)}{1 - 0.6} = \frac{110}{0.4}$$
$$= 140 + 0.6(Y - 50) = 140 + 0.6Y - 30 \qquad\qquad\qquad = 275$$
$$Y - 0.6Y = 110$$
$$0.4Y = 110$$
$$Y = 275$$

The graph of this function is given in Problem 2.9.

c) As seen in part a), incorporation of a lump-sum tax into the model leaves the multiplier at $1/(1-b)$. Only the aggregate value of the exogenous variables is reduced by an amount equal to $-bT$. Incorporation of other autonomous variables such as G_0, X_0, or Z_0 will not affect the value of the multiplier either.

2.21. Find (*a*) the reduced form, (*b*) the numerical value of Y_e, and (*c*) the effect on the multiplier if a proportional income tax (*t*) is incorporated into the model.

$$Y = C + I \qquad C = C_0 + bYd \qquad T = T_0 + tY \qquad Yd = Y - T$$

where $I = I_0 = 30$, $C_0 = 85$, $b = 0.75$, $t = 0.2$, and $T_0 = 20$.

a)

$$Y = C + I = C_0 + bYd + I_0$$
$$= C_0 + b(Y - T) + I_0 = C_0 + b(Y - T_0 - tY) + I_0$$
$$= C_0 + bY - bT_0 - btY + I_0$$
$$Y - bY + btY = C_0 + I_0 - bT_0$$
$$(1 - b + bt)Y = C_0 + I_0 - bT_0$$
$$Y = \frac{C_0 + I_0 - bT_0}{1 - b + bt}$$

b) Once the reduced form is found, its use speeds the solution. But sometimes the reduced form is not available, making it necessary to be familiar with the other method.

$$Y = C + I = 85 + 0.75Yd + 30 = 115 + 0.75(Y - T)$$
$$= 115 + 0.75(Y - 20 - 0.2Y) = 115 + 0.75Y - 15 - 0.15Y$$
$$Y - 0.75Y + 0.15Y = 100$$
$$0.4Y = 100$$
$$Y = 250$$

The graph of this function is given in Problem 2.10.

c) The multiplier is changed from $1/(1 - b)$ to $1/(1 - b + bt)$. This reduces the size of the multiplier because it makes the denominator larger and the fraction smaller:

$$\frac{1}{1 - b} = \frac{1}{1 - 0.75} = \frac{1}{0.25} = 4$$

$$\frac{1}{1 - b + bt} = \frac{1}{1 - 0.75 + 0.75(0.2)} = \frac{1}{1 - 0.75 + 0.15} = \frac{1}{0.4} = 2.5$$

2.22. If the foreign sector is added to the model and there is a positive marginal propensity to import (*z*), find (*a*) the reduced form, (*b*) the equilibrium level of income, and (*c*) the effect on the multiplier.

$$Y = C + I + G + (X - Z) \qquad C = C_0 + bY \qquad Z = Z_0 + zY$$

where $I = I_0 = 90$, $G = G_0 = 65$, $X = X_0 = 80$, $C_0 = 70$, $Z_0 = 40$, $b = 0.9$, and $z = 0.15$.

a)

$$Y = C + I + G + (X - Z) = C_0 + bY + I_0 + G_0 + X_0 - Z_0 - zY$$
$$Y - bY + zY = C_0 + I_0 + G_0 + X_0 - Z_0$$
$$(1 - b + z)Y = C_0 + I_0 + G_0 + X_0 - Z_0$$
$$Y = \frac{C_0 + I_0 + G_0 + X_0 - Z_0}{1 - b + z}$$

b) Using the reduced form above,

$$Y = \frac{70 + 90 + 65 + 80 - 40}{1 - 0.9 + 0.15} - \frac{265}{0.25} = 1060$$

c) Introduction of the marginal propensity to import (z) into the model reduces the size of the multiplier. It makes the denominator larger and the fraction smaller:

$$\frac{1}{1-b} = \frac{1}{1-0.9} = \frac{1}{0.1} = 10$$

$$\frac{1}{1-b+z} = \frac{1}{1-0.9+0.15} = \frac{1}{0.25} = 4$$

IS-LM EQUATIONS

2.23. Given: $C = 102 + 0.7Y$, $I = 150 - 100i$, $M_s = 300$, $M_t = 0.25Y$, and $M_z = 124 - 200i$. Find (a) the equilibrium level of income and the equilibrium rate of interest and (b) the level of C, I, M_t, and M_z when the economy is in equilibrium.

a) Commodity market equilibrium (*IS*) exists where

$$Y = C + I$$
$$= 102 + 0.7Y + 150 - 100i$$
$$Y - 0.7Y = 252 - 100i$$
$$0.3Y + 100i - 252 = 0$$

Monetary equilibrium (*LM*) exists where

$$M_s = M_t + M_z$$
$$300 = 0.25Y + 124 - 200i$$
$$0.25Y - 200i - 176 = 0$$

Simultaneous equilibrium in both markets requires that

$$0.3Y + 100i - 252 = 0 \tag{2.19}$$
$$0.25Y - 200i - 176 = 0 \tag{2.20}$$

Multiply (*2.19*) by 2, and add the result to (*2.20*) to eliminate i:

$$\begin{aligned}
0.6Y + 200i - 504 &= 0 \\
\underline{0.25Y - 200i - 176 = 0} \\
0.85Y \qquad\qquad\quad &= 680 \\
Y &= 800
\end{aligned}$$

Substitute $Y = 800$ in (*2.19*) or (*2.20*):

$$0.25Y - 200i - 176 = 0$$
$$0.25(800) - 200i - 176 = 0$$
$$-200i = -24$$
$$i = 0.12$$

b) At $Y = 800$ and $i = 0.12$,

$$C = 102 + 0.7(800) = 662 \qquad M_t = 0.25(800) = 200$$
$$I = 150 - 100(0.12) = 138 \qquad M_z = 124 - 200(0.12) = 100$$

and

$$C + I = Y \qquad\qquad M_t + M_z = M_s$$
$$662 + 138 = 800 \qquad 200 + 100 = 300$$

2.24. Find (a) the equilibrium income level and interest rate and (b) the levels of C, I, M_t, and M_z in equilibrium when

$$C = 89 + 0.6Y \qquad I = 120 - 150i \qquad M_s = 275 \qquad M_t = 0.1Y \qquad M_z = 240 - 250i$$

a)   For *IS*:

$$Y = 89 + 0.6Y + 120 - 150i$$
$$Y - 0.6Y = 209 - 150i$$
$$0.4Y + 150i - 209 = 0$$

For *LM*:

$$M_s = M_t + M_z$$
$$275 = 0.1Y + 240 - 250i$$
$$0.1Y - 250i - 35 = 0$$

In equilibrium,

$$0.4Y + 150i - 209 = 0 \qquad (2.21)$$
$$0.1Y - 250i - 35 = 0 \qquad (2.22)$$

Multiply (*2.22*) by 4, and subtract the result from (*2.21*) to eliminate *Y*.

$$0.4Y + 150i - 209 = 0$$
$$\underline{-(0.4Y - 1000i - 140 = 0)}$$
$$1150i \qquad = 69$$
$$i = 0.06$$

Substitute $i = 0.06$ in (*2.21*) or (*2.22*).

$$0.4Y + 150(0.06) - 209 = 0$$
$$0.4Y = 200$$
$$Y = 500$$

b)   At $Y = 500$ and $i = 0.06$,

$$C = 89 + 0.6(500) = 389 \qquad M_t = 0.1(500) = 50$$
$$I = 120 - 150(0.06) = 111 \qquad M_z = 240 - 250(0.06) = 225$$

and

$$C + I = Y \qquad M_t + M_z = M_s$$
$$389 + 111 = 500 \qquad 50 + 225 = 275$$

CHAPTER 3

The Derivative and the Rules of Differentiation

3.1 LIMITS

If the functional values $f(x)$ of a function f draw closer to one and only one finite real number L for all values of x as x draws closer to a from both sides, but does not equal a, L is defined as the *limit* of $f(x)$ as x approaches a and is written

$$\lim_{x \to a} f(x) = L$$

Assuming that $\lim_{x \to a} f(x)$ and $\lim_{x \to a} g(x)$ both exist, the *rules of limits* are given below, explained in Example 2, and treated in Problems 3.1 to 3.4.

1. $\lim_{x \to a} k = k \qquad (k = \text{a constant})$

2. $\lim_{x \to a} x^n = a^n \qquad (n = \text{a positive integer})$

3. $\lim_{x \to a} k f(x) = k \lim_{x \to a} f(x) \qquad (k = \text{a constant})$

4. $\lim_{x \to a} [f(x) \pm g(x)] = \lim_{x \to a} f(x) \pm \lim_{x \to a} g(x)$

5. $\lim_{x \to a} [f(x) \cdot g(x)] = \lim_{x \to a} f(x) \cdot \lim_{x \to a} g(x)$

6. $\lim_{x \to a} [f(x) \div g(x)] = \lim_{x \to a} f(x) \div \lim_{x \to a} g(x) \qquad \left[\lim_{x \to a} g(x) \neq 0 \right]$

7. $\lim_{x \to a} [f(x)]^n = \left[\lim_{x \to a} f(x) \right]^n \qquad (n > 0)$

EXAMPLE 1.   *a*) From the graph of the function $f(x)$ in Fig. 3-1, it is clear that as the value of x approaches 3 from either side, the value of $f(x)$ approaches 2. This means that the limit of $f(x)$ as x approaches 3 is the number 2, which is written

$$\lim_{x \to 3} f(x) = 2$$

As x approaches 7 from either side in Fig. 3-1, where the open circle in the graph of $f(x)$ signifies there is a gap in the function at that point, the value of $f(x)$ approaches 4 even though the function is not defined at that point.

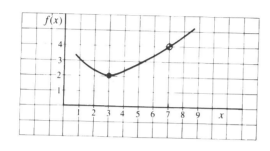
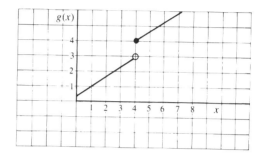

Fig. 3-1   **Fig. 3-2**

Since the limit of a function as x approaches a number depends only on the values of x *close to* that number, the limit exists and is written

$$\lim_{x \to 7} f(x) = 4$$

b) In Fig. 3-2, as x approaches 4 from the left (from values less than 4), written $x \to 4^-$, $g(x)$ approaches 3, called a *one-sided limit*; as x approaches 4 from the right (from values greater than 4), written $x \to 4^+$, $g(x)$ approaches 4. The limit does not exist, therefore, since $g(x)$ does not approach a *single* number as x approaches 4 from *both* sides.

EXAMPLE 2.   In the absence of a graph, limits can be found by using the rules of limits enumerated above.

a) $\lim\limits_{x \to 5} 9 = 9$ Rule 1

b) $\lim\limits_{x \to 6} x^2 = (6)^2 = 36$ Rule 2

c) $\lim\limits_{x \to 3} 2x^3 = 2 \lim\limits_{x \to 3} x^3 = 2(3)^3 = 54$ Rules 2 and 3

d) $\lim\limits_{x \to 2} (x^4 + 3x) = \lim\limits_{x \to 2} x^4 + 3 \lim\limits_{x \to 2} x$ Rule 4
$$= (2)^4 + 3(2) = 22$$

e) $\lim\limits_{x \to 4} [(x + 8)(x - 5)] = \lim\limits_{x \to 4} (x + 8) \cdot \lim\limits_{x \to 4} (x - 5)$ Rule 5
$$= (4 + 8) \cdot (4 - 5) = -12$$

3.2   CONTINUITY

A *continuous* function is one which has no breaks in its curve. It can be drawn without lifting the pencil from the paper. A function f is continuous at $x = a$ if:

1. $f(x)$ is defined, i.e., exists, at $x = a$

2. $\lim\limits_{x \to a} f(x)$ exists, *and*

3. $\lim\limits_{x \to a} f(x) = f(a)$

All polynomial functions are continuous, as are all rational functions, except where undefined, i.e., where their denominators equal zero. See Problem 3.5.

EXAMPLE 3.   Given that the graph of a continuous function can be sketched without ever removing pencil from paper and that an open circle means a gap in the function, it is clear that $f(x)$ is discontinuous at $x = 4$ in Fig. 3-3(a) and $g(x)$ is discontinuous at $x = 5$ in Fig. 3-3(b), even though $\lim_{x \to 5} g(x)$ exists.

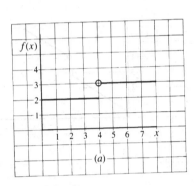
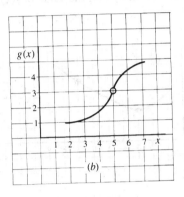

(a) (b)

Fig. 3-3

3.3   THE SLOPE OF A CURVILINEAR FUNCTION

The slope of a curvilinear function is not constant. It differs at different points on the curve. In geometry, the slope of a curvilinear function at a given point is measured by the slope of a line drawn tangent to the function at that point. A *tangent line* is a straight line that touches a curve at only one point. Measuring the slope of a curvilinear function at different points requires separate tangent lines, as in Fig. 3-4(a).

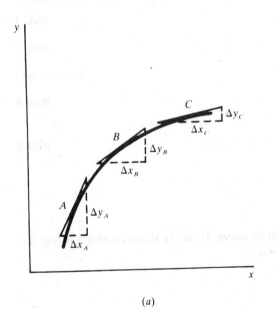
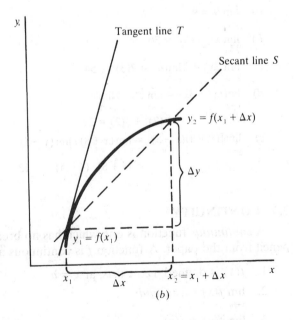

(a) (b)

Fig. 3-4

The slope of a tangent line is derived from the slopes of a family of secant lines. A *secant line S* is a straight line that intersects a curve at two points, as in Fig. 3-4(*b*), where

$$\text{Slope } S = \frac{y_2 - y_1}{x_2 - x_1}$$

By letting $x_2 = x_1 + \Delta x$ and $y_2 = f(x_1 + \Delta x)$, the slope of the secant line can also be expressed by a *difference quotient*:

$$\text{Slope } S = \frac{f(x_1 + \Delta x) - f(x_1)}{(x_1 + \Delta x) - x_1}$$

$$= \frac{f(x_1 + \Delta x) - f(x_1)}{\Delta x}$$

If the distance between x_2 and x_1 is made smaller and smaller, i.e., if $\Delta x \to 0$, the secant line pivots back to the left and draws progressively closer to the tangent line. If the slope of the secant line approaches a limit as $\Delta x \to 0$, the limit is the slope of the tangent line T, which is also the slope of the function at the point. It is written

$$\text{Slope } T = \lim_{\Delta x \to 0} \frac{f(x_1 + \Delta x) - f(x_1)}{\Delta x} \tag{3.1}$$

Note: In many texts h is used in place of Δx, giving

$$\text{Slope } T = \lim_{h \to 0} \frac{f(x_1 + h) - f(x_1)}{h} \tag{3.1a}$$

EXAMPLE 4. To find the slope of a curvilinear function, such as $f(x) = 2x^2$, (1) employ the specific function in the algebraic formula (*3.1*) or (*3.1a*) and substitute the arguments $x_1 + \Delta x$ (or $x_1 + h$) and x_1, respectively, (2) simplify the function, and (3) evaluate the limit of the function in its simplified form. From (*3.1*),

$$\text{Slope } T = \lim_{\Delta x \to 0} \frac{f(x + \Delta x) - f(x)}{\Delta x}$$

1) Employ the function $f(x) = 2x^2$ and substitute the arguments.

$$\text{Slope } T = \lim_{\Delta x \to 0} \frac{2(x + \Delta x)^2 - 2x^2}{\Delta x}$$

2) Simplify the result.

$$\text{Slope } T = \lim_{\Delta x \to 0} \frac{2[x^2 + 2x(\Delta x) + (\Delta x)^2] - 2x^2}{\Delta x}$$

$$= \lim_{\Delta x \to 0} \frac{4x(\Delta x) + 2(\Delta x)^2}{\Delta x}$$

Divide through by Δx.

$$\text{Slope } T = \lim_{\Delta x \to 0} (4x + 2\Delta x)$$

3) Take the limit of the simplified expression.

$$\text{Slope } T = 4x$$

Note: The value of the slope depends on the value of x chosen. At $x = 1$, slope $T = 4(1) = 4$; at $x = 2$, slope $T = 4(2) = 8$.

3.4 THE DERIVATIVE

Given a function $y = f(x)$, the *derivative* of the function f at x, written $f'(x)$ or dy/dx, is defined as

$$f'(x) = \lim_{\Delta x \to 0} \frac{f(x + \Delta x) - f(x)}{\Delta x} \qquad \text{if the limit exists} \qquad (3.2)$$

or from (3.1a),

$$f'(x) = \lim_{h \to 0} \frac{f(x_1 + h) - f(x_1)}{h} \qquad (3.2a)$$

where $f'(x)$ is read "the derivative of f with respect to x" or "f prime of x."

The derivative of a function $f'(x)$, or simply f', is itself a function which measures both the slope and the instantaneous rate of change of the original function $f(x)$ at a given point.

3.5 DIFFERENTIABILITY AND CONTINUITY

A function is *differentiable* at a point if the derivative exists (may be taken) at that point. To be differentiable at a point, a function must (1) be continuous at that point and (2) have a unique tangent at that point. In Fig. 3-5, $f(x)$ is not differentiable at a and c because gaps exist in the function at those points and the derivative cannot be taken at any point where the function is discontinuous.

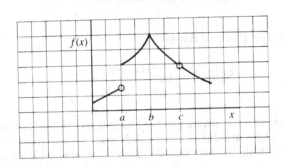

Fig. 3-5

Continuity alone, however, does not ensure (is not a sufficient condition for) differentiability. In Fig. 3-5, $f(x)$ is continuous at b, but it is not differentiable at b because at a sharp point or kink, called a *cusp*, an infinite number of tangent lines (and no one unique tangent line) can be drawn.

3.6 DERIVATIVE NOTATION

The derivative of a function can be written in many different ways. If $y = f(x)$, the derivatives can be expressed as

$$f'(x) \qquad y' \qquad \frac{dy}{dx} \qquad \frac{df}{dx} \qquad \frac{d}{dx}[f(x)] \qquad \text{or} \qquad D_x[f(x)]$$

If $y = \phi(t)$, the derivative can be written

$$\phi'(t) \qquad y' \qquad \frac{dy}{dt} \qquad \frac{d\phi}{dt} \qquad \frac{d}{dt}[\phi(t)] \qquad \text{or} \qquad D_t[\phi(t)]$$

If the derivative of $y = f(x)$ is evaluated at $x = a$, proper notation includes $f'(a)$ and $\left.\dfrac{dy}{dx}\right|_a$.

EXAMPLE 5.   If $y = 5x^2 + 7x + 12$, the derivative can be written

$$y' \qquad \frac{dy}{dx} \qquad \frac{d}{dx}(5x^2 + 7x + 12) \qquad \text{or} \qquad D_x(5x^2 + 7x + 12)$$

If $z = \sqrt{8t - 3}$, the derivative can be expressed as

$$z' \qquad \frac{dz}{dt} \qquad \frac{d}{dt}(\sqrt{8t - 3}) \qquad \text{or} \qquad D_t(\sqrt{8t - 3})$$

See Problems 3.6 to 3.8.

3.7 RULES OF DIFFERENTIATION

Differentiation is the process of finding the derivative of a function. It involves nothing more complicated than applying a few basic rules or formulas to a given function. In explaining the rules of differentiation for a function such as $y = f(x)$, other functions such as $g(x)$ and $h(x)$ are commonly used, where g and h are both unspecified functions of x. The rules of differentiation are listed below and treated in Problems 3.6 to 3.21. Selected proofs are found in Problems 3.24 to 3.26.

3.7.1 The Constant Function Rule

The derivative of a constant function $f(x) = k$, where k is a constant, is zero.

Given $f(x) = k$, $\qquad\qquad\qquad\qquad\qquad f'(x) = 0$

EXAMPLE 6.   Given $f(x) = 8$, $\qquad\qquad\qquad f'(x) = 0$
Given $f(x) = -6$, $\qquad\qquad\qquad\qquad f'(x) = 0$

3.7.2 The Linear Function Rule

The derivative of a linear function $f(x) = mx + b$ is equal to m, the coefficient of x. The derivative of a variable raised to the first power is always equal to the coefficient of the variable, while the derivative of a constant is simply zero.

Given $f(x) = mx + b$, $\qquad\qquad\qquad\qquad f'(x) = m$

EXAMPLE 7.   Given $f(x) = 3x + 2$, $\qquad\qquad f'(x) = 3$
Given $f(x) = 5 - \frac{1}{4}x$, $\qquad\qquad\qquad\qquad f'(x) = -\frac{1}{4}$
Given $f(x) = 12x$, $\qquad\qquad\qquad\qquad\qquad f'(x) = 12$

3.7.3 The Power Function Rule

The derivative of a power function $f(x) = kx^n$, where k is a constant and n is any real number, is equal to the coefficient k times the exponent n, multiplied by the variable x raised to the $n - 1$ power.

Given $f(x) = kx^n$ $\qquad\qquad\qquad\qquad f'(x) = k \cdot n \cdot x^{n-1}$

EXAMPLE 8.   Given $f(x) = 4x^3$ $\qquad\quad f'(x) = 4 \cdot 3 \cdot x^{3-1} = 12x^2$
Given $f(x) = 5x^2$, $\qquad\qquad\qquad f'(x) = 5 \cdot 2 \cdot x^{2-1} = 10x$
Given $f(x) = x^4$, $\qquad\qquad\qquad\quad f'(x) = 1 \cdot 4 \cdot x^{4-1} = 4x^3$

See also Problem 3.7.

3.7.4   The Rules for Sums and Differences

The derivative of a sum of two functions $f(x) = g(x) + h(x)$, where $g(x)$ and $h(x)$ are both differentiable functions, is equal to the sum of the derivatives of the individual functions. Similarly, the derivative of the difference of two functions is equal to the difference of the derivatives of the two functions.

Given $f(x) = g(x) \pm h(x)$, $f'(x) = g'(x) \pm h'(x)$

EXAMPLE 9.   Given $f(x) = 12x^5 - 4x^4$, $f'(x) = 60x^4 - 16x^3$
Given $f(x) = 9x^2 + 2x - 3$, $f'(x) = 18x + 2$

See Problem 3.8. For derivation of the rule, see Problem 3.24.

3.7.5   The Product Rule

The derivative of a product $f(x) = g(x) \cdot h(x)$, where $g(x)$ and $h(x)$ are both differentiable functions, is equal to the first function multiplied by the derivative of the second plus the second function multiplied by the derivative of the first. Given $f(x) = g(x) \cdot h(x)$,

$$f'(x) = g(x) \cdot h'(x) + h(x) \cdot g'(x) \tag{3.3}$$

EXAMPLE 10.   Given $f(x) = 3x^4(2x - 5)$, let $g(x) = 3x^4$ and $h(x) = 2x - 5$. Taking the individual derivatives, $g'(x) = 12x^3$ and $h'(x) = 2$. Then by substituting these values in the product-rule formula (3.3),

$$f'(x) = 3x^4(2) + (2x - 5)(12x^3)$$

and simplifying algebraically gives

$$f'(x) = 6x^4 + 24x^4 - 60x^3 = 30x^4 - 60x^3$$

See Problems 3.9 to 3.11; for the derivation of the rule, see Problem 3.25.

3.7.6   The Quotient Rule

The derivative of a quotient $f(x) = g(x) \div h(x)$, where $g(x)$ and $h(x)$ are both differentiable functions and $h(x) \neq 0$, is equal to the denominator times the derivative of the numerator, minus the numerator times the derivative of the denominator, all divided by the denominator squared. Given $f(x) = g(x)/h(x)$,

$$f'(x) = \frac{h(x) \cdot g'(x) - g(x) \cdot h'(x)}{[h(x)]^2} \tag{3.4}$$

EXAMPLE 11.   Given

$$f(x) = \frac{5x^3}{4x + 3}$$

where $g(x) = 5x^3$ and $h(x) = 4x + 3$, we know that $g'(x) = 15x^2$ and $h'(x) = 4$. Substituting these values in the quotient—rule formula (3.4),

$$f'(x) = \frac{(4x + 3)(15x^2) - 5x^3(4)}{(4x + 3)^2}$$

Simplifying algebraically,

$$f'(x) = \frac{60x^3 + 45x^2 - 20x^3}{(4x + 3)^2} = \frac{40x^3 + 45x^2}{(4x + 3)^2} = \frac{5x^2(8x + 9)}{(4x + 3)^2}$$

See Problems 3.12 and 3.13; for the derivation of the rule, see Problem 3.26.

3.7.7 *The Generalized Power Function Rule*

The derivative of a function raised to a power, $f(x) = [g(x)]^n$, where $g(x)$ is a differentiable function and n is any real number, is equal to the exponent n times the function $g(x)$ raised to the $n - 1$ power, multiplied in turn by the derivative of the function itself $g'(x)$. Given $f(x) = [g(x)]^n$,

$$f'(x) = n[g(x)]^{n-1} \cdot g'(x) \qquad (3.5)$$

EXAMPLE 12. Given $f(x) = (x^3 + 6)^5$, let $g(x) = x^3 + 6$, then $g'(x) = 3x^2$. Substituting these values in the generalized power function formula (3.5) gives

$$f'(x) = 5(x^3 + 6)^{5-1} \cdot 3x^2$$

Simplifying algebraically,

$$f'(x) = 5(x^3 + 6)^4 \cdot 3x^2 = 15x^2(x^3 + 6)^4$$

Note: The generalized power function rule is derived from the chain rule which follows below. See Problems 3.14 and 3.15.

3.7.8 *The Chain Rule*

Given a *composite function*, also called a *function of a function*, in which y is a function of u and u in turn is a function of x, that is, $y = f(u)$ and $u = g(x)$, then $y = f[g(x)]$ and the derivative of y with respect to x is equal to the derivative of the first function with respect to u times the derivative of the second function with respect to x:

$$\frac{dy}{dx} = \frac{dy}{du} \cdot \frac{du}{dx} \qquad (3.6)$$

See Problems 3.16 and 3.17.

EXAMPLE 13. Consider the function $y = (5x^2 + 3)^4$. To use the chain rule, let $y = u^4$ and $u = 5x^2 + 3$. Then $dy/du = 4u^3$ and $du/dx = 10x$. Substitute these values in (3.6):

$$\frac{dy}{dx} = 4u^3 \cdot 10x = 40xu^3$$

Then to express the derivative in terms of a single variable, substitute $5x^2 + 3$ for u.

$$\frac{dy}{dx} = 40x(5x^2 + 3)^3$$

For more complicated functions, different combinations of the basic rules must be used. See Problems 3.18 and 3.19.

3.8 HIGHER-ORDER DERIVATIVES

The second-order derivative, written $f''(x)$, measures the slope and the rate of change of the first derivative, just as the first derivative measures the slope and the rate of change of the original or *primitive function*. The third-order derivative $f'''(x)$ measures the slope and rate of change of the second-order derivative, etc. Higher-order derivatives are found by applying the rules of differentiation to lower-order derivatives, as illustrated in Example 14 and treated in Problems 3.20 and 3.21.

EXAMPLE 14. Given $y = f(x)$, common notation for the second-order derivative includes $f''(x)$, d^2y/dx^2, y'', and D^2y; for the third-order derivative, $f'''(x)$, d^3y/dx^3, y''', and D^3y; for the fourth-order derivatives, $f^{(4)}(x)$, d^4y/dx^4, $y^{(4)}$, and D^4y; etc.

Higher-order derivatives are found by successively applying the rules of differentiation to derivatives of the previous order. Thus, if $f(x) = 2x^4 + 5x^3 + 3x^2$,

$$f'(x) = 8x^3 + 15x^2 + 6x$$
$$f''(x) = 24x^2 + 30x + 6$$
$$f'''(x) = 48x + 30$$
$$f^{(4)}(x) = 48 \qquad f^{(5)}(x) = 0$$

See Problems 3.20 and 3.21.

3.9 IMPLICIT DIFFERENTIATION

Introductory economics deals most often with *explicit functions* in which the dependent variable appears to the left of the equal sign and the independent variable appears to the right. Frequently encountered in more advanced economics courses, however, are *implicit functions* in which both variables and constants are to the left of the equal sign. Some implicit functions can be easily converted to explicit functions by solving for the dependent variable in terms of the independent variable; others cannot. For those not readily convertible, the derivative may be found by *implicit differentiation*. See Example 16 and Problems 3.22, 3.23, 4.25, and 4.26; also see Section 5.10 and Problems 5.20, 5.21, 6.51, and 6.52.

EXAMPLE 15.   Samples of explicit and implicit functions include:

Explicit: $\qquad\qquad\qquad y = 4x \qquad y = x^2 + 6x - 7 \qquad y = \dfrac{x^4 - 9x^3}{x^2 - 13}$

Implicit: $\qquad\qquad 8x + 5y - 21 = 0 \qquad 3x^2 - 8xy - 5y - 49 = 0 \qquad 35x^3 y^7 - 106 = 0$

EXAMPLE 16.   Given $3x^4 - 7y^5 - 86 = 0$, the derivative dy/dx is found by means of implicit differentiation in two easy steps.

1)   Differentiate both sides of the equation with respect to x while treating y as a function of x,

$$\frac{d}{dx}(3x^4 - 7y^5 - 86) = \frac{d}{dx}(0) \qquad\qquad (3.7)$$

$$\frac{d}{dx}(3x^4) - \frac{d}{dx}(7y^5) - \frac{d}{dx}(86) = \frac{d}{dx}(0)$$

where $\dfrac{d}{dx}(3x^4) = 12x^3, \dfrac{d}{dx}(86) = 0, \dfrac{d}{dx}(0) = 0$. Using the generalized power function rule for $\dfrac{d}{dx}(7y^5)$ and noting that $\dfrac{d}{dx}(y) = \dfrac{dy}{dx}$, we get

$$\frac{d}{dx}(7y^5) = 7 \cdot 5 \cdot y^{5-1} \cdot \frac{d}{dx}(y) = 35y^4 \frac{dy}{dx}$$

Substitute the above values in (*3.7*).

$$12x^3 - 35y^4 \frac{dy}{dx} = 0 \qquad\qquad (3.8)$$

(2)   Now simply solve (*3.8*) algebraically for dy/dx:

$$-35y^4 \frac{dy}{dx} = -12x^3$$

$$\frac{dy}{dx} = \frac{12x^3}{35y^4}$$

Compare this answer to that in Example 16 of Chapter 5.

Solved Problems

LIMITS AND CONTINUITY

3.1. Use the rules of limits to find the limits for the following functions:

a) $\lim_{x \to 2} [x^3(x + 4)]$

$$\lim_{x \to 2} [x^3(x + 4)] = \lim_{x \to 2} x^3 \cdot \lim_{x \to 2} (x + 4)$$ Rule 5
$$= (2)^3 \cdot (2 + 4) = 8 \cdot 6 = 48$$

b) $\lim_{x \to 4} \dfrac{3x^2 - 5x}{x + 6}$

$$\lim_{x \to 4} \frac{3x^2 - 5x}{x + 6} = \frac{\lim_{x \to 4} (3x^2 - 5x)}{\lim_{x \to 4} (x + 6)}$$ Rule 6

$$= \frac{3(4)^2 - 5(4)}{4 + 6} = \frac{48 - 20}{10}$$
$$= 2.8$$

c) $\lim_{x \to 2} \sqrt{6x^3 + 1}$

$$\lim_{x \to 2} \sqrt{6x^3 + 1} = \lim_{x \to 2} (6x^3 + 1)^{1/2}$$

$$= [\lim_{x \to 2} (6x^3 + 1)]^{1/2}$$ Rule 7

$$= [6(2)^3 + 1]^{1/2} = (49)^{1/2} = \pm 7$$

3.2. Find the limits for the following polynomial and rational functions.

a) $\lim_{x \to 3} (5x^2 - 4x + 9)$

From the properties of limits it can be shown that for all polynomial functions and all rational functions, where defined, $\lim_{x \to a} f(x) = f(a)$. The limits can be taken, therefore, by simply evaluating the functions at the given level of a.

$$\lim_{x \to 3} (5x^2 - 4x + 9) = 5(3)^2 - 4(3) + 9 = 42$$

b) $\lim_{x \to -4} (3x^2 + 7x - 12)$

$$\lim_{x \to -4} (3x^2 + 7x - 12) = 3(-4)^2 + 7(-4) - 12 = 8$$

c) $\lim_{x \to 6} \dfrac{4x^2 - 2x - 8}{5x^2 + 6}$

$$\lim_{x \to 6} \frac{4x^2 - 2x - 8}{5x^2 + 6} = \frac{4(6)^2 - 2(6) - 8}{5(6)^2 + 6} = \frac{124}{186} = \frac{2}{3}$$

3.3. Find the limits of the following rational functions. If the limit of the denominator equals zero, neither Rule 6 nor the generalized rule for rational functions used above applies.

a) $\lim_{x \to 7} \dfrac{x - 7}{x^2 - 49}$

The limit of the denominator is zero, so Rule 6 cannot be used. Since we are only interested in the function as x draws *near* to 7, however, the limit can be found if by factoring and canceling, the problem of zero in the denominator is removed.

$$\lim_{x \to 7} \frac{x-7}{x^2-49} = \lim_{x \to 7} \frac{x-7}{(x+7)(x-7)}$$

$$= \lim_{x \to 7} \frac{1}{x+7} = \frac{1}{14}$$

b) $\lim_{x \to -7} \dfrac{x-7}{x^2-49}$

$$\lim_{x \to -7} \frac{x-7}{x^2-49} = \lim_{x \to -7} \frac{x-7}{(x+7)(x-7)}$$

$$= \lim_{x \to -7} \frac{1}{x+7}$$

The limit does not exist.

c) $\lim_{x \to 6} \dfrac{x^2-2x-24}{x-6}$

With the limit of the denominator equal to zero, factor.

$$\lim_{x \to 6} \frac{x^2-2x-24}{x-6} = \lim_{x \to 6} \frac{(x+4)(x-6)}{x-6} = \lim_{x \to 6}(x+4) = 10$$

3.4.   Find the limits of the following functions, noting the role that infinity plays.

a) $\lim_{x \to 0} \dfrac{2}{x} \quad (x \neq 0)$

As seen in Fig. 1-3(b), as x approaches 0 from the right ($x \to 0^+$), $f(x)$ approaches positive infinity; as x approaches 0 from the left ($x \to 0^-$), $f(x)$ approaches negative infinity. If a limit approaches either positive or negative infinity, the limit does *not* exist and is written

$$\lim_{x \to 0^+} \frac{2}{x} = \infty \qquad \lim_{x \to 0^-} \frac{2}{x} = -\infty \qquad \text{The limit does not exist.}$$

b) $\lim_{x \to \infty} \dfrac{2}{x} \qquad \lim_{x \to -\infty} \dfrac{2}{x}$

As also seen in Fig. 1-3(b), as x approaches ∞, $f(x)$ approaches 0; as x approaches $-\infty$, $f(x)$ also approaches 0. The limit exists in both cases and is written

$$\lim_{x \to \infty} \frac{2}{x} = 0 \qquad \lim_{x \to -\infty} \frac{2}{x} = 0$$

c) $\lim_{x \to \infty} \dfrac{3x^2-7x}{4x^2-21}$

As $x \to \infty$, both numerator and denominator become infinite, leaving matters unclear. A trick in such circumstances is to divide all terms by the highest power of x which appears in the function. Here dividing all terms by x^2 leaves

$$\lim_{x \to \infty} \frac{3x^2-7x}{4x^2-21} = \lim_{x \to \infty} \frac{3-(7/x)}{4-(21/x^2)} = \frac{3-0}{4-0} = \frac{3}{4}$$

3.5. Indicate whether the following functions are continuous at the specified points by determining whether at the given point all the following conditions from Section 3.2 hold: (1) $f(x)$ is defined, (2) $\lim_{x \to a} f(x)$ exists, and (3) $\lim_{x \to a} f(x) = f(a)$.

    a) $f(x) = 5x^2 - 8x + 9$ at $x = 3$

        1) $f(3) = 5(3)^2 - 8(3) + 9 = 30$

        2) $\lim_{x \to 3} (5x^2 - 8x + 9) = 5(3)^2 - 8(3) + 9 = 30$

        3) $\lim_{x \to 3} f(x) = 30 = f(3)$ $f(x)$ is continuous.

    b) $f(x) = \dfrac{x^2 + 3x + 12}{x - 3}$ at $x = 4$

        1) $f(4) = \dfrac{(4)^2 + 3(4) + 12}{4 - 3} = \dfrac{40}{1} = 40$

        2) $\lim_{x \to 4} \dfrac{x^2 + 3x + 12}{x - 3} = 40$

        3) $\lim_{x \to 4} f(x) = 40 = f(4)$ $f(x)$ is continuous.

    c) $f(x) = \dfrac{x - 3}{x^2 - 9}$ at $x = 3$

        1) $f(3) = \dfrac{3 - 3}{(3)^2 - 9}$

            With the denominator equal to zero, $f(x)$ is not defined at $x = 3$ and so cannot be continuous at $x = 3$ even though the limit exists at $x = 3$. See steps 2 and 3.

        2) $\lim_{x \to 3} \dfrac{x - 3}{x^2 - 9} = \lim_{x \to 3} \dfrac{x - 3}{(x + 3)(x - 3)} = \lim_{x \to 3} \dfrac{1}{x + 3} = \dfrac{1}{6}$

        3) $\lim_{x \to 3} f(x) = \frac{1}{6} \neq f(3)$. So $f(x)$ is discontinuous at $x = 3$.

DERIVATIVE NOTATION AND SIMPLE DERIVATIVES

3.6. Differentiate each of the following functions and practice the use of the different notations for a derivative.

    a) $f(x) = 17$ *b)* $y = -12$

         $f'(x) = 0$ (constant rule) $\dfrac{dy}{dx} = 0$

    c) $y = 5x + 12$ *d)* $f(x) = 9x - 6$

         $y' = 5$ (linear function rule) $f' = 9$

3.7. Differentiate each of the following functions using the power function rule. Continue to use the different notations.

    a) $y = 8x^3$ *b)* $f(x) = -6x^5$

         $\dfrac{d}{dx}(8x^3) = 24x^2$ $f' = -30x^4$

c) $f(x) = 5x^{-2}$

$$f'(x) = 5(-2) \cdot x^{[-2-(1)]} = -10x^{-3} = -\frac{10}{x^3}$$

d) $y = -9x^{-4}$

$$\frac{dy}{dx} = -9(-4) \cdot x^{[-4-(1)]} = 36x^{-5} = \frac{36}{x^5}$$

e) $y = \dfrac{7}{x} = 7x^{-1}$

$$D_x(7x^{-1}) = 7(-1)x^{-2} = -7x^{-2} = -\frac{7}{x^2}$$

f) $f(x) = 18\sqrt{x} = 18x^{1/2}$

$$\frac{df}{dx} = 18\left(\frac{1}{2}\right) \cdot x^{1/2-1} = 9x^{-1/2} = \frac{9}{\sqrt{x}}$$

3.8. Use the rule for sums and differences to differentiate the following functions. Treat the dependent variable on the left as y and the independent variable on the right as x.

a) $R = 8t^2 + 5t - 6$ b) $C = 4t^3 - 9t^2 + 28t - 68$

$$\frac{dR}{dt} = 16t + 5 \qquad\qquad C' = 12t^2 - 18t + 28$$

c) $p = 6q^5 - 3q^3$ d) $q = 7p^4 + 15p^{-3}$

$$\frac{dp}{dq} = 30q^4 - 9q^2 \qquad D_p(7p^4 + 15p^{-3}) = 28p^3 - 45p^{-4}$$

THE PRODUCT RULE

3.9. Given $y = f(x) = 5x^4(3x - 7)$, (a) use the product rule to find the derivative. (b) Simplify the original function first and then find the derivative. (c) Compare the two derivatives.

a) Recalling the formula for the product rule from (3.3),

$$f'(x) = g(x) \cdot h'(x) + h(x) \cdot g'(x)$$

let $g(x) = 5x^4$ and $h(x) = 3x - 7$. Then $g'(x) = 20x^3$ and $h'(x) = 3$. Substitute these values in the product − rule formula.

$$y' = f'(x) = 5x^4(3) + (3x - 7)(20x^3)$$

Simplify algebraically.

$$y' = 15x^4 + 60x^4 - 140x^3 = 75x^4 - 140x^3$$

b) Simplify the original function by multiplication.

$$y = 5x^4(3x - 7) = 15x^5 - 35x^4$$

Take the derivative.

$$y' = 75x^4 - 140x^3$$

c) The derivatives found in parts (a) and (b) are identical. The derivative of a product can be found by either method, but as the functions grow more complicated, the product rule becomes more useful. Knowledge of another method helps to check answers.

3.10. Redo Problem 3.9, given $y = f(x) = (x^8 + 8)(x^6 + 11)$.

a) Let $g(x) = x^8 + 8$ and $h(x) = x^6 + 11$. Then $g'(x) = 8x^7$ and $h'(x) = 6x^5$. Substituting these values in *(3.3)*,

$$y' = f'(x) = (x^8 + 8)(6x^5) + (x^6 + 11)(8x^7)$$
$$= 6x^{13} + 48x^5 + 8x^{13} + 88x^7 = 14x^{13} + 88x^7 + 48x^5$$

b) Simplifying first through multiplication,

$$y = (x^8 + 8)(x^6 + 11) = x^{14} + 11x^8 + 8x^6 + 88$$

Then
$$y' = 14x^{13} + 88x^7 + 48x^5$$

c) The derivatives are identical.

3.11. Differentiate each of the following functions using the product rule. *Note:* The choice of problems is purposely kept simple in this and other sections of the book to enable students to see how various rules work. While it is proper and often easier to simplify a function algebraically before taking the derivative, applying the rules to the problems as given in the long run will help the student to master the rules more efficiently.

a) $y = (4x^2 - 3)(2x^5)$

$$\frac{dy}{dx}(4x^2 - 3)(10x^4) + 2x^5(8x) = 40x^6 - 30x^4 + 16x^6 = 56x^6 - 30x^4$$

b) $y = 7x^9(3x^2 - 12)$

$$\frac{dy}{dx} = 7x^9(6x) + (3x^2 - 12)(63x^8) = 42x^{10} + 189x^{10} - 756x^8 = 231x^{10} - 756x^8$$

c) $y = (2x^4 + 5)(3x^5 - 8)$

$$\frac{dy}{dx} = (2x^4 + 5)(15x^5) + (3x^5 - 8)(8x^3) = 30x^8 + 75x^4 + 24x^8 - 64x^3 = 54x^8 + 75x^4 - 64x^2$$

d) $z = (3 - 12t^3)(5 + 4t^6)$

$$\frac{dz}{dt} = (3 - 12t^3)(24t^5) + (5 + 4t^6)(-36t^2) = 72t^5 - 288t^8 - 180t^2 - 144t^8 = -432t^8 + 72t^5 - 180t^2$$

QUOTIENT RULE

3.12. Given

$$y = \frac{10x^8 - 6x^7}{2x}$$

(a) Find the derivative directly, using the quotient rule. *(b)* Simplify the function by division and then take its derivative. *(c)* Compare the two derivatives.

a) From *(3.4)*, the formula for the quotient rule is

$$f'(x) = \frac{h(x) \cdot g'(x) - g(x) \cdot h'(x)}{[h(x)]^2}$$

where $g(x) =$ the numerator $= 10x^8 - 6x^7$ and $h(x) =$ the denominator $= 2x$. Take the individual derivatives.

$$g'(x) = 80x^7 - 42x^6 \qquad h'(x) = 2$$

Substitute in the formula,

$$y' = \frac{2x(80x^7 - 42x^6) - (10x^8 - 6x^7)(2)}{(2x)^2}$$

$$= \frac{160x^8 - 84x^7 - 20x^8 + 12x^7}{4x^2} = \frac{140x^8 - 72x^7}{4x^2}$$

$$= 35x^6 - 18x^5$$

b) Simplifying the original function first by division,

$$y = \frac{10x^8 - 6x^7}{2x} = 5x^7 - 3x^6$$

$$y' = 35x^6 - 18x^5$$

c) The derivatives will always be the same if done correctly, but as functions grow in complexity, the quotient rule becomes more important. A second method is also a way to check answers.

3.13. Differentiate each of the following functions by means of the quotient rule. Continue to apply the rules to the functions as given. Later, when all the rules have been mastered, the functions can be simplified first and the easiest rule applied.

a) $y = \dfrac{3x^8 - 4x^7}{4x^3}$

Here $g(x) = 3x^8 - 4x^7$ and $h(x) = 4x^3$. Thus, $g'(x) = 24x^7 - 28x^6$ and $h'(x) = 12x^2$. Substituting in the quotient formula,

$$y' = \frac{4x^3(24x^7 - 28x^6) - (3x^8 - 4x^7)(12x^2)}{(4x^3)^2}$$

$$= \frac{96x^{10} - 112x^9 - 36x^{10} + 48x^9}{16x^6} = \frac{60x^{10} - 64x^9}{16x^6} = 3.75x^4 - 4x^3$$

b) $y = \dfrac{4x^5}{1 - 3x}$ $(x \neq \tfrac{1}{3})$

(*Note*: The qualifying statement is added because if $x = \tfrac{1}{3}$, the denominator would equal zero and the function would be undefined.)

$$\frac{dy}{dx} = \frac{(1 - 3x)(20x^4) - 4x^5(-3)}{(1 - 3x)^2} = \frac{20x^4 - 60x^5 + 12x^5}{(1 - 3x)^2} = \frac{20x^4 - 48x^5}{(1 - 3x)^2}$$

c) $y = \dfrac{15x^2}{2x^2 + 7x - 3}$

$$\frac{dy}{dx} = \frac{(2x^2 + 7x - 3)(30x) - 15x^2(4x + 7)}{(2x^2 + 7x - 3)^2}$$

$$= \frac{60x^3 + 210x^2 - 90x - 60x^3 - 105x^2}{(2x^2 + 7x - 3)^2} = \frac{105x^2 - 90x}{(2x^2 + 7x - 3)^2}$$

d) $y = \dfrac{6x - 7}{8x - 5}$ $(x \neq \tfrac{5}{8})$

$$\frac{dy}{dx} = \frac{(8x - 5)(6) - (6x - 7)(8)}{(8x - 5)^2} = \frac{48x - 30 - 48x + 56}{(8x - 5)^2} = \frac{26}{(8x - 5)^2}$$

e) $y = \dfrac{5x^2 - 9x + 8}{x^2 + 1}$

$$\frac{dy}{dx} = \frac{(x^2 + 1)(10x - 9) - (5x^2 - 9x + 8)(2x)}{(x^2 + 1)^2}$$

$$= \frac{10x^3 - 9x^2 + 10x - 9 - 10x^3 + 18x^2 - 16x}{(x^2 + 1)^2} = \frac{9x^2 - 6x - 9}{(x^2 + 1)^2}$$

THE GENERALIZED POWER FUNCTION RULE

3.14. Given $y = (5x + 8)^2$, (*a*) use the generalized power function rule to find the derivative; (*b*) simplify the function first by squaring it and then take the derivative; (*c*) compare answers.

a) From the generalized power function rule in (*3.5*), if $f(x) = [g(x)]^n$,

$$f'(x) = n[g(x)]^{n-1} \cdot g'(x)$$

Here $g(x) = 5x + 8$, $g'(x) = 5$, and $n = 2$. Substitute these values in the generalized power function rule,

$$y' = 2(5x + 8)^{2-1} \cdot 5 = 10(5x + 8) = 50x + 80$$

b) Square the function first and then take the derivative,

$$y = (5x + 8)(5x + 8) = 25x^2 + 80x + 64$$
$$y' = 50x + 80$$

c) The derivatives are identical. But for higher, negative, and fractional values of n, the generalized power function rule is faster and more practical.

3.15. Find the derivative for each of the following functions with the help of the generalized power function rule.

a) $y = (6x^3 + 9)^4$

Here $g(x) = 6x^3 + 9$, $g'(x) = 18x^2$, and $n = 4$. Substitute in the generalized power function rule,

$$y' = 4(6x^3 + 9)^{4-1} \cdot 18x^2$$
$$= 4(6x^3 + 9)^3 \cdot 18x^2 = 72x^2(6x^3 + 9)^3$$

b) $y = (2x^2 - 5x + 7)^3$

$$y' = 3(2x^2 - 5x + 7)^2 \cdot (4x - 5)$$
$$= (12x - 15)(2x^2 - 5x + 7)^2$$

c) $y = \dfrac{1}{7x^3 + 13x + 3}$

First convert the function to an easier equivalent form,

$$y = (7x^3 + 13x + 3)^{-1}$$

then use the generalized power function rule,

$$y' = -1(7x^3 + 13x + 3)^{-2} \cdot (21x^2 + 13)$$
$$= -(21x^2 + 13)(7x^3 + 13x + 3)^{-2}$$
$$= \frac{-(21x^2 + 13)}{(7x^3 + 13x + 3)^2}$$

d) $y = \sqrt{34 - 6x^2}$

Convert the radical to a power function, then differentiate.

$$y = (34 - 6x^2)^{1/2}$$
$$y' = \tfrac{1}{2}(34 - 6x^2)^{-1/2} \cdot (-12x)$$
$$= -6x(34 - 6x^2)^{-1/2} = \frac{-6x}{\sqrt{34 - 6x^2}}$$

e) $y = \dfrac{1}{\sqrt{4x^3 + 94}}$

Convert to an equivalent form; then take the derivative.

$$y = (4x^3 + 94)^{-1/2}$$
$$y' = -\tfrac{1}{2}(4x^3 + 94)^{-3/2} \cdot (12x^2) = -6x^2(4x^3 + 94)^{-3/2}$$
$$= \frac{-6x^2}{(4x^3 + 94)^{3/2}} = \frac{-6x^2}{\sqrt{(4x^3 + 94)^3}}$$

CHAIN RULE

3.16. Use the chain rule to find the derivative dy/dx for each of the following functions of a function. Check each answer on your own with the generalized power function rule, noting that the generalized power function rule is simply a specialized use of the chain rule.

a) $y = (3x^4 + 5)^6$

Let $y = u^6$ and $u = 3x^4 + 5$. Then $dy/du = 6u^5$ and $du/dx = 12x^3$. From the chain rule in (*3.6*),

$$\frac{dy}{dx} = \frac{dy}{du}\frac{du}{dx}$$

Substituting,
$$\frac{dy}{dx} = 6u^5 \cdot 12x^3 = 72x^3 u^5$$

But $u = 3x^4 + 5$. Substituting again,

$$\frac{dy}{dx} = 72x^3(3x^4 + 5)^5$$

b) $y = (7x + 9)^2$

Let $y = u^2$ and $u = 7x + 9$, then $dy/du = 2u$ and $du/dx = 7$. Substitute these values in the chain rule,

$$\frac{dy}{dx} = 2u \cdot 7 = 14u$$

Then substitute $7x + 9$ for u.

$$\frac{dy}{dx} = 14(7x + 9) = 98x + 126$$

c) $y = (4x^5 - 1)^7$

Let $y = u^7$ and $u = 4x^5 - 1$; then $dy/du = 7u^6$, $du/dx = 20x^4$, and

$$\frac{dy}{dx} = 7u^6 \cdot 20x^4 = 140x^4 u^6$$

Substitute $u = 4x^5 - 1$.

$$\frac{dy}{dx} = 140x^4(4x^5 - 1)^6$$

3.17. Redo Problem 3.16, given:

  a) $y = (x^2 + 3x - 1)^5$

   Let $y = u^5$ and $u = x^2 + 3x - 1$, then $dy/du = 5u^4$ and $du/dx = 2x + 3$. Substitute in *(3.6)*.

$$\frac{dy}{dx} = 5u^4(2x + 3) = (10x + 15)u^4$$

   But $u = x^2 + 3x - 1$. Therefore,

$$\frac{dy}{dx} = (10x + 15)(x^2 + 3x - 1)^4$$

  b) $y = -3(x^2 - 8x + 7)^4$

   Let $y = -3u^4$ and $u = x^2 - 8x + 7$. Then $dy/du = -12u^3$, $du/dx = 2x - 8$, and

$$\frac{dy}{dx} = -12u^3(2x - 8) = (-24x + 96)u^3$$

$$= (-24x + 96)(x^2 - 8x + 7)^3$$

COMBINATION OF RULES

3.18. Use whatever combination of rules is necessary to find the derivatives of the following functions. Do not simplify the original functions first. They are deliberately kept simple to facilitate the practice of the rules.

  a) $y = \dfrac{3x(2x - 1)}{5x - 2}$

   The function involves a quotient with a product in the numerator. Hence both the quotient rule and the product rule are required. Start with the quotient rule from *(3.4)*.

$$y' = \frac{h(x) \cdot g'(x) - g(x) \cdot h'(x)}{[h(x)]^2}$$

   where $g(x) = 3x(2x - 1)$, $h(x) = 5x - 2$, and $h'(x) = 5$. Then use the product rule from *(3.3)* for $g'(x)$.

$$g'(x) = 3x \cdot 2 + (2x - 1) \cdot 3 = 12x - 3$$

   Substitute the appropriate values in the quotient rule.

$$y' = \frac{(5x - 2)(12x - 3) - [3x(2x - 1)] \cdot 5}{(5x - 2)^2}$$

   Simplify algebraically.

$$y' = \frac{60x^2 - 15x - 24x + 6 - 30x^2 + 15x}{(5x - 2)^2} = \frac{30x^2 - 24x + 6}{(5x - 2)^2}$$

   Note: To check this answer one could let

$$y = 3x \cdot \frac{2x - 1}{5x - 2} \qquad \text{or} \qquad y = \frac{3x}{5x - 2} \cdot (2x - 1)$$

   and use the product rule involving a quotient.

b) $y = 3x(4x - 5)^2$

    The function involves a product in which one function is raised to a power. Both the product rule and the generalized power function rule are needed. Starting with the product rule,

$$y' = g(x) \cdot h'(x) + h(x) \cdot g'(x)$$

where $\qquad\qquad g(x) = 3x \qquad h(x) = (4x - 5)^2 \qquad$ and $\qquad g'(x) = 3$

Use the generalized power function rule for $h'(x)$.

$$h'(x) = 2(4x - 5) \cdot 4 = 8(4x - 5) = 32x - 40$$

Substitute the appropriate values in the product rule,

$$y' = 3x \cdot (32x - 40) + (4x - 5)^2 \cdot 3$$

and simplify algebraically,

$$y' = 96x^2 - 120x + 3(16x^2 - 40x + 25) = 144x^2 - 240x + 75$$

c) $y = (3x - 4) \cdot \dfrac{5x + 1}{2x + 7}$

    Here we have a product involving a quotient. Both the product rule and the quotient rule are needed. Start with the product rule,

$$y' = g(x) \cdot h'(x) + h(x) \cdot g'(x)$$

where $\qquad\qquad g(x) = 3x - 4 \qquad h(x) = \dfrac{5x + 1}{2x + 7} \qquad$ and $\qquad g'(x) = 3$

and use the quotient rule for $h'(x)$.

$$h'(x) = \frac{(2x + 7)(5) - (5x + 1)(2)}{(2x + 7)^2} = \frac{33}{(2x + 7)^2}$$

Substitute the appropriate values in the product rule,

$$y' = (3x - 4) \cdot \frac{33}{(2x + 7)^2} + \frac{5x + 1}{2x + 7} \cdot 3 = \frac{99x - 132}{(2x + 7)^2} + \frac{15x + 3}{2x + 7}$$

$$= \frac{99x - 132 + (15x + 3)(2x + 7)}{(2x + 7)^2} = \frac{30x^2 + 210x - 111}{(2x + 7)^2}$$

One could check this answer by letting $y = (3x - 4)(5x + 1)/(2x + 7)$ and using the quotient rule involving a product.

d) $y = \dfrac{(8x - 5)^3}{(7x + 4)}$

    Start with the quotient rule, where

$$g(x) = (8x - 5)^3 \qquad h(x) = 7x + 4 \qquad h'(x) = 7$$

and use the generalized power function rule for $g'(x)$,

$$g'(x) = 3(8x - 5)^2 \cdot 8 = 24(8x - 5)^2$$

Substitute these values in the quotient rule,

$$y' = \frac{(7x + 4) \cdot 24(8x - 5)^2 - (8x - 5)^3 \cdot 7}{(7x + 4)^2}$$

$$= \frac{(168x + 96)(8x - 5)^2 - 7(8x - 5)^3}{(7x + 4)^2}$$

To check this answer, one could let $y = (8x - 5)^3 \cdot (7x + 4)^{-1}$ and use the product rule involving the generalized power function rule twice.

e) $y = \left(\dfrac{3x + 4}{2x + 5} \right)^2$

Start with the generalized power function rule,

$$y' = 2\left(\frac{3x + 4}{2x + 5} \right) \cdot \frac{d}{dx}\left(\frac{3x + 4}{2x + 5} \right) \tag{3.9}$$

Then use the quotient rule,

$$\frac{d}{dx}\left(\frac{3x + 4}{2x + 5} \right) = \frac{(2x + 5)(3) - (3x + 4)(2)}{(2x + 5)^2} = \frac{7}{(2x + 5)^2}$$

and substitute this value in (3.9),

$$y' = 2\left(\frac{3x + 4}{2x + 5} \right) \cdot \frac{7}{(2x + 5)^2} = \frac{14(3x + 4)}{(2x + 5)^3} = \frac{42x + 56}{(2x + 5)^3}$$

To check this answer, let $y = (3x + 4)^2 \cdot (2x + 5)^{-2}$, and use the product rule involving the generalized power function rule twice.

3.19.   Differentiate each of the following, using whatever rules are necesary:

a) $y = (5x - 1)(3x + 4)^3$

Using the product rule together with the generalized power function rule,

$$\frac{dy}{dx} = (5x - 1)[3(3x + 4)^2(3)] + (3x + 4)^3(5)$$

Simplifying algebraically,

$$\frac{dy}{dx} = (5x - 1)(9)(3x + 4)^2 + 5(3x + 4)^3 = (45x - 9)(3x + 4)^2 + 5(3x + 4)^3$$

b) $y = \dfrac{(9x^2 - 2)(7x + 3)}{5x}$

Using the quotient rule along with the product rule,

$$y' = \frac{5x[(9x^2 - 2)(7) + (7x + 3)(18x)] - (9x^2 - 2)(7x + 3)(5)}{(5x)^2}$$

Simplifying algebraically,

$$y' = \frac{5x(63x^2 - 14 + 126x^2 + 54x) - 5(63x^3 + 27x^2 - 14x - 6)}{25x^2} = \frac{630x^3 + 135x^2 + 30}{25x^2}$$

c) $y = \dfrac{15x + 23}{(3x + 1)^2}$

Using the quotient rule plus the generalized power function rule,

$$y' = \frac{(3x + 1)^2(15) - (15x + 23)[2(3x + 1)(3)]}{(3x + 1)^4}$$

Simplifying algebraically,

$$y' = \frac{15(3x + 1)^2 - (15x + 23)(18x + 6)}{(3x + 1)^4} = \frac{-135x^2 - 414x - 123}{(3x + 1)^4}$$

d) $y = (6x + 1)\dfrac{4x}{9x - 1}$

Using the product rule and the quotient rule,

$$D_x = (6x + 1)\frac{(9x - 1)(4) - 4x(9)}{(9x - 1)^2} + \frac{4x}{9x - 1} \quad (6)$$

Simplifying algebraically,

$$D_x = \frac{(6x + 1)(36x - 4 - 36x)}{(9x - 1)^2} + \frac{24x}{9x - 1} = \frac{216x^2 - 48x - 4}{(9x - 1)^2}$$

e) $y = \left(\dfrac{3x - 1}{2x + 5}\right)^3$

Using the generalized power function rule and the quotient rule,

$$y' = 3\left(\frac{3x - 1}{2x + 5}\right)^2 \frac{(2x + 5)(3) - (3x - 1)(2)}{(2x + 5)^2}$$

Simplifying algebraically,

$$y' = \frac{3(3x - 1)^2}{(2x + 5)^2} \frac{17}{(2x + 5)^2} = \frac{51(3x - 1)^2}{(2x + 5)^4}$$

HIGHER-ORDER DERIVATIVES

3.20. For each of the following functions, (1) find the second-order derivative and (2) evaluate it at $x = 2$. Practice the use of the different second-order notations.

a) $y = 7x^3 + 5x^2 + 12$

1) $\dfrac{dy}{dx} = 21x^2 + 10x$

$\dfrac{d^2y}{dx^2} = 42x + 10$

2) At $x = 2$, $\dfrac{d^2y}{dx^2} = 42(2) + 10$

$= 94$

b) $f(x) = x^6 + 3x^4 + x$

1) $f'(x) = 6x^5 + 12x^3 + 1$

$f''(x) = 30x^4 + 36x^2$

2) At $x = 2$, $f''(x) = 30(2)^4 + 36(2)^2$

$= 624$

c) $y = (2x + 3)(8x^2 - 6)$

1) $Dy = (2x + 3)(16x) + (8x^2 - 6)(2)$

$= 32x^2 + 48x + 16x^2 - 12$

$= 48x^2 + 48x - 12$

$D^2y = 96x + 48$

2) At $x = 2$, $D^2y = 96(2) + 48$

$= 240$

d) $f(x) = (x^4 - 3)(x^3 - 2)$

1) $f' = (x^4 - 3)(3x^2) + (x^3 - 2)(4x^3)$

$= 3x^6 - 9x^2 + 4x^6 - 8x^3$

$= 7x^6 - 8x^3 - 9x^2$

$f'' = 42x^5 - 24x^2 - 18x$

2) At $x = 2$, $f'' = 42(2)^5 - 24(2)^2 - 18(2)$

$= 1212$

e) $y = \dfrac{5x}{1 - 3x}$

$\quad$ 1) $y' = \dfrac{(1 - 3x)(5) - 5x(-3)}{(1 - 3x)^2}$

$\qquad\qquad = \dfrac{5 - 15x + 15x}{(1 - 3x)^2} = \dfrac{5}{(1 - 3x)^2}$

$\qquad y'' = \dfrac{(1 - 3x)^2(0) - 5[2(1 - 3x)(-3)]}{(1 - 3x)^4}$

$\qquad\qquad = \dfrac{-5(-6 + 18x)}{(1 - 3x)^4} = \dfrac{30 - 90x}{(1 - 3x)^4} = \dfrac{30}{(1 - 3x)^3}$

$\quad$ 2) At $x = 2$, $y'' = \dfrac{30 - 90(2)}{[1 - 3(2)]^4}$

$\qquad\qquad\qquad\qquad = \dfrac{-150}{(-5)^4}$

$\qquad\qquad\qquad\qquad = -\dfrac{6}{25}$

f) $y = \dfrac{7x^2}{x - 1}$

$\quad$ 1) $y' = \dfrac{(x - 1)(14x) - 7x^2(1)}{(x - 1)^2}$

$\qquad\qquad = \dfrac{14x^2 - 14x - 7x^2}{(x - 1)^2} = \dfrac{7x^2 - 14x}{(x - 1)^2}$

$\qquad y'' = \dfrac{(x - 1)^2(14x - 14) - (7x^2 - 14x)[2(x - 1)(1)]}{(x - 1)^4}$

$\qquad\qquad = \dfrac{(x^2 - 2x + 1)(14x - 14) - (7x^2 - 14x)(2x - 2)}{(x - 1)^4}$

$\qquad\qquad = \dfrac{14(x - 1)}{(x - 1)^4} = \dfrac{14}{(x - 1)^3}$

$\quad$ 2) At $x = 2$, $y'' = \dfrac{14}{(2 - 1)^3}$

$\qquad\qquad\qquad\qquad = 14$

g) $f(x) = (8x - 4)^3$

$\quad$ 1) $f' = 3(8x - 4)^2(8)$

$\qquad\qquad = 24(8x - 4)^2$

$\qquad f'' = 2(24)(8x - 4)(8)$

$\qquad\qquad = 384(8x - 4)$

$\quad$ 2) At $x = 2$, $f'' = 384[8(2) - 4]$

$\qquad\qquad\qquad\qquad = 4608$

h) $y = (5x^3 - 7x^2)^2$

$\quad$ 1) $Dy = 2(5x^3 - 7x^2)(15x^2 - 14x)$

$\qquad\qquad = 150x^5 - 350x^4 + 196x^3$

$\qquad D^2y = 750x^4 - 1400x^3 + 588x^2$

$\quad$ 2) At $x = 2$, $D^2y = 750(2)^4 - 1400(2)^3 + 588(2)^2$

$\qquad\qquad\qquad\qquad = 3152$

3.21. For each of the following functions, (1) investigate the successive derivatives and (2) evaluate them at $x = 3$.

a) $y = x^3 + 3x^2 + 9x - 7$

$\quad$ 1) $y' = 3x^2 + 6x + 9$

$\qquad y'' = 6x + 6$

$\qquad y''' = 6$

$\qquad y^{(4)} = 0$

$\quad$ 2) At $x = 3$, $y' = 3(3)^2 + 6(3) + 9 = 54$

$\qquad\qquad\qquad y'' = 6(3) + 6 = 24$

$\qquad\qquad\qquad y''' = 6$

$\qquad\qquad\qquad y^{(4)} = 0$

b) $y = (4x - 7)(9x + 2)$

$\quad$ 1) $y' = (4x - 7)(9) + (9x + 2)(4)$

$\qquad\qquad = 36x - 63 + 36x + 8 = 72x - 55$

$\qquad y'' = 72$

$\qquad y''' = 0$

$\quad$ 2) At $x = 3$, $y' = 72(3) - 55 = 161$

$\qquad\qquad\qquad y'' = 72$

$\qquad\qquad\qquad y''' = 0$

c) $y = (5 - x)^4$

1) $D_x = 4(5 - x)^3(-1) = -4(5 - x)^3$
    $D_x^2 = -12(5 - x)^2(-1) = 12(5 - x)^2$
    $D_x^3 = 24(5 - x)(-1)$
    $\quad = -24(5 - x) = 24x - 120$
    $D_x^4 = 24$
    $D_x^5 = 0$

2) At $x = 3$, $D_x = -4(5 - 3)^3 = -32$
    $D_x^2 = 12(5 - 3)^2 = 48$
    $D_x^3 = 24(3) - 120 = -48$
    $D_x^4 = 24$
    $D_x^5 = 0$

IMPLICIT DIFFERENTIATION

3.22. Use implicit differentiation to find the derivative dy/dx for each of the following equations.

a) $4x^2 - y^3 = 97$

Take the derivative with respect to x of both sides,

$$\frac{d}{dx}(4x^2) - \frac{d}{dx}(y^3) = \frac{d}{dx}(97) \tag{3.10}$$

where $\frac{d}{dx}(4x^2) = 8x$, $\frac{d}{dx}(97) = 0$, and use the generalized power function rule because y is considered a function of x,

$$\frac{d}{dx}(y^3) = 3 \cdot y^2 \cdot \frac{d}{dx}(y)$$

Set these values in (*3.10*) and recall that $\frac{d}{dx}(y) = \frac{dy}{dx}$.

$$8x - 3y^2\left(\frac{dy}{dx}\right) = 0$$

$$-3y^2\left(\frac{dy}{dx}\right) = -8x$$

$$\frac{dy}{dx} = \frac{8x}{3y^2}$$

b) $3y^5 - 6y^4 + 5x^6 = 243$

Taking the derivative with respect to x of both sides,

$$\frac{d}{dx}(3y^5) - \frac{d}{dx}(6y^4) + \frac{d}{dx}(5x^6) = \frac{d}{dx}(243)$$

$$15y^4\left(\frac{dy}{dx}\right) - 24y^3\left(\frac{dy}{dx}\right) + 30x^5 = 0$$

Solve for dy/dx,

$$(15y^4 - 24y^3)\left(\frac{dy}{dx}\right) = -30x^5$$

$$\frac{dy}{dx} = \frac{-30x^5}{15y^4 - 24y^3}$$

c) $2x^4 + 7x^3 + 8y^5 = 136$

$$\frac{d}{dx}(2x^4) + \frac{d}{dx}(7x^3) + \frac{d}{dx}(8y^5) = \frac{d}{dx}(136)$$

$$8x^3 + 21x^2 + 40y^4\left(\frac{dy}{dx}\right) = 0$$

$$40y^4\left(\frac{dy}{dx}\right) = -(8x^3 + 21x^2)$$

$$\frac{dy}{dx} = \frac{-(8x^3 + 21x^2)}{40y^4}$$

3.23. Use the different rules of differentiation in implicit differentiation to find dy/dx for each of the following:

a) $x^4 y^6 = 89$

$$\frac{d}{dx}(x^4 y^6) = \frac{d}{dx}(89)$$

Use the product rule and the generalized power function rule.

$$x^4 \cdot \frac{d}{dx}(y^6) + y^6 \cdot \frac{d}{dx}(x^4) = \frac{d}{dx}(89)$$

$$x^4 \cdot 6y^5\frac{dy}{dx} + y^6 \cdot 4x^3 = 0$$

Solve algebraically for dy/dx.

$$6x^4 y^5\frac{dy}{dx} = -4x^3 y^6$$

$$\frac{dy}{dx} = \frac{-4x^3 y^6}{6x^4 y^5} = \frac{-2y}{3x}$$

b) $2x^3 + 5xy + 6y^2 = 87$

$$\frac{d}{dx}(2x^3 + 5xy + 6y^2) = \frac{d}{dx}(87)$$

Note that the derivative of $5xy$ requires the product rule.

$$6x^2 + \left[5x \cdot \left(\frac{dy}{dx}\right) + y \cdot (5)\right] + 12y\left(\frac{dy}{dx}\right) = 0$$

Solving algebraically for dy/dx

$$(5x + 12y)\left(\frac{dy}{dx}\right) = -6x^2 - 5y$$

$$\frac{dy}{dx} = \frac{-(6x^2 + 5y)}{5x + 12y}$$

c) $7x^4 + 3x^3 y + 9xy^2 = 496$

$$28x^3 + \left[3x^3 \cdot \left(\frac{dy}{dx}\right) + y \cdot 9x^2\right] + \left[9x \cdot 2y\left(\frac{dy}{dx}\right) + y^2 \cdot 9\right] = 0$$

$$28x^3 + 3x^3\left(\frac{dy}{dx}\right) + 9x^2 y + 18xy\left(\frac{dy}{dx}\right) + 9y^2 = 0$$

$$(3x^3 + 18xy)\left(\frac{dy}{dx}\right) = -28x^3 - 9x^2 y - 9y^2$$

$$\frac{dy}{dx} = \frac{-(28x^3 + 9x^2 y + 9y^2)}{3x^3 + 18xy}$$

d) $(5y - 21)^3 = 6x^5$

$$\frac{d}{dx}[(5y - 21)^3] = \frac{d}{dx}(6x^5)$$

Use the generalized power function rule.

$$3(5y - 21)^2 \cdot 5\left(\frac{dy}{dx}\right) = 30x^4$$

$$15(5y - 21)^2\left(\frac{dy}{dx}\right) = 30x^4$$

$$\frac{dy}{dx} = \frac{30x^4}{15(5y - 21)^2}$$

e) $(2x^3 + 7y)^2 = x^5$

$$\frac{d}{dx}(2x^3 + 7y)^2 = \frac{d}{dx}(x^5)$$

$$2(2x^3 + 7y) \cdot \frac{d}{dx}(2x^3 + 7y) = 5x^4$$

$$(4x^3 + 14y)\left[6x^2 + 7\left(\frac{dy}{dx}\right)\right] = 5x^4$$

$$24x^5 + 28x^3\left(\frac{dy}{dx}\right) + 84x^2y + 98y\left(\frac{dy}{dx}\right) = 5x^4$$

$$(28x^3 + 98y)\frac{dy}{dx} = 5x^4 - 24x^5 - 84x^2y$$

$$\frac{dy}{dx} = \frac{5x^4 - 24x^5 - 84x^2y}{28x^3 + 98y}$$

See also Problems 4.24, 4.25, 5.20, 5.21, 6.51, and 6.52.

DERIVATION OF THE RULES OF DIFFERENTIATION

3.24. Given $f(x) = g(x) + h(x)$, where $g(x)$ and $h(x)$ are both differentiable functions, prove the rule of sums by demonstrating that $f'(x) = g'(x) + h'(x)$.

From (3.2) the derivative of $f(x)$ is

$$f'(x) = \lim_{\Delta x \to 0} \frac{f(x + \Delta x) - f(x)}{\Delta x}$$

Substituting $f(x) = g(x) + h(x)$,

$$f'(x) = \lim_{\Delta x \to 0} \frac{[g(x + \Delta x) + h(x + \Delta x)] - [g(x) + h(x)]}{\Delta x}$$

Rearrange terms.

$$f'(x) = \lim_{\Delta x \to 0} \frac{g(x + \Delta x) - g(x) + h(x + \Delta x) - h(x)}{\Delta x}$$

Separate terms, and take the limits.

$$f'(x) = \lim_{\Delta x \to 0}\left[\frac{g(x + \Delta x) - g(x)}{\Delta x} + \frac{h(x + \Delta x) - h(x)}{\Delta x}\right]$$

$$= \lim_{\Delta x \to 0}\frac{g(x + \Delta x) - g(x)}{\Delta x} + \lim_{\Delta x \to 0}\frac{h(x + \Delta x) - h(x)}{\Delta x}$$

$$= g'(x) + h'(x)$$

3.25. Given $f(x) = g(x) \cdot h(x)$, where $g'(x)$ and $h'(x)$ both exist, prove the product rule by demonstrating that $f'(x) = g(x) \cdot h'(x) + h(x) \cdot g'(x)$.

$$f'(x) = \lim_{\Delta x \to 0} \frac{f(x + \Delta x) - f(x)}{\Delta x}$$

Substitute $f(x) = g(x) \cdot h(x)$.

$$f'(x) = \lim_{\Delta x \to 0} \frac{g(x + \Delta x) \cdot h(x + \Delta x) - g(x) \cdot h(x)}{\Delta x}$$

Add and subtract $g(x + \Delta x) \cdot h(x)$,

$$f'(x) = \lim_{\Delta x \to 0} \frac{g(x + \Delta x)h(x + \Delta x) - g(x + \Delta x)h(x) + g(x + \Delta x)h(x) - g(x)h(x)}{\Delta x}$$

Partially factor out $g(x + \Delta x)$ and $h(x)$.

$$f'(x) = \lim_{\Delta x \to 0} \frac{g(x + \Delta x)[h(x + \Delta x) - h(x)] + h(x)[g(x + \Delta x) - g(x)]}{\Delta x}$$

$$= \lim_{\Delta x \to 0} \frac{g(x + \Delta x)[h(x + \Delta x) - h(x)]}{\Delta x} + \lim_{\Delta x \to 0} \frac{h(x)[g(x + \Delta x) - g(x)]}{\Delta x}$$

$$= \lim_{\Delta x \to 0} g(x + \Delta x) \cdot \lim_{\Delta x \to 0} \frac{h(x + \Delta x) - h(x)}{\Delta x} + \lim_{\Delta x \to 0} h(x) \cdot \lim_{\Delta x \to 0} \frac{g(x + \Delta x) - g(x)}{\Delta x}$$

$$= g(x) \cdot h'(x) + h(x) \cdot g'(x)$$

3.26. Given $f(x) = g(x)/h(x)$, where $g'(x)$ and $h'(x)$ both exist and $h(x) \neq 0$, prove the quotient rule by demonstrating

$$f'(x) = \frac{h(x) \cdot g'(x) - g(x) \cdot h'(x)}{[h(x)]^2}$$

Start with $f(x) = g(x)/h(x)$ and solve for $g(x)$,

$$g(x) = f(x) \cdot h(x)$$

Then take the derivative of $g(x)$, using the product rule,

$$g'(x) = f(x) \cdot h'(x) + h(x) \cdot f'(x)$$

and solve algebraically for $f'(x)$.

$$h(x) \cdot f'(x) = g'(x) - f(x) \cdot h'(x)$$

$$f'(x) = \frac{g'(x) - f(x) \cdot h'(x)}{h(x)}$$

Substitute $g(x)/h(x)$ for $f(x)$.

$$f'(x) = \frac{g'(x) - \dfrac{g(x) \cdot h'(x)}{h(x)}}{h(x)}$$

Now multiply both numerator and denominator by $h(x)$,

$$f'(x) = \frac{h(x) \cdot g'(x) - g(x) \cdot h'(x)}{[h(x)]^2}$$

CHAPTER 4

Uses of the Derivative in Mathematics and Economics

4.1 INCREASING AND DECREASING FUNCTIONS

A function $f(x)$ is said to be *increasing* (*decreasing*) at $x = a$ if in the immediate vicinity of the point $[a, f(a)]$ the graph of the function rises (falls) as it moves from left to right. Since the first derivative measures the rate of change and slope of a function, a positive first derivative at $x = a$ indicates the function is increasing at s; a negative first derivative indicates it is decreasing. In short, as seen in Fig. 4-1,

$$f'(a) > 0: \quad \text{increasing function at } x = a$$
$$f'(a) < 0: \quad \text{decreasing function at } x = a$$

A function that increases (or decreases) over its entire domain is called a *monotonic function*. It is said to increase (decrease) *monotonically*. See Problems 4.1 to 4.3.

4.2 CONCAVITY AND CONVEXITY

A function $f(x)$ is *concave* at $x = a$ if in some small region close to the point $[a, f(a)]$ the graph of the function lies completely below its tangent line. A function is *convex* at $x = a$ if in an area very close to $[a, f(a)]$ the graph of the function lies completely above its tangent line. A positive second derivative at $x = a$ denotes the function is convex at $x = a$; a negative second derivative at $x = a$ denotes the function is concave at a. The sign of the first derivative is irrelevant for concavity. In brief, as seen in Fig. 4-2 and Problems 4.1 to 4.4,

$$f''(a) > 0: \quad f(x) \text{ is convex at } x = a$$
$$f''(a) < 0: \quad f(x) \text{ is concave at } x = a$$

58

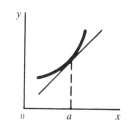

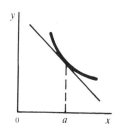

Slope > 0
Increasing function at $x = a$

(a)

Slope < 0
Decreasing function at $x = a$

(b)

Fig. 4-1

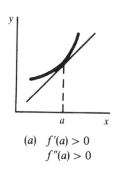

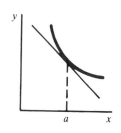

(a) $f'(a) > 0$
      $f''(a) > 0$

(b) $f'(a) < 0$
      $f''(a) > 0$

Convex at $x = a$

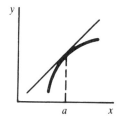

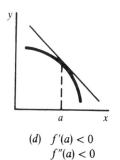

(d) $f'(a) < 0$
      $f''(a) < 0$

Concave at $x = a$

Fig. 4-2

If $f''(x) > 0$ for all x in the domain, $f(x)$ is *strictly convex*. If $f''(x) < 0$ for all x in the domain, $f(x)$ is *strictly concave*.

4.3 RELATIVE EXTREMA

A *relative extremum* is a point at which a function is at a relative maximum or minimum. To be at a relative maximum or minimum at a point a, the function must be at a relative *plateau*, i.e., neither increasing nor decreasing at a. If the function is neither increasing nor decreasing at a, the first derivative of the function at a must equal zero or be undefined. A point in the domain of a function where the derivative equals zero or is undefined is called a *critical point* or *value*.

To distinguish mathematically between a relative maximum and minimum, the *second-derivative test* is used. Assuming $f'(a) = 0$,

1. If $f''(a) > 0$, indicating that the function is convex and the graph of the function lies completely above its tangent line at $x = a$, the function is at a relative minimum at $x = a$.
2. If $f''(a) < 0$, denoting that the function is concave and the graph of the function lies completely below its tangent line at $x = a$, the function is at a relative maximum at $x = a$.
3. If $f''(a) = 0$, the test is inconclusive.

For functions which are differentiable at all values of x, called *differentiable or smooth functions*, one need only consider cases where $f'(x) = 0$ in looking for critical points. To summarize,

$$f'(a) = 0 \qquad f''(a) > 0: \qquad \text{relative minimum at } x = a$$
$$f'(a) = 0 \qquad f''(a) < 0: \qquad \text{relative maximum at } x = a$$

See Fig. 4-3 and Problems 4.5 and 4.6.

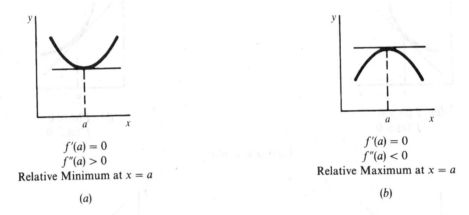

$f'(a) = 0$
$f''(a) > 0$
Relative Minimum at $x = a$

(a)

$f'(a) = 0$
$f''(a) < 0$
Relative Maximum at $x = a$

(b)

Fig. 4-3

4.4 INFLECTION POINTS

An *inflection point* is a point on the graph where the function crosses its tangent line and changes from concave to convex or vice versa. Inflection points occur only where the *second* derivative equals zero or is undefined. The sign of the first derivative is immaterial. In sum, for an inflection point at a, as seen in Fig. 4-4 and Problems 4.6 and 4.7(c),

1. $f''(a) = 0$ or is undefined.
2. Concavity changes at $x = a$.
3. Graph crosses its tangent line at $x = a$.

4.5 OPTIMIZATION OF FUNCTIONS

Optimization is the process of finding the relative maximum or minimum of a function. Without the aid of a graph, this is done with the techniques developed in Sections 4.3 through 4.4 and outlined below. Given the usual differentiable function,

1. Take the first derivative, set it equal to zero, and solve for the critical point(s). This step represents a necessary condition known as the *first-order condition*. It identifies all the points at which the function is neither increasing nor decreasing, but at a plateau. All such points are candidates for a possible relative maximum or minimum.

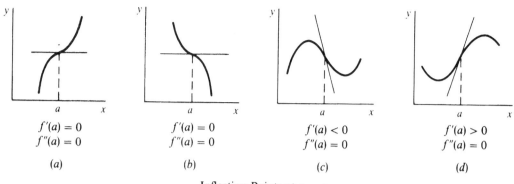

$f'(a) = 0$ $f'(a) = 0$ $f'(a) < 0$ $f'(a) > 0$
$f''(a) = 0$ $f''(a) = 0$ $f''(a) = 0$ $f''(a) = 0$

(a)   (b)   (c)   (d)

Inflection Points at $x = a$

Fig. 4-4

2. Take the second derivative, evaluate it at the critical point(s), and check the sign(s). If at a critical point a,

$f''(a) < 0$, the function is concave at a, and hence at a relative maximum.

$f''(a) > 0$, the function is convex at a, and hence at a relative minimum.

$f''(a) = 0$, the test is inconclusive. See Section 4.6.

Assuming the necessary first-order condition is met, this step, known as the *second-order derivative test*, or simply the *second-order condition*, represents a sufficiency condition. In sum,

Relative maximum	Relative minimum
$f'(a) = 0$	$f'(a) = 0$
$f''(a) < 0$	$f''(a) > 0$

Note that if the function is *strictly* concave (convex), there will be only one maximum (minimum), called a *global* maximum (minimum). See Example 1 and Problems 4.7 to 4.9.

EXAMPLE 1.   Optimize $f(x) = 2x^3 - 30x^2 + 126x + 59$.

a)   Find the critical points by taking the first derivative, setting it equal to zero, and solving for x.

$$f'(x) = 6x^2 - 60x + 126 = 0$$
$$6(x - 3)(x - 7) = 0$$
$$x = 3 \quad x = 7 \quad \text{critical points}$$

(b)   Test for concavity by taking the second derivative, evaluating it at the critical points, and checking the signs to distinguish betewen a relative maximum and minimum.

$$f''(x) = 12x - 60$$
$$f''(3) = 12(3) - 60 = -24 < 0 \quad \text{concave, relative maximum}$$
$$f''(7) = 12(7) - 60 = 24 > 0 \quad \text{convex, relative minimum}$$

The function is maximized at $x = 3$ and minimized at $x = 7$.

4.6   SUCCESSIVE-DERIVATIVE TEST FOR OPTIMIZATION

If $f''(a) = 0$, as in Fig. 4-4(a) through (d), the second-derivative test is inconclusive. In such cases, without a graph for guidance, the successive-derivative test is helpful:

1. If the first nonzero value of a higher-order derivative, when evaluated at a critical point, is an odd-numbered derivative (third, fifth, etc.), the function is at an inflection point. See Problems 4.6(b) and (d) and 4.7(c).

2. If the first nonzero value of a higher-order derivative, when evaluated at a critical point a, is an even-numbered derivative, the function is at a relative extremum at a, with a negative value of the derivative indicating that the function is concave and at a relative maximum and a positive value signifying the function is convex and at a relative minimum. See Problems 4.6(a) and (c), 4.7(d), and 4.9(c) and (d).

4.7 MARGINAL CONCEPTS

Marginal cost in economics is defined as the change in total cost incurred from the production of an additional unit. *Marginal revenue* is defined as the change in total revenue brought about by the sale of an extra good. Since total cost (TC) and total revenue (TR) are both functions of the level of output (Q), marginal cost (MC) and marginal revenue (MR) can each be expressed mathematically as derivatives of their respective total functions. Thus,

$$\text{if} \quad TC = TC(Q), \quad \text{then} \qquad MC = \frac{dTC}{dQ}$$

$$\text{and if} \quad TR = TR(Q), \quad \text{then} \qquad MR = \frac{dTR}{dQ}$$

In short, the marginal concept of any economic function can be expressed as the derivative of its total function. See Examples 2 and 3 and Problems 4.10 to 4.16.

EXAMPLE 2.

1. If $TR = 75Q - 4Q^2$, then $MR = dTR/dQ = 75 - 8Q$.
2. If $TC = Q^2 + 7Q + 23$, then $MC = dTC/dQ = 2Q + 7$.

EXAMPLE 3. Given the demand function $P = 30 - 2Q$, the marginal revenue function can be found by first finding the total revenue function and then taking the derivative of that function with respect to Q. Thus,

$$TR = PQ = (30 - 2Q)Q = 30Q - 2Q^2$$

Then
$$MR = \frac{dTR}{dQ} = 30 - 4Q$$

If $Q = 4$, $MR = 30 - 4(4) = 14$; if $Q = 5$, $MR = 30 - 4(5) = 10$.

4.8 OPTIMIZING ECONOMIC FUNCTIONS

The economist is frequently called upon to help a firm maximize profits and levels of physical output and productivity, as well as to minimize costs, levels of pollution, and the use of scarce natural resources. This is done with the help of techniques developed earlier and illustrated in Example 4 and Problems 4.17 to 4.23.

EXAMPLE 4. Maximize profits π for a firm, given total revenue $R = 4000Q - 33Q^2$ and total cost $C = 2Q^3 - 3Q^2 + 400Q + 5000$, assuming $Q > 0$.

a) Set up the profit function: $\pi = R - C$.

$$\pi = 4000Q - 33Q^2 - (2Q^3 - 3Q^2 + 400Q + 5000)$$
$$= -2Q^3 - 30Q^2 + 3600Q - 5000$$

b) Take the first derivative, set it equal to zero, and solve for Q to find the critical points.

$$\pi' = -6Q^2 - 60Q + 3600 = 0$$
$$= -6(Q^2 + 10Q - 600) = 0$$
$$= -6(Q + 30)(Q - 20) = 0$$
$$Q = -30 \qquad Q = 20 \qquad \text{critical points}$$

c) Take the second derivative; evaluate it at the positive critical point and ignore the negative critical point, which has no economic significance and will prove mathematically to be a relative minimum. Then check the sign for concavity to be sure of a relative maximum.

$$\pi'' = -12Q - 60$$
$$\pi''(20) = -12(20) - 60 = -300 < 0 \qquad \text{concave, relative maximum}$$

Profit is maximized at $Q = 20$ where

$$\pi(20) = -2(20)^3 - 30(20)^2 + 3600(20) - 5000 = 39{,}000$$

4.9 RELATIONSHIP AMONG TOTAL, MARGINAL, AND AVERAGE CONCEPTS

A *total product* (TP) *curve* of an input is derived from a production function by allowing the amounts of one input (say, capital) to vary while holding the other inputs (labor and land) constant. A graph showing the relationship between the total, average, and marginal products of an input can easily be sketched by using now familiar methods, as demonstrated in Example 5.

EXAMPLE 5. Given $\text{TP} = 90K^2 - K^3$, the relationship among the total, average, and marginal products can be illustrated graphically as follows.

1. Test the first-order condition to find the critical values.

$$\text{TP}' = 180K - 3K^2 = 0$$
$$3K(60 - K) = 0$$
$$K = 0 \qquad K = 60 \qquad \text{critical values}$$

Check the second-order conditions.

$$\text{TP}'' = 180 - 6K$$
$$\text{TP}''(0) = 180 > 0 \qquad \text{convex, relative minimum}$$
$$\text{TP}''(60) = -180 < 0 \qquad \text{concave, relative maximum}$$

Check for inflection points.

$$\text{TP}'' = 180 - 6K = 0$$
$$K = 30$$
$$K < 30 \qquad \text{TP}'' > 0 \qquad \text{convex}$$
$$K > 30 \qquad \text{TP}'' < 0 \qquad \text{concave}$$

Since, at $K = 30$, $\text{TP}'' = 0$ and concavity changes, there is an inflection point at $K = 30$.

2. Find and maximize the average product of capital AP_K.

$$\text{AP}_K = \frac{\text{TP}}{K} = 90K - K^2$$
$$\text{AP}_K' = 90 - 2K = 0$$
$$K = 45 \qquad \text{critical value}$$
$$\text{AP}_K'' = -2 < 0 \qquad \text{concave, relative maximum}$$

3. Find and maximize the marginal product of capital MP_K, recalling that $MP_K = TP' = 180K - 3K^2$:

$$MP'_K = 180 - 6K = 0$$
$$K = 30 \quad \text{critical value}$$
$$MP''_K = -6 < 0 \quad \text{concave, relative maximum}$$

4. Sketch the graphs, as in Fig. 4-5.

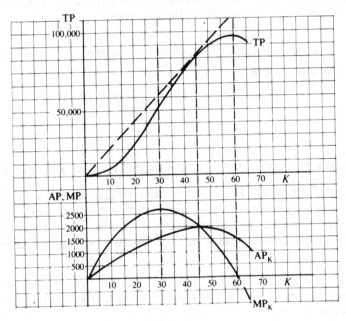

Fig. 4-5

Note that (a) MP_K increases when TP is convex and increasing at an increasing rate, is at a maximum where TP is at an inflection point, and decreases when TP is concave and increasing at a decreasing rate; (b) TP increases over the whole range where MP_K is positive, is at a maximum where $MP_K = 0$, and declines when MP_K is negative; (c) AP_K is at a maximum where the slope of a line from the origin to the TP curve is tangent to the TP curve, i.e., where $MP_K = AP_K$; (d) $MP_K > AP_K$ when AP_K is increasing, $MP_K = AP_K$ when AP_K is at a maximum, and $MP_K < AP_K$ when AP_K decreases; and (e) MP_K is negative when TP declines. See also Problem 4.26.

Solved Problems

INCREASING AND DECREASING FUNCTIONS, CONCAVITY AND CONVEXITY

4.1. From the graphs in Fig. 4-6, indicate which graphs (1) are increasing for all x, (2) are decreasing for all x, (3) are convex for all x, (4) are concave for all x, (5) have relative maxima or minima, and (6) have inflection points.

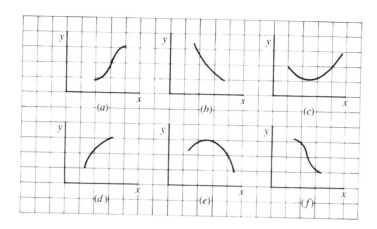

Fig. 4-6

1) *a*, *d*: increasing for all *x*.

2) *b*, *f*: decreasing for all *x*.

3) *b*, *c*: convex for all *x*.

4) *d*, *e*: concave for all *x*.

5) *c*, *e*: exhibit a relative maximum or minimum.

6) *a*, *f*: have an inflection point.

4.2. Indicate with respect to the graphs in Fig. 4-7 which functions have (1) positive first derivatives for all *x*, (2) negative first derivatives for all *x*, (3) positive second derivatives for all *x*, (4) negative second derivatives for all *x*, (5) first derivatives equal to zero or undefined at some point, and (6) second derivatives equal to zero or undefined at some point.

1) *a*, *b*, *h*: the graphs all move up from left to right.

2) *d*, *f*, *g*: the graphs all move down from left to right.

3) *d*, *e*, *h*: the graphs are all convex.

4) *a*, *c*, *f*: the graphs are all concave.

5) *c*, *e*: the graphs reach a plateau (at an extreme point).

6) *b*, *g*: the graphs have inflection points.

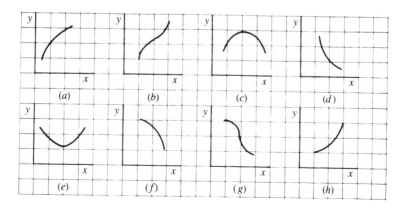

Fig. 4-7

4.3. Test to see whether the following functions are increasing, decreasing, or stationary at $x = 4$.

a) $y = 3x^2 - 14x + 5$

$$y' = 6x - 14$$
$$y'(4) = 6(4) - 14 = 10 > 0 \qquad \text{Function is increasing.}$$

b) $y = x^3 - 7x^2 + 6x - 2$

$$y' = 3x^2 - 14x + 6$$
$$y'(4) = 3(4)^2 - 14(4) + 6 = -2 < 0 \qquad \text{Function is decreasing.}$$

c) $y = x^4 - 6x^3 + 4x^2 - 13$

$$y' = 4x^3 - 18x^2 + 8x$$
$$y'(4) = 4(4)^3 - 18(4)^2 + 8(4) = 0 \qquad \text{Function is stationary.}$$

4.4. Test to see if the following functions are concave or convex at $x = 3$.

a) $y = -2x^3 + 4x^2 + 9x - 15$

$$y' = -6x^2 + 8x + 9$$
$$y'' = 12x + 8$$
$$y''(3) = -12(3) + 8 = -28 < 0 \qquad \text{concave}$$

b) $y = (5x^2 - 8)^2$

$$y' = 2(5x^2 - 8)(10x) = 20x(5x^2 - 8) = 100x^3 - 160x$$
$$y'' = 300x^2 - 160$$
$$y''(3) = 300(3)^2 - 160 = 2540 > 0 \qquad \text{convex}$$

RELATIVE EXTREMA

4.5. Find the relative extrema for the following functions by (1) finding the critical value(s) and (2) determining if at the critical value(s) the function is at a relative maximum or minimum.

a) $f(x) = -7x^2 + 126x - 23$

1) Take the first derivative, set it equal to zero, and solve for x to find the critical value(s).

$$f'(x) = -14x + 126 = 0$$
$$x = 9 \qquad \text{critical value}$$

2) Take the second derivative, evaluate it at the critical value(s), and check for concavity to distinguish between a relative maximum and minimum.

$$f''(x) = -14$$
$$f''(9) = -14 < 0 \qquad \text{concave, relative maximum}$$

b) $f(x) = 3x^3 - 36x^2 + 135x - 13$

1)
$$f'(x) = 9x^2 - 72x + 135 = 0$$
$$= 9(x^2 - 8x + 15) = 0$$
$$= 9(x - 3)(x - 5) = 0$$
$$x = 3 \qquad x = 5 \qquad \text{critical values}$$

2)
$$f''(x) = 18x - 72$$
$$f''(3) = 18(3) - 72 = -18 < 0 \qquad \text{concave, relative maximum}$$
$$f''(5) = 18(5) - 72 = 18 > 0 \qquad \text{convex, relative minimum}$$

c) $f(x) = 2x^4 - 16x^3 + 32x^2 + 5$

1)

$$f'(x) = 8x^3 - 48x^2 + 64x = 0$$
$$= 8x(x^2 - 6x + 8) = 0$$
$$= 8x(x - 2)(x - 4) = 0$$

$x = 0 \quad x = 2 \quad x = 4 \quad$ critical values

2) $f''(x) = 24x^2 - 96x + 64$

$f''(0) = 24(0)^2 - 96(0) + 64 = 64 > 0 \qquad$ convex, relative minimum

$f''(2) = 24(2)^2 - 96(2) + 64 = -32 < 0 \qquad$ concave, relative maximum

$f''(4) = 24(4)^2 - 96(4) + 64 = 64 > 0 \qquad$ convex, relative minimum

4.6. For the following functions, (1) find the critical values and (2) test to see if at the critical values the function is at a relative maximum, minimum, or possible inflection point.

a) $y = -(x - 8)^4$

1) Take the first derivative, set it equal to zero, and solve for x to obtain the critical value(s).

$$y' = -4(x - 8)^3 = 0$$
$$x - 8 = 0$$

$x = 8 \qquad$ critical value

2) Take the second derivative, evaluate it at the critical value(s), and check the sign for concavity to distinguish between a relative maximum, minimum, or inflection point.

$$y'' = -12(x - 8)^2$$
$$y''(8) = -12(8 - 8)^2 = 0 \qquad \text{test inconclusive}$$

If the second-derivative test is inconclusive, continue to take successively higher derivatives and evaluate them at the critical values until you come to the first higher-order derivative that is nonzero:

$$y''' = -24(x - 8)$$
$$y'''(8) = 24(8 - 8) = 0 \qquad \text{test inconclusive}$$
$$y^{(4)} = -24$$
$$y^{(4)}(8) = -24 < 0$$

As explained in Section 4.6, with the first nonzero higher-order derivative an even-numbered derivative, y is at a relative extremum. With that derivative negative, y is concave and at a relative maximum. See Fig. 4-8(a).

b) $y = (5 - x)^3$

1)

$$y' = 3(5 - x)^2(-1) = -3(5 - x)^2 = 0$$
$$x = 5 \qquad \text{critical value}$$

2)

$$y'' = 6(5 - x)$$
$$y''(5) = 6(5 - 5) = 0 \qquad \text{test inconclusive}$$

Continuing to take successively higher-order derivatives and evaluating them at the critical value(s) in search of the first higher-order derivative that does not equal zero, we get

$$y''' = -6$$
$$y'''(5) = -6 < 0$$

As explained in Section 4.6, with the first nonzero higher-order derivative an odd-numbered derivative, y is at an inflection point and not at an extreme point. See Fig. 4-8(b).

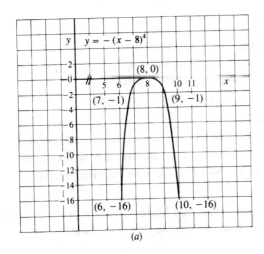

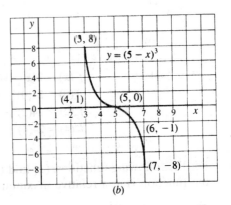

Fig. 4-8

c) $y = -2(x-6)^6$

    1)
$$y' = -12(x-6)^5 = 0$$
$$x = 6 \quad \text{critical value}$$

    2)
$$y'' = -60(x-6)^4$$
$$y''(6) = -60(0)^4 = 0 \quad \text{test inconclusive}$$

Continuing, we get

$$y''' = -240(x-6)^3 \quad y'''(6) = 0 \quad \text{test inconclusive}$$
$$y^{(4)} = -720(x-6)^2 \quad y^{(4)}(6) = 0 \quad \text{test inconclusive}$$
$$y^{(5)} = -1440(x-6) \quad y^{(5)}(6) = 0 \quad \text{test inconclusive}$$
$$y^{(6)} = -1440 \quad\quad y^{(6)}(6) = -1440 < 0$$

With the first nonzero higher-order derivative an even-numbered derivative, y is at an extreme point; with $y^{(6)}(6) < 0$, y is concave and at a relative maximum.

d) $y = (4-x)^5$

    1)
$$y' = 5(4-x)^4(-1) = -5(4-x)^4 = 0$$
$$x = 4 \quad \text{critical value}$$

    2)
$$y'' = 20(4-x)^3$$
$$y''(4) = 20(0)^3 = 0 \quad \text{test inconclusive}$$

Moving on to the third- and higher-order derivatives, we get

$$y''' = -60(4-x)^2 \quad y'''(4) = 0 \quad \text{test inconclusive}$$
$$y^{(4)} = 120(4-x) \quad y^{(4)}(4) = 0 \quad \text{test inconclusive}$$
$$y^{(5)} = -120 \quad\quad y^{(5)}(4) = -120 < 0$$

With the first nonzero higher-order derivative an odd-numbered derivative, y is at an inflection point.

OPTIMIZATION

4.7. For the following functions, (1) find the critical values, (2) test for concavity to determine relative maxima or minima, (3) check for inflection points, and (4) evaluate the function at the critical values and inflection points.

a) $f(x) = x^3 - 18x^2 + 96x - 80$

1)
$$f'(x) = 3x^2 - 36x + 96 = 0$$
$$= 3(x - 4)(x - 8) = 0$$
$$x = 4 \qquad x = 8 \qquad \text{critical values}$$

2)
$$f''(x) = 6x - 36$$
$$f''(4) = 6(4) - 36 = -12 < 0 \qquad \text{concave, relative maximum}$$
$$f''(8) = 6(8) - 36 = 12 > 0 \qquad \text{convex, relative minimum}$$

3)
$$f'' = 6x - 36 = 0$$
$$x = 6$$

With $f''(6) = 0$ and concavity changing between $x = 4$ and $x = 8$, as seen in step 2, there is an inflection point at $x = 6$.

4)
$$f(4) = (4)^3 - 18(4)^2 + 96(4) - 80 = 80 \qquad (4, 80) \qquad \text{relative maximum}$$
$$f(6) = (6)^3 - 18(6)^2 + 96(6) - 80 = 64 \qquad (6, 64) \qquad \text{inflection point}$$
$$f(8) = (8)^3 - 18(8)^2 + 96(8) - 80 = 48 \qquad (8, 48) \qquad \text{relative minimum}$$

b) $f(x) = -x^3 + 6x^2 + 15x - 32$

1)
$$f'(x) = -3x^2 + 12x + 15 = 0$$
$$= -3(x + 1)(x - 5) = 0$$
$$x = -1 \qquad x = 5 \qquad \text{critical values}$$

2)
$$f''(x) = -6x + 12$$
$$f''(-1) = -6(-1) + 12 = 18 > 0 \qquad \text{convex, relative minimum}$$
$$f''(5) = -6(5) + 12 = -18 < 0 \qquad \text{concave, relative maximum}$$

3)
$$f''(x) = -6x + 12 = 0$$
$$x = 2 \qquad \text{inflection point at } x = 2$$

4)
$$f(-1) = -40 \qquad (-1, -40) \qquad \text{relative minimum}$$
$$f(2) = 14 \qquad (2, 14) \qquad \text{inflection point}$$
$$f(5) = 68 \qquad (5, 68) \qquad \text{relative maximum}$$

c) $f(x) = (2x - 7)^3$

1)
$$f'(x) = 3(2x - 7)^2(2) = 6(2x - 7)^2 = 0$$
$$x = 3.5 \qquad \text{critical value}$$

2)
$$f''(x) = 12(2x - 7)(2) = 24(2x - 7)$$
$$f''(3.5) = 24[2(3.5) - 7] = 0 \qquad \text{test inconclusive}$$

Continuing on to successively higher-order derivatives, we find

$$f''' = 48$$
$$f'''(3.5) = 48 > 0$$

3) As explained in Section 4.6, with the first nonzero higher-order derivative an odd-numbered derivative, the function is at an inflection point at $x = 3.5$. With an inflection point at the only critical value, there is no relative maximum or minimum.

4)
$$f(3.5) = 0 \qquad (3.5, 0) \qquad \text{inflection point}$$

Testing for concavity to the left ($x = 3$) and right ($x = 4$) of $x = 3.5$ gives

$$f''(3) = 24[2(3) - 7] = -24 < 0 \qquad \text{concave}$$
$$f''(4) = 24[2(4) - 7] = 24 > 0 \qquad \text{convex}$$

d) $f(x) = (x+2)^4$

1)
$$f'(x) = 4(x+2)^3 = 0$$
$$x = -2 \quad \text{critical value}$$

2)
$$f''(x) = 12(x+2)^2$$
$$f''(-2) = 12(-2+2)^2 = 0 \quad \text{test inconclusive}$$

Continuing, as explained in Section 4.6, we get

$$f'''(x) = 24(x+2)$$
$$f'''(-2) = 24(-2+2) = 0 \quad \text{test inconclusive}$$
$$f^{(4)}(x) = 24$$
$$f^{(4)}(-2) = 24 > 0 \quad \text{relative minimum}$$

With the first nonzero higher-order derivative even-numbered and greater than 0, $f(x)$ is minimized at $x = -2$.

3) There is no inflection point.

4)
$$f(-2) = 0 \quad (-2, 0) \quad \text{relative minimum}$$

4.8. Optimize the following quadratic and cubic functions by (1) finding the critical value(s) at which the function is optimized and (2) testing the second-order condition to distinguish between a relative maximum or minimum.

a) $y = 7x^2 + 112x - 54$

1) Take the first derivative, set it equal to zero, and solve for x to find the critical value(s).

$$y' = 14x + 112 = 0$$
$$x = -8 \quad \text{critical value}$$

2) Take the second derivative, evaluate it at the critical value, and check the sign for a relative maximum and minimum.

$$y'' = 14$$
$$y''(-8) = 14 > 0 \quad \text{convex, relative minimum}$$

Here, with $y'' = $ a constant greater than zero, y is strictly convex and so we can draw the further conclusion that y is at a global minimum at $x = -8$.

b) $y = -9x^2 + 72x - 13$

1)
$$y' = -18x + 72 = 0$$
$$x = 4 \quad \text{critical value}$$

2)
$$y'' = -18$$
$$y''(4) = -18 < 0 \quad \text{concave, relative maximum}$$

Here, with $y'' = $ a constant less than zero, y is strictly concave and so we can also conclude that y is at a global maximum at $x = 4$.

c) $y = x^3 - 6x^2 - 135x + 4$

1)
$$y' = 3x^2 - 12x - 135 = 0$$
$$= 3(x^2 - 4x - 45) = 0$$
$$= 3(x+5)(x-9) = 0$$
$$x = -5 \quad x = 9 \quad \text{critical values}$$

2)
$$y'' = 6x - 12$$
$$y''(-5) = 6(-5) - 12 = -42 < 0 \quad \text{concave, relative maximum}$$
$$y''(9) = 6(9) - 12 = 42 > 0 \quad \text{convex, relative minimum}$$

d) $y = -2x^3 + 15x^2 + 84x - 25$

    1)

$$y' = -6x^2 + 30x + 84 = 0$$
$$= -6(x + 2)(x - 7) = 0$$
$$x = -2 \quad x = 7 \quad \text{critical values}$$

    2)

$$y'' = -12x + 30$$
$$y''(-2) = -12(-2) + 30 = 54 > 0 \quad \text{convex, relative minimum}$$
$$y''(7) = -12(7) + 30 = -54 < 0 \quad \text{concave, relative maximum}$$

4.9. Optimize the following higher-order polynomial functions, using the same procedure as in Problem 4.8.

a) $y = x^4 - 8x^3 - 80x^2 + 15$

    1)

$$y' = 4x^3 - 24x^2 - 160x = 0$$
$$= 4x(x + 4)(x - 10) = 0$$
$$x = 0 \quad x = -4 \quad x = 10 \quad \text{critical values}$$

    2)

$$y'' = 12x^2 - 48x - 160$$
$$y''(-4) = 12(-4)^2 - 48(-4) - 160 = 224 > 0 \quad \text{convex, relative minimum}$$
$$y''(0) = 12(0)^2 - 48(0) - 160 = -160 < 0 \quad \text{concave, relative maximum}$$
$$y''(10) = 12(10)^2 - 48(10) - 160 = 560 > 0 \quad \text{convex, relative minimum}$$

b) $y = -3x^4 - 20x^3 + 144x^2 + 17$

    1)

$$y' = -12x^3 - 60x^2 + 288x = 0$$
$$= -12x(x - 3)(x + 8) = 0$$
$$x = 0 \quad x = 3 \quad x = -8 \quad \text{critical values}$$

    2)

$$y'' = -36x^2 - 120x + 288$$
$$y''(-8) = -36(-8)^2 - 120(-8) + 288 = -1056 < 0 \quad \text{concave, relative maximum}$$
$$y''(0) = -36(0)^2 - 120(0) + 288 = 288 > 0 \quad \text{convex, relative minimum}$$
$$y''(3) = -36(3)^2 - 120(3) + 288 = -396 < 0 \quad \text{concave, relative maximum}$$

c) $y = -(x + 13)^4$

    1)

$$y' = -4(x + 13)^3 = 0$$
$$x + 13 = 0 \quad x = -13 \quad \text{critical value}$$

    2)

$$y'' = -12(x + 13)^2$$
$$y''(-13) = -12(-13 + 13)^2 = 0 \quad \text{test inconclusive}$$

Continuing as explained in Section 4.6 and Problem 4.6, we get

$$y''' = -24(x + 13)$$
$$y'''(-13) = -24(0) = 0 \quad \text{test inconclusive}$$
$$y^{(4)} = -24$$
$$y^{(4)}(-13) = -24 < 0 \quad \text{concave, relative maximum}$$

d) $y = (9 - 4x)^4$

    1)

$$y' = 4(9 - 4x)^3(-4) = -16(9 - 4x)^3 = 0$$
$$9 - 4x = 0 \quad x = 2\tfrac{1}{4} \quad \text{critical value}$$

2)
$$y'' = -48(9 - 4x)^2(-4) = 192(9 - 4x)^2$$
$$y''(2\tfrac{1}{4}) = 192(0)^2 = 0 \quad \text{test inconclusive}$$
$$y''' = 384(9 - 4x)(-4) = -1536(9 - 4x)$$
$$y'''(2\tfrac{1}{4}) = -1536(0) = 0 \quad \text{test inconclusive}$$
$$y^{(4)} = 6144$$
$$y^{(4)}(2\tfrac{1}{4}) = 6144 > 0 \quad \text{convex, relative minimum}$$

MARGINAL, AVERAGE, AND TOTAL CONCEPTS

4.10. Find (1) the marginal and (2) the average functions for each of the following total functions. Evaluate them at $Q = 3$ and $Q = 5$.

a) $TC = 3Q^2 + 7Q + 12$

1) $MC = \dfrac{dTC}{dQ} = 6Q + 7$

  At $Q = 3$, $MC = 6(3) + 7 = 25$
  At $Q = 5$, $MC = 6(5) + 7 = 37$

2) $AC = \dfrac{TC}{Q} = 3Q + 7 + \dfrac{12}{Q}$

  At $Q = 3$, $AC = 3(3) + 7 + \frac{12}{3} = 20$
  At $Q = 5$, $AC = 3(5) + 7 + \frac{12}{5} = 24.4$

Note: When finding the average function, be sure to divide the constant term by Q.

b) $\pi = Q^2 - 13Q + 78$

1) $\dfrac{d\pi}{dQ} = 2Q - 13$

  At $Q = 3$, $\dfrac{d\pi}{dQ} = 2(3) - 13 = -7$

  At $Q = 5$, $\dfrac{d\pi}{dQ} = 2(5) - 13 = -3$

2) $A\pi = \dfrac{\pi}{Q} = Q - 13 + \dfrac{78}{Q}$

  At $Q = 3$, $A\pi = 3 - 13 + \frac{78}{3} = 16$
  At $Q = 5$, $A\pi = 5 - 13 + \frac{78}{5} = 7.6$

c) $TR = 12Q - Q^2$

1) $MR = \dfrac{dTR}{dQ} = 12 - 2Q$

  At $Q = 3$, $MR = 12 - 2(3) = 6$
  At $Q = 5$, $MR = 12 - 2(5) = 2$

2) $AR = \dfrac{TR}{Q} = 12 - Q$

  At $Q = 3$, $AR = 12 - 3 = 9$
  At $Q = 5$, $AR = 12 - 5 = 7$

d) $TC = 35 + 5Q - 2Q^2 + 2Q^3$

1) $MC = \dfrac{dTC}{dQ} = 5 - 4Q + 6Q^2$

  At $Q = 3$, $MC = 5 - 4(3) + 6(3)^2 = 47$
  At $Q = 5$, $MC = 5 - 4(5) + 6(5)^2 = 135$

2) $AC = \dfrac{TC}{Q} = \dfrac{35}{Q} + 5 - 2Q + 2Q^2$

  At $Q = 3$, $AC = \frac{35}{3} + 5 - 2(3) + 2(3)^2 = 28.67$
  At $Q = 5$, $AC = \frac{35}{5} + 5 - 2(5) + 2(5)^2 = 52$

4.11. Find the marginal expenditure (ME) functions associated with each of the following supply functions. Evaluate them at $Q = 4$ and $Q = 10$.

a) $P = Q^2 + 2Q + 1$

  To find the ME function, given a simple supply function, find the total expenditure (TE) function and take its derivative with respect to Q.

$$TE = PQ = (Q^2 + 2Q + 1)Q = Q^3 + 2Q^2 + Q$$

$$ME = \dfrac{dTE}{dQ} = 3Q^2 + 4Q + 1$$

At $Q = 4$, $ME = 3(4)^2 + 4(4) + 1 = 65$. At $Q = 10$, $ME = 3(10)^2 + 4(10) + 1 = 341$.

b) $P = Q^2 + 0.5Q + 3$

$$\text{TE} = PQ = (Q^2 + 0.5Q + 3)Q = Q^3 + 0.5Q^2 + 3Q$$
$$\text{ME} = 3Q^2 + Q + 3$$

At $Q = 4$, ME $= 3(4)^2 + 4 + 3 = 55$. At $Q = 10$, ME $= 3(10)^2 + 10 + 3 = 313$.

4.12. Find the MR functions for each of the following demand functions and evaluate them at $Q = 4$ and $Q = 10$.

a) $Q = 36 - 2P$

$$P = 18 - 0.5Q$$
$$\text{TR} = (18 - 0.5Q)Q = 18Q - 0.5Q^2$$
$$\text{MR} = \frac{d\text{TR}}{dQ} = 18 - Q$$

At $Q = 4$, MR $= 18 - 4 = 14$
At $Q = 10$, MR $= 18 - 10 = 8$

b) $44 - 4P - Q = 0$

$$P = 11 - 0.25Q$$
$$\text{TR} = (11 - 0.25Q)Q = 11Q - 0.25Q^2$$
$$\text{MR} = \frac{d\text{TR}}{dQ} = 11 - 0.5Q$$

At $Q = 4$, MR $= 11 - 0.5(4) = 9$
At $Q = 10$, MR $= 11 - 0.5(10) = 6$

4.13. For each of the following consumption functions, use the derivative to find the marginal propensity to consume MPC $= dC/dY$.

a) $C = C_0 + bY$

$$\text{MPC} = \frac{dC}{dY} = b$$

b) $C = 1500 + 0.75Y$

$$\text{MPC} = \frac{dC}{dY} = 0.75$$

4.14. Given $C = 1200 + 0.8Yd$, where $Yd = Y - T$ and $T = 100$, use the derivative to find the MPC.

When $C = f(Yd)$, make $C = f(Y)$ before taking the derivative. Thus,

$$C = 1200 + 0.8(Y - 100) = 1120 + 0.8Y$$
$$\text{MPC} = \frac{dC}{dY} = 0.8$$

Note that the introduction of a lump-sum tax into the income determination model *does not* affect the value of the MPC (or the multiplier).

4.15. Given $C = 2000 + 0.9Yd$, where $Yd = Y - T$ and $T = 300 + 0.2Y$, use the derivative to find the MPC.

$$C = 2000 + 0.9(Y - 300 - 0.2Y) = 2000 + 0.9Y - 270 - 0.18Y = 1730 + 0.72Y$$
$$\text{MPC} = \frac{dC}{dY} = 0.72$$

The introduction of a proportional tax into the income determination model *does* affect the value of the MPC and hence the multiplier.

4.16. Find the marginal cost functions for each of the following average cost functions.

a) $AC = 1.5Q + 4 + \dfrac{46}{Q}$

Given the average cost function, the marginal cost function is determined by first finding the total cost function and then taking its derivative, as follows:

$$\text{TC} = (\text{AC})Q = \left(1.5Q + 4 + \frac{46}{Q}\right)Q = 1.5Q^2 + 4Q + 46$$

$$\text{MC} = \frac{d\text{TC}}{dQ} = 3Q + 4$$

b) $\text{AC} = \dfrac{160}{Q} + 5 - 3Q + 2Q^2$

$$\text{TC} = \left(\frac{160}{Q} + 5 - 3Q + 2Q^2\right)Q = 160 + 5Q - 3Q^2 + 2Q^3$$

$$\text{MC} = \frac{d\text{TC}}{dQ} = 5 - 6Q + 6Q^2$$

OPTIMIZING ECONOMIC FUNCTIONS

4.17. Maximize the following total revenue TR and total profit π functions by (1) finding the critical value(s), (2) testing the second-order conditions, and (3) calculating the maximum TR or π.

a) $\text{TR} = 32Q - Q^2$

1) $\text{TR}' = 32 - 2Q = 0$
$Q = 16$ critical value

2) $\text{TR}'' = -2 < 0$ concave, relative maximum

3) $\text{TR} = 32(16) - (16)^2 = 256$

Note that whenever the value of the second derivative is negative over the whole domain of the function, as in (2) above, we can also conclude that the function is strictly concave and at a global maximum.

b) $\pi = -Q^2 + 11Q - 24$

1) $\pi' = -2Q + 11 = 0$
$Q = 5.5$ critical value

2) $\pi'' = -2 < 0$ concave, relative maximum

3) $\pi = -(5.5)^2 + 11(5.5) - 24 = 6.25$

c) $\pi = -\frac{1}{3}Q^3 - 5Q^2 + 2000Q - 326$

1) $\pi' = -Q^2 - 10Q + 2000 = 0$ (4.1)
$-1(Q^2 + 10Q - 2000) = 0$ (4.2)
$(Q + 50)(Q - 40) = 0$
$Q = -50$ $Q = 40$ critical values

2) $\pi'' = -2Q - 10$
$\pi''(40) = -2(40) - 10 = -90 < 0$ concave, relative maximum
$\pi''(-50) = -2(-50) - 10 = 90 > 0$ convex, relative minimum

Negative critical values will subsequently be ignored as having no economic significance.

3) $\pi = -\frac{1}{3}(40)^3 - 5(40)^2 + 2000(40) - 326 = 50,340.67$

Note: In testing the second-order conditions, as in step 2, always take the second derivative from the original first derivative (*4.1*) before any *negative* number has been factored out. Taking the second derivative from the first derivative after a negative has been factored out, as in (*4.2*), will reverse the second-order conditions and suggest that the function is maximized at $Q = -50$ and minimized at $Q = 40$. Test it yourself.

d) $\pi = -Q^3 - 6Q^2 + 1440Q - 545$

   1) $\pi' = -3Q^2 - 12Q + 1440 = 0$

       $-3(Q - 20)(Q + 24) = 0$

                  $Q = 20$ $Q = -24$ critical values

   2) $\pi'' = -6Q - 12$

       $\pi''(20) = -6(20) - 12 = -132 < 0$ concave, relative maximum

   3) $\pi = -(20)^3 - 6(20)^2 + 1440(20) - 545 = 17,855$

4.18. From each of the following total cost TC functions, find (1) the average cost AC function, (2) the critical value at which AC is minimized, and (3) the minimum average cost.

a) $TC = Q^3 - 5Q^2 + 60Q$

   1) $AC = \dfrac{TC}{Q} = \dfrac{Q^3 - 5Q^2 + 60Q}{Q} = Q^2 - 5Q + 60$

   2) $AC' = 2Q - 5 = 0$ $Q = 2.5$

       $AC'' = 2 > 0$ convex, relative minimum

   3) $AC(2.5) = (2.5)^2 - 5(2.5) + 60 = 53.75$

Note that whenever the value of the second derivative is positive over the whole domain of the function, as in (2) above, we can also conclude that the function is strictly convex and at a global minimum.

b) $TC = Q^3 - 21Q^2 + 500Q$

   1) $AC = \dfrac{Q^3 - 21Q^2 + 500Q}{Q} = Q^2 - 21Q + 500$

   2) $AC' = 2Q - 21 = 0$ $Q = 10.5$

       $AC'' = 2 > 0$ convex, relative minimum

   3) $AC = (10.5)^2 - 21(10.5) + 500 = 389.75$

4.19. Given the following total revenue and total cost functions for different firms, maximize profit π for the firms as follows: (1) Set up the profit function $\pi = TR - TC$, (2) find the critical value(s) where π is at a relative extremum and test the second-order condition, and (3) calculate the maximum profit.

a) $TR = 1400Q - 6Q^2$ $TC = 1500 + 80Q$

   1) $\pi = 1400Q - 6Q^2 - (1500 + 80Q)$

       $= -6Q^2 + 1320Q - 1500$

   2) $\pi' = -12Q + 1320 = 0$

            $Q = 110$ critical value

       $\pi'' = -12 < 0$ concave, relative maximum

   3) $\pi = -6(110)^2 + 1320(110) - 1500 = 71,000$

b) $TR = 1400Q - 7.5Q^2$ $TC = Q^3 - 6Q^2 + 140Q + 750$

   1) $\pi = 1400Q - 7.5Q^2 - (Q^3 - 6Q^2 + 140Q + 750)$

           $= -Q^3 - 1.5Q^2 + 1260Q - 750$ (*4.3*)

   2) $\pi' = -3Q^2 - 3Q + 1260 = 0$

           $= -3(Q^2 + Q - 420) = 0$

           $= -3(Q + 21)(Q - 20) = 0$

               $Q = -21$ $Q = 20$ critical values

   Take the second derivative directly from (*4.3*), as explained in Problem 4.17(*c*), and ignore all negative critical values.

$$\pi'' = -6Q - 3$$
$$\pi''(20) = -6(20) - 3 = -123 < 0 \qquad \text{concave, relative maximum}$$

   3) $\pi = -(20)^3 - 1.5(20)^2 + 1260(20) - 750 = 15{,}850$

c) $TR = 4350Q - 13Q^2$ $TC = Q^3 - 5.5Q^2 + 150Q + 675$

   1) $\pi = 4350Q - 13Q^2 - (Q^3 - 5.5Q^2 + 150Q + 675)$

           $= -Q^3 - 7.5Q^2 + 4200Q - 675$

   2) $\pi' = -3Q^2 - 15Q + 4200 = 0$

           $= -3(Q^2 + 5Q - 1400) = 0$

           $= -3(Q + 40)(Q - 35) = 0$

               $Q = -40$ $Q = 35$ critical values

$$\pi'' = -6Q - 15$$
$$\pi''(35) = -6(35) - 15 = -225 < 0 \qquad \text{concave, relative maximum}$$

   3) $\pi = -(35)^3 - 7.5(35)^2 + 4200(35) - 675 = 94{,}262.50$

d) $TR = 5900Q - 10Q^2$ $TC = 2Q^3 - 4Q^2 + 140Q + 845$

   1) $\pi = 5900Q - 10Q^2 - (2Q^3 - 4Q^2 + 140Q + 845)$

           $= -2Q^3 - 6Q^2 + 5760Q - 845$

   2) $\pi' = -6Q^2 - 12Q + 5760 = 0$

           $= -6(Q^2 + 2Q - 960) = 0$

           $= -6(Q + 32)(Q - 30) = 0$

               $Q = -32$ $Q = 30$ critical values

$$\pi'' = -12Q - 12$$
$$\pi''(30) = -12(30) - 12 = -372 < 0 \qquad \text{concave, relative maximum}$$

   3) $\pi = -2(30)^3 - 6(30)^2 + 5760(30) - 845 = 112{,}555$

4.20.   Prove that marginal cost (MC) must equal marginal revenue (MR) at the profit-maximizing level of output.

$$\pi = TR - TC$$

To maximize π, $d\pi/dQ$ must equal zero.

$$\frac{d\pi}{dQ} = \frac{dTR}{dQ} - \frac{dTC}{dQ} = 0$$

$$\frac{dTR}{dQ} = \frac{dTC}{dQ}$$

$$MR = MC \qquad \text{Q.E.D.}$$

4.21. A producer has the possibility of discriminating between the domestic and foreign markets for a product where the demands, respectively, are

$$Q_1 = 21 - 0.1P_1 \qquad (4.4)$$
$$Q_2 = 50 - 0.4P_2 \qquad (4.5)$$

Total cost $= 2000 + 10Q$ where $Q = Q_1 + Q_2$. What price will the producer charge in order to maximize profits (a) with discrimination between markets and (b) without discrimination? (c) Compare the profit differential between discrimination and nondiscrimination.

a) To maximize profits under price discrimination, the producer will set prices so that MC = MR in each market. Thus, $MC = MR_1 = MR_2$. With $TC = 2000 + 10Q$,

$$MC = \frac{dTC}{dQ} = 10$$

Hence MC will be the same at all levels of output. In the domestic market,

$$Q_1 = 21 - 0.1P_1$$

Hence,
$$P_1 = 210 - 10Q_1$$

$$TR_1 = (210 - 10Q_1)Q_1 = 210Q_1 - 10Q_1^2$$

and
$$MR_1 = \frac{dTR_1}{dQ_1} = 210 - 20Q_1$$

When $MR_1 = MC$,
$$210 - 20Q_1 = 10 \qquad Q_1 = 10$$

When $Q_1 = 10$,
$$P_1 = 210 - 10(10) = 110$$

In the foreign market,
$$Q_2 = 50 - 0.4P_2$$

Hence,
$$P_2 = 125 - 2.5Q_2$$

$$TR_2 = (125 - 2.5Q_2)Q_2 = 125Q_2 - 2.5Q_2^2$$

Thus,
$$MR_2 = \frac{dTR_2}{dQ_2} = 125 - 5Q_2$$

When $MR_2 = MC$,
$$125 - 5Q_2 = 10 \qquad Q_2 = 23$$

When $Q_2 = 23$,
$$P_2 = 125 - 2.5(23) = 67.5$$

The discriminating producer charges a lower price in the foreign market where the demand is relatively more elastic and a higher price ($P_1 = 110$) in the domestic market where the demand is relatively less elastic.

b) If the producer does not discriminate, $P_1 = P_2$ and the two demand functions (4.4) and (4.5) may simply be aggregated. Thus,

$$Q = Q_1 + Q_2 = 21 - 0.1P + 50 - 0.4P = 71 - 0.5P$$

Hence,
$$P = 142 - 2Q$$

$$TR = (142 - 2Q)Q = 142Q - 2Q^2$$

and
$$MR = \frac{dTR}{dQ} = 142 - 4Q$$

When $MR = MC$,
$$142 - 4Q = 10 \qquad Q = 33$$

When $Q = 33$,
$$P = 142 - 2(33) = 76$$

When no discrimination takes place, the price falls somewhere between the relatively high price of the domestic market and the relatively low price of the foreign market. Notice, however, that the quantity sold remains the same: at $P = 76$, $Q_1 = 13.4$, $Q_2 = 19.6$, and $Q = 33$.

c)   With discrimination,

$$TR = TR_1 + TR_2 = P_1Q_1 + P_2Q_2 = 110(10) + 67.5(23) = 2652.50$$

$TC = 2000 + 10Q$, where $Q = Q_1 + Q_2$.

$$TC = 2000 + 10(10 + 23) = 2330$$

Thus, $\pi = TR - TC = 2652.50 - 2330 = 322.50$

Without discrimination,

$$TR = PQ = 76(33) = 2508$$

$TC = 2330$ since costs do not change with or without discrimination. Thus, $\pi = 2508 - 2330 = 178$. Profits are higher with discrimination (322.50) than without discrimination.

4.22.   Faced with two distinct demand functions

$$Q_1 = 24 - 0.2P_1 \qquad Q_2 = 10 - 0.05P_2$$

where $TC = 35 + 40Q$, what price will the firm charge (a) with discrimination and (b) without discrimination?

a)   With $Q_1 = 24 - 0.2P_1$,

$$P_1 = 120 - 5Q_1$$
$$TR_1 = (120 - 5Q_1)Q_1 = 120Q_1 - 5Q_1^2$$
$$MR_1 = 120 - 10Q_1$$

The firm will maximize profits where $MC = MR_1 = MR_2$

$$TC = 35 + 40Q$$
$$MC = 40$$

When $MC = MR_1$, $40 = 120 - 10Q_1 \qquad Q_1 = 8$

When $Q_1 = 8$, $P_1 = 120 - 5(8) = 80$

In the second market, with $Q_2 = 10 - 0.05P_2$,

$$P_2 = 200 - 20Q_2$$
$$TR_2 = (200 - 20Q_2)Q_2 = 200Q_2 - 20Q_2^2$$
$$MR_2 = 200 - 40Q_2$$

When $MC = MR_2$, $40 = 200 - 40Q_2 \qquad Q_2 = 4$

When $Q_2 = 4$, $P_2 = 200 - 20(4) = 120$

b)   If the producer does not discriminate, $P_1 = P_2 = P$ and the two demand functions can be combined, as follows:

$$Q = Q_1 + Q_2 = 24 - 0.2P + 10 - 0.05P = 34 - 0.25P$$

Thus,

$$P = 136 - 4Q$$
$$TR = (136 - 4Q)Q = 136Q - 4Q^2$$
$$MR = 136 - 8Q$$

At the profit-maximizing level, $MC = MR$.

$$40 = 136 - 8Q \qquad Q = 12$$

At $Q = 12$, $P = 136 - 4(12) = 88$

4.23. Use the MR = MC method to (*a*) maximize profit π and (*b*) check the second-order conditions, given

$$\text{TR} = 1400Q - 7.5Q^2 \qquad \text{TC} = Q^3 - 6Q^2 + 140Q + 750$$

a)   MR = TR$'$ = $1400 - 15Q$, MC = TC$'$ = $3Q^2 - 12Q + 140$

Equate MR = MC. $1400 - 15Q = 3Q^2 - 12Q + 140$

Solve for Q by moving everything to the right.

$$3Q^2 + 3Q - 1260 = 0$$
$$3(Q + 21)(Q - 20) = 0$$
$$Q = -21 \qquad Q = 20 \qquad \text{critical values}$$

b) TR$''$ = -15 TC$''$ = $6Q - 12$

Since $\pi = \text{TR} - \text{TC}$ and the objective is to maximize π, be sure to *subtract* TC$''$ *from* TR$''$, or you will reverse the second-order conditions and select the wrong critical value.

$$\pi'' = \text{TR}'' - \text{TC}''$$
$$= -15 - 6Q + 12 = -6Q - 3$$
$$\pi''(20) = -6(20) - 3 = -123 < 0 \qquad \text{concave, relative maximum}$$

Compare these results with Problem 4.19(*b*).

THE MARGINAL RATE OF TECHNICAL SUBSTITUTION

4.24.   An *isoquant* depicts the different combinations of inputs K and L that can be used to produce a specific level of output Q. One such isoquant for the output level $Q = 2144$ is

$$16K^{1/4}L^{3/4} = 2144$$

(*a*) Use implicit differentiation from Section 3.9 to find the slope of the isoquant dK/dL which in economics is called the *marginal rate of technical substitution* (MRTS). (*b*) Evaluate the marginal rate of technical substitution at $K = 256$, $L = 108$.

a)   Take the derivative of each term with respect to L and treat K as a function of L.

$$\frac{d}{dL}(16K^{1/4}L^{3/4}) = \frac{d}{dL}(2144)$$

Use the product rule since K is being treated as a function of L.

$$16K^{1/4} \cdot \frac{d}{dL}(L^{3/4}) + L^{3/4} \cdot \frac{d}{dL}(16K^{1/4}) = \frac{d}{dL}(2144)$$

$$\left(16K^{1/4} \cdot \frac{3}{4}L^{-1/4}\right) + \left(L^{3/4} \cdot 16 \cdot \frac{1}{4}K^{-3/4} \cdot \frac{dK}{dL}\right) = 0$$

$$12K^{1/4}L^{-1/4} + 4K^{-3/4}L^{3/4} \cdot \frac{dK}{dL} = 0$$

Solve algebraically for dK/dL.

$$\frac{dK}{dL} = \frac{-12K^{1/4}L^{-1/4}}{4K^{-3/4}L^{3/4}} = \frac{-3K}{L}$$

b)   At $K = 256$ and $L = 108$.

$$\text{MRTS} = \frac{dK}{dL} = \frac{-3(256)}{108} = -7.11$$

This means that if L is increased by 1 relatively small unit, K must decrease by 7.11 units in order to remain on the production isoquant where the production level is constant. See also Problem 6.51.

4.25. The equation for the production isoquant is

$$25K^{3/5}L^{2/5} = 5400$$

(a) Find the MRTS and (b) evaluate it at $K = 243$, $L = 181$.

a) Treat K as a function of L and use the product rule to find dK/dL, which is the MRTS.

$$25K^{3/5} \cdot \frac{2}{5}L^{-3/5} + L^{2/5} \cdot 25 \cdot \frac{3}{5}K^{-2/5} \cdot \frac{dK}{dL} = 0$$

$$10K^{3/5}L^{-3/5} + 15K^{-2/5}L^{2/5} \cdot \frac{dK}{dL} = 0$$

Solve algebraically for dK/dL.

$$\frac{dK}{dL} = \frac{-10K^{3/5}L^{-3/5}}{15K^{-2/5}L^{2/5}} = \frac{-2K}{3L} = \text{MRTS}$$

b) At $K = 243$ and $K = 181$,

$$\text{MRTS} = \frac{dK}{dL} = \frac{-2(243)}{3(181)} = -0.895$$

This means that if L is increased by 1 relatively small unit, K must decrease by 0.895 unit in order to remain on the production isoquant where the production level is constant. See Problem 6.52.

RELATIONSHIP BETWEEN FUNCTIONS AND GRAPHS

4.26. Given the total cost function $C = Q^3 - 18Q^2 + 750Q$, use your knowledge of calculus to help sketch a graph showing the relationship between total, average, and marginal costs.

a) Take the first and second derivatives of the total cost function

$$C' = 3Q^2 - 36Q + 750$$
$$C'' = 6Q - 36$$

and check for (1) concavity and (2) inflection points.

1) For $Q < 6$, $C'' < 0$ concave
   For $Q > 6$, $C'' > 0$ convex

2) $6Q - 36 = 0$
                                 $Q = 6$
   $$C(6) = (6)^3 - 18(6)^2 + 750(6) = 4068$$

With $C(Q)$ changing from concave to convex at $Q = 6$,

$$(6, 4068) \quad \text{inflection point}$$

b) Find the average cost function AC and the relative extrema.

$$\text{AC} = \frac{\text{TC}}{Q} = Q^2 - 18Q + 750$$

$$\text{AC}' = 2Q - 18 = 0$$

$$Q = 9 \quad \text{critical value}$$

$$\text{AC}'' = 2 > 0 \quad \text{convex, relative minimum}$$

c) Do the same thing for the marginal cost function

$$MC = C' = 3Q^2 - 36Q + 750$$
$$MC' = 6Q - 36 = 0$$
$$Q = 6 \qquad \text{critical value}$$
$$MC'' = 6 > 0 \qquad \text{convex, relative minimum}$$

d) Sketch the graph as in Fig. 4-9, noting that (1) MC decreases when TC is concave and increasing at a decreasing rate, increases when TC is convex and increasing at an increasing rate, and is at a minimum when TC is at an inflection point and changing concavity; and (2) AC decreases over the whole region where MC < AC, is at a minimum when MC = AC, and increases when MC > AC.

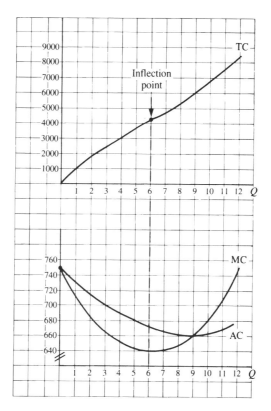

Fig. 4-9

CHAPTER 5

Calculus of Multivariable Functions

5.1 FUNCTIONS OF SEVERAL VARIABLES AND PARTIAL DERIVATIVES

Study of the derivative in Chapter 4 was limited to functions of a single independent variable such as $y = f(x)$. Many economic activities, however, involve functions of more than one independent variable. $z = f(x, y)$ is defined as a *function of two independent variables* if there exists one and only one value of z in the range of f for each ordered pair of real numbers (x, y) in the domain of f. By convention, z is the *dependent variable*; x and y are the *independent variables*.

To measure the effect of a change in a single independent variable (x or y) on the dependent variable (z) in a multivariable function, the partial derivative is needed. The *partial derivative of z with respect to x* measures the instantaneous rate of change of z with respect to x while y is held constant. It is written $\partial z/\partial x$, $\partial f/\partial x$, $f_x(x, y)$, f_x, or z_x. The *partial derivative of z with respect to y* measures the rate of change of z with respect to y while x is held constant. It is written $\partial z/\partial y$, $\partial f/\partial y$, $f_y(x, y)$, f_y, or z_y. Expressed mathematically,

$$\frac{\partial z}{\partial x} = \lim_{\Delta x \to 0} \frac{f(x + \Delta x, y) - f(x, y)}{\Delta x} \qquad (5.1a)$$

$$\frac{\partial z}{\partial y} = \lim_{\Delta y \to 0} \frac{f(x, y + \Delta y) - f(x, y)}{\Delta y} \qquad (5.1b)$$

Partial differentiation with respect to one of the independent variables follows the same rules as ordinary differentiation while the other independent variables are treated as constant. See Examples 1 and 2 and Problems 5.1 and 5.23.

EXAMPLE 1. The partial derivatives of a multivariable function such as $z = 3x^2 y^3$ are found as follows:

a) When differentiating with respect to x, treat the y term as a constant by mentally bracketing it with the coefficient:

$$z = [3y^3] \cdot x^2$$

Then take the derivative of the x term, holding the y term constant,

$$\frac{\partial z}{\partial x} = z_x = [3y^3] \cdot \frac{d}{dx}(x^2)$$

$$= [3y^3] \cdot 2x$$

Recalling that a multiplicative constant remains in the process of differentiation, simply multiply and rearrange terms to obtain

$$\frac{\partial z}{\partial x} = z_x = 6xy^3$$

b) When differentiating with respect to y, treat the x term as a constant by bracketing it with the coefficient; then take the derivative as was done above:

$$z = [3x^2] \cdot y^3$$

$$\frac{\partial z}{\partial y} = z_y = [3x^2] \cdot \frac{d}{dy}(y^3)$$

$$= [3x^2] \cdot 3y^2 = 9x^2 y^2$$

EXAMPLE 2.   To find the partial derivatives for $z = 5x^3 - 3x^2 y^2 + 7y^5$:

a) When differentiating with respect to x, mentally bracket all y terms to remember to treat them as constants:

$$z = 5x^3 - [3y^2]x^2 + [7y^5]$$

Then take the derivative of each term, remembering that in differentiation multiplicative constants remain but additive constants drop out, because the derivative of a constant is zero.

$$\frac{\partial z}{\partial x} = \frac{d}{dx}(5x^3) - [3y^2] \cdot \frac{d}{dx}(x^2) + \frac{d}{dx}[7y^5]$$

$$= 15x^2 - [3y^2] \cdot 2x + 0$$

$$= 15x^2 - 6xy^2$$

b) When differentiating with respect to y, block off all the x terms and differentiate as above.

$$z = [5x^3] - [3x^2]y^2 + 7y^5$$

$$\frac{\partial z}{\partial y} = \frac{d}{dy}[5x^3] - [3x^2] \cdot \frac{d}{dy}(y^2) + \frac{d}{dy}(7y^5)$$

$$= 0 - [3x^2] \cdot 2y + 35y^4$$

$$= -6x^2 y + 35y^4$$

See Problem 5.1.

5.2 RULES OF PARTIAL DIFFERENTIATION

Partial derivatives follow the same basic patterns as the rules of differentiation in Section 3.7. A few key rules are given below, illustrated in Examples 3 to 5, treated in Problems 5.2 to 5.5, and verified in Problem 5.23.

5.2.1 Product Rule

Given $z = g(x, y) \cdot h(x, y)$,

$$\frac{\partial z}{\partial x} = g(x, y) \cdot \frac{\partial h}{\partial x} + h(x, y) \cdot \frac{\partial g}{\partial x} \qquad (5.2a)$$

$$\frac{\partial z}{\partial y} = g(x, y) \cdot \frac{\partial h}{\partial y} + h(x, y) \cdot \frac{\partial g}{\partial y} \qquad (5.2b)$$

EXAMPLE 3.   Given $z = (3x + 5)(2x + 6y)$, by the product rule,

$$\frac{\partial z}{\partial x} = (3x + 5)(2) + (2x + 6y)(3) = 12x + 10 + 18y$$

$$\frac{\partial z}{\partial y} = (3x + 5)(6) + (2x + 6y)(0) = 18x + 30$$

5.2.2 Quotient Rule

Given $z = g(x, y)/h(x, y)$ and $h(x, y) \neq 0$,

$$\frac{\partial z}{\partial x} = \frac{h(x, y) \cdot \partial g/\partial x - g(x, y) \cdot \partial h/\partial x}{[h(x, y)]^2} \qquad (5.3a)$$

$$\frac{\partial z}{\partial y} = \frac{h(x, y) \cdot \partial g/\partial y - g(x, y) \cdot \partial h/\partial y}{[h(x, y)]^2} \qquad (5.3b)$$

EXAMPLE 4.   Given $z = (6x + 7y)/(5x + 3y)$, by the quotient rule,

$$\frac{\partial z}{\partial x} = \frac{(5x + 3y)(6) - (6x + 7y)(5)}{(5x + 3y)^2}$$

$$= \frac{30x + 18y - 30x - 35y}{(5x + 3y)^2} = \frac{-17y}{(5x + 3y)^2}$$

$$\frac{\partial z}{\partial y} = \frac{(5x + 3y)(7) - (6x + 7y)(3)}{(5x + 3y)^2}$$

$$= \frac{35x + 21y - 18x - 21y}{(5x + 3y)^2} = \frac{17x}{(5x + 3y)^2}$$

5.2.3 Generalized Power Function Rule

Given $z = [g(x, y)]^n$,

$$\frac{\partial z}{\partial x} = n[g(x, y)]^{n-1} \cdot \frac{\partial g}{\partial x} \qquad (5.4a)$$

$$\frac{\partial z}{\partial y} = n[g(x, y)]^{n-1} \cdot \frac{\partial g}{\partial y} \qquad (5.4b)$$

EXAMPLE 5.   Given $z = (x^3 + 7y^2)^4$, by the generalized power function rule,

$$\frac{\partial z}{\partial x} = 4(x^3 + 7y^2)^3 \cdot (3x^2) = 12x^2(x^3 + 7y^2)^3$$

$$\frac{\partial z}{\partial y} = 4(x^3 + 7y^2)^3 \cdot (14y) = 56y(x^3 + 7y^2)^3$$

5.3 SECOND-ORDER PARTIAL DERIVATIVES

Given a function $z = f(x, y)$, *the second-order (direct) partial derivative* signifies that the function has been differentiated partially with respect to one of the independent variables twice while the other independent variable has been held constant:

$$f_{xx} = (f_x)_x = \frac{\partial}{\partial x}\left(\frac{\partial z}{\partial x}\right) = \frac{\partial^2 z}{\partial x^2} \qquad f_{yy} = (f_y)_y = \frac{\partial}{\partial y}\left(\frac{\partial z}{\partial y}\right) = \frac{\partial^2 z}{\partial y^2}$$

In effect, f_{xx} measures the rate of change of the first-order partial derivative f_x with respect to x while y is held constant. And f_{yy} is exactly parallel. See Problems 5.6 and 5.8.

The *cross* (or *mixed*) *partial derivatives* f_{xy} and f_{yx} indicate that first the primitive function has been partially differentiated with respect to one independent variable and then that partial derivative has in turn been partially differentiated with respect to the other independent variable:

$$f_{xy} = (f_x)_y = \frac{\partial}{\partial y}\left(\frac{\partial z}{\partial x}\right) = \frac{\partial^2 z}{\partial y\,\partial x} \qquad f_{yx} = (f_y)_x = \frac{\partial}{\partial x}\left(\frac{\partial z}{\partial y}\right) = \frac{\partial^2 z}{\partial x\,\partial y}$$

In brief, a cross partial measures the rate of change of a first-order partial derivative with respect to the other independent variable. Notice how the order of independent variables changes in the different forms of notation. See Problems 5.7 and 5.9.

EXAMPLE 6. The (*a*) first, (*b*) second, and (*c*) cross partial derivatives for $z = 7x^3 + 9xy + 2y^5$ are taken as shown below.

a) $\dfrac{\partial z}{\partial x} = z_x = 21x^2 + 9y \qquad \dfrac{\partial z}{\partial y} = z_y = 9x + 10y^4$

b) $\dfrac{\partial^2 z}{\partial x^2} = z_{xx} = 42x \qquad\qquad \dfrac{\partial^2 z}{\partial y^2} = z_{yy} = 40y^3$

c) $\dfrac{\partial^2 z}{\partial y\,\partial x} = \dfrac{\partial}{\partial y}\left(\dfrac{\partial z}{\partial x}\right) = \dfrac{\partial}{\partial y}(21x^2 + 9y) = z_{xy} = 9$

$\dfrac{\partial^2 z}{\partial x\,\partial y} = \dfrac{\partial}{\partial x}\left(\dfrac{\partial z}{\partial y}\right) = \dfrac{\partial}{\partial x}(9x + 10y^4) = z_{yx} = 9$

EXAMPLE 7. The (*a*) first, (*b*) second, and (*c*) cross partial derivatives for $z = 3x^2 y^3$ are evaluated below at $x = 4, y = 1$.

a) $z_x = 6xy^3 \qquad\qquad\qquad z_y = 9x^2 y^2$

$z_x(4, 1) = 6(4)(1)^3 = 24 \qquad z_y(4, 1) = 9(4)^2(1)^2 = 144$

b) $z_{xx} = 6y^3 \qquad\qquad\qquad z_{yy} = 18x^2 y$

$z_{xx}(4, 1) = 6(1)^3 = 6 \qquad z_{yy}(4, 1) = 18(4)^2(1) = 288$

c) $z_{xy} = \dfrac{\partial}{\partial y}(6xy^3) = 18xy^2 \qquad z_{yx} = \dfrac{\partial}{\partial x}(9x^2 y^2) = 18xy^2$

$z_{xy}(4, 1) = 18(4)(1)^2 = 72 \qquad z_{yx}(4, 1) = 18(4)(1)^2 = 72$

By *Young's theorem*, if both cross partial derivatives are continuous, they will be identical. See Problems 5.7 to 5.9.

5.4 OPTIMIZATION OF MULTIVARIABLE FUNCTIONS

For a multivariable function such as $z = f(x, y)$ to be at a relative minimum or maximum, three conditions must be met:

1. The first-order partial derivatives must equal zero simultaneously. This indicates that at the

given point (a, b), called a *critical point*, the function is neither increasing nor decreasing with respect to the principal axes but is at a relative plateau.

2.   The second-order direct partial derivatives, when evaluated at the critical point (a, b), must both be negative for a relative maximum and positive for a relative minimum. This ensures that from a relative plateau at (a, b) the function is concave and moving downward in relation to the principal axes in the case of a maximum and convex and moving upward in relation to the principal axes in the case of a minimum.

3.   The product of the second-order direct partial derivatives evaluated at the critical point must exceed the product of the cross partial derivatives also evaluated at the critical point. This added condition is needed to preclude an inflection point or saddle point.

In sum, as seen in Fig. 5-1, when evaluated at a critical point (a, b),

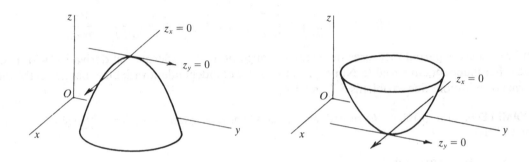

Fig. 5-1

Relative maximum   *Relative minimum*
1. $f_x, f_y = 0$ 1. $f_x, f_y = 0$
2. $f_{xx}, f_{yy} < 0$ 2. $f_{xx}, f_{yy} > 0$
3. $f_{xx} \cdot f_{yy} > (f_{xy})^2$ 3. $f_{xx} \cdot f_{yy} > (f_{xy})^2$

Note the following:

1)   Since $f_{xy} = f_{yx}$ by Young's theorem, $f_{xy} \cdot f_{yx} = (f_{xy})^2$. Step 3 can also be written $f_{xx} \cdot f_{yy} - (f_{xy})^2 > 0$.

2)   If $f_{xx} \cdot f_{yy} < (f_{xy})^2$, when f_{xx} and f_{yy} have the same signs, the function is at an *inflection point*; when f_{xx} and f_{yy} have different signs, the function is at a *saddle point*, as seen in Fig. 5-2, where the function is at a maximum when viewed from one axis but at a minimum when viewed from the other axis.

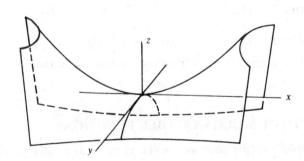

Fig. 5-2

3) If $f_{xx} \cdot f_{yy} = (f_{xy})^2$, the test is inconclusive. See Example 8 and Problems 5.10 and 5.11; for inflection points, see Problems 5.10(c) and 5.11(b) and (c); for saddle points see Problems 5.10(d) and 5.11(a) and (d).

4) If the function is *strictly* concave (convex) in x and y, as in Fig. 5-1, there will be only one maximum (minimum), called an absolute or *global* maximum (minimum). If the function is simply concave (convex) in x and y on an interval, the critical point is a relative or *local* maximum (minimum).

EXAMPLE 8. (a) Find the critical points. (b) Test whether the function is at a relative maximum or minimum, given

$$z = 2y^3 - x^3 + 147x - 54y + 12$$

a)   Take the first-order partial derivatives, set them equal to zero, and solve for x and y:

$$z_x = -3x^2 + 147 = 0 \qquad z_y = 6y^2 - 54 = 0 \qquad\qquad (5.5)$$
$$x^2 = 49 \qquad\qquad y^2 = 9$$
$$x = \pm 7 \qquad\qquad y = \pm 3$$

With $x = \pm 7$, $y = \pm 3$, there are four distinct sets of critical points: $(7,3)$, $(7,-3)$, $(-7,3)$, and $(-7,-3)$.

b)   Take the second-order direct partials from (5.5), evaluate them at each of the critical points, and check the signs:

$$z_{xx} = -6x \qquad\qquad\qquad z_{yy} = 12y$$

1) $z_{xx}(7,3) = -6(7) = -42 < 0$ $\qquad z_{yy}(7,3) = 12(3) = 36 > 0$
2) $z_{xx}(7,-3) = -6(7) = -42 < 0$ $\qquad z_{yy}(7,-3) = 12(-3) = -36 < 0$
3) $z_{xx}(-7,3) = -6(-7) = 42 > 0$ $\qquad z_{yy}(-7,3) = 12(3) = 36 > 0$
4) $z_{xx}(-7,-3) = -6(-7) = 42 > 0$ $\qquad z_{yy}(-7,-3) = 12(-3) = -36 < 0$

Since there are different signs for each of the second direct partials in (1) and (4), the function cannot be at a relative maximum or minimum at $(7,3)$ or $(-7,-3)$. When f_{xx} and f_{yy} are of different signs, $f_{xx} \cdot f_{yy}$ cannot be greater than $(f_{xy})^2$, and the function is at a saddle point.

With both signs of the second direct partials negative in (2) and positive in (3), the function *may be* at a relative maximum at $(7,-3)$ and at a relative minimum at $(-7,3)$, but the third condition must be tested first to ensure against the possibility of an inflection point.

c)   From (5.5). take the cross partial derivatives and check to make sure that $z_{xx}(a,b) \cdot z_{yy}(a,b) > [z_{xy}(a,b)]^2$.

$$z_{xy} = 0 \qquad z_{yx} = 0$$
$$z_{xx}(a,b) \cdot z_{yy}(a,b) \qquad > [z_{xy}(a,b)]^2$$

From (2), $(-42) \cdot (-36) \qquad > \qquad (0)^2$
$$1512 > 0$$
From (3), $(42) \quad \cdot \quad (36) \qquad > \qquad (0)^2$
$$1512 > 0$$

The function is maximized at $(7,-3)$ and minimized at $(-7,3)$; for inflection points, see Problems 5.10(c) and 5.11(b) and (c).

5.5   CONSTRAINED OPTIMIZATION WITH LAGRANGE MULTIPLIERS

Differential calculus is also used to maximize or minimize a function subject to constraint. Given a function $f(x,y)$ subject to a constraint $g(x,y) = k$ (a constant), a new function F can be formed by

(1) setting the constraint equal to zero, (2) multiplying it by λ (the *Lagrange multiplier*), and (3) adding the product to the original function:

$$F(x, y, \lambda) = f(x, y) + \lambda[k - g(x, y)] \tag{5.6}$$

Here $F(x, y, \lambda)$ is the *Lagrangian function*, $f(x, y)$ is the original or *objective function*, and $g(x, y)$ is the *constraint*. Since the constraint is always set equal to zero, the product $\lambda[k - g(x, y)]$ also equals zero, and the addition of the term does not change the value of the objective function. Critical values x_0, y_0, and λ_0, at which the function is optimized, are found by taking the partial derivatives of F with respect to *all three* independent variables, setting them equal to zero, and solving simultaneously:

$$F_x(x, y, \lambda) = 0 \qquad F_y(x, y, \lambda) = 0 \qquad F_\lambda(x, y, \lambda) = 0$$

Second-order conditions differ from those of unconstrained optimization and are treated in Section 12.5. See Example 9; Problems 5.12 to 5.14; Sections 6.6, 6.9, and 6.10; and Problems 6.28 to 6.39 and 6.41 to 6.44.

For constraints involving inequalities, see Chapter 13 for concave programming.

EXAMPLE 9.   Optimize the function

$$z = 4x^2 + 3xy + 6y^2$$

subject to the constraint $x + y = 56$.

1.   Set the constraint equal to zero by subtracting the variables from the constant as in (5.6), for reasons to be explained in Section 5.6.

$$56 - x - y = 0$$

Multiply this difference by λ and add the product of the two to the objective function in order to form the Lagrangian function Z.

$$Z = 4x^2 + 3xy + 6y^2 + \lambda(56 - x - y) \tag{5.7}$$

2.   Take the first-order partials, set them equal to zero, and solve simultaneously.

$$Z_x = 8x + 3y - \lambda = 0 \tag{5.8}$$

$$Z_y = 3x + 12y - \lambda = 0 \tag{5.9}$$

$$Z_\lambda = 56 - x - y = 0 \tag{5.10}$$

Subtracting (5.9) from (5.8) to eliminate λ gives

$$5x - 9y = 0 \qquad x = 1.8y$$

Substitute $x = 1.8y$ in (5.10),

$$56 - 1.8y - y = 0 \qquad y_0 = 20$$

From which we find

$$x_0 = 36 \qquad \lambda_0 = 348$$

Substitute the critical values in (5.7),

$$Z = 4(36)^2 + 3(36)(20) + 6(20)^2 + (348)(56 - 36 - 20)$$
$$= 4(1296) + 3(720) + 6(400) + 348(0) = 9744$$

In Chapter 12, Example 5, it will be shown that Z is at a minimum. Notice that at the critical values, the Lagrangian function Z equals the objective function z because the constraint equals zero. See Problems 5.12 to 5.14 and Sections 6.6, 6.9, and 6.10.

5.6 SIGNIFICANCE OF THE LAGRANGE MULTIPLIER

The Lagrange multiplier λ *approximates* the marginal impact on the objective function caused by a small change in the constant of the constraint. With $\lambda = 348$ in Example 9, for instance, a 1-unit increase (decrease) in the constant of the constraint would cause Z to increase (decrease) by approximately 348 units, as is demonstrated in Example 10. Lagrange multipliers are often referred to as *shadow prices*. In utility maximization subject to a budget constraint, for example, λ will estimate the marginal utility of an extra dollar of income. See Problem 6.36.

Note: Since in (5.6) above $\lambda[k - g(x, y)] = \lambda[g(x, y) - k] = 0$, either form can be added to *or* subtracted from the objective function without changing the critical values of x and y. Only the sign of λ will be affected. For the interpretation of λ given in Section 5.6 to be valid, however, the precise form used in Equation (5.6) should be adhered to. See Problems 5.12 to 5.14.

EXAMPLE 10. To verify that a 1-unit change in the constant of the constraint will cause a change of approximately 348 units in Z from Example 9, take the original objective function $z = 4x^2 + 3xy + 6y^2$ and optimize it subject to a new constraint $x + y = 57$ in which the constant of the constraint is 1 unit larger.

$$Z = 4x^2 + 3xy + 6y^2 + \lambda(57 - x - y)$$
$$Z_x = 8x + 3y - \lambda = 0$$
$$Z_y = 3x + 12y - \lambda = 0$$
$$Z_\lambda = 57 - x - y = 0$$

When solved simultaneously this gives

$$x_0 = 36.64 \qquad y_0 = 20.36 \qquad \lambda_0 = 354.2$$

Substituting these values in the Lagrangian function gives $Z = 10{,}095$ which is 351 larger than the old constrained optimum of 9744, close to the approximation of the 348 increment suggested by λ.

5.7 DIFFERENTIALS

In Section 3.4 the derivative dy/dx was presented as a single symbol denoting the limit of $\Delta y/\Delta x$ as Δx approaches zero. The derivative dy/dx may also be treated as a ratio of differentials in which dy is the differential of y and dx the differential of x. Given a function of a single independent variable $y = f(x)$, the *differential of* y, dy, measures the change in y resulting from a small change in x, written dx.

Given $y = 2x^2 + 5x + 4$, the differential of y is found by first taking the derivative of y with respect to x, which measures the rate at which y changes for a small change in x,

$$\frac{dy}{dx} = 4x + 5 \qquad \text{a derivative or rate of change}$$

and then multiplying that rate at which y changes for a small change in x by a specific change in $x(dx)$ to find the resulting change in $y(dy)$.

$$dy = (4x + 5)\, dx \qquad \text{a differential or simple change}$$

Change in y = rate at which y changes for a small change in $x \cdot$ a small change in x.

EXAMPLE 11.

1. If $y = 4x^3 + 5x^2 - 7$, then $dy/dx = 12x^2 + 10x$ and the differential is

$$dy = (12x^2 + 10x)\, dx$$

2. If $y = (2x - 5)^2$, then $dy/dx = 2(2x - 5)(2) = 8x - 20$ and the differential is

$$dy = (8x - 20)\, dx$$

See Problem 5.15.

5.8 TOTAL AND PARTIAL DIFFERENTIALS

For a function of two or more independent variables, the *total differential* measures the change in the dependent variable brought about by a small change in each of the independent variables. If $z = f(x, y)$, the total differential dz is expressed mathematically as

$$dz = z_x \, dx + z_y \, dy \tag{5.11}$$

where z_x and z_y are the partial derivatives of z with respect to x and y respectively, and dx and dy are small changes in x and y. The total differential can thus be found by taking the partial derivatives of the function with respect to each independent variable and substituting these values in the formula above.

EXAMPLE 12. The total differential is found as follows:

1. Given: $z = x^4 + 8xy + 3y^3$

$$z_x = 4x^3 + 8y \qquad z_y = 8x + 9y^2$$

which, when substituted in the total differential formula, gives

$$dz = (4x^3 + 8y) \, dx + (8x + 9y^2) \, dy$$

2. Given: $z = (x - y)/(x + 1)$

$$z_x = \frac{(x+1)(1) - (x-y)(1)}{(x+1)^2} = \frac{y+1}{(x+1)^2}$$

$$z_y = \frac{(x+1)(-1) - (x-y)(0)}{(x+1)^2} = \frac{-1(x+1)}{(x+1)^2} = \frac{-1}{x+1}$$

The total differential is $dz = \dfrac{y+1}{(x+1)^2} \, dx - \left(\dfrac{1}{x+1} \right) dy$

If one of the independent variables is held constant, for example, $dy = 0$, we then have a partial differential:

$$dz = z_x \, dx$$

A *partial differential* measures the change in the dependent variable of a multivariate function resulting from a small change in one of the independent variables and assumes the other independent variables are constant. See Problems 5.16 and 5.17 and 6.45 to 6.52.

5.9 TOTAL DERIVATIVES

Given a case where $z = f(x, y)$ and $y = g(x)$, that is, when x and y are not independent, a change in x will affect z directly through the function f and indirectly through the function g. This is illustrated in the channel map in Fig. 5-3. To measure the effect of a change in x on z when x and y are not independent, the total derivative must be found. The *total derivative* measures the *direct* effect of x

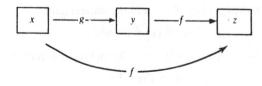

Fig. 5-3

on z, $\partial z/\partial x$, *plus* the *indirect* effect of x on z through y, $\dfrac{\partial z}{\partial y}\dfrac{dy}{dx}$. In brief, the total derivative is

$$\frac{dz}{dx} = z_x + z_y \frac{dy}{dx} \qquad\qquad (5.12)$$

See Examples 13 to 15 and Problems 5.18 and 5.19.

EXAMPLE 13.   An alternative method of finding the total derivative is to take the total differential of z

$$dz = z_x\, dx + z_y\, dy$$

and divide through mentally by dx. Thus,

$$\frac{dz}{dx} = z_x \frac{dx}{dx} + z_y \frac{dy}{dx}$$

Since $dx/dx = 1$, $\dfrac{dz}{dx} = z_x + z_y \dfrac{dy}{dx}$

EXAMPLE 14.   Given

$$z = f(x, y) = 6x^3 + 7y$$

where $y = g(x) = 4x^2 + 3x + 8$, the total derivative dz/dx with respect to x is

$$\frac{dz}{dx} = z_x + z_y \frac{dy}{dx}$$

where $z_x = 18x^2$, $z_y = 7$, and $dy/dx = 8x + 3$. Substituting above,

$$\frac{dz}{dx} = 18x^2 + 7(8x + 3) = 18x^2 + 56x + 21$$

To check the answer, substitute $y = 4x^2 + 3x + 8$ in the original function to make z a function of x alone and then take the derivative as follows:

$$z = 6x^3 + 7(4x^2 + 3x + 8) = 6x^3 + 28x^2 + 21x + 56$$

Thus, $\dfrac{dz}{dx} = 18x^2 + 56x + 21$

EXAMPLE 15.   The total derivative can be expanded to accommodate other interconnections as well. Given

$$z = 8x^2 + 3y^2 \qquad x = 4t \qquad y = 5t$$

the total derivative of z with respect to t then becomes

$$\frac{dz}{dt} = z_x \frac{dx}{dt} + z_y \frac{dy}{dt}$$

where $z_x = 16x$, $z_y = 6y$, $dx/dt = 4$, and $dy/dt = 5$. Substituting above,

$$\frac{dz}{dt} = 16x(4) + 6y(5) = 64x + 30y$$

Then substituting $x = 4t$ and $y = 5t$ immediately above,

$$\frac{dz}{dt} = 64(4t) + 30(5t) = 406t$$

5.10 IMPLICIT AND INVERSE FUNCTION RULES

As seen in Section 3.9, functions of the form $y = f(x)$ express y explicitly in terms of x and are called *explicit functions*. Functions of the form $f(x, y) = 0$ do not express y explicitly in terms of x and are called *implicit functions*. If an implicit function $f(x, y) = 0$ exists and $f_y \neq 0$ at the point around which the implicit function is defined, the total differential is simply $f_x\, dx + f_y\, dy = 0$.

Recalling that a derivative is a ratio of differentials, we can then rearrange the terms to get the *implicit function rule*:

$$\frac{dy}{dx} = \frac{-f_x}{f_y} \tag{5.13}$$

Notice that the derivative dy/dx is the *negative* of the *reciprocal* of the corresponding partials.

$$\frac{dy}{dx} = \frac{-f_x}{f_y} = -\frac{1}{f_y/f_x}$$

Given a function $y = f(x)$, an *inverse function* $x = f^{-1}(y)$ exists if each value of y yields one and only one value of x. Assuming the inverse function exists, the *inverse function rule* states that the derivative of the inverse function is the reciprocal of the derivative of the original function. Thus, if $Q = f(P)$ is the original function, the derivative of the original function is dQ/dP, the derivative of the inverse function $[P = f^{-1}(Q)]$ is dP/dQ, and

$$\frac{dP}{dQ} = \frac{1}{dQ/dP} \qquad \text{provided } \frac{dQ}{dP} \neq 0 \tag{5.14}$$

See Examples 16 and 17 and Problems 5.20 to 5.22, 6.51, and 6.52.

EXAMPLE 16. Given the implicit functions:

(*a*) $7x^2 - y = 0$ (*b*) $3x^4 - 7y^5 - 86 = 0$

the derivative dy/dx is found as follows:

a) From (*5.13*),

$$\frac{dy}{dx} = -\frac{f_x}{f_y}$$

Here $f_x = 14x$ and $f_y = -1$. Substituting above,

$$\frac{dy}{dx} = -\frac{14x}{(-1)} = 14x$$

The function in this case was deliberately kept simple so the answer could easily be checked by solving for y in terms of x and then taking the derivative directly. Since $y = 7x^2$, $dy/dx = 14x$.

b)

$$\frac{dy}{dx} = -\frac{f_x}{f_y} = -\frac{12x^3}{-35y^4} = \frac{12x^3}{35y^4}$$

Compare this answer with that in Example 16 of Chapter 3.

EXAMPLE 17. Find the derivative for the inverse of the following functions:

1. Given $Q = 20 - 2P$,

$$\frac{dP}{dQ} = \frac{1}{dQ/dP}$$

where $dQ/dP = -2$. Thus,

$$\frac{dP}{dQ} = \frac{1}{-2} = -\frac{1}{2}$$

2. Given $Q = 25 + 3P^3$,

$$\frac{dP}{dQ} = \frac{1}{dQ/dP} = \frac{1}{9P^2} \qquad (P \neq 0)$$

Solved Problems

FIRST-ORDER PARTIAL DERIVATIVES

5.1.   Find the first-order partial derivatives for each of the following functions:

a) $z = 8x^2 + 14xy + 5y^2$

  $z_x = 16x + 14y$

  $z_y = 14x + 10y$

b) $z = 4x^3 + 2x^2 y - 7y^5$

  $z_x = 12x^2 + 4xy$

  $z_y = 2x^2 - 35y^4$

c) $z = 6w^3 + 4wx + 3x^2 - 7xy - 8y^2$

  $z_w = 18w^2 + 4x$

  $z_x = 4w + 6x - 7y$

  $z_y = -7x - 16y$

d) $z = 2w^2 + 8wxy - x^2 + y^3$

  $z_w = 4w + 8xy$

  $z_x = 8wy - 2x$

  $z_y = 8wx + 3y^2$

5.2.   Use the product rule from (5.2) to find the first-order partials for each of the following functions:

a) $z = 3x^2(5x + 7y)$

  $z_x = 3x^2(5) + (5x + 7y)(6x)$

    $= 45x^2 + 42xy$

  and $z_y = 3x^2(7) + (5x + 7y)(0)$

    $= 21x^2$

b) $z = (9x - 4y)(12x + 2y)$

  $z_x = (9x - 4y)(12) + (12x + 2y)(9)$

    $= 108x - 48y + 108x + 18y = 216x - 30y$

  and $z_y = (9x - 4y)(2) + (12x + 2y)(-4)$

    $= 18x - 8y - 48x - 8y = -30x - 16y$

c) $z = (2x^2 + 6y)(5x - 3y^3)$

  $z_x = (2x^2 + 6y)(5) + (5x - 3y^3)(4x)$

    $= 10x^2 + 30y + 20x^2 - 12xy^3$

    $= 30x^2 + 30y - 12xy^3$

  and $z_y = (2x^2 + 6y)(-9y^2) + (5x - 3y^3)(6)$

    $= -18x^2 y^2 - 54y^3 + 30x - 18y^3$

    $= -72y^3 - 18x^2 y^2 + 30x$

d) $z = (w - x - y)(3w + 2x - 4y)$

  $z_w = (w - x - y)(3) + (3w + 2x - 4y)(1)$

    $= 6w - x - 7y$

  $z_x = (w - x - y)(2) + (3w + 2x - 4y)(-1)$

    $= -w - 4x + 2y$

  and $z_y = (w - x - y)(-4) + (3w + 2x - 4y)(-1)$

    $= -7w + 2x + 8y$

5.3. Use the quotient rule from (5.3) to find the first-order partials of the following functions:

a) $z = \dfrac{5x}{6x - 7y}$

$$z_x = \frac{(6x - 7y)(5) - (5x)(6)}{(6x - 7y)^2}$$

$$= \frac{-35y}{(6x - 7y)^2}$$

and $z_y = \dfrac{(6x - 7y)(0) - (5x)(-7)}{(6x - 7y)^2}$

$$= \frac{35x}{(6x - 7y)^2}$$

b) $z = \dfrac{x + y}{3y}$

$$z_x = \frac{3y(1) - (x + y)(0)}{(3y)^2}$$

$$= \frac{1}{3y}$$

and $z_y = \dfrac{3y(1) - (x + y)(3)}{(3y)^2}$

$$= \frac{-3x}{(3y)^2} = \frac{-x}{3y^2}$$

c) $z = \dfrac{4x - 9y}{5x + 2y}$

$$z_x = \frac{(5x + 2y)(4) - (4x - 9y)(5)}{(5x + 2y)^2}$$

$$= \frac{53y}{(5x + 2y)^2}$$

and $z_y = \dfrac{(5x + 2y)(-9) - (4x - 9y)(2)}{(5x + 2y)^2}$

$$= \frac{-53x}{(5x + 2y)^2}$$

d) $z = \dfrac{x^2 - y^2}{3x + 2y}$

$$z_x = \frac{(3x + 2y)(2x) - (x^2 - y^2)(3)}{(3x + 2y)^2}$$

$$= \frac{3x^2 + 4xy + 3y^2}{(3x + 2y)^2}$$

and $z_y = \dfrac{(3x + 2y)(-2y) - (x^2 - y^2)(2)}{(3x + 2y)^2}$

$$= \frac{-2x^2 - 6xy - 2y^2}{(3x + 2y)^2}$$

5.4. Find the first-order partial derivatives for each of the following functions by using the generalized power function rule from (5.4):

a) $z = (x + y)^2$

$\quad z_x = 2(x + y)(1)$

$\quad\quad = 2(x + y)$

and $z_y = 2(x + y)(1)$

$\quad\quad = 2(x + y)$

b) $z = (2x - 5y)^3$

$\quad z_x = 3(2x - 5y)^2(2)$

$\quad\quad = 6(2x - 5y)^2$

and $z_y = 3(2x - 5y)^2(-5)$

$\quad\quad = -15(2x - 5y)^2$

c) $z = (7x^2 + 4y^3)^5$

$\quad z_x = 5(7x^2 + 4y^3)^4(14x)$

$\quad\quad = 70x(7x^2 + 4y^3)^4$

and $z_y = 5(7x^2 + 4y^3)^4(12y^2)$

$\quad\quad = 60y^2(7x^2 + 4y^3)^4$

d) $z = (5w + 4x + 7y)^3$

$\quad z_w = 3(5w + 4x + 7y)^2(5)$

$\quad\quad = 15(5w + 4x + 7y)^2$

$\quad z_x = 3(5w + 4x + 7y)^2(4)$

$\quad\quad = 12(5w + 4x + 7y)^2$

and $z_y = 3(5w + 4x + 7y)^2(7)$

$\quad\quad = 21(5w + 4x + 7y)^2$

5.5. Use whatever combination of rules is necessary to find the first-order partials for the following functions:

a) $z = \dfrac{(5x^2 - 7y)(3x^2 + 8y)}{4x + 2y}$

Using the quotient rule and product rule,

$$z_x = \frac{(4x+2y)[(5x^2-7y)(6x)+(3x^2+8y)(10x)]-(5x^2-7y)(3x^2+8y)(4)}{(4x+2y)^2}$$

$$= \frac{(4x+2y)(30x^3-42xy+30x^3+80xy)-(5x^2-7y)(12x^2+32y)}{(4x+2y)^2}$$

$$= \frac{(4x+2y)(60x^3+38xy)-(5x^2-7y)(12x^2+32y)}{(4x+2y)^2}$$

and $$z_y = \frac{(4x+2y)[(5x^2-7y)(8)+(3x^2+8y)(-7)]-(5x^2-7y)(3x^2+8y)(2)}{(4x+2y)^2}$$

$$= \frac{(4x+2y)(40x^2-56y-21x^2-56y)-(5x^2-7y)(6x^2+16y)}{(4x+2y)^2}$$

$$= \frac{(4x+2y)(19x^2-112y)-(5x^2-7y)(6x^2+16y)}{(4x+2y)^2}$$

b) $z = (5x^2-4y)^2(2x+7y^3)$

Using the product rule and the generalized power function rule,

$$z_x = (5x^2-4y)^2(2)+(2x+7y^3)[2(5x^2-4y)(10x)]$$
$$= 2(5x^2-4y)^2+(2x+7y^3)(100x^3-80xy)$$
and
$$z_y = (5x^2-4y)^2(21y^2)+(2x+7y^3)[2(5x^2-4y)(-4)]$$
$$= 21y^2(5x^2-4y)^2+(2x+7y^3)(-40x^2+32y)$$

c) $z = \dfrac{(3x+11y)^3}{2x+6y}$

Using the quotient rule and the generalized power function rule,

$$z_x = \frac{(2x+6y)[3(3x+11y)^2(3)]-(3x+11y)^3(2)}{(2x+6y)^2}$$

$$= \frac{(18x+54y)(3x+11y)^2-2(3x+11y)^3}{(2x+6y)^2}$$

and $$z_y = \frac{(2x+6y)[3(3x+11y)^2(11)]-(3x+11y)^3(6)}{(2x+6y)^2}$$

$$= \frac{(66x+198y)(3x+11y)^2-6(3x+11y)^3}{(2x+6y)^2}$$

d) $z = \left(\dfrac{8x+7y}{5x+2y}\right)^2$

Using the generalized power function rule and the quotient rule,

$$z_x = 2\left(\frac{8x+7y}{5x+2y}\right)\left[\frac{(5x+2y)(8)-(8x+7y)(5)}{(5x+2y)^2}\right]$$

$$= \frac{16x+14y}{5x+2y}\left[\frac{-19y}{(5x+2y)^2}\right] = \frac{-(266y^2+304xy)}{(5x+2y)^3}$$

and $$z_y = 2\left(\frac{8x+7y}{5x+2y}\right)\left[\frac{(5x+2y)(7)-(8x+7y)(2)}{(5x+2y)^2}\right]$$

$$= \frac{16x+14y}{5x+2y}\left[\frac{19x}{(5x+2y)^2}\right] = \frac{304x^2+266xy}{(5x+2y)^3}$$

SECOND-ORDER PARTIAL DERIVATIVES

5.6.   Find the second-order direct partial derivatives z_{xx} and z_{yy} for each of the following functions:

a) $z = x^2 + 2xy + y^2$

          $z_x = 2x + 2y$ $z_y = 2x + 2y$

          $z_{xx} = 2$ $z_{yy} = 2$

b) $z = x^3 - 9xy - 3y^3$

          $z_x = 3x^2 - 9y$ $z_y = -9x - 9y^2$

          $z_{xx} = 6x$ $z_{yy} = -18y$

c) $z = 2xy^4 + 7x^3y$

          $z_x = 2y^4 + 21x^2y$ $z_y = 8xy^3 + 7x^3$

          $z_{xx} = 42xy$ $z_{yy} = 24xy^2$

d) $z = x^4 + x^3y^2 - 3xy^3 - 2y^3$

          $z_x = 4x^3 + 3x^2y^2 - 3y^3$ $z_y = 2x^3y - 9xy^2 - 6y^2$

          $z_{xx} = 12x^2 + 6xy^2$ $z_{yy} = 2x^3 - 18xy - 12y$

e) $z = (12x - 7y)^2$

          $z_x = 2(12x - 7y)(12)$ $z_y = 2(12x - 7y)(-7)$

          $= 288x - 168y$ $= -168x + 98y$

          $z_{xx} = 288$ $z_{yy} = 98$

f) $z = (7x + 3y)^3$

          $z_x = 3(7x + 3y)^2(7)$ $z_y = 3(7x + 3y)^2(3)$

          $= 21(7x + 3y)^2$ $= 9(7x + 3y)^2$

          $z_{xx} = 42(7x + 3y)(7)$ $z_{yy} = 18(7x + 3y)(3)$

          $= 2058x + 882y$ $= 378x + 162y$

g) $z = (x^2 + 2y)^4$

          $z_x = 4(x^2 + 2y)^3(2x) = 8x(x^2 + 2y)^3$ $z_y = 4(x^2 + 2y)^3(2) = 8(x^2 + 2y)^3$

          $^*z_{xx} = 8x[3(x^2 + 2y)^2(2x)] + (x^2 + 2y)^3(8)$ $z_{yy} = 24(x^2 + 2y)^2(2) = 48(x^2 + 2y)^2$

          $= 48x^2(x^2 + 2y)^2 + 8(x^2 + 2y)^3$

5.7.   Find the cross partial derivatives z_{xy} and z_{yx} for each of the following functions:

a) $z = 3x^2 + 12xy + 5y^2$

          $z_x = 6x + 12y$ $z_y = 12x + 10y$

          $z_{xy} = 12$ $z_{yx} = 12$

b) $z = x^3 - xy - 2y^3$

          $z_x = 3x^2 - y$ $z_y = -x - 6y^2$

          $z_{xy} = -1$ $z_{yx} = -1$

c) $z = 8x^2y - 11xy^3$

          $z_x = 16xy - 11y^3$ $z_y = 8x^2 - 33xy^2$

          $z_{xy} = 16x - 33y^2$ $z_{yx} = 16x - 33y^2$

*By the product rule.

d) $z = (8x - 4y)^5$

$$z_x = 5(8x - 4y)^4(8) \qquad\qquad z_y = 5(8x - 4y)^4(-4)$$
$$= 40(8x - 4y)^4 \qquad\qquad = -20(8x - 4y)^4$$
$$z_{xy} = 160(8x - 4y)^3(-4) \qquad z_{yx} = -80(8x - 4y)^3(8)$$
$$= -640(8x - 4y)^3 \qquad\qquad = -640(8x - 4y)^3$$

In items (*a*) through (*d*) above, notice how, in accord with Young's theorem, $z_{xy} = z_{yx}$ no matter which first partial is taken initially.

5.8.   Find the first-order and second-order direct partial derivatives for the following functions:

a) $z = x^{0.4}y^{0.6}$

$$z_x = 0.4x^{-0.6}y^{0.6} \qquad\qquad z_y = 0.6x^{0.4}y^{-0.4}$$
$$z_{xx} = -0.24x^{-1.6}y^{0.6} \qquad z_{yy} = -0.24x^{0.4}y^{-1.4}$$

b) $f(x, y) = x^{0.7}y^{0.2}$

$$f_x = 0.7x^{-0.3}y^{0.2} \qquad\qquad f_y = 0.2x^{0.7}y^{-0.8}$$
$$f_{xx} = -0.21x^{-1.3}y^{0.2} \qquad f_{yy} = -0.16x^{0.7}y^{-1.8}$$

c) $z = 2w^6x^5y^3$

$$z_w = 12w^5x^5y^3 \qquad z_x = 10w^6x^4y^3 \qquad z_y = 6w^6x^5y^2$$
$$z_{ww} = 60w^4x^5y^3 \qquad z_{xx} = 40w^6x^3y^3 \qquad z_{yy} = 12w^6x^5y$$

d) $f(x, y, z) = 10x^3y^2z^4$

$$f_x = 30x^2y^2z^4 \qquad f_y = 20x^3yz^4 \qquad f_z = 40x^3y^2z^3$$
$$f_{xx} = 60xy^2z^4 \qquad f_{yy} = 20x^3z^4 \qquad f_{zz} = 120x^3y^2z^2$$

5.9.   Find the cross partials for each of the following functions:

a) $z = x^{0.3}y^{0.5}$

$$z_x = 0.3x^{-0.7}y^{0.5} \qquad\qquad z_y = 0.5x^{0.3}y^{-0.5}$$
$$z_{xy} = 0.15x^{-0.7}y^{-0.5} \qquad z_{yx} = 0.15x^{-0.7}y^{-0.5}$$

b) $f(x, y) = x^{0.1}y^{0.8}$

$$f_x = 0.1x^{-0.9}y^{0.8} \qquad\qquad f_y = 0.8x^{0.1}y^{-0.2}$$
$$f_{xy} = 0.08x^{-0.9}y^{-0.2} \qquad f_{yx} = 0.08x^{-0.9}y^{-0.2}$$

c) $z = w^3x^4y^3$

$$z_w = 3w^2x^4y^3 \qquad z_x = 4w^3x^3y^3 \qquad z_y = 3w^3x^4y^2$$
$$z_{wx} = 12w^2x^3y^3 \qquad z_{xw} = 12w^2x^3y^3 \qquad z_{yw} = 9w^2x^4y^2$$
$$z_{wy} = 9w^2x^4y^2 \qquad z_{xy} = 12w^3x^3y^2 \qquad z_{yx} = 12w^3x^3y^2$$

d) $f(x, y, z) = x^3y^{-4}z^{-5}$

$$f_x = 3x^2y^{-4}z^{-5} \qquad f_y = -4x^3y^{-5}z^{-5} \qquad f_z = -5x^3y^{-4}z^{-6}$$
$$f_{xy} = -12x^2y^{-5}z^{-5} \qquad f_{yx} = -12x^2y^{-5}z^{-5} \qquad f_{zx} = -15x^2y^{-4}z^{-6}$$
$$f_{xz} = -15x^2y^{-4}z^{-6} \qquad f_{yz} = 20x^3y^{-5}z^{-6} \qquad f_{zy} = 20x^3y^{-5}z^{-6}$$

Note how by Young's theorem in (*c*) $z_{wx} = z_{xw}$, $z_{yw} = z_{wy}$, and $z_{xy} = z_{yx}$ and in (*d*) $f_{xy} = f_{yx}$, $f_{xz} = f_{zx}$, and $f_{yz} = f_{zy}$.

OPTIMIZING MULTIVARIABLE FUNCTIONS

5.10. For each of the following quadratic functions, (1) find the critical points at which the function may be optimized and (2) determine whether at these points the function is maximized, is minimized, is at an inflection point, or is at a saddle point.

a) $z = 3x^2 - xy + 2y^2 - 4x - 7y + 12$

1) Take the first-order partial derivatives, set them equal to zero, and solve simultaneously, using the methods of Section 1.4.

$$z_x = 6x - y - 4 = 0 \qquad\qquad (5.15)$$
$$z_y = -x + 4y - 7 = 0 \qquad\qquad (5.16)$$
$$z = 1 \quad\quad y = 2 \quad\quad (1,2) \quad\quad \text{critical point}$$

2) Take the second-order direct partial derivatives from (5.15) and (5.16), evaluate them at the critical point, and check signs.

$$z_{xx} = 6 \qquad\qquad z_{yy} = 4$$
$$z_{xx}(1,2) = 6 > 0 \qquad z_{yy}(1,2) = 4 > 0$$

With both second-order direct partial derivatives everywhere positive, the function is possibly at a global minimum. Now take the cross partial from (5.15) or (5.16),

$$z_{xy} = -1 = z_{yx}$$

evaluate it at the critical point and test the third condition:

$$z_{xy}(1,2) = -1 = z_{yx}(1,2)$$
$$z_{xx}(1,2) \quad\cdot\quad z_{yy}(1,2) > [z_{xy}(1,2)]^2$$
$$6 \quad\quad\cdot\quad\quad 4 \;>\;\; (-1)^2$$

With $z_{xx}z_{yy} > (z_{xy})^2$ and $z_{xx}, z_{yy} > 0$, the function is at a global minimum at $(1,2)$.

b) $f(x,y) = 60x + 34y - 4xy - 6x^2 - 3y^2 + 5$

1) Take the first-order partials, set them equal to zero, and solve.

$$f_x = 60 - 4y - 12x = 0 \qquad\qquad (5.17)$$
$$f_y = 34 - 6y - 4x = 0 \qquad\qquad (5.18)$$
$$x = 4 \quad\quad y = 3 \quad\quad (4,3) \quad\quad \text{critical point}$$

2) Take the second-order direct partials, evaluate them at the critical point, and check their signs.

$$f_{xx} = -12 \qquad\qquad f_{yy} = -6$$
$$f_{xx}(4,3) = -12 < 0 \qquad f_{yy}(4,3) = -6 < 0$$

Take the cross partial from (5.17) or (5.18),

$$f_{xy} = -4 = f_{yx}$$

evaluate it at the critical point and test the third condition:

$$f_{xy}(4,3) = -4 = f_{yx}(4,3)$$
$$f_{xx}(4,3) \cdot f_{yy}(4,3) > [f_{xy}(4,3)]^2$$
$$-12 \quad\cdot\quad -6 \;>\;\; (-4)^2$$

With $f_{xx}f_{yy} > (f_{xy})^2$ and $f_{xx}, f_{yy} < 0$, the function is at a global maximum at $(4,3)$.

c) $z = 48y - 3x^2 - 6xy - 2y^2 + 72x$

   1)

$$z_x = -6x - 6y + 72 = 0$$
$$z_y = -6x - 4y + 48 = 0$$
$$x = 0 \quad y = 12 \quad (0, 12) \quad \text{critical point}$$

   2)   Test the second-order direct partials at the critical point.

$$z_{xx} = -6 \qquad\qquad z_{yy} = -4$$
$$z_{xx}(0, 12) = -6 < 0 \qquad z_{yy}(0, 12) = -4 < 0$$

With z_{xx} and $z_{yy} < 0$ for all values, the function may be at a global maximum. Test the cross partials to be sure.

$$z_{xy} = -6 = z_{yx}$$
$$z_{xy}(0, 12) = -6 = z_{yx}(0, 12)$$
$$z_{xx}(0, 12) \cdot z_{yy}(0, 12) > [z_{xy}(0, 12)]^2$$
$$-6 \quad \cdot \quad -4 \quad < (-6)^2$$

With z_{xx} and z_{yy} of the same sign and $z_{xx}z_{yy} < (z_{xy})^2$, the function is at an inflection point at $(0, 12)$.

d) $f(x, y) = 5x^2 - 3y^2 - 30x + 7y + 4xy$

   1)
$$f_x = 10x + 4y - 30 = 0$$
$$f_y = 4x - 6y + 7 = 0$$
$$x = 2 \quad y = 2.5 \quad (2, 2.5) \quad \text{critical point}$$

   2)
$$f_{xx} = 10 \qquad\qquad f_{yy} = -6$$
$$f_{xx}(2, 2.5) = 10 > 0 \qquad f_{yy}(2, 2.5) = -6 < 0$$

Testing the cross partials,

$$f_{xy} = 4 = f_{yx}$$
$$f_{xx}(2, 2.5) \quad \cdot \quad f_{yy}(2, 2.5) > [f_{xy}(2, 2.5)]^2$$
$$10 \quad \cdot \quad -6 \quad < \quad 4^2$$

Whenever f_{xx} and f_{yy} are of different signs, $f_{xx}f_{yy}$ cannot be greater than $(f_{xy})^2$, and the function will be at a saddle point.

5.11.   For the following cubic functions, (1) find the critical points and (2) determine if at these points the function is at a relative maximum, relative minimum, inflection point, or saddle point.

a) $z(x, y) = 3x^3 - 5y^2 - 225x + 70y + 23$

   1)   Take the first-order partials and set them equal to zero.

$$z_x = 9x^2 - 225 = 0 \tag{5.19}$$
$$z_y = -10y + 70 = 0 \tag{5.20}$$

Solve for the critical points.

$$9x^2 = 225 \qquad -10y = -70$$
$$x^2 = 25 \qquad\qquad y = 7$$
$$x = \pm 5$$
$$(5, 7) \quad (-5, 7) \quad \text{critical points}$$

2) From (*5.19*) and (*5.20*), take the second-order direct partials,

$$z_{xx} = 18x \qquad z_{yy} = -10$$

evaluate them at the critical points and note the signs.

$$z_{xx}(5,7) = 18(5) = 90 > 0 \qquad\qquad z_{yy}(5,7) = -10 < 0$$
$$z_{xx}(-5,7) = 18(-5) = -90 < 0 \qquad z_{yy}(-5,7) = -10 < 0$$

Then take the cross partial from (*5.19*) or (*5.20*),

$$z_{xy} = 0 = z_{yx}$$

evaluate it at the critical points and test the third condition.

	$z_{xx}(a,b) \cdot z_{yy}(a,b)$		$> [z_{xy}(a,b)]^2$
At (5,7),	90 $\cdot$ -10	<	0
At (−5,7),	-90 $\cdot$ -10	>	0

With $z_{xx}z_{yy} > (z_{xy})^2$ and z_{xx}, $z_{yy} < 0$ at $(-5,7)$, $z(-5,7)$ is a **relative maximum.** With $z_{xx}z_{yy} < (z_{xy})^2$ and z_{xx} and z_{yy} of different signs at $(5,7)$, $z(5,7)$ is a **saddle point.**

b) $f(x,y) = 3x^3 + 1.5y^2 - 18xy + 17$

1) Set the first-order partial derivatives equal to zero,

$$f_x = 9x^2 - 18y = 0 \tag{5.21}$$
$$f_y = 3y - 18x = 0 \tag{5.22}$$

and solve for the critical values:

$$18y = 9x^2 \qquad 3y = 18x$$
$$y = \tfrac{1}{2}x^2 \qquad y = 6x \tag{5.23}$$

Setting *y* equal to *y*,

$$\tfrac{1}{2}x^2 = 6x$$
$$x^2 - 12x = 0$$
$$x(x - 12) = 0$$
$$x = 0 \qquad x = 12$$

Substituting $x = 0$ and $x = 12$ in $y = 6x$ from (*5.23*),

$$y = 6(0) = 0$$
$$y = 6(12) = 72$$

Therefore, $(0,0)$ $(12,72)$ critical points

2) Take the second-order direct partials from (*5.21*) and (*5.22*),

$$f_{xx} = 18x \qquad f_{yy} = 3$$

evaluate them at the critical points and note the signs.

$$f_{xx}(0,0) = 18(0) = 0 \qquad\qquad f_{yy}(0,0) = 3 > 0$$
$$f_{xx}(12,72) = 18(12) = 216 > 0 \qquad f_{yy}(12,72) = 3 > 0$$

Then take the cross partial from (*5.21*) or (*5.22*),

$$f_{xy} = -18 = f_{yx}$$

evaluate it at the critical points and test the third conditon.

	$f_{xx}(a,b)$ $\cdot$ $f_{yy}(a,b) > [f_{xy}(a,b)]^2$		
At (0,0),	0 $\cdot$ 3	<	$(-18)^2$
At (12,72),	216 $\cdot$ 3	>	$(-18)^2$
	$648 > 324$		

With $f_{xx}f_{yy} > (f_{xy})^2$ and $f_{xx}, f_{yy} > 0$ at $(12, 72)$, $f(12, 72)$ is a relative minimum. With $f_{xx}f_{yy} < (f_{xy})^2$ and f_{xx} and f_{yy} of the same sign at $(0, 0)$, $f(0, 0)$ is an inflection point.

c) $f = 3x^3 - 9xy + 3y^3$

1)
$$f_x = 9x^2 - 9y = 0 \tag{5.24}$$
$$f_y = 9y^2 - 9x = 0 \tag{5.25}$$

From (5.24),
$$9y = 9x^2 \qquad y = x^2$$

Substitute $y = x^2$ in (5.25),

$$9(x^2)^2 - 9x = 0$$
$$9x^4 - 9x = 0$$
$$9x(x^3 - 1) = 0$$
$$9x = 0 \qquad \text{or} \qquad x^3 - 1 = 0$$
$$x = 0 \qquad\qquad\qquad x^3 = 1$$
$$x = 1$$

Substituting these values in (5.24), we find that if $x = 0$, $y = 0$, and if $x = 1$, $y = 1$. Therefore,

$$(0, 0) \qquad (1, 1) \qquad \text{critical points}$$

2)   Test the second-order conditons from (5.24) and (5.25).

$$f_{xx} = 18x \qquad\qquad\qquad f_{yy} = 18y$$
$$f_{xx}(0, 0) = 18(0) = 0 \qquad f_{yy}(0, 0) = 18(0) = 0$$
$$f_{xx}(1, 1) = 18(1) = 18 > 0 \qquad f_{yy}(1, 1) = 18(1) = 18 > 0$$
$$f_{xy} = -9 = f_{yx}$$

$$f_{xx}(a, b) \cdot f_{yy}(a, b) > [f_{xy}(a, b)]^2$$

At $(0, 0)$, $0 \quad \cdot \quad 0 \quad < \quad (-9)^2$
At $(1, 1)$, $18 \quad \cdot \quad 18 \quad > \quad (-9)^2$

With f_{xx} and $f_{yy} > 0$ and $f_{xx}f_{yy} > (f_{xy})^2$ at $(1, 1)$, the function is at a relative minimum at $(1, 1)$. With f_{xx} and f_{yy} of the same sign at $(0, 0)$ and $f_{xx}f_{yy} < (f_{xy})^2$, the function is at an inflection point at $(0, 0)$.

d) $f(x, y) = x^3 - 6x^2 + 2y^3 + 9y^2 - 63x - 60y$

1)
$$f_x = 3x^2 - 12x - 63 = 0 \qquad f_y = 6y^2 + 18y - 60 = 0 \tag{5.26}$$
$$3(x^2 - 4x - 21) = 0 \qquad\qquad 6(y^2 + 3y - 10) = 0$$
$$(x + 3)(x - 7) = 0 \qquad\qquad (y - 2)(y + 5) = 0$$
$$x = -3 \quad x = 7 \qquad\qquad\qquad y = 2 \quad y = -5$$

Hence $(-3, 2) \qquad (-3, -5) \qquad (7, 2) \qquad (7, -5) \qquad \text{critical points}$

2)   Test the second-order direct partials at each of the critical points. From (5.26),

$$f_{xx} = 6x - 12 \qquad\qquad f_{yy} = 12y + 18$$

(i) $f_{xx}(-3, 2) = -30 < 0 \qquad f_{yy}(-3, 2) = 42 > 0$
(ii) $f_{xx}(-3, -5) = -30 < 0 \qquad f_{yy}(-3, -5) = -42 < 0$
(iii) $f_{xx}(7, 2) = 30 > 0 \qquad f_{yy}(7, 2) = 42 > 0$
(iv) $f_{xx}(7, -5) = 30 > 0 \qquad f_{yy}(7, -5) = -42 < 0$

With different signs in (i) and (iv), $(-3,2)$ and $(7,-5)$ can be ignored, if desired, as saddle points. Now take the cross partial from (5.26) and test the third condition.

$$f_{xy} = 0 = f_{yx}$$
$$f_{xx}(a,b) \quad \cdot \quad f_{yy}(a,b) > [f_{xy}(a,b)]^2$$

From (ii), $(-30) \quad \cdot \quad (-42) > \quad (0)^2$

From (iii), $(30) \quad \cdot \quad (42) > \quad (0)^2$

The function is at a relative maximum at $(-3,-5)$, at a relative minimum at $(7,2)$, and at a saddle point at $(-3,2)$ and $(7,-5)$.

CONSTRAINED OPTIMIZATION AND LAGRANGE MULTIPLIERS

5.12.   (1) Use Lagrange multipliers to optimize the following functions subject to the given constraint, and (2) estimate the effect on the value of the objective function from a 1-unit change in the constant of the constraint.

a) $z = 4x^2 - 2xy + 6y^2$ subject to $x + y = 72$

1)   Set the constraint equal to zero, multiply it by λ, and add it to the objective function, to obtain

$$Z = 4x^2 - 2xy + 6y^2 + \lambda(72 - x - y)$$

The first-order conditions are

$$Z_x = 8x - 2y - \lambda = 0 \tag{5.27}$$

$$Z_y = -2x + 12y - \lambda = 0 \tag{5.28}$$

$$Z_\lambda = 72 - x - y = 0 \tag{5.29}$$

Subtract (5.28) from (5.27) to eliminate λ.

$$10x - 14y = 0 \qquad x = 1.4y$$

Substitute $x = 1.4y$ in (5.29) and rearrange.

$$1.4y + y = 72 \qquad y_0 = 30$$

Substitute $y_0 = 30$ in the previous equations to find that at the critical point

$$x_0 = 42 \qquad y_0 = 30 \qquad \lambda_0 = 276$$

Thus, $Z = 4(42)^2 - 2(42)(30) + 6(30)^2 + 276(72 - 42 - 30) = 9936$.

2)   With $\lambda = 276$, a 1-unit increase in the constant of the constraint will lead to an increase of approximately 276 in the value of the objective function, and $Z \approx 10{,}212$.

b) $f(x,y) = 26x - 3x^2 + 5xy - 6y^2 + 12y$ subject to $3x + y = 170$

1)   The Lagrangian function is

$$F = 26x - 3x^2 + 5xy - 6y^2 + 12y + \lambda(170 - 3x - y)$$

Thus, $F_x = 26 - 6x + 5y - 3\lambda = 0 \tag{5.30}$

$$F_y = 5x - 12y + 12 - \lambda = 0 \tag{5.31}$$

$$F_\lambda = 170 - 3x - y = 0 \tag{5.32}$$

Multiply (5.31) by 3 and subtract from (5.30) to eliminate λ.

$$-21x + 41y - 10 = 0 \tag{5.33}$$

Multiply (5.32) by 7 and subtract from (5.33) to eliminate x.

$$48y - 1200 = 0 \qquad y_0 = 25$$

Then substituting $y_0 = 25$ into the previous equations shows that at the critical point

$$x_0 = \frac{145}{3} = 48\frac{1}{3} \qquad y_0 = 25 \qquad \lambda_0 = \frac{-139}{3} = -46\frac{1}{3}$$

Using $x_0 = \dfrac{145}{3}$, $y_0 = 25$, and $\lambda_0 = \dfrac{-139}{3}$, $F = -3160$.

2) With $\lambda = -46\frac{1}{3}$, a 1-unit increase in the constant of the constraint will lead to a *decrease* of approximately $46\frac{1}{3}$ in the value of the objective function, and $F \approx -3206.33$.

c) $f(x, y, z) = 4xyz^2$ subject to $x + y + z = 56$

    1)
$$F = 4xyz^2 + \lambda(56 - x - y - z)$$

$$F_x = 4yz^2 - \lambda = 0 \tag{5.34}$$

$$F_y = 4xz^2 - \lambda = 0 \tag{5.35}$$

$$F_z = 8xyz - \lambda = 0 \tag{5.36}$$

$$F_\lambda = 56 - x - y - z = 0 \tag{5.37}$$

Equate λ's from (5.34) and (5.35).

$$4yz^2 = 4xz^2 \qquad y = x$$

Equate λ's from (5.34) and (5.36)

$$4yz^2 = 8xyz \qquad z = 2x$$

Substitute $y = x$ and $z = 2x$ in (5.37).

$$56 - x - x - 2x = 0 \qquad 4x = 56 \qquad x_0 = 14$$

Then substituting $x_0 = 14$ in the previous equations gives

$$x_0 = 14 \quad y_0 = 14 \qquad z_0 = 28 \qquad \lambda_0 = 43{,}904$$
$$F_0 = 614{,}656$$

    2) $\qquad\qquad\qquad F_1 \approx F_0 + \lambda_2 \approx 614{,}656 + 43{,}904 \approx 658{,}560$

See Problem 12.28 for the second-order conditions.

d) $f(x, y, z) = 5xy + 8xz + 3yz$ subject to $2xyz = 1920$

    1)
$$F = 5xy + 8xz + 3yz + \lambda(1920 - 2xyz)$$

$$F_x = 5y + 8z - 2\lambda yz = 0 \tag{5.38}$$

$$F_y = 5x + 3z - 2\lambda xz = 0 \tag{5.39}$$

$$F_z = 8x + 3y - 2\lambda xy = 0 \tag{5.40}$$

$$F_\lambda = 1920 - 2xyz = 0 \tag{5.41}$$

Solve (5.38), (5.39), and (5.40) for λ.

$$\lambda = \frac{5y + 8z}{2yz} = \frac{2.5}{z} + \frac{4}{y} \tag{5.42}$$

$$\lambda = \frac{5x + 3z}{2xz} = \frac{2.5}{z} + \frac{1.5}{x} \tag{5.43}$$

$$\lambda = \frac{8x + 3y}{2xy} = \frac{4}{y} + \frac{1.5}{x} \tag{5.44}$$

Equate λ's in (5.42) and (5.43) to eliminate $2.5/z$,

$$\frac{4}{y} = \frac{1.5}{x} \qquad 4x = 1.5y \qquad x = \frac{1.5}{4}y$$

and λ's in (5.43) and (5.44) to eliminate $1.5/x$.

$$\frac{2.5}{z} = \frac{4}{y} \qquad 4z = 2.5y \qquad z = \frac{2.5}{4}y$$

Then substitute $x = (1.5/4)y$ and $z = (2.5/4)y$ in (5.41).

$$1920 = 2 \cdot \left(\frac{1.5}{4}y\right) \cdot y \cdot \left(\frac{2.5}{4}y\right)$$

$$y^3 = 1920 \cdot \frac{16}{7.5} = 4096$$

$$y_0 = 16$$

and the critical values are $x_0 = 6$, $y_0 = 16$, $z_0 = 10$, and $\lambda_0 = 0.5$.

$$F_0 = 1440$$

2)
$$F_1 \approx F_0 + \lambda_0 \approx 1440 + 0.5 \approx 1440.5$$

5.13. In Problem 5.12(a) it was estimated that if the constant of the constraint were increased by 1 unit, the constrained optimum would increase by approximately 276, from 9936 to 10,212. Check the accuracy of the estimate by optimizing the original function $z = 4x^2 - 2xy + 6y^2$ subject to a new constraint $x + y = 73$.

$$Z = 4x^2 - 2xy + 6y^2 + \lambda(73 - x - y)$$
$$Z_x = 8x - 2y - \lambda = 0$$
$$Z_y = -2x + 12y - \lambda = 0$$
$$Z_\lambda = 73 - x - y = 0$$

Simultaneous solution gives $x_0 = 42.58$, $y_0 = 30.42$, $\lambda_0 = 279.8$. Thus, $Z_0 = 10,213.9$, compared to the 10,212 estimate from the original λ, a difference of 1.9 units or 0.02 percent.

5.14. Constraints can also be used simply to ensure that the two independent variables will always be in constant proportion, for example, $x = 3y$. In this case measuring the effect of λ has no economic significance since a 1-unit increase in the constant of the constraint would alter the constant proportion between the independent variables. With this in mind, optimize the following functions subject to the constant proportion constraint:

$a)$ $z = 4x^2 - 3x + 5xy - 8y + 2y^2$ subject to $x = 2y$

With $x - 2y = 0$, the Lagrangian function is

$$Z = 4x^2 - 3x + 5xy - 8y + 2y^2 + \lambda(x - 2y)$$
$$Z_x = 8x - 3 + 5y + \lambda = 0$$
$$Z_y = 5x - 8 + 4y - 2\lambda = 0$$
$$Z_\lambda = x - 2y = 0$$

When solved simultaneously, $x_0 = 0.5$, $y_0 = 0.25$, and $\lambda_0 = -2.25$. Thus, $Z_0 = -1.75$.

b) $z = -5x^2 + 7x + 10xy + 9y - 2y^2$ subject to $y = 5x$

The Lagrangian function is $Z = -5x^2 + 7x + 10xy + 9y - 2y^2 + \lambda(5x - y)$

$$Z_x = -10x + 7 + 10y + 5\lambda = 0$$
$$Z_y = 10x + 9 - 4y - \lambda = 0$$
$$Z_\lambda = 5x - y = 0$$

Solving simultaneously, $x_0 = 5.2$, $y_0 = 26$, and $\lambda_0 = -43$. Thus, $Z_0 = 135.2$.

See also Problems 12.19 to 12.28.

DIFFERENTIALS

5.15. Find the differential dy for each of the following functions:

a) $y = 7x^3 - 5x^2 + 6x - 3$

$$\frac{dy}{dx} = 21x^2 - 10x + 6$$

Thus, $dy = (21x^2 - 10x + 6)\,dx$

b) $y = (4x + 3)(3x - 8)$

$$\frac{dy}{dx} = (4x + 3)(3) + (3x - 8)(4) = 24x - 23$$

Thus, $dy = (24x - 23)\,dx$

c) $y = \dfrac{9x - 4}{5x}$

$$\frac{dy}{dx} = \frac{5x(9) - (9x - 4)(5)}{(5x)^2} = \frac{20}{25x^2}$$

$$dy = \frac{4}{5x^2}\,dx$$

d) $y = (11x + 9)^3$

$$\frac{dy}{dx} = 3(11x + 9)^2(11)$$

$$dy = 33(11x + 9)^2\,dx$$

5.16. Find the total differential $dz = z_x\,dx + z_y\,dy$ for each of the following functions:

a) $z = 5x^3 - 12xy - 6y^5$

$$z_x = 15x^2 - 12y \qquad z_y = -12x - 30y^4$$
$$dz = (15x^2 - 12y)\,dx - (12x + 30y^4)\,dy$$

b) $z = 7x^2 y^3$

$$z_x = 14xy^3 \qquad z_y = 21x^2 y^2$$
$$dz = 14xy^3\,dx + 21x^2 y^2\,dy$$

c) $z = 3x^2(8x - 7y)$

$$z_x = 3x^2(8) + (8x - 7y)(6x) \qquad z_y = 3x^2(-7) + (8x - 7y)(0)$$
$$dz = (72x^2 - 42xy)\,dx - 21x^2\,dy$$

d) $z = (5x^2 + 7y)(2x - 4y^3)$

$$z_x = (5x^2 + 7y)(2) + (2x - 4y^3)(10x) \qquad z_y = (5x^2 + 7y)(-12y^2) + (2x - 4y^3)(7)$$
$$dz = (30x^2 - 40xy^3 + 14y)\,dx - (112y^3 + 60x^2 y^2 - 14x)\,dy$$

e) $z = \dfrac{9y^3}{x-y}$

$$z_x = \frac{(x-y)(0) - 9y^3(1)}{(x-y)^2} \qquad z_y = \frac{(x-y)(27y^2) - 9y^3(-1)}{(x-y)^2}$$

$$dz = \frac{-9y^3}{(x-y)^2}\,dx + \frac{27xy^2 - 18y^3}{(x-y)^2}\,dy$$

f) $z = (x - 3y)^3$

$$z_x = 3(x-3y)^2(1) \qquad z_y = 3(x-3y)^2(-3)$$
$$dz = 3(x-3y)^2\,dx - 9(x-3y)^2\,dy$$

5.17. Find the partial differential for a small change in x for each of the functions given in Problem 5.16, assuming $dy = 0$.

a) $dz = (15x^2 - 12y)\,dx$

b) $dz = 14xy^3\,dx$

c) $dz = (72x^2 - 42xy)\,dx$

d) $dz = (30x^2 - 40xy^3 + 14y)\,dx$

e) $dz = \dfrac{-9y^3}{(x-y)^2}\,dx$

f) $dz = 3(x-3y)^2\,dx$

TOTAL DERIVATIVES

5.18. Find the total derivative dz/dx for each of the following functions:

a) $z = 6x^2 + 15xy + 3y^2 \qquad$ where $y = 7x^2$

$$\frac{dz}{dx} = z_x + z_y\frac{dy}{dx}$$
$$= (12x + 15y) + (15x + 6y)(14x)$$
$$= 210x^2 + 84xy + 12x + 15y$$

b) $z = (13x - 18y)^2 \qquad$ where $y = x + 6$

$$\frac{dz}{dx} = z_x + z_y\frac{dy}{dx}$$
$$= 26(13x - 18y) - 36(13x - 18y)(1)$$
$$= -10(13x - 18y)$$

c) $z = \dfrac{9x - 7y}{2x + 5y} \qquad$ where $y = 3x - 4$

$$\frac{dz}{dx} = z_x + z_y\frac{dy}{dx}$$
$$= \frac{59y}{(2x + 5y)^2} - \frac{59x}{(2x + 5y)^2}(3) = \frac{59(y - 3x)}{(2x + 5y)^2}$$

d) $z = 8x - 12y \qquad$ where $y = (x + 1)/x^2$

$$\frac{dz}{dx} = z_x + z_y\frac{dy}{dx}$$
$$= 8 - \frac{12(-x^2 - 2x)}{x^4}$$
$$= 8 + \frac{12(x + 2)}{x^3}$$

5.19. Find the total derivative dz/dw for each of the following functions:

a) $z = 7x^2 + 4y^2$ where $x = 5w$ and $y = 4w$

$$\frac{dz}{dw} = z_x \frac{dx}{dw} + z_y \frac{dy}{dw} = 14x(5) + 8y(4) = 70x + 32y$$

b) $z = 10x^2 - 6xy - 12y^2$ where $x = 2w$ and $y = 3w$

$$\frac{dz}{dw} = z_x \frac{dx}{dw} + z_y \frac{dy}{dw} = (20x - 6y)(2) + (-6x - 24y)(3) = 22x - 84y$$

IMPLICIT AND INVERSE FUNCTION RULES

5.20. Find the derivatives dy/dx and dx/dy for each of the following implicit functions:

a) $y - 6x + 7 = 0$

$$\frac{dy}{dx} = \frac{-f_x}{f_y} = \frac{-(-6)}{1} = 6 \qquad \frac{dx}{dy} = \frac{-f_y}{f_x} = \frac{-(1)}{-6} = \frac{1}{6}$$

b) $3y - 12x + 17 = 0$

$$\frac{dy}{dx} = \frac{-f_x}{f_y} = \frac{-(-12)}{3} = 4 \qquad \frac{dx}{dy} = \frac{-f_y}{f_x} = \frac{-(3)}{-12} = \frac{1}{4}$$

c) $x^2 + 6x - 13 - y = 0$

$$\frac{dy}{dx} = \frac{-f_x}{f_y} = \frac{-(2x+6)}{-1} = 2x + 6 \qquad \frac{dx}{dy} = \frac{-f_y}{f_x} = \frac{-(-1)}{2x+6} = \frac{1}{2x+6} \qquad (x \neq -3)$$

Notice that in each of the above cases, one derivative is the inverse of the other.

5.21. Use the implicit function rule to find dy/dx and, where applicable, dy/dz.

a) $f(x,y) = 3x^2 + 2xy + 4y^3$

$$\frac{dy}{dx} = \frac{-f_x}{f_y} = -\frac{6x + 2y}{12y^2 + 2x}$$

b) $f(x,y) = 12x^5 - 2y$

$$\frac{dy}{dx} = \frac{-f_x}{f_y} = \frac{-60x^4}{-2} = 30x^4$$

c) $f(x,y) = 7x^2 + 2xy^2 + 9y^4$

$$\frac{dy}{dx} = \frac{-f_x}{f_y} = -\frac{14x + 2y^2}{36y^3 + 4xy}$$

d) $f(x,y) = 6x^3 - 5y$

$$\frac{dy}{dx} = \frac{-f_x}{f_y} = \frac{18x^2}{-5} = 3.6x^2$$

e) $f(x,y,z) = x^2y^3 + z^2 + xyz$

$$\frac{dy}{dx} = \frac{-f_x}{f_y} = -\frac{2xy^3 + yz}{3x^2y^2 + xz}$$

$$\frac{dy}{dz} = \frac{-f_z}{f_y} = -\frac{2z + xy}{3x^2y^2 + xz}$$

f) $f(x,y,z) = x^3z^2 + y^3 + 4xyz$

$$\frac{dy}{dx} = \frac{-f_x}{f_y} = -\frac{3x^2z^2 + 4yz}{3y^2 + 4xz}$$

$$\frac{dy}{dz} = \frac{-f_z}{f_y} = -\frac{2x^3z + 4xy}{3y^2 + 4xz}$$

5.22. Find the derivative for the inverse function dP/dQ.

a) $Q = 210 - 3P$

$$\frac{dP}{dQ} = \frac{1}{dQ/dP} = -\frac{1}{3}$$

b) $Q = 35 - 0.25P$

$$\frac{dP}{dQ} = \frac{1}{-0.25} = -4$$

c) $Q = 14 + P^2$

$$\frac{dP}{dQ} = \frac{1}{2P} \quad (P \neq 0)$$

d) $Q = P^3 + 2P^2 + 7P$

$$\frac{dP}{dQ} = \frac{1}{3P^2 + 4P + 7}$$

VERIFICATION OF RULES

5.23.   For each of the following functions, use (1) the definition in (5.1a) to find $\partial z / \partial x$ and (2) the definition in (5.1b) to find $\partial z / \partial y$ in order to confirm the rules of differentiation.

a) $z = 38 + 7x - 4y$

    1)   From (5.1a),
$$\frac{\partial z}{\partial x} = \lim_{\Delta x \to 0} \frac{f(x + \Delta x, y) - f(x, y)}{\Delta x}$$

        Substituting,
$$\frac{\partial z}{\partial x} = \lim_{\Delta x \to 0} \frac{[38 + 7(x + \Delta x) - 4y] - (38 + 7x - 4y)}{\Delta x}$$

$$= \lim_{\Delta x \to 0} \frac{38 + 7x + 7\Delta x - 4y - 38 - 7x + 4y}{\Delta x}$$

$$= \lim_{\Delta x \to 0} \frac{7\Delta x}{\Delta x} = \lim_{\Delta x \to 0} 7 = 7$$

    2)   From (5.1b),
$$\frac{\partial z}{\partial y} = \lim_{\Delta y \to 0} \frac{f(x, y + \Delta y) - f(x, y)}{\Delta y}$$

        Substituting,
$$\frac{\partial z}{\partial y} = \lim_{\Delta y \to 0} \frac{[38 + 7x - 4(y + \Delta y)] - (38 + 7x - 4y)}{\Delta y}$$

$$= \lim_{\Delta y \to 0} \frac{38 + 7x - 4y - 4\Delta y - 38 - 7x + 4y}{\Delta y}$$

$$= \lim_{\Delta y \to 0} \frac{-4\Delta y}{\Delta y} = \lim_{\Delta y \to 0} (-4) = -4$$

b) $z = 18x - 5xy + 14y$

    1)
$$\frac{\partial z}{\partial x} = \lim_{\Delta x \to 0} \frac{[18(x + \Delta x) - 5(x + \Delta x)y + 14y] - (18x - 5xy + 14y)}{\Delta x}$$

$$= \lim_{\Delta x \to 0} \frac{18x + 18\Delta x - 5xy - 5\Delta xy + 14y - 18x + 5xy - 14y}{\Delta x}$$

$$= \lim_{\Delta x \to 0} \frac{18\Delta x - 5\Delta xy}{\Delta x} = \lim_{\Delta x \to 0} (18 - 5y) = 18 - 5y$$

2)
$$\frac{\partial z}{\partial y} = \lim_{\Delta y \to 0} \frac{[18x - 5x(y + \Delta y) + 14(y + \Delta y)] - (18x - 5xy + 14y)}{\Delta y}$$

$$= \lim_{\Delta y \to 0} \frac{18x - 5xy - 5x\,\Delta y + 14y + 14\Delta y - 18x + 5xy - 14y}{\Delta y}$$

$$= \lim_{\Delta y \to 0} \frac{-5x\,\Delta y + 14\Delta y}{\Delta y} = \lim_{\Delta y \to 0} (-5x + 14) = -5x + 14$$

c) $z = 3x^2 y$

1)
$$\frac{\partial z}{\partial x} = \lim_{\Delta x \to 0} \frac{[3(x + \Delta x)^2 y] - 3x^2 y}{\Delta x}$$

$$= \lim_{\Delta x \to 0} \frac{3x^2 y + 6x\,\Delta xy + 3(\Delta x)^2 y - 3x^2 y}{\Delta x}$$

$$= \lim_{\Delta x \to 0} \frac{6x\,\Delta xy + 3(\Delta x)^2 y}{\Delta x}$$

$$= \lim_{\Delta x \to 0} (6xy + 3\Delta xy) = 6xy$$

2)
$$\frac{\partial z}{\partial y} = \lim_{\Delta y \to 0} \frac{[3x^2(y + \Delta y)] - 3x^2 y}{\Delta y}$$

$$= \lim_{\Delta y \to 0} \frac{3x^2 y + 3x^2\,\Delta y - 3x^2 y}{\Delta y} = \lim_{\Delta y \to 0} \frac{3x^2\,\Delta y}{\Delta y} = \lim_{\Delta y \to 0} 3x^2 = 3x^2$$

d) $z = 4x^2 y^2$

1)
$$\frac{\partial z}{\partial x} = \lim_{\Delta x \to 0} \frac{[4(x + \Delta x)^2 y^2] - 4x^2 y^2}{\Delta x}$$

$$= \lim_{\Delta x \to 0} \frac{4x^2 y^2 + 8x\,\Delta xy^2 + 4(\Delta x)^2 y^2 - 4x^2 y^2}{\Delta x}$$

$$= \lim_{\Delta x \to 0} \frac{8x\,\Delta xy^2 + 4(\Delta x)^2 y^2}{\Delta x}$$

$$= \lim_{\Delta x \to 0} (8xy^2 + 4\Delta xy^2) = 8xy^2$$

2)
$$\frac{\partial z}{\partial y} = \lim_{\Delta y \to 0} \frac{[4x^2(y + \Delta y)^2] - 4x^2 y^2}{\Delta y}$$

$$= \lim_{\Delta y \to 0} \frac{4x^2 y^2 + 8x^2 y\,\Delta y + 4x^2(\Delta y)^2 - 4x^2 y^2}{\Delta y}$$

$$= \lim_{\Delta y \to 0} \frac{8x^2 y\,\Delta y + 4x^2(\Delta y)^2}{\Delta y} = \lim_{\Delta y \to 0} (8x^2 y + 4x^2\,\Delta y) = 8x^2 y$$

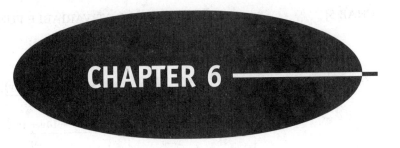

CHAPTER 6

Calculus of Multivariable Functions in Economics

6.1 MARGINAL PRODUCTIVITY

The *marginal product* of capital (MP_K) is defined as the change in output brought about by a small change in capital when all the other factors of production are held constant. Given a production function such as

$$Q = 36KL - 2K^2 - 3L^2$$

the MP_K is measured by taking the partial derivative $\partial Q/\partial K$. Thus,

$$MP_K = \frac{\partial Q}{\partial K} = 36L - 4K$$

Similarly, for labor, $MP_L = \partial Q/\partial L = 36K - 6L$. See Problems 6.1 to 6.3

6.2 INCOME DETERMINATION MULTIPLIERS AND COMPARATIVE STATICS

The partial derivative can also be used to derive the various multipliers of an income determination model. In calculating how the equilibrium level of the endogenous variable can be expected to change in response to a change in any of the exogenous variables or parameters, income determination multipliers provide an elementary exercise in what is called *comparative static analysis* or, more simply, *comparative statics*, which we shall study later in greater detail in Chapter 13. Given

$$Y = C + I + G + (X - Z)$$

where

$$C = C_0 + bY \qquad G = G_0 \qquad Z = Z_0$$
$$I = I_0 + aY \qquad X = X_0$$

Using simple substitution as in Problem 2.19, the equilibrium level of income is

$$\bar{Y} = \frac{1}{1 - b - a}(C_0 + I_0 + G_0 + X_0 - Z_0) \tag{6.1}$$

Taking the partial derivative of (6.1) with respect to any of the variables or parameters gives the multiplier for that variable or parameter. Thus, the government multiplier is given by

$$\frac{\partial \bar{Y}}{\partial G_0} = \frac{1}{1 - b - a}$$

The import multiplier is given by

$$\frac{\partial \bar{Y}}{\partial Z_0} = -\frac{1}{1 - b - a}$$

And the multiplier for a change in the marginal propensity to invest is given by $\partial \bar{Y}/\partial a$, where, by means of the quotient rule,

$$\frac{\partial \bar{Y}}{\partial a} = \frac{(1 - b - a)(0) - (C_0 + I_0 + G_0 + X_0 - Z_0)(-1)}{(1 - b - a)^2} = \frac{C_0 + I_0 + G_0 + X_0 - Z_0}{(1 - b - a)^2}$$

This can alternately be expressed as

$$\frac{\partial \bar{Y}}{\partial a} = \frac{1}{1 - b - a}(C_0 + I_0 + G_0 + X_0 - Z_0)\left(\frac{1}{1 - b - a}\right)$$

which from (6.1) reduces to

$$\frac{\partial \bar{Y}}{\partial a} = \frac{\bar{Y}}{1 - b - a}$$

See Problems 6.4 to 6.8.

6.3 INCOME AND CROSS PRICE ELASTICITIES OF DEMAND

Income elasticity of demand ϵ_Y measures the percentage change in the demand for a good resulting from a small percentage change in income, when all other variables are held constant. *Cross price elasticity of demand* ϵ_c measures the relative responsiveness of the demand for one product to changes in the price of another, when all other variables are held constant. Given the demand function

$$Q_1 = a - bP_1 + cP_2 + mY$$

where Y = income and P_2 = the price of a substitute good, the income elasticity of demand is

$$\epsilon_Y = \frac{\partial Q_1}{Q_1} \div \frac{\partial Y}{Y} = \frac{\partial Q_1}{\partial Y}\left(\frac{Y}{Q_1}\right)$$

and the cross price elasticity of demand is

$$\epsilon_c = \frac{\partial Q_1}{Q_1} \div \frac{\partial P_2}{P_2} = \frac{\partial Q_1}{\partial P_2}\left(\frac{P_2}{Q_1}\right)$$

See Examples 1 and 2 and Problems 6.18 to 6.21.

EXAMPLE 1. Given the demand for beef

$$Q_b = 4850 - 5P_b + 1.5P_p + 0.1Y \tag{6.2}$$

with $Y = 10,000$, $P_b = 200$, and the price of pork $P_p = 100$. The calculations for (1) the income elasticity and (2) the cross price elasticity of demand for beef are given below.

1)
$$\epsilon_Y = \frac{\partial Q_b}{Q_b} \div \frac{\partial Y}{Y} = \frac{\partial Q_b}{\partial Y}\left(\frac{Y}{Q_b}\right) \tag{6.3}$$

From (6.2),
$$\frac{\partial Q_b}{\partial Y} = 0.1$$

and
$$Q_b = 4850 - 5(200) + 1.5(100) + 0.1(10,000) = 5000 \tag{6.4}$$

Substituting in (6.3), $\epsilon_Y = 0.1(10,000/5000) = 0.2$.

With $\epsilon_Y < 1$, the good is income-inelastic. For any given percentage increase in national income, demand for the good will increase less than proportionately. Hence the relative market share of the good will decline as the economy expands. Since the income elasticity of demand suggests the growth potential of a market, the growth potential in this case is limited.

2)
$$\epsilon_c = \frac{\partial Q_b}{Q_b} \div \frac{\partial P_p}{P_p} = \frac{\partial Q_b}{\partial P_p}\left(\frac{P_p}{Q_b}\right)$$

From (6.2), $\partial Q_b/\partial P_p = 1.5$; from (6.4), $Q_b = 5000$. Thus,

$$\epsilon_c = 1.5\left(\frac{100}{5000}\right) = 0.03$$

For *substitute goods*, such as beef and pork, $\partial Q_1/\partial P_2 > 0$ and the cross price elasticity will be positive. For *complementary goods*, $\partial Q_1/\partial P_2 < 0$ and the cross price elasticity will be negative. If $\partial Q_1/\partial P_2 = 0$, the goods are unrelated.

EXAMPLE 2. Continuing with Example 1, the percentage change in the demand for beef resulting from a 10 percent increase in the price of pork is estimated as follows:

$$\epsilon_c = \frac{\partial Q_b}{Q_b} \div \frac{\partial P_p}{P_p}$$

Rearranging terms and substituting the known parameters,

$$\frac{\partial Q_b}{Q_b} = \epsilon_c\frac{\partial P_p}{P_p} = (0.03)(0.10) = 0.003$$

The percentage change in the demand for beef $\partial Q_b/Q_b$ will be 0.3 percent.

6.4 DIFFERENTIALS AND INCREMENTAL CHANGES

Frequently in economics we want to measure the effect on the dependent variable (costs, revenue, profit) of a change in an independent variable (labor hired, capital used, items sold). If the change is a relatively small one, the differential will measure the effect. Thus, if $z = f(x, y)$, the effect on z of a small change in x is given by the partial differential

$$dz = z_x\, dx$$

The effect of larger changes can be approximated by multiplying the partial derivative by the proposed change. Thus,

$$\Delta z \approx z_x \Delta x$$

If the original function $z = f(x, y)$ is linear,

$$\frac{dz}{dx} = \frac{\Delta z}{\Delta x}$$

and the effect of the change will be measured exactly:

$$\Delta z = z_x \Delta x$$

See Examples 3 and 4 and Problems 6.9 to 6.17.

EXAMPLE 3.   A firm's costs are related to its output of two goods x and y. The functional relationship is

$$TC = x^2 - 0.5xy + y^2$$

The additional cost of a slight increment in output x will be given by the differential

$$d\text{TC} = (2x - 0.5y)\, dx$$

The costs of larger increments can be approximated by multiplying the partial derivative with respect to x by the change in x. Mathematically,

$$\Delta \text{TC} \approx \frac{\partial \text{TC}}{\partial x} \Delta x \tag{6.5}$$

Since $\partial \text{TC}/\partial x =$ the marginal cost (MC_x) of x, we can also write (6.5) as

$$\Delta \text{TC} \approx \text{MC}_x \Delta x$$

If initially $x = 100$, $y = 60$, and $\Delta x = 3$, then

$$\Delta \text{TC} \approx [2(100) - 0.5(60)] \cong 510$$

EXAMPLE 4.   Assume in Section 6.2 that $b = 0.7$, $a = 0.1$, and $Y = 1200$. The differential can then be used to calculate the effect of an increase in any of the independent variables. Given the partial derivative

$$\frac{\partial \bar{Y}}{\partial G_0} = \frac{1}{1 - b - a}$$

the partial differential is

$$d\bar{Y} = \frac{1}{1 - b - a} dG_0$$

In a linear model such as this, where the slope is everywhere constant,

$$\frac{\partial \bar{Y}}{\partial G_0} = \frac{\Delta \bar{Y}}{\Delta G_0}$$

Hence

$$\Delta \bar{Y} = \frac{1}{1 - b - a} \Delta G_0$$

If the government increases expenditures by $100,

$$\Delta \bar{Y} = \frac{1}{1 - 0.7 - 0.1}(100) = 500$$

6.5   OPTIMIZATION OF MULTIVARIABLE FUNCTIONS IN ECONOMICS

Food processors frequently sell different grades of the same product: quality, standard, economy; some, too, sell part of their output under their own brand name and part under the brand name of a large chain store. Clothing manufacturers and designers frequently have a top brand and cheaper imitations for discount department stores. Maximizing profits or minimizing costs under these conditions involve functions of more than one variable. Thus, the basic rules for optimization of multivariate functions (see Section 5.4) are required. See Examples 5 and 6 and Problems 6.22 to 6.27.

EXAMPLE 5.   A firm producing two goods x and y has the profit function

$$\pi = 64x - 2x^2 + 4xy - 4y^2 + 32y - 14$$

To find the profit-maximizing level of output for each of the two goods and test to be sure profits are maximized:

1. Take the first-order partial derivatives, set them equal to zero, and solve for x and y simultaneously.

$$\pi_x = 64 - 4x + 4y = 0 \qquad (6.6)$$

$$\pi_y = 4x - 8y + 32 = 0 \qquad (6.7)$$

   When solved simultaneously, $\bar{x} = 40$ and $\bar{y} = 24$.

2. Take the second-order direct partial derivatives and make sure both are negative, as is required for a relative maximum. From (6.6) and (6.7),

$$\pi_{xx} = -4 \qquad \pi_{yy} = -8$$

3. Take the cross partials to make sure $\pi_{xx}\pi_{yy} > (\pi_{xy})^2$. From (6.6) and (6.7), $\pi_{xy} = 4 = \pi_{yx}$. Thus,

$$\pi_{xx}\pi_{yy} > (\pi_{xy})^2$$
$$(-4)(-8) > (4)^2$$
$$32 > 16$$

   Profits are indeed maximized at $\bar{x} = 40$ and $\bar{y} = 24$. At that point, $\pi = 1650$.

EXAMPLE 6.   In monopolistic competition producers must determine the price that will maximize their profit. Assume that a producer offers two different brands of a product, for which the demand functions are

$$Q_1 = 14 - 0.25P_1 \qquad (6.8)$$

$$Q_2 = 24 - 0.5P_2 \qquad (6.9)$$

and the joint cost function is

$$TC = Q_1^2 + 5Q_1Q_2 + Q_2^2 \qquad (6.10)$$

The profit-maximizing level of output, the price that should be charged for each brand, and the profits are determined as follows:

First, establish the profit function π in terms of Q_1 and Q_2. Since $\pi = $ total revenue (TR) minus total cost (TC) and the total revenue for the firm is $P_1Q_1 + P_2Q_2$, the firm's profit is

$$\pi = P_1Q_1 + P_2Q_2 - TC$$

Substituting from (6.10),

$$\pi = P_1Q_1 + P_2Q_2 - (Q_1^2 + 5Q_1Q_2 + Q_2^2) \qquad (6.11)$$

Next find the inverse functions of (6.8) and (6.9) by solving for P in terms of Q. Thus, from (6.8),

$$P_1 = 56 - 4Q_1 \qquad (6.12)$$

and from (6.9),

$$P_2 = 48 - 2Q_2 \qquad (6.13)$$

Substituting in (6.11),

$$\pi = (56 - 4Q_1)Q_1 + (48 - 2Q_2)Q_2 - Q_1^2 - 5Q_1Q_2 - Q_2^2$$
$$= 56Q_1 - 5Q_1^2 + 48Q_2 - 3Q_2^2 - 5Q_1Q_2 \qquad (6.14)$$

Then maximize (6.14) by the familiar rules:

$$\pi_1 = 56 - 10Q_1 - 5Q_2 = 0 \qquad \pi_2 = 48 - 6Q_2 - 5Q_1 = 0$$

which, when solved simultaneously, give $\bar{Q}_1 = 2.75$ and $\bar{Q}_2 = 5.7$.

Take the second derivatives to be sure π is maximized:

$$\pi_{11} = -10 \qquad \pi_{22} = -6 \qquad \pi_{12} = -5 = \pi_{21}$$

With both second direct partials negative and $\pi_{11}\pi_{22} > (\pi_{12})^2$, the function is maximized at the critical values.

Finally, substitute $\bar{Q}_1 = 2.75$ and $\bar{Q}_2 = 5.7$ in (6.12) and (6.13), respectively, to find the profit-maximizing price.

$$P_1 = 56 - 4(2.75) = 45 \qquad P_2 = 48 - 2(5.7) = 36.6$$

Prices should be set at $45 for brand 1 and $36.60 for brand 2, leading to sales of 2.75 of brand 1 and 5.7 of brand 2. From (6.11) or (6.14), the maximum profit is

$$\pi = 45(2.75) + 36.6(5.7) - (2.75)^2 - 5(2.75)(5.7) - (5.7)^2 = 213.94$$

6.6 CONSTRAINED OPTIMIZATION OF MULTIVARIABLE FUNCTIONS IN ECONOMICS

Solutions to economic problems frequently have to be found under constraints (e.g., maximizing utility subject to a budget constraint or minimizing costs subject to some such minimal requirement of output as a production quota). Use of the Lagrangian function (see Section 5.5) greatly facilitates this task. See Example 7 and Problems 6.28 to 6.39. For inequality constraints, see concave programming (Section 13.7) in Chapter 13.

EXAMPLE 7.   Find the critical values for minimizing the costs of a firm producing two goods x and y when the total cost function is $c = 8x^2 - xy + 12y^2$ and the firm is bound by contract to produce a minimum combination of goods totaling 42, that is, subject to the constraint $x + y = 42$.

Set the constraint equal to zero, multiply it by λ, and form the Lagrangian function,

$$C = 8x^2 - xy + 12y^2 + \lambda(42 - x - y)$$

Take the first-order partials,

$$C_x = 16x - y - \lambda = 0$$
$$C_y = -x + 24y - \lambda = 0$$
$$C_\lambda = 42 - x - y = 0$$

Solving simultaneously, $\bar{x} = 25$, $\bar{y} = 17$, and $\bar{\lambda} = 383$. With $\bar{\lambda} = 383$, a 1-unit increase in the constraint or production quota will lead to an increase in cost of approximately $383. For second-order conditions, see Section 12.5 and Problem 12.27(a).

6.7 HOMOGENEOUS PRODUCTION FUNCTIONS

A production function is said to be homogeneous if when each input factor is multiplied by a positive real constant k, the constant can be completely factored out. If the exponent of the factor is 1, the function is homogeneous of degree 1; if the exponent of the factor is greater than 1, the function is homogeneous of degree greater than 1; and if the exponent of the factor is less than 1, the function is homogeneous of degree less than 1. Mathematically, a function $z = f(x, y)$ is homogeneous of degree n if for all positive real values of k, $f(kx, ky) = k^n f(x, y)$. See Example 8 and Problem 6.40.

EXAMPLE 8.   The degree of homogeneity of a function is illustrated below.

1. $z = 8x + 9y$ is homogeneous of degree 1 because

$$f(kx, ky) = 8kx + 9ky = k(8x + 9y)$$

2. $z = x^2 + xy + y^2$ is homogeneous of degree 2 because

$$f(kx, ky) = (kx)^2 + (kx)(ky) + (ky)^2 = k^2(x^2 + xy + y^2)$$

3. $z = x^{0.3}y^{0.4}$ is homogeneous of degree less than 1 because

$$f(kx, ky) = (kx)^{0.3}(ky)^{0.4} = k^{0.3+0.4}(x^{0.3}y^{0.4}) = k^{0.7}(x^{0.3}y^{0.4})$$

4. $z = 2x/y$ is homogeneous of degree 0 because

$$f(kx, ky) = \frac{2kx}{ky} = 1\left(\frac{2x}{y}\right) \qquad \text{since } \frac{k}{k} = k^0 = 1$$

5. $z = x^3 + 2xy + y^3$ is not homogeneous because k cannot be completely factored out:

$$f(kx, ky) = (kx)^3 + 2(kx)(ky) + (ky)^3$$
$$= k^3x^3 + 2k^2xy + k^3y^3 = k^2(kx^3 + 2xy + ky^3)$$

6. $Q = AK^\alpha L^\beta$ is homogeneous of degree $\alpha + \beta$ because

$$Q(kK, kL) = A(kK)^\alpha(kL)^\beta = Ak^\alpha K^\alpha k^\beta L^\beta = k^{\alpha+\beta}(AK^\alpha L^\beta)$$

6.8 RETURNS TO SCALE

A production function exhibits *constant returns to scale* if when all inputs are increased by a given proportion k, output increases by the same proportion. If output increases by a proportion greater than k, there are *increasing returns to scale*; and if output increases by a proportion smaller than k, there are *diminishing returns to scale*. In other words, if the production function is homogeneous of degree greater than, equal to, or less than 1, returns to scale are increasing, constant, or diminishing. See Problem 6.40.

6.9 OPTIMIZATION OF COBB-DOUGLAS PRODUCTION FUNCTIONS

Economic analysis frequently employs the *Cobb-Douglas production function* $q = AK^\alpha L^\beta$ ($A > 0$; $0 < \alpha, \beta < 1$), where q is the quantity of output in physical units, K is the quantity of capital, and L is the quantity of labor. Here α (the *output elasticity of capital*) measures the percentage change in q for a 1 percent change in K while L is held constant; β (the *output elasticity of labor*) is exactly parallel; and A is an *efficiency parameter* reflecting the level of technology.

A *strict* Cobb-Douglas function, in which $\alpha + \beta = 1$, exhibits *constant returns to scale*. A *generalized* Cobb-Douglas function, in which $\alpha + \beta \neq 1$, exhibits *increasing returns to scale* if $\alpha + \beta > 1$ and *decreasing returns to scale* if $\alpha + \beta < 1$. A Cobb-Douglas function is optimized subject to a budget constraint in Example 10 and Problems 6.41 and 6.42; second-order condtions are explained in Section 12.5. Selected properties of Cobb-Douglas functions are demonstrated and proved in Problems 6.53 to 6.58.

EXAMPLE 9.   The first and second partial derivatives for (*a*) $q = AK^\alpha L^\beta$ and (*b*) $q = 5K^{0.4}L^{0.6}$ are illustrated below.

a) $q_K = \alpha AK^{\alpha-1}L^\beta$ $q_L = \beta AK^\alpha L^{\beta-1}$

$q_{KK} = \alpha(\alpha - 1)AK^{\alpha-2}L^\beta$ $q_{LL} = \beta(\beta - 1)AK^\alpha L^{\beta-2}$

$q_{KL} = \alpha\beta AK^{\alpha-1}L^{\beta-1}$ $q_{LK} = \alpha\beta AK^{\alpha-1}L^{\beta-1}$

b) $q_K = 2K^{-0.6}L^{0.6}$ $q_L = 3K^{0.4}L^{-0.4}$

$q_{KK} = -1.2K^{-1.6}L^{0.6}$ $q_{LL} = -1.2K^{0.4}L^{-1.4}$

$q_{KL} = 1.2K^{-0.6}L^{-0.4}$ $q_{LK} = 1.2K^{-0.6}L^{-0.4}$

EXAMPLE 10.   Given a budget constraint of \$108 when $P_K = 3$ and $P_L = 4$, the generalized Cobb-Douglas production function $q = K^{0.4}L^{0.5}$ is optimized as follows:

1. Set up the Lagrangian function.

$$Q = K^{0.4}L^{0.5} + \lambda(108 - 3K - 4L)$$

2. Using the simple power function rule, take the first-order partial derivatives, set them equal to zero, and solve simultaneously for K_0 and L_0 (and λ_0, if desired).

$$\frac{\partial Q}{\partial K} = Q_K = 0.4K^{-0.6}L^{0.5} - 3\lambda = 0 \qquad (6.15)$$

$$\frac{\partial Q}{\partial L} = Q_L = 0.5K^{0.4}L^{-0.5} - 4\lambda = 0 \qquad (6.16)$$

$$\frac{\partial Q}{\partial \lambda} = Q_\lambda = 108 - 3K - 4L = 0 \qquad (6.17)$$

Rearrange, then divide (6.15) by (6.16) to eliminate λ.

$$\frac{0.4K^{-0.6}L^{0.5}}{0.5K^{0.4}L^{-0.5}} = \frac{3\lambda}{4\lambda}$$

Remembering to subtract exponents in division,

$$0.8K^{-1}L^1 = 0.75$$

$$\frac{L}{K} = \frac{0.75}{0.8} \qquad L = 0.9375K$$

Substitute $L = 0.9375K$ in (6.17).

$$108 - 3K - 4(0.9375K) = 0 \qquad K_0 = 16$$

Then by substituting $K_0 = 16$ in (6.17),

$$L_0 = 15$$

EXAMPLE 11.   The problem in Example 10 where $q = K^{0.4}L^{0.5}$, $P_K = 3$, $P_L = 4$, and $B = 108$ can also be solved using the familiar condition from microeconomic theory for output maximization

$$\frac{MU_K}{MU_L} = \frac{P_K}{P_L}$$

as demonstrated in (a) below and illustrated in (b).

a) $$MU_K = \frac{\partial q}{\partial K} = 0.4K^{-0.6}L^{0.5} \qquad MU_L = \frac{\partial q}{\partial L} = 0.5K^{0.4}L^{-0.5}$$

Substituting in the equality of ratios above,

$$\frac{0.4K^{-0.6}L^{0.5}}{0.5K^{0.4}L^{-0.5}} = \frac{3}{4}$$

and solving as in Example 10,

$$0.8K^{-1}L^1 = 0.75$$
$$L = 0.9375K$$

Then substituting in the budget constraint,

$$3K + 4L = 108$$
$$3K + 4(0.9375K) = 108$$
$$K_0 = 16 \qquad L_0 = 15$$

which is exactly what we found in Example 10 using the calculus. See Fig. 6-1.

b)

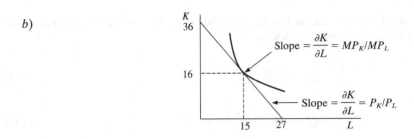

Fig. 6-1

6.10 OPTIMIZATION OF CONSTANT ELASTICITY OF SUBSTITUTION PRODUCTION FUNCTIONS

The *elasticity of substitution* σ measures the percentage change in the least-cost (K/L) input ratio resulting from a small percentage change in the input-price ratio (P_L/P_K).

$$\sigma = \frac{\dfrac{d(K/L)}{K/L}}{\dfrac{d(P_L/P_K)}{P_L/P_K}} = \frac{d(K/L)}{d(P_L/P_K)} \cdot \frac{P_L/P_K}{K/L} \tag{6.18}$$

where $0 \leq \sigma \leq \infty$. If $\sigma = 0$, there is no substitutability; the two inputs are complements and must be used together in fixed proportions. If $\sigma = \infty$, the two goods are perfect substitutes. A Cobb-Douglas production function, as shown in Problem 6.57, has a constant elasticity of substitution equal to 1. A *constant elasticity of substitution (CES) production function*, of which a Cobb-Douglas function is but one example, has an elasticity of substitution that is constant but not necessarily equal to 1.

A CES production function is typically expressed in the form

$$q = A[\alpha K^{-\beta} + (1 - \alpha)L^{-\beta}]^{-1/\beta} \tag{6.19}$$

where A is the efficiency parameter, α is the *distribution parameter* denoting relative factor shares, β is the *substitution parameter* determining the value of the elasticity of substitution, and the parameters are restricted so that $A > 0$, $0 < \alpha < 1$, and $\beta > -1$. CES production functions are optimized subject to budget constraints in Example 12 and Problems 6.43 and 6.44. Various important properties of the CES production function are demonstrated and proved in Problems 6.59 to 6.69.

EXAMPLE 12. The CES production function

$$q = 75[0.3K^{-0.4} + (1 - 0.3)L^{-0.4}]^{-1/0.4}$$

is maximized subject to the constraint $4K + 3L = 120$ as follows:

1. Set up the Lagrangian function.

$$Q = 75(0.3K^{-0.4} + 0.7L^{-0.4})^{-2.5} + \lambda(120 - 4K - 3L)$$

2. Test the first-order conditions, using the generalized power function rule for Q_K and Q_L.

$$\begin{aligned}
Q_K &= -187.5(0.3K^{-0.4} + 0.7L^{-0.4})^{-3.5}(-0.12K^{-1.4}) - 4\lambda = 0 \\
&= 22.5K^{-1.4}(0.3K^{-0.4} + 0.7L^{-0.4})^{-3.5} - 4\lambda = 0
\end{aligned} \tag{6.20}$$

$$\begin{aligned}
Q_L &= -187.5(0.3K^{-0.4} + 0.7L^{-0.4})^{-3.5}(-0.28L^{-1.4}) - 3\lambda = 0 \\
&= 52.5L^{-1.4}(0.3K^{-0.4} + 0.7L^{-0.4})^{-3.5} - 3\lambda = 0
\end{aligned} \tag{6.21}$$

$$Q_\lambda = 120 - 4K - 3L = 0 \tag{6.22}$$

Rearrange, then divide (6.20) by (6.21) to eliminate λ.

$$\frac{22.5K^{-1.4}(0.3K^{-0.4} + 0.7L^{-0.4})^{-3.5}}{52.5L^{-1.4}(0.3K^{-0.4} + 0.7L^{-0.4})^{-3.5}} = \frac{4\lambda}{3\lambda}$$

$$\frac{22.5K^{-1.4}}{52.5L^{-1.4}} = \frac{4}{3}$$

Cross multiply. $67.5K^{-1.4} = 210L^{-1.4}$

$$K^{-1.4} = 3.11L^{-1.4}$$

Take the -1.4 root, $K = (3.11)^{-1/1.4}L = (3.11)^{-0.71}L$

and use a calculator. $K \approx 0.45L$

Substitute in (6.22).

$$120 - 4(0.45L) - 3L = 0 \qquad L_0 = 25 \qquad K_0 = 11.25$$

Note: To find $(3.11)^{-0.71}$ with a calculator, enter 3.11, press the $\boxed{y^x}$ key, then enter 0.71 followed by the $\boxed{+/-}$ key to make it negative, and hit the $\boxed{=}$ key to find $(3.11)^{-0.71} = 0.44683$.

Solved Problems

MARGINAL CONCEPTS

6.1. Find the marginal productivity of the different inputs or factors of production for each of the following production functions Q:

a) $Q = 6x^2 + 3xy + 2y^2$

$$\text{MP}_x = \frac{\partial Q}{\partial x} = 12x + 3y$$

$$\text{MP}_y = \frac{\partial Q}{\partial y} = 3x + 4y$$

b) $Q = 0.5K^2 - 2KL + L^2$

$$\text{MP}_K = K - 2L$$
$$\text{MP}_L = 2L - 2K$$

c) $Q = 20 + 8x + 3x^2 - 0.25x^3 + 5y + 2y^2 - 0.5y^3$

$$\text{MP}_x = 8 + 6x - 0.75x^2$$
$$\text{MP}_y = 5 + 4y - 1.5y^2$$

d) $Q = x^2 + 2xy + 3y^2 + 1.5yz + 0.2z^2$

$$\text{MP}_x = 2x + 2y$$
$$\text{MP}_y = 2x + 6y + 1.5z$$
$$\text{MP}_z = 1.5y + 0.4z$$

6.2. (a) Assume $\bar{y} = 4$ in Problem 6.1(a) and find the MP_x for $x = 5$ and $x = 8$. (b) If the marginal revenue at $\bar{x} = 5$, $\bar{y} = 4$ is \$3, compute the marginal revenue product for the fifth unit of x.

a) $MP_x = 12x + 3y$

At $x = 5$, $\bar{y} = 4$, $MP_x = 12(5) + 3(4) = 72$.

At $x = 8$, $\bar{y} = 4$, $MP_x = 12(8) + 3(4) = 108$.

b) $MRP_x = MP_x(MR)$

At $\bar{x} = 5$, $\bar{y} = 4$, $MRP_x = (72)(3) = 216$.

6.3. (a) Find the marginal cost of a firm's different products when the total cost function is $c = 3x^2 + 7x + 1.5xy + 6y + 2y^2$. (b) Determine the marginal cost of x when $x = 5$, $\bar{y} = 3$.

a)
$$MC_x = 6x + 7 + 1.5y$$
$$MC_y = 1.5x + 6 + 4y$$

b) The marginal cost of x when $x = 5$ and y is held constant at 3 is

$$MC_x = 6(5) + 7 + 1.5(3) = 41.5$$

INCOME DETERMINATION MULTIPLIERS AND COMPARATIVE STATICS

6.4. Given a three-sector income determination model in which

$$Y = C + I_0 + G_0 \qquad Yd = Y - T \qquad C_0, I_0, G_0, T_0 > 0 \qquad 0 < b, t < 1$$
$$C = C_0 + bYd \qquad T = T_0 + tY$$

determine the magnitude and direction of a 1-unit change in (a) government spending, (b) lump-sum taxation, and (c) the tax rate on the equilibrium level of income. In short, perform the comparative-static exercise of determining the *government multiplier*, the *autonomous tax multiplier*, and the *tax rate multiplier*.

To find the different multipliers, first solve for the equilibrium level of income, as follows:

$$Y = C_0 + bY - bT_0 - btY + I_0 + G_0$$

$$\bar{Y} = \frac{1}{1 - b + bt}(C_0 - bT_0 + I_0 + G_0) \tag{6.23}$$

Then take the appropriate partial derivatives.

a)
$$\frac{\partial \bar{Y}}{\partial G_0} = \frac{1}{1 - b + bt}$$

Since $0 < b < 1$, $\partial \bar{Y}/\partial G_0 > 0$. A 1-unit increase in government spending will increase the equilibrium level of income by $1/(1 - b + bt)$.

b)
$$\frac{\partial \bar{Y}}{\partial T_0} = \frac{-b}{1 - b + bt} < 0$$

A 1-unit increase in autonomous taxation will cause national income to fall by $b/(1 - b + bt)$.

c) Since t appears in the denominator in (6.23), the quotient rule is necessary.

$$\frac{\partial \bar{Y}}{\partial t} = \frac{(1 - b + bt)(0) - (C_0 - bT_0 + I_0 + G_0)(b)}{(1 - b + bt)^2}$$

$$= \frac{-b(C_0 - bT_0 + I_0 + G_0)}{(1 - b + bt)^2} = \frac{-b}{1 - b + bt}\left(\frac{C_0 - bT_0 + I_0 + G_0}{1 - b + bt}\right)$$

Thus, from (6.23),
$$\frac{\partial \bar{Y}}{\partial t} = \frac{-b\bar{Y}}{1 - b + bt} < 0$$

A 1-unit increase in the tax rate will cause national income to fall by an amount equal to the tax rate multiplier.

6.5. Given a simple model

$$Y = C + I_0 + G_0 \qquad Yd = Y - T$$
$$C = C_0 + bYd \qquad T = T_0$$

where taxation does *not* depend on income, calculate the effect on the equilibrium level of income of a 1-unit change in government expenditure exactly offset by a 1-unit change in taxation. That is, do the comparative-static analysis of finding the balanced-budget multiplier for an economy in which there is only autonomous taxation.

$$Y = C_0 + b(Y - T_0) + I_0 + G_0$$

$$\bar{Y} = \frac{1}{1 - b}(C_0 - bT_0 + I_0 + G_0)$$

Thus, the government multiplier is

$$\frac{\partial \bar{Y}}{\partial G_0} = \frac{1}{1 - b} \tag{6.24}$$

and the tax multiplier is

$$\frac{\partial \bar{Y}}{\partial T_0} = \frac{-b}{1 - b} \tag{6.25}$$

The balanced-budget effect of a 1-unit increase in government spending matched by a 1-unit increase in taxation is the sum of (*6.24*) and (*6.25*). Therefore,

$$\Delta \bar{Y} = \frac{1}{1 - b} + \left(\frac{-b}{1 - b} \right) = \frac{1}{1 - b} - \frac{b}{1 - b} = \frac{1 - b}{1 - b} = 1$$

A change in government expenditure matched by an equal change in government taxation will have a *positive* effect on the equilibrium level of income exactly equal to the change in government expenditure and taxation. The multiplier in this case is +1.

6.6. Given

$$Y = C + I_0 + G_0 \qquad Yd = Y - T$$
$$C = C_0 + bYd \qquad T = T_0 + tY$$

where taxation is now a function of income, demonstrate the effect on the equilibrium level of income of a 1-unit change in government expenditure offset by a 1-unit change in *autonomous* taxation T_0. That is, demonstrate the effect of the *balanced-budget multiplier* in an economy in which taxes are a positive function of income.

From (*6.23*), $\bar{Y} = [1/(1 - b + bt)](C_0 - bT_0 + I_0 + G_0)$. Thus,

$$\frac{\partial \bar{Y}}{\partial G_0} = \frac{1}{1 - b + bt} \tag{6.26}$$

and

$$\frac{\partial \bar{Y}}{\partial T_0} = \frac{-b}{1 - b + bt} \tag{6.27}$$

The combined effect on $\bar{Y}$ of a 1-unit increase in government spending and an equal increase in autonomous taxation is the sum of (*6.26*) and (*6.27*). Thus,

$$\Delta \bar{Y} = \frac{1}{1 - b + bt} + \left(\frac{-b}{1 - b + bt} \right) = \frac{1 - b}{1 - b + bt}$$

which is positive but less than 1 because $1 - b < 1 - b + bt$. A change in government expenditures equaled by a change in autonomous taxes when taxes are positively related to income in the model, will have a positive effect on the equilibrium level of income, but the effect is smaller than the initial change in government expenditure. Here the multiplier is less than 1 because the total change in taxes ($\Delta T = \Delta T_0 + t\,\Delta Y$) is greater than the change in G_0.

6.7. Given

$$Y = C + I_0 + G_0 + X_0 - Z \qquad T = T_0 + tY$$
$$C = C_0 + bYd \qquad\qquad Z = Z_0 + zYd$$

where all the independent variables are positve and $0 < b, z, t < 1$. Determine the effect on the equilibrium level of income of a 1-unit change in (a) exports, (b) autonomous imports, and (c) autonomous taxation. In short, perform the comparative-static analysis of finding the *export*, *autonomous import*, and *autonomous taxation multipliers*. [Note that $Z = f(Y_d)$].

From the equilibrium level of income,

$$Y = C_0 + b(Y - T_0 - tY) + I_0 + G_0 + X_0 - Z_0 - z(Y - T_0 - tY)$$

$$\bar{Y} = \frac{1}{1 - b + bt + z - zt}(C_0 - bT_0 + I_0 + G_0 + X_0 - Z_0 + zT_0)$$

a)
$$\frac{\partial \bar{Y}}{\partial X_0} = \frac{1}{1 - b + bt + z - zt} > 0$$

because $0 < b, z < 1$. A 1-unit increase in exports will have a positive effect on $\bar{Y}$, which is given by the multiplier.

b)
$$\frac{\partial \bar{Y}}{\partial Z_0} = \frac{-1}{1 - b + bt + z - zt} < 0$$

An increase in autonomous imports will lead to a decrease in $\bar{Y}$.

c)
$$\frac{\partial \bar{Y}}{\partial T_0} = \frac{z - b}{1 - b + bt + z - zt} < 0$$

because a country's marginal propensity to import z is usually smaller than its marginal propensity to consume b. With $z < b$, $z - b < 0$. An increase in autonomous taxes will lead to a decrease in national income, as in (6.27), but the presence of z in the numerator has a mitigating effect on the decrease in income. When there is a positive marginal propensity to import, increased taxes will reduce cash outflows for imports and thus reduce the negative effect of increased taxes on the equilibrium level of income.

6.8. Determine the effect on $\bar{Y}$ of a 1-unit change in the marginal propensity to import z in Problem 6.7.

$$\frac{\partial \bar{Y}}{\partial z} = \frac{(1 - b + bt + z - zt)(T_0) - (C_0 - bT_0 + I_0 + G_0 + X_0 - Z_0 + zT_0)(1 - t)}{(1 - b + bt + z - zt)^2}$$

$$= \frac{T_0}{1 - b + bt + z - zt} - \frac{\bar{Y}(1 - t)}{1 - b + bt + z - zt}$$

$$= \frac{-[\bar{Y} - (T_0 + t\bar{Y})]}{1 - b + bt + z - zt} = \frac{-\bar{Y}d}{1 - b + bt + z - zt} < 0$$

DIFFERENTIALS AND COMPARATIVE STATICS

6.9. $Y = C + I_0 + G_0$ $Yd = Y - T$ $C_0 = 100$ $I_0 = 90$ $b = 0.75$
     $C = C_0 + bYd$ $T = T_0 + tY$ $G_0 = 330$ $T_0 = 240$ $t = 0.20$

(a) What is the equilibrium level of income $\bar{Y}$? What is the effect on $\bar{Y}$ of a \$50 increase in (b) government spending and (c) autonomous taxation T_0?

a) From (6.23), $\bar{Y} = \dfrac{1}{1-b+bt}(C_0 - bT_0 + I_0 + G_0)$

$$= \dfrac{1}{1-0.75+0.75(0.20)}[100 - 0.75(240) + 90 + 330]$$

$$= \dfrac{1}{0.40}(100 - 180 + 90 + 330) = 2.5(340) = 850$$

b) If government increases spending by 50,

$$\Delta\bar{Y} = \dfrac{\partial\bar{Y}}{\partial G_0}\Delta G_0 = \dfrac{1}{1-b+bt}(50) = 2.5(50) = 125$$

c) If autonomous taxation T_0 increases by 50,

$$\Delta\bar{Y} = \dfrac{\partial\bar{Y}}{\partial T_0}\Delta T_0 = \dfrac{-b}{1-b+bt}(50) = \dfrac{-0.75}{1-0.75+0.75(0.20)}(50) = -1.875(50) = -93.75$$

6.10. If the full-employment level of income Y_{fe} in Problem 6.9(a) is 1000 and the government wishes to achieve it, by how much should it change (a) government spending or (b) autonomous taxation?

a) The desired increase in economic activity is the difference between the full-employment level of income (1000) and the present level (850). Thus, the desired $\Delta\bar{Y} = 150$. Substituting in the formula from Problem 6.9(b),

$$\Delta\bar{Y} = \dfrac{\partial\bar{Y}}{\partial G_0}\Delta G_0$$

$$150 = 2.5\,\Delta G_0 \qquad \Delta G_0 = 60$$

Increased government expenditure of 60 will increase $\bar{Y}$ by 150.

b) If the government wishes to alter autonomous taxes to achieve full employment, from Problem 6.9(c),

$$\Delta\bar{Y} = \dfrac{\partial\bar{Y}}{\partial T_0}\Delta T_0$$

$$150 = -1.875\,\Delta T_0 \qquad \Delta T_0 = -80$$

The government should cut autonomous taxes by 80.

6.11. Explain the effect on the government deficit (a) if policy a in Problem 6.10 is adopted and (b) if policy b is adopted instead.

a) The government's financial condition is given by the difference between receipts T and expenditures G. At the initial 850 level of income,

$$T = 240 + 0.2(850) = 410 \qquad G_0 = 330 \qquad T - G_0 = 410 - 330 = 80$$

The government has a surplus of 80.

   If the government increases spending by 60, expenditures rise by 60. But tax revenues also increase as a result of the increase in income. With $\Delta\bar{Y} = 150$, $\Delta T = 0.2(150) = 30$. With expenditures rising by 60 and receipts increasing by 30, the net cost to the government of stimulating the economy to full employment is only \$30. At the new $\bar{Y} = 1000$,

$$T = 240 + 0.2(1000) = 440 \qquad G_0 = 330 + 60 = 390 \qquad T - G_0 = 440 - 390 = 50$$

The government surplus is reduced to \$50 from the previous \$80 surplus.

b)   If the government reduces T_0 by 80, tax revenue falls initially by 80. But the \$150 stimulatory effect on income has a positive effect on total tax collections, since $\Delta T = 0.2(150) = 30$. Thus, the net cost of reducing autonomous taxation to stimulate the economy to full employment is \$50. The government surplus is reduced to \$30:

$$T = 160 + 0.2(1000) = 360 \qquad G_0 = 330 \qquad T - G_0 = 360 - 330 = 30$$

6.12. (a) If the proportional tax in Problem 6.9 is increased by 10 percent, what is the effect on $\bar{Y}$? (b) If the government wants to alter the original marginal tax rate of 20 percent to achieve $Y_{fe} = 1000$, by how much should it change t?

a)   If the proportional tax is increased by 10 percent,

$$\Delta t = 0.10(0.20) = 0.02$$

The resulting change in income is

$$\Delta \bar{Y} = \frac{\partial \bar{Y}}{\partial t} \Delta t$$

Substituting from Problem 6.4(c),

$$\Delta \bar{Y} \approx \frac{-b\bar{Y}}{1 - b + bt}(0.02)$$

Since a change in one of the parameters, unlike a change in one of the independent variables, will alter the value of the multiplier, the multiplier will only approximate the effect of the change.

$$\Delta \bar{Y} \approx \frac{-0.75(850)}{0.4}(0.02) = -31.88$$

b)   The government wants to raise $\bar{Y}$ by 150. Substituting $\Delta \bar{Y} = 150$ in the equation above,

$$150 \approx \frac{-0.75(850)}{0.4} \Delta t$$

$$\Delta t \approx -0.09$$

The tax rate should be reduced by approximately 0.09. The new tax rate should be around 11 percent $(0.20 - 0.09 = 0.11)$.

6.13.   Given
$$Y = C + I_0 + G_0 + X_0 - Z \qquad T = T_0 + tY$$
$$C = C_0 + bYd \qquad\qquad Z = Z_0 + zYd$$

with
$$b = 0.9 \qquad t = 0.2 \qquad C_0 = 125$$
$$X_0 = 150 \qquad Z_0 = 55 \qquad I_0 = 92.5$$
$$z = 0.15 \qquad T_0 = 150 \qquad G_0 = 600$$

Calculate (a) the equilibrium level of income, (b) the effect on $\bar{Y}$ of an increase of 60 in autonomous exports X_0, and (c) the effect on $\bar{Y}$ of an increase of 30 in autonomous imports Z_0.

a)   From Problem 6.7,

$$\bar{Y} = \frac{1}{1 - b + bt + z - zt}(C_0 - bT_0 + I_0 + G_0 + X_0 - Z_0 + zT_0)$$

$$= \frac{1}{1 - 0.9 + 0.9(0.2) + 0.15 - 0.15(0.2)}[125 - 0.9(150) + 92.5 + 600 + 150 - 55 + 22.5]$$

$$= 2.5(800) = 2000$$

b)

$$\Delta \bar{Y} = \frac{\partial \bar{Y}}{\partial X_0} \Delta X_0 = \frac{1}{1 - b + bt + z - zt}(60) = 2.5(60) = 150$$

c)

$$\Delta \bar{Y} = \frac{\partial \bar{Y}}{\partial Z_0} \Delta Z_0 = \frac{-1}{1 - b + bt + z - zt}(30) = -2.5(30) = -75$$

6.14. If the full-employment level of income in Problem 6.13 is 2075, (*a*) by how much should the government increase expenditures to achieve it? (*b*) By how much should it cut autonomous taxes to have the same effect?

a)   The effect of government spending on national income is

$$\Delta \bar{Y} = \frac{\partial \bar{Y}}{\partial G_0} \Delta G_0$$

substituting $\Delta \bar{Y} = 75$,

$$75 = \frac{1}{1 - b + bt + z - zt} \Delta G_0 = 2.5 \, \Delta G_0 \qquad \Delta G_0 = 30$$

b)

$$\Delta \bar{Y} = \frac{\partial \bar{Y}}{\partial T_0} \Delta T_0$$

$$75 = \frac{z - b}{1 - b + bt + z - zt} \Delta T_0 = -1.875 \, \Delta T_0 \qquad \Delta T_0 = -40$$

The government should cut autonomous taxation by 40.

6.15. Calculate the effect on the government deficit if the government in Problem 6.14 achieves full employment through (*a*) increased expenditures or (*b*) a tax cut.

a)   If the government increases expenditures by 30, the government deficit increases initially by 30. However, income is stimulated by 75. With $\Delta \bar{Y} = 75$, $\Delta T = 0.2(75) = 15$. Tax revenue increases by 15. Thus the net cost to the government from this policy, and the effect on the deficit, is $15 ($30 - 15 = 15$).

b)   If the government cuts autonomous taxation by 40, tax revenues fall initially by 40. But income increases by 75, causing tax revenues to increase by 15. Thus the net cost to the government of this policy is 25 ($40 - 15 = 25$), and the government deficit worsens by 25.

6.16. Calculate the effect on the balance of payments (B/P) from (*a*) government spending and (*b*) the tax reduction in Problem 6.14.

a)   Since B/P = $X - Z$, substituting from Problem 6.13,

$$\text{B/P} = X_0 - (Z_0 + zYd) = X_0 - Z_0 - zY - zT_0 + ztY \qquad (6.28)$$

With an increase of 30 in government spending, $\Delta \bar{Y} = 75$. Since Y is the only variable on the right-hand side of (*6.28*) to change,

$$\Delta(\text{B/P}) = -z(75) + zt(75)$$

Substituting $z = 0.15$, $zt = 0.03$, $\Delta(\text{B/P}) = -9$.

b)   When the government cuts autonomous taxes by 40, $\Delta \bar{Y} = 75$. Adjusting (*6.28*),

$$\Delta(\text{B/P}) = -z(75) + z(-40) + zt(75) = -15$$

The reduction in taxes leads to a greater increase in disposable income than the increased government spending, resulting in a higher level of imports and a more serious balance of payments deficit.

6.17. Estimate the effect on $\bar{Y}$ of a one-percentage-point decrease in the marginal propensity to import from Problem 6.13.

$$\Delta \bar{Y} \approx \frac{\partial \bar{Y}}{\partial z} \Delta z$$

Substituting from Problem 6.8,

$$\Delta \bar{Y} = \frac{-\bar{Y}d}{1 - b + bt + z - zt} \Delta z$$

where

$$\bar{Y}d = \bar{Y} - T_0 - t\bar{Y} = 2000 - 150 - 0.2(2000) = 1450$$

Thus,

$$\Delta \bar{Y} = \frac{-1450}{0.4}(-0.01) = +36.25$$

PARTIAL ELASTICITIES

6.18. Given $Q = 700 - 2P + 0.02Y$, where $P = 25$ and $Y = 5000$. Find (a) the price elasticity of demand and (b) the income elasticity of demand.

a)

$$\epsilon_d = \frac{\partial Q}{\partial P}\left(\frac{P}{Q}\right)$$

where $\partial Q / \partial P = -2$ and $Q = 700 - 2(25) + 0.02(5000) = 750$. Thus,

$$\epsilon_d = -2\left(\frac{25}{750}\right) = -0.067$$

b)

$$\epsilon_Y = \frac{\partial Q}{\partial Y}\left(\frac{Y}{Q}\right) = 0.02\left(\frac{5000}{750}\right) = 0.133$$

6.19. Given $Q = 400 - 8P + 0.05Y$, where $P = 15$ and $Y = 12,000$. Find (a) the income elasticity of demand and (b) the growth potential of the product, if income is expanding by 5 percent a year. (c) Comment on the growth potential of the product.

a) $Q = 400 - 8(15) + 0.05(12,000) = 880$ and $\partial Q / \partial Y = 0.05$. Thus,

$$\epsilon_Y = \frac{\partial Q}{\partial Y}\left(\frac{Y}{Q}\right) = 0.05\left(\frac{12,000}{880}\right) = 0.68$$

b)

$$\epsilon_Y = \frac{\partial Q}{Q} \div \frac{\partial Y}{Y}$$

Rearranging terms and substituting the known parameters,

$$\frac{\partial Q}{Q} = \epsilon_Y \frac{\partial Y}{Y} = 0.68(0.05) = 0.034$$

The demand for the good will increase by 3.4 percent.

c)   Since $0 < \epsilon_Y < 1$, it can be expected that demand for the good will increase with national income, but the increase will be less than proportionate. Thus, while demand grows absolutely, the relative market share of the good will decline in an expanding economy. If $\epsilon_Y > 1$, the demand for the product would grow faster than the rate of expansion in the economy, and increase its relative market share. And if $\epsilon_Y < 0$, demand for the good would decline as income increases.

6.20.   Given $Q_1 = 100 - P_1 + 0.75P_2 - 0.25P_3 + 0.0075Y$.   At $P_1 = 10$, $P_2 = 20$, $P_3 = 40$,   and $Y = 10,000$, $Q_1 = 170$. Find the different cross price elasticities of demand.

$$\epsilon_{12} = \frac{\partial Q_1}{\partial P_2}\left(\frac{P_2}{Q_1}\right) = 0.75\left(\frac{20}{170}\right) = 0.088$$

$$\epsilon_{13} = \frac{\partial Q_1}{\partial P_3}\left(\frac{P_3}{Q_1}\right) = -0.25\left(\frac{40}{170}\right) = -0.059$$

6.21.   Given $Q_1 = 50 - 4P_1 - 3P_2 + 2P_3 + 0.001Y$. At $P_1 = 5$, $P_2 = 7$, $P_3 = 3$, and $Y = 11,000$, $Q_1 = 26$. (a) Use cross price elasticities to determine the relationship between good 1 and the other two goods. (b) Determine the effect on Q_1 of a 10 percent price increase for each of the other goods individually.

a)
$$\epsilon_{12} = -3(\tfrac{7}{26}) = -0.81 \qquad \epsilon_{13} = 2(\tfrac{3}{26}) = 0.23$$

With ϵ_{12} negative, goods 1 and 2 are complements. An increase in P_2 will lead to a decrease in Q_1. With ϵ_{13} positive, goods 1 and 3 are substitutes. An increase in P_3 will increase Q_1.

b)
$$\epsilon_{12} = \frac{\partial Q_1}{Q_1} \div \frac{\partial P_2}{P_2}$$

Rearranging terms and substituting the known parameters,

$$\frac{\partial Q_1}{Q_1} = \epsilon_{12}\frac{\partial P_2}{P_2} = -0.81(0.10) = -0.081$$

If P_2 increases by 10 percent, Q_1 decreases by 8.1 percent.

$$\epsilon_{13} = \frac{\partial Q_1}{Q_1} \div \frac{\partial P_3}{P_3}$$

$$\frac{\partial Q_1}{Q_1} = \epsilon_{13}\frac{\partial P_3}{P_3} = 0.23(0.10) = 0.023$$

If P_3 increases by 10 percent, Q_1 increases by 2.3 percent.

OPTIMIZING ECONOMIC FUNCTIONS

6.22.   Given the profit function $\pi = 160x - 3x^2 - 2xy - 2y^2 + 120y - 18$ for a firm producing two goods x and y, (a) maximize profits, (b) test the second-order condition, and (c) evaluate the function at the critical values $\bar{x}$ and $\bar{y}$.

a)
$$\pi_x = 160 - 6x - 2y = 0 \qquad \pi_y = -2x - 4y + 120 = 0$$

When solved simultaneously, $\bar{x} = 20$ and $\bar{y} = 20$.

b)   Taking the second partials,

$$\pi_{xx} = -6 \qquad \pi_{yy} = -4 \qquad \pi_{xy} = -2$$

With both direct second partials negative, and $\pi_{xx}\pi_{yy} > (\pi_{xy})^2$, π is maximized at $\bar{x} = \bar{y} = 20$.

c)
$$\pi = 2782$$

6.23.   Redo Problem 6.22, given $\pi = 25x - x^2 - xy - 2y^2 + 30y - 28$.

a)
$$\pi_x = 25 - 2x - y = 0 \qquad \pi_y = -x - 4y + 30 = 0$$

Thus, $\bar{x} = 10$ and $\bar{y} = 5$.

b)
$$\pi_{xx} = -2 \qquad \pi_{yy} = -4 \qquad \pi_{xy} = -1$$

With π_{xx} and π_{yy} both negative and $\pi_{xx}\pi_{yy} > (\pi_{xy})^2$, π is maximized.

c)
$$\pi = 172$$

6.24. A monopolist sells two products x and y for which the demand functions are

$$x = 25 - 0.5P_x \tag{6.29}$$

$$y = 30 - P_y \tag{6.30}$$

and the combined cost function is

$$c = x^2 + 2xy + y^2 + 20 \tag{6.31}$$

Find (a) the profit-maximizing level of output for each product, (b) the profit-maximizing price for each product, and (c) the maximum profit.

a)   Since $\pi = \text{TR}_x + \text{TR}_y - \text{TC}$, in this case,

$$\pi = P_x x + P_y y - c \tag{6.32}$$

From (6.29) and (6.30),

$$P_x = 50 - 2x \tag{6.33}$$

$$P_y = 30 - y \tag{6.34}$$

Substituting in (6.32),

$$\pi = (50 - 2x)x + (30 - y)y - (x^2 + 2xy + y^2 + 20)$$
$$= 50x - 3x^2 + 30y - 2y^2 - 2xy - 20 \tag{6.35}$$

The first-order condition for maximizing (6.35) is

$$\pi_x = 50 - 6x - 2y = 0 \qquad \pi_y = 30 - 4y - 2x = 0$$

Solving simultaneously, $\bar{x} = 7$ and $\bar{y} = 4$. Testing the second-order conditon, $\pi_{xx} = -6$, $\pi_{yy} = -4$, and $\pi_{xy} = -2$. With both direct partials negative and $\pi_{xx}\pi_{yy} > (\pi_{xy})^2$, π is maximized.

b)   Substituting $\bar{x} = 7$, $\bar{y} = 4$ in (6.33) and (6.34),

$$P_x = 50 - 2(7) = 36 \qquad P_y = 30 - 4 = 26$$

c)   Substituting $\bar{x} = 7$, $\bar{y} = 4$ in (6.35), $\pi = 215$.

6.25. Find the profit-maximizing level of (a) output, (b) price, and (c) profit for a monopolist with the demand functions

$$x = 50 - 0.5P_x \tag{6.36}$$

$$y = 76 - P_y \tag{6.37}$$

and the total cost function $c = 3x^2 + 2xy + 2y^2 + 55$.

a)   From (6.36) and (6.37),

$$P_x = 100 - 2x \tag{6.38}$$

$$P_y = 76 - y \tag{6.39}$$

Substituting in $\pi = P_x x + P_y y - c$,

$$\pi = (100 - 2x)x + (76 - y)y - (3x^2 + 2xy + 2y^2 + 55)$$
$$= 100x - 5x^2 + 76y - 3y^2 - 2xy - 55 \tag{6.40}$$

Maximizing (6.40),

$$\pi_x = 100 - 10x - 2y = 0 \qquad \pi_y = 76 - 6y - 2x = 0$$

Thus, $\bar{x} = 8$ and $\bar{y} = 10$. Checking the second-order condition, $\pi_{xx} = -10$, $\pi_{yy} = -6$, and $\pi_{xy} = -2$. Since $\pi_{xx}, \pi_{yy} < 0$ and $\pi_{xx}\pi_{yy} > (\pi_{xy})^2$, π is maximized at the critical values.

b) Substituting $\bar{x} = 8$, $\bar{y} = 10$ in (6.38) and (6.39),

$$P_x = 100 - 2(8) = 84 \qquad P_y = 76 - 10 = 66$$

c) From (6.40), $\pi = 725$.

6.26. Find the profit-maximizing level of (a) output, (b) price, and (c) profit for the monopolistic producer with the demand functions

$$Q_1 = 49\tfrac{1}{3} - \tfrac{2}{3}P_1 \tag{6.41}$$
$$Q_2 = 36 - \tfrac{1}{2}P_2 \tag{6.42}$$

and the joint cost function $c = Q_1^2 + 2Q_1 Q_2 + Q_2^2 + 120$.

a) From (6.41) and (6.42),

$$P_1 = 74 - 1.5Q_1 \tag{6.43}$$
$$P_2 = 72 - 2Q_2 \tag{6.44}$$

Substituting in $\pi = P_1 Q_1 + P_2 Q_2 - c$,

$$\pi = (74 - 1.5Q_1)Q_1 + (72 - 2Q_2)Q_2 - (Q_1^2 + 2Q_1 Q_2 + Q_2^2 + 120)$$
$$= 74Q_1 - 2.5Q_1^2 + 72Q_2 - 3Q_2^2 - 2Q_1 Q_2 - 120 \tag{6.45}$$

The first-order condition for maximizing (6.45) is

$$\pi_1 = 74 - 5Q_1 - 2Q_2 = 0 \qquad \pi_2 = 72 - 6Q_2 - 2Q_1 = 0$$

Thus, $\bar{Q}_1 = 11.54$ and $\bar{Q}_2 = 8.15$. Testing the second-order condition, $\pi_{11} = -5$, $\pi_{22} = -6$, and $\pi_{12} = -2$. Thus, $\pi_{11}, \pi_{22} < 0$; $\pi_{11}\pi_{22} > (\pi_{12})^2$, and π is maximized.

b) Substituting the critical values in (6.43) and (6.44),

$$P_1 = 74 - 1.5(11.54) = 56.69 \qquad P_2 = 72 - 2(8.15) = 55.70$$

c)
$$\pi = 600.46$$

6.27. Find the profit-maximizing level of (a) output, (b) price, and (c) profit when

$$Q_1 = 5200 - 10P_1 \tag{6.46}$$
$$Q_2 = 8200 - 20P_2 \tag{6.47}$$

and
$$c = 0.1Q_1^2 + 0.1Q_1 Q_2 + 0.2Q_2^2 + 325$$

a) From (6.46) and (6.47)

$$P_1 = 520 - 0.1Q_1 \tag{6.48}$$
$$P_2 = 410 - 0.05Q_2 \tag{6.49}$$

Thus, $\pi = (520 - 0.1Q_1)Q_1 + (410 - 0.05Q_2)Q_2 - (0.1Q_1^2 + 0.1Q_1Q_2 + 0.2Q_2^2 + 325)$

$$= 520Q_1 - 0.2Q_1^2 + 410Q_2 - 0.25Q_2^2 - 0.1Q_1Q_2 - 325 \tag{6.50}$$

Maximizing (6.50),

$$\pi_1 = 520 - 0.4Q_1 - 0.1Q_2 = 0 \qquad \pi_2 = 410 - 0.5Q_2 - 0.1Q_1 = 0$$

Thus, $\bar{Q}_1 = 1152.63$ and $\bar{Q}_2 = 589.47$. Checking the second-order condition, $\pi_{11} = -0.4$, $\pi_{22} = -0.5$, and $\pi_{12} = -0.1 = \pi_{21}$. Since π_{11}, $\pi_{22} < 0$ and $\pi_{11}\pi_{22} > (\pi_{12})^2$, π is maximized at $\bar{Q}_1 = 1152.63$ and $\bar{Q}_2 = 589.47$.

b)   Substituting in (6.48) and (6.49),

$$P_1 = 520 - 0.1(1152.63) = 404.74 \qquad P_2 = 410 - 0.05(589.47) = 380.53$$

c) $\pi = 420{,}201.32$

CONSTRAINED OPTIMIZATION IN ECONOMICS

6.28.   (a) What combination of goods x and y should a firm produce to minimize costs when the joint cost function is $c = 6x^2 + 10y^2 - xy + 30$ and the firm has a production quota of $x + y = 34$? (b) Estimate the effect on costs if the production quota is reduced by 1 unit.

a)   Form a new function by setting the constraint equal to zero, multiplying it by λ, and adding it to the original or objective function. Thus,

$$C = 6x^2 + 10y^2 - xy + 30 + \lambda(34 - x - y)$$
$$C_x = 12x - y - \lambda = 0$$
$$C_y = 20y - x - \lambda = 0$$
$$C_\lambda = 34 - x - y = 0$$

Solving simultaneously, $\bar{x} = 21$, $\bar{y} = 13$, and $\bar{\lambda} = 239$. Thus, $C = 4093$. Second-order conditions are discussed in Section 12.5.

b)   With $\lambda = 239$, a decrease in the constant of the constraint (the production quota) will lead to a cost reduction of approximately 239.

6.29.   (a) What output mix should a profit-maximizing firm produce when its total profit function is $\pi = 80x - 2x^2 - xy - 3y^2 + 100y$ and its maximum output capacity is $x + y = 12$? (b) Estimate the effect on profits if output capacity is expanded by 1 unit.

a) $\Pi = 80x - 2x^2 - xy - 3y^2 + 100y + \lambda(12 - x - y)$
$$\Pi_x = 80 - 4x - y - \lambda = 0$$
$$\Pi_y = -x - 6y + 100 - \lambda = 0$$
$$\Pi_\lambda = 12 - x - y = 0$$

When solved simultaneously, $\bar{x} = 5$, $\bar{y} = 7$, and $\bar{\lambda} = 53$. Thus, $\pi = 868$.

b)   With $\bar{\lambda} = 53$, an increase in output capacity should lead to increased profits of approximately 53.

6.30.   A rancher faces the profit function

$$\pi = 110x - 3x^2 - 2xy - 2y^2 + 140y$$

where x = sides of beef and y = hides. Since there are two sides of beef for every hide, it follows that output must be in the proportion

$$\frac{x}{2} = y \qquad x = 2y$$

At what level of output will the rancher maximize profits?

$$\Pi = 110x - 3x^2 - 2xy - 2y^2 + 140y + \lambda(x - 2y)$$
$$\Pi_x = 110 - 6x - 2y + \lambda = 0$$
$$\Pi_y = -2x - 4y + 140 - 2\lambda = 0$$
$$\Pi_\lambda = x - 2y = 0$$

Solving simultaneously, $\bar{x} = 20$, $\bar{y} = 10$, $\bar{\lambda} = 30$, and $\pi = 1800$.

6.31. (a) Minimize costs for a firm with the cost function $c = 5x^2 + 2xy + 3y^2 + 800$ subject to the production quota $x + y = 39$. (b) Estimate additional costs if the production quota is increased to 40.

a)
$$C = 5x^2 + 2xy + 3y^2 + 800 + \lambda(39 - x - y)$$
$$C_x = 10x + 2y - \lambda = 0$$
$$C_y = 2x + 6y - \lambda = 0$$
$$C_\lambda = 39 - x - y = 0$$

When solved simultaneously, $\bar{x} = 13$, $\bar{y} = 26$, $\bar{\lambda} = 182$, and $c = 4349$.

b) Since $\bar{\lambda} = 182$, an increased production quota will lead to additional costs of approximately 182.

6.32. A monopolistic firm has the following demand functions for each of its products x and y:

$$x = 72 - 0.5P_x \qquad (6.51)$$

$$x = 120 - P_y \qquad (6.52)$$

The combined cost function is $c = x^2 + xy + y^2 + 35$, and maximum joint production is 40. Thus $x + y = 40$. Find the profit-maximizing level of (a) output, (b) price, and (c) profit.

a) From (6.51) and (6.52),

$$P_x = 144 - 2x \qquad (6.53)$$

$$P_y = 120 - y \qquad (6.54)$$

Thus, $\pi = (144 - 2x)x + (120 - y)y - (x^2 + xy + y^2 + 35) = 144x - 3x^2 - xy - 2y^2 + 120y - 35$. Incorporating the constraint,

$$\Pi = 144x - 3x^2 - xy - 2y^2 + 120y - 35 + \lambda(40 - x - y)$$

Thus,
$$\Pi_x = 144 - 6x - y - \lambda = 0$$
$$\Pi_y = -x - 4y + 120 - \lambda = 0$$
$$\Pi_\lambda = 40 - x - y = 0$$

and, $\bar{x} = 18$, $\bar{y} = 22$, and $\bar{\lambda} = 14$.

b) Substituting in (6.53) and (6.54),

$$P_x = 144 - 2(18) = 108 \qquad P_y = 120 - 22 = 98$$

c)
$$\pi = 2861$$

6.33. A manufacturer of parts for the tricycle industry sells three tires (x) for every frame (y). Thus,

$$\frac{x}{3} = y \qquad x = 3y$$

If the demand functions are

$$x = 63 - 0.25P_x \qquad (6.55)$$

$$y = 60 - \tfrac{1}{3}P_y \qquad (6.56)$$

and costs are

$$c = x^2 + xy + y^2 + 190$$

find the profit-maximizing level of (a) output, (b) price, and (c) profit.

a) From (6.55) and (6.56),

$$P_x = 252 - 4x \qquad (6.57)$$

$$P_y = 180 - 3y \qquad (6.58)$$

Thus, $\pi = (252 - 4x)x + (180 - 3y)y - (x^2 + xy + y^2 + 190) = 252x - 5x^2 - xy + 180y - 190 - 4y^2$

Forming a new, constrained function,

$$\Pi = 252x - 5x^2 - xy - 4y^2 + 180y - 190 + \lambda(x - 3y)$$

Hence, $\Pi_x = 252 - 10x - y + \lambda = 0 \qquad \Pi_y = -x - 8y + 180 - 3\lambda = 0 \qquad \Pi_\lambda = x - 3y = 0$

and $\bar{x} = 27$, $\bar{y} = 9$, and $\bar{\lambda} = 27$.

b) From (6.57) and (6.58), $P_x = 144$ and $P_y = 153$.

c) $\pi = 4022$

6.34. Problem 4.22 dealt with the profit-maximizing level of output for a firm producing a single product that is sold in two distinct markets when it does and does not discriminate. The functions given were

$$Q_1 = 21 - 0.1P_1 \qquad (6.59)$$

$$Q_2 = 50 - 0.4P_2 \qquad (6.60)$$

$$c = 2000 + 10Q \qquad \text{where } Q = Q_1 + Q_2 \qquad (6.61)$$

Use multivariable calculus to check your solution to Problem 4.22.

From (6.59), (6.60), and (6.61),

$$P_1 = 210 - 10Q_1 \qquad (6.62)$$

$$P_2 = 125 - 2.5Q_2 \qquad (6.63)$$

$$c = 2000 + 10Q_1 + 10Q_2$$

With discrimination $P_1 \neq P_2$ since different prices are charged in different markets, and therefore

$$\pi = (210 - 10Q_1)Q_1 + (125 - 2.5Q_2)Q_2 - (2000 + 10Q_1 + 10Q_2)$$
$$= 200Q_1 - 10Q_1^2 + 115Q_2 - 2.5Q_2^2 - 2000$$

Taking the first partials,

$$\pi_1 = 200 - 20Q_1 = 0 \qquad \pi_2 = 115 - 5Q_2 = 0$$

Thus, $\bar{Q}_1 = 10$ and $\bar{Q}_2 = 23$. Substituting in (6.62) and (6.63), $\bar{P}_1 = 110$ and $\bar{P}_2 = 67.5$.

If there is no discrimination, the same price must be charged in both markets. Hence $P_1 = P_2$. Substituting from (6.62) and (6.63),

$$210 - 10Q_1 = 125 - 2.5Q_2$$
$$2.5Q_2 - 10Q_1 = -85$$

Rearranging this as a constraint and forming a new function,

$$\Pi = 200Q_1 - 10Q_1^2 + 115Q_2 - 2.5Q_2^2 - 2000 + \lambda(85 - 10Q_1 + 2.5Q_2)$$

Thus, $\Pi_1 = 200 - 20Q_1 - 10\lambda = 0$ $\Pi_2 = 115 - 5Q_2 + 2.5\lambda = 0$ $\Pi_\lambda = 85 - 10Q_1 + 2.5Q_2 = 0$

and $\bar{Q}_1 = 13.4$, $\bar{Q}_2 = 19.6$, and $\bar{\lambda} = -6.8$. Substituting in (6.62) and (6.63),

$$P_1 = 210 - 10(13.4) = 76$$
$$P_2 = 125 - 2.5(19.6) = 76$$
$$Q = 13.4 + 19.6 = 33$$

6.35. Check your answers to Problem 4.23, given

$$Q_1 = 24 - 0.2P_1 \qquad Q_2 = 10 - 0.05P_2$$
$$c = 35 + 40Q \qquad \text{where } Q = Q_1 + Q_2$$

From the information given,

$$P_1 = 120 - 5Q_1 \tag{6.64}$$

$$P_2 = 200 - 20Q_2 \tag{6.65}$$

$$c = 35 + 40Q_1 + 40Q_2$$

With price discrimination,

$$\pi = (120 - 5Q_1)Q_1 + (200 - 20Q_2)Q_2 - (35 + 40Q_1 + 40Q_2) = 80Q_1 - 5Q_1^2 + 160Q_2 - 20Q_2^2 - 35$$

Thus, $\pi_1 = 80 - 10Q_1 = 0$ $\pi_2 = 160 - 40Q_2 = 0$

and $\bar{Q}_1 = 8$, $\bar{Q}_2 = 4$, $P_1 = 80$, and $P_2 = 120$.

If there is no price discrimination, $P_1 = P_2$. Substituting from (6.64) and (6.65),

$$120 - 5Q_1 = 200 - 20Q_2$$
$$20Q_2 - 5Q_1 = 80 \tag{6.66}$$

Forming a new function with (6.66) as a constraint,

$$\Pi = 80Q_1 - 5Q_1^2 + 160Q_2 - 20Q_2^2 - 35 + \lambda(80 + 5Q_1 - 20Q_2)$$

Thus, $\Pi_1 = 80 - 10Q_1 + 5\lambda = 0$ $\Pi_2 = 160 - 40Q_2 - 20\lambda = 0$ $\Pi_\lambda = 80 + 5Q_1 - 20Q_2 = 0$

and $\bar{Q}_1 = 6.4$, $\bar{Q}_2 = 5.6$, and $\bar{\lambda} = -3.2$. Substituting in (6.64) and (6.65),

$$P_1 = 120 - 5(6.4) = 88$$
$$P_2 = 200 - 20(5.6) = 88$$
$$Q = 6.4 + 5.6 = 12$$

6.36. (a) Maximize utility $u = Q_1Q_2$ when $P_1 = 1$, $P_2 = 4$, and one's budget $B = 120$. (b) Estimate the effect of a 1-unit increase in the budget.

a)   The budget constraint is $Q_1 + 4Q_2 = 120$. Forming a new function to incorporate the constraint,

$$U = Q_1Q_2 + \lambda(120 - Q_1 - 4Q_2)$$

Thus, $U_1 = Q_2 - \lambda = 0$ $U_2 = Q_1 - 4\lambda = 0$ $U_\lambda = 120 - Q_1 - 4Q_2 = 0$

and $\bar{Q}_1 = 60$, $\bar{Q}_2 = 15$, and $\bar{\lambda} = 15$.

b)   With $\bar{\lambda} = 15$, a \$1 increase in the budget will lead to an increase in the utility function of approximately 15. Thus, the marginal utility of money (or income) at $\bar{Q}_1 = 60$ and $\bar{Q}_2 = 15$ is approximately 15.

6.37. (*a*) Maximize utility $u = Q_1 Q_2$, subject to $P_1 = 10$, $P_2 = 2$, and $B = 240$. (*b*) What is the marginal utility of money?

a) Form the Lagrangian function $U = Q_1 Q_2 + \lambda(240 - 10Q_1 - 2Q_2)$.

$$U_1 = Q_2 - 10\lambda = 0 \qquad U_2 = Q_1 - 2\lambda = 0 \qquad U_\lambda = 240 - 10Q_1 - 2Q_2 = 0$$

Thus, $\bar{Q}_1 = 12$, $\bar{Q}_2 = 60$, and $\bar{\lambda} = 6$.

b) The marginal utility of money at $\bar{Q}_1 = 12$ and $\bar{Q}_2 = 60$ is approximately 6.

6.38. Maximize utility $u = Q_1 Q_2 + Q_1 + 2Q_2$, subject to $P_1 = 2$, $P_2 = 5$, and $B = 51$.

Form the Lagrangian function $U = Q_1 Q_2 + Q_1 + 2Q_2 + \lambda(51 - 2Q_1 - 5Q_2)$.

$$U_1 = Q_2 + 1 - 2\lambda = 0 \qquad U_2 = Q_1 + 2 - 5\lambda = 0 \qquad U_\lambda = 51 - 2Q_1 - 5Q_2 = 0$$

Thus, $\bar{Q}_1 = 13$, $\bar{Q}_2 = 5$, and $\bar{\lambda} = 3$.

6.39. Maximize utility $u = xy + 3x + y$ subject to $P_x = 8$, $P_y = 12$, and $B = 212$.

The Lagrangian function is $U = xy + 3x + y + \lambda(212 - 8x - 12y)$.

$$U_x = y + 3 - 8\lambda = 0 \qquad U_y = x + 1 - 12\lambda = 0 \qquad U_\lambda = 212 - 8x - 12y = 0$$

Thus, $\bar{x} = 15$, $\bar{y} = 7\frac{2}{3}$, and $\bar{\lambda} = 1\frac{1}{3}$.

HOMOGENEITY AND RETURNS TO SCALE

6.40. Determine the level of homogeneity and returns to scale for each of the following production functions:

a) $Q = x^2 + 6xy + 7y^2$

here Q is homogeneous of degree 2, and returns to scale are increasing because

$$f(kx, ky) = (kx)^2 + 6(kx)(ky) + 7(ky)^2 = k^2(x^2 + 6xy + 7y^2)$$

b) $Q = x^3 - xy^2 + 3y^3 + x^2 y$

here Q is homogeneous of degree 3, and returns to scale are increasing because

$$f(kx, ky) = (kx)^3 - (kx)(ky)^2 + 3(ky)^3 + (kx)^2(ky) = k^3(x^3 - xy^2 + 3y^3 + x^2 y)$$

c) $Q = \dfrac{3x^2}{5y^2}$

here Q is homogeneous of degree 0, and returns to scale are decreasing because

$$f(kx, ky) = \frac{3(kx)^2}{5(ky)^2} = \frac{3x^2}{5y^2} \quad \text{and} \quad k^0 = 1$$

d) $Q = 0.9K^{0.2}L^{0.6}$

here Q is homogeneous of degree 0.8 and returns to scale are decreasing because

$$Q(kK, kL) = 0.9(kK)^{0.2}(kL)^{0.6} = Ak^{0.2}K^{0.2}k^{0.6}L^{0.6}$$
$$= k^{0.2+0.6}(0.9K^{0.2}L^{0.6}) = k^{0.8}(0.9K^{0.2}L^{0.6})$$

Note that the returns to scale of a Cobb-Douglas function will always equal the sum of the exponents $\alpha + \beta$, as is illustrated in part 6 of Example 8.

CONSTRAINED OPTIMIZATION OF COBB-DOUGLAS FUNCTIONS

6.41. Optimize the following Cobb-Douglas production functions subject to the given constraints by (1) forming the Lagrange function and (2) finding the critical values as in Example 10.

a) $q = K^{0.3}L^{0.5}$ subject to $6K + 2L = 384$

1) $Q = K^{0.3}L^{0.5} + \lambda(384 - 6K - 2L)$

2) $Q_K = 0.3K^{-0.7}L^{0.5} - 6\lambda = 0$ (6.67)

$Q_L = 0.5K^{0.3}L^{-0.5} - 2\lambda = 0$ (6.68)

$Q_\lambda = 384 - 6K - 2L = 0$ (6.69)

Rearrange, then divide (6.67) by (6.68) to eliminate λ.

$$\frac{0.3K^{-0.7}L^{0.5}}{0.5K^{0.3}L^{-0.5}} = \frac{6\lambda}{2\lambda}$$

Subtracting exponents in division,

$$0.6K^{-1}L^{1} = 3$$

$$\frac{L}{K} = \frac{3}{0.6} \qquad L = 5K$$

Substitute $L = 5K$ in (6.69).

$$384 - 6K - 2(5K) = 0 \qquad K_0 = 24 \qquad L_0 = 120$$

Second-order conditions are tested in Problem 12.27(b).

b) $q = 10K^{0.7}L^{0.1}$, given $P_K = 28$, $P_L = 10$, and $B = 4000$

1) $Q = 10K^{0.7}L^{0.1} + \lambda(4000 - 28K - 10L)$

2) $Q_K = 7K^{-0.3}L^{0.1} - 28\lambda = 0$ (6.70)

$Q_L = 1K^{0.7}L^{-0.9} - 10\lambda = 0$ (6.71)

$Q_\lambda = 4000 - 28K - 10L = 0$ (6.72)

Divide (6.70) by (6.71) to eliminate λ.

$$\frac{7K^{-0.3}L^{0.1}}{1K^{0.7}L^{-0.9}} = \frac{28\lambda}{10\lambda}$$

$$7K^{-1}L^{1} = 2.8$$

$$\frac{L}{K} = \frac{2.8}{7} \qquad L = 0.4K$$

Substituting in (6.72), $K_0 = 125 \qquad L_0 = 50$

See Problem 12.27(c) for the second-order conditions.

6.42. Maximize the following utility functions subject to the given budget constraints, using the same steps as above.

a) $u = x^{0.6}y^{0.25}$, given $P_x = 8$, $P_y = 5$, and $B = 680$

1) $U = x^{0.6}y^{0.25} + \lambda(680 - 8x - 5y)$

2) $U_x = 0.6x^{-0.4}y^{0.25} - 8\lambda = 0$ (6.73)

$U_y = 0.25x^{0.6}y^{-0.75} - 5\lambda = 0$ (6.74)

$U_\lambda = 680 - 8x - 5y = 0$ (6.75)

Divide (6.73) by (6.74).

$$\frac{0.6x^{-0.4}y^{0.25}}{0.25x^{0.6}y^{-0.75}} = \frac{8\lambda}{5\lambda}$$

$$2.4x^{-1}y^{1} = 1.6$$

$$y = \tfrac{2}{3}x$$

Substitute in (6.75). $x_0 = 60$ $y_0 = 40$

b) $u = x^{0.8}y^{0.2}$, given $P_x = 5$, $P_y = 3$, and $B = 75$

1) $U = x^{0.8}y^{0.2} + \lambda(75 - 5x - 3y)$

2) $U_x = 0.8x^{-0.2}y^{0.2} - 5\lambda = 0$ (6.76)

                         $U_y = 0.2x^{0.8}y^{-0.8} - 3\lambda = 0$ (6.77)

                         $U_\lambda = 75 - 5x - 3y = 0$ (6.78)

Divide (6.76) by (6.77).

$$\frac{0.8x^{-0.2}y^{0.2}}{0.2x^{0.8}y^{-0.8}} = \frac{5\lambda}{3\lambda}$$

$$4x^{-1}y^{1} = \tfrac{5}{3}$$

$$y = \tfrac{5}{12}x$$

Substitute in (6.78). $x_0 = 12$ $y_0 = 5$

CONSTRAINED OPTIMIZATION OF CES PRODUCTION FUNCTIONS

6.43.   Optimize the following CES production function subject to the given constraint by (1) forming the Lagrange function and (2) finding the critical values as in Example 12:

$$q = 80[0.4K^{-0.25} + (1 - 0.4)L^{-0.25}]^{-1/0.25}   \text{subject to } 5K + 2L = 150$$

1) $Q = 80(0.4K^{-0.25} + 0.6L^{-0.25})^{-4} + \lambda(150 - 5K - 2L)$

2)   Using the generalized power function rule for Q_K and Q_L,

$$Q_K = -320(0.4K^{-0.25} + 0.6L^{-0.25})^{-5}(-0.1K^{-1.25}) - 5\lambda = 0$$
$$= 32K^{-1.25}(0.4K^{-0.25} + 0.6L^{-0.25})^{-5} - 5\lambda = 0   (6.79)$$

$$Q_L = -320(0.4K^{-0.25} + 0.6L^{-0.25})^{-5}(-0.15L^{-1.25}) - 2\lambda = 0$$
$$= 48L^{-1.25}(0.4K^{-0.25} + 0.6L^{-0.25})^{-5} - 2\lambda = 0   (6.80)$$

$$Q_\lambda = 150 - 5K - 2L = 0   (6.81)$$

Rearrange, then divide (6.79) by (6.80) to eliminate λ.

$$\frac{32K^{-1.25}(0.4K^{-0.25} + 0.6L^{-0.25})^{-5}}{48L^{-1.25}(0.4K^{-0.25} + 0.6L^{-0.25})^{-5}} = \frac{5\lambda}{2\lambda}$$

$$\frac{32K^{-1.25}}{48L^{-1.25}} = 2.5$$

$$K^{-1.25} = 3.75L^{-1.25}$$

Take the -1.25 root.

$$K = (3.75)^{-1/1.25}L = (3.75)^{-0.8}L$$

To find $(3.75)^{-0.8}$, enter 3.75 on a calculator, press the ⌐ y^x ⌐ key, then enter 0.8 followed by the ⌐ $+/-$ ⌐ key to make it negative, and hit the ⌐ $=$ ⌐ key to find $(3.75)^{-0.8} = 0.34736$.

Thus,

$$K \approx 0.35L$$

Substitute in (6.81).

$$150 - 5(0.35L) - 2L = 0 \qquad L_0 = 40 \qquad K_0 = 14$$

6.44. Optimize the CES production function

$$q = 100[0.2K^{-(-0.5)} + (1 - 0.2)L^{-(-0.5)}]^{-1/(-0.5)}$$

subject to the constraint $10K + 4L = 4100$, as in Problem 6.43.

1) $$Q = 100(0.2K^{0.5} + 0.8L^{0.5})^2 + \lambda(4100 - 10K - 4L)$$

2) $$Q_K = 200(0.2K^{0.5} + 0.8L^{0.5})(0.1K^{-0.5}) - 10\lambda = 0$$
$$= 20K^{-0.5}(0.2K^{0.5} + 0.8L^{0.5}) - 10\lambda = 0 \qquad (6.82)$$

$$Q_L = 200(0.2K^{0.5} + 0.8L^{0.5})(0.4L^{-0.5}) - 4\lambda = 0$$
$$= 80L^{-0.5}(0.2K^{0.5} + 0.8L^{0.5}) - 4\lambda = 0 \qquad (6.83)$$

$$Q_\lambda = 4100 - 10K - 4L = 0 \qquad (6.84)$$

Divide (6.82) by (6.83) to eliminate λ.

$$\frac{20K^{-0.5}(0.2K^{0.5} + 0.8L^{0.5})}{80L^{-0.5}(0.2K^{0.5} + 0.8L^{0.5})} = \frac{10\lambda}{4\lambda}$$

$$\frac{20K^{-0.5}}{80L^{-0.5}} = 2.5$$

$$K^{-0.5} = 10L^{-0.5}$$

Take the -0.5 root. $$K = (10)^{-1/0.5}L = (10)^{-2}L$$
$$K = 0.01L$$

Substitute in (6.84). $$L_0 = 1000 \qquad K_0 = 10$$

PARTIAL DERIVATIVES AND DIFFERENTIALS

6.45. Given $Q = 10K^{0.4}L^{0.6}$, (a) find the marginal productivity of capital and labor and (b) determine the effect on output of an additional unit of capital and labor at $K = 8$, $L = 20$.

a) $$\text{MP}_K = \frac{\partial Q}{\partial K} = 0.4(10)K^{-0.6}L^{0.6} = 4K^{-0.6}L^{0.6} \qquad \text{MP}_L = \frac{\partial Q}{\partial L} = 0.6(10)K^{0.4}L^{-0.4} = 6K^{0.4}L^{-0.4}$$

b) $\Delta Q \approx (\partial Q/\partial K)\,\Delta K$. For a 1-unit change in K, at $K = 8$, $L = 20$, $\Delta Q \approx 4K^{-0.6}L^{0.6} = 4(8)^{-0.6}(20)^{0.6}$. Using a calculator,

$$\Delta Q \approx 4K^{-0.6}L^{0.6} \approx 4(8)^{-0.6}(20)^{0.6} \approx 4(0.28717)(6.03418) \approx 6.93$$

Note: to find $(8)^{-0.6}$ on a calculator, enter 8, press the $\boxed{y^x}$ key, then enter 0.6 followed by the $\boxed{+/-}$ key to make it negative, and hit the $\boxed{=}$ key to find $(8)^{-0.6} = 0.28717$. To find $(20)^{0.6}$, enter 20, press the $\boxed{y^x}$ key, then enter 0.6, and hit the $\boxed{=}$ key to find $(20)^{0.6} = 6.03418$.

For a 1-unit change in L,

$$\Delta Q \approx 6K^{0.4}L^{-0.4} \approx 6(8)^{0.4}(20)^{-0.4} \approx 6(2.29740)(0.30171) \approx 4.16$$

6.46. Redo Problem 6.45, given $Q = 12K^{0.3}L^{0.5}$ at $K = 10$, $L = 15$.

a) $$\text{MP}_K = 3.6K^{-0.7}L^{0.5} \qquad \text{MP}_L = 6K^{0.3}L^{-0.5}$$

b)   For a 1-unit change in K, at $K = 10$, $L = 15$, $\Delta Q \approx 3.6 K^{-0.7} L^{0.5}$.

$$\Delta Q \approx 3.6(10)^{-0.7}(15)^{0.5} \approx 3.6(0.19953)(3.87298) \approx 2.78$$

For a 1-unit change in L,

$$\Delta Q \approx 6(10)^{0.3}(15)^{-0.5} \approx 6(1.99526)(0.25820) \approx 3.09$$

6.47.   Given $Q = 4\sqrt{KL}$, find *(a)* MP_K and MP_L, and *(b)* determine the effect on Q of a 1-unit change in K and L, when $K = 50$ and $L = 600$.

a) $Q = 4\sqrt{KL} = 4(KL)^{1/2}$. By the generalized power function rule,

$$MP_K = Q_K = 2(KL)^{-1/2}(L) = \frac{2L}{\sqrt{KL}} \qquad MP_L = Q_L = 2(KL)^{-1/2}(K) = \frac{2K}{\sqrt{KL}}$$

b)   For a 1-unit change in K at $K = 50$, $L = 600$,

$$\Delta Q \approx 2[50(600)]^{-1/2}(600) \approx 2(0.00577)(600) \approx 6.93$$

For a 1-unit change in L,

$$\Delta Q \approx 2[50(600)]^{-1/2}(50) \approx 2(0.00577)(50) \approx 0.58$$

6.48.   Redo Problem 6.47, given $Q = 2\sqrt{KL}$, where $K = 100$ and $L = 1000$.

a)
$$Q = 2(KL)^{1/2}$$

$$MP_K = (KL)^{-1/2}(L) = \frac{L}{\sqrt{KL}} \qquad MP_L = (KL)^{-1/2}(K) = \frac{K}{\sqrt{KL}}$$

b)   For a 1-unit change in K at $K = 100$, $L = 1000$,

$$\Delta Q \approx [100(1000)]^{-1/2}(1000) \approx (0.00316)(1000) \approx 3.16$$

For a 1-unit change in L,

$$\Delta Q \approx [100(1000)]^{-1/2}(100) \approx (0.00316)(100) \approx 0.316$$

6.49.   A company's sales s have been found to depend on price P, advertising A, and the number of field representatives r it maintains.

$$s = (12{,}000 - 900P)A^{1/2} r^{1/2}$$

Find the change in sales associated with *(a)* hiring another field representative, *(b)* an extra \$1 of advertising, *(c)* a \$0.10 reduction in price, at $P = \$6$, $r = 49$, and $A = \$8100$.

a)
$$\Delta s \approx \frac{\partial s}{\partial r} \Delta r = \frac{1}{2}(12{,}000 - 900P)A^{1/2} r^{-1/2} \Delta r$$

$$= \tfrac{1}{2}[12{,}000 - 900(6)](8100)^{1/2}(49)^{-1/2}(1) = \tfrac{1}{2}(6600)(90)(\tfrac{1}{7}) = 42{,}429$$

b)
$$\Delta s \approx \frac{\partial s}{\partial A} \Delta A = \frac{1}{2}(12{,}000 - 900P)A^{-1/2} r^{1/2} \Delta A = \frac{1}{2}(6600)\left(\frac{1}{90}\right)(7)(1) = 256.67$$

c)
$$\Delta s \approx \frac{\partial s}{\partial P} \Delta P = -900 A^{1/2} r^{1/2} \Delta P = -900(90)(7)(-0.10) = 56{,}700$$

6.50.   Given the sales function for a firm similar to the one in Problem 6.49: $s = (15{,}000 - 1000P)A^{2/3} r^{1/4}$, estimate the change in sales from *(a)* hiring an extra field representative, *(b)* a \$1 increase in advertising, and *(c)* a \$0.01 reduction in price, when $P = 4$, $A = \$6000$, and $r = 24$.

a)
$$\Delta s \approx \tfrac{1}{4}(15{,}000 - 1000P)A^{2/3}r^{-3/4}\Delta r$$
$$\approx \tfrac{1}{4}(11{,}000)(6000)^{2/3}(24)^{-3/4}(1)$$
$$\approx 2750(330.19)(0.09222) \approx 83{,}740$$

b)
$$\Delta s \approx \tfrac{2}{3}(15{,}000 - 1000P)A^{-1/3}r^{1/4}\Delta A$$
$$\approx \tfrac{2}{3}(11{,}000)(6000)^{-1/3}(24)^{1/4}(1)$$
$$\approx (7333.33)(0.05503)(2.21336) \approx 893$$

c)
$$\Delta s \approx -1000A^{2/3}r^{1/4}\Delta P$$
$$\approx -1000(6000)^{2/3}(24)^{1/4}(-0.01)$$
$$\approx 10(330.19)(2.21336) \approx 7308$$

6.51. Given the equation for a *production isoquant*

$$16K^{1/4}L^{3/4} = 2144$$

use the implicit function rule from Section 5.10 to find the slope of the isoquant dK/dL which is the *marginal rate of technical substitution* (MRTS).

Set the equation equal to zero to get

$$F(K, L) = 16K^{1/4}L^{3/4} - 2144 = 0$$

Then from the implicit function rule in Equation (*5.13*),

$$\frac{dK}{dL} = \frac{-F_L}{F_K} = \frac{-12K^{1/4}L^{-1/4}}{4K^{-3/4}L^{3/4}} = \frac{-3K}{L} = \text{MRTS}$$

Compare this answer with that in Problem 4.24.

6.52. Given the equation for the production isoquant

$$25K^{3/5}L^{2/5} = 5400$$

find the MRTS, using the implicit function rule.

Set up the implicit function,

$$F(K, L) = 25K^{3/5}L^{2/5} - 5400 = 0$$

and use (*5.13*).

$$\frac{dK}{dL} = \frac{-F_L}{F_K} = \frac{-10K^{3/5}L^{-3/5}}{15K^{-2/5}L^{2/5}} = \frac{-2K}{3L} = \text{MRTS}$$

Compare this answer with that in Problem 4.25.

PROOFS

6.53. Use the properties of homogeneity to show that a strict Cobb-Douglas production function $q = AK^{\alpha}L^{\beta}$, where $\alpha + \beta = 1$, exhibits constant returns to scale.

Multiply each of the inputs by a constant k and factor.

$$q(kK, kL) = A(kK)^{\alpha}(kL)^{\beta} = Ak^{\alpha}K^{\alpha}k^{\beta}L^{\beta}$$
$$= k^{\alpha+\beta}(AK^{\alpha}L^{\beta}) = k^{\alpha+\beta}(q)$$

As explained in Section 6.9, if $\alpha + \beta = 1$, returns to scale are constant; if $\alpha + \beta > 1$, returns to scale are increasing; and if $\alpha + \beta < 1$, returns to scale are decreasing.

6.54. Given the utility function $u = Ax^a y^b$ subject to the budget constraint $P_x x + P_y y = B$, prove that at the point of constrained utility maximization the ratio of prices P_x/P_y must equal the ratio of marginal utilities MU_x/MU_y.

$$U = Ax^a y^b + \lambda(B - P_x x - P_y y)$$

$$U_x = aAx^{l-1} y^b - \lambda P_x = 0 \tag{6.85}$$

$$U_y = bAx^a y^{b-1} - \lambda P_y = 0 \tag{6.86}$$

$$U_\lambda = B - P_x x - P_y y = 0$$

where in (6.85) $aAx^{a-1} y^b = u_x = MU_x$, and in (6.86) $bAx^a y^{b-1} = u_y = MU_y$.

From (6.85),
$$\lambda = \frac{aAx^{a-1} y^b}{P_x} = \frac{MU_x}{P_x}$$

From (6.86),
$$\lambda = \frac{bAx^a y^{b-1}}{P_y} = \frac{MU_y}{P_y}$$

Equating λ's,
$$\frac{MU_x}{P_x} = \frac{MU_y}{P_y} \qquad \frac{MU_x}{MU_y} = \frac{P_x}{P_y} \qquad \text{Q.E.D.}$$

6.55. Given a generalized Cobb-Douglas production function $q = AK^\alpha L^\beta$ subject to the budget constraint $P_K K + P_L L = B$, prove that for constrained optimization the least-cost input ratio is

$$\frac{K}{L} = \frac{\alpha P_L}{\beta P_K}$$

Using the Lagrangian method,

$$Q = AK^\alpha L^\beta + \lambda(B - P_K K - P_L L)$$

$$Q_K = \alpha AK^{\alpha-1} L^\beta - \lambda P_K = 0 \tag{6.87}$$

$$Q_L = \beta AK^\alpha L^{\beta-1} - \lambda P_L = 0 \tag{6.88}$$

$$Q_\lambda = B - P_K K - P_L L = 0$$

From (6.87) and (6.88),

$$\frac{\alpha AK^{\alpha-1} L^\beta}{P_K} = \lambda = \frac{\beta AK^\alpha L^{\beta-1}}{P_L}$$

Rearranging terms,
$$\frac{P_L}{P_K} = \frac{\beta AK^\alpha L^{\beta-1}}{\alpha AK^{\alpha-1} L^\beta}$$

where $L^{\beta-1} = L^\beta/L$ and $1/K^{\alpha-1} = K/K^\alpha$. Thus,

$$\frac{P_L}{P_K} = \frac{\beta K}{\alpha L} \qquad \frac{K}{L} = \frac{\alpha P_L}{\beta P_K} \qquad \text{Q.E.D.} \tag{6.89}$$

6.56. Prove that for a linearly homogeneous Cobb-Douglas production function $Q = AK^\alpha L^\beta$, α = the output elasticity of capital (ϵ_{QK}) and β = the output elasticity of labor (ϵ_{QL}).

From the definition of output elasticity,

$$\epsilon_{QK} = \frac{\partial Q/\partial K}{Q/K} \qquad \text{and} \qquad \epsilon_{QL} = \frac{\partial Q/\partial L}{Q/L}$$

Since $\alpha + \beta = 1$, let $\beta = 1 - \alpha$ and let $k = K/L$. Then

$$Q = AK^\alpha L^{1-\alpha} = A\left(\frac{K}{L}\right)^\alpha L = Ak^\alpha L$$

Find the marginal functions.

$$\frac{\partial Q}{\partial K} = \alpha AK^{\alpha-1}L^{1-\alpha} = \alpha AK^{\alpha-1}L^{-(\alpha-1)} = \alpha A\left(\frac{K}{L}\right)^{\alpha-1} = \alpha Ak^{\alpha-1}$$

$$\frac{\partial Q}{\partial L} = (1-\alpha)AK^\alpha L^{-\alpha} = (1-\alpha)A\left(\frac{K}{L}\right)^\alpha = (1-\alpha)Ak^\alpha$$

Find the average functions.

$$\frac{Q}{K} = \frac{Ak^\alpha L}{K} = \frac{Ak^\alpha}{K} = Ak^{\alpha-1}$$

$$\frac{Q}{L} = \frac{Ak^\alpha L}{L} = Ak^\alpha$$

Then divide the marginal functions by their respective average functions to obtain ϵ.

$$\epsilon_{QK} = \frac{\partial Q/\partial K}{Q/K} = \frac{\alpha Ak^{\alpha-1}}{Ak^{\alpha-1}} = \alpha \qquad \text{Q.E.D.}$$

$$\epsilon_{QL} = \frac{\partial Q/\partial L}{Q/L} = \frac{(1-\alpha)Ak^\alpha}{Ak^\alpha} = 1 - \alpha = \beta \qquad \text{Q.E.D.}$$

6.57. Equation (6.89) gave the least-cost input ratio for a generalized Cobb-Douglas production function. Prove that the elasticity of substitution σ of any generalized Cobb-Douglas production function is unitary, i.e., that $\sigma = 1$.

In Section 6.10, the elasticity of substitution is defined as the percentage change in the least-cost K/L ratio resulting from a small percentage change in the input-price ratio P_L/P_K.

$$\sigma = \frac{\dfrac{d(K/L)}{K/L}}{\dfrac{d(P_L/P_K)}{P_L/P_K}} = \frac{d(K/L)}{d(P_L/P_K)} \cdot \frac{K/L}{P_L/P_K} \tag{6.90}$$

Since α and β are constants in (6.89) and P_K and P_L are independent variables, K/L can be considered a function of P_L/P_K. Noting that in the second ratio of (6.90), $\sigma = $ the marginal function divided by the average function, first find the marginal function of (6.89).

$$\frac{d(K/L)}{d(P_L/P_K)} = \frac{\alpha}{\beta}$$

Then find the average function by dividing both sides of (6.89) by P_L/P_K.

$$\frac{K/L}{P_L/P_K} = \frac{\alpha}{\beta}$$

Substituting in (6.90),

$$\sigma = \frac{\dfrac{d(K/L)}{d(P_L/P_K)}}{\dfrac{K/L}{P_L/P_K}} = \frac{\alpha/\beta}{\alpha/\beta} = 1 \qquad \text{Q.E.D.}$$

6.58. Use the least-cost input ratio for a Cobb-Douglas function given in (6.89) to check the answer to Example 10, where $q = K^{0.4}L^{0.5}$, $P_K = 3$, and $P_L = 4$.

With $\alpha = 0.4$ and $\beta = 0.5$, from (6.89),

$$\frac{K}{L} = \frac{0.4(4)}{0.5(3)} = \frac{1.6}{1.5}$$

Capital and labor must be used in the ratio of $16K:15L$. This confirms the answer found in Example 10 of $K_0 = 16$, $L_0 = 15$.

6.59. Given the CES production function

$$q = A[\alpha K^{-\beta} + (1 - \alpha)L^{-\beta}]^{-1/\beta} \qquad (6.91)$$

and bearing in mind from Problem 6.54 that the ratio of prices must equal the ratios of marginal products if a function is to be optimized, (a) prove that the elasticity of substitution σ of a CES production is constant and (b) demonstrate the range that σ may assume.

a)   First-order conditions require that

$$\frac{\partial Q/\partial L}{\partial Q/\partial K} = \frac{P_L}{P_K} \qquad (6.92)$$

Using the generalized power function rule to take the first-order partials of (6.91),

$$\frac{\partial Q}{\partial L} = -\frac{1}{\beta}A[\alpha K^{-\beta} + (1 - \alpha)L^{-\beta}]^{-(1/\beta)-1}(-\beta)(1 - \alpha)L^{-\beta-1}$$

Canceling $-\beta$'s, rearranging $1 - \alpha$, and adding the exponents $-(1/\beta) - 1$, we get

$$\frac{\partial Q}{\partial L} = (1 - \alpha)A[\alpha K^{-\beta} + (1 - \alpha)L^{-\beta}]^{-(1+\beta)/\beta}L^{-(1+\beta)}$$

Substituting $A^{1+\beta}/A^\beta = A$ for A,

$$\frac{\partial Q}{\partial L} = (1 - \alpha)\frac{A^{1+\beta}}{A^\beta}[\alpha K^{-\beta} + (1 - \alpha)L^{-\beta}]^{-(1+\beta)/\beta}L^{-(1+\beta)}$$

From (6.91), $A^{1+\beta}[\alpha K^{-\beta} + (1 - \alpha)L^{-\beta}]^{-(1+\beta)/\beta} = Q^{1+\beta}$ and $L^{-(1+\beta)} = 1/L^{1+\beta}$. Thus,

$$\frac{\partial Q}{\partial L} = \frac{1 - \alpha}{A^\beta}\left(\frac{Q}{L}\right)^{1+\beta} \qquad (6.93)$$

Similarly,

$$\frac{\partial Q}{\partial K} = \frac{\alpha}{A^\beta}\left(\frac{Q}{K}\right)^{1+\beta} \qquad (6.94)$$

Substituting (6.93) and (6.94) in (6.92), which leads to the cancellation of A^β and Q,

$$\frac{1 - \alpha}{\alpha}\left(\frac{K}{L}\right)^{1+\beta} = \frac{P_L}{P_K}$$

$$\left(\frac{K}{L}\right)^{1+\beta} = \frac{\alpha}{1 - \alpha}\frac{P_L}{P_K}$$

$$\frac{\bar{K}}{\bar{L}} = \left(\frac{\alpha}{1 - \alpha}\right)^{1/(1+\beta)}\left(\frac{P_L}{P_K}\right)^{1/(1+\beta)} \qquad (6.95)$$

Since α and β are constants, by considering $\bar{K}/\bar{L}$ a function of P_L/P_K, as in Problem 6.57, we can find the elasticity of substitution as the ratio of the marginal and average functions. Simplifying first by letting

$$h = \left(\frac{\alpha}{1-\alpha}\right)^{1/(1+\beta)}$$

$$\frac{\bar{K}}{\bar{L}} = h\left(\frac{P_L}{P_K}\right)^{1/(1+\beta)} \tag{6.96}$$

The marginal function is

$$\frac{d(\bar{K}/\bar{L})}{d(P_L/P_K)} = \frac{h}{1+\beta}\left(\frac{P_L}{P_K}\right)^{1/(1+\beta)-1} \tag{6.97}$$

and the average function is

$$\frac{\bar{K}/\bar{L}}{P_L/P_K} = \frac{h(P_L/P_K)^{1/(1+\beta)}}{P_L/P_K} = h\left(\frac{P_L}{P_K}\right)^{1/(1+\beta)-1} \tag{6.98}$$

By dividing the marginal function in (6.97) by the average function in (6.98), the elasticity of substitution is

$$\sigma = \frac{\dfrac{d(K/L)}{d(P_L/P_K)}}{\dfrac{K/L}{P_L/P_K}} = \frac{\dfrac{h}{1+\beta}(P_L/P_K)^{1/(1+\beta)-1}}{h(P_L/P_K)^{1/(1+\beta)-1}} = \frac{1}{1+\beta} \tag{6.99}$$

Since β is a given parameter, $\sigma = 1/(1+\beta)$ is a constant.

    b)   If $-1 < \beta < 0$, $\sigma > 1$. If $\beta = 0$, $\sigma = 1$. If $0 < \beta < \infty$, $\sigma < 1$.

6.60. Prove that the CES production function is homogeneous of degree 1 and thus has constant returns to scale.

    From (6.91), $Q = A[\alpha K^{-\beta} + (1-\alpha)L^{-\beta}]^{-1/\beta}$

Multiplying inputs K and L by k, as in Section 6.7,

$$\begin{aligned}
f(kK, kL) &= A[\alpha(kK)^{-\beta} + (1-\alpha)(kL)^{-\beta}]^{-1/\beta} \\
&= A\{k^{-\beta}[\alpha K^{-\beta} + (1-\alpha)L^{-\beta}]\}^{-1/\beta} \\
&= A(k^{-\beta})^{-1/\beta}[\alpha K^{-\beta} + (1-\alpha)L^{-\beta}]^{-1/\beta} \\
&= kA[\alpha K^{-\beta} + (1-\alpha)L^{-\beta}]^{-1/\beta} = kQ \qquad \text{Q.E.D.}
\end{aligned}$$

6.61. Find the elasticity of substitution for the CES production function, $q = 75(0.3K^{-0.4} + 0.7L^{-0.4})^{-2.5}$, given in Example 12.

    From (6.99), $\sigma = \dfrac{1}{1+\beta}$

where $\beta = 0.4$. Thus, $\sigma = 1/(1+0.4) = 0.71$.

6.62. Use the optimal K/L ratio in (6.95) to check the answer in Example 12 where $q = 75(0.3K^{-0.4} + 0.7L^{-0.4})^{-2.5}$ was optimized under the constraint $4K + 3L = 120$, giving $\bar{K} = 11.25$ and $\bar{L} = 25$.

    From (6.95), $\dfrac{\bar{K}}{\bar{L}} = \left(\dfrac{\alpha}{1-\alpha}\dfrac{P_L}{P_K}\right)^{1/(1+\beta)}$

Substituting $\alpha = 0.3$, $1 - \alpha = 0.7$, and $\beta = 0.4$,

$$\frac{\bar{K}}{\bar{L}} = \left(\frac{0.3}{0.7}\frac{3}{4}\right)^{1/1.4} = \left(\frac{0.9}{2.8}\right)^{0.71} = (0.32)^{0.71} \approx 0.45$$

With $\bar{K} = 11.25$ and $\bar{L} = 25$, $\bar{K}/\bar{L} = 11.25/25 = 0.45$.

6.63. Use (6.95) to check the answer to Problem 6.43 where $q = 80(0.4K^{-0.25} + 0.6L^{-0.25})^{-4}$ was optimized subject to the constraint $5K + 2L = 150$ at $\bar{K} = 14$ and $\bar{L} = 40$.

Substituting $\alpha = 0.4$, $1 - \alpha = 0.6$, and $\beta = 0.25$ in (6.95),

$$\frac{\bar{K}}{\bar{L}} = \left(\frac{0.4}{0.6}\frac{2}{5}\right)^{1/1.25} = \left(\frac{0.8}{3}\right)^{0.8} \approx 0.35$$

Substituting $\bar{K} = 14$ and $\bar{L} = 40$, $\frac{14}{40} = 0.35$.

6.64. Find the elasticity of substitution from Problem 6.63.

From (6.99), $$\sigma = \frac{1}{1 + \beta} = \frac{1}{1 + 0.25} = 0.8$$

6.65. Use (6.95) to check the answer to Problem 6.44 where $q = 100(0.2K^{0.5} + 0.8L^{0.5})^2$ was optimized subject to the constraint $10K + 4L = 4100$ at $\bar{K} = 10$ and $\bar{L} = 1000$.

With $\alpha = 0.2$, $1 - \alpha = 0.8$, and $\beta = -0.5$,

$$\frac{\bar{K}}{\bar{L}} = \left[\left(\frac{0.2}{0.8}\right)\left(\frac{4}{10}\right)\right]^{1/(1-0.5)} = \left(\frac{0.8}{8}\right)^2 = (0.1)^2 = 0.01$$

Substituting $\bar{K} = 10$ and $\bar{L} = 1000$, $\frac{10}{1000} = 0.01$.

6.66. Find the elasticity of substitution from Problem 6.65.

From (6.99), $$\sigma = \frac{1}{1 + \beta} = \frac{1}{1 - 0.5} = 2$$

6.67. (a) Use the elasticity of substitution found in Problem 6.64 to estimate the effect on the least-cost $(\bar{K}/\bar{L})$ ratio in Problem 6.43 if P_L increases by 25 percent. (b) Check your answer by substituting the new P_L in (6.95).

a) The elasticity of substitution measures the relative change in the $\bar{K}/\bar{L}$ ratio brought about by a relative change in the price ratio P_L/P_K. If P_L increases by 25 percent, $P_L = 1.25(2) = 2.5$. Thus, $P_L/P_K = 2.5/5 = 0.5$ vs. $\frac{2}{5} = 0.4$ in Problem 6.43.

  The percentage increase in the price ratio, therefore, is $(0.5 - 0.4)/0.4 = 0.25$. With the elasticity of substitution $= 0.8$ from Problem 6.64, the expected percentage change in the $\bar{K}/\bar{L}$ ratio is

$$\frac{\Delta(\bar{K}/\bar{L})}{\bar{K}/\bar{L}} \approx 0.8(0.25) = 0.2 \quad \text{or} \quad 20\%$$

With $(\bar{K}/\bar{L})_1 = 0.35$, $(\bar{K}/\bar{L})_2 \approx 1.2(0.35) \approx 0.42$.

b) Substituting $P_L = 2.5$ in (6.95),

$$\frac{\bar{K}}{\bar{L}} = \left(\frac{0.4}{0.6}\frac{2.5}{5}\right)^{0.8} = \left(\frac{1}{3}\right)^{0.8} \approx 0.42$$

6.68. (*a*) Use the elasticity of substitution to estimate the new $\bar{K}/\bar{L}$ ratio if the price of capital decreases by 20 percent. Assume the initial data of Problem 6.43. (*b*) Check your answer.

a) If P_K decreases by 20 percent, $P_K = 0.8(5) = 4$. Thus, $P_L/P_K = \frac{2}{4} = 0.5$ which is a 25 percent increase in the P_L/P_K ratio, as seen above. Therefore,

$$\frac{\Delta(\bar{K}/\bar{L})}{\bar{K}/\bar{L}} = 0.8(0.25) = 0.2 \qquad \text{or} \qquad 20\%$$

and $(\bar{K}/\bar{L})_2 \approx 1.2(0.35) = 0.42$.

b) Substituting $P_K = 4$ in (6.95),

$$\frac{\bar{K}}{\bar{L}} = \left(\frac{0.4}{0.6}\frac{2}{4}\right)^{0.8} = \left(\frac{0.8}{2.4}\right)^{0.8} = 0.42$$

6.69. (*a*) If the price of labor decreases by 10 percent in Problem 6.44, use the elasticity of substitution to estimate the effect on the least-cost $\bar{K}/\bar{L}$ ratio. (*b*) Check your answer.

a) If P_L decreases by 10 percent, $P_L = 0.9(4) = 3.6$, and the P_L/P_K ratio also decreases by 10 percent. With a 10 percent decrease in P_L/P_K and an elasticity of substitution $= 2$,

$$\frac{\Delta(\bar{K}/\bar{L})}{\bar{K}/\bar{L}} \approx 2(-0.10) = -0.20 \qquad \text{or} \qquad -20\%$$

With the old $\bar{K}/\bar{L} = 0.01$, $(\bar{K}/\bar{L})_2 \approx (1-0.2)(0.01) = 0.8(0.01) = 0.008$.

b) Substituting $P_L = 3.6$ in (6.95),

$$\frac{\bar{K}}{\bar{L}} = \left(\frac{0.2}{0.8}\frac{3.6}{10}\right)^2 = \left(\frac{0.72}{8}\right)^2 = (0.09)^2 = 0.0081$$

Exponential and Logarithmic Functions

7.1 EXPONENTIAL FUNCTIONS

Previous chapters dealt mainly with *power functions*, such as $y = x^a$, in which a variable base x is raised to a constant exponent a. In this chapter we introduce an important new function in which a constant base a is raised to a variable exponent x. It is called an *exponential function* and is defined as

$$y = a^x \qquad a > 0 \qquad \text{and} \qquad a \neq 1$$

Commonly used to express rates of growth and decay, such as interest compounding and depreciation, exponential functions have the following general properties. Given $y = a^x$, $a > 0$, and $a \neq 1$:

1. The domain of the function is the set of all real numbers; the range of the function is the set of all positive real numbers, i.e., for all x, even $x < 0$, $y > 0$.
2. For $a > 1$, the function is increasing and convex; for $0 < a < 1$, the function is decreasing and convex.
3. At $x = 0$, $y = 1$, independently of the base.

See Example 1 and Problems 7.1 and 7.2; for a review of exponents, see Section 1.1 and Problem 1.1.

EXAMPLE 1. Given (a) $y = 2^x$ and (b) $y = 2^{-x} = (\frac{1}{2})^x$, the above properties of exponential functions can readily be seen from the tables and graphs of the functions in Fig. 7-1. More complicated exponential functions are estimated with the help of the $\boxed{y^x}$ key on pocket calculators.

146

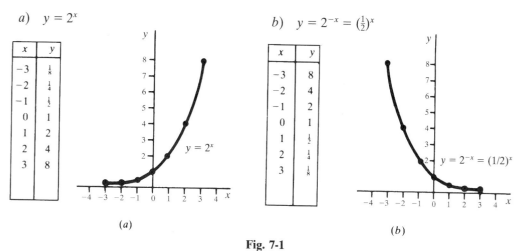

Fig. 7-1

7.2 LOGARITHMIC FUNCTIONS

Interchanging the variables of an exponential function f defined by $y = a^x$ gives rise to a new function g defined by $x = a^y$ such that any ordered pair of numbers in f will also be found in g in reverse order. For example, if $f(2) = 4$, then $g(4) = 2$; if $f(3) = 8$, then $g(8) = 3$. The new function g, the *inverse* of the exponential function f, is called a *logarithmic function with base a*. Instead of $x = a^y$, the logarithmic function with base a is more commonly written

$$y = \log_a x \qquad a > 0, a \ne 1$$

$\text{Log}_a x$ is the exponent to which a must be raised to get x. Any positive number except 1 may serve as the base for a logarithm. The *common logarithm of x*, written $\log_{10} x$ or simply $\log x$, is the exponent to which 10 must be raised to get x. Logarithms have the following properties. Given $y = \log_a x, a > 0, a \ne 1$:

1. The domain of the function is the set of all positive real numbers; the range is the set of all real numbers—the exact opposite of its inverse function, the exponential function.
2. For base $a > 1, f(x)$ is increasing and concave. For $0 < a < 1, f(x)$ is decreasing and convex.
3. At $x = 1, y = 0$ independent of the base.

See Examples 2 to 4 and Problems 7.5 and 7.6.

EXAMPLE 2.   A graph of two functions f and g in which x and y are interchanged, such as $y = 2^x$ and $x = 2^y$ in Fig. 7-2, reveals that one function is a *mirror image* of the other along the 45° line $y = x$, such that if $f(x) = y$, then $g(y) = x$. Recall that $x = 2^y$ is equivalent to and more commonly expressed as $y = \log_2 x$.

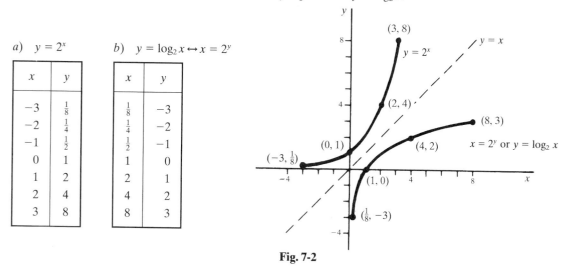

Fig. 7-2

EXAMPLE 3. Knowing that the common logarithm of x is the power to which 10 must be raised to get x, it follows that

$$\log 10 = 1 \quad \text{since } 10^1 = 10 \qquad \log 1 = 0 \quad \text{since } 10^0 = 1$$
$$\log 100 = 2 \quad \text{since } 10^2 = 100 \qquad \log 0.1 = -1 \quad \text{since } 10^{-1} = 0.1$$
$$\log 1000 = 3 \quad \text{since } 10^3 = 1000 \qquad \log 0.01 = -2 \quad \text{since } 10^{-2} = 0.01$$

EXAMPLE 4. For numbers that are exact powers of the base, logs are easily calculated without the aid of calculators.

$$\log_7 49 = 2 \quad \text{since } 7^2 = 49 \qquad \log_2 16 = 4 \quad \text{since } 2^4 = 16$$
$$\log_{36} 6 = \tfrac{1}{2} \quad \text{since } 36^{1/2} = 6 \qquad \log_{16} 2 = \tfrac{1}{4} \quad \text{since } 16^{1/4} = 2$$
$$\log_3 \tfrac{1}{9} = -2 \quad \text{since } 3^{-2} = \tfrac{1}{9} \qquad \log_2 \tfrac{1}{8} = -3 \quad \text{since } 2^{-3} = \tfrac{1}{8}$$

For numbers that are not exact powers of the base, log tables or calculators are needed.

7.3 PROPERTIES OF EXPONENTS AND LOGARITHMS

Assuming $a, b > 0$; $a, b \neq 1$; and x and y are any real numbers:

1. $a^x \cdot a^y = a^{x+y}$ 4. $(a^x)^y = a^{xy}$

2. $\dfrac{1}{a^x} = a^{-x}$ 5. $a^x \cdot b^x = (ab)^x$

3. $\dfrac{a^x}{a^y} = a^{x-y}$ 6. $\dfrac{a^x}{b^x} = \left(\dfrac{a}{b}\right)^x$

For a, x, and y positive real numbers, n a real number, and $a \neq 1$:

1. $\log_a xy = \log_a x + \log_a y$ 3. $\log_a x^n = n \log_a x$

2. $\log_a \dfrac{x}{y} = \log_a x - \log_a y$ 4. $\log_a \sqrt[n]{x} = \dfrac{1}{n} \log_a x$

Properties of exponents were treated in Section 1.1 and Problem 1.1. Properties of logarithms are treated in Example 5 and Problems 7.12 to 7.16.

Table 7.1

x	$\log x$	x	$\log x$	x	$\log x$	x	$\log x$
1	0.0000	6	0.7782	11	1.0414	16	1.2041
2	0.3010	7	0.8451	12	1.0792	17	1.2304
3	0.4771	8	0.9031	13	1.1139	18	1.2553
4	0.6021	9	0.9542	14	1.1461	19	1.2788
5	0.6990	10	1.0000	15	1.1761	20	1.3010

EXAMPLE 5. The problems below are kept simple and solved by means of logarithms to illustrate the properties of logarithms.

a) $x = 7 \cdot 2$
$$\log x = \log 7 + \log 2$$
$$\log x = 0.8451 + 0.3010$$
$$\log x = 1.1461$$
$$x = 14$$

b) $x = 18 \div 3$
$$\log x = \log 18 - \log 3$$
$$\log x = 1.2553 - 0.4771$$
$$\log x = 0.7782$$
$$x = 6$$

c) $x = 3^2$

$\log x = 2 \log 3$

$\log x = 2(0.4771)$

$\log x = 0.9542$

$x = 9$

d) $x = \sqrt[3]{8}$

$\log x = \frac{1}{3} \log 8$

$\log x = \frac{1}{3}(0.9031)$

$\log x = 0.3010$

$x = 2$

7.4   NATURAL EXPONENTIAL AND LOGARITHMIC FUNCTIONS

The most commonly used base for exponential and logarithmic functions is the irrational number e. Expressed mathematically,

$$e = \lim_{n \to \infty} \left(1 + \frac{1}{n}\right)^n \approx 2.71828 \qquad (7.1)$$

Exponential functions to base e are called *natural exponential functions* and are written $y = e^x$; logarithmic functions to base e are termed *natural logarithmic functions* and are expressed as $y = \log_e x$ or, more commonly, $\ln x$. Thus $\ln x$ is simply the exponent or power to which e must be raised to get x.

As with other exponential and logarithmic functions to a common base, one function is the inverse of the other, such that the ordered pair (a, b) will belong to the set of e^x if and only if (b, a) belongs to the set of $\ln x$. Natural exponential and logarithmic functions follow the same rules as other exponential and logarithmic functions and are estimated with the help of tables or the $\boxed{e^x}$ and $\boxed{\ln x}$ keys on pocket calculators. See Problems 7.3, 7.4, and 7.6.

7.5   SOLVING NATURAL EXPONENTIAL AND LOGARITHMIC FUNCTIONS

Since natural exponential functions and natural logarithmic functions are inverses of each other, one is generally helpful in solving the other. Mindful that $\ln x$ signifies the power to which e must be raised to get x, it follows that:

1. e raised to the natural log of a constant $(a > 0)$, a variable $(x > 0)$, or a function of a variable $[f(x) > 0]$ must equal that constant, variable, or function of the variable:

$$e^{\ln a} = a \qquad e^{\ln x} = x \qquad e^{\ln f(x)} = f(x) \qquad (7.2)$$

2.   Conversely, the natural log of e raised to the power of a constant, variable, or function of a variable must also equal that constant, variable, or function of the variable:

$$\ln e^a = a \qquad \ln e^x = x \qquad \ln e^{f(x)} = f(x) \qquad (7.3)$$

See Example 6 and Problems 7.18 to 7.22.

EXAMPLE 6.   The principles of (7.2) and (7.3) are used below to solve the given equations for x.

a) $5e^{x+2} = 120$

1)   Solve algebraically for e^{x+2},

$$5e^{x+2} = 120$$
$$e^{x+2} = 24$$

2)   Take the natural log of both sides to eliminate e.

$$\ln e^{x+2} = \ln 24$$

From (7.3),
$$x + 2 = \ln 24$$
$$x = \ln 24 - 2$$

Enter 24 on your calculator and press the $\boxed{\ln x}$ key to find $\ln 24 = 3.17805$. Then substitute and solve.

$$x = 3.17805 - 2 = 1.17805$$

b) $6 \ln x - 7 = 12.2$

  1) Solve algebraically for $\ln x$,

$$6 \ln x = 19.2$$
$$\ln x = 3.2$$

  2) Set both sides of the equation as exponents of e to eliminate the natural log expression,

$$e^{\ln x} = e^{3.2}$$

From (7.2), $x = e^{3.2}$

Enter 3.2 on your calculator and press the $\boxed{e^x}$ key to find $e^{3.2} = 24.53253$ and substitute.

$$x = 24.53253$$

Note: On many calculators the $\boxed{e^x}$ key is the inverse (shift, or second function) of the $\boxed{\ln x}$ key, and to activate the $\boxed{e^x}$ key, one must first press the $\boxed{\text{INV}}$ ($\boxed{\text{Shift}}$, or $\boxed{\text{2ndF}}$) key followed by the $\boxed{\ln x}$ key.

7.6 LOGARITHMIC TRANSFORMATION OF NONLINEAR FUNCTIONS

Linear algebra and regression analysis involving ordinary or two-stage least squares which are common tools in economic analysis assume linear functions or equations. Some nonlinear functions, such as Cobb-Douglas production functions, can easily be converted to linear functions through simple logarithmic transformation; others, such as CES production functions, cannot. For example, from the properties of logarithms, it is clear that given a generalized Cobb-Douglas production function

$$q = AK^{\alpha}L^{\beta}$$
$$\ln q = \ln A + \alpha \ln K + \beta \ln L \qquad (7.4)$$

which is log-linear. But given the CES production function,

$$q = A[\alpha K^{-\beta} + (1 - \alpha)L^{-\beta}]^{-1/\beta}$$
$$\ln q = \ln A - \frac{1}{\beta} \ln [\alpha K^{-\beta} + (1 - \alpha)L^{-\beta}]$$

which is not linear even in logarithms because of $K^{-\beta}$ and $L^{-\beta}$. Ordinary least-square estimation of the coefficients in a log transformation of a Cobb-Douglas production function, such as in (7.4), has the nice added feature that estimates for α and β provide direct measures of the *output elasticity* of K and L, respectively, as was proved in Problem 6.56.

Solved Problems

GRAPHS

7.1. Make a schedule for each of the following exponential functions with base $a > 1$ and then sketch them on the same graph to convince yourself that (1) the functions never equal zero; (2) they all pass through $(0, 1)$; and (3) they are all positively sloped and convex.

a) $y = 3^x$ b) $y = 4^x$ c) $y = 5^x$

a)

x	y
-3	$\frac{1}{27}$
-2	$\frac{1}{9}$
-1	$\frac{1}{3}$
0	1
1	3
2	9
3	27

b)

x	y
-3	$\frac{1}{64}$
-2	$\frac{1}{16}$
-1	$\frac{1}{4}$
0	1
1	4
2	16
3	64

c)

x	y
-3	$\frac{1}{125}$
-2	$\frac{1}{25}$
-1	$\frac{1}{5}$
0	1
1	5
2	25
3	125

Fig. 7-3

7.2. Make a schedule for each of the following exponential functions with $0 < a < 1$ and then sketch them on the same graph to convince yourself that (1) the functions never equal zero; (2) they all pass through $(0, 1)$, and (3) they are all negatively sloped and convex.

a) $y = (\frac{1}{3})^x = 3^{-x}$ b) $y = (\frac{1}{4})^x = 4^{-x}$ c) $y = (\frac{1}{5})^x = 5^{-x}$

a)

x	y
-3	27
-2	9
-1	3
0	1
1	$\frac{1}{3}$
2	$\frac{1}{9}$
3	$\frac{1}{27}$

b)

x	y
-3	64
-2	16
-1	4
0	1
1	$\frac{1}{4}$
2	$\frac{1}{16}$
3	$\frac{1}{64}$

c)

x	y
-3	125
-2	25
-1	5
0	1
1	$\frac{1}{5}$
2	$\frac{1}{25}$
3	$\frac{1}{125}$

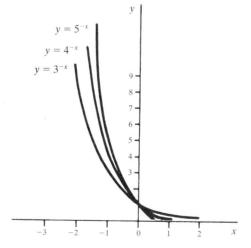

Fig. 7-4

7.3. Using a calculator or tables, set up a schedule for each of the following natural exponential functions $y = e^{kx}$ where $k > 0$, noting (1) the functions never equal zero; (2) they all pass through $(0, 1)$, and (3) they are all positively sloped and convex.

a) $y = e^{0.5x}$ b) $y = e^x$ c) $y = e^{2x}$

a)

x	y
−2	0.37
−1	0.61
0	1.00
1	1.65
2	2.72

b)

r	y
−2	0.14
−1	0.37
0	1.00
1	2.72
2	7.39

c)

x	y
−2	0.02
−1	0.14
0	1.00
1	7.39
2	54.60

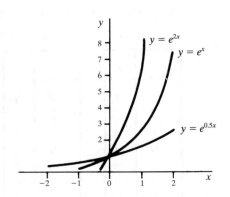

Fig. 7-5

7.4. Set up a schedule, rounding to two decimal places, for the following natural exponential functions $y = e^{kx}$ where $k < 0$, noting (1) the functions never equal zero; (2) they all pass through $(0, 1)$; and (3) they are all negatively sloped and convex.

a) $y = e^{-0.5x}$ b) $y = e^{-x}$ c) $y = e^{-2x}$

a)

x	y
−2	2.72
−1	1.65
0	1.00
1	0.61
2	0.37

b)

x	y
−2	7.39
−1	2.72
0	1.00
1	0.37
2	0.14

c)

x	y
−2	54.60
−1	7.39
0	1.00
1	0.14
2	0.02

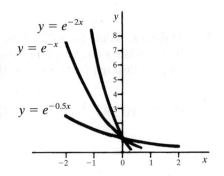

Fig. 7-6

7.5. Construct a schedule and draw a graph for the following functions to show that one is the mirror image and hence the inverse of the other, noting that (1) the domain of (a) is the range of (b) and the range of (a) is the domain of (b); and (2) a logarithmic function with $0 < a < 1$ is a decreasing function and convex.

a) $y = (\frac{1}{2})^x = 2^{-x}$ b) $x = (\frac{1}{2})^y$ or $y = \log_{1/2} x$

a)

x	y
−3	8
−2	4
−1	2
0	1
1	$\frac{1}{2}$
2	$\frac{1}{4}$
3	$\frac{1}{8}$

b)

x	y
8	−3
4	−2
2	−1
1	0
$\frac{1}{2}$	1
$\frac{1}{4}$	2
$\frac{1}{8}$	3

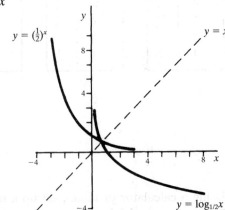

Fig. 7-7

7.6. Given (*a*) $y = e^x$ and (*b*) $y = \ln x$, and using a calculator or tables, construct a schedule and draw a graph for each of the functions to show that one function is the mirror image or inverse of the other, noting that (1) the domain of (*a*) is the range of (*b*) while the range of (*a*) is the domain of (*b*), (2) $\ln x$ is negative for $0 < x < 1$ and positive for $x > 1$; and (3) $\ln x$ is an increasing function and concave.

a) $y = e^x$

x	y
-2	0.13534
-1	0.36788
0	1.00000
1	2.71828
2	7.38906

b) $y = \ln x$

x	y
0.13534	-2
0.36788	-1
1.00000	0
2.71828	1
7.38906	2

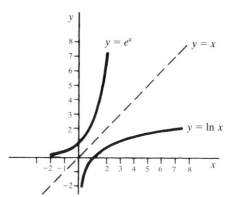

Fig. 7-8

EXPONENTIAL-LOGARITHMIC CONVERSION

7.7. Change the following logarithms to their equivalent exponential forms:

a) $\log_8 64 = 2$

$$64 = 8^2$$

b) $\log_5 125 = 3$

$$125 = 5^3$$

c) $\log_7 \frac{1}{7} = -1$

$$\frac{1}{7} = 7^{-1}$$

d) $\log_3 \frac{1}{81} = -4$

$$\frac{1}{81} = 3^{-4}$$

e) $\log_{36} 6 = \frac{1}{2}$

$$6 = 36^{1/2}$$

f) $\log_{16} 2 = \frac{1}{4}$

$$2 = 16^{1/4}$$

g) $\log_a y = 6x$

$$y = a^{6x}$$

h) $\log_2 y = 7x$

$$y = 2^{7x}$$

7.8. Convert the following natural logarithms to natural exponential functions:

a) $\ln 32 = 3.46574$

$$32 = e^{3.46574}$$

b) $\ln 0.8 = -0.22314$

$$0.8 = e^{-0.22314}$$

c) $\ln 20 = 2.99573$

$$20 = e^{2.99573}$$

d) $\ln 2.5 = 0.91629$

$$2.5 = e^{0.91629}$$

e) $\ln y = -4x$

$$y = e^{-4x}$$

f) $\ln y = 2t + 1$

$$y = e^{2t+1}$$

7.9. Change the following exponential forms to logarithmic forms:

a) $81 = 9^2$

$$\log_9 81 = 2$$

b) $32 = 2^5$

$$\log_2 32 = 5$$

c) $\frac{1}{9} = 3^{-2}$

$\log_3 \frac{1}{9} = -2$

(d) $\frac{1}{16} = 2^{-4}$

$\log_2 \frac{1}{16} = -4$

e) $5 = 125^{1/3}$

$\log_{125} 5 = \frac{1}{3}$

f) $11 = 121^{1/2}$

$\log_{121} 11 = \frac{1}{2}$

g) $27 = 9^{3/2}$

$\log_9 27 = \frac{3}{2}$

h) $64 = 256^{3/4}$

$\log_{256} 64 = \frac{3}{4}$

7.10. Convert the following natural exponential expressions to equivalent natural logarithmic forms:

a) $4.8 = e^{1.56862}$

$\ln 4.8 = 1.56862$

b) $15 = e^{2.70805}$

$\ln 15 = 2.70805$

c) $0.6 = e^{-0.51083}$

$\ln 0.6 = -0.51083$

d) $130 = e^{4.86753}$

$\ln 130 = 4.86753$

e) $y = e^{(1/2)t}$

$\ln y = \frac{1}{2}t$

f) $y = e^{t-5}$

$\ln y = t - 5$

7.11. Solve the following for x, y, or a by finding the equivalent expression:

a) $y = \log_{30} 900$

$900 = 30^y$

$y = 2$

b) $y = \log_2 \frac{1}{32}$

$\frac{1}{32} = 2^y$

$y = -5$

c) $\log_4 x = 3$

$x = 4^3$

$x = 64$

d) $\log_{81} x = \frac{3}{4}$

$x = 81^{3/4}$

$x = 27$

e) $\log_a 27 = 3$

$27 = a^3$

$a = 27^{1/3}$

$a = 3$

f) $\log_a 4 = \frac{2}{3}$

$4 = a^{2/3}$

$a = 4^{3/2}$

$a = 8$

g) $\log_a 125 = \frac{3}{2}$

$125 = a^{3/2}$

$a = 125^{2/3}$

$a = 25$

h) $\log_a 8 = \frac{3}{4}$

$8 = a^{3/4}$

$a = 8^{4/3}$

$a = 16$

PROPERTIES OF LOGARITHMS AND EXPONENTS

7.12. Use the properties of logarithms to write the following expressions as sums, differences, or products:

a) $\log_a 56x$

$\log_a 56x = \log_a 56 + \log_a x$

b) $\log_a 33x^4$

$\log_a 33x^4 = \log_a 33 + 4\log_a x$

c) $\log_a x^2 y^3$

$\log_a x^2 y^3 = 2\log_a x + 3\log_a y$

d) $\log_a u^5 v^{-4}$

$\log_a u^5 v^{-4} = 5\log_a u - 4\log_a v$

$e)$ $\log_a \dfrac{6x}{7y}$

$$\log_a \dfrac{6x}{7y} = \log_a 6x - \log_a 7y$$
$$= \log_a 6 + \log_a x - (\log_a 7 + \log_a y)$$
$$= \log_a 6 + \log_a x - \log_a 7 - \log_a y$$

$f)$ $\log_a \dfrac{x^7}{y^4}$

$$\log_a \dfrac{x^7}{y^4} = 7\log_a x - 4\log_a y$$

$g)$ $\log_a \sqrt[3]{x}$

$$\log_a \sqrt[3]{x} = \tfrac{1}{3}\log_a x$$

7.13. Use of the properties of logarithms to write the following natural logarithmic forms as sums, differences, or products:

$a)$ $\ln 76x^3$

$$\ln 76^3 = \ln 76 + 3\ln x$$

$b)$ $\ln x^5 y^2$

$$\ln x^5 y^2 = 5\ln x + 2\ln y$$

$c)$ $\ln \dfrac{x^4}{y^6}$

$$\ln \dfrac{x^4}{y^6} = 4\ln x - 6\ln y$$

$d)$ $\ln \dfrac{8x}{9y}$

$$\ln \dfrac{8x}{9y} = \ln 8x - \ln 9y$$
$$= \ln 8 + \ln x - (\ln 9 + \ln y)$$
$$= \ln 8 + \ln x - \ln 9 - \ln y$$

$e)$ $\ln \sqrt[4]{x}$

$$\ln \sqrt[4]{x} = \tfrac{1}{4}\ln x$$

$f)$ $\ln(x\sqrt[5]{y})$

$$\ln(x\sqrt[5]{y}) = 5\ln x + \tfrac{1}{2}\ln y$$

$g)$ $\ln \dfrac{3\sqrt[5]{x}}{\sqrt{y}}$

$$\ln \dfrac{3\sqrt[5]{x}}{\sqrt{y}} = \ln 3 + \tfrac{1}{5}\ln x - \tfrac{1}{2}\ln y$$

$h)$ $\ln \sqrt{\dfrac{x^7}{y^4}}$

$$\ln \sqrt{\dfrac{x^7}{y^4}} = \tfrac{1}{2}(7\ln x - 4\ln y)$$

7.14. Use the properties of exponents to simplify the following exponential expressions, assuming a, $b > 0$ and $a \neq b$:

$a)$ $a^x \cdot a^y$

$$a^x \cdot a^y = a^{x+y}$$

$b)$ $a^{4x} \cdot a^{5y}$

$$a^{4x} \cdot a^{5y} = a^{4x+5y}$$

$c)$ $\dfrac{a^{2x}}{a^{3y}}$

$$\dfrac{a^{2x}}{a^{3y}} = a^{2x-3y}$$

$d)$ $\dfrac{a^x}{b^x}$

$$\dfrac{a^x}{b^x} = \left(\dfrac{a}{b}\right)^x$$

$e)$ $\sqrt{a^{7x}}$

$$\sqrt{a^{7x}} = (a^{7x})^{1/2} = a^{(7/2)x}$$

$f)$ $(a^x)^{4y}$

$$(a^x)^{4y} = a^{4xy}$$

7.15. Simplify the following natural exponential expressions:

$a)$ $e^{5x} \cdot e^{2y}$

$$e^{5x} \cdot e^{2y} = e^{5x+2y}$$

$b)$ $(e^{3x})^5$

$$(e^{3x})^5 = e^{15x}$$

c) $\dfrac{e^{8x}}{e^{6x}}$

$$\frac{e^{8x}}{e^{6x}} = e^{8x-6x} = e^{2x}$$

d) $\dfrac{e^{4x}}{e^{7x}}$

$$\frac{e^{4x}}{e^{7x}} = e^{4x-7x} = e^{-3x} = \frac{1}{e^{3x}}$$

7.16. Simplify the following natural logarithmic expressions:

a) $\ln 8 + \ln x$

$$\ln 8 + \ln x = \ln 8x$$

b $\ln x^5 - \ln x^3$

$$\ln x^5 - \ln x^3 = \ln \frac{x^5}{x^3} = \ln x^2 = 2\ln x$$

c) $\ln 12 + \ln 5 - \ln 6$

$$\ln 12 + \ln 5 - \ln 6 = \ln \frac{12 \cdot 5}{6} = \ln 10$$

d) $\ln 7 - \ln x + \ln 9$

$$\ln 7 - \ln x + \ln 9 = \ln \frac{7 \cdot 9}{x} = \ln \frac{63}{x}$$

e) $\frac{1}{2}\ln 81$

$$\tfrac{1}{2}\ln 81 = \ln 81^{1/2} = \ln 9$$

f) $5\ln \frac{1}{2}$

$$5\ln \tfrac{1}{2} = \ln \left(\tfrac{1}{2}\right)^5 = \ln \tfrac{1}{32}$$

g) $\frac{1}{3}\ln 27 + 4\ln 2$

$$\tfrac{1}{3}\ln 27 + 4\ln 2 = \ln 27^{1/3} + \ln 2^4 = \ln (3 \cdot 16) = \ln 48$$

h) $2\ln 4 - \frac{1}{3}\ln 8$

$$2\ln 4 - \tfrac{1}{3}\ln 8 = \ln 4^2 - \ln 8^{1/3} = \ln \tfrac{16}{2} = \ln 8$$

7.17. Simplify each of the following exponential expressions:

a) $e^{3\ln x}$

$$e^{3\ln x} = e^{\ln x^3} \qquad \text{But from (7.2), } e^{\ln f(x)} = f(x), \text{ so}$$
$$e^{3\ln x} = x^3$$

b) $e^{4\ln x + 5\ln y}$

$$e^{4\ln x + 5\ln y} = e^{\ln x^4} \cdot e^{\ln y^5} = x^4 y^5$$

c) $e^{(1/2)\ln 6x}$

$$e^{(1/2)\ln 6x} = e^{\ln (6x)1/2} = (6x)^{1/2} = \sqrt{6x}$$

d) $e^{4\ln x - 9\ln y}$

$$e^{4\ln x - 9\ln y} = \frac{e^{\ln x^4}}{e^{\ln y^9}} = \frac{x^4}{y^9}$$

SOLVING EXPONENTIAL AND LOGARITHMIC FUNCTIONS

7.18. Use the techniques from Section 7.5 to solve the following natural exponential functions for x:

a) $3e^{5x} = 8943$

1) Solve algebraically for e^{5x}.

$$e^{5x} = 2981$$

2)   Then take the natural log of both sides to eliminate e.

$$\ln e^{5x} = \ln 2981$$

From (7.3), $5x = \ln 2981$

To find the value of $\ln 2981$, enter 2981 on a calculator and press the $\boxed{\ln x}$ key to find $\ln 2981 = 8.00001 \approx 8$. Then substitute and solve algebraically.

$$5x = 8$$
$$x = 1.6$$

b) $4e^{3x-1.5} = 360$

    1)   Solve for $e^{3x-1.5}$. $e^{3x-1.5} = 90$

    2)   Take the natural log. $\ln e^{3x-1.5} = \ln 90$

    From (7.3), $3x - 1.5 = \ln 90$

For $\ln 90$, enter 90 on a calculator and press the $\boxed{\ln x}$ key to find $\ln 90 = 4.49981 \approx 4.5$. Substitute and solve.

$$3x - 1.5 = 4.5 \qquad x = 2$$

c) $\frac{1}{2}e^{x^2} = 259$

    1)   Solve for e^{x^2} $e^{x^2} = 518$

    2)   Take the natural log. $\ln e^{x^2} = \ln 518$

    From (7.3), $x^2 = 6.24998 \approx 6.25 \qquad x = \pm 2.5$

7.19.   Using the techniques of Section 7.5, solve the following natural logarithmic functions for x:

a) $5 \ln x + 8 = 14$

    1)   Solve algebraically for $\ln x$.

$$5 \ln x = 6 \qquad \ln x = 1.2$$

    2)   Set both sides of the equation as exponents of e to eliminate the natural log.

$$e^{\ln x} = e^{1.2}$$

From (7.2), $x = e^{1.2}$

To find the value of $e^{1.2}$, enter 1.2 on a calculator, press the $\boxed{e^x}$ key to find $e^{1.2} = 3.32012$, and substitute. If the $\boxed{e^x}$ key is the inverse of the $\boxed{\ln x}$ key, enter 1.2, then press the $\boxed{\text{INV}}$ key followed by the $\boxed{\ln x}$ key.

$$x = 3.32012 \approx 3.32$$

b) $\ln (x + 4)^2 = 3$

    1)   Simplify with the laws of logs, then solve for $\ln x$.

$$2 \ln (x + 4) = 3$$
$$\ln (x + 4) = 1.5$$

    2) $e^{\ln(x+4)} = e^{1.5}$

    From (7.2), $x + 4 = e^{1.5}$

    Using a calculator, $x + 4 = 4.48169$

$$x = 4.48169 - 4 = 0.48169$$

c) In $\sqrt{x + 34} = 2.55$

   1) Simplify and solve, $\frac{1}{2}\ln(x + 34) = 2.55$

                 $\ln(x + 34) = 5.1$

   2) From (7.2) $x + 34 = e^{5.1} \approx 164$

                  $x = 130$

7.20. Solve each of the following equations for x in terms of y:

a) $\log_a x = y^3$ *b)* $\log_a x = \log_a 3 + \log_a y$

   $x = a^{y^3}$ $x = 3y$ since addition in logs = multiplication in algebra

c) $\ln x = 3y$ *d)* $\ln x = \log_a y$ *e)* $\log_a x = \ln y$

   $x = e^{3y}$ $x = e^{\log_a y}$ $x = a^{\ln y}$

f) $y = ge^{hx}$

  To solve for x when x is an exponent in a natural exponential function, take the natural log of both sides and solve algebraically, as follows:

$$\ln y = \ln g + hx \ln e = \ln g + hx$$

$$x = \frac{\ln y - \ln g}{h}$$

g) $y = ae^{x+1}$

$$\ln y = \ln a + (x + 1)\ln e = \ln a + x + 1$$
$$x = \ln y - \ln a - 1$$

h) $y = p(1 + i)^x$

  When x is an exponent in an exponential function with a base other than e, take the common log of both sides and solve algebraically.

$$\log y = \log p + x \log(1 + i)$$

$$x = \frac{\log y - \log p}{\log(1 + i)}$$

7.21. Use common logs to solve each of the following equations:

a) $y = 625(0.8)$

   1) Take the common log of both sides of the equation, using the properties of logarithms from Section 7.3.

$$\log y = \log 625 + \log 0.8$$

   To find the logs of 625 and 0.8, enter each number individually on the calculator, press the $\boxed{\log x}$ key to get the common log of each number, and perform the required arithmetic.

$$\log y = 2.79588 + (-0.09691) = 2.69897$$

   2) Since $\log y = 2.69897$ indicates that 10 must be raised to the 2.69897 power to get y, to find the *antilogarithm* of 2.69897 and solve for y, enter 2.69897 on a calculator, press the $\boxed{10^x}$ key to find that $10^{2.69897} = 500$, and substitute. If the $\boxed{10^x}$ key is the inverse of the $\boxed{\log x}$ key, enter 2.69897 and press the $\boxed{\text{INV}}$ key followed by the $\boxed{\log x}$ key.

$$y = \text{antilog } 2.69897 = 10^{2.69897} = 500$$

b) $y = \frac{40}{100}$

  1) $\log y = \log 40 - \log 100$

  2) $\log y = 1.60206 - 2 = -0.39794$

                    $y = \text{antilog}\,(-0.39794) = 10^{-0.39794} = 0.4$

c) $y = \dfrac{130}{0.25}$

  1) $\log y = \log 130 - \log 0.25$

  2) $\log y = 2.11394 - (-0.60206) = 2.71600$

            $y = \text{antilog}\,2.71600 = 10^{2.71600} = 519.996 \approx 520$

d) $y = (1.06)^{10}$

  1) $\log y = 10 \log 1.06$

  2) $\log y = 10(0.02531) = 0.2531$

              $y = \text{antilog}\,0.2531 = 10^{0.2531} = 1.791$

e) $y = 1024^{0.2}$

  1) $\log y = 0.2 \log 1024$

  2) $\log y = 0.2(3.0103) = 0.60206$

              $y = \text{antilog}\,0.60206 = 10^{0.60206} = 4$

f) $y = \sqrt[5]{1024}$

  1) $\log y = \tfrac{1}{5} \log 1024$

  2) $\log y = \tfrac{1}{5}(3.0103) = 0.60206$

              $y = \text{antilog}\,0.60206 = 10^{0.60206} = 4$

The answer is the same as in part (*e*) because $y = 1024^{1/5} = 1024^{0.2} = 4$. Taking the fifth root is the same thing as raising to the 0.2 or one-fifth power. in one case the log is divided by 5; in the other, it is multiplied by 0.2.

7.22. Use natural logs to solve the following equations:

a) $y = 12.5^3$

  1) $\ln y = 3 \ln 12.5$

To find the natural log of 12.5, enter 12.5 on a calculator and press the $\boxed{\ln x}$ key to find $\ln 12.5 = 2.52573$. Then substitute.

  2) $\ln y = 3(2.52573) = 7.57719$

Since $\ln y = 7.57719$ indicates that e must be raised to the 7.57719 power to get y, to find the *antilogarithm$_e$* of 7.57719 and solve for y, enter 7.57719 on a calculator press the $\boxed{e^x}$ key to find that $e^{7.57719} \approx 1953.1$ and substitute. If the $\boxed{e^x}$ key is the inverse of the $\boxed{\ln x}$ key, enter 7.57719 and press the $\boxed{\text{INV}}$ key followed by the $\boxed{\ln x}$ key.

                $y = \text{antilog}_e\,7.57719 = e^{7.57719} \approx 1953.1$

b) $y = \sqrt[4]{28{,}561}$

  1) $\ln y = \tfrac{1}{4} \ln 28{,}561$

  2) $\ln y = \tfrac{1}{4}(10.25980) = 2.56495$

                $y = \text{antilog}_e\,2.56495 = e^{2.56495} = 13$

CHAPTER 8

Exponential and Logarithmic Functions in Economics

8.1 INTEREST COMPOUNDING

A given principal P compounded annually at an interest rate i for a given number of years t will have a value S at the end of that time given by the exponential function

$$S = P(1 + i)^t \tag{8.1}$$

If compounded m times a year for t years,

$$S = P\left(1 + \frac{i}{m}\right)^{mt} \tag{8.2}$$

If compounded continuously at 100 percent interest for 1 year,

$$S = P \lim_{m \to \infty} \left(1 + \frac{1}{m}\right)^m = P(2.71828) = Pe$$

For interest rates r other than 100 percent and time periods t other than 1 year,

$$S = Pe^{rt} \tag{8.3}$$

For negative growth rates, such as depreciation or deflation, the same formulas apply, but i and r are negative. See Example 1 and Problems 8.1 to 8.6 and 8.9 to 8.17.

EXAMPLE 1. Find the value of $100 at 10 percent interest for 2 years compounded:

1. Annually, $S = P(1 + i)^t$.

$$S = 100(1 + 0.10)^2 = 121$$

160

2. Semiannually, $S = P[1 + (i/m)]^{mt}$, where $m = 2$ and $t = 2$.

$$S = 100\left(1 + \frac{0.10}{2}\right)^{2(2)} = 100(1 + 0.05)^4$$

To find the value of $(1.05)^4$, enter 1.05 on a calculator, press the $\boxed{y^x}$ key, enter 4, and hit the $\boxed{=}$ key to find $(1.05)^4 = 1.2155$. Then substitute.

$$S = 100(1.2155) = 121.55$$

3. Continuously, $S = Pe^{rt}$.

$$S = 100e^{0.1(2)} = 100e^{0.2}$$

For $e^{0.2}$, enter 0.2 on a calculator, press the $\boxed{e^x}$ key to find that $e^{0.2} = 1.2214$, and substitute. If the $\boxed{e^x}$ key is the inverse (shift, or second function) of the $\boxed{\ln x}$ key, enter 0.2 and press the $\boxed{\text{INV}}$ ($\boxed{\text{Shift}}$, or $\boxed{\text{2nd F}}$) key followed by the $\boxed{\ln x}$ key.

$$S = 100(1.2214) = 122.14$$

8.2   EFFECTIVE VS. NOMINAL RATES OF INTEREST

As seen in Example 1, a given principal set out at the same *nominal rate of interest* will earn different *effective rates of interest* which depend on the type of compounding. When compounded annually for 2 years, \$100 will be worth \$121; when compounded semiannually, $S = \$121.55$; when compounded continuously, $S = \$122.14$.

To find the effective annual rate of interest i_e for multiple compounding:

$$P(1 + i_e)^t = P\left(1 + \frac{i}{m}\right)^{mt}$$

Dividing by P and taking the tth root of each side,

$$1 + i_e = \left(1 + \frac{i}{m}\right)^m$$

$$i_e = \left(1 + \frac{i}{m}\right)^m - 1 \tag{8.4}$$

To find the effective annual rate of interest for continuous compounding:

$$1 + i_e = e^r$$
$$i_e = e^r - 1 \tag{8.5}$$

See Example 2 and Problems 8.7 and 8.8.

EXAMPLE 2. Find the effective annual rate of interest for a nominal interest rate of 10 percent when compounded for 2 years (1) semiannually and (2) continuously.

1. Semiannually, $\quad i_e = \left(1 + \dfrac{i}{m}\right)^m - 1 = (1.05)^2 - 1$

For $(1.05)^2$, enter 1.05 on a calculator, press the $\boxed{x^2}$ key, or press the $\boxed{y^x}$ key, and then enter 2 followed by the $\boxed{=}$ key, to find $(1.05)^2 = 1.1025$, and substitute.

$$i_e = 1.1025 - 1 = 0.1025 = 10.25\%$$

2. Continuously, $i_e = e^r - 1 = e^{0.1} - 1$

To find the value of $e^{0.1}$, enter 0.1 on a calculator, press the $\boxed{e^x}$ key to learn $e^{0.1} = 1.10517$, and substitute.

$$i_e = 1.10517 - 1 = 0.10517 \approx 10.52\%$$

8.3 DISCOUNTING

A sum of money to be received in the future is not worth as much as an equivalent amount of money in the present, because the money on hand can be lent at interest to grow to an even larger sum by the end of the year. If present market conditions will enable a person to earn 8 percent interest compounded annually, $100 will grow to $108 by the end of the year. And $108 1 year from now, therefore, is worth (has a *present value* of) only $100 today.

Discounting is the process of determining the present value P of a future sum of money S. If under annual compounding

$$S = P(1 + i)^t$$

then $$P = \frac{S}{(1 + i)^t} = S(1 + i)^{-t} \qquad (8.6)$$

Similarly, under multiple compoundings $P = S[1 + (i/m)]^{-mt}$, and under continuous compounding $P = Se^{-rt}$. When finding the present value, the interest rate is called the *discount rate*. See Example 3 and Problems 8.18 to 8.22.

EXAMPLE 3.   The present value of a 5-year bond with a face value of $1000 and no coupons is calculated below. It is assumed that comparable opportunities offer interest rates of 9 percent under annual compounding.

$$P = S(1 + i)^{-t} = 1000(1 + 0.09)^{-5}$$

To find the value of $(1.09)^{-5}$, enter 1.09 on a calculator, press the $\boxed{y^x}$ key, and enter -5 by first entering 5 and then pressing the $\boxed{+/-}$ key followed by the $\boxed{=}$ key to find $(1.09)^{-5} = 0.64993$, and substitute.

$$P = 1000(0.64993) = 649.93$$

Thus, a bond with no coupons promising to pay $1000 5 years from now is worth approximately $649.93 today since $649.93 at 9 percent interest will grow to $1000 in 5 years.

8.4 CONVERTING EXPONENTIAL TO NATURAL EXPONENTIAL FUNCTIONS

In Section 8.1 we saw that (1) exponential functions are used to measure rates of *discrete growth*, i.e., growth that takes place at discrete intervals of time, such as the end of the year or the end of the quarter as in ordinary interest compounding or discounting; and (2) natural exponential functions are used to measure rates of *continuous growth*, i.e., growth that takes place constantly rather than at discrete intervals, as in continuous compounding, animal development, or population growth. An exponential function $S = P(1 + i/m)^{mt}$ expressing discrete growth can be converted to an equivalent natural exponential function $S = Pe^{rt}$ measuring continuous growth, by setting the two expressions equal to each other, and solving for r, as follows:

$$P\left(1 + \frac{i}{m}\right)^{mt} = Pe^{rt}$$

By canceling P's,

$$\left(1 + \frac{i}{m}\right)^{mt} = e^{rt}$$

Taking the natural log of each side,

$$\ln\left(1 + \frac{i}{m}\right)^{mt} = \ln e^{rt}$$

$$mt \ln\left(1 + \frac{i}{m}\right) = rt$$

Dividing both sides by t,

$$r = m \ln\left(1 + \frac{i}{m}\right)$$

Thus,
$$S = P\left(1 + \frac{i}{m}\right)^{mt} = Pe^{m \ln (1+i/m)t} \tag{8.7}$$

See Examples 4 and 5 and Problems 8.37 to 8.42.

EXAMPLE 4. A natural exponential function can be used to determine the value of $100 at 10 percent interest compounded semiannually for 2 years, as shown below.

$$S = Pe^{rt}$$

where $r = m \ln(1 + i/m)$. Thus,

$$r = 2 \ln\left(1 + \frac{0.10}{2}\right) = 2 \ln 1.05 = 2(0.04879) = 0.09758$$

Substituting above,

$$S = 100e^{(0.09758)2} = 100e^{0.19516}$$

Using a calculator here and throughout,

$$S = 100(1.2155) = 121.55$$

as was found in Example 1.

Note that with natural exponential functions, the continuous growth is given by r in Pe^{rt}. Thus, the continuous growth rate of $100 at 10 percent interest compounded semiannually is 0.09758, or 9.758 percent a year. That is to say, 9.758 percent interest at continuous compounding is equivalent to 10 percent interest when compounded semiannually.

EXAMPLE 5. A small firm with current annual sales of $10,000 projects a 12 percent growth in sales annually. Its projected sales in 4 years are calculated below in terms of an *ordinary exponential function*.

$$S = 10,000(1 + 0.12)^4$$
$$= 10,000(1.5735) = 15,735$$

EXAMPLE 6. The sales projections specified in Example 5 are recalculated below, using a natural exponential function with $r = m \ln(1 + i/m)$ and $m = 1$.

$$r = \ln 1.12 = 0.11333$$
$$S = 10,000e^{0.11333(4)} = 10,000(1.5735) = 15,735$$

8.5 ESTIMATING GROWTH RATES FROM DATA POINTS

Given two sets of data for a function—sales, costs, profits—growing consistently over time, annual growth rates can be measured and a natural exponential function estimated through a system of simultaneous equations. For example, if sales volume equals 2.74 million in 1996 and 4.19 million in

2001, let $t = 0$ for the base year 1996, then $t = 5$ for 2001. Express the two sets of data points in terms of a natural exponential function $S = Pe^{rt}$, recalling that $e^0 = 1$.

$$2.74 = Pe^{r(0)} = P \tag{8.8}$$

$$4.19 = Pe^{r(5)} \tag{8.9}$$

Substitute $P = 2.74$ from (8.8) in (8.9) and simplify algebraically.

$$4.19 = 2.74e^{5r}$$
$$1.53 = e^{5r}$$

Take the natural log of both sides.

$$\ln 1.53 = \ln e^{5r} = 5r$$
$$0.42527 = 5r$$
$$r = 0.08505 \approx 8.5\%$$

Substituting,
$$S = 2.74e^{0.085t}$$

With $r = 0.085$, the rate of continuous growth per year is 8.5 percent. To find the rate of discrete growth i, recall that

$$r = m \ln\left(1 + \frac{i}{m}\right)$$

Thus, for annual compounding with $m = 1$,

$$0.085 = \ln(1 + i)$$
$$1 + i = \text{antilog}_e\, 0.085 = e^{0.085} = 1.08872$$
$$i = 1.08872 - 1 = 0.08872 \approx 8.9\%$$

See Example 7 and Problems 8.43 to 8.45.

EXAMPLE 7. Given the original information above, an ordinary exponential function for growth in terms of $S = P(1 + i)^t$ can also be estimated directly from the data.

Set the data in ordinary exponential form.

$$2.74 = P(1 + i)^0 = P \tag{8.10}$$

$$4.19 = P(1 + i)^5 \tag{8.11}$$

Substitute $P = 2.74$ from (8.10) in (8.11) and simplify.

$$4.19 = 2.74(1 + i)^5$$
$$1.53 = (1 + i)^5$$

Take the common log of both sides.

$$\log 1.53 = 5 \log(1 + i)$$
$$\tfrac{1}{5}(0.18469) = \log(1 + i)$$
$$\log(1 + i) = 0.03694$$

$$1 + i = \text{antilog}\, 0.03694 = 10^{0.03694} = 1.08878$$
$$i = 1.08878 - 1 = 0.08878 \approx 8.9\%$$

Substituting,
$$S = 2.74(1 + 0.089)^t$$

Solved Problems

COMPOUNDING INTEREST

8.1. Given a principal P of \$1000 at 6 percent interest i for 3 years, find the future value S when the principal is compounded (a) annually, (b) semiannually, and (c) quarterly.

    a)   From (8.1),
$$S = P(1 + i)^t = 1000(1 + 0.06)^3$$

For $(1.06)^3$, enter 1.06 on a calculator, press the $\boxed{y^x}$ key, enter 3 followed by the $\boxed{=}$ key to find $(1.06)^3 = 1.19102$, then substitute.
$$S = 1000(1.19102) \approx 1191.02$$

    b)   From (8.2),
$$S = P\left(1 + \frac{i}{m}\right)^{mt} = 1000\left(1 + \frac{0.06}{2}\right)^{2(3)} = 1000(1.03)^6$$

For $(1.03)^6$, enter 1.03, hit the $\boxed{y^x}$ key and then 6, and substitute.
$$S = 1000(1.19405) \approx 1194.05$$

    c)
$$S = 1000\left(1 + \frac{0.06}{4}\right)^{4(3)} = 1000(1.015)^{12}$$

Enter 1.015, hit the $\boxed{y^x}$ key then 12, and substitute.
$$S = 1000(1.19562) \approx 1195.62$$

8.2. Redo Problem 8.1, given a principal of \$100 at 8 percent for 5 years.

    a)
$$S = 100(1.08)^5$$
$$= 100(1.46933) \approx 146.93$$

    b)
$$S = 100\left(1 + \frac{0.08}{2}\right)^{2(5)} = 100(1.04)^{10}$$
$$= 100(1.48024) \approx 148.02$$

    c)
$$S = 100\left(1 + \frac{0.08}{4}\right)^{4(5)} = 100(1.02)^{20}$$
$$= 100(1.48595) \approx 148.60$$

8.3. Redo Problem 8.1, given a principal of \$1250 at 12 percent for 4 years.

    a) $S = 1250(1.12)^4 = 1250(1.57352) \approx 1966.90$

    b) $S = 1250(1.06)^8 = 1250(1.59385) \approx 1992.31$

    c) $S = 1250(1.03)^{16} = 1250(1.60471) \approx 2005.89$

8.4. Find the future value of a principal of \$100 at 5 percent for 6 years when compounded (a) annually and (b) continually.

    a) $S = 100(1.05)^6 = 100(1.34010) \approx 134.01$

    b)   From (8.3), $S = Pe^{rt} = 100e^{0.05(6)} = 100e^{0.3}$

For $e^{0.3}$, enter 0.3, hit the $\boxed{e^x}$ key, and substitute.
$$S = 100(1.34986) \approx 134.99$$

8.5. Redo Problem 8.4, given a principal of $150 at 7 percent for 4 years.

a) $$S = 150(1.07)^4 = 150(1.31080) \approx 196.62$$

b) $$S = 150e^{0.07(4)} = 150e^{0.28} = 150(1.32313) \approx 198.47$$

8.6. From Problems 8.4 and 8.5, use natural logs to find (*a*) $S = 100e^{0.3}$ and (*b*) $S = 150e^{0.28}$.

a)
$$\ln S = \ln 100 + 0.3 \ln e = 4.60517 + 0.3(1) = 4.90517$$
$$S = \text{antilog}_e\, 4.90517 = e^{4.90517} \approx 134.99$$

b)
$$\ln S = \ln 150 + 0.28 \ln e = 5.01064 + 0.28 = 5.29064$$
$$S = \text{antilog}_e\, 5.29064 = e^{5.29064} \approx 198.47$$

8.7. Find the effective annual interest rate on $100 at 6 percent compounded (*a*) semiannually and (*b*) continuously.

a) From (*8.4*),
$$i_e = \left(1 + \frac{i}{m}\right)^m - 1$$
$$= 1.0609 - 1 = 0.06090 \approx 6.09\%$$

b) From (*8.5*),
$$i_e = e^r - 1 = e^{0.06} - 1$$
$$= 1.06184 - 1 = 0.06184 \approx 6.18\%$$

8.8. Calculate the rate of effective annual interest on $1000 at 12 percent compounded (*a*) quarterly and (*b*) continuously.

a)
$$i_e = \left(1 + \frac{i}{m}\right)^m - 1 = (1.03)^4 - 1$$
$$= 1.12551 - 1 = 0.12551 \approx 12.55\%$$

b)
$$i_e = e^r - 1 = e^{0.12} - 1$$
$$= 1.12750 - 1 = 0.12750 \approx 12.75\%$$

TIMING

8.9. Determine the interest rate needed to have money double in 10 years under annual compounding.

$$S = P(1 + i)^t$$

If money doubles, $S = 2P$. Thus, $2P = P(1 + i)^{10}$.

Dividing by P, and taking the tenth root of each side,

$$2 = (1 + i)^{10} \qquad (1 + i) = \sqrt[10]{2}$$

For $\sqrt[10]{2}$, enter 2, press the $\boxed{\sqrt[x]{y}}$ key, then 10 followed by the $\boxed{=}$ key, and substitute. If the $\boxed{\sqrt[x]{y}}$ is the inverse (shift, or second function) of the $\boxed{y^x}$ key, enter 2, hit the $\boxed{\text{INV}}$ ($\boxed{\text{Shift}}$, or $\boxed{\text{2ndF}}$) key followed by the $\boxed{y^x}$ key, and then enter 10 and hit the $\boxed{=}$ key.

$$1 + i = 1.07177$$
$$i = 1.07177 - 1 = 0.07177 \approx 7.18\%$$

Note that since $\sqrt[10]{2} = 2^{1/10} = 2^{0.1}$, $\sqrt[10]{2}$ or any root can also be found with the $\boxed{y^x}$ key.

8.10. Determine the interest rate needed to have money double in 6 years when compounded semiannually.

$$S = P\left(1 + \frac{i}{m}\right)^{mt}$$

$$2P = P\left(1 + \frac{i}{2}\right)^{2(6)}$$

$$2 = 1(1 + 0.5i)^{12}$$

$$1 + 0.5i = \sqrt[12]{2}$$

$$\ln(1 + 0.5i) = \tfrac{1}{12}\ln 2 = \tfrac{1}{12}(0.69315) = 0.05776$$

$$1 + 0.5i = e^{0.05776} = 1.05946$$

$$0.5i = 1.05946 - 1 = 0.05946$$

$$i = 0.11892 \approx 11.89\%$$

8.11. What interest rate is needed to have money treble in 10 years when compounded quarterly?

$$S = P\left(1 + \frac{i}{4}\right)^{4(10)}$$

If money trebles,

$$3P = P\left(1 + \frac{i}{4}\right)^{40}$$

$$3 = (1 + 0.25i)^{40}$$

$$1 + 0.25i = \sqrt[40]{3}$$

$$\ln(1 + 0.25i) = \frac{1}{40}\ln 3 = 0.02747$$

$$1 + 0.25i = e^{0.02747} = 1.02785$$

$$i = 0.1114 \approx 11.14\%$$

8.12. At what interest rate will money treble if compounded continuously for 8 years?

$$S = Pe^{rt}$$

$$3P = Pe^{r(8)}$$

$$\ln 3 = \ln e^{8r}$$

$$1.09861 = 8r \qquad r = 0.1373 = 13.73\%$$

8.13. At what interest rate will money quintuple if compounded continuously for 25 years?

$$S = Pe^{rt}$$

$$5 = e^{25r}$$

$$\ln 5 = 25r$$

$$1.60944 = 25r \qquad r = 0.0644 = 6.44\%$$

8.14. How long will it take money to double at 12 percent interest under annual compounding? Round answers to two decimal places.

$$S = P(1 + i)^t \qquad 2 = (1 + 0.12)^t$$

$$\ln 2 = t \ln 1.12 \qquad 0.69315 = 0.11333t$$

$$t \approx 6.12 \text{ years}$$

8.15. How long will it take money to increase to $2\frac{1}{2}$ times its present value when compounded semiannually at 8 percent?

$$S = P\left(1 + \frac{0.08}{2}\right)^{2t} \qquad 2.5 = (1.04)^{2t}$$

$$\ln 2.5 = 2t \ln 1.04 \qquad 0.91629 = 2(0.03922)t$$

$$t \approx 11.68 \text{ years}$$

8.16. How long will it take money to double at 5 percent interest when compounded quarterly?

$$S = P\left(1 + \frac{0.05}{4}\right)^{4t} \qquad 2 = (1.0125)^{4t}$$

$$\ln 2 = 4t \ln 1.0125 \qquad 0.69315 = 4(0.01242)t$$

$$t \approx 13.95 \text{ years}$$

8.17. How long will it take money (a) to quadruple when compounded continuously at 9 percent and (b) to treble at 12 percent?

a) $\quad S = Pe^{rt} \qquad 4 = e^{0.09t}$ b) $\quad S = Pe^{rt} \qquad 3 = e^{0.12t}$

$\quad\quad \ln 4 = 0.09t \qquad 1.38629 = 0.09t$ $\ln 3 = 0.12t \qquad 1.09861 = 0.12t$

$\quad\quad t \approx 15.4 \text{ years}$ $t \approx 9.16 \text{ years}$

DISCOUNTING

8.18. Find the present value of $750 to be paid 4 years from now when the prevailing interest rate is 10 percent if interest is compounded (a) annually and (b) semiannually.

a)   Using (8.6) and its modifications throughout,

$$P = S(1 + i)^{-t} = 750(1.10)^{-4}$$

For $(1.10)^{-4}$, enter 1.10, hit the $\boxed{y^x}$ key, enter 4, then press the $\boxed{+/-}$ key to find $(1.10)^{-4} = 0.68301$, and substitute.

$$P = 750(0.68301) \approx 512.26$$

b) $$P = S\left(1 + \frac{i}{m}\right)^{-mt} = 750(1.05)^{-8}$$

$$= 750(0.67684) \approx 507.63$$

8.19. Redo Problem 8.18, for $600 to be paid 7 years hence at a prevailing interest rate of 4 percent.

a) $\quad P = 600(1.04)^{-7}$ b) $\quad P = 600(1.02)^{-14}$

$\quad\quad = 600(0.75992) \approx 455.95$ $= 600(0.75788) \approx 454.73$

8.20. Find the present value of $500 in 3 years at 8 percent when interest is compounded (a) annually and (b) continuously.

a) $\quad P = 500(1.08)^{-3}$ b) $\quad P = Se^{-rt} = 500e^{-0.08(3)} = 500e^{-0.24}$

$\quad\quad = 500(0.79383) \approx 396.92$ $= 500(0.78663) \approx 393.32$

8.21. Redo Problem 8.20, for \$120 in 5 years at 9 percent.

a) $P = 120(1.09)^{-5}$
     $= 120(0.64993) \approx 77.99$

b) $P = 120e^{-0.09(5)} = 120e^{-0.45}$
     $= 120(0.63763) \approx 76.52$

8.22. Use natural logs to solve Problem 8.21(b).

$$P = 120e^{-0.45}$$
$$\ln P = \ln 120 + (-0.45) = 4.78749 - 0.45 = 4.33749$$
$$P = \text{antilog}_e\, 4.33749 = e^{4.33749} \approx 76.52$$

EXPONENTIAL GROWTH FUNCTIONS

8.23. A firm with sales of 150,000 a year expects to grow by 8 percent a year. Determine the expected level of sales in 6 years.

$$S = 150{,}000(1.08)^6$$
$$= 150{,}000(1.58687) \approx 238{,}031$$

8.24. Profits are projected to rise by 9 percent a year over the decade. With current profits of 240,000, what will the level of profits be at the end of the decade?

$$\pi = 240{,}000(1.09)^{10}$$
$$= 240{,}000(2.36736) \approx 568{,}166$$

8.25. The cost of food has been increasing by 3.6 percent a year. What can a family with current food expenditures of \$200 a month be expected to pay for food each month in 5 years?

$$F = 200(1.036)^5$$
$$= 200(1.19344) \approx 238.69$$

8.26. If the cost of living had continued to increase by 12.5 percent a year from a base of 100 in 1993, what would the cost-of-living index be in 2000?

$$C = 100(1.125)^7$$
$$= 100(2.28070) \approx 228.07$$

8.27. A discount clothing store reduces prices by 10 percent each day until the goods are sold. What will a \$175 suit sell for in 5 days?

$$P = 175(1 - 0.10)^5$$
$$= 175(0.9)^5 = 175(0.59049) \approx 103.34$$

8.28. A new car depreciates in value by 3 percent a month for the first year. What is the book value of a \$6000 car at the end of the first year?

$$B = 6000(0.97)^{12}$$
$$= 6000(0.69384) \approx 4163.04$$

8.29. If the dollar depreciates at 2.6 percent a year, what will a dollar be worth in real terms 25 years from now?

$$D = 1.00(0.974)^{25}$$
$$= 1.00(0.51758) \approx 0.5176 \text{ or } 51.76\text{¢}$$

8.30. The cost of an average hospital stay was $500 at the end of 1989. The average cost in 1999 was $1500. What was the annual rate of increase?

$$1500 = 500(1 + i)^{10}$$
$$3 = (1 + i)^{10}$$
$$1 + i = \sqrt[10]{3}$$

For $\sqrt[10]{3}$, enter 3, press the $\boxed{\sqrt[x]{y}}$ key then 10, and substitute.

$$1 + i = 1.11612 \qquad i = 0.11612 \approx 11.6\%$$

8.31. A 5-year development plan calls for boosting investment from 2.6 million a year to 4.2 million. What average annual increase in investment is needed each year?

$$4.2 = 2.6(1 + i)^5$$
$$1.615 = (1 + i)^5$$
$$1 + i = \sqrt[5]{1.615} = 1.10061$$
$$i = 0.10061 \approx 10\%$$

8.32. A developing country wishes to increase savings from a present level of 5.6 million to 12 million. How long will it take if it can increase savings by 15 percent a year?

$$12 = 5.6(1.15)^t$$

To solve for an exponent, use a logarithmic transformation

$$\ln 12 = \ln 5.6 + t \ln 1.15$$
$$2.48491 = 1.72277 + 0.13976t$$
$$0.13976t = 0.76214 \qquad t \approx 5.45 \text{ years}$$

8.33. Population in many third-world countries is growing at 3.2 percent. Calculate the population 20 years from now for a country with 1,000,000 people.

Since population increases continually over time, a natural exponential function is needed.

$$P = 1,000,000e^{0.032(20)} = 1,000,000e^{0.64}$$
$$= 1,000,000(1.89648) \approx 1,896,480$$

8.34. If the country in Problem 8.33 reduces its population increase to 2.4 percent, what will the population be in 20 years?

$$P = 1,000,000e^{0.024(20)} = 1,000,000e^{0.48}$$
$$= 1,000,000(1.61607) \approx 1,616,070$$

8.35. If world population grows at 2.6 percent, how long will it take to double?

$$2 = e^{0.026t}$$
$$\ln 2 = 0.026t$$
$$0.69315 = 0.026t \qquad t = 26.66 \text{ years}$$

8.36. If arable land in the Sahel is eroding by 3.5 percent a year because of climatic conditions, how much of the present arable land A will be left in 12 years?

$$P = Ae^{-0.035(12)} = Ae^{-0.42}$$
$$= 0.657047A \qquad \text{or} \qquad 66\%$$

CONVERTING EXPONENTIAL FUNCTIONS

8.37. Find the future value of a principal of $2000 compounded semiannually at 12 percent for 3 years, using (a) an exponential function and (b) the equivalent natural exponential function.

a)
$$S = P\left(1 + \frac{i}{m}\right)^{mt} = 2000\left(1 + \frac{0.12}{2}\right)^{2(3)} = 2000(1.06)^6 = 2000(1.41852) \approx 2837.04$$

b)
$$S = Pe^{rt}$$

where $r = m \ln(1 + i/m) = 2 \ln 1.06 = 2(0.05827) = 0.11654$.

Thus,
$$S = 2000e^{0.11654(3)} = 2000e^{0.34962} = 2000(1.41853) \approx 2837.06^*$$

8.38. Redo Problem 8.37 for a principal of $600 compounded annually at 9 percent for 5 years.

a)
$$S = 600(1.09)^5 = 600(1.53862) \approx 923.17$$

b) $S = Pe^{rt}$, where $r = \ln 1.09 = 0.08618$. Thus,

$$S = 600e^{0.08618(5)} = 600e^{0.4309} = 600(1.53864) \approx 923.18^*$$

8.39. Redo Problem 8.37 for a principal of $1800 compounded quarterly at 8 percent interest for $2\frac{1}{2}$ years.

a) $\quad S = 1800(1.02)^{10}$
$$= 1800(1.21899) \approx 2194.18$$

b) $\quad r = 4 \ln 1.02 = 4(0.01980) = 0.07920$
$$S = 1800e^{0.07920(2.5)} = 1800e^{0.19800}$$
$$= 1800(1.21896) = 2194.13^*$$

8.40. Find the equivalent form under annual discrete compounding for $S = Pe^{0.07696t}$.

$$r = m \ln\left(1 + \frac{i}{m}\right)$$

Since compounding is annual, $m = 1$

$$0.07696 = \ln(1 + i)$$
$$1 + i = \text{antilog}_e 0.07696 = 1.08$$
$$i = 0.08$$

Thus,
$$S = P(1.08)^t$$

8.41. Find the equivalent form under semiannual discrete compounding for $Pe^{0.09758t}$.

$$r = 2 \ln(1 + 0.5i)$$
$$0.09758 = 2 \ln(1 + 0.5i)$$
$$1 + 0.5i = \text{antilog}_e 0.04879 = 1.05$$
$$0.5i = 0.05 \qquad i = 0.10$$

Thus,
$$S = P(1.05)^{2t}$$

8.42. Find the equivalent form for $S = Pe^{0.15688t}$ under quarterly compounding.

$$r = 4 \ln(1 + 0.25i)$$
$$\tfrac{1}{4}(0.15688) = \ln(1 + 0.25i)$$
$$1 + 0.25i = \text{antilog}_e 0.03922 = 1.04$$
$$i = 0.16$$
$$S = P(1.04)^{4t}$$

---

* Slight discrepancy is due to earlier rounding.

ESTABLISHING EXPONENTIAL FUNCTIONS FROM DATA

8.43. An animal population goes from 3.5 million in 1997 to 4.97 million in 2001. Express population growth P in terms of a natural exponential function and determine the rate of growth.

$$3.50 = P_0 e^{r(0)} = P_0 \qquad (8.12)$$

$$4.97 = P_0 e^{r(4)} \qquad (8.13)$$

Substitute $P_0 = 3.50$ from (8.12) in (8.13) and simplify.

$$4.97 = 3.50 e^{4r}$$
$$1.42 = e^{4r}$$

Take the natural log of both sides.

$$\ln 1.42 = \ln e^{4r} = 4r$$
$$0.35066 = 4r$$
$$r = 0.08767 \approx 8.8\%$$

Thus, $$P = 3.50 e^{0.088t} \qquad r = 8.8\%$$

8.44. Costs C of a government program escalate from 5.39 billion in 1995 to 10.64 billion in 2001. Express costs in terms of an ordinary exponential function, and find the annual rate of growth.

$$5.39 = C_0 (1 + i)^0 = C_0 \qquad (8.14)$$

$$10.64 = C_0 (1 + i)^6 \qquad (8.15)$$

Substitute $C_0 = 5.39$ from (8.14) in (8.15) and simplify.

$$10.64 = 5.39 (1 + i)^6$$
$$1.974 = (1 + i)^6$$

Take the common log of both sides.

$$\log 1.974 = 6 \log (1 + i)$$
$$\tfrac{1}{6}(0.29535) = \log (1 + i)$$
$$\log (1 + i) = 0.04923$$
$$1 + i = \text{antilog } 0.04923 = 10^{0.04923} = 1.12$$
$$i = 1.12 - 1 = 0.12 = 12\%$$

Hence, $$C = 5.39 (1 + 0.12)^t \qquad i = 12\%$$

8.45. Redo problem 8.44, given $C = 2.80$ in 1991 and $C = 5.77$ in 2001.

$$2.80 = C_0 (1 + i)^0 = C_0 \qquad (8.16)$$

$$5.77 = C_0 (1 + i)^{10} \qquad (8.17)$$

Substitute $C_0 = 2.80$ in (8.17) and simplify.

$$5.77 = 2.80 (1 + i)^{10}$$
$$2.06 = (1 + i)^{10}$$

Take the logs. $$\log 2.06 = 10 \log (1 + i)$$
$$\tfrac{1}{10}(0.31387) = \log (1 + i)$$
$$\log (1 + i) = 0.03139$$
$$1 + i = \text{antilog } 0.03139 = 10^{0.03139} = 1.07495$$
$$i = 1.07495 - 1 = 0.07495 \approx 7.5\%$$

Hence, $$C = 2.80 (1 + 0.075)^t \qquad i = 7.5\%$$

CHAPTER 9

Differentiation of Exponential and Logarithmic Functions

9.1 RULES OF DIFFERENTIATION

The rules of exponential and logarithmic differentiation are presented below, illustrated in Examples 1 to 4, and treated in Problems 9.1 to 9.8. Selected proofs for the rules are offered in Problems 9.35 to 9.40.

9.1.1 The Natural Exponential Function Rule

Given $f(x) = e^{g(x)}$, where $g(x)$ is a differentiable function of x, the derivative is

$$f'(x) = e^{g(x)} \cdot g'(x) \tag{9.1}$$

In short, the derivative of a natural exponential function is equal to the original natural exponential function times the derivative of the exponent.

EXAMPLE 1. The derivatives of each of the natural exponential functions below are found as follows:

1. $f(x) = e^x$
    Let $g(x) = x$, then $g'(x) = 1$. Substituting in (9.1),

    $$f'(x) = e^x \cdot 1 = e^x$$

    The derivative of e^x is simply e^x, the original function itself.

2. $f(x) = e^{x^2}$
    Since $g(x) = x^2$, then $g'(x) = 2x$. Substituting in (9.1),

    $$f'(x) = e^{x^2} \cdot 2x = 2xe^{x^2}$$

173

3. $f(x) = 3e^{7-2x}$

Here $g(x) = 7 - 2x$, so $g'(x) = -2$. From (9.1),

$$f'(x) = 3e^{7-2x} \cdot -2 = -6e^{7-2x}$$

Evaluating the slope of this function at $x = 4$,

$$f'(4) = -6e^{7-2(4)} = -6e^{-1} = -6\left(\frac{1}{2.71828}\right) = -2.2$$

See also Problem 9.1.

9.1.2   The Exponential Function Rule for Base a Other Than e

Given $f(x) = a^{g(x)}$, where $a > 0$, $a \neq 1$, and $g(x)$ is a differentiable function of x, the derivative is

$$f'(x) = a^{g(x)} \cdot g'(x) \cdot \ln a \tag{9.2}$$

The derivative is simply the original function times the derivative of the exponent times the natural log of the base.

EXAMPLE 2.   The exponential function rule for base a is demonstrated in the following cases:

1. $f(x) = a^{1-2x}$. Let $g(x) = 1 - 2x$, then $g'(x) = -2$. Substituting in (9.2),

$$f'(x) = a^{1-2x} \cdot -2 \cdot \ln a = -2a^{1-2x} \ln a$$

2. $y = a^x$. Here $g(x) = x$ and $g'(x) = 1$. From (9.2),

$$y' = a^x \cdot 1 \cdot \ln a = a^x \ln a$$

Remember that a may also assume a numerical value. See Problem 9.2(c) through (g).

3. $y = x^2 a^{3x}$. With y a product of x^2 and a^{3x}, the product rule is necessary.

$$y' = x^2(a^{3x} \cdot 3 \cdot \ln a) + a^{3x}(2x)$$
$$= xa^{3x}(3x \ln a + 2)$$

See also Problem 9.2.

9.1.3   The Natural Logarithmic Function Rule

Given $f(x) = \ln g(x)$, where $g(x)$ is positive and differentiable, the derivative is

$$f'(x) = \frac{1}{g(x)} \cdot g'(x) = \frac{g'(x)}{g(x)} \tag{9.3}$$

See Example 3 and Problems 9.3 to 9.5.

EXAMPLE 3.   Finding the derivative of a natural logarithmic function is demonstrated below:

1. $f(x) = \ln 6x^2$. Let $g(x) = 6x^2$, then $g'(x) = 12x$. Substituting in (9.3),

$$f'(x) = \frac{1}{6x^2} \cdot 12x = \frac{2}{x}$$

2. $y = \ln x$. Since $g(x) = x$, $g'(x) = 1$. From (9.3),

$$y' = \frac{1}{x} \cdot 1 = \frac{1}{x}$$

3. $y = \ln(x^2 + 6x + 2)$. The derivative is

$$y' = \frac{1}{x^2 + 6x + 2} \cdot (2x + 6) = \frac{2x + 6}{x^2 + 6x + 2}$$

Evaluating the slope of this function at $x = 4$,

$$y'(4) = \tfrac{14}{42} = \tfrac{1}{3}$$

9.1.4   The Logarithmic Function Rule for Base a Other Than e

Given $f(x) = \log_a g(x)$, where $a > 0$, $a \neq 1$, and $g(x)$ is positive and differentiable, the derivative is

$$f'(x) = \frac{1}{g(x)} \cdot g'(x) \cdot \log_a e \qquad \text{or} \qquad f'(x) = \frac{1}{g(x)} \cdot g'(x) \cdot \frac{1}{\ln a} \qquad (9.4)$$

since $\log_a e = 1/\ln a$. See Example 4 and Problems 9.6 and 9.40.

EXAMPLE 4.   Derivatives of logarithmic functions to base a are found as shown below.

1. $f(x) = \log_a(2x^2 + 1)$. Let $g(x) = 2x^2 + 1$; then $g'(x) = 4x$. Substituting in (9.4),

$$f'(x) = \frac{1}{2x^2 + 1} \cdot 4x \cdot \log_a e = \frac{4x}{2x^2 + 1} \log_a e$$

or, from (9.4),

$$f'(x) = \frac{4x}{(2x^2 + 1)\ln a}$$

2. $y = \log_a x$. Here $g(x) = x$, and $g'(x) = 1$. From (9.4),

$$y' = \frac{1}{x} \cdot 1 \cdot \log_a e = \frac{\log_a e}{x}$$

or

$$y' = \frac{1}{x \ln a}$$

9.2   HIGHER-ORDER DERIVATIVES

Higher-order derivatives are found by taking the derivative of the previous derivative, as illustrated in Example 5 and Problems 9.9 and 9.10.

EXAMPLE 5.   Finding the first and second derivatives of exponential and logarithmic functions is illustrated below:

1.   Given $y = e^{5x}$. The first and second derivatives are

$$\frac{dy}{dx} = e^{5x}(5) = 5e^{5x}$$

$$\frac{d^2 y}{dx^2} = 5e^{5x}(5) = 25e^{5x}$$

2.   Given $y = a^x$. The first derivative is

$$\frac{dy}{dx} = a^x(1)\ln a = a^x \ln a$$

where $\ln a$ is a constant. Thus, the second derivative is

$$\frac{d^2y}{dx^2} = a^x(\ln a)(1)(\ln a) = a^x(\ln a)^2 = a^x \ln^2 a$$

3. Given $y = \ln 2x$. The first derivative is

$$\frac{dy}{dx} = \frac{1}{2x}(2) = \frac{1}{x} \quad \text{or} \quad x^{-1}$$

By the simple power function rule,

$$\frac{d^2y}{dx^2} = -x^{-2} \quad \text{or} \quad -\frac{1}{x^2}$$

4. Given $y = \log_a 3x$. The first derivative is

$$\frac{dy}{dx} = \frac{1}{3x}(3)\frac{1}{\ln a} = \frac{1}{x \ln a}$$

By the quotient rule, where $\ln a$ is a constant, the second derivative is

$$\frac{d^2y}{dx^2} = \frac{x \ln a(0) - 1 \ln a}{(x \ln a)^2} = \frac{-\ln a}{x^2 \ln^2 a} = -\frac{1}{x^2 \ln a}$$

9.3 PARTIAL DERIVATIVES

Partial derivatives are found by differentiating the function with respect to one variable, while keeping the other independent variables constant. See Example 6 and Problem 9.11.

EXAMPLE 6. Finding all the first and second partial derivatives for a function is illustrated below:

1. Given $z = e^{(3x+2y)}$. The first and second partials are

$$z_x = e^{(3x+2y)}(3) = 3e^{(3x+2y)} \qquad z_y = e^{(3x+2y)}(2) = 2e^{(3x+2y)}$$
$$z_{xx} = 3e^{(3x+2y)}(3) = 9e^{(3x+2y)} \qquad z_{yy} = 2e^{(3x+2y)}(2) = 4e^{(3x+2y)}$$
$$z_{xy} = 6e^{(3x+2y)} = z_{yx}$$

2. Given $z = \ln(5x + 9y)$, the partial derivatives are

$$z_x = \frac{5}{5x + 9y} \qquad z_y = \frac{9}{5x + 9y}$$

By the simple quotient rule,

$$z_{xx} = \frac{(5x+9y)(0) - 5(5)}{(5x+9y)^2} \qquad z_{yy} = \frac{(5x+9y)(0) - 9(9)}{(5x+9y)^2}$$
$$= \frac{-25}{(5x+9y)^2} \qquad = \frac{-81}{(5x+9y)^2}$$
$$z_{xy} = \frac{-45}{(5x+9y)^2} = z_{yx}$$

9.4 OPTIMIZATION OF EXPONENTIAL AND LOGARITHMIC FUNCTIONS

Exponential and logarithmic functions follow the general rules for optimization presented in Sections 4.5 and 5.4. The method is demonstrated in Example 7 and treated in Problems 9.12 to 9.21.

EXAMPLE 7. The procedure for finding critical values and determining whether exponential and logarithmic functions are maximized or minimized is illustrated below:

1. Given $y = 2xe^{4x}$. Using the product rule and setting the derivative equal to zero, we get

$$\frac{dy}{dx} = 2x(4e^{4x}) + 2(e^{4x}) = 0$$

$$= 2e^{4x}(4x + 1) = 0$$

For the derivative to equal zero, either $2e^{4x} = 0$ or $4x + 1 = 0$. Since $2e^{4x} \neq 0$ for any value of x,

$$4x + 1 = 0 \qquad \bar{x} = -\tfrac{1}{4}$$

Testing the second-order condition,

$$\frac{d^2y}{dx^2} = 2e^{4x}(4) + (4x + 1)(2e^{4x})(4) = 8e^{4x}(4x + 2)$$

Evaluated at the critical value, $\bar{x} = -\tfrac{1}{4}$, $d^2y/dx^2 = 8e^{-1}(-1 + 2) = 8/e > 0$. The function is thus at a minimum, since the second derivative is positive.

2. Given $y = \ln(x^2 - 6x + 10)$. By the natural log rule,

$$\frac{dy}{dx} = \frac{2x - 6}{x^2 - 6x + 10} = 0$$

Multiplying both sides by the denominator $x^2 - 6x + 10$,

$$2x - 6 = 0 \qquad \bar{x} = 3$$

Using the simple quotient rule for the second derivative,

$$\frac{d^2y}{dx^2} = \frac{(x^2 - 6x + 10)(2) - (2x - 6)(2x - 6)}{(x^2 - 6x + 10)^2}$$

Evaluating the second derivative at $\bar{x} = 3$, $d^2y/dx^2 = 2 > 0$. The function is minimized.

3. Given $z = e^{(x^2 - 2x + y^2 - 6y)}$.

$$z_x = (2x - 2)e^{(x^2 - 2x + y^2 - 6y)} = 0 \qquad z_y = (2y - 6)e^{(x^2 - 2x + y^2 - 6y)} = 0$$

Since $e^{(x^2 - 2x + y^2 - 6x)} \neq 0$ for any value of x or y,

$$2x - 2 = 0 \qquad 2y - 6 = 0$$
$$\bar{x} = 1 \qquad\quad \bar{y} = 3$$

Testing the second-order conditions, using the product rule,

$$z_{xx} = (2x - 2)(2x - 2)e^{(x^2 - 2x + y^2 - 6y)} + e^{(x^2 - 2x + y^2 - 6y)}(2)$$
$$z_{yy} = (2y - 6)(2y - 6)e^{(x^2 - 2x + y^2 - 6y)} + e^{(x^2 - 2x + y^2 - 6y)}(2)$$

When evaluated at $\bar{x} = 1$, $\bar{y} = 3$,

$$z_{xx} = 0 + 2e^{-10} > 0 \qquad z_{yy} = 0 + 2e^{-10} > 0$$

since e to any power is positive. Then testing the mixed partials,

$$z_{xy} = (2x - 2)(2y - 6)e^{(x^2 - 2x + y^2 - 6y)} = z_{yx}$$

Evaluated at $\bar{x} = 1$, $\bar{y} = 3$, $z_{xy} = 0 = z_{yx}$. Thus, the function is at a minimum at $\bar{x} = 1$ and $\bar{y} = 3$ since z_{xx} and $z_{yy} > 0$ and $z_{xx}z_{yy} > (z_{xy})^2$.

9.5 LOGARITHMIC DIFFERENTIATION

The natural logarithm function and its derivative are frequently used to facilitate the differentiation of products and quotients involving multiple terms. The process is called *logarithmic differentiation* and is demonstrated in Example 8 and Problem 9.22.

EXAMPLE 8.   To find the derivative of a function such as

$$g(x) = \frac{(5x^3 - 8)(3x^4 + 7)}{(9x^5 - 2)} \tag{9.5}$$

use logarithmic differentiation as follows:

a)   Take the natural logarithm of both sides.

$$\ln g(x) = \ln \frac{(5x^3 - 8)(3x^4 + 7)}{(9x^5 - 2)}$$
$$= \ln (5x^3 - 8) + \ln (3x^4 + 7) - \ln (9x^5 - 2)$$

b)   Take the derivative of $\ln g(x)$.

$$\frac{d}{dx}[\ln g(x)] = \frac{g'(x)}{g(x)} = \frac{15x^2}{5x^3 - 8} + \frac{12x^3}{3x^4 + 7} - \frac{45x^4}{9x^5 - 2} \tag{9.6}$$

c)   Solve algebraically for $g'(x)$ in (9.6).

$$g'(x) = \left(\frac{15x^2}{5x^3 - 8} + \frac{12x^3}{3x^4 + 7} - \frac{45x^4}{9x^5 - 2} \right) \cdot g(x) \tag{9.7}$$

d)   Then substitute (9.5) for $g(x)$ in (9.7).

$$g'(x) = \left(\frac{15x^2}{5x^3 - 8} + \frac{12x^3}{3x^4 + 7} - \frac{45x^4}{9x^5 - 2} \right) \cdot \frac{(5x^3 - 8)(3x^4 + 7)}{(9x^5 - 2)}$$

9.6   ALTERNATIVE MEASURES OF GROWTH

Growth G of a function $y = f(t)$ is defined as

$$G = \frac{dy/dt}{y} = \frac{f'(t)}{f(t)} = \frac{y'}{y}$$

From Section 9.1.3 this is exactly equivalent to the derivative of $\ln y$. The growth of a function, therefore, can be measured (1) by dividing the derivative of the function by the function itself or (2) by taking the natural log of the function and then simply differentiating the natural log function. This latter method is sometimes helpful with more complicated functions. See Example 9 and Problems 9.23 to 9.30.

EXAMPLE 9.   Finding the growth rate of $V = Pe^{rt}$, where P is a constant, is illustrated below by using the two methods outlined above.

1.   By the first method,
$$G = \frac{V'}{V}$$

where $V' = Pe^{rt}(r) = rPe^{rt}$. Thus,

$$G = \frac{rPe^{rt}}{Pe^{rt}} = r$$

2.   For the second method, take the natural log of the function.

$$\ln V = \ln P + \ln e^{rt} = \ln P + rt$$

and then take the derivative of the natural log function with respect to t.

$$G = \frac{1}{V}\frac{dV}{dt} = \frac{d}{dt}(\ln V) = \frac{d}{dt}(\ln P + rt) = 0 + r = r$$

9.7   OPTIMAL TIMING

Exponential functions are used to express the value of goods that appreciate or depreciate over time. Such goods include wine, cheese, and land. Since a dollar in the future is worth less than a dollar today, its future value must be discounted to a present value. Investors and speculators seek to maximize the present value of their assets, as is illustrated in Example 10 and Problems 9.31 to 9.34.

EXAMPLE 10.   The value of cheese that improves with age is given by $V = 1400(1.25)^{\sqrt{t}}$. If the cost of capital under continuous compounding is 9 percent a year and there is no storage cost for aging the cheese in company caves, how long should the company store the cheese?

The company wants to maximize the present value of the cheese: $P = Ve^{-rt}$. Substituting the given values of V and r, $P = 1400(1.25)^{\sqrt{t}}e^{-0.09t}$. Taking the natural log,

$$\ln P = \ln 1400 + t^{1/2}\ln 1.25 - 0.09t$$

Then taking the derivative and setting it equal to zero to maximize P,

$$\frac{1}{P}\frac{dP}{dt} = 0 + \frac{1}{2}(\ln 1.25)t^{-1/2} - 0.09 = 0$$

$$\frac{dP}{dt} = P\left[\frac{1}{2}(\ln 1.25)t^{-1/2} - 0.09\right] = 0 \qquad (9.8)$$

Since $P \neq 0$,

$$\tfrac{1}{2}(\ln 1.25)t^{-1/2} - 0.09 = 0$$

$$t^{-1/2} = \frac{0.18}{\ln 1.25}$$

$$t = \left(\frac{\ln 1.25}{0.18}\right)^2 = \left(\frac{0.22314}{0.18}\right)^2 \approx 1.54 \text{ years}$$

Using the product rule when taking the second derivative from (9.8), because $P = f(t)$, we get

$$\frac{d^2P}{dt^2} = P\left[-\frac{1}{4}(\ln 1.25)t^{-3/2}\right] + \left[\frac{1}{2}(\ln 1.25)t^{-1/2} - 0.09\right]\frac{dP}{dt}$$

Since $dP/dt = 0$ at the critical point,

$$\frac{d^2P}{dt^2} = P\left[-\frac{1}{4}(\ln 1.25)t^{-3/2}\right] = -P(0.05579t^{-3/2})$$

With $P, t > 0$, $d^2P/dt^2 < 0$ and the function is at a maximum.

9.8   DERIVATION OF A COBB-DOUGLAS DEMAND FUNCTION USING A LOGARITHMIC TRANSFORMATION

A demand function expresses the amount of a good a consumer will purchase as a function of commodity prices and consumer income. A Cobb-Douglas demand function is derived by maximizing a Cobb-Douglas utility function subject to the consumer's income. Given $u = x^{\alpha}y^{\beta}$ and the budget constraint $p_x x + p_y y = M$, begin with a logarithmic transformation of the utility function

$$\ln u = \alpha \ln x + \beta \ln y$$

Then set up the Lagrangian function and maximize.

$$U = \alpha \ln x + \beta \ln y + \lambda(M - p_x x - p_y y)$$

$$U_x = \alpha \cdot \frac{1}{x} - \lambda p_x = 0 \qquad \alpha = \lambda p_x x$$

$$U_y = \beta \cdot \frac{1}{y} - \lambda p_y = 0 \qquad \beta = \lambda p_y y$$

$$U_\lambda = M - p_x x - p_y y = 0$$

Add $\alpha + \beta$ from U_x and U_y, recalling that $p_x x + p_y y = M$.

$$\alpha + \beta = \lambda(p_x x + p_y y) = \lambda M$$

Thus, $$\lambda = \frac{\alpha + \beta}{M}$$

Now substitute $\lambda = (\alpha + \beta)/M$ back in U_x and U_y to get

$$\frac{\alpha}{x} - \left(\frac{\alpha + \beta}{M}\right)p_x = 0 \qquad \bar{x} = \left(\frac{\alpha}{\alpha + \beta}\right)\left(\frac{M}{p_x}\right) \tag{9.9a}$$

$$\frac{\beta}{y} - \left(\frac{\alpha + \beta}{M}\right)p_y = 0 \qquad \bar{y} = \left(\frac{\beta}{\alpha + \beta}\right)\left(\frac{M}{p_y}\right) \tag{9.9b}$$

For a strict Cobb-Douglas function where $\alpha + \beta = 1$,

$$\bar{x} = \frac{\alpha M}{p_x} \quad \text{and} \quad \bar{y} = \frac{\beta M}{p_y} \tag{9.9c}$$

EXAMPLE 11. Given the utility function $u = x^{0.3} y^{0.7}$ and the income constraint $M = 200$, from the information derived in Section 9.8, the demand functions for x and y are (a) derived and (b) evaluated at $p_x = 5$, $p_y = 8$ and $p_x = 6$, $p_y = 10$, as follows:

a)   From (9.9c),
$$\bar{x} = \frac{\alpha M}{p_x} \quad \text{and} \quad \bar{y} = \frac{\beta M}{p_y}$$

b)   At $p_x = 5$, $p_y = 8$,
$$\bar{x} = \frac{0.3(200)}{5} = 12 \quad \text{and} \quad \bar{y} = \frac{0.7(200)}{8} = 17.5$$

At $p_x = 6$, $p_y = 10$,
$$\bar{x} = \frac{0.3(200)}{6} = 10 \quad \text{and} \quad \bar{y} = \frac{0.7(200)}{10} = 14$$

Solved Problems

DERIVATIVES OF NATURAL EXPONENTIAL FUNCTIONS

9.1. Differentiate each of the following natural exponential functions according to the rule $d/dx[e^{g(x)}] = e^{g(x)} \cdot g'(x)$:

a) $y = e^{2x}$

Letting $g(x) = 2x$, then $g'(x) = 2$,
and $y' = e^{2x}(2) = 2e^{2x}$

b) $y = e^{(-1/3)x}$

$g(x) = -\frac{1}{3}x$, $g'(x) = -\frac{1}{3}$, and
$y' = e^{-(1/3)x}(-\frac{1}{3}) = -\frac{1}{3}e^{(-1/3)x}$

c) $y = e^{x^3}$

$$y' = e^{x^3}(3x^2) = 3x^2 e^{x^3}$$

d) $y = 3e^{x^2}$

$$y' = 3e^{x^2}(2x) = 6xe^{x^2}$$

e) $y = e^{2x+1}$

$$y' = e^{2x+1}(2) = 2e^{x+1}$$

f) $y = e^{1-4x}$

$$y' = -4e^{1-4x}$$

g) $y = 5e^{1-x^2}$

$$y' = -10xe^{1-x^2}$$

h) $y = 2xe^x$

By the product rule,
$$y' = 2x(e^x) + e^x(2) = 2e^x(x+1)$$

i) $y = 3xe^{2x}$

$$y' = 3x(2e^{2x}) + e^{2x}(3) = 3e^{2x}(2x+1)$$

j) $y = x^2 e^{5x}$

$$y' = x^2(5e^{5x}) + e^{5x}(2x) = xe^{5x}(5x+2)$$

k) $y = \dfrac{e^{5x} - 1}{e^{5x} + 1}$

By the quotient rule,

$$y' = \frac{(e^{5x} + 1)(5e^{5x}) - (e^{5x} - 1)(5e^{5x})}{(e^{5x} + 1)^2} = \frac{10e^{5x}}{(e^{5x} + 1)^2}$$

l) $y = \dfrac{e^{2x} + 1}{e^{2x} - 1}$

$$y' = \frac{(e^{2x} - 1)(2e^{2x}) - (e^{2x} + 1)(2e^{2x})}{(e^{2x} - 1)^2} = \frac{-4e^{2x}}{(e^{2x} - 1)^2}$$

DIFFERENTIATION OF EXPONENTIAL FUNCTIONS WITH BASES OTHER THAN e

9.2.   Differentiate each of the following exponential functions according to the rule
$d/dx[a^{g(x)}] = a^{g(x)} \cdot g'(x) \cdot \ln a$:

a) $y = a^{2x}$

Letting $g(x) = 2x$, then $g'(x) = 2$, and
$$y' = a^{2x}(2)\ln a = 2a^{2x}\ln a$$

b) $y = a^{5x^2}$

$$y' = a^{5x^2}(10x)\ln a = 10xa^{5x^2}\ln a$$

c) $y = 4^{2x+7}$

$$y' = 4^{2x+7}(2)\ln 4 = 2(4)^{2x+7}\ln 4$$

Using a calculator, $y' = 2(1.38629)(4)^{2x+7} = 2.77258(4)^{2x+7}$

d) $y = 2^x$

$$y' = 2^x(1)\ln 2 = 2^x\ln 2 = 0.69315(2)^x$$

e) $y = 7^{x^2}$

$$y' = 7^{x^2}(2x)\ln 7 = 2x(7)^{x^2}\ln 7 = 2x(7)^{x^2}(1.94591) = 3.89182x(7)^{x^2}$$

f) $y = x^3 2^x$

By the product rule, recalling that x^3 is a power function and 2^x is an exponential function,
$$y' = x^3[2^x(1)\ln 2] + 2^x(3x^2) = x^2 2^x(x\ln 2 + 3)$$

g) $y = x^2 2^{5x}$

$$y' = x^2[2^{5x}(5)\ln 2] + 2^{5x}(2x) = x2^{5x}(5x\ln 2 + 2)$$

DERIVATIVES OF NATURAL LOGARITHMIC FUNCTIONS

9.3. Differentiate each of the following natural log functions according to the rule $d/dx[\ln g(x)] = 1/[g(x)] \cdot g'(x)$:

a) $y = \ln 2x^3$

   Let $g(x) = 2x^3$, then $g'(x) = 6x^2$, and

   $$y' = \frac{1}{2x^3}(6x^2) = \frac{3}{x}$$

b) $y = \ln 7x^2$

   $$y' = \frac{1}{7x^2}(14x) = \frac{2}{x}$$

c) $y = \ln(1 + x)$

   $$y' = \frac{1}{1 + x}$$

d) $y = \ln(4x + 7)$

   $$y' = \frac{1}{4x + 7}(4) = \frac{4}{4x + 7}$$

e) $y = \ln 6x$

   $$y' = \frac{1}{6x}(6) = \frac{1}{x}$$

f) $y = 6\ln x$

   $$y' = 6\left(\frac{1}{x}\right) = \frac{6}{x}$$

   Notice how a multiplicative constant within the log expression in part (*e*) drops out in differentiation, whereas a multiplicative constant outside the log expression in part (*f*) remains.

9.4. Redo Problem 9.3 for each of the following functions:

a) $y = \ln^2 x = (\ln x)^2$

   By the generalized power function rule,

   $$y' = 2\ln x \frac{d}{dx}(\ln x) = 2\ln x\left(\frac{1}{x}\right) = \frac{2\ln x}{x}$$

b) $y = \ln^2 8x = (\ln 8x)^2$

   $$y' = 2\ln 8x\left(\frac{1}{8x}\right)(8) = \frac{2\ln 8x}{x}$$

c) $y = \ln^2(3x + 1) = [\ln(3x + 1)]^2$

   $$y' = 2\ln(3x + 1)\left(\frac{1}{3x + 1}\right)(3) = \frac{6\ln(3x + 1)}{3x + 1}$$

d) $y = \ln^2(5x + 6)$

   $$y' = 2\ln(5x + 6)\left(\frac{1}{5x + 6}\right)(5) = \frac{10\ln(5x + 6)}{5x + 6}$$

e) $y = \ln^3(4x + 13)$

   $$y' = 3[\ln(4x + 13)]^2\left(\frac{1}{4x + 13}\right)(4) = 3\ln^2(4x + 13)\left(\frac{4}{4x + 13}\right) = \frac{12}{4x + 13}\ln^2(4x + 13)$$

f) $y = \ln(x + 5)^2 \neq [\ln(x + 5)]^2$

   Letting $g(x) = (x + 5)^2$, then $g'(x) = 2(x + 5)$, and

   $$y' = \frac{1}{(x + 5)^2}[2(x + 5)] = \frac{2}{x + 5}$$

g) $y = \ln(x - 8)^2$

$$y' = \frac{1}{(x-8)^2}[2(x-8)] = \frac{2}{x-8}$$

h) $y = 3\ln(1 + x)^2$

$$y' = 3\left[\frac{1}{(1+x)^2}\right][2(1+x)] = \frac{6}{1+x}$$

9.5. Use the laws of logarithms in Section 7.3 to simplify the differentiation of each of the following natural log functions:

a) $y = \ln(x + 5)^2$

From the rules for logs, $\ln(x + 5)^2 = 2\ln(x + 5)$. Thus, as in Problem 9.4(f),

$$y' = 2\left(\frac{1}{x+5}\right)(1) = \frac{2}{x+5}$$

b) $y = \ln(2x + 7)^2$

$$y = 2\ln(2x + 7)$$

$$y' = 2\left(\frac{1}{2x+7}\right)(2) = \frac{4}{2x+7}$$

c) $y = \ln[(3x + 7)(4x + 2)]$

$$y = \ln(3x + 7) + \ln(4x + 2)$$

$$y' = \frac{3}{3x+7} + \frac{4}{4x+2}$$

d) $y = \ln[5x^2(3x^3 - 7)]$

$$y = \ln 5x^2 + \ln(3x^3 - 7)$$

$$y' = \frac{10x}{5x^2} + \frac{9x^2}{3x^3 - 7}$$

$$= \frac{2}{x} + \frac{9x^2}{3x^3 - 7}$$

e) $y = \ln\dfrac{3x^2}{x^2 - 1}$

$$y = \ln 3x^2 - \ln(x^2 - 1)$$

$$y' = \frac{2}{x} - \frac{2x}{x^2 - 1}$$

f) $y = \ln\dfrac{x^3}{(2x + 5)^2}$

$$y = \ln x^3 - \ln(2x + 5)^2$$

$$y' = \frac{1}{x^3}(3x^2) - 2\left(\frac{1}{2x+5}\right)(2)$$

$$= \frac{3}{x} - \frac{4}{2x+5}$$

g) $y = \ln\sqrt{\dfrac{2x^2 + 3}{x^2 + 9}}$

$$y = \tfrac{1}{2}[\ln(2x^2 + 3) - \ln(x^2 + 9)]$$

$$y' = \frac{1}{2}\left(\frac{4x}{2x^2 + 3} - \frac{2x}{x^2 + 9}\right)$$

$$= \frac{2x}{2x^2 + 3} - \frac{x}{x^2 + 9}$$

DERIVATIVES OF LOGARITHMIC FUNCTIONS WITH BASES OTHER THAN e

9.6. Differentiate each of the following logarithmic functions according to the rule

$$\frac{d}{dx}[\log_a g(x)] = \frac{1}{g(x)} \cdot g'(x) \cdot \log_a e = \frac{1}{g(x)} \cdot g'(x) \cdot \frac{1}{\ln a}$$

a) $y = \log_a(4x^2 - 3)$

$$y' = \frac{1}{4x^2 - 3}(8x)\left(\frac{1}{\ln a}\right)$$

$$= \frac{8x}{(4x^2 - 3)\ln a}$$

b) $y = \log_4 9x^3$

$$y' = \frac{1}{9x^3}(27x^2)\left(\frac{1}{\ln 4}\right)$$

$$= \frac{3}{x\ln 4}$$

c) $y = \log_2 (8 - x)$

$$y' = \frac{1}{(8-x)}(-1)\left(\frac{1}{\ln 2}\right)$$

$$= -\frac{1}{(8-x)\ln 2}$$

d) $y = x^3 \log_6 x$

By the product rule,

$$y' = x^3\left[\frac{1}{x}(1)\left(\frac{1}{\ln 6}\right)\right] + \log_6 x(3x^2)$$

$$= \frac{x^2}{\ln 6} + 3x^2 \log_6 x$$

e) $y = \log_a \sqrt{x^2 - 7}$

From the law of logs, $y = \frac{1}{2}\log_a (x^2 - 7)$. Thus,

$$y' = \frac{1}{2}\left[\frac{1}{x^2 - 7}(2x)\left(\frac{1}{\ln a}\right)\right] = \frac{x}{(x^2 - 7)\ln a}$$

COMBINATION OF RULES

9.7.   Use whatever combination of rules are necessary to differentiate the following functions:

a) $y = x^2 \ln x^3$

By the product rule,

$$y' = x^2\left(\frac{1}{x^3}\right)(3x^2) + \ln x^3(2x) = 3x + 2x\ln x^3 = 3x + 6x\ln x = 3x(1 + 2\ln x)$$

b) $y = x^3 \ln x^2$

$$y' = x^3\left(\frac{1}{x^2}\right)(2x) + \ln x^2(3x^2) = 2x^2 + 6x^2\ln x = 2x^2(1 + 3\ln x)$$

c) $y = e^x \ln x$

By the product rule,

$$y' = e^x\left(\frac{1}{x}\right) + (\ln x)(e^x) = e^x\left(\frac{1}{x} + \ln x\right)$$

d) $y = e^{-2x} \ln 2x$

$$y' = e^{-2x}\left(\frac{1}{2x}\right)(2) + \ln 2x(-2e^{-2x}) = e^{-2x}\left(\frac{1}{x} - 2\ln 2x\right)$$

e) $y = \ln e^{3x+2}$

$$y' = \frac{1}{e^{3x+2}}(3e^{3x+2}) = 3$$

since $\ln e^{3x+2} = 3x + 2$, $d/dx(\ln e^{3x+2}) = d/dx(3x + 2) = 3$.

f) $y = e^{\ln x}$

$$y' = e^{\ln x}\left(\frac{1}{x}\right) = x\left(\frac{1}{x}\right) = 1$$

since $e^{\ln x} = x$.

g) $y = e^{\ln (2x+1)}$

$$y' = e^{\ln (2x+1)}\left(\frac{1}{2x + 1}\right)(2) = 2$$

since $e^{\ln (2x+1)} = 2x + 1$.

h) $y = e^{x \ln x}$

$$y' = e^{x \ln x} \frac{d}{dx}(x \ln x)$$

Then using the product rule for $d/dx(x \ln x)$,

$$y' = e^{x \ln x} \left[x\left(\frac{1}{x}\right) + (\ln x)(1) \right] = e^{x \ln x}(1 + \ln x)$$

i) $y = e^{x^2 \ln 3x}$

$$y' = e^{x^2 \ln 3x} \left[x^2\left(\frac{1}{3x}\right)(3) + \ln 3x(2x) \right]$$
$$= e^{x^2 \ln 3x}(x + 2x \ln 3x) = xe^{x^2 \ln 3x}(1 + 2 \ln 3x)$$

SLOPES OF EXPONENTIAL AND LOGARITHMIC FUNCTIONS

9.8.   Evaluate the slope of each of the following functions at the point indicated:

a) $y = 3e^{0.2x}$ at $x = 5$.

$$y' = 0.6e^{0.2x}$$

At $x = 5$, $y' = 0.6e^{0.2(5)} = 0.6(2.71828) \approx 1.63097$.

b) $y = 2e^{-1.5x}$ at $x = 4$.

$$y' = -3e^{-1.5x}$$

At $x = 4$, $y' = -3e^{-1.5(4)} = -3e^{-6} = -3(0.00248) \approx -0.00744$.

c) $y = \ln(x^2 + 8x + 4)$ at $x = 2$.

$$y' = \frac{2x + 8}{x^2 + 8x + 4}$$

At $x = 2$, $y' = \frac{12}{24} = 0.5$.

d) $y = \ln^2(x + 4)$ at $x = 6$.

$$y' = [2 \ln(x+4)]\left(\frac{1}{x+4}\right)(1) = \frac{2 \ln(x+4)}{x+4}$$

At $x = 6$,

$$y' = \frac{2 \ln 10}{10} = \frac{2(2.30259)}{10} \approx 0.46052$$

SECOND DERIVATIVES

9.9.   Find the first and second derivatives of the following functions:

a) $y = e^{3x}$

$$y' = 3e^{3x}$$
$$y'' = 9e^{3x}$$

b) $y = e^{-(1/2)x}$

$$y' = -\tfrac{1}{2}e^{-(1/2)x}$$
$$y'' = \tfrac{1}{4}e^{-(1/2)x}$$

c) $y = 3e^{5x+1}$

$$y' = 15e^{5x+1}$$
$$y'' = 75e^{5x+1}$$

d) $y = 2xe^x$

By the product rule,

$$y' = 2x(e^x) + e^x(2) = 2e^x(x + 1)$$
$$y'' = 2e^x(1) + (x + 1)(2e^x) = 2e^x(x + 2)$$

e) $y = \ln 2x^5$

$$y' = \frac{1}{2x^5}(10x^4) = \frac{5}{x} = 5x^{-1}$$

$$y'' = -5x^{-2} = \frac{-5}{x^2}$$

f) $y = 4\ln x$

$$y' = 4\left(\frac{1}{x}\right)(1) = \frac{4}{x} = 4x^{-1}$$

$$y'' = -4x^{-2}$$

9.10. Take the first and second derivatives of each of the following functions:

a) $y = a^{3x}$

$$y' = a^{3x}(3)\ln a = 3a^{3x}\ln a$$

where $\ln a = $ a constant. Thus,

$$y'' = (3a^{3x}\ln a)(3)\ln a = 9a^{3x}(\ln a)^2$$

b) $y = a^{5x+1}$

$$y' = a^{5x+1}(5)\ln a = 5a^{5x+1}\ln a$$
$$y'' = (5a^{5x+1}\ln a)(5)\ln a = 25a^{5x+1}(\ln a)^2$$

c) $y = \log_a 5x$

$$y' = \frac{1}{5x}(5)\frac{1}{\ln a} = \frac{1}{x\ln a} = (x\ln a)^{-1}$$

Using the generalized power function rule,

$$y'' = -1(x\ln a)^{-2}\ln a = \frac{-\ln a}{x^2\ln^2 a} = -\frac{1}{x^2\ln a}$$

d) $y = \log_3 6x$

$$y' = \frac{1}{6x}(6)\left(\frac{1}{\ln 3}\right) = \frac{1}{x\ln 3} = (x\ln 3)^{-1}$$

$$y'' = -1(x\ln 3)^{-2}(\ln 3) = \frac{-\ln 3}{x^2\ln^2 3} = -\frac{1}{x^2\ln 3}$$

e) $y = 3xe^x$

By the product rule,

$$y' = 3x(e^x) + e^x(3) = 3e^x(x+1)$$
$$y'' = 3e^x(1) + (x+1)(3e^x) = 3e^x(x+2)$$

f) $y = \dfrac{4x}{3\ln x}$

By the quotient rule,

$$y' = \frac{(3\ln x)(4) - 4x(3)(1/x)}{9\ln^2 x} = \frac{12\ln x - 12}{9\ln^2 x} = \frac{12(\ln x - 1)}{9\ln^2 x}$$

$$y'' = \frac{(9\ln^2 x)[12(1/x)] - 12(\ln x - 1)\{[9(2)\ln x](1/x)\}}{81\ln^4 x} = \frac{(108/x)(\ln^2 x) - (216/x)(\ln x - 1)(\ln x)}{81\ln^4 x}$$

$$= \frac{4\ln x - 8(\ln x - 1)}{3x\ln^3 x} = \frac{-4\ln x + 8}{3x\ln^3 x} = \frac{4(2 - \ln x)}{3x\ln^3 x}$$

PARTIAL DERIVATIVES

9.11. Find all the first and second partial derivatives for each of the following functions:

a) $z = e^{x^2+y^2}$

$$z_x = 2xe^{x^2+y^2} \qquad z_y = 2ye^{x^2+y^2}$$

By the product rule,

$$z_{xx} = 2x(2xe^{x^2+y^2}) + e^{x^2+y^2}(2) \qquad z_{yy} = 2y(2ye^{x^2+y^2}) + e^{x^2+y^2}(2)$$
$$= 2e^{x^2+y^2}(2x^2 + 1) \qquad\qquad = 2e^{x^2+y^2}(2y^2 + 1)$$
$$z_{xy} = 4xye^{x^2+y^2} = z_{yx}$$

b) $z = e^{2x^2+3y}$

$$z_x = 4xe^{2x^2+3y} \qquad\qquad z_y = 3e^{2x^2+3y}$$
$$z_{xx} = 4x(4xe^{2x^2+3y}) + e^{2x^2+3y}(4) \qquad z_{yy} = 9e^{2x^2+3y}$$
$$= 4e^{2x^2+3y}(4x^2 + 1)$$
$$z_{xy} = 12xe^{2x^2+3y} = z_{yx}$$

c) $z = a^{2x+3y}$

$$z_x = a^{2x+3y}(2) \ln a \qquad\qquad z_y = a^{2x+3y}(3) \ln a$$
$$= 2a^{2x+3y} \ln a \qquad\qquad = 3a^{2x+3y} \ln a$$
$$z_{xx} = 2a^{2x+3y}(\ln a)(2)(\ln a) \qquad z_{yy} = 3a^{2x+3y}(\ln a)(3)(\ln a)$$
$$= 4a^{2x+3y} \ln^2 a \qquad\qquad = 9a^{2x+3y} \ln^2 a$$
$$z_{xy} = 6a^{2x+3y} \ln^2 a = z_{yx}$$

d) $z = 4^{3x+5y}$

$$z_x = 4^{3x+5y}(3) \ln 4 \qquad\qquad z_y = 4^{3x+5y}(5) \ln 4$$
$$= 3(4)^{3x+5y} \ln 4 \qquad\qquad = 5(4)^{3x+5y} \ln 4$$
$$z_{xx} = 3(4)^{3x+5y}(\ln 4)(3)(\ln 4) \qquad z_{yy} = 5(4)^{3x+5y}(\ln 4)(5)(\ln 4)$$
$$= 9(4)^{3x+5y} \ln^2 4 \qquad\qquad = 25(4)^{3x+5y} \ln^2 4$$
$$z_{xy} = 15(4)^{3x+5y} \ln^2 4 = z_{yx}$$

e) $z = \ln (7x + 2y)$

$$z_x = \frac{7}{7x + 2y} \qquad z_y = \frac{2}{7x + 2y}$$

By the quotient rule,

$$z_{xx} = \frac{(7x + 2y)(0) - 7(7)}{(7x + 2y)^2} = \frac{-49}{(7x + 2y)^2} \qquad z_{yy} = \frac{(7x + 2y)(0) - 2(2)}{(7x + 2y)^2} = \frac{-4}{(7x + 2y)^2}$$

$$z_{xy} = \frac{-14}{(7x + 2y)^2} = z_{yx}$$

f) $z = \ln (x^2 + 4y^2)$

$$z_x = \frac{2x}{x^2 + 4y^2} \qquad\qquad z_y = \frac{8y}{x^2 + 4y^2}$$

$$z_{xx} = \frac{(x^2 + 4y^2)(2) - 2x(2x)}{(x^2 + 4y^2)^2} = \frac{8y^2 - 2x^2}{(x^2 + 4y^2)^2} \qquad z_{yy} = \frac{(x^2 + 4y^2)(8) - 8y(8y)}{(x^2 + 4y^2)^2} = \frac{8x^2 - 32y^2}{(x^2 + 4y^2)^2}$$

$$z_{xy} = \frac{-16xy}{(x^2 + 4y^2)^2} = z_{yx}$$

g) $z = \log_a (x - 2y)$

$$z_x = \frac{1}{(x - 2y) \ln a} \qquad\qquad z_y = \frac{-2}{(x - 2y) \ln a}$$

$$z_{xx} = \frac{-1 \ln a}{(x - 2y)^2 \ln^2 a} = \frac{-1}{(x - 2y)^2 \ln a} \qquad z_{yy} = \frac{4 \ln a}{(x - 2y)^2 \ln^2 a} = \frac{4}{(x - 2y)^2 \ln a}$$

$$z_{xy} = \frac{2}{(x - 2y)^2 \ln a} = z_{yx}$$

h) $z = \log_a (3x^2 + y^2)$

$$z_x = \frac{6x}{(3x^2 + y^2) \ln a} \qquad\qquad z_y = \frac{2y}{(3x^2 + y^2) \ln a}$$

$$z_{xx} = \frac{(3x^2 + y^2)(\ln a)(6) - 6x(6x \ln a)}{(3x^2 + y^2)^2 \ln^2 a} \qquad z_{yy} = \frac{(3x^2 + y^2)(\ln a)(2) - 2y(2y \ln a)}{(3x^2 + y^2)^2 \ln^2 a}$$

$$= \frac{6y^2 - 18x^2}{(3x^2 + y^2)^2 \ln a} \qquad\qquad = \frac{6x^2 - 2y^2}{(3x^2 + y^2)^2 \ln a}$$

$$z_{xy} = \frac{-12xy}{(3x^2 + y^2)^2 \ln a} = z_{yx}$$

OPTIMIZATION OF EXPONENTIAL AND LOGARITHMIC FUNCTIONS

9.12. Given $y = 4xe^{3x}$, (a) find the critical values and (b) determine whether the function is maximized or minimized.

a) By the product rule,

$$y' = 4x(3e^{3x}) + e^{3x}(4) = 0$$
$$4e^{3x}(3x + 1) = 0$$

Since there is no value of x for which $4e^{3x} = 0$, or for which $e^x = 0$,

$$3x + 1 = 0 \qquad \bar{x} = -\tfrac{1}{3}$$

b) $$y'' = 4e^{3x}(3) + (3x + 1)(12e^{3x}) = 12e^{3x}(3x + 2)$$

At $\bar{x} = -\tfrac{1}{3}$, $y'' = 12e^{-1}(1)$. And $y'' = 12(0.36788) > 0$. The function is minimized.

9.13. Redo Problem 9.12, given $y = 5xe^{-0.2x}$

a) $$y' = 5x(-0.2e^{-0.2x}) + e^{-0.2x}(5) = 0$$
$$5e^{-0.2x}(1 - 0.2x) = 0$$

Since $5e^{-0.2x} \neq 0$, $(1 - 0.2x) = 0 \qquad \bar{x} = 5$

b) $$y'' = 5e^{-0.2x}(-0.2) + (1 - 0.2x)(-1e^{-0.2x}) = e^{-0.2x}(0.2x - 2)$$

At $\bar{x} = 5$, $y'' = e^{-1}(1 - 2)$. And $y'' = (0.36788)(-1) < 0$. The function is at a maximum.

9.14. Redo Problem 9.12, given $y = \ln (x^2 - 8x + 20)$.

a) $$y' = \frac{2x - 8}{x^2 - 8x + 20} = 0$$

Multiplying both sides by $x^2 - 8x + 20$ gives $2x - 8 = 0$ and $\bar{x} = 4$.

b) $$y'' = \frac{(x^2 - 8x + 20)(2) - (2x - 8)(2x - 8)}{(x^2 - 8x + 20)^2}$$

At $\bar{x} = 4$, $y'' = \tfrac{8}{16} > 0$. The function is at a minimum.

9.15. Redo Problem 9.12, given $y = \ln(2x^2 - 20x + 5)$.

a)
$$y' = \frac{4x - 20}{2x^2 - 20x + 5} = 0$$
$$4x - 20 = 0 \qquad \bar{x} = 5$$

b)
$$y'' = \frac{(2x^2 - 20x + 5)(4) - (4x - 20)(4x - 20)}{(2x^2 - 20x + 5)^2}$$

At $\bar{x} = 5$, $y'' = -180/2025 < 0$. The function is at a maximum.

9.16. Given the function $z = \ln(2x^2 - 12x + y^2 - 10y)$, (a) find the critical values and (b) indicate whether the function is at a maximum or minimum.

a)
$$z_x = \frac{4x - 12}{2x^2 - 12x + y^2 - 10y} = 0 \qquad z_y = \frac{2y - 10}{2x^2 - 12x + y^2 - 10y} = 0$$
$$4x - 12 = 0 \qquad \bar{x} = 3 \qquad 2y - 10 = 0 \qquad \bar{y} = 5$$

b)
$$z_{xx} = \frac{(2x^2 - 12x + y^2 - 10y)(4) - (4x - 12)(4x - 12)}{(2x^2 - 12x + y^2 - 10y)^2}$$
$$z_{yy} = \frac{(2x^2 - 12x + y^2 - 10y)(2) - (2y - 10)(2y - 10)}{(2x^2 - 12x + y^2 - 10y)^2}$$

Evaluated at $\bar{x} = 3$, $\bar{y} = 5$,

$$z_{xx} = \frac{(-43)(4) - 0}{(-43)^2} = \frac{-172}{1849} < 0 \qquad z_{yy} = \frac{(-43)(2) - 0}{(-43)^2} = \frac{-86}{1849} < 0$$

$$z_{xy} = \frac{-(4x - 12)(2y - 10)}{(2x^2 - 12x + y^2 - 10y)^2} = z_{yx}$$

At $\bar{x} = 3$, $\bar{y} = 5$, $z_{xy} = 0 = z_{yx}$. With $z_{xx}, z_{yy} < 0$ and $z_{xx}z_{yy} > (z_{xy})^2$, the function is at a maximum.

9.17. Redo Problem 9.16, given $z = \ln(x^2 - 4x + 3y^2 - 6y)$.

a)
$$z_x = \frac{2x - 4}{x^2 - 4x + 3y^2 - 6y} = 0 \qquad z_y = \frac{6y - 6}{x^2 - 4x + 3y^2 - 6y} = 0$$
$$2x - 4 = 0 \qquad \bar{x} = 2 \qquad 6y - 6 = 0 \qquad \bar{y} = 1$$

b)
$$z_{xx} = \frac{(x^2 - 4x + 3y^2 - 6y)(2) - (2x - 4)(2x - 4)}{(x^2 - 4x + 3y^2 - 6y)^2}$$
$$z_{yy} = \frac{(x^2 - 4x + 3y^2 - 6y)(6) - (6y - 6)(6y - 6)}{(x^2 - 4x + 3y^2 - 6y)^2}$$

At $\bar{x} = 2$, $\bar{y} = 1$,

$$z_{xx} = \frac{(-7)(2) - 0}{(-7)^2} = -\frac{14}{49} < 0 \qquad z_{yy} = \frac{(-7)(6) - 0}{(-7)^2} = -\frac{42}{49} < 0$$

$$z_{xy} = \frac{-(2x - 4)(6y - 6)}{(x^2 - 4x + 3y^2 - 6y)^2} = z_{yx}$$

At $\bar{x} = 2$, $\bar{y} = 1$, $z_{xy} = 0 = z_{yx}$. With $z_{xx}, z_{yy} < 0$ and $z_{xx}z_{yy} > (z_{xy})^2$, the function is at a maximum.

9.18. Redo Problem 9.16, given $z = e^{(3x^2 - 6x + y^2 - 8y)}$.

a)
$$z_x = (6x - 6)e^{(3x^2 - 6x + y^2 - 8y)} = 0 \qquad z_y = (2y - 8)e^{(3x^2 - 6x + y^2 - 8y)} = 0$$
$$6x - 6 = 0 \qquad \bar{x} = 1 \qquad 2y - 8 = 0 \qquad \bar{y} = 4$$

b) Using the product rule,

$$z_{xx} = (6x - 6)(6x - 6)e^{(3x^2 - 6x + y^2 - 8y)} + e^{(3x^2 - 6x + y^2 - 8y)}(6)$$
$$z_{yy} = (2y - 8)(2y - 8)e^{(3x^2 - 6x + y^2 - 8y)} + e^{(3x^2 - 6x + y^2 - 8y)}(2)$$

Evaluated at $\bar{x} = 1, \bar{y} = 4$,

$$z_{xx} = 0 + 6e^{-19} > 0 \qquad z_{yy} = 0 + 2e^{-19} > 0$$

Then testing the cross partials,

$$z_{xy} = (6x - 6)(2y - 8)e^{(3x^2 - 6x + y^2 - 8y)} = z_{yx}$$

At $\bar{x} = 1, \bar{y} = 4, z_{xy} = 0 = z_{yx}$. The function is at a minimum since $z_{xx}, z_{yy} > 0$ and $z_{xx}z_{yy} > (z_{xy})^2$.

9.19. Given $z = e^{(2x^2 - 12x - 2xy + y^2 - 4y)}$, redo Problem 9.16.

a)
$$z_x = (4x - 12 - 2y)e^{(2x^2 - 12x - 2xy + y^2 - 4y)} = 0$$
$$4x - 2y - 12 = 0 \qquad\qquad (9.10)$$
$$z_y = (-2x + 2y - 4)e^{(2x^2 - 12x - 2xy + y^2 - 4y)} = 0$$
$$-2x + 2y - 4 = 0 \qquad\qquad (9.11)$$

Solving (9.10) and (9.11) simultaneously, $\bar{x} = 8, \bar{y} = 10$.

b)
$$z_{xx} = (4x - 12 - 2y)(4x - 12 - 2y)e^{(2x^2 - 12x - 2xy + y^2 - 4y)} + e^{(2x^2 - 12x - 2xy + y^2 - 4y)}(4)$$
$$z_{yy} = (-2x + 2y - 4)(-2x + 2y - 4)e^{(2x^2 - 12x - 2xy + y^2 - 4y)} + e^{(2x^2 - 12x - 2xy + y^2 - 4y)}(2)$$

Evaluated at $\bar{x} = 8, \bar{y} = 10$,

$$z_{xx} = 0 + 4e^{-68} > 0 \qquad z_{yy} = 0 + 2e^{-68} > 0$$

Testing the mixed partials, by the product rule,

$$z_{xy} = (4x - 12 - 2y)(-2x + 2y - 4)e^{(2x^2 - 12x - 2xy + y^2 - 4y)} + e^{(2x^2 - 12x - 2xy + y^2 - 4y)}(-2) = z_{yx}$$

Evaluated at $\bar{x} = 8, \bar{y} = 10, z_{xy} = 0 - 2e^{-68} = z_{yx}$. Since $z_{xx}, z_{yy} > 0$ and $z_{xx}z_{yy} > (z_{xy})^2$, the function is at a minimum.

9.20. Given the demand function

$$P = 8.25e^{-0.02Q} \qquad\qquad (9.12)$$

(a) Determine the quantity and price at which total revenue will be maximized and (b) test the second-order condition.

a)
$$\text{TR} = PQ = (8.25e^{-0.02Q})Q$$

By the product rule,

$$\frac{d\text{TR}}{dQ} = (8.25e^{-0.02Q})(1) + Q(-0.02)(8.25e^{-0.02Q}) = 0$$
$$(8.25e^{-0.02Q})(1 - 0.02Q) = 0$$

Since $(8.25e^{-0.02Q}) \neq 0$ for any value of Q, $1 - 0.02Q = 0$; $\bar{Q} = 50$.
Substituting $\bar{Q} = 50$ in (9.12), $P = 8.25e^{-0.02(50)} = 8.25e^{-1}$. And $P = 8.25(0.36788) = 3.04$.

b) By the product rule,

$$\frac{d^2\text{TR}}{dQ^2} = (8.25e^{-0.02Q})(-0.02) + (1 - 0.02Q)(-0.02)(8.25e^{-0.02Q}) = (-0.02)(8.25e^{-0.02Q})(2 - 0.02Q)$$

Evaluated at $\bar{Q} = 50$, $d^2\text{TR}/dQ^2 = (-0.02)(8.25e^{-1})(1) = -0.165(0.36788) < 0$. TR is at a maximum.

9.21. (*a*) Find the price and quantity that will maximize total revenue, given the demand function $P = 12.50e^{-0.005Q}$. (*b*) Check the second-order condition.

a)
$$\text{TR} = (12.50e^{-0.005Q})Q$$

$$\frac{d\text{TR}}{dQ} = (12.50e^{-0.005Q})(1) + Q(-0.005)(12.50e^{-0.005Q})$$

$$= (12.50e^{-0.005Q})(1 - 0.005Q) = 0$$

$$1 - 0.005Q = 0 \qquad \bar{Q} = 200$$

Thus,
$$P = 12.50e^{-0.005(200)} = 12.50e^{-1} = 12.50(0.36788) = 4.60$$

b)
$$\frac{d^2\text{TR}}{dQ^2} = (12.50e^{-0.005Q})(-0.005) + (1 - 0.005Q)(-0.005)(12.50e^{-0.005Q})$$

$$= (-0.005)(12.50e^{-0.005Q})(2 - 0.005Q)$$

Evaluated at $\bar{Q} = 200$, $d^2\text{TR}/dQ^2 = (-0.005)(12.50e^{-1})(1) = -0.0625(0.36788) < 0$. The function is maximized.

LOGARITHMIC DIFFERENTIATION

9.22. Use logarithmic differentiation to find the derivatives for the following functions:

a)
$$g(x) = (x^3 - 2)(x^2 - 3)(8x - 5) \tag{9.13}$$

1) Take the natural logarithm of both sides.
$$\ln g(x) = \ln (x^3 - 2) + \ln (x^2 - 3) + \ln (8x - 5)$$

2) Take the derivative of $\ln g(x)$.
$$\frac{d}{dx}[\ln g(x)] = \frac{g'(x)}{g(x)} = \frac{3x^2}{x^3 - 2} + \frac{2x}{x^2 - 3} + \frac{8}{8x - 5} \tag{9.14}$$

3) Solve algebraically for $g'(x)$ in (*9.14*).
$$g'(x) = \left(\frac{3x^2}{x^3 - 2} + \frac{2x}{x^2 - 3} + \frac{8}{8x - 5}\right) \cdot g(x) \tag{9.15}$$

4) Then substitute (*9.13*) for $g(x)$ in (*9.15*).
$$g'(x) = \left(\frac{3x^2}{x^3 - 2} + \frac{2x}{x^2 - 3} + \frac{8}{8x - 5}\right)[(x^3 - 2)(x^2 - 3)(8x - 5)]$$

b)
$$g(x) = (x^4 + 7)(x^5 + 6)(x^3 + 2) \tag{9.16}$$

1)
$$\ln g(x) = \ln (x^4 + 7) + \ln (x^5 + 6) + \ln (x^3 + 2)$$

2)
$$\frac{d}{dx}[\ln g(x)] = \frac{g'(x)}{g(x)} = \frac{4x^3}{x^4 + 7} + \frac{5x^4}{x^5 + 6} + \frac{3x^2}{x^3 + 2}$$

3)
$$g'(x) = \left(\frac{4x^3}{x^4 + 7} + \frac{5x^4}{x^5 + 6} + \frac{3x^2}{x^3 + 2}\right) \cdot g(x) \tag{9.17}$$

4) Finally, substituting (*9.16*) for $g(x)$ in (*9.17*),
$$g'(x) = \left(\frac{4x^3}{x^4 + 7} + \frac{5x^4}{x^5 + 6} + \frac{3x^2}{x^3 + 2}\right) \cdot (x^4 + 7)(x^5 + 6)(x^3 + 2)$$

c)
$$g(x) = \frac{(3x^5 - 4)(2x^3 + 9)}{(7x^4 - 5)}$$

1)
$$\ln g(x) = \ln (3x^5 - 4) + \ln (2x^3 + 9) - \ln (7x^4 - 5)$$

2)

$$\frac{d}{dx}[\ln g(x)] = \frac{g'(x)}{g(x)} = \frac{15x^4}{3x^5 - 4} + \frac{6x^2}{2x^3 + 9} - \frac{28x^3}{7x^4 - 5}$$

3)

$$g'(x) = \left(\frac{15x^4}{3x^5 - 4} + \frac{6x^2}{2x^3 + 9} - \frac{28x^3}{7x^4 - 5}\right) \cdot g(x)$$

4)

$$g'(x) = \left(\frac{15x^4}{3x^5 - 4} + \frac{6x^2}{2x^3 + 9} - \frac{28x^3}{7x^4 - 5}\right) \cdot \frac{(3x^5 - 4)(2x^3 + 9)}{(7x^4 - 5)}$$

GROWTH

9.23. The price of agricultural goods is going up by 4 percent each year, the quantity by 2 percent. What is the annual rate of growth of revenue R derived from the agricultural sector?

Converting the revenue formula $R = PQ$ to natural logs,

$$\ln R = \ln P + \ln Q$$

The derivative of the natural log function equals the instantaneous rate of growth G of the function (see Section 9.6). Thus,

$$G = \frac{d}{dt}(\ln R) = \frac{d}{dt}(\ln P) + \frac{d}{dt}(\ln Q)$$

But

$$\frac{d}{dt}(\ln P) = \text{growth of } P = 4\% \qquad \frac{d}{dt}(\ln Q) = \text{growth of } Q = 2\%$$

Thus,

$$G = \frac{d}{dt}(\ln R) = 0.04 + 0.02 = 0.06$$

The rate of growth of a function involving a product is the sum of the rates of growth of the individual components.

9.24. A firm experiences a 10 percent increase in the use of inputs at a time when input costs are rising by 3 percent. What is the rate of increase in total input costs?

$$C = PQ$$
$$\ln C = \ln P + \ln Q$$
$$G = \frac{d}{dt}(\ln C) = \frac{d}{dt}(\ln P) + \frac{d}{dt}(\ln Q) = 0.03 + 0.10 = 0.13$$

9.25. Employment opportunities E are increasing by 4 percent a year and population P by 2.5 percent. What is the rate of growth of per capita employment PCE?

$$\text{PCE} = \frac{E}{P}$$

$$\ln \text{PCE} = \ln E - \ln P$$

Taking the derivative to find the growth rate,

$$G = \frac{d}{dt}(\ln \text{PCE}) = \frac{d}{dt}(\ln E) - \frac{d}{dt}(\ln P) = 0.04 - 0.025 = 0.015 = 1.5\%$$

The rate of growth of a function involving a quotient is the difference between the rate of growth of the numerator and denominator.

9.26. National income Y is increasing by 1.5 percent a year and population P by 2.5 percent a year. What is the rate of growth of per capita income PCY?

$$PCY = \frac{Y}{P}$$

$$\ln PCY = \ln Y - \ln P$$

$$G = \frac{d}{dt}(\ln PCY) = \frac{d}{dt}(\ln Y) - \frac{d}{dt}(\ln P) = 0.015 - 0.025 = -0.01 = -1\%$$

Per capita income is falling by 1 percent a year.

9.27. A country exports two goods, copper c and bananas b, where earnings in terms of million dollars are

$$c = c(t_0) = 4 \qquad b = b(t_0) = 1$$

If c grows by 10 percent and b by 20 percent, what is the rate of growth of export earnings E?

$$E = c + b$$

$$\ln E = \ln(c + b)$$

$$G = \frac{d}{dt}(\ln E) = \frac{d}{dt}\ln(c + b)$$

From the rules of derivatives in Section 9.1.3,

$$G_E = \frac{1}{c + b}[c'(t) + b'(t)] \qquad\qquad (9.18)$$

From Section 9.6,

$$G_c = \frac{c'(t)}{c(t)} \qquad G_b = \frac{b'(t)}{b(t)}$$

Thus,

$$c'(t) = G_c c(t) \qquad b'(t) = G_b b(t)$$

Substituting in (9.18),

$$G_E = \frac{1}{c + b}[G_c c(t) + G_b b(t)]$$

Rearranging terms,

$$G_E = \frac{c(t)}{c + b}G_c + \frac{b(t)}{c + b}G_b$$

Then substituting the given values,

$$G_E = \frac{4}{4 + 1}(0.10) + \frac{1}{4 + 1}(0.20) = \frac{4}{5}(0.10) + \frac{1}{5}(0.20) = 0.12 \qquad \text{or} \qquad 12\%$$

The growth rate of a function involving the sum of other functions is the sum of the weighted average of the growth of the other functions.

9.28. A company derives 70 percent of its revenue from bathing suits, 20 percent from bathing caps, and 10 percent from bathing slippers. If revenues from bathing suits increase by 15 percent, from caps by 5 percent, and from slippers by 4 percent, what is the rate of growth of total revenue?

$$G_R = 0.70(0.15) + 0.20(0.05) + 0.10(0.04) = 0.105 + 0.01 + 0.004 = 0.119 \qquad \text{or} \qquad 11.9\%$$

9.29. Find the relative growth rate G of sales at $t = 4$, given $S(t) = 100,000e^{0.5\sqrt{t}}$

$$\ln S(t) = \ln 100,000 + \ln e^{0.5\sqrt{t}} = \ln 100,000 + 0.5\sqrt{t}$$

Take the derivative and recall that $0.5\sqrt{t} = 0.5t^{1/2}$,

$$G = \frac{d}{dt}\ln S(t) = \frac{S'(t)}{S(t)} = 0.5\left(\frac{1}{2}\right)t^{-1/2} = \frac{0.25}{\sqrt{t}}$$

At $t = 4$,

$$G = \frac{0.25}{\sqrt{4}} = 0.125 = 12.5\%$$

9.30. Find the relative growth of profits at $t = 8$, given $\pi(t) = 250,000e^{1.2t^{1/3}}$.

$$\ln \pi(t) = \ln 250,000 + 1.2t^{1/3}$$

and

$$G = \frac{d}{dt}\ln \pi(t) = \frac{\pi'(t)}{\pi(t)} = 1.2\left(\frac{1}{3}\right)t^{-2/3} = \frac{0.4}{t^{2/3}}$$

At $t = 8$,

$$G = \frac{0.4}{8^{2/3}} = \frac{0.4}{4} = 0.1 = 10\%$$

OPTIMAL TIMING

9.31. Cut glass currently worth \$100 is appreciating in value according to the formula

$$V = 100e^{\sqrt{t}} = 100e^{t^{1/2}}$$

How long should the cut glass be kept to maximize its present value if under continuous compounding (*a*) $r = 0.08$ and (*b*) $r = 0.12$?

a)   The present value P is $P = Ve^{-rt}$. Substituting for V and r,

$$P = 100e^{\sqrt{t}}e^{-0.08t} = 100e^{\sqrt{t}-0.08t}$$

Converting to natural logs, $\ln P = \ln 100 + \ln e^{\sqrt{t}-0.08t} = \ln 100 + t^{1/2} - 0.08t$. Taking the derivative, setting it equal to zero, and recalling that $\ln 100$ is a constant,

$$\frac{d}{dt}(\ln P) = \frac{1}{P}\frac{dP}{dt} = \frac{1}{2}t^{-1/2} - 0.08$$

$$\frac{dP}{dt} = P\left(\frac{1}{2}t^{-1/2} - 0.08\right) = 0 \qquad\qquad (9.19)$$

Since $P \neq 0$,

$$\tfrac{1}{2}t^{-1/2} = 0.08$$
$$t^{-1/2} = 0.16$$

$$t = (0.16)^{-2} = \frac{1}{0.0256} = 39.06$$

Testing the second-order condition, and using the product rule since $P = f(t)$,

$$\frac{d^2P}{dt^2} = P\left(-\frac{1}{4}t^{-3/2}\right) + \left(\frac{1}{2}t^{-1/2} - 0.08\right)\frac{dP}{dt}$$

Since $dP/dt = 0$ at the critical value,

$$\frac{d^2P}{dt^2} = \frac{-P}{4\sqrt{t^3}}$$

which is negative, since P and t must both be positive. Thus, $t = 39.06$ maximizes the function.

b)   If $r = 0.12$, substituting 0.12 for 0.08 in (*9.19*) above,

$$\frac{dP}{dt} = P\left(\frac{1}{2}t^{-1/2} - 0.12\right) = 0$$

$$\tfrac{1}{2}t^{-1/2} = 0.12$$

$$t = (0.24)^{-2} = \frac{1}{0.0576} = 17.36$$

The second-order condition is unchanged. Note that the higher the rate of discount r, the shorter the period of storage.

9.32.   Land bought for speculation is increasing in value according to the formula

$$V = 1000e^{\sqrt[3]{t}}$$

The discount rate under continuous compounding is 0.09. How long should the land be held to maximize the present value?

$$P = 1000e^{\sqrt[3]{t}}e^{-0.09t} = 1000e^{\sqrt[3]{t}-0.09t}$$

Convert to natural logs. $$\ln P = \ln 1000 + t^{1/3} - 0.09t$$

Take the derivative. $$\frac{d}{dt}(\ln P) = \frac{1}{P}\frac{dP}{dt} = \frac{1}{3}t^{-2/3} - 0.09 = 0$$

$$\frac{dP}{dt} = P\left(\frac{1}{3}t^{-2/3} - 0.09\right) = 0$$

$$\tfrac{1}{3}t^{-2/3} = 0.09 \qquad t = 0.27^{-3/2} \approx 7.13 \text{ years}$$

The second-order condition, recalling $dP/dt = 0$ at the critical value, is

$$\frac{d^2P}{dt^2} = P\left(-\frac{2}{9}t^{-5/3}\right) + \left(\frac{1}{3}t^{-2/3} - 0.09\right)\frac{dP}{dt} = -\frac{2P}{9\sqrt[3]{t^5}} < 0$$

9.33.   The art collection of a recently deceased painter has an estimated value of

$$V = 200{,}000(1.25)^{\sqrt[3]{t^2}}$$

How long should the executor of the estate hold on to the collection before putting it up for sale if the discount rate under continuous compounding is 6 percent?

Substituting the value of V in $P = Ve^{-rt}$,

$$P = 200{,}000(1.25)^{t^{2/3}}e^{-0.06t}$$

$$\ln P = \ln 200{,}000 + t^{2/3}\ln 1.25 - 0.06t$$

$$\frac{d}{dt}(\ln P) = \frac{1}{P}\frac{dP}{dt} = \frac{2}{3}(\ln 1.25)t^{-1/3} - 0.06 = 0$$

$$\frac{dP}{dt} = P\left[\frac{2}{3}(\ln 1.25)t^{-1/3} - 0.06\right] = 0$$

$$t^{-1/3} = \frac{3(0.06)}{2\ln 1.25} \qquad t = \left[\frac{0.18}{2(0.22314)}\right]^{-3} = (0.403)^{-3} \approx 15.3 \text{ years}$$

9.34.   The estimated value of a diamond bought for investment purposes is

$$V = 250{,}000(1.75)^{\sqrt[3]{t}}$$

If the discount rate under continuous compounding is 7 percent, how long should the diamond be held?

$$P = 250{,}000(1.75)^{t^{1/4}} e^{-0.07t}$$

$$\ln P = \ln 250{,}000 + t^{1/4} (\ln 1.75) - 0.07t$$

$$\frac{d}{dt}(\ln P) = \frac{1}{P}\frac{dP}{dt} = \frac{1}{4}(\ln 1.75)t^{-3/4} - 0.07 = 0$$

$$\frac{dP}{dt} = P\left[\frac{1}{4}(\ln 1.75)t^{-3/4} - 0.07\right] = 0$$

$$\frac{1}{4}(\ln 1.75)t^{-3/4} - 0.07 = 0 \qquad t = \left(\frac{0.28}{\ln 1.75}\right)^{-4/3} = (0.50)^{-4/3} \approx 2.52 \text{ years}$$

SELECTED PROOFS

9.35. Derive the derivative for $\ln x$.

From the definition of a derivative in Equation (3.2),

$$f'(x) = \lim_{\Delta x \to 0} \frac{f(x + \Delta x) - f(x)}{\Delta x}$$

Specifying $f(x) = \ln x$,

$$\frac{d}{dx}(\ln x) = \lim_{\Delta x \to 0} \frac{\ln(x + \Delta x) - \ln x}{\Delta x}$$

From the properties of logs, where $\ln a - \ln b = \ln(a/b)$,

$$\frac{d}{dx}(\ln x) = \lim_{\Delta x \to 0} \frac{\ln[(x + \Delta x)/x]}{\Delta x}$$

Rearranging first the denominator and then the numerator,

$$\frac{d}{dx}(\ln x) = \lim_{\Delta x \to 0}\left(\frac{1}{\Delta x}\ln\frac{x + \Delta x}{x}\right) = \lim_{\Delta x \to 0}\left[\frac{1}{\Delta x}\ln\left(1 + \frac{\Delta x}{x}\right)\right]$$

Multiply by x/x.

$$\frac{d}{dx}(\ln x) = \lim_{\Delta x \to 0}\left[\frac{1}{x}\cdot\frac{x}{\Delta x}\ln\left(1 + \frac{\Delta x}{x}\right)\right]$$

From the properties of logs, where $a\ln x = \ln x^a$,

$$\frac{d}{dx}(\ln x) = \lim_{\Delta x \to 0}\left[\frac{1}{x}\ln\left(1 + \frac{\Delta x}{x}\right)^{x/\Delta x}\right]$$

$$= \frac{1}{x}\lim_{\Delta x \to 0}\left[\ln\left(1 + \frac{\Delta x}{x}\right)^{x/\Delta x}\right]$$

Since the logarithmic function is continuous,

$$\frac{d}{dx}(\ln x) = \frac{1}{x}\ln\left[\lim_{\Delta x \to 0}\left(1 + \frac{\Delta x}{x}\right)^{x/\Delta x}\right]$$

Let $n = x/\Delta x$ and note that as $\Delta x \to 0$, $n \to \infty$. Then

$$\frac{d}{dx}(\ln x) = \frac{1}{x}\ln\left[\lim_{n \to \infty}\left(1 + \frac{1}{n}\right)^n\right]$$

But from Section 7.4, $e =$ the limit as $n \to \infty$ of $(1 + 1/n)^n$, so

$$\frac{d}{dx}(\ln x) = \frac{1}{x}\ln e = \frac{1}{x}\cdot 1 = \frac{1}{x}$$

9.36. Derive the derivative for $y = \ln g(x)$, assuming $g(x)$ is positive and differentiable.

Using the chain rule notation from (3.6), let $y = \ln u$ and $u = g(x)$. Then

$$\frac{dy}{dx} = \frac{dy}{du} \cdot \frac{du}{dx}$$

where

$$\frac{dy}{du} = \frac{1}{u} \quad \text{and} \quad \frac{du}{dx} = g'(x)$$

Substituting,

$$\frac{dy}{dx} = \frac{1}{u} \cdot g'(x)$$

Then replacing u with $g(x)$,

$$\frac{dy}{dx} = \frac{1}{g(x)} \cdot g'(x)$$

Hence,

$$\frac{d}{dx}[\ln g(x)] = \frac{1}{g(x)} \cdot g'(x) = \frac{g'(x)}{g(x)}$$

9.37. Show that the derivative of the function $y = e^x$ is e^x.

Take the natural logarithm of both sides,

$$\ln y = \ln e^x$$

From Equation (7.3),

$$\ln y = x$$

Use implicit differentiation and recall that y is a function of x and so requires the chain rule,

$$\frac{1}{y} \cdot \frac{dy}{dx} = 1$$

$$\frac{dy}{dx} = y$$

Replace y with e^x,

$$\frac{d}{dx}(e^x) = e^x$$

9.38. Given $y = e^{g(x)}$, prove that $dy/dx = e^{g(x)} \cdot g'(x)$.

Use the chain rule, letting $y = e^u$ and $u = g(x)$; then

$$\frac{dy}{dx} = \frac{dy}{du} \cdot \frac{du}{dx}$$

where

$$\frac{dy}{du} = e^u \quad \text{and} \quad \frac{du}{dx} = g'(x)$$

Substituting,

$$\frac{dy}{dx} = e^u \cdot g'(x)$$

Then replace y with $e^{g(x)}$ on the left-hand side and u with $g(x)$ on the right-hand side,

$$\frac{d}{dx}(e^{g(x)}) = e^{g(x)} \cdot g'(x)$$

9.39. Prove that $\log_a (x/y) = \log_a x - \log_a y$.

Let $s = \log_a x$ and $t = \log_a y$. Then from the definition of a logarithm in Section 7.2

$$a^s = x \quad \text{and} \quad a^t = y$$

and

$$\frac{a^s}{a^t} = \frac{x}{y}$$

Substituting from the property of exponents, where $a^s/a^t = a^{s-t}$,

$$a^{s-t} = \frac{x}{y}$$

Again using the definition of a logarithm,

$$\log_a \frac{x}{y} = s - t$$

But $s = \log_a x$ and $t = \log_a y$, so

$$\log_a \frac{x}{y} = \log_a x - \log_a y$$

9.40. From Equation (9.4), prove that $\log_a e = 1/\ln a$.

Set each side of the given equation as an exponent of a.

$$a^{\log_a e} = a^{1/\ln a}$$

But $a^{\log_a e} = e$. Substituting,

$$e = a^{1/\ln a}$$

Then taking the log of both sides,

$$\log_a e = \log_a a^{1/\ln a} = \frac{1}{\ln a}$$

CHAPTER 10

The Fundamentals of Linear (or Matrix) Algebra

10.1 THE ROLE OF LINEAR ALGEBRA

Linear algebra (1) permits expression of a complicated system of equations in a succinct, simplified way, (2) provides a shorthand method to determine whether a solution exists before it is attempted, and (3) furnishes the means of solving the equation system. Linear algebra however, can be applied *only* to systems of linear equations. Since many economic relationships can be approximated by linear equations and others can be converted to linear relationships, this limitation can in part be averted. See Example 2 and Section 7.6.

EXAMPLE 1. For a company with several different outlets selling several different products, a matrix provides a concise way of keeping track of stock.

Outlet	Skis	Poles	Bindings	Outfits
1	120	110	90	150
2	200	180	210	110
3	175	190	160	80
4	140	170	180	140

By reading across a row of the matrix, the firm can determine the level of stock in any of its outlets. By reading down a column of the matrix, the firm can determine the stock of any line of its products.

EXAMPLE 2. A nonlinear function, such as the rational function $z = x^{0.3}/y^{0.6}$, can be easily converted to a linear function by a simple rearrangement

$$z = \frac{x^{0.3}}{y^{0.6}} = x^{0.3}y^{-0.6}$$

followed by a logarithmic transformation

$$\ln z = 0.3 \ln x - 0.6 \ln y$$

which is log-linear. In similar fashion, many exponential and power functions are readily convertible to linear functions and then handled by linear algebra. See Section 7.6.

10.2 DEFINITIONS AND TERMS

A *matrix* is a rectangular array of numbers, parameters, or variables, each of which has a carefully ordered place within the matrix. The numbers (parameters, or variables) are referred to as *elements* of the matrix. The numbers in a horizontal line are called *rows*; the numbers in a vertical line are called *columns*. The number of rows r and columns c defines the *dimensions* of the matrix ($r \times c$), which is read "r by c." The row number always precedes the column number. In a *square matrix*, the number of rows equals the number of columns (that is, $r = c$). If the matrix is composed of a single column, such that its dimensions are $r \times 1$, it is a *column vector*; if a matrix is a single row, with dimensions $1 \times c$, it is a *row vector*. A matrix which converts the rows of A to columns and the columns of A to rows is called the *transpose* of $\mathbf{A}$ and is designated by $\mathbf{A}'$ (or $\mathbf{A}^T$). See Example 3 and Problems 10.1 to 10.3.

EXAMPLE 3. Given

$$\mathbf{A} = \begin{bmatrix} a_{11} & a_{12} & a_{13} \\ a_{21} & a_{22} & a_{23} \\ a_{31} & a_{32} & a_{33} \end{bmatrix}_{3\times3} \quad \mathbf{B} = \begin{bmatrix} 3 & 9 & 8 \\ 4 & 2 & 7 \end{bmatrix}_{2\times3} \quad \mathbf{C} = \begin{bmatrix} 7 \\ 4 \\ 5 \end{bmatrix}_{3\times1} \quad \mathbf{D} = \begin{bmatrix} 3 & 0 & 1 \end{bmatrix}_{1\times3}$$

Here $\mathbf{A}$ is a general matrix composed of $3 \times 3 = 9$ elements, arranged in three rows and three columns. It is thus a square matrix. Note that no punctuation separates the elements of a matrix. The elements all have double subscripts which give the *address* or placement of the element in the matrix; the first subscript identifies the row in which the element appears, and the second identifies the column. Positioning is precise within a matrix. Thus, a_{23} is the element which appears in the second row, third column; a_{32} is the element which appears in the third row, second column. Since row always precedes column in matrix notation, it might be helpful to think of the subscripts in terms of RC cola or some other mnemonic device. To determine the number of rows, always count down; to find the number of columns, count across.

Here $\mathbf{B}$ is a 2×3 matrix. Its b_{12} element is 9, its b_{21} element is 4. And $\mathbf{C}$ is a column vector with dimensions 3×1; $\mathbf{D}$ is a row vector with dimensions 1×3.

The transpose of $\mathbf{A}$ is

$$\mathbf{A}' = \begin{bmatrix} a_{11} & a_{21} & a_{31} \\ a_{12} & a_{22} & a_{32} \\ a_{13} & a_{23} & a_{33} \end{bmatrix}$$

and the transpose of $\mathbf{C}$ is

$$\mathbf{C}' = \begin{bmatrix} 7 & 4 & 5 \end{bmatrix}$$

10.3 ADDITION AND SUBTRACTION OF MATRICES

Addition (and subtraction) of two matrices $\mathbf{A} + \mathbf{B}$ (or $\mathbf{A} - \mathbf{B}$) requires that the matrices be of equal dimensions. Each element of one matrix is then added to (subtracted from) the corresponding element of the other matrix. Thus, a_{11} in $\mathbf{A}$ will be added to (subtracted from) b_{11} in $\mathbf{B}$; a_{12} to b_{12}; etc. See Examples 4 and 5 and Problems 10.4 to 10.8.

EXAMPLE 4. The sum $\mathbf{A} + \mathbf{B}$ is calculated below, given matrices $\mathbf{A}$ and $\mathbf{B}$:

$$\mathbf{A} = \begin{bmatrix} 8 & 9 & 7 \\ 3 & 6 & 2 \\ 4 & 5 & 10 \end{bmatrix}_{3\times3} \quad \mathbf{B} = \begin{bmatrix} 1 & 3 & 6 \\ 5 & 2 & 4 \\ 7 & 9 & 2 \end{bmatrix}_{3\times3} \quad \mathbf{A}+\mathbf{B} = \begin{bmatrix} 8+1 & 9+3 & 7+6 \\ 3+5 & 6+2 & 2+4 \\ 4+7 & 5+9 & 10+2 \end{bmatrix}_{3\times3} = \begin{bmatrix} 9 & 12 & 13 \\ 8 & 8 & 6 \\ 11 & 14 & 12 \end{bmatrix}_{3\times3}$$

The difference $\mathbf{C} - \mathbf{D}$, given matrices $\mathbf{C}$ and $\mathbf{D}$, is found as follows:

$$\mathbf{C} = \begin{bmatrix} 4 & 9 \\ 2 & 6 \end{bmatrix}_{2 \times 2} \qquad \mathbf{D} = \begin{bmatrix} 1 & 7 \\ 5 & 4 \end{bmatrix}_{2 \times 2} \qquad \mathbf{C} - \mathbf{D} = \begin{bmatrix} 4-1 & 9-7 \\ 2-5 & 6-4 \end{bmatrix}_{2 \times 2} = \begin{bmatrix} 3 & 2 \\ -3 & 2 \end{bmatrix}_{2 \times 2}$$

EXAMPLE 5. Suppose that deliveries $\mathbf{D}$ are made to the outlets of the firm in Example 1. What is the new level of stock?

$$\mathbf{D} = \begin{bmatrix} 40 & 20 & 50 & 10 \\ 25 & 30 & 10 & 60 \\ 15 & 0 & 40 & 70 \\ 60 & 40 & 10 & 50 \end{bmatrix}$$

To find the new level of stock, label the initial matrix $\mathbf{S}$ and solve for $\mathbf{S} + \mathbf{D}$. Adding the corresponding elements of each matrix,

$$\mathbf{S} + \mathbf{D} = \begin{bmatrix} 120+40 & 110+20 & 90+50 & 150+10 \\ 200+25 & 180+30 & 210+10 & 110+60 \\ 175+15 & 190+0 & 160+40 & 80+70 \\ 140+60 & 170+40 & 180+10 & 140+50 \end{bmatrix} = \begin{bmatrix} 160 & 130 & 140 & 160 \\ 225 & 210 & 220 & 170 \\ 190 & 190 & 200 & 150 \\ 200 & 210 & 190 & 190 \end{bmatrix}$$

10.4 SCALAR MULTIPLICATION

In matrix algebra, a simple number such as 12, -2, or 0.07 is called a *scalar*. Multiplication of a matrix by a number or scalar involves multiplication of every element of the matrix by the number. The process is called *scalar multiplication* because it scales the matrix up or down according to the size of the number. See Example 6 and Problems 10.10 to 10.12.

EXAMPLE 6. The result of scalar multiplication $k\mathbf{A}$, given $k = 8$ and

$$\mathbf{A} = \begin{bmatrix} 6 & 9 \\ 2 & 7 \\ 8 & 4 \end{bmatrix}_{3 \times 2}$$

is shown below

$$k\mathbf{A} = \begin{bmatrix} 8(6) & 8(9) \\ 8(2) & 8(7) \\ 8(8) & 8(4) \end{bmatrix}_{3 \times 2} = \begin{bmatrix} 48 & 72 \\ 16 & 56 \\ 64 & 32 \end{bmatrix}_{3 \times 2}$$

10.5 VECTOR MULTIPLICATION

Multiplication of a row vector $\mathbf{A}$ by a column vector $\mathbf{B}$ requires as a precondition that each vector have precisely the same number of elements. The product is then found by multiplying the individual elements of the row vector by their corresponding elements in the column vector and summing the products:

$$\mathbf{AB} = (a_{11} \times b_{11}) + (a_{12} \times b_{21}) + (a_{13} \times b_{31}) \qquad \text{etc.}$$

The product of row-column multiplication will thus be a single number or scalar. Row-column vector multiplication is of paramount importance. It serves as the basis for all matrix multiplication. See Example 7 and Problems 10.13 to 10.18.

EXAMPLE 7. The product **AB** of the row vector **A** and the column vector **B**, given

$$A = [4 \quad 7 \quad 2 \quad 9]_{1\times4} \qquad B = \begin{bmatrix} 12 \\ 1 \\ 5 \\ 6 \end{bmatrix}_{4\times1}$$

is calculated as follows:

$$AB = 4(12) + 7(1) + 2(5) + 9(6) = 48 + 7 + 10 + 54 = 119$$

The product of vectors

$$C = [3 \quad 6 \quad 8]_{1\times3} \qquad D = \begin{bmatrix} 2 \\ 4 \\ 5 \end{bmatrix}_{3\times1}$$

is

$$CD = (3 \times 2) + (6 \times 4) + (8 \times 5) = 6 + 24 + 40 = 70$$

Note that since each set of vectors above has the same number of elements, multiplication is possible.

Reversing the order of multiplication in either of the above and having column-row vector multiplication (**BA** or **DC**) will give a totally different answer. See Problem 10.36.

EXAMPLE 8. The meaning of vector multiplication is perhaps easiest understood in terms of the following example. Assume **Q** is a row vector of the physical quantities of hamburgers, fries, and sodas sold, respectively, on a given day and **P** is a column vector of the corresponding prices of hamburgers, fries, and sodas.

$$Q = [12 \quad 8 \quad 10] \qquad P = \begin{bmatrix} 1.25 \\ 0.75 \\ 0.50 \end{bmatrix}$$

Then by vector multiplication the total value of sales (TVS) for the day is

$$TVS = QP = [12(1.25) + 8(0.75) + 10(0.50)] = 26.00$$

10.6 MULTIPLICATION OF MATRICES

Multiplication of two matrices with dimensions $(r \times c)_1$ and $(r \times c)_2$ requires that the matrices be *conformable*, i.e., that $c_1 = r_2$, or the number of columns in 1, the *lead matrix*, equal the number of rows in 2, the *lag matrix*. Each row vector in the lead matrix is then multiplied by each column vector of the lag matrix, according to the rules for multiplying row and column vectors discussed in Section 10.5. The row-column products, called *inner products* or *dot products*, are then used as elements in the formation of the product matrix, such that each element c_{ij} of the product matrix **C** is a scalar derived from the multiplication of the *i*th row of the lead matrix and the *j*th column of the lag matrix. See Examples 9 to 11 and Problems 10.19 to 10.33.

EXAMPLE 9. Given

$$A = \begin{bmatrix} 3 & 6 & 7 \\ 12 & 9 & 11 \end{bmatrix}_{2\times3} \qquad B = \begin{bmatrix} 6 & 12 \\ 5 & 10 \\ 13 & 2 \end{bmatrix}_{3\times2} \qquad C = \begin{bmatrix} 1 & 7 & 8 \\ 2 & 4 & 3 \end{bmatrix}_{2\times3}$$

A shorthand test for conformability, which should be applied before undertaking any matrix multiplication, is to place the two sets of dimensions in the order in which the matrices are to be multiplied, then mentally circle the last number of the first set and the first number of the second set. If the two numbers are equal, the number of columns in the lead matrix will equal the number of rows in the lag matrix, and the two matrices will be conformable for multiplication in the given order. Moreover, the numbers outside the circle will provide, in proper order, the dimensions of the resulting product matrix. Thus, for **AB**

The number of columns in the lead matrix equals the number of rows in the lag matrix, 3 = 3; the matrices are conformable for multiplication; and the dimensions of the product matrix **AB** will be 2 × 2. When two matrices such as **A** and **B** are conformable for multiplication, the product **AB** is said to be *defined*.
    For **BC**,

The number of columns in the lead matrix equals the number of rows in the lag matrix, 2 = 2; hence **B** and **C** are conformable. The product **BC** is defined, and **BC** will be a 3 × 3 matrix.
    For **AC**,

$$2 \times 3 \quad \neq \quad 2 \times 3$$

A and **C** are not conformable for multiplication. Thus, **AC** is not defined.

EXAMPLE 10.   Having determined that **A** and **B** in Example 9 are conformable, the product **AB** = **D** can be found. Remembering to use only rows R from the lead matrix and only columns **C** from the lag matrix, multiply the first row R_1 of the lead matrix by the first column C_1 of the lag matrix to find the first element d_{11} ($= R_1 C_1$) of the product matrix **D**. Then multiply the first row R_1 of the lead matrix by the second column C_2 of the lag matrix to get d_{12} ($= R_1 C_2$). Since there are no more columns left in the lag matrix to be multiplied by the first row of the lead matrix, move to the second row of the lead matrix. Multiply the second row R_2 of the lead matrix by the first column C_1 of the lag matrix to get d_{21} ($= R_2 C_1$). Finally, multiply the second row R_2 of the lead matrix by the second column C_2 of the lag matrix to get d_{22} ($= R_2 C_2$). Thus,

$$\mathbf{AB} = \mathbf{D} = \begin{bmatrix} R_1 C_1 & R_1 C_2 \\ R_2 C_1 & R_2 C_2 \end{bmatrix} = \begin{bmatrix} 3(6) + 6(5) + 7(13) & 3(12) + 6(10) + 7(2) \\ 12(6) + 9(5) + 11(13) & 12(12) + 9(10) + 11(2) \end{bmatrix}_{2\times2} = \begin{bmatrix} 139 & 110 \\ 260 & 256 \end{bmatrix}_{2\times2}$$

The product of **BC** is calculated below, using the same method:

$$\mathbf{BC} = \mathbf{E} = \begin{bmatrix} R_1 C_1 & R_1 C_2 & R_1 C_3 \\ R_2 C_1 & R_2 C_2 & R_2 C_3 \\ R_3 C_1 & R_3 C_2 & R_3 C_3 \end{bmatrix} = \begin{bmatrix} 6(1) + 12(2) & 6(7) + 12(4) & 6(8) + 12(3) \\ 5(1) + 10(2) & 5(7) + 10(4) & 5(8) + 10(3) \\ 13(1) + 2(2) & 13(7) + 2(4) & 13(8) + 2(3) \end{bmatrix}_{3\times3} = \begin{bmatrix} 30 & 90 & 84 \\ 25 & 75 & 70 \\ 17 & 99 & 110 \end{bmatrix}_{3\times3}$$

EXAMPLE 11.   Referring to Example 1, suppose that the price of skis is $200, poles $50, bindings $100, and outfits $150. To find the value **V** of the stock in the different outlets, express the prices as a column vector **P**, and multiply **S** by **P**.

$$\mathbf{V} = \mathbf{SP} = \begin{bmatrix} 120 & 110 & 90 & 150 \\ 200 & 180 & 210 & 110 \\ 175 & 190 & 160 & 80 \\ 140 & 170 & 180 & 140 \end{bmatrix}_{4\times4} \begin{bmatrix} 200 \\ 50 \\ 100 \\ 150 \end{bmatrix}_{4\times1}$$

The matrices are conformable, and the product matrix will be 4×1 since

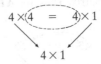

Thus,

$$
\mathbf{V} = \begin{bmatrix} R_1 C_1 \\ R_2 C_1 \\ R_3 C_1 \\ R_4 C_1 \end{bmatrix} = \begin{bmatrix} 120(200) + 110(50) + 90(100) + 150(150) \\ 200(200) + 180(50) + 210(100) + 110(150) \\ 175(200) + 190(50) + 160(100) + 80(150) \\ 140(200) + 170(50) + 180(100) + 140(150) \end{bmatrix}_{4 \times 1} = \begin{bmatrix} 61{,}000 \\ 86{,}500 \\ 72{,}500 \\ 75{,}500 \end{bmatrix}_{4 \times 1}
$$

10.7 COMMUTATIVE, ASSOCIATIVE, AND DISTRIBUTIVE LAWS IN MATRIX ALGEBRA

Matrix addition is commutative (that is, $\mathbf{A} + \mathbf{B} = \mathbf{B} + \mathbf{A}$) since matrix addition merely involves the summing of corresponding elements of two matrices and the order in which the addition takes place is inconsequential. For the same reason, matrix addition is also associative, $(\mathbf{A} + \mathbf{B}) + \mathbf{C} = \mathbf{A} + (\mathbf{B} + \mathbf{C})$. The same is true of matrix subtraction. Since matrix subtraction $\mathbf{A} - \mathbf{B}$ can be converted to matrix addition $\mathbf{A} + (-\mathbf{B})$, matrix subtraction is also commutative and associative.

Matrix multiplication, with few exceptions, is not commutative (that is, $\mathbf{AB} \neq \mathbf{BA}$). Scalar multiplication, however, is commutative (that is, $k\mathbf{A} = \mathbf{A}k$). If three or more matrices are conformable, that is, $\mathbf{X}_{a \times b}$, $\mathbf{Y}_{c \times d}$, $\mathbf{Z}_{e \times f}$, where $b = c$ and $d = e$, the associative law will apply as long as the matrices are multiplied in the order of conformability. Thus $(\mathbf{XY})\mathbf{Z} = \mathbf{X}(\mathbf{YZ})$. Subject to these same conditions, matrix multiplication is also distributive: $\mathbf{A}(\mathbf{B} + \mathbf{C}) = \mathbf{AB} + \mathbf{AC}$. See Examples 12 to 14 and Problems 10.34 to 10.48.

EXAMPLE 12. Given

$$
\mathbf{A} = \begin{bmatrix} 4 & 11 \\ 17 & 6 \end{bmatrix} \qquad \mathbf{B} = \begin{bmatrix} 3 & 7 \\ 6 & 2 \end{bmatrix}
$$

To show that matrix addition and matrix subtraction are commutative, demonstrate that (1) $\mathbf{A} + \mathbf{B} = \mathbf{B} + \mathbf{A}$ and (2) $\mathbf{A} - \mathbf{B} = -\mathbf{B} + \mathbf{A}$. The calculations are shown below.

1)
$$
\mathbf{A} + \mathbf{B} = \begin{bmatrix} 4 + 3 & 11 + 7 \\ 17 + 6 & 6 + 2 \end{bmatrix} = \begin{bmatrix} 7 & 18 \\ 23 & 8 \end{bmatrix} = \mathbf{B} + \mathbf{A} = \begin{bmatrix} 3 + 4 & 7 + 11 \\ 6 + 17 & 2 + 6 \end{bmatrix} = \begin{bmatrix} 7 & 18 \\ 23 & 8 \end{bmatrix}
$$

2)
$$
\mathbf{A} - \mathbf{B} = \begin{bmatrix} 4 - 3 & 11 - 7 \\ 17 - 6 & 6 - 2 \end{bmatrix} = \begin{bmatrix} 1 & 4 \\ 11 & 4 \end{bmatrix} = -\mathbf{B} + \mathbf{A} = \begin{bmatrix} -3 + 4 & -7 + 11 \\ -6 + 17 & -2 + 6 \end{bmatrix} = \begin{bmatrix} 1 & 4 \\ 11 & 4 \end{bmatrix}
$$

EXAMPLE 13. Given

$$
\mathbf{A} = \begin{bmatrix} 3 & 6 & 7 \\ 12 & 9 & 11 \end{bmatrix}_{2 \times 3} \qquad \mathbf{B} = \begin{bmatrix} 6 & 12 \\ 5 & 10 \\ 13 & 2 \end{bmatrix}_{3 \times 2}
$$

It can be demonstrated that matrix multiplication is not commutative, by showing $\mathbf{AB} \neq \mathbf{BA}$, as follows:

Matrix $\mathbf{AB}$ is conformable. $2 \times 3 \ = \ 3 \times 2$ $\mathbf{AB}$ will be 2×2

$$
\mathbf{AB} = \begin{bmatrix} 3(6) + 6(5) + 7(13) & 3(12) + 6(10) + 7(2) \\ 12(6) + 9(5) + 11(13) & 12(12) + 9(10) + 11(2) \end{bmatrix}_{2 \times 2} = \begin{bmatrix} 139 & 110 \\ 260 & 256 \end{bmatrix}_{2 \times 2}
$$

:

Matrix **BA** is conformable. $3 \times 2 = 2 \times 3$ **BA** will be 3×3

$$\mathbf{BA} = \begin{bmatrix} 6(3)+12(12) & 6(6)+12(9) & 6(7)+12(11) \\ 5(3)+10(12) & 5(6)+10(9) & 5(7)+10(11) \\ 13(3)+2(12) & 13(6)+2(9) & 13(7)+2(11) \end{bmatrix}_{3\times3} = \begin{bmatrix} 162 & 144 & 174 \\ 135 & 120 & 145 \\ 63 & 96 & 113 \end{bmatrix}_{3\times3}$$

Hence **AB** ≠ **BA**. Frequently matrices will not even be conformable in two directions.

EXAMPLE 14.   Given

$$\mathbf{A} = \begin{bmatrix} 7 & 5 \\ 1 & 3 \\ 8 & 6 \end{bmatrix}_{3\times2} \qquad \mathbf{B} = \begin{bmatrix} 4 & 9 & 10 \\ 2 & 6 & 5 \end{bmatrix}_{2\times3} \qquad \mathbf{C} = \begin{bmatrix} 2 \\ 6 \\ 7 \end{bmatrix}_{3\times1}$$

To illustrate that matrix multiplication is associative, that is, $(\mathbf{AB})\mathbf{C} = \mathbf{A}(\mathbf{BC})$, the calculations are as follows:

$$\mathbf{AB} = \begin{bmatrix} 7(4)+5(2) & 7(9)+5(6) & 7(10)+5(5) \\ 1(4)+3(2) & 1(9)+3(6) & 1(10)+3(5) \\ 8(4)+6(2) & 8(9)+6(6) & 8(10)+6(5) \end{bmatrix}_{3\times3} = \begin{bmatrix} 38 & 93 & 95 \\ 10 & 27 & 25 \\ 44 & 108 & 110 \end{bmatrix}_{3\times3}$$

$$(\mathbf{AB})\mathbf{C} = \begin{bmatrix} 38 & 93 & 95 \\ 10 & 27 & 25 \\ 44 & 108 & 110 \end{bmatrix}_{3\times3} \begin{bmatrix} 2 \\ 6 \\ 7 \end{bmatrix}_{3\times1} = \begin{bmatrix} 38(2)+93(6)+95(7) \\ 10(2)+27(6)+25(7) \\ 44(2)+108(6)+110(7) \end{bmatrix}_{3\times1} = \begin{bmatrix} 1299 \\ 357 \\ 1506 \end{bmatrix}_{3+1}$$

$$\mathbf{BC} = \begin{bmatrix} 4(2)+9(6)+10(7) \\ 2(2)+6(6)+5(7) \end{bmatrix}_{2\times1} = \begin{bmatrix} 132 \\ 75 \end{bmatrix}_{2\times1}$$

$$\mathbf{A}(\mathbf{BC}) = \begin{bmatrix} 7 & 5 \\ 1 & 3 \\ 8 & 6 \end{bmatrix}_{3\times2} \begin{bmatrix} 132 \\ 75 \end{bmatrix}_{2\times1} = \begin{bmatrix} 7(132)+5(75) \\ 1(132)+3(75) \\ 8(132)+6(75) \end{bmatrix}_{3\times1} = \begin{bmatrix} 1299 \\ 357 \\ 1506 \end{bmatrix}_{3\times1} \qquad \text{Q.E.D.}$$

10.8   IDENTITY AND NULL MATRICES

An identity matrix **I** is a square matrix which has 1 for every element on the principal diagonal from left to right and 0 everywhere else. See Example 15. When a subscript is used, as in $\mathbf{I}_n$, n denotes the dimensions of the matrix $(n \times n)$. The identity matrix is similar to the number 1 in algebra since multiplication of a matrix by an identity matrix leaves the original matrix unchanged (that is, $\mathbf{AI} = \mathbf{IA} = \mathbf{A}$). Multiplication of an identity matrix by itself leaves the identity matrix unchanged: $\mathbf{I} \times \mathbf{I} = \mathbf{I}^2 = \mathbf{I}$. Any matrix for which $\mathbf{A} = \mathbf{A}'$ is a *symmetric matrix*. A symmetric matrix for which $\mathbf{A} \times \mathbf{A} = \mathbf{A}$ is an *idempotent matrix*. The identity matrix is symmetric and idempotent.

A *null matrix* is composed of all 0s and can be of any dimension; it is not necessarily square. Addition or subtraction of the null matrix leaves the original matrix unchanged; multiplication by a null matrix produces a null matrix. See Example 15 and Problems 10.49 to 10.51.

EXAMPLE 15.   Given

$$\mathbf{A} = \begin{bmatrix} 7 & 10 & 14 \\ 9 & 2 & 6 \\ 1 & 3 & 7 \end{bmatrix} \qquad \mathbf{B} = \begin{bmatrix} 5 & 12 \\ 20 & 4 \end{bmatrix} \qquad \mathbf{N} = \begin{bmatrix} 0 & 0 \\ 0 & 0 \end{bmatrix} \qquad \mathbf{I} = \begin{bmatrix} 1 & 0 & 0 \\ 0 & 1 & 0 \\ 0 & 0 & 1 \end{bmatrix}$$

it is possible to show that (1) multiplication by an identity matrix leaves the original matrix unchanged, that is, $\mathbf{AI} = \mathbf{A}$, (2) multiplication by a null matrix produces a null matrix, that is, $\mathbf{BN} = \mathbf{N}$, and (3) addition or subtraction of a null matrix leaves the original matrix unchanged, that is, $\mathbf{B} + \mathbf{N} = \mathbf{B}$. The calculations are shown below.

1) $\mathbf{AI} = \begin{bmatrix} 7 & 10 & 14 \\ 9 & 2 & 6 \\ 1 & 3 & 7 \end{bmatrix} \begin{bmatrix} 1 & 0 & 0 \\ 0 & 1 & 0 \\ 0 & 0 & 1 \end{bmatrix} = \begin{bmatrix} 7(1)+10(0)+14(0) & 7(0)+10(1)+14(0) & 7(0)+10(0)+14(1) \\ 9(1)+2(0)+6(0) & 9(0)+2(1)+6(0) & 9(0)+2(0)+6(1) \\ 1(1)+3(0)+7(0) & 1(0)+3(1)+7(0) & 1(0)+3(0)+7(1) \end{bmatrix}$

$= \begin{bmatrix} 7 & 10 & 14 \\ 9 & 2 & 6 \\ 1 & 3 & 7 \end{bmatrix}$ Q.E.D.

2) $\mathbf{BN} = \begin{bmatrix} 5(0)+12(0) & 5(0)+12(0) \\ 20(0)+4(0) & 20(0)+4(0) \end{bmatrix} = \begin{bmatrix} 0 & 0 \\ 0 & 0 \end{bmatrix}$ Q.E.D.

3) $\mathbf{B}+\mathbf{N} = \begin{bmatrix} 5+0 & 12+0 \\ 20+0 & 4+0 \end{bmatrix} = \begin{bmatrix} 5 & 12 \\ 20 & 4 \end{bmatrix}$ Q.E.D.

10.9 MATRIX EXPRESSION OF A SYSTEM OF LINEAR EQUATIONS

Matrix algebra permits the concise expression of a system of linear equations. As a simple illustration, note that the system of linear equations

$$7x_1 + 3x_2 = 45$$
$$4x_1 + 5x_2 = 29$$

can be expressed in matrix form

$$\mathbf{AX} = \mathbf{B}$$

where $\mathbf{A} = \begin{bmatrix} 7 & 3 \\ 4 & 5 \end{bmatrix}$ $\mathbf{X} = \begin{bmatrix} x_1 \\ x_2 \end{bmatrix}$ and $\mathbf{B} = \begin{bmatrix} 45 \\ 29 \end{bmatrix}$

Here $\mathbf{A}$ is the *coefficient matrix*, $\mathbf{X}$ is the *solution vector*, and $\mathbf{B}$ is the *vector of constant terms*. And $\mathbf{X}$ and $\mathbf{B}$ will always be column vectors. See Examples 16 and 17.

EXAMPLE 16.   To show that $\mathbf{AX} = \mathbf{B}$ accurately represents the given system of equations above, find the product $\mathbf{AX}$. Multiplication is possible since $\mathbf{AX}$ is conformable, and the product matrix will be 2×1.

$$2 \times 2 = 2 \times 1$$

$$(2 \times 1)$$

Thus, $\mathbf{AX} = \begin{bmatrix} 7 & 3 \\ 4 & 5 \end{bmatrix} \begin{bmatrix} x_1 \\ x_2 \end{bmatrix} = \begin{bmatrix} 7x_1 + 3x_2 \\ 4x_1 + 5x_2 \end{bmatrix}_{2 \times 1}$

and $\mathbf{AX} = \mathbf{B}$: $\begin{bmatrix} 7x_1 + 3x_2 \\ 4x_1 + 5x_2 \end{bmatrix} = \begin{bmatrix} 45 \\ 29 \end{bmatrix}$ Q.E.D.

Here, despite appearances, $\mathbf{AX}$ is a 2×1 column vector since each row is composed of a single element which cannot be simplified further through addition.

EXAMPLE 17.   Given

$$8w + 12x - 7y + 2z = 139$$
$$3w - 13x + 4y + 9z = 242$$

To express this system of equations in matrix notation, mentally reverse the order of matrix multiplication:

$$\begin{bmatrix} 8 & 12 & -7 & 2 \\ 3 & -13 & 4 & 9 \end{bmatrix}_{2\times 4} \begin{bmatrix} w \\ x \\ y \\ z \end{bmatrix}_{4\times 1} = \begin{bmatrix} 139 \\ 242 \end{bmatrix}_{2\times 1}$$

Then, letting **A** = matrix of coefficients, **W** = the column vector of variables, and **B** = the column vector of constants, the given system of equations can be expressed in matrix form

$$\mathbf{A}_{2\times 4}\mathbf{W}_{4\times 1} = \mathbf{B}_{2\times 1}$$

Solved Problems

MATRIX FORMAT

10.1. (a) Give the dimensions of each of the following matrices. (b) Give their transposes and indicate the new dimensions.

$$\mathbf{A} = \begin{bmatrix} 6 & 7 & 9 \\ 2 & 8 & 4 \end{bmatrix} \qquad \mathbf{B} = \begin{bmatrix} 12 & 9 & 2 & 6 \\ 7 & 5 & 8 & 3 \\ 9 & 1 & 0 & 4 \end{bmatrix} \qquad \mathbf{C} = \begin{bmatrix} 12 \\ 19 \\ 25 \end{bmatrix}$$

$$\mathbf{D} = \begin{bmatrix} 2 & 1 \\ 7 & 8 \\ 3 & 0 \\ 9 & 5 \end{bmatrix} \qquad \mathbf{E} = \begin{bmatrix} 10 & 2 & 9 & 6 & 8 & 1 \end{bmatrix} \qquad \mathbf{F} = \begin{bmatrix} 1 & 2 & 5 \\ 5 & 9 & 3 \\ 6 & 7 & 6 \\ 3 & 8 & 9 \end{bmatrix}$$

a) Recalling that dimensions are always listed row by column or rc, $\mathbf{A} = 2\times 3$, $\mathbf{B} = 3\times 4$, $\mathbf{C} = 3\times 1$, $\mathbf{D} = 4\times 2$, $\mathbf{E} = 1\times 6$, and $\mathbf{F} = 4\times 3$. $\mathbf{C}$ is also called a column vector; $\mathbf{E}$, a row vector.

b) The transpose of **A** converts the rows of **A** to columns and the columns of **A** to rows.

$$\mathbf{A}' = \begin{bmatrix} 6 & 2 \\ 7 & 8 \\ 9 & 4 \end{bmatrix}_{3\times 2} \qquad \mathbf{B}' = \begin{bmatrix} 12 & 7 & 9 \\ 9 & 5 & 1 \\ 2 & 8 & 0 \\ 6 & 3 & 4 \end{bmatrix}_{4\times 3} \qquad \mathbf{C}' = \begin{bmatrix} 12 & 19 & 25 \end{bmatrix}_{1\times 3}$$

$$\mathbf{D}' = \begin{bmatrix} 2 & 7 & 3 & 9 \\ 1 & 8 & 0 & 5 \end{bmatrix}_{2\times 4} \qquad \mathbf{E}' = \begin{bmatrix} 10 \\ 2 \\ 9 \\ 6 \\ 8 \\ 1 \end{bmatrix}_{6\times 1} \qquad \mathbf{F}' = \begin{bmatrix} 1 & 5 & 6 & 3 \\ 2 & 9 & 7 & 8 \\ 5 & 3 & 6 & 9 \end{bmatrix}_{3\times 4}$$

10.2. Given $a_{21} = 4$, $a_{32} = 5$, $a_{13} = 3$, $a_{23} = 6$, $a_{12} = 10$, and $a_{31} = -5$, use your knowledge of subscripts to complete the following matrix:

$$\mathbf{A} = \begin{bmatrix} 6 & - & - \\ - & 7 & - \\ - & - & 9 \end{bmatrix}$$

Since the subscripts are always given in row-column order, $a_{21} = 4$ means that 4 is located in the second row, first column; $a_{32} = 5$ means that 5 appears in the third row, second column; etc. Thus,

$$A = \begin{bmatrix} 6 & 10 & 3 \\ 4 & 7 & 6 \\ -5 & 5 & 9 \end{bmatrix}$$

10.3. A firm with five retail stores has 10 TVs t, 15 stereos s, 9 tape decks d, and 12 recorders r in store 1; 20t, 14s, 8d, and 5r in store 2; 16t, 8s, 15d, and 6r in store 3; 25t, 15s, 7d, and 16r in store 4; and 5t, 12s, 20d, and 18r in store 5. Express present inventory in matrix form.

$$\begin{array}{c} \text{Retail} \\ \text{store} \end{array} \quad \begin{array}{cccc} t & s & d & r \end{array}$$

$$\begin{array}{c} 1 \\ 2 \\ 3 \\ 4 \\ 5 \end{array} \begin{bmatrix} 10 & 15 & 9 & 12 \\ 20 & 14 & 8 & 5 \\ 16 & 8 & 15 & 6 \\ 25 & 15 & 7 & 16 \\ 5 & 12 & 20 & 18 \end{bmatrix}$$

MATRIX ADDITION AND SUBTRACTION

10.4. Find the sums $A + B$ of the following matrices:

a) $A = \begin{bmatrix} 8 & 9 \\ 12 & 7 \end{bmatrix}$ $B = \begin{bmatrix} 13 & 4 \\ 2 & 6 \end{bmatrix}$

$$A + B = \begin{bmatrix} 8+13 & 9+4 \\ 12+2 & 7+6 \end{bmatrix} = \begin{bmatrix} 21 & 13 \\ 14 & 13 \end{bmatrix}$$

b) $A = \begin{bmatrix} 7 & -10 \\ -8 & 2 \end{bmatrix}$ $B = \begin{bmatrix} -8 & 4 \\ 12 & -6 \end{bmatrix}$

$$A + B = \begin{bmatrix} 7+(-8) & -10+4 \\ -8+12 & 2+(-6) \end{bmatrix} = \begin{bmatrix} -1 & -6 \\ 4 & -4 \end{bmatrix}$$

c) $A = \begin{bmatrix} 12 & 16 & 2 & 7 & 8 \end{bmatrix}$ $B = \begin{bmatrix} 0 & 1 & 9 & 5 & 6 \end{bmatrix}$

$$A + B = \begin{bmatrix} 12 & 17 & 11 & 12 & 14 \end{bmatrix}$$

d) $A = \begin{bmatrix} 9 & 4 \\ 2 & 7 \\ 3 & 5 \\ 8 & 6 \end{bmatrix}$ $B = \begin{bmatrix} 1 & 3 \\ 6 & 5 \\ 2 & 8 \\ 9 & 2 \end{bmatrix}$

$$A + B = \begin{bmatrix} 10 & 7 \\ 8 & 12 \\ 5 & 13 \\ 17 & 8 \end{bmatrix}$$

10.5. Redo Problem 10.4, given

$$A = \begin{bmatrix} 0 & 1 & -6 & 2 \\ -3 & 5 & 8 & 7 \\ 2 & 9 & -1 & 6 \end{bmatrix} \qquad B = \begin{bmatrix} 7 & 2 & 12 & 6 & 5 \\ 4 & 3 & 8 & 10 & 6 \\ 1 & 0 & 5 & 11 & 9 \end{bmatrix}$$

Matrices **A** and **B** are not conformable for addition because they are not of equal dimensions; $\mathbf{A} = 3 \times 4$, $\mathbf{B} = 3 \times 5$.

10.6. The parent company in Problem 10.3 sends out deliveries **D** to its stores:

$$\mathbf{D} = \begin{bmatrix} 4 & 3 & 5 & 2 \\ 0 & 9 & 6 & 1 \\ 5 & 7 & 2 & 6 \\ 12 & 2 & 4 & 8 \\ 9 & 6 & 3 & 5 \end{bmatrix}$$

What is the new level of inventory?

$$\mathbf{I}_2 = \mathbf{I}_1 + \mathbf{D} = \begin{bmatrix} 10 & 15 & 9 & 12 \\ 20 & 14 & 8 & 5 \\ 16 & 8 & 15 & 6 \\ 25 & 15 & 7 & 16 \\ 5 & 12 & 20 & 18 \end{bmatrix} + \begin{bmatrix} 4 & 3 & 5 & 2 \\ 0 & 9 & 6 & 1 \\ 5 & 7 & 2 & 6 \\ 12 & 2 & 4 & 8 \\ 9 & 6 & 3 & 5 \end{bmatrix} = \begin{bmatrix} 14 & 18 & 14 & 14 \\ 20 & 23 & 14 & 6 \\ 21 & 15 & 17 & 12 \\ 37 & 17 & 11 & 24 \\ 14 & 18 & 23 & 23 \end{bmatrix}$$

10.7. Find the difference $\mathbf{A} - \mathbf{B}$ for each of the following:

a) $\mathbf{A} = \begin{bmatrix} 3 & 7 & 11 \\ 12 & 9 & 2 \end{bmatrix} \qquad \mathbf{B} = \begin{bmatrix} 6 & 8 & 1 \\ 9 & 5 & 8 \end{bmatrix}$

$$\mathbf{A} - \mathbf{B} = \begin{bmatrix} 3-6 & 7-8 & 11-1 \\ 12-9 & 9-5 & 2-8 \end{bmatrix} = \begin{bmatrix} -3 & -1 & 10 \\ 3 & 4 & -6 \end{bmatrix}$$

b) $\mathbf{A} = \begin{bmatrix} 16 \\ 2 \\ 15 \\ 9 \end{bmatrix} \qquad \mathbf{B} = \begin{bmatrix} 7 \\ 11 \\ 3 \\ 8 \end{bmatrix}$

$$\mathbf{A} - \mathbf{B} = \begin{bmatrix} 16-7 \\ 2-11 \\ 15-3 \\ 9-8 \end{bmatrix} = \begin{bmatrix} 9 \\ -9 \\ 12 \\ 1 \end{bmatrix}$$

c) $\mathbf{A} = \begin{bmatrix} 13 & -5 & 8 \\ 4 & 9 & 1 \\ 10 & 6 & -2 \end{bmatrix} \qquad \mathbf{B} = \begin{bmatrix} 14 & 2 & -5 \\ 9 & 6 & 8 \\ -3 & 13 & 11 \end{bmatrix}$

$$\mathbf{A} - \mathbf{B} = \begin{bmatrix} -1 & -7 & 13 \\ -5 & 3 & -7 \\ 13 & -7 & -13 \end{bmatrix}$$

10.8. A monthly report **R** on sales for the company in Problem 10.6 indicates

$$\mathbf{R} = \begin{bmatrix} 8 & 12 & 6 & 9 \\ 10 & 11 & 8 & 3 \\ 15 & 6 & 9 & 7 \\ 21 & 14 & 5 & 18 \\ 6 & 11 & 13 & 9 \end{bmatrix}$$

What is the inventory left at the end of the month?

$$\mathbf{I_2 - R} - \begin{bmatrix} 14 & 18 & 14 & 14 \\ 20 & 23 & 14 & 6 \\ 21 & 15 & 17 & 12 \\ 37 & 17 & 11 & 24 \\ 14 & 18 & 23 & 23 \end{bmatrix} \begin{bmatrix} 8 & 12 & 6 & 9 \\ 10 & 11 & 8 & 3 \\ 15 & 6 & 9 & 7 \\ 21 & 14 & 5 & 18 \\ 6 & 11 & 13 & 9 \end{bmatrix} - \begin{bmatrix} 6 & 6 & 8 & 5 \\ 10 & 12 & 6 & 3 \\ 6 & 9 & 8 & 5 \\ 16 & 3 & 6 & 6 \\ 8 & 7 & 10 & 14 \end{bmatrix}$$

CONFORMABILITY

10.9.   Given

$$\mathbf{A} = \begin{bmatrix} 7 & 2 & 6 \\ 5 & 4 & 8 \\ 3 & 1 & 9 \end{bmatrix} \qquad \mathbf{B} = \begin{bmatrix} 6 & 2 \\ 5 & 0 \end{bmatrix} \qquad \mathbf{C} = \begin{bmatrix} 11 \\ 4 \\ 13 \end{bmatrix}$$

$$\mathbf{D} = \begin{bmatrix} 14 \\ 4 \end{bmatrix} \qquad \mathbf{E} = [8 \quad 1 \quad 10] \qquad \mathbf{F} = [13 \quad 3]$$

Determine for each of the following whether the products are defined, i.e., conformable for multiplication. If so, indicate the dimensions of the product matrix. (*a*) **AC**, (*b*) **BD**, (*c*) **EC**, (*d*) **DF**, (*e*) **CA**, (*f*) **DE**, (*g*) **DB**, (*h*) **CF**, (*i*) **EF**.

a)   The dimensions of **AC**, in the order of multiplication, are $3 \times (3 = 3) \times 1$. Matrix **AC** is defined since the numbers within the dashed circle indicate that the number of columns in **A** equals the number of rows in **C**. The numbers outside the circle indicate that the product matrix will be 3×1.

b)   The dimensions of **BD** are $2 \times (2 = 2) \times 1$. Matrix **BD** is defined; the product matrix will be 2×1.

c)   The dimensions of **EC** are $1 \times (3 = 3) \times 1$. Matrix **EC** is defined; the product matrix will be 1×1, or a scalar.

d)   The dimensions of **DF** are $2 \times (1 = 1) \times 2$. Matrix **DF** is defined; the product matrix will be 2×2.

e)   The dimensions of **CA** are $3 \times (1 \neq 3) \times 3$. Matrix **CA** is undefined. The matrices are not conformable for multiplication in that order. [Note that **AC** in part (*a*) is defined. This illustrates that matrix multiplication is not commutative: $\mathbf{AC} \neq \mathbf{CA}$.]

f)   The dimensions of **DE** are $2 \times (1 = 1) \times 3$. Matrix **DE** is defined; the product matrix will be 2×3.

g)   The dimensions of **DB** are $2 \times (1 \neq 2) \times 2$. The matrices are not conformable for multiplication. Matrix **DB** is not defined.

h)   The dimensions of **CF** are $3 \times (1 = 1) \times 2$. The matrices are conformable; the product matrix will be 3×2.

i)   The dimensions of **EF** are $1 \times (3 \neq 1) \times 2$. The matrices are not conformable, and **EF** is not defined.

SCALAR AND VECTOR MULTIPLICATION

10.10. Determine $\mathbf{A}k$, given

$$\mathbf{A} = \begin{bmatrix} 3 & 2 \\ 9 & 5 \\ 6 & 7 \end{bmatrix} \qquad k = 4$$

Here k is a scalar, and scalar multiplication is possible with a matrix of any dimension. Hence the product is defined.

$$\mathbf{A}k = \begin{bmatrix} 3(4) & 2(4) \\ 9(4) & 5(4) \\ 6(4) & 7(4) \end{bmatrix} = \begin{bmatrix} 12 & 8 \\ 36 & 20 \\ 24 & 28 \end{bmatrix}$$

10.11. Find $k\mathbf{A}$, given

$$k = -2 \qquad \mathbf{A} = \begin{bmatrix} 7 & -3 & 2 \\ -5 & 6 & 8 \\ 2 & -7 & -9 \end{bmatrix}$$

$$k\mathbf{A} = \begin{bmatrix} -2(7) & -2(-3) & -2(2) \\ -2(-5) & -2(6) & -2(8) \\ -2(2) & -2(-7) & -2(-9) \end{bmatrix} = \begin{bmatrix} -14 & 6 & -4 \\ 10 & -12 & -16 \\ -4 & 14 & 18 \end{bmatrix}$$

10.12. A clothing store discounts all its slacks, jackets, and suits by 20 percent at the end of the year. If $\mathbf{V}_1$ is the value of stock in its three branches prior to the discount, find the value $\mathbf{V}_2$ after the discount, when

$$\mathbf{V}_1 = \begin{bmatrix} 5{,}000 & 4{,}500 & 6{,}000 \\ 10{,}000 & 12{,}000 & 7{,}500 \\ 8{,}000 & 9{,}000 & 11{,}000 \end{bmatrix}$$

A 20 percent reduction means that the clothing is selling for 80 percent of its original value. Hence $\mathbf{V}_2 = 0.8\mathbf{V}_1$,

$$\mathbf{V}_2 = 0.8 \begin{bmatrix} 5{,}000 & 4{,}500 & 6{,}000 \\ 10{,}000 & 12{,}000 & 7{,}500 \\ 8{,}000 & 9{,}000 & 11{,}000 \end{bmatrix} = \begin{bmatrix} 4{,}000 & 3{,}600 & 4{,}800 \\ 8{,}000 & 9{,}600 & 6{,}000 \\ 6{,}400 & 7{,}200 & 8{,}800 \end{bmatrix}$$

10.13. Find $\mathbf{AB}$, given

$$\mathbf{A} = \begin{bmatrix} 9 & 11 & 3 \end{bmatrix} \qquad \mathbf{B} = \begin{bmatrix} 2 \\ 6 \\ 7 \end{bmatrix}$$

Matrix $\mathbf{AB}$ is defined; $1 \times \underline{3} = \underline{3} \times 1$; the product will be a scalar, derived by multiplying each element of the row vector by its corresponding element in the column vector and then summing the products.

$$\mathbf{AB} = 9(2) + 11(6) + 3(7) = 18 + 66 + 21 = 105$$

10.14. Find **AB**, given

$$\mathbf{A} = \begin{bmatrix} 12 & -5 & 6 & 11 \end{bmatrix} \quad \mathbf{B} = \begin{bmatrix} 3 \\ 2 \\ -8 \\ 6 \end{bmatrix}$$

Matrix **AB** is defined; $1 \times (4 = 4) \times 1$.

$$\mathbf{AB} = 12(3) + (-5)(2) + 6(-8) + 11(6) = 44$$

10.15. Find **AB**, given

$$\mathbf{A} = \begin{bmatrix} 9 & 6 & 2 & 0 & -5 \end{bmatrix} \quad \mathbf{B} = \begin{bmatrix} 2 \\ 13 \\ 5 \\ 8 \\ 1 \end{bmatrix}$$

Matrix **AB** is defined; $1 \times (5 = 5) \times 1$.

$$\mathbf{AB} = 9(2) + 6(13) + 2(5) + 0(8) + (-5)(1) = 101$$

10.16. Find **AB**, given

$$\mathbf{A} = \begin{bmatrix} 12 & 9 & 2 & 4 \end{bmatrix} \quad \mathbf{B} = \begin{bmatrix} 6 \\ 1 \\ 2 \end{bmatrix}$$

Matrix **AB** is undefined; $1 \times (4 \neq 3) \times 1$. Multiplication is not possible.

10.17. If the price of a TV is \$300, the price of a stereo is \$250, the price of a tape deck is \$175, and the price of a recorder is \$125, use vectors to determine the value of stock for outlet 2 in Problem 10.3.

The value of stock is $\mathbf{V} = \mathbf{QP}$. The physical volume of stock in outlet 2 in vector form is $\mathbf{Q} = \begin{bmatrix} 20 & 14 & 8 & 5 \end{bmatrix}$. The price vector **P** can be written

$$\mathbf{P} = \begin{bmatrix} 300 \\ 250 \\ 175 \\ 125 \end{bmatrix}$$

Matrix **QP** is defined; $1 \times (4 = 4) \times 1$. Thus

$$\mathbf{V} = \mathbf{QP}$$
$$= 20(300) + 14(250) + 8(175) + 5(125) = 11,525$$

10.18. Redo Problem 10.17 for outlet 5 in Problem 10.3.

Here $\mathbf{Q} = \begin{bmatrix} 5 & 12 & 20 & 18 \end{bmatrix}$, **P** remains the same. Matrix **QP** is defined. Thus,

$$\mathbf{V} = 5(300) + 12(250) + 20(175) + 18(125) = 10,250$$

MATRIX MULTIPLICATION

10.19. Determine whether **AB** is defined, indicate what the dimensions of the product matrix will be, and find the product matrix **AB**, given

$$\mathbf{A} = \begin{bmatrix} 12 & 14 \\ 20 & 5 \end{bmatrix} \quad \mathbf{B} = \begin{bmatrix} 3 & 9 \\ 0 & 2 \end{bmatrix}$$

Matrix $\mathbf{AB}$ is defined; $2 \times (2 = 2) \times 2$; the product matrix will be 2×2. Matrix multiplication is nothing but a series of row-column vector multiplications in which the a_{11} element of the product matrix is determined by the product of the first row R_1 of the lead matrix and the first column C_1 of the lag matrix; the a_{12} element of the product matrix is determined by the product of the first row R_1 of the lead matrix and the second column C_2 of the lag matrix; the a_{ij} element of the product matrix is determined by the product of the ith row R_i of the lead matrix and the jth column C_j of the lag matrix, etc. Thus,

$$\mathbf{AB} = \begin{bmatrix} R_1 C_1 & R_1 C_2 \\ R_2 C_1 & R_2 C_2 \end{bmatrix} = \begin{bmatrix} 12(3) + 14(0) & 12(9) + 14(2) \\ 20(3) + 5(0) & 20(9) + 5(2) \end{bmatrix} = \begin{bmatrix} 36 & 136 \\ 60 & 190 \end{bmatrix}$$

10.20. Redo Problem 10.19, given

$$\mathbf{A} = \begin{bmatrix} 4 & 7 \\ 9 & 1 \end{bmatrix} \qquad \mathbf{B} = \begin{bmatrix} 3 & 8 & 5 \\ 2 & 6 & 7 \end{bmatrix}$$

Matrix $\mathbf{AB}$ is defined; $2 \times (2 = 2) \times 3$; the product matrix will be 2×3.

$$\mathbf{AB} = \begin{bmatrix} R_1 C_1 & R_1 C_2 & R_1 C_3 \\ R_2 C_1 & R_2 C_2 & R_2 C_3 \end{bmatrix} = \begin{bmatrix} 4(3) + 7(2) & 4(8) + 7(6) & 4(5) + 7(7) \\ 9(3) + 1(2) & 9(8) + 1(6) & 9(5) + 1(7) \end{bmatrix} = \begin{bmatrix} 26 & 74 & 69 \\ 29 & 78 & 52 \end{bmatrix}$$

10.21. Redo Problem 10.19, given

$$\mathbf{A} = \begin{bmatrix} 3 & 1 \\ 8 & 2 \end{bmatrix} \qquad \mathbf{B} = \begin{bmatrix} 2 & 9 \\ 4 & 6 \\ 7 & 5 \end{bmatrix}$$

Matrix $\mathbf{AB}$ is not defined; $2 \times (2 \neq 3) \times 2$. The matrices cannot be multiplied because they are not conformable in the given order. The number of columns (2) in $\mathbf{A}$ does not equal the number of rows (3) in $\mathbf{B}$.

10.22. Redo Problem 10.19 for $\mathbf{BA}$ in Problem 10.21.

Matrix $\mathbf{BA}$ is defined; $3 \times (2 = 2) \times 2$; the product matrix will be 3×2.

$$\mathbf{BA} = \begin{bmatrix} 2 & 9 \\ 4 & 6 \\ 7 & 5 \end{bmatrix} \begin{bmatrix} 3 & 1 \\ 8 & 2 \end{bmatrix} = \begin{bmatrix} R_1 C_1 & R_1 C_2 \\ R_2 C_1 & R_2 C_2 \\ R_3 C_1 & R_3 C_2 \end{bmatrix} = \begin{bmatrix} 2(3) + 9(8) & 2(1) + 9(2) \\ 4(3) + 6(8) & 4(1) + 6(2) \\ 7(3) + 5(8) & 7(1) + 5(2) \end{bmatrix} = \begin{bmatrix} 78 & 20 \\ 60 & 16 \\ 61 & 17 \end{bmatrix}$$

10.23. Redo Problem 10.19 for $\mathbf{AB'}$ in Problem 10.21, where $\mathbf{B'}$ is the transpose of $\mathbf{B}$:

$$\mathbf{B'} = \begin{bmatrix} 2 & 4 & 7 \\ 9 & 6 & 5 \end{bmatrix}$$

Matrix $\mathbf{AB'}$ is defined: $2 \times (2 = 2) \times 3$; the product will be a 2×3 matrix.

$$\mathbf{AB'} = \begin{bmatrix} 3 & 1 \\ 8 & 2 \end{bmatrix} \begin{bmatrix} 2 & 4 & 7 \\ 9 & 6 & 5 \end{bmatrix} = \begin{bmatrix} 3(2) + 1(9) & 3(4) + 1(6) & 3(7) + 1(5) \\ 8(2) + 2(9) & 8(4) + 2(6) & 8(7) + 2(5) \end{bmatrix} = \begin{bmatrix} 15 & 18 & 26 \\ 34 & 44 & 66 \end{bmatrix}$$

(Note from Problems 10.21 to 10.23 that $\mathbf{AB} \neq \mathbf{BA} \neq \mathbf{AB'}$. The noncommutative aspects of matrix multiplication are treated in Problems 10.36 to 10.41.)

10.24. Redo Problem 10.19, given

$$\mathbf{A} = \begin{bmatrix} 7 & 11 \\ 2 & 9 \\ 10 & 6 \end{bmatrix} \qquad \mathbf{B} = \begin{bmatrix} 12 & 4 & 5 \\ 3 & 6 & 1 \end{bmatrix}$$

Matrix **AB** is defined; $3 \times (2 = 2) \times 3$; the product matrix will be 3×3.

$$\mathbf{AB} = \begin{bmatrix} R_1C_1 & R_1C_2 & R_1C_3 \\ R_2C_1 & R_2C_2 & R_2C_3 \\ R_3C_1 & R_3C_2 & R_3C_3 \end{bmatrix} = \begin{bmatrix} 7(12)+11(3) & 7(4)+11(6) & 7(5)+11(1) \\ 2(12)+9(3) & 2(4)+9(6) & 2(5)+9(1) \\ 10(12)+6(3) & 10(4)+6(6) & 10(5)+6(1) \end{bmatrix} = \begin{bmatrix} 117 & 94 & 46 \\ 51 & 62 & 19 \\ 138 & 76 & 56 \end{bmatrix}$$

10.25. Redo Problem 10.19, given

$$\mathbf{A} = \begin{bmatrix} 6 & 2 & 5 \\ 7 & 9 & 4 \end{bmatrix} \qquad \mathbf{B} = \begin{bmatrix} 10 & 1 \\ 11 & 3 \\ 2 & 9 \end{bmatrix}$$

Matrix **AB** is defined; $2 \times (3 = 3) \times 2$; the product matrix will be 2×2.

$$\mathbf{AB} = \begin{bmatrix} R_1C_1 & R_1C_2 \\ R_2C_1 & R_2C_2 \end{bmatrix} = \begin{bmatrix} 6(10)+2(11)+5(2) & 6(1)+2(3)+5(9) \\ 7(10)+9(11)+4(2) & 7(1)+9(3)+4(9) \end{bmatrix} = \begin{bmatrix} 92 & 57 \\ 177 & 70 \end{bmatrix}$$

10.26. Redo Problem 10.19, given

$$\mathbf{A} = \begin{bmatrix} 2 & 3 & 5 \end{bmatrix} \qquad \mathbf{B} = \begin{bmatrix} 7 & 1 & 6 \\ 5 & 2 & 4 \\ 9 & 2 & 7 \end{bmatrix}$$

Matrix **AB** is defined; $1 \times (3 = 3) \times 3$. The product matrix will be 1×3.

$$\mathbf{AB} = \begin{bmatrix} R_1C_1 & R_1C_2 & R_1C_3 \end{bmatrix} = \begin{bmatrix} 2(7)+3(5)+5(9) & 2(1)+3(2)+5(2) & 2(6)+3(4)+5(7) \end{bmatrix} = \begin{bmatrix} 74 & 18 & 59 \end{bmatrix}$$

10.27. Redo Problem 10.19, given

$$\mathbf{A} = \begin{bmatrix} 5 \\ 1 \\ 10 \end{bmatrix} \qquad \mathbf{B} = \begin{bmatrix} 3 & 9 & 4 \\ 2 & 1 & 8 \\ 5 & 6 & 1 \end{bmatrix}$$

Matrix **AB** is not defined; $3 \times (1 \neq 3) \times 3$. Multiplication is impossible in the given order.

10.28. Find **BA** from Problem 10.27.

Matrix **BA** is defined; $3 \times (3 = 3) \times 1$. The product matrix will be 3×1.

$$\mathbf{BA} = \begin{bmatrix} 3 & 9 & 4 \\ 2 & 1 & 8 \\ 5 & 6 & 1 \end{bmatrix}\begin{bmatrix} 5 \\ 1 \\ 10 \end{bmatrix} = \begin{bmatrix} R_1C_1 \\ R_2C_1 \\ R_3C_1 \end{bmatrix} = \begin{bmatrix} 3(5)+9(1)+4(10) \\ 2(5)+1(1)+8(10) \\ 5(5)+6(1)+1(10) \end{bmatrix} = \begin{bmatrix} 64 \\ 91 \\ 41 \end{bmatrix}$$

10.29. Redo Problem 10.19, given

$$\mathbf{A} = \begin{bmatrix} 2 & 1 & 5 \\ 3 & 2 & 6 \\ 1 & 4 & 3 \end{bmatrix} \qquad \mathbf{B} = \begin{bmatrix} 10 & 1 & 2 \\ 5 & 3 & 6 \\ 2 & 1 & 2 \end{bmatrix}$$

Matrix **AB** is defined; $3 \times (3 = 3) \times 3$. The product matrix will be 3×3.

$$\mathbf{AB} = \begin{bmatrix} R_1C_1 & R_1C_2 & R_1C_3 \\ R_2C_1 & R_2C_2 & R_2C_3 \\ R_3C_1 & R_3C_2 & R_3C_3 \end{bmatrix} = \begin{bmatrix} 2(10)+1(5)+5(2) & 2(1)+1(3)+5(1) & 2(2)+1(6)+5(2) \\ 3(10)+2(5)+6(2) & 3(1)+2(3)+6(1) & 3(2)+2(6)+6(2) \\ 1(10)+4(5)+3(2) & 1(1)+4(3)+3(1) & 1(2)+4(6)+3(2) \end{bmatrix}$$

$$= \begin{bmatrix} 35 & 10 & 20 \\ 52 & 15 & 30 \\ 36 & 16 & 32 \end{bmatrix}$$

10.30. Redo Problem 10.19, given

$$\mathbf{A} = \begin{bmatrix} 3 \\ 1 \\ 4 \\ 5 \end{bmatrix} \qquad \mathbf{B} = \begin{bmatrix} 2 & 6 & 5 & 3 \end{bmatrix}$$

Matrix **AB** is defined; $4 \times (1 = 1) \times 4$. The product matrix will be 4×4.

$$\mathbf{AB} = \begin{bmatrix} R_1C_1 & R_1C_2 & R_1C_3 & R_1C_4 \\ R_2C_1 & R_2C_2 & R_2C_3 & R_2C_4 \\ R_3C_1 & R_3C_2 & R_3C_3 & R_3C_4 \\ R_4C_1 & R_4C_2 & R_4C_3 & R_4C_4 \end{bmatrix} = \begin{bmatrix} 3(2) & 3(6) & 3(5) & 3(3) \\ 1(2) & 1(6) & 1(5) & 1(3) \\ 4(2) & 4(6) & 4(5) & 4(3) \\ 5(2) & 5(6) & 5(5) & 5(3) \end{bmatrix} = \begin{bmatrix} 6 & 18 & 15 & 9 \\ 2 & 6 & 5 & 3 \\ 8 & 24 & 20 & 12 \\ 10 & 30 & 25 & 15 \end{bmatrix}$$

10.31. Find **AB** when

$$\mathbf{A} = \begin{bmatrix} 3 & 9 & 8 & 7 \end{bmatrix} \qquad \mathbf{B} = \begin{bmatrix} 2 \\ 5 \\ 3 \end{bmatrix}$$

Matrix **AB** is undefined and cannot be multiplied as given; $1 \times (4 \neq 3) \times 1$.

10.32. Find **BA** from Problem 10.31.

Matrix **BA** is defined; $3 \times (1 = 1) \times 4$. The product matrix will be 3×4.

$$\mathbf{BA} = \begin{bmatrix} R_1C_1 & R_1C_2 & R_1C_3 & R_1C_4 \\ R_2C_1 & R_2C_2 & R_2C_3 & R_2C_4 \\ R_3C_1 & R_3C_2 & R_3C_3 & R_3C_4 \end{bmatrix} = \begin{bmatrix} 2 \\ 5 \\ 3 \end{bmatrix} \begin{bmatrix} 3 & 9 & 8 & 7 \end{bmatrix}$$

$$= \begin{bmatrix} 2(3) & 2(9) & 2(8) & 2(7) \\ 5(3) & 5(9) & 5(8) & 5(7) \\ 3(3) & 3(9) & 3(8) & 3(7) \end{bmatrix} = \begin{bmatrix} 6 & 18 & 16 & 14 \\ 15 & 45 & 40 & 35 \\ 9 & 27 & 24 & 21 \end{bmatrix}$$

10.33. Use the inventory matrix for the company in Problem 10.3 and the price vector from Problem 10.17 to determine the value of inventory in all five of the company's outlets.

$\mathbf{V} = \mathbf{QP}.$ $\mathbf{QP}$ is defined; $5 \times (4 \ = \ 4) \times 1$; $\mathbf{V}$ will be 5×1.

$$\mathbf{V} = \begin{bmatrix} 10 & 15 & 9 & 12 \\ 20 & 14 & 8 & 5 \\ 16 & 8 & 15 & 6 \\ 25 & 15 & 7 & 16 \\ 5 & 12 & 20 & 18 \end{bmatrix} \begin{bmatrix} 300 \\ 250 \\ 175 \\ 125 \end{bmatrix} = \begin{bmatrix} R_1 C_1 \\ R_2 C_1 \\ R_3 C_1 \\ R_4 C_1 \\ R_5 C_1 \end{bmatrix} = \begin{bmatrix} 10(300) + 15(250) + 9(175) + 12(125) \\ 20(300) + 14(250) + 8(175) + 5(125) \\ 16(300) + 8(250) + 15(175) + 6(125) \\ 25(300) + 15(250) + 7(175) + 16(125) \\ 5(300) + 12(250) + 20(175) + 18(125) \end{bmatrix} = \begin{bmatrix} 9,825 \\ 11,525 \\ 10,175 \\ 14,475 \\ 10,250 \end{bmatrix}$$

THE COMMUTATIVE LAW AND MATRIX OPERATIONS

10.34. To illustrate the commutative or noncommutative aspects of matrix operations (that is, $\mathbf{A} \pm \mathbf{B} = \mathbf{B} \pm \mathbf{A}$, but in general, $\mathbf{AB} \neq \mathbf{BA}$), find (a) $\mathbf{A} + \mathbf{B}$ and (b) $\mathbf{B} + \mathbf{A}$, given

$$\mathbf{A} = \begin{bmatrix} 7 & 3 & 2 \\ 1 & 4 & 6 \\ 2 & 5 & 4 \end{bmatrix} \qquad \mathbf{B} = \begin{bmatrix} 2 & 0 & 5 \\ 3 & 4 & 1 \\ 7 & 9 & 6 \end{bmatrix}$$

a) $\mathbf{A} + \mathbf{B} = \begin{bmatrix} 7+2 & 3+0 & 2+5 \\ 1+3 & 4+4 & 6+1 \\ 2+7 & 5+9 & 4+6 \end{bmatrix} = \begin{bmatrix} 9 & 3 & 7 \\ 4 & 8 & 7 \\ 9 & 14 & 10 \end{bmatrix}$

b) $\mathbf{B} + \mathbf{A} = \begin{bmatrix} 2+7 & 0+3 & 5+2 \\ 3+1 & 4+4 & 1+6 \\ 7+2 & 9+5 & 6+4 \end{bmatrix} = \begin{bmatrix} 9 & 3 & 7 \\ 4 & 8 & 7 \\ 9 & 14 & 10 \end{bmatrix}$

$\mathbf{A} + \mathbf{B} = \mathbf{B} + \mathbf{A}$. This illustrates that the commutative law does apply to matrix addition.

Problems 10.35 to 10.42 illustrate the application of the commutative law to other matrix operations.

10.35. Find (a) $\mathbf{A} - \mathbf{B}$ and (b) $-\mathbf{B} + \mathbf{A}$ given

$$\mathbf{A} = \begin{bmatrix} 5 & 3 \\ 4 & 9 \\ 10 & 8 \\ 6 & 12 \end{bmatrix} \qquad \mathbf{B} = \begin{bmatrix} 3 & 13 \\ 7 & 9 \\ 2 & 1 \\ 8 & 6 \end{bmatrix}$$

a) $\mathbf{A} - \mathbf{B} = \begin{bmatrix} 5-3 & 3-13 \\ 4-7 & 9-9 \\ 10-2 & 8-1 \\ 6-8 & 12-6 \end{bmatrix} = \begin{bmatrix} 2 & -10 \\ -3 & 0 \\ 8 & 7 \\ -2 & 6 \end{bmatrix}$

b) $-\mathbf{B} + \mathbf{A} = \begin{bmatrix} -3+5 & -13+3 \\ -7+4 & -9+9 \\ -2+10 & -1+8 \\ -8+6 & -6+12 \end{bmatrix} = \begin{bmatrix} 2 & -10 \\ -3 & 0 \\ 8 & 7 \\ -2 & 6 \end{bmatrix}$

$\mathbf{A} - \mathbf{B} = -\mathbf{B} + \mathbf{A}$. This illustrates that matrix subtraction is commutative.

10.36. Find (*a*) **AB** and (*b*) **BA**, given

$$\mathbf{A} = [4 \quad 12 \quad 9 \quad 6] \qquad \mathbf{B} = \begin{bmatrix} 13 \\ 5 \\ -2 \\ 7 \end{bmatrix}$$

Check for conformability first and indicate the dimensions of the product matrix.

a) Matrix **AB** is defined; $1 \times 4 = 4 \times 1$. The product will be a 1×1 matrix or scalar.

$$\mathbf{AB} = [4(13) + 12(5) + 9(-2) + 6(7)] = 136$$

b) Matrix **BA** is also defined; $4 \times 1 = 1 \times 4$; the product will be a 4×4 matrix.

$$\mathbf{BA} = \begin{bmatrix} 13(4) & 13(12) & 13(9) & 13(6) \\ 5(4) & 5(12) & 5(9) & 5(6) \\ -2(4) & -2(12) & -2(9) & -2(6) \\ 7(4) & 7(12) & 7(9) & 7(6) \end{bmatrix} = \begin{bmatrix} 52 & 156 & 117 & 78 \\ 20 & 60 & 45 & 30 \\ -8 & -24 & -18 & -12 \\ 28 & 84 & 63 & 42 \end{bmatrix}$$

$\mathbf{AB} \neq \mathbf{BA}$. This illustrates the noncommutative aspect of matrix multiplication. Products generally differ in dimensions and elements if the order of multiplication is reversed.

10.37. Find (*a*) **AB** and (*b*) **BA**, given

$$\mathbf{A} = \begin{bmatrix} 7 & 4 \\ 6 & 2 \\ 1 & 8 \end{bmatrix} \qquad \mathbf{B} = \begin{bmatrix} -3 & 9 & 1 \\ 2 & 12 & 7 \end{bmatrix}$$

a) Matrix **AB** is defined; $3 \times 2 = 2 \times 3$; the product will be 3×3.

$$\mathbf{AB} = \begin{bmatrix} 7(-3) + 4(2) & 7(9) + 4(12) & 7(1) + 4(7) \\ 6(-3) + 2(2) & 6(9) + 2(12) & 6(1) + 2(7) \\ 1(-3) + 8(2) & 1(9) + 8(12) & 1(1) + 8(7) \end{bmatrix} = \begin{bmatrix} -13 & 111 & 35 \\ -14 & 78 & 20 \\ 13 & 105 & 57 \end{bmatrix}$$

b) Matrix **BA** is also defined; $2 \times 3 = 3 \times 2$; the product will be 2×2.

$$\mathbf{BA} = \begin{bmatrix} -3(7) + 9(6) + 1(1) & -3(4) + 9(2) + 1(8) \\ 2(7) + 12(6) + 7(1) & 2(4) + 12(2) + 7(8) \end{bmatrix} = \begin{bmatrix} 34 & 14 \\ 93 & 88 \end{bmatrix}$$

$\mathbf{AB} \neq \mathbf{BA}$. Matrix multiplication is not commutative. Here the products again differ in dimensions and elements.

10.38. Find (*a*) **AB** and (*b*) **BA**, given

$$\mathbf{A} = \begin{bmatrix} 4 & 9 & 8 \\ 7 & 6 & 2 \\ 1 & 5 & 3 \end{bmatrix} \qquad \mathbf{B} = \begin{bmatrix} 1 & 2 & 0 \\ 5 & 3 & 1 \\ 0 & 2 & 4 \end{bmatrix}$$

a) Matrix **AB** is defined; $3 \times 3 = 3 \times 3$; the product will be 3×3.

$$\mathbf{AB} = \begin{bmatrix} 4(1) + 9(5) + 8(0) & 4(2) + 9(3) + 8(2) & 4(0) + 9(1) + 8(4) \\ 7(1) + 6(5) + 2(0) & 7(2) + 6(3) + 2(2) & 7(0) + 6(1) + 2(4) \\ 1(1) + 5(5) + 3(0) & 1(2) + 5(3) + 3(2) & 1(0) + 5(1) + 3(4) \end{bmatrix} = \begin{bmatrix} 49 & 51 & 41 \\ 37 & 36 & 14 \\ 26 & 23 & 17 \end{bmatrix}$$

b) Matrix **BA** is also defined and will result in a 3 × 3 matrix.

$$\mathbf{BA} = \begin{bmatrix} 1(4)+2(7)+0(1) & 1(9)+2(6)+0(5) & 1(8)+2(2)+0(3) \\ 5(4)+3(7)+1(1) & 5(9)+3(6)+1(5) & 5(8)+3(2)+1(3) \\ 0(4)+2(7)+4(1) & 0(9)+2(6)+4(5) & 0(8)+2(2)+4(3) \end{bmatrix} = \begin{bmatrix} 18 & 21 & 12 \\ 42 & 68 & 49 \\ 18 & 32 & 16 \end{bmatrix}$$

AB ≠ **BA**. The dimensions are the same but the elements differ.

10.39. Find (a) **AB** and (b) **BA**, given

$$\mathbf{A} = \begin{bmatrix} 7 & 5 & 2 & 6 \\ 1 & 3 & 9 & 4 \end{bmatrix} \qquad \mathbf{B} = \begin{bmatrix} 1 \\ 0 \\ -1 \\ 3 \end{bmatrix}$$

a) Matrix **AB** is defined; 2 × (4 = 4) × 1; the product will be 2 × 1.

$$\mathbf{AB} = \begin{bmatrix} 7(1)+5(0)+2(-1)+6(3) \\ 1(1)+3(0)+9(-1)+4(3) \end{bmatrix} = \begin{bmatrix} 23 \\ 4 \end{bmatrix}$$

b) Matrix **BA** is not defined; 4 × (1 ≠ 2) × 4. Multiplication is impossible. This is but another way in which matrix multiplication is noncommutative.

10.40. Find (a) **AB** and (b) **BA**, given

$$\mathbf{A} = \begin{bmatrix} 11 & 14 \\ 2 & 6 \end{bmatrix} \qquad \mathbf{B} = \begin{bmatrix} 7 & 6 \\ 4 & 5 \\ 1 & 3 \end{bmatrix}$$

a) Matrix **AB** is not defined; 2 × (2 ≠ 3) × 2 and so cannot be multiplied.

b) Matrix **BA** is defined; 3 × (2 = 2) × 2 and will produce a 3 × 2 matrix.

$$\mathbf{BA} = \begin{bmatrix} 7(11)+6(2) & 7(14)+6(6) \\ 4(11)+5(2) & 4(14)+5(6) \\ 1(11)+3(2) & 1(14)+3(6) \end{bmatrix} = \begin{bmatrix} 89 & 134 \\ 54 & 86 \\ 17 & 32 \end{bmatrix}$$

BA ≠ **AB**, because **AB** does not exist.

10.41. Find (a) **AB** and (b) **BA**, given

$$\mathbf{A} = \begin{bmatrix} -2 \\ 4 \\ 7 \end{bmatrix} \qquad \mathbf{B} = \begin{bmatrix} 3 & 6 & -2 \end{bmatrix}$$

a) Matrix **AB** is defined; 3 × (1 = 1) × 3; the product will be a 3 × 3 matrix.

$$\mathbf{AB} = \begin{bmatrix} -2(3) & -2(6) & -2(-2) \\ 4(3) & 4(6) & 4(-2) \\ 7(3) & 7(6) & 7(-2) \end{bmatrix} = \begin{bmatrix} -6 & -12 & 4 \\ 12 & 24 & -8 \\ 21 & 42 & -14 \end{bmatrix}$$

b) Matrix **BA** is also defined; 1 × (3 = 3) × 1, producing a 1 × 1 matrix or scalar.

$$\mathbf{BA} = [3(-2)+6(4)+(-2)(7)] = 4$$

Since matrix multiplication is not commutative, reversing the order of multiplication can lead to widely different answers. Matrix **AB** results in a 3 × 3 matrix, **BA** results in a scalar.

10.42. Find (*a*) **AB** and (*b*) **BA**, for a case where **B** is an identity matrix, given

$$\mathbf{A} = \begin{bmatrix} 23 & 6 & 14 \\ 18 & 12 & 9 \\ 24 & 2 & 6 \end{bmatrix} \qquad \mathbf{B} = \begin{bmatrix} 1 & 0 & 0 \\ 0 & 1 & 0 \\ 0 & 0 & 1 \end{bmatrix}$$

a)   Matrix **AB** is defined; 3 × ③ = ③ × 3. The product matrix will also be 3 × 3.

$$\mathbf{AB} = \begin{bmatrix} 23(1)+6(0)+14(0) & 23(0)+6(1)+14(0) & 23(0)+6(0)+14(1) \\ 18(1)+12(0)+9(0) & 18(0)+12(1)+9(0) & 18(0)+12(0)+9(1) \\ 24(1)+2(0)+6(0) & 24(0)+2(1)+6(0) & 24(0)+2(0)+6(1) \end{bmatrix} = \begin{bmatrix} 23 & 6 & 14 \\ 18 & 12 & 9 \\ 24 & 2 & 6 \end{bmatrix}$$

b)   Matrix **BA** is also defined; 3 × ③ = ③ × 3. The product matrix will also be 3 × 3.

$$\mathbf{BA} = \begin{bmatrix} 1(23)+0(18)+0(24) & 1(6)+0(12)+0(2) & 1(14)+0(9)+0(6) \\ 0(23)+1(18)+0(24) & 0(6)+1(12)+0(2) & 0(14)+1(9)+0(6) \\ 0(23)+0(18)+1(24) & 0(6)+0(12)+1(2) & 0(14)+0(9)+1(6) \end{bmatrix} = \begin{bmatrix} 23 & 6 & 14 \\ 18 & 12 & 9 \\ 24 & 2 & 6 \end{bmatrix}$$

Here **AB = BA**. Premultiplication or postmultiplication by an identity matrix gives the original matrix. Thus in the case of an identity matrix, matrix multiplication is commutative. This will also be true of a matrix and its inverse. See Section 11.7.

ASSOCIATIVE AND DISTRIBUTIVE LAWS

10.43. To illustrate whether the associative and distributive laws apply to matrix operations [that is, (**A** + **B**) + **C** = **A** + (**B** + **C**), (**AB**)**C** = **A**(**BC**), and **A**(**B** + **C**) = **AB** + **AC**, subject to the conditions in Section 10.7], find (*a*) (**A** + **B**) + **C** and (*b*) **A** + (**B** + **C**), given

$$\mathbf{A} = \begin{bmatrix} 6 & 2 & 7 \\ 9 & 5 & 3 \end{bmatrix} \qquad \mathbf{B} = \begin{bmatrix} 9 & 1 & 3 \\ 4 & 2 & 6 \end{bmatrix} \qquad \mathbf{C} = \begin{bmatrix} 7 & 5 & 1 \\ 10 & 3 & 8 \end{bmatrix}$$

a)
$$\mathbf{A} + \mathbf{B} = \begin{bmatrix} 6+9 & 2+1 & 7+3 \\ 9+4 & 5+2 & 3+6 \end{bmatrix} = \begin{bmatrix} 15 & 3 & 10 \\ 13 & 7 & 9 \end{bmatrix}$$

$$(\mathbf{A} + \mathbf{B}) + \mathbf{C} = \begin{bmatrix} 15+7 & 3+5 & 10+1 \\ 13+10 & 7+3 & 9+8 \end{bmatrix} = \begin{bmatrix} 22 & 8 & 11 \\ 23 & 10 & 17 \end{bmatrix}$$

b)
$$\mathbf{B} + \mathbf{C} = \begin{bmatrix} 9+7 & 1+5 & 3+1 \\ 4+10 & 2+3 & 6+8 \end{bmatrix} = \begin{bmatrix} 16 & 6 & 4 \\ 14 & 5 & 14 \end{bmatrix}$$

$$\mathbf{A} + (\mathbf{B} + \mathbf{C}) = \begin{bmatrix} 6+16 & 2+6 & 7+4 \\ 9+14 & 5+5 & 3+14 \end{bmatrix} = \begin{bmatrix} 22 & 8 & 11 \\ 23 & 10 & 17 \end{bmatrix}$$

Thus, (**A** + **B**) + **C** = **A** + (**B** + **C**). This illustrates that matrix addition is associative. Other aspects of these laws are demonstrated in Problems 10.44 to 10.47.

10.44. Find (*a*) (**A** − **B**) + **C** and (*b*) **A** + (−**B** + **C**), given

$$\mathbf{A} = \begin{bmatrix} 7 \\ 6 \\ 12 \end{bmatrix} \qquad \mathbf{B} = \begin{bmatrix} 3 \\ 8 \\ 5 \end{bmatrix} \qquad \mathbf{C} = \begin{bmatrix} 13 \\ 2 \\ 6 \end{bmatrix}$$

a) $\mathbf{A} - \mathbf{B} = \begin{bmatrix} 7-3 \\ 6-8 \\ 12-5 \end{bmatrix} = \begin{bmatrix} 4 \\ -2 \\ 7 \end{bmatrix}$ *b*) $-\mathbf{B} + \mathbf{C} = \begin{bmatrix} -3+13 \\ -8+2 \\ -5+6 \end{bmatrix} = \begin{bmatrix} 10 \\ -6 \\ 1 \end{bmatrix}$

$(\mathbf{A} - \mathbf{B}) + \mathbf{C} = \begin{bmatrix} 4+13 \\ -2+2 \\ 7+6 \end{bmatrix} = \begin{bmatrix} 17 \\ 0 \\ 13 \end{bmatrix}$ $\mathbf{A} + (-\mathbf{B} + \mathbf{C}) = \begin{bmatrix} 7+10 \\ 6+(-6) \\ 12+1 \end{bmatrix} = \begin{bmatrix} 17 \\ 0 \\ 13 \end{bmatrix}$

Matrix subtraction is also associative.

10.45. Find (*a*) **(AB)C** and (*b*) **A(BC)**, given

$$\mathbf{A} = [7 \quad 1 \quad 5] \quad \mathbf{B} = \begin{bmatrix} 6 & 5 \\ 2 & 4 \\ 3 & 8 \end{bmatrix} \quad \mathbf{C} = \begin{bmatrix} 9 & 4 \\ 3 & 10 \end{bmatrix}$$

a) Matrix **AB** is defined; $1 \times 3 = 3 \times 2$, producing a 1×2 matrix.

$$\mathbf{AB} = [7(6) + 1(2) + 5(3) \quad 7(5) + 1(4) + 5(8)] = [59 \quad 79]$$

Matrix **(AB)C** is defined; $1 \times 2 = 2 \times 2$, leaving a 1×2 matrix.

$$\mathbf{(AB)C} = [59(9) + 79(3) \quad 59(4) + 79(10)] = [768 \quad 1026]$$

Matrix **BC** is defined; $3 \times 2 = 2 \times 2$, creating a 3×2 matrix.

$$\mathbf{BC} = \begin{bmatrix} 6(9) + 5(3) & 6(4) + 5(10) \\ 2(9) + 4(3) & 2(4) + 4(10) \\ 3(9) + 8(3) & 3(4) + 8(10) \end{bmatrix} = \begin{bmatrix} 69 & 74 \\ 30 & 48 \\ 51 & 92 \end{bmatrix}$$

Matrix **A(BC)** is also defined; $1 \times 3 = 3 \times 2$, producing a 1×2 matrix.

$$\mathbf{A(BC)} = [7(69) + 1(30) + 5(51) \quad 7(74) + 1(48) + 5(92)] = [768 \quad 1026]$$

Matrix multiplication is associative, provided the proper order of multiplication is maintained.

10.46. Find (*a*) **A(B + C)** and (*b*) **AB + AC**, given

$$\mathbf{A} = [4 \quad 7 \quad 2] \quad \mathbf{B} = \begin{bmatrix} 6 \\ 5 \\ 1 \end{bmatrix} \quad \mathbf{C} = \begin{bmatrix} 9 \\ 5 \\ 8 \end{bmatrix}$$

a)
$$\mathbf{B + C} = \begin{bmatrix} 6 + 9 \\ 5 + 5 \\ 1 + 8 \end{bmatrix} = \begin{bmatrix} 15 \\ 10 \\ 9 \end{bmatrix}$$

Matrix **A(B + C)** is defined; $1 \times 3 = 3 \times 1$. The product matrix will be 1×1.

$$\mathbf{A(B + C)} = [4(15) + 7(10) + 2(9)] = 148$$

b) Matrix **AB** is defined; $1 \times 3 = 3 \times 1$, producing a 1×1 matrix.

$$\mathbf{AB} = [4(6) + 7(5) + 2(1)] = 61$$

Matrix **AC** is defined; $1 \times 3 = 3 \times 1$, also producing a 1×1 matrix.

$$\mathbf{AC} = [4(9) + 7(5) + 2(8)] = 87$$

Thus, **AB + AC** = 61 + 87 = 148. This illustrates the distributive law of matrix multiplication.

10.47. A hamburger chain sells 1000 hamburgers, 600 cheeseburgers, and 1200 milk shakes in a week. The price of a hamburger is 45¢, a cheeseburger 60¢, and a milk shake 50¢. The cost to the chain of a hamburger is 38¢, a cheeseburger 42¢, and a milk shake 32¢. Find the firm's profit for the week, using (*a*) total concepts and (*b*) per-unit analysis to prove that matrix multiplication is distributive.

a) The quantity of goods sold **Q**, the selling price of the goods **P**, and the cost of goods **C**, can all be represented in matrix form:

$$\mathbf{Q} = \begin{bmatrix} 1000 \\ 600 \\ 1200 \end{bmatrix} \qquad \mathbf{P} = \begin{bmatrix} 0.45 \\ 0.60 \\ 0.50 \end{bmatrix} \qquad \mathbf{C} = \begin{bmatrix} 0.38 \\ 0.42 \\ 0.32 \end{bmatrix}$$

Total revenue TR is

$$\text{TR} = \mathbf{PQ} = \begin{bmatrix} 0.45 \\ 0.60 \\ 0.50 \end{bmatrix} \begin{bmatrix} 1000 \\ 600 \\ 1200 \end{bmatrix}$$

which is not defined as given. Taking the transpose of **P** or **Q** will render the vectors conformable for multiplication. Note that the order of multiplication is all-important. Row-vector multiplication (**P′Q** or **Q′P**) will produce the scalar required; vector-row multiplication (**PQ′** or **QP′**) will produce a 3 × 3 matrix that has no economic meaning. Thus, taking the transpose of **P** and premultiplying, we get

$$\text{TR} = \mathbf{P'Q} = \begin{bmatrix} 0.45 & 0.60 & 0.50 \end{bmatrix} \begin{bmatrix} 1000 \\ 600 \\ 1200 \end{bmatrix}$$

where **P′Q** is defined; $1 \times (3 = 3) \times 1$, producing a 1 × 1 matrix or scalar.

$$\text{TR} = [0.45(1000) + 0.60(600) + 0.50(1200)] = 1410$$

Similarly, total cost TC is TC = **C′Q**:

$$\text{TC} = \begin{bmatrix} 0.38 & 0.42 & 0.32 \end{bmatrix} \begin{bmatrix} 1000 \\ 600 \\ 1200 \end{bmatrix} = [0.38(1000) + 0.42(600) + 0.32(1200)] = 1016$$

Profits, therefore, are

$$\Pi = \text{TR} - \text{TC} = 1410 - 1016 = 394$$

b) Using per-unit analysis, the per-unit profit **U** is

$$\mathbf{U} = \mathbf{P} - \mathbf{C} = \begin{bmatrix} 0.45 \\ 0.60 \\ 0.50 \end{bmatrix} - \begin{bmatrix} 0.38 \\ 0.42 \\ 0.32 \end{bmatrix} = \begin{bmatrix} 0.07 \\ 0.18 \\ 0.18 \end{bmatrix}$$

Total profit Π is per-unit profit times the number of items sold

$$\Pi = \mathbf{UQ} = \begin{bmatrix} 0.07 \\ 0.18 \\ 0.18 \end{bmatrix} \begin{bmatrix} 1000 \\ 600 \\ 1200 \end{bmatrix}$$

which is undefined. Taking the transpose of **U**,

$$\Pi = \mathbf{U'P} = \begin{bmatrix} 0.07 & 0.18 & 0.18 \end{bmatrix} \begin{bmatrix} 1000 \\ 600 \\ 1200 \end{bmatrix}$$

$$= [0.07(1000) + 0.18(600) + 0.18(1200)] = 394 \qquad \text{Q.E.D.}$$

10.48. Crazy Teddie's sells 700 CDs, 400 cassettes, and 200 CD players each week. The selling price of CDs is \$4, cassettes \$6, and CD players \$150. The cost to the shop is \$3.25 for a CD, \$4.75 for a cassette, and \$125 for a CD player. Find weekly profits by using (a) total and (b) per-unit concepts.

a)
$$\mathbf{Q} = \begin{bmatrix} 700 \\ 400 \\ 200 \end{bmatrix} \quad \mathbf{P} = \begin{bmatrix} 4 \\ 6 \\ 150 \end{bmatrix} \quad \mathbf{C} = \begin{bmatrix} 3.25 \\ 4.75 \\ 125.00 \end{bmatrix}$$

$$\mathrm{TR} = \mathbf{P'Q} = \begin{bmatrix} 4 & 6 & 150 \end{bmatrix} \begin{bmatrix} 700 \\ 400 \\ 200 \end{bmatrix} = [4(700) + 6(400) + 150(200)] = 35{,}200$$

$$\mathrm{TC} = \mathbf{C'Q} = \begin{bmatrix} 3.25 & 4.75 & 125 \end{bmatrix} \begin{bmatrix} 700 \\ 400 \\ 200 \end{bmatrix} = [3.25(700) + 4.75(400) + 125(200)] = 29{,}175$$

$$\Pi = \mathrm{TR} - \mathrm{TC} = 35{,}200 - 29{,}175 = 6025$$

b) Per-unit profit $\mathbf{U}$ is

$$\mathbf{U} = \mathbf{P} - \mathbf{C} = \begin{bmatrix} 4 \\ 6 \\ 150 \end{bmatrix} - \begin{bmatrix} 3.25 \\ 4.75 \\ 125.00 \end{bmatrix} = \begin{bmatrix} 0.75 \\ 1.25 \\ 25.00 \end{bmatrix}$$

Total profit Π is

$$\Pi = \mathbf{U'Q} = \begin{bmatrix} 0.75 & 1.25 & 25 \end{bmatrix} \begin{bmatrix} 700 \\ 400 \\ 200 \end{bmatrix} = [0.75(700) + 1.25(400) + 25(200)] = 6025$$

UNIQUE PROPERTIES OF MATRICES

10.49. Given

$$\mathbf{A} = \begin{bmatrix} 6 & -12 \\ -3 & 6 \end{bmatrix} \quad \mathbf{B} = \begin{bmatrix} 12 & 6 \\ 6 & 3 \end{bmatrix}$$

(a) Find $\mathbf{AB}$. (b) Why is the product unique?

a)
$$\mathbf{AB} = \begin{bmatrix} 6(12) - 12(6) & 6(6) - 12(3) \\ -3(12) + 6(6) & -3(6) + 6(3) \end{bmatrix} = \begin{bmatrix} 0 & 0 \\ 0 & 0 \end{bmatrix}$$

b) The product $\mathbf{AB}$ is unique to matrix algebra in that, unlike ordinary algebra in which the product of two nonzero numbers can never equal zero, the product of two non-null matrices may produce a null matrix. The reason for this is that the two original matrices are singular. A *singular matrix* is one in which a row or column is a multiple of another row or column (see Section 11.1). In this problem, row 1 of $\mathbf{A}$ is -2 times row 2, and column 2 is -2 times column 1. In $\mathbf{B}$, row 1 is 2 times row 2, and column 1 is 2 times column 2. Thus, in matrix algebra, multiplication involving singular matrices may, but need not, produce a null matrix as a solution. See Problem 10.50.

10.50. (a) Find $\mathbf{AB}$ and (b) comment on the solution, given

$$\mathbf{A} = \begin{bmatrix} 6 & 12 \\ 3 & 6 \end{bmatrix} \quad \mathbf{B} = \begin{bmatrix} 12 & 6 \\ 6 & 3 \end{bmatrix}$$

a)
$$\mathbf{AB} = \begin{bmatrix} 6(12) + 12(6) & 6(6) + 12(3) \\ 3(12) + 6(6) & 3(6) + 6(3) \end{bmatrix} = \begin{bmatrix} 144 & 72 \\ 72 & 36 \end{bmatrix}$$

b) While both $\mathbf{A}$ and $\mathbf{B}$ are singular, they do not produce a null matrix. The product $\mathbf{AB}$, however, is also singular.

10.51. Given

$$\mathbf{A} = \begin{bmatrix} 4 & 8 \\ 1 & 2 \end{bmatrix} \quad \mathbf{B} = \begin{bmatrix} 2 & 1 \\ 2 & 2 \end{bmatrix} \quad \mathbf{C} = \begin{bmatrix} -2 & 1 \\ 4 & 2 \end{bmatrix}$$

a) Find **AB** and **AC**. (*b*) Comment on the unusual property of the solutions.

a)
$$\mathbf{AB} = \begin{bmatrix} 4(2) + 8(2) & 4(1) + 8(2) \\ 1(2) + 2(2) & 1(1) + 2(2) \end{bmatrix} = \begin{bmatrix} 24 & 20 \\ 6 & 5 \end{bmatrix}$$

$$\mathbf{AC} = \begin{bmatrix} 4(-2) + 8(4) & 4(1) + 8(2) \\ 1(-2) + 2(4) & 1(1) + 2(2) \end{bmatrix} = \begin{bmatrix} 24 & 20 \\ 6 & 5 \end{bmatrix}$$

b) Even though **B** ≠ **C**, **AB** = **AC**. Unlike algebra, where multiplication of one number by two different numbers cannot give the same product, in matrix algebra multiplication of one matrix by two different matrices may, but need not, produce identical matrices. In this case, **A** is a singular matrix.

CHAPTER 11

Matrix Inversion

11.1 DETERMINANTS AND NONSINGULARITY

The determinant $|\mathbf{A}|$ of a 2×2 matrix, called a *second-order determinant*, is derived by taking the product of the two elements on the principal diagonal and subtracting from it the product of the two elements off the principal diagonal. Given a general 2×2 matrix

$$\mathbf{A} = \begin{bmatrix} a_{11} & a_{12} \\ a_{21} & a_{22} \end{bmatrix}$$

the determinant is

$$|\mathbf{A}| = \begin{vmatrix} a_{11} & a_{12} \\ a_{21} & a_{22} \end{vmatrix} \begin{matrix} (-) \\ (+) \end{matrix} = a_{11}a_{22} - a_{12}a_{21}$$

The determinant is a single number or scalar and is found only for square matrices. If the determinant of a matrix is equal to zero, the determinant is said to *vanish* and the matrix is termed *singular*. A *singular matrix* is one in which there exists linear dependence between at least two rows or columns. If $|\mathbf{A}| \neq 0$, matrix $\mathbf{A}$ is *nonsingular* and all its rows and columns are linearly independent.

If linear dependence exists in a system of equations, the system as a whole will have an infinite number of possible solutions, making a unique solution impossible. Hence we want to preclude linearly dependent equations from our models and will generally fall back on the following simple determinant test to spot potential problems. Given a system of equations with coefficient matrix $\mathbf{A}$,

If $|\mathbf{A}| = 0$, the matrix is singular and there is linear dependence among the equations. No unique solution is possible.

If $|\mathbf{A}| \neq 0$, the matrix is nonsingular and there is no linear dependence among the equations. A unique solution can be found.

The *rank* ρ of a matrix is defined as the maximum number of linearly independent rows or columns in the matrix. The rank of a matrix also allows for a simple test of linear dependence which follows immediately. Assuming a square matrix of order n,

If $\rho(\mathbf{A}) = n$, $\mathbf{A}$ is nonsingular and there is no linear dependence.

If $\rho(\mathbf{A}) < n$, $\mathbf{A}$ is singular and there is linear dependence.

See Example 1 and Problems 11.1, 11.3, and 11.17. For proof of nonsingularity and linear independence, see Problem 11.16.

224

EXAMPLE 1. Determinants are calculated as follows, given

$$\mathbf{A} = \begin{bmatrix} 6 & 4 \\ 7 & 9 \end{bmatrix} \qquad \mathbf{B} = \begin{bmatrix} 4 & 6 \\ 6 & 9 \end{bmatrix}$$

From the rules stated above,

$$|\mathbf{A}| = 6(9) - 4(7) = 26$$

Since $|\mathbf{A}| \neq 0$, the matrix is nonsingular, i.e., there is no linear dependence between any of its rows or columns. The rank of $\mathbf{A}$ is 2, written $\rho(\mathbf{A}) = 2$. By way of contrast,

$$|\mathbf{B}| = 4(9) - 6(6) = 0$$

With $|\mathbf{B}| = 0$, $\mathbf{B}$ is singular and linear dependence exists between its rows and columns. Closer inspection reveals that row 2 and column 2 are equal to 1.5 times row 1 and column 1, respectively. Hence $\rho(\mathbf{B}) = 1$.

11.2 THIRD-ORDER DETERMINANTS

The determinant of a 3×3 matrix

$$\mathbf{A} = \begin{bmatrix} a_{11} & a_{12} & a_{13} \\ a_{21} & a_{22} & a_{23} \\ a_{31} & a_{32} & a_{33} \end{bmatrix}$$

is called a *third-order determinant* and is the summation of three products. To derive the three products:

1. Take the first element of the first row, a_{11}, and mentally delete the row and column in which it appears. See (a) below. Then multiply a_{11} by the determinant of the remaining elements.
2. Take the second element of the first row, a_{12}, and mentally delete the row and column in which it appears. See (b) below. Then multiply a_{12} by -1 times the determinant of the remaining elements.
3. Take the third element of the first row, a_{13}, and mentally delete the row and column in which it appears. See (c) below. Then multiply a_{13} by the determinant of the remaining elements.

$$\begin{bmatrix} \textcircled{a_{11}} & a_{12} & a_{13} \\ a_{21} & a_{22} & a_{23} \\ a_{31} & a_{32} & a_{33} \end{bmatrix} \qquad \begin{bmatrix} a_{11} & \textcircled{a_{12}} & a_{13} \\ a_{21} & a_{22} & a_{23} \\ a_{31} & a_{32} & a_{33} \end{bmatrix} \qquad \begin{bmatrix} a_{11} & a_{12} & \textcircled{a_{13}} \\ a_{21} & a_{22} & a_{23} \\ a_{31} & a_{32} & a_{33} \end{bmatrix}$$

$$(a) \qquad\qquad\qquad\qquad (b) \qquad\qquad\qquad\qquad (c)$$

Thus, the calculations for the determinant are as follows:

$$\begin{aligned} |\mathbf{A}| &= a_{11} \begin{vmatrix} a_{22} & a_{23} \\ a_{32} & a_{33} \end{vmatrix} + a_{12}(-1) \begin{vmatrix} a_{21} & a_{23} \\ a_{31} & a_{33} \end{vmatrix} + a_{13} \begin{vmatrix} a_{21} & a_{22} \\ a_{31} & a_{32} \end{vmatrix} \\ &= a_{11}(a_{22}a_{33} - a_{23}a_{32}) - a_{12}(a_{21}a_{33} - a_{23}a_{31}) + a_{13}(a_{21}a_{32} - a_{22}a_{31}) \qquad (11.1) \\ &= \text{a scalar} \end{aligned}$$

See Examples 2 and 3 and Problems 11.2, 11.3, and 11.17.

In like manner, the determinant of a 4×4 matrix is the sum of four products; the determinant of a 5×5 matrix is the sum of five products; etc. See Section 11.4 and Example 5.

EXAMPLE 2. Given

$$\mathbf{A} = \begin{bmatrix} 8 & 3 & 2 \\ 6 & 4 & 7 \\ 5 & 1 & 3 \end{bmatrix}$$

the determinant $|\mathbf{A}|$ is calculated as follows:

$$|\mathbf{A}| = 8\begin{vmatrix} 4 & 7 \\ 1 & 3 \end{vmatrix} + 3(-1)\begin{vmatrix} 6 & 7 \\ 5 & 3 \end{vmatrix} + 2\begin{vmatrix} 6 & 4 \\ 5 & 1 \end{vmatrix}$$

$$= 8[4(3) - 7(1)] - 3[6(3) - 7(5)] + 2[6(1) - 4(5)]$$

$$= 8(5) - 3(-17) + 2(-14) = 63$$

With $|\mathbf{A}| \neq 0$, $\mathbf{A}$ is nonsingular and $\rho(\mathbf{A}) = 3$.

11.3 MINORS AND COFACTORS

The elements of a matrix remaining after the deletion process described in Section 11.2 form a subdeterminant of the matrix called a *minor*. Thus, a *minor* $|M_{ij}|$ is the determinant of the submatrix formed by deleting the *i*th row and *j*th column of the matrix. Using the matrix from Section 11.2,

$$|M_{11}| = \begin{vmatrix} a_{22} & a_{23} \\ a_{32} & a_{33} \end{vmatrix} \qquad |M_{12}| = \begin{vmatrix} a_{21} & a_{23} \\ a_{31} & a_{33} \end{vmatrix} \qquad |M_{13}| = \begin{vmatrix} a_{21} & a_{22} \\ a_{31} & a_{32} \end{vmatrix}$$

where $|M_{11}|$ is the minor of a_{11}, $|M_{12}|$ the minor of a_{12}, and $|M_{13}|$ the minor of a_{13}. Thus, the determinant in (*11.1*) can be written

$$|\mathbf{A}| = a_{11}|M_{11}| + a_{12}(-1)|M_{12}| + a_{13}|M_{13}| \qquad\qquad (11.2)$$

A *cofactor* $|C_{ij}|$ is a minor with a prescribed sign. The rule for the sign of a cofactor is

$$|C_{ij}| = (-1)^{i+j}|M_{ij}|$$

Thus if the sum of the subscripts is an even number, $|C_{ij}| = |M_{ij}|$, since -1 raised to an even power is positive. If $i + j$ is equal to an odd number, $|C_{ij}| = -|M_{ij}|$, since -1 raised to an odd power is negative. See Example 3 and Problems 11.18 to 11.24.

EXAMPLE 3. The cofactors (1) $|C_{11}|$, (2) $|C_{12}|$, and (3) $|C_{13}|$ for the matrix in Section 11.2 are found as follows:

1)
$$|C_{11}| = (-1)^{1+1}|M_{11}|$$

Since $(-1)^{1+1} = (-1)^2 = 1$,

$$|C_{11}| = |M_{11}| = \begin{vmatrix} a_{22} & a_{23} \\ a_{32} & a_{33} \end{vmatrix}$$

2) $\qquad |C_{12}| = (-1)^{1+2}|M_{12}|$

Since $(-1)^{1+2} = (-1)^3 = -1$,

$$|C_{12}| = -|M_{12}| = -\begin{vmatrix} a_{21} & a_{23} \\ a_{31} & a_{33} \end{vmatrix}$$

3) $\qquad |C_{13}| = (-1)^{1+3}|M_{13}|$

Since $(-1)^{1+3} = (-1)^4 = 1$,

$$|C_{13}| = |M_{13}| = \begin{vmatrix} a_{21} & a_{22} \\ a_{31} & a_{32} \end{vmatrix}$$

11.4 LAPLACE EXPANSION AND HIGHER-ORDER DETERMINANTS

Laplace expansion is a method for evaluating determinants in terms of cofactors. It thus simplifies matters by permitting higher-order determinants to be established in terms of lower-order determinants. Laplace expansion of a third-order determinant can be expressed as

$$|\mathbf{A}| = a_{11}|C_{11}| + a_{12}|C_{12}| + a_{13}|C_{13}| \tag{11.3}$$

where $|C_{ij}|$ is a cofactor based on a second-order determinant. Here, unlike (11.1) and (11.2), a_{12} is not explicitly multiplied by -1, since by the rule of cofactors $|C_{12}|$ will automatically be multiplied by -1.

Laplace expansion permits evaluation of a determinant along any row or column. Selection of a row or column with more zeros than others simplifies evaluation of the determinant by eliminating terms. Laplace expansion also serves as the basis for evaluating determinants of orders higher than three. See Examples 4 and 5 and Problem 11.25.

EXAMPLE 4.   Given

$$\mathbf{A} = \begin{bmatrix} 12 & 7 & 0 \\ 5 & 8 & 3 \\ 6 & 7 & 0 \end{bmatrix}$$

the determinant is found by Laplace expansion along the third column, as demonstrated below:

$$|\mathbf{A}| = a_{13}|C_{13}| + a_{23}|C_{23}| + a_{33}|C_{33}|$$

Since a_{13} and $a_{33} = 0$

$$|\mathbf{A}| = a_{23}|C_{23}| \tag{11.4}$$

Deleting row 2 and column 3 to find $|C_{23}|$,

$$|C_{23}| = (-1)^{2+3} \begin{vmatrix} 12 & 7 \\ 6 & 7 \end{vmatrix}$$
$$= (-1)[12(7) - 7(6)] = -42$$

Then substituting in (11.4) where $a_{23} = 3$, $|\mathbf{A}| = 3(-42) = -126$. So $\mathbf{A}$ is nonsingular and $\rho(\mathbf{A}) = 3$.

The accuracy of this answer can be readily checked by expanding along the first row and solving for $|\mathbf{A}|$.

EXAMPLE 5.   Laplace expansion for a fourth-order determinant is

$$|\mathbf{A}| = a_{11}|C_{11}| + a_{12}|C_{12}| + a_{13}|C_{13}| + a_{14}|C_{14}|$$

where the cofactors are third-order subdeterminants which in turn can be reduced to second-order subdeterminants, as above. Fifth-order determinants and higher are treated in similar fashion. See Problem 11.25(d) to (e).

11.5 PROPERTIES OF A DETERMINANT

The following seven properties of determinants provide the ways in which a matrix can be manipulated to simplify its elements or reduce part of them to zero, before evaluating the determinant:

1. Adding or subtracting any nonzero multiple of one row (or column) from another row (or column) will have no effect on the determinant.
2. Interchanging any two rows or columns of a matrix will change the sign, but not the absolute value, of the determinant.
3. Multiplying the elements of any row or column by a constant will cause the determinant to be multiplied by the constant.

4. The determinant of a *triangular matrix*, i.e., a matrix with zero elements everywhere above *or* below the principal diagonal, is equal to the product of the elements on the principal diagonal.
5. The determinant of a matrix equals the determinant of its transpose: $|\mathbf{A}| = |\mathbf{A}'|$.
6. If all the elements of any row or column are zero, the determinant is zero.
7. If two rows or columns are identical or proportional, i.e., linearly dependent, the determinant is zero.

These properties and their use in matrix manipulation are treated in Problems 11.4 to 11.15.

11.6 COFACTOR AND ADJOINT MATRICES

A *cofactor matrix* is a matrix in which every element a_{ij} is replaced with its cofactor $|C_{ij}|$. An *adjoint matrix* is the transpose of a cofactor matrix. Thus,

$$\mathbf{C} = \begin{bmatrix} |C_{11}| & |C_{12}| & |C_{13}| \\ |C_{21}| & |C_{22}| & |C_{23}| \\ |C_{31}| & |C_{32}| & |C_{33}| \end{bmatrix} \qquad \text{Adj } \mathbf{A} = \mathbf{C}' = \begin{bmatrix} |C_{11}| & |C_{21}| & |C_{31}| \\ |C_{12}| & |C_{22}| & |C_{32}| \\ |C_{13}| & |C_{23}| & |C_{33}| \end{bmatrix}$$

EXAMPLE 6.   The cofactor matrix $\mathbf{C}$ and the adjoint matrix Adj $\mathbf{A}$ are found below, given

$$\mathbf{A} = \begin{bmatrix} 2 & 3 & 1 \\ 4 & 1 & 2 \\ 5 & 3 & 4 \end{bmatrix}$$

Replacing the elements a_{ij} with their cofactors $|C_{ij}|$ according to the laws of cofactors,

$$\mathbf{C} = \begin{bmatrix} \begin{vmatrix} 1 & 2 \\ 3 & 4 \end{vmatrix} & -\begin{vmatrix} 4 & 2 \\ 5 & 4 \end{vmatrix} & \begin{vmatrix} 4 & 1 \\ 5 & 3 \end{vmatrix} \\ -\begin{vmatrix} 3 & 1 \\ 3 & 4 \end{vmatrix} & \begin{vmatrix} 2 & 1 \\ 5 & 4 \end{vmatrix} & -\begin{vmatrix} 2 & 3 \\ 5 & 3 \end{vmatrix} \\ \begin{vmatrix} 3 & 1 \\ 1 & 2 \end{vmatrix} & -\begin{vmatrix} 2 & 1 \\ 4 & 2 \end{vmatrix} & \begin{vmatrix} 2 & 3 \\ 4 & 1 \end{vmatrix} \end{bmatrix} = \begin{bmatrix} -2 & -6 & 7 \\ -9 & 3 & 9 \\ 5 & 0 & -10 \end{bmatrix}$$

The adjoint matrix Adj $\mathbf{A}$ is the transpose of $\mathbf{C}$,

$$\text{Adj } \mathbf{A} = \mathbf{C}' = \begin{bmatrix} -2 & -9 & 5 \\ -6 & 3 & 0 \\ 7 & 9 & -10 \end{bmatrix}$$

11.7 INVERSE MATRICES

An *inverse matrix* $\mathbf{A}^{-1}$, which can be found only for a square, nonsingular matrix $\mathbf{A}$, is a unique matrix satisfying the relationship

$$\mathbf{A}\mathbf{A}^{-1} = \mathbf{I} = \mathbf{A}^{-1}\mathbf{A}$$

Multiplying a matrix by its inverse reduces it to an identity matrix. Thus, the inverse matrix in linear algebra performs much the same function as the reciprocal in ordinary algebra. The formula for deriving the inverse is

$$\mathbf{A}^{-1} = \frac{1}{|\mathbf{A}|} \text{Adj } \mathbf{A}$$

See Example 7 and Problem 11.25.

EXAMPLE 7.　Find the inverse for

$$\mathbf{A} = \begin{bmatrix} 4 & 1 & -5 \\ -2 & 3 & 1 \\ 3 & -1 & 4 \end{bmatrix}$$

1. Check that it is a square matrix, here 3×3, since only square matrices can have inverses.
2. Evaluate the determinant to be sure $|\mathbf{A}| \neq 0$, since only nonsingular matrices can have inverses.

$$|\mathbf{A}| = 4[3(4) - 1(-1)] - 1[(-2)(4) - 1(3)] + (-5)[(-2)(-1) - 3(3)]$$
$$= 52 + 11 + 35 = 98 \neq 0$$

Matrix $\mathbf{A}$ is nonsingular; $\rho(\mathbf{A}) = 3$.

3. Find the cofactor matrix of $\mathbf{A}$,

$$\mathbf{C} = \begin{bmatrix} \begin{vmatrix} 3 & 1 \\ -1 & 4 \end{vmatrix} & -\begin{vmatrix} -2 & 1 \\ 3 & 4 \end{vmatrix} & \begin{vmatrix} -2 & 3 \\ 3 & -1 \end{vmatrix} \\ -\begin{vmatrix} 1 & -5 \\ -1 & 4 \end{vmatrix} & \begin{vmatrix} 4 & -5 \\ 3 & 4 \end{vmatrix} & -\begin{vmatrix} 4 & 1 \\ 3 & -1 \end{vmatrix} \\ \begin{vmatrix} 1 & -5 \\ 3 & 1 \end{vmatrix} & -\begin{vmatrix} 4 & -5 \\ -2 & 1 \end{vmatrix} & \begin{vmatrix} 4 & 1 \\ -2 & 3 \end{vmatrix} \end{bmatrix} = \begin{bmatrix} 13 & 11 & -7 \\ 1 & 31 & 7 \\ 16 & 6 & 14 \end{bmatrix}$$

Then transpose the cofactor matrix to get the adjoint matrix.

$$\text{Adj } \mathbf{A} = \mathbf{C}' = \begin{bmatrix} 13 & 1 & 16 \\ 11 & 31 & 6 \\ -7 & 7 & 14 \end{bmatrix}$$

4. Multiply the adjoint matrix by $1/|\mathbf{A}| = \frac{1}{98}$ to get $\mathbf{A}^{-1}$.

$$\mathbf{A}^{-1} = \frac{1}{98}\begin{bmatrix} 13 & 1 & 16 \\ 11 & 31 & 6 \\ -7 & 7 & 14 \end{bmatrix} = \begin{bmatrix} \frac{13}{98} & \frac{1}{98} & \frac{16}{98} \\ \frac{11}{98} & \frac{31}{98} & \frac{6}{98} \\ -\frac{1}{14} & \frac{1}{14} & \frac{1}{7} \end{bmatrix} = \begin{bmatrix} 0.1327 & 0.0102 & 0.1633 \\ 0.1122 & 0.3163 & 0.0612 \\ -0.0714 & 0.0714 & 0.1429 \end{bmatrix}$$

5. To check your answer, multiply $\mathbf{AA}^{-1}$ or $\mathbf{A}^{-1}\mathbf{A}$. Both products will equal $\mathbf{I}$ if the answer is correct. An inverse is checked in Problem 11.26(a).

11.8　SOLVING LINEAR EQUATIONS WITH THE INVERSE

An inverse matrix can be used to solve matrix equations. If

$$\mathbf{A}_{n \times n} \mathbf{X}_{n \times 1} = \mathbf{B}_{n \times 1}$$

and the inverse $\mathbf{A}^{-1}$ exists, multiplication of both sides of the equation by $\mathbf{A}^{-1}$, following the laws of conformability, gives

$$\mathbf{A}_{n \times n}^{-1} \mathbf{A}_{n \times n} \mathbf{X}_{n \times 1} = \mathbf{A}_{n \times n}^{-1} \mathbf{B}_{n \times 1}$$

From Section 11.7,　　$\mathbf{A}^{-1}\mathbf{A} = \mathbf{I}$. Thus,

$$\mathbf{I}_{n \times n} \mathbf{X}_{n \times 1} = \mathbf{A}_{n \times n}^{-1} \mathbf{B}_{n \times 1}$$

From Section 10.8,　　$\mathbf{IX} = \mathbf{X}$. Therefore,

$$\mathbf{X}_{n \times 1} = (\mathbf{A}^{-1}\mathbf{B})_{n \times 1}$$

The solution of the equation is given by the product of the inverse of the coefficient matrix $\mathbf{A}^{-1}$ and the column vector of constants $\mathbf{B}$.　　See Problems 11.27 to 11.33.

EXAMPLE 8. Matrix equations and the inverse are used below to solve for x_1, x_2, and x_3, given

$$4x_1 + x_2 - 5x_3 = 8$$
$$-2x_1 + 3x_2 + x_3 = 12$$
$$3x_1 - x_2 + 4x_3 = 5$$

First, express the system of equations in matrix form,

$$\mathbf{AX} = \mathbf{B}$$

$$\begin{bmatrix} 4 & 1 & -5 \\ -2 & 3 & 1 \\ 3 & -1 & 4 \end{bmatrix} \begin{bmatrix} x_1 \\ x_2 \\ x_3 \end{bmatrix} = \begin{bmatrix} 8 \\ 12 \\ 5 \end{bmatrix}$$

From Section 11.8,

$$\mathbf{X} = \mathbf{A}^{-1}\mathbf{B}$$

Substituting $\mathbf{A}^{-1}$ from Example 7 and multiplying,

$$\mathbf{X} = \begin{bmatrix} \frac{13}{98} & \frac{1}{98} & \frac{16}{98} \\ \frac{11}{98} & \frac{31}{98} & \frac{6}{98} \\ -\frac{1}{14} & \frac{1}{14} & \frac{1}{7} \end{bmatrix} \begin{bmatrix} 8 \\ 12 \\ 5 \end{bmatrix} = \begin{bmatrix} \frac{104}{98} + \frac{12}{98} + \frac{80}{98} \\ \frac{88}{98} + \frac{372}{98} + \frac{30}{98} \\ -\frac{8}{14} + \frac{12}{14} + \frac{5}{7} \end{bmatrix} = \begin{bmatrix} \frac{196}{98} \\ \frac{490}{98} \\ \frac{14}{14} \end{bmatrix} = \begin{bmatrix} 2 \\ 5 \\ 1 \end{bmatrix}$$

Thus, $\bar{x}_1 = 2$, $\bar{x}_2 = 5$, and $\bar{x}_3 = 1$.

11.9 CRAMER'S RULE FOR MATRIX SOLUTIONS

Cramer's rule provides a simplified method of solving a system of linear equations through the use of determinants. Cramer's rule states

$$\bar{x}_i = \frac{|\mathbf{A}_i|}{|\mathbf{A}|}$$

where x_i is the ith unknown variable in a series of equations, $|\mathbf{A}|$ is the determinant of the coefficient matrix, and $|\mathbf{A}_i|$ is the determinant of a special matrix formed from the original coefficient matrix by replacing the column of coefficients of x_i with the column vector of constants. See Example 9 and Problems 11.34 to 11.37. Proof for Cramer's rule is given in Problem 11.38.

EXAMPLE 9. Cramer's rule is used below to solve the system of equations

$$6x_1 + 5x_2 = 49$$
$$3x_1 + 4x_2 = 32$$

1. Express the equations in matrix form.

$$\mathbf{AX} = \mathbf{B}$$

$$\begin{bmatrix} 6 & 5 \\ 3 & 4 \end{bmatrix} \begin{bmatrix} x_1 \\ x_2 \end{bmatrix} = \begin{bmatrix} 49 \\ 32 \end{bmatrix}$$

2. Find the determinant of $\mathbf{A}$

$$|\mathbf{A}| = 6(4) - 5(3) = 9$$

3. Then to solve for x_1, replace column 1, the coefficients of x_1, with the vector of constants $\mathbf{B}$, forming a new matrix $\mathbf{A}_1$.

$$\mathbf{A}_1 = \begin{bmatrix} 49 & 5 \\ 32 & 4 \end{bmatrix}$$

Find the determinant of $\mathbf{A}_1$,

$$|\mathbf{A}_1| = 49(4) - 5(32) = 36$$

and use the formula for Cramer's rule,

$$\bar{x}_1 = \frac{|\mathbf{A}_1|}{|\mathbf{A}|} = \frac{36}{9} = 4$$

4.   To solve for x_2, replace column 2, the coefficients of x_2, from the *original* matrix, with the column vector of constants $\mathbf{B}$, forming a new matrix $\mathbf{A}_2$.

$$\mathbf{A}_2 = \begin{bmatrix} 6 & 49 \\ 3 & 32 \end{bmatrix}$$

Take the determinant,

$$|\mathbf{A}_2| = 6(32) - 49(3) = 45$$

and use the formula

$$\bar{x}_2 = \frac{|\mathbf{A}_2|}{|\mathbf{A}|} = \frac{45}{9} = 5$$

For a system of three linear equations, see Problem 11.35(b) to (e).

Solved Problems

DETERMINANTS

11.1.   Find the determinant $|\mathbf{A}|$ for the following matrices:

a) $\mathbf{A} = \begin{bmatrix} 9 & 13 \\ 15 & 18 \end{bmatrix}$

$|\mathbf{A}| = 9(18) - 13(15) = -33$

b) $\mathbf{A} = \begin{bmatrix} 40 & -10 \\ 25 & -5 \end{bmatrix}$

$|\mathbf{A}| = 40(-5) - (-10)(25) = 50$

c) $\mathbf{A} = \begin{bmatrix} 7 & 6 \\ 9 & 5 \\ 2 & 12 \end{bmatrix}$

The determinant does not exist because $\mathbf{A}$ is a 3×2 matrix and only a square matrix can have a determinant.

11.2.   Find the determinant $|\mathbf{A}|$ for the following matrices. Notice how the presence of zeros simplifies the task of evaluating a determinant.

a) $\mathbf{A} = \begin{bmatrix} 3 & 6 & 5 \\ 2 & 1 & 8 \\ 7 & 9 & 1 \end{bmatrix}$

$$|\mathbf{A}| = 3\begin{vmatrix} 1 & 8 \\ 9 & 1 \end{vmatrix} - 6\begin{vmatrix} 2 & 8 \\ 7 & 1 \end{vmatrix} + 5\begin{vmatrix} 2 & 1 \\ 7 & 9 \end{vmatrix}$$

$$= 3[1(1) - 8(9)] - 6[2(1) - 8(7)] + 5[2(9) - 1(7)]$$

$$= 3(-71) - 6(-54) + 5(11) = 166$$

b) $\mathbf{A} = \begin{bmatrix} 12 & 0 & 3 \\ 9 & 2 & 5 \\ 4 & 6 & 1 \end{bmatrix}$

$$|\mathbf{A}| = 12 \begin{vmatrix} 2 & 5 \\ 6 & 1 \end{vmatrix} - 0 \begin{vmatrix} 9 & 5 \\ 4 & 1 \end{vmatrix} + 3 \begin{vmatrix} 9 & 2 \\ 4 & 6 \end{vmatrix}$$

$$= 12(2 - 30) - 0 + 3(54 - 8) = -198$$

c) $\mathbf{A} = \begin{bmatrix} 0 & 6 & 0 \\ 3 & 5 & 2 \\ 7 & 6 & 9 \end{bmatrix}$

$$|\mathbf{A}| = 0 \begin{vmatrix} 5 & 2 \\ 6 & 9 \end{vmatrix} - 6 \begin{vmatrix} 3 & 2 \\ 7 & 9 \end{vmatrix} + 0 \begin{vmatrix} 3 & 5 \\ 7 & 6 \end{vmatrix}$$

$$= 0 - 6(27 - 14) + 0 = -78$$

RANK OF A MATRIX

11.3.   Determine the rank ρ of the following matrices:

a) $\mathbf{A} = \begin{bmatrix} -3 & 6 & 2 \\ 1 & 5 & 4 \\ 4 & -8 & 2 \end{bmatrix}$

$$|\mathbf{A}| = -3 \begin{vmatrix} 5 & 4 \\ -8 & 2 \end{vmatrix} + 6(-1) \begin{vmatrix} 1 & 4 \\ 4 & 2 \end{vmatrix} + 2 \begin{vmatrix} 1 & 5 \\ 4 & -8 \end{vmatrix}$$

$$= -3[10 - (-32)] - 6(2 - 16) + 2(-8 - 20) = -98$$

With $|\mathbf{A}| \neq 0$, $\mathbf{A}$ is nonsingular and the three rows and columns are linearly independent. Hence, $\rho(\mathbf{A}) = 3$.

b) $\mathbf{B} = \begin{bmatrix} 5 & -9 & 3 \\ 2 & 12 & -4 \\ -3 & -18 & 6 \end{bmatrix}$

$$|\mathbf{B}| = 5 \begin{vmatrix} 12 & -4 \\ -18 & 6 \end{vmatrix} - 9(-1) \begin{vmatrix} 2 & -4 \\ -3 & 6 \end{vmatrix} + 3 \begin{vmatrix} 2 & 12 \\ -3 & -18 \end{vmatrix}$$

$$= 5[72 - (+72)] + 9[12 - (+12)] + 3[-36 - (-36)]$$

$$= 5(0) + 9(0) + 3(0) = 0$$

With $|\mathbf{B}| = 0$, $\mathbf{B}$ is singular and the three rows and columns are not linearly independent. Hence, $\rho(\mathbf{B}) \neq 3$. Now test to see if any two rows or columns are independent. Starting with the submatrix in the upper left corner, take the 2×2 determinant.

$$\begin{vmatrix} 5 & -9 \\ 2 & 12 \end{vmatrix} = 60 - (-18) = 78 \neq 0$$

Thus, $\rho(\mathbf{B}) = 2$. There are only two linearly independent rows and columns in $\mathbf{B}$. Row 3 is -1.5 times row 2, and column 3 is $-\frac{1}{3}$ times column 2.

c) $\mathbf{C} = \begin{bmatrix} -8 & 2 & -6 \\ 10 & -2.5 & 7.5 \\ 24 & -6 & 18 \end{bmatrix}$

$$|\mathbf{C}| = -8 \begin{vmatrix} -2.5 & 7.5 \\ -6 & 18 \end{vmatrix} + 2(-1) \begin{vmatrix} 10 & 7.5 \\ 24 & 18 \end{vmatrix} - 6 \begin{vmatrix} 10 & -2.5 \\ 24 & -6 \end{vmatrix}$$

$$= -8[-45 - (-45)] - 2(180 - 180) - 6[-60 - (-60)] = 0$$

With $|\mathbf{C}| = 0$, $\rho(\mathbf{C}) \neq 3$. Trying various 2×2 submatrices,

$$\begin{vmatrix} -8 & 2 \\ 10 & -2.5 \end{vmatrix} = 20 - 20 = 0 \qquad\qquad \begin{vmatrix} 2 & -6 \\ -2.5 & 7.5 \end{vmatrix} = 15 - 15 = 0$$

$$\begin{vmatrix} 10 & -2.5 \\ 24 & -6 \end{vmatrix} = -60 - (-60) = 0 \qquad\qquad \begin{vmatrix} -2.5 & 7.5 \\ -6 & 18 \end{vmatrix} = -45 - (-45) = 0$$

With all the determinants of the different 2×2 submatrices equal to zero, no two rows or columns of $\mathbf{C}$ are linearly independent. So $\rho(\mathbf{C}) \neq 2$ and $\rho(\mathbf{C}) = 1$. Row 2 is -1.25 times row 1, row 3 is -3 times row 1, column 2 is $-\frac{1}{4}$ times column 1, and column 3 is $\frac{3}{4}$ times column 1.

d) $\mathbf{D} = \begin{bmatrix} 2 & 5 \\ 7 & 11 \\ 3 & 1 \end{bmatrix}$

Since the maximum number of linearly independent rows (columns) must equal the maximum number of linearly independent columns (rows), the rank of $\mathbf{D}$ cannot exceed 2. Testing a submatrix,

$$\begin{vmatrix} 2 & 5 \\ 7 & 11 \end{vmatrix} = 22 - 35 = -13 \neq 0 \qquad \rho(\mathbf{D}) = 2$$

While it is clear that there are only two linearly independent columns, there are also only two linearly independent rows because row 2 = 2 times row 1 plus row 3.

PROPERTIES OF DETERMINANTS

11.4.   Given

$$\mathbf{A} = \begin{bmatrix} 2 & 5 & 1 \\ 3 & 2 & 4 \\ 1 & 4 & 2 \end{bmatrix}$$

Compare (a) the determinant of $\mathbf{A}$ and (b) the determinant of the transpose of $\mathbf{A}$. (c) Specify which property of determinants the comparison illustrates.

a)
$$|\mathbf{A}| = 2(4 - 16) - 5(6 - 4) + 1(12 - 2) = -24$$

b)
$$\mathbf{A}' = \begin{bmatrix} 2 & 3 & 1 \\ 5 & 2 & 4 \\ 1 & 4 & 2 \end{bmatrix}$$

$$|\mathbf{A}'| = 2(4 - 16) - 3(10 - 4) + 1(20 - 2) = -24$$

c)   This illustrates that the determinant of a matrix equals the determinant of its transpose. See Section 11.5.

11.5.   Compare (a) the determinant of $\mathbf{A}$ and (b) the determinant of $\mathbf{A}'$, given

$$\mathbf{A} = \begin{bmatrix} a_{11} & a_{12} \\ a_{21} & a_{22} \end{bmatrix}$$

$$a) \quad |\mathbf{A}| = a_{11}a_{22} - a_{12}a_{21} \qquad b) \quad \mathbf{A}' = \begin{bmatrix} a_{11} & a_{21} \\ a_{12} & a_{22} \end{bmatrix} \qquad |\mathbf{A}| = a_{11}a_{22} - a_{21}a_{12}$$

11.6. Given
$$\mathbf{A} = \begin{bmatrix} 1 & 4 & 2 \\ 3 & 5 & 4 \\ 2 & 3 & 2 \end{bmatrix}$$

(*a*) Find the determinant of **A**. (*b*) Form a new matrix **B** by interchanging row 1 and row 2 of **A**, and find $|\mathbf{B}|$. (*c*) Form another matrix **C** by interchanging column 1 and column 3 of **A**, and find $|\mathbf{C}|$. (*d*) Compare determinants and specify which property of determinants is illustrated.

a)
$$|\mathbf{A}| = 1(10 - 12) - 4(6 - 8) + 2(9 - 10) = 4$$

b)
$$\mathbf{B} = \begin{bmatrix} 3 & 5 & 4 \\ 1 & 4 & 2 \\ 2 & 3 & 2 \end{bmatrix}$$
$$|\mathbf{B}| = 3(8 - 6) - 5(2 - 4) + 4(3 - 8) = -4$$

c)
$$\mathbf{C} = \begin{bmatrix} 2 & 4 & 1 \\ 4 & 5 & 3 \\ 2 & 3 & 2 \end{bmatrix}$$
$$|\mathbf{C}| = 2(10 - 9) - 4(8 - 6) + 1(12 - 10) = -4$$

d) $|\mathbf{B}| = -|\mathbf{A}|$. Interchanging any two rows or columns will affect the sign of the determinant, but not the absolute value of the determinant.

11.7. Given
$$\mathbf{W} = \begin{bmatrix} w & x \\ y & z \end{bmatrix}$$

(*a*) Find the determinant of **W**. (*b*) Interchange row 1 and row 2 of **W**, forming a new matrix **Y**, and compare the determinant of **Y** with that of **W**.

a)
$$|\mathbf{W}| = wz - yx$$

b)
$$\mathbf{Y} = \begin{bmatrix} y & z \\ w & x \end{bmatrix} \qquad |\mathbf{Y}| = yx - wz = -(wz - yx) = -|\mathbf{W}|$$

11.8. Given
$$\mathbf{A} = \begin{bmatrix} 3 & 5 & 7 \\ 2 & 1 & 4 \\ 4 & 2 & 3 \end{bmatrix}$$

(*a*) Find the determinant of **A**. (*b*) Form a new matrix **B** by multiplying the first row of **A** by 2, and find the determinant of **B**. (*c*) Compare determinants and indicate which property of determinants this illustrates.

a)
$$|\mathbf{A}| = 3(3 - 8) - 5(6 - 16) + 7(4 - 4) = 35$$

b)
$$\mathbf{B} = \begin{bmatrix} 6 & 10 & 14 \\ 2 & 1 & 4 \\ 4 & 2 & 3 \end{bmatrix} \qquad |\mathbf{B}| = 6(3 - 8) - 10(6 - 16) + 14(4 - 4) = 70$$

c) $|\mathbf{B}| = 2|\mathbf{A}|$. Multiplying a single row or column of a matrix by a scalar will cause the value of the determinant to be multiplied by the scalar. Here doubling row 1 doubles the determinant.

11.9. Given

$$\mathbf{A} = \begin{bmatrix} 2 & 5 & 8 \\ 3 & 10 & 1 \\ 1 & 15 & 4 \end{bmatrix}$$

(a) Find $|\mathbf{A}|$. (b) Form a new matrix $\mathbf{B}$ by multiplying column 2 by $\frac{1}{5}$ and find $|\mathbf{B}|$. (c) Compare determinants.

a)
$$|\mathbf{A}| = 2(40 - 15) - 5(12 - 1) + 8(45 - 10) = 275$$

b)   Recalling that multiplying by $\frac{1}{5}$ is the same thing as dividing by or factoring out 5,

$$\mathbf{B} = \begin{bmatrix} 2 & 1 & 8 \\ 3 & 2 & 1 \\ 1 & 3 & 4 \end{bmatrix} \qquad |\mathbf{B}| = 2(8 - 3) - 1(12 - 1) + 8(9 - 2) = 55$$

c)
$$|\mathbf{B}| = \tfrac{1}{5}|\mathbf{A}|$$

11.10. Given

$$\mathbf{A} = \begin{bmatrix} a_{11} & a_{12} \\ a_{21} & a_{22} \end{bmatrix} \qquad \mathbf{B} = \begin{bmatrix} a_{11} & ka_{12} \\ a_{21} & ka_{22} \end{bmatrix}$$

Compare (a) the determinant of $\mathbf{A}$ and (b) the determinant of $\mathbf{B}$.

a) $|\mathbf{A}| = a_{11}a_{22} - a_{12}a_{21}$

b) $|\mathbf{B}| = a_{11}ka_{22} - ka_{12}a_{21} = k(a_{11}a_{22}) - k(a_{12}a_{21})$
$$= k(a_{11}a_{22} - a_{12}a_{21}) = k|\mathbf{A}|$$

11.11. Given

$$\mathbf{A} = \begin{bmatrix} 5 & 1 & 4 \\ 3 & 2 & 5 \\ 4 & 1 & 6 \end{bmatrix}$$

(a) Find $|\mathbf{A}|$. (b) Subtract 5 times column 2 from column 1, forming a new matrix $\mathbf{B}$, and find $|\mathbf{B}|$. (c) Compare determinants and indicate which property of determinants is illustrated.

a)
$$|\mathbf{A}| = 5(12 - 5) - 1(18 - 20) + 4(3 - 8) = 17$$

b)
$$\mathbf{B} = \begin{bmatrix} 0 & 1 & 4 \\ -7 & 2 & 5 \\ -1 & 1 & 6 \end{bmatrix} \qquad |\mathbf{B}| = 0 - 1(-42 + 5) + 4(-7 + 2) = 17$$

c) $|\mathbf{B}| = |\mathbf{A}|$. Addition or subtraction of a nonzero multiple of any row or column to or from another row or column does not change the value of the determinant.

11.12. (a) Subtract row 3 from row 1 in $\mathbf{A}$ of Problem 11.11, forming a new matrix $\mathbf{C}$, and (b) find $|\mathbf{C}|$.

a)
$$\mathbf{C} = \begin{bmatrix} 1 & 0 & -2 \\ 3 & 2 & 5 \\ 4 & 1 & 6 \end{bmatrix}$$

b)
$$|\mathbf{C}| = 1(12 - 5) - 0 + (-2)(3 - 8) = 17$$

11.13. Given the *upper-triangular matrix*

$$\mathbf{A} = \begin{bmatrix} -3 & 0 & 0 \\ 2 & -5 & 0 \\ 6 & 1 & 4 \end{bmatrix}$$

which has zero elements everywhere above the principal diagonal, (*a*) find $|\mathbf{A}|$. (*b*) Find the product of the elements along the principal diagonal and (*c*) specify which property of determinants this illustrates.

a)
$$|\mathbf{A}| = -3(-20 - 0) - 0 + 0 = 60$$

b)   Multiplying the elements along the principal diagonal, $(-3)(-5)(4) = 60$.

c)   The determinant of a triangular matrix is equal to the product of the elements along the principal diagonal.

11.14. Given the *lower-triangular matrix*

$$\mathbf{A} = \begin{bmatrix} 2 & -5 & -1 \\ 0 & 3 & 6 \\ 0 & 0 & -7 \end{bmatrix}$$

which has zero elements everywhere below the principal diagonal, find (*a*) $|\mathbf{A}|$ and (*b*) the product of the diagonal elements.

a)
$$|\mathbf{A}| = 2(-21 - 0) - (-5)(0 - 0) - 1(0 - 0) = -42$$

b)
$$2(3)(-7) = -42$$

11.15. Given
$$\mathbf{A} = \begin{bmatrix} 12 & 16 & 13 \\ 0 & 0 & 0 \\ -15 & 20 & -9 \end{bmatrix}$$

(*a*) Find $|\mathbf{A}|$. (*b*) What property of determinants is illustrated?

a)
$$|\mathbf{A}| = 12(0 - 0) - 16(0 - 0) + 13(0 - 0) = 0$$

b)   If all the elements of a row or column equal zero, the determinant will equal zero. With all the elements of row 2 in $\mathbf{A}$ equal to zero, the matrix is, in effect, a 2×3 matrix, not a 3×3 matrix. Only square matrices have determinants.

SINGULAR AND NONSINGULAR MATRICES

11.16. Using a 2×2 coefficient matrix $\mathbf{A}$, prove that if $|\mathbf{A}| \neq 0$, there is linear independence between the rows and columns of $\mathbf{A}$ and a unique solution exists for the system of equations.

Start with two linear equations in two unknowns

$$a_{11}x + a_{12}y = b_1 \qquad (11.5)$$

$$a_{21}x + a_{22}y = b_2 \qquad (11.6)$$

and solve for x by multiplying (*11.5*) by a_{22} and (*11.6*) by $-a_{12}$ and then adding to eliminate y.

$$a_{11}a_{22}x + a_{12}a_{22}y = a_{22}b_1$$
$$\underline{-a_{12}a_{21}x - a_{12}a_{22}y = -a_{12}b_2}$$
$$(a_{11}a_{22} - a_{12}a_{21})x = a_{22}b_1 - a_{12}b_2$$
$$x = \frac{a_{22}b_1 - a_{12}b_2}{a_{11}a_{22} - a_{12}a_{21}} \qquad (11.7)$$

where $a_{11}a_{22} - a_{12}a_{21} = |\mathbf{A}|$. If, in (*11.7*), $|\mathbf{A}| = a_{11}a_{22} - a_{12}a_{21} = 0$, x has no unique solution, indicating linear dependence between the equations; if $|\mathbf{A}| = a_{11}a_{22} - a_{12}a_{21} \neq 0$, x has a unique solution and the equations must be linearly independent.

11.17. Use determinants to determine whether a unique solution exists for each of the following systems of equations:

a) $12x_1 + 7x_2 = 147$
    $15x_1 + 19x_2 = 168$

To determine whether a unique solution exists, find the coefficient matrix **A** and take the determinant $|\mathbf{A}|$. If $|\mathbf{A}| \neq 0$, the matrix is nonsingular and a unique solution exists. If $|\mathbf{A}| = 0$, the matrix is singular and there is no unique solution. Thus,

$$\mathbf{A} = \begin{bmatrix} 12 & 7 \\ 15 & 19 \end{bmatrix}$$
$$|\mathbf{A}| = 12(19) - (7)15 = 123$$

Since $|\mathbf{A}| \neq 0$, **A** is nonsingular and a unique solution exists.

b) $2x_1 + 3x_2 = 27$
    $6x_1 + 9x_2 = 81$

$$\mathbf{A} = \begin{bmatrix} 2 & 3 \\ 6 & 9 \end{bmatrix} \qquad |\mathbf{A}| = 2(9) - 6(3) = 0$$

There is no unique solution. The equations are linearly dependent. The second equation is 3 times the first equation.

c) $72x_1 - 54x_2 = 216$
    $64x_1 - 48x_2 = 192$

$$\mathbf{A} = \begin{bmatrix} 72 & -54 \\ 64 & -48 \end{bmatrix} \qquad |\mathbf{A}| = 72(-48) - (-54)(64) = -3456 + 3456 = 0$$

A unique solution does not exist because the equations are linearly dependent. Closer inspection reveals the second equation is $\frac{8}{9}$ times the first equation.

d) $4x_1 + 3x_2 + 5x_3 = 27$
    $x_1 + 6x_2 + 2x_3 = 19$
    $3x_1 + x_2 + 3x_3 = 15$

$$\mathbf{A} = \begin{bmatrix} 4 & 3 & 5 \\ 1 & 6 & 2 \\ 3 & 1 & 3 \end{bmatrix} \qquad |\mathbf{A}| = 4(18 - 2) - 3(3 - 6) + 5(1 - 18) = -12$$

A unique solution exists.

e) $4x_1 + 2x_2 + 6x_3 = 28$
    $3x_1 + x_2 + 2x_3 = 20$
    $10x_1 + 5x_2 + 15x_3 = 70$

$$\mathbf{A} = \begin{bmatrix} 4 & 2 & 6 \\ 3 & 1 & 2 \\ 10 & 5 & 15 \end{bmatrix} \qquad |\mathbf{A}| = 4(15 - 10) - 2(45 - 20) + 6(15 - 10) = 0$$

There is no unique solution because the equations are linearly dependent. Closer examination reveals the third equation is 2.5 times the first equation.

f) $56x_1 + 47x_2 + 8x_3 = 365$
        $84x_1 - 39x_2 + 12x_3 = 249$
        $28x_1 - 81x_2 + 4x_3 = 168$

$$\mathbf{A} = \begin{bmatrix} 56 & 47 & 8 \\ 84 & -39 & 12 \\ 28 & -81 & 4 \end{bmatrix}$$

Factoring out 28 from column 1 and 4 from column 3 before taking the determinant,

$$|\mathbf{A}| = 28(4) \begin{vmatrix} 2 & 47 & 2 \\ 3 & -39 & 3 \\ 1 & -81 & 1 \end{vmatrix}$$

The linear dependence between column 1 and column 3 is now evident. The determinant will therefore be zero, and no unique solution exists.

$$|\mathbf{A}| = 112[2(-39 + 243) - 47(0) + 2(-243 + 39)] = 112(0) = 0$$

MINORS AND COFACTORS

11.18. Find (a) the minor $|M_{ij}|$ and (b) the cofactor $|C_{ij}|$ for each of the elements in the first row, given

$$\mathbf{A} = \begin{bmatrix} a_{11} & a_{12} \\ a_{21} & a_{22} \end{bmatrix}$$

a)   To find the minor of a_{11}, mentally delete the row and column in which it appears. The remaining element is the minor. Thus, $|M_{11}| = a_{22}$. Similarly, $|M_{12}| = a_{21}$.

b)   From the rule of cofactors,

$$|C_{11}| = (-1)^{1+1}|M_{11}| = +1(a_{22}) = a_{22}$$
$$|C_{12}| = (-1)^{1+2}|M_{12}| = -1(a_{21}) = -a_{21}$$

11.19. Find (a) the minors and (b) the cofactors for the elements of the second row, given

$$\mathbf{A} = \begin{bmatrix} 13 & 17 \\ 19 & 15 \end{bmatrix}$$

a)
$$|M_{21}| = 17 \qquad |M_{22}| = 13$$

b)
$$|C_{21}| = (-1)^{2+1}|M_{21}| = -1(17) = -17$$
$$|C_{22}| = (-1)^{2+2}|M_{22}| = +1(13) = 13$$

11.20. Find (a) the minors and (b) the cofactors for the elements of the second column, given

$$\mathbf{A} = \begin{bmatrix} 6 & 7 \\ 12 & 9 \end{bmatrix}$$

a)
$$|M_{12}| = 12 \qquad |M_{22}| = 6$$

b)
$$|C_{12}| = (-1)^{1+2}|M_{12}| = -12$$
$$|C_{22}| = (-1)^{2+2}|M_{22}| = 6$$

11.21. Find (*a*) the minors and (*b*) the cofactors for the elements of the first row, given

$$\mathbf{A} = \begin{bmatrix} 5 & 2 & -4 \\ 6 & -3 & 7 \\ 1 & 2 & 4 \end{bmatrix}$$

a)   Deleting row 1 and column 1,

$$|M_{11}| = \begin{vmatrix} -3 & 7 \\ 2 & 4 \end{vmatrix} = -26$$

Similarly,

$$|M_{12}| = \begin{vmatrix} 6 & 7 \\ 1 & 4 \end{vmatrix} = 17$$

$$|M_{13}| = \begin{vmatrix} 6 & -3 \\ 1 & 2 \end{vmatrix} = 15$$

b)

$$|C_{11}| = (-1)^2 |M_{11}| = -26$$
$$|C_{12}| = (-1)^3 |M_{12}| = -17$$
$$|C_{13}| = (-1)^4 |M_{13}| = 15$$

11.22. Find (*a*) the minors and (*b*) the cofactors for the elements of the third row, given

$$\mathbf{A} = \begin{bmatrix} 9 & 11 & 4 \\ 3 & 2 & 7 \\ 6 & 10 & 4 \end{bmatrix}$$

a)   Deleting row 3 and column 1,

$$|M_{31}| = \begin{vmatrix} 11 & 4 \\ 2 & 7 \end{vmatrix} = 69$$

Similarly,

$$|M_{32}| = \begin{vmatrix} 9 & 4 \\ 3 & 7 \end{vmatrix} = 51$$

$$|M_{33}| = \begin{vmatrix} 9 & 11 \\ 3 & 2 \end{vmatrix} = -15$$

b)

$$|C_{31}| = (-1)^4 |M_{31}| = 69$$
$$|C_{32}| = (-1)^5 |M_{32}| = -51$$
$$|C_{33}| = (-1)^6 |M_{33}| = -15$$

11.23. Find (*a*) the minors and (*b*) the cofactors for the elements in the second column, given

$$\mathbf{A} = \begin{bmatrix} 13 & 6 & 11 \\ 12 & 9 & 4 \\ 7 & 10 & 2 \end{bmatrix}$$

a)

$$|M_{12}| = \begin{vmatrix} 12 & 4 \\ 7 & 2 \end{vmatrix} = -4$$

$$|M_{22}| = \begin{vmatrix} 13 & 11 \\ 7 & 2 \end{vmatrix} = -51$$

$$|M_{32}| = \begin{vmatrix} 13 & 11 \\ 12 & 4 \end{vmatrix} = -80$$

b)

$$|C_{12}| = (-1)^3 |M_{12}| = -1(-4) = 4$$
$$|C_{22}| = (-1)^4 |M_{22}| = -51$$
$$|C_{32}| = (-1)^5 |M_{32}| = -1(-80) = 80$$

11.24. Find (1) the cofactor matrix $\mathbf{C}$ and (2) the adjoint matrix Adj $\mathbf{A}$ for each of the following:

a) $\mathbf{A} = \begin{bmatrix} 7 & 12 \\ 4 & 3 \end{bmatrix}$

1) $\mathbf{C} = \begin{bmatrix} |C_{11}| & |C_{12}| \\ |C_{21}| & |C_{22}| \end{bmatrix} = \begin{bmatrix} |M_{11}| & -|M_{12}| \\ -|M_{21}| & |M_{22}| \end{bmatrix} = \begin{bmatrix} 3 & -4 \\ -12 & 7 \end{bmatrix}$

2) Adj $\mathbf{A} = \mathbf{C}' = \begin{bmatrix} 3 & -12 \\ -4 & 7 \end{bmatrix}$

b) $\mathbf{A} = \begin{bmatrix} -2 & 5 \\ 13 & 6 \end{bmatrix}$

1) $\mathbf{C} = \begin{bmatrix} 6 & -13 \\ -5 & -2 \end{bmatrix}$ 2) Adj $\mathbf{A} = \begin{bmatrix} 6 & -5 \\ -13 & -2 \end{bmatrix}$

c) $\mathbf{A} = \begin{bmatrix} 9 & -16 \\ -20 & 7 \end{bmatrix}$

1) $\mathbf{C} = \begin{bmatrix} 7 & 20 \\ 16 & 9 \end{bmatrix}$ 2) Adj $\mathbf{A} = \begin{bmatrix} 7 & 16 \\ 20 & 9 \end{bmatrix}$

d) $\mathbf{A} = \begin{bmatrix} 6 & 2 & 7 \\ 5 & 4 & 9 \\ 3 & 3 & 1 \end{bmatrix}$

1) $\mathbf{C} = \begin{bmatrix} |C_{11}| & |C_{12}| & |C_{13}| \\ |C_{21}| & |C_{22}| & |C_{23}| \\ |C_{31}| & |C_{32}| & |C_{33}| \end{bmatrix} = \begin{bmatrix} \begin{vmatrix} 4 & 9 \\ 3 & 1 \end{vmatrix} & -\begin{vmatrix} 5 & 9 \\ 3 & 1 \end{vmatrix} & \begin{vmatrix} 5 & 4 \\ 3 & 3 \end{vmatrix} \\ -\begin{vmatrix} 2 & 7 \\ 3 & 1 \end{vmatrix} & \begin{vmatrix} 6 & 7 \\ 3 & 1 \end{vmatrix} & -\begin{vmatrix} 6 & 2 \\ 3 & 3 \end{vmatrix} \\ \begin{vmatrix} 2 & 7 \\ 4 & 9 \end{vmatrix} & -\begin{vmatrix} 6 & 7 \\ 5 & 9 \end{vmatrix} & \begin{vmatrix} 6 & 2 \\ 5 & 4 \end{vmatrix} \end{bmatrix} = \begin{bmatrix} -23 & 22 & 3 \\ 19 & -15 & -12 \\ -10 & -19 & 14 \end{bmatrix}$

2) Adj $\mathbf{A} = \mathbf{C}' = \begin{bmatrix} -23 & 19 & -10 \\ 22 & -15 & -19 \\ 3 & -12 & 14 \end{bmatrix}$

e) $\mathbf{A} = \begin{bmatrix} 13 & -2 & 8 \\ -9 & 6 & -4 \\ -3 & 2 & -1 \end{bmatrix}$

1) $\mathbf{C} = \begin{bmatrix} \begin{vmatrix} 6 & -4 \\ 2 & -1 \end{vmatrix} & -\begin{vmatrix} -9 & -4 \\ -3 & -1 \end{vmatrix} & \begin{vmatrix} -9 & 6 \\ -3 & 2 \end{vmatrix} \\ -\begin{vmatrix} -2 & 8 \\ 2 & -1 \end{vmatrix} & \begin{vmatrix} 13 & 8 \\ -3 & -1 \end{vmatrix} & -\begin{vmatrix} 13 & -2 \\ -3 & 2 \end{vmatrix} \\ \begin{vmatrix} -2 & 8 \\ 6 & -4 \end{vmatrix} & -\begin{vmatrix} 13 & 8 \\ -9 & -4 \end{vmatrix} & \begin{vmatrix} 13 & -2 \\ -9 & 6 \end{vmatrix} \end{bmatrix} = \begin{bmatrix} 2 & 3 & 0 \\ 14 & 11 & -20 \\ -40 & -20 & 60 \end{bmatrix}$

2) Adj $\mathbf{A} = \mathbf{C}' = \begin{bmatrix} 2 & 14 & -40 \\ 3 & 11 & -20 \\ 0 & -20 & 60 \end{bmatrix}$

LAPLACE EXPANSION

11.25. Use Laplace expansion to find the determinants for each of the following, using whatever row or column is easiest:

a) $\mathbf{A} = \begin{bmatrix} 15 & 7 & 9 \\ 2 & 5 & 6 \\ 9 & 0 & 12 \end{bmatrix}$

Expanding along the second column,

$$|\mathbf{A}| = a_{12}|C_{12}| + a_{22}|C_{22}| + a_{32}|C_{32}| = 7(-1)\begin{vmatrix} 2 & 6 \\ 9 & 12 \end{vmatrix} + 5\begin{vmatrix} 15 & 9 \\ 9 & 12 \end{vmatrix} + 0$$

$$= -7(-30) + 5(99) = 705$$

b) $\mathbf{A} = \begin{bmatrix} 23 & 35 & 0 \\ 72 & 46 & 10 \\ 15 & 29 & 0 \end{bmatrix}$

Expanding along the third column,

$$|\mathbf{A}| = a_{13}|C_{13}| + a_{23}|C_{23}| + a_{33}|C_{33}|$$

$$= 0 + 10(-1)\begin{vmatrix} 23 & 35 \\ 15 & 29 \end{vmatrix} + 0 = -10(142) = -1420$$

c) $\mathbf{A} = \begin{bmatrix} 12 & 98 & 15 \\ 0 & 25 & 0 \\ 21 & 84 & 19 \end{bmatrix}$

Expanding along the second row,

$$|\mathbf{A}| = a_{21}|C_{21}| + a_{22}|C_{22}| + a_{23}|C_{23}| = 0 + 25\begin{vmatrix} 12 & 15 \\ 21 & 19 \end{vmatrix} + 0 = 25(-87) = -2175$$

d) $\mathbf{A} = \begin{bmatrix} 2 & 4 & 1 & 5 \\ 3 & 2 & 5 & 1 \\ 1 & 2 & 1 & 4 \\ 3 & 4 & 3 & 2 \end{bmatrix}$

Expanding along the first row,

$$|\mathbf{A}| = a_{11}|C_{11}| + a_{12}|C_{12}| + a_{13}|C_{13}| + a_{14}|C_{14}|$$

$$= 2(-1)^{1+1}\begin{vmatrix} 2 & 5 & 1 \\ 2 & 1 & 4 \\ 4 & 3 & 2 \end{vmatrix} + 4(-1)^{1+2}\begin{vmatrix} 3 & 5 & 1 \\ 1 & 1 & 4 \\ 3 & 3 & 2 \end{vmatrix} + 1(-1)^{1+3}\begin{vmatrix} 3 & 2 & 1 \\ 1 & 2 & 4 \\ 3 & 4 & 2 \end{vmatrix} + 5(-1)^{1+4}\begin{vmatrix} 3 & 2 & 5 \\ 1 & 2 & 1 \\ 3 & 4 & 3 \end{vmatrix}$$

Then expanding each of the 3×3 subdeterminants along the first row,

$$|\mathbf{A}| = 2\left[2\begin{vmatrix} 1 & 4 \\ 3 & 2 \end{vmatrix} - 5\begin{vmatrix} 2 & 4 \\ 4 & 2 \end{vmatrix} + 1\begin{vmatrix} 2 & 1 \\ 4 & 3 \end{vmatrix}\right] - 4\left[3\begin{vmatrix} 1 & 4 \\ 3 & 2 \end{vmatrix} - 5\begin{vmatrix} 1 & 4 \\ 3 & 2 \end{vmatrix} + 1\begin{vmatrix} 1 & 1 \\ 3 & 3 \end{vmatrix}\right]$$

$$+ 1\left[3\begin{vmatrix} 2 & 4 \\ 4 & 2 \end{vmatrix} - 2\begin{vmatrix} 1 & 4 \\ 3 & 2 \end{vmatrix} + 1\begin{vmatrix} 1 & 2 \\ 3 & 4 \end{vmatrix}\right] - 5\left[3\begin{vmatrix} 2 & 1 \\ 4 & 3 \end{vmatrix} - 2\begin{vmatrix} 1 & 1 \\ 3 & 3 \end{vmatrix} + 5\begin{vmatrix} 1 & 2 \\ 3 & 4 \end{vmatrix}\right]$$

$$= 2[2(-10) - 5(-12) + 1(2)] - 4[3(-10) - 5(-10) + 1(0)]$$

$$+ 1[3(-12) - 2(-10) + 1(-2)] - 5[3(2) - 2(0) + 5(-2)]$$

$$= 2(42) - 4(20) + 1(-18) - 5(-4) = 6$$

$$e) \quad \mathbf{A} = \begin{bmatrix} 5 & 0 & 1 & 3 \\ 4 & 2 & 6 & 0 \\ 3 & 0 & 1 & 5 \\ 0 & 1 & 4 & 2 \end{bmatrix}$$

Expanding along the second column,

$$|\mathbf{A}| = a_{12}|C_{12}| + a_{22}|C_{22}| + a_{32}|C_{32}| + a_{42}|C_{42}|$$

$$= 0 + 2(-1)^{2+2} \begin{vmatrix} 5 & 1 & 3 \\ 3 & 1 & 5 \\ 0 & 4 & 2 \end{vmatrix} + 0 + 1(-1)^{4+2} \begin{vmatrix} 5 & 1 & 3 \\ 4 & 6 & 0 \\ 3 & 1 & 5 \end{vmatrix}$$

Then substituting the values for the 3×3 subdeterminants,

$$|\mathbf{A}| = 2(-60) + 1(88) = -32$$

INVERTING A MATRIX

11.26. Find the inverse $\mathbf{A}^{-1}$ for the following matrices. Check your answer to part (a).

$$a) \quad \mathbf{A} = \begin{bmatrix} 24 & 15 \\ 8 & 7 \end{bmatrix}$$

$$\mathbf{A}^{-1} = \frac{1}{|\mathbf{A}|} \operatorname{Adj} \mathbf{A}$$

Evaluating the determinant, $\qquad |\mathbf{A}| = 24(7) - 15(8) = 48$

Then finding the cofactor matrix to get the adjoint,

$$\mathbf{C} = \begin{bmatrix} 7 & -8 \\ -15 & 24 \end{bmatrix}$$

and $\qquad \operatorname{Adj} \mathbf{A} = \mathbf{C}' = \begin{bmatrix} 7 & -15 \\ -8 & 24 \end{bmatrix}$

Thus, $\qquad \mathbf{A}^{-1} = \frac{1}{48} \begin{bmatrix} 7 & -15 \\ -8 & 24 \end{bmatrix} = \begin{bmatrix} \frac{7}{48} & -\frac{5}{16} \\ -\frac{1}{6} & \frac{1}{2} \end{bmatrix} \qquad (11.8)$

Checking to make sure $\mathbf{A}^{-1}\mathbf{A} = \mathbf{I}$, and using the unreduced form of $\mathbf{A}^{-1}$ from (11.8) for easier computation,

$$\mathbf{A}^{-1}\mathbf{A} = \frac{1}{48} \begin{bmatrix} 7 & -15 \\ -8 & 24 \end{bmatrix} \begin{bmatrix} 24 & 15 \\ 8 & 7 \end{bmatrix} = \frac{1}{48} \begin{bmatrix} 7(24) - 15(8) & 7(15) - 15(7) \\ -8(24) + 24(8) & -8(15) + 24(7) \end{bmatrix}$$

$$= \frac{1}{48} \begin{bmatrix} 48 & 0 \\ 0 & 48 \end{bmatrix} = \begin{bmatrix} 1 & 0 \\ 0 & 1 \end{bmatrix}$$

$$b) \quad \mathbf{A} = \begin{bmatrix} 7 & 9 \\ 6 & 12 \end{bmatrix}$$

$$|\mathbf{A}| = 7(12) - 9(6) = 30$$

The cofactor matrix is

$$\mathbf{C} = \begin{bmatrix} 12 & -6 \\ -9 & 7 \end{bmatrix}$$

and $\qquad \operatorname{Adj} \mathbf{A} = \mathbf{C}' = \begin{bmatrix} 12 & -9 \\ -6 & 7 \end{bmatrix}$

Thus,

$$\mathbf{A}^{-1} = \frac{1}{30} \begin{bmatrix} 12 & -9 \\ -6 & 7 \end{bmatrix} = \begin{bmatrix} \frac{2}{5} & -\frac{3}{10} \\ -\frac{1}{5} & \frac{7}{30} \end{bmatrix}$$

c) $\mathbf{A} = \begin{bmatrix} -7 & 16 \\ -9 & 13 \end{bmatrix}$

$$|\mathbf{A}| = -7(13) - 16(-9) = 53$$

$$\mathbf{C} = \begin{bmatrix} 13 & 9 \\ -16 & -7 \end{bmatrix}$$

$$\text{Adj}\,\mathbf{A} = \mathbf{C}' = \begin{bmatrix} 13 & -16 \\ 9 & -7 \end{bmatrix}$$

$$\mathbf{A}^{-1} = \frac{1}{53} \begin{bmatrix} 13 & -16 \\ 9 & -7 \end{bmatrix} = \begin{bmatrix} \frac{13}{53} & -\frac{16}{53} \\ \frac{9}{53} & -\frac{7}{53} \end{bmatrix}$$

d) $\mathbf{A} = \begin{bmatrix} 4 & 2 & 5 \\ 3 & 1 & 8 \\ 9 & 6 & 7 \end{bmatrix}$

$$|\mathbf{A}| = 4(7 - 48) - 2(21 - 72) + 5(18 - 9) = -17$$

The cofactor matrix is

$$\mathbf{C} = \begin{bmatrix} \begin{vmatrix} 1 & 8 \\ 6 & 7 \end{vmatrix} & -\begin{vmatrix} 3 & 8 \\ 9 & 7 \end{vmatrix} & \begin{vmatrix} 3 & 1 \\ 9 & 6 \end{vmatrix} \\ -\begin{vmatrix} 2 & 5 \\ 6 & 7 \end{vmatrix} & \begin{vmatrix} 4 & 5 \\ 9 & 7 \end{vmatrix} & -\begin{vmatrix} 4 & 2 \\ 9 & 6 \end{vmatrix} \\ \begin{vmatrix} 2 & 5 \\ 1 & 8 \end{vmatrix} & -\begin{vmatrix} 4 & 5 \\ 3 & 8 \end{vmatrix} & \begin{vmatrix} 4 & 2 \\ 3 & 1 \end{vmatrix} \end{bmatrix} = \begin{bmatrix} -41 & 51 & 9 \\ 16 & -17 & -6 \\ 11 & -17 & -2 \end{bmatrix}$$

and

$$\text{Adj}\,\mathbf{A} = \mathbf{C}' = \begin{bmatrix} -41 & 16 & 11 \\ 51 & -17 & -17 \\ 9 & -6 & -2 \end{bmatrix}$$

Thus,

$$\mathbf{A}^{-1} = -\frac{1}{17} \begin{bmatrix} -41 & 16 & 11 \\ 51 & -17 & -17 \\ 9 & -6 & -2 \end{bmatrix} = \begin{bmatrix} \frac{41}{17} & -\frac{16}{17} & -\frac{11}{17} \\ -3 & 1 & 1 \\ -\frac{9}{17} & \frac{6}{17} & \frac{2}{17} \end{bmatrix}$$

e) $\mathbf{A} = \begin{bmatrix} 14 & 0 & 6 \\ 9 & 5 & 0 \\ 0 & 11 & 8 \end{bmatrix}$

$$|\mathbf{A}| = 14(40) - 0 + 6(99) = 1154$$

The cofactor matrix is

$$\mathbf{C} = \begin{bmatrix} 40 & -72 & 99 \\ 66 & 112 & -154 \\ -30 & 54 & 70 \end{bmatrix}$$

The adjoint is

$$\text{Adj}\,\mathbf{A} = \begin{bmatrix} 40 & 66 & -30 \\ -72 & 112 & 54 \\ 99 & -154 & 70 \end{bmatrix}$$

Then
$$\mathbf{A}^{-1} = \frac{1}{1154} \begin{bmatrix} 40 & 66 & -30 \\ -72 & 112 & 54 \\ 99 & -154 & 70 \end{bmatrix} = \begin{bmatrix} \frac{20}{577} & \frac{33}{577} & -\frac{15}{577} \\ -\frac{36}{577} & \frac{56}{577} & \frac{27}{577} \\ \frac{99}{1154} & -\frac{77}{577} & \frac{35}{577} \end{bmatrix}$$

MATRIX INVERSION IN EQUATION SOLUTIONS

11.27. Use matrix inversion to solve the following systems of linear equations. Check your answers on your own by substituting into the original equations.

a) $4x_1 + 3x_2 = 28$
     $2x_1 + 5x_2 = 42$

$$\begin{bmatrix} 4 & 3 \\ 2 & 5 \end{bmatrix} \begin{bmatrix} x_1 \\ x_2 \end{bmatrix} = \begin{bmatrix} 28 \\ 42 \end{bmatrix}$$

where from Section 11.8, $\mathbf{X} = \mathbf{A}^{-1}\mathbf{B}$. Find first the inverse of $\mathbf{A}$, where $|\mathbf{A}| = 4(5) - 3(2) = 14$. The cofactor matrix of $\mathbf{A}$ is

$$\mathbf{C} = \begin{bmatrix} 5 & -2 \\ -3 & 4 \end{bmatrix}$$

and
$$\text{Adj } \mathbf{A} = \mathbf{C}' = \begin{bmatrix} 5 & -3 \\ -2 & 4 \end{bmatrix}$$

Thus,
$$\mathbf{A}^{-1} = \frac{1}{14} \begin{bmatrix} 5 & -3 \\ -2 & 4 \end{bmatrix} = \begin{bmatrix} \frac{5}{14} & -\frac{3}{14} \\ -\frac{1}{7} & \frac{2}{7} \end{bmatrix}$$

Then substituting in $\mathbf{X} = \mathbf{A}^{-1}\mathbf{B}$ and simply multiplying matrices,

$$\mathbf{X} = \begin{bmatrix} \frac{5}{14} & -\frac{3}{14} \\ -\frac{1}{7} & \frac{2}{7} \end{bmatrix}_{2\times2} \begin{bmatrix} 28 \\ 42 \end{bmatrix}_{2\times1} = \begin{bmatrix} 10 - 9 \\ -4 + 12 \end{bmatrix}_{2\times1} = \begin{bmatrix} 1 \\ 8 \end{bmatrix}_{2\times1}$$

Thus, $\bar{x} = 1$ and $\bar{x}_2 = 8$.

b) $6x_1 + 7x_2 = 56$
     $2x_1 + 3x_2 = 44$

$$\begin{bmatrix} 6 & 7 \\ 2 & 3 \end{bmatrix} \begin{bmatrix} x_1 \\ x_2 \end{bmatrix} = \begin{bmatrix} 56 \\ 44 \end{bmatrix}$$

where $|\mathbf{A}| = 6(3) - 7(2) = 4$.

$$\mathbf{C} = \begin{bmatrix} 3 & -2 \\ -7 & 6 \end{bmatrix} \qquad \text{Adj } \mathbf{A} = \mathbf{C}' = \begin{bmatrix} 3 & -7 \\ -2 & 6 \end{bmatrix}$$

and
$$\mathbf{A}^{-1} = \frac{1}{4} \begin{bmatrix} 3 & -7 \\ -2 & 6 \end{bmatrix} = \begin{bmatrix} \frac{3}{4} & -\frac{7}{4} \\ -\frac{1}{2} & \frac{3}{2} \end{bmatrix}$$

Thus,
$$\mathbf{X} = \begin{bmatrix} \frac{3}{4} & -\frac{7}{4} \\ -\frac{1}{2} & \frac{3}{2} \end{bmatrix}_{2\times2} \begin{bmatrix} 56 \\ 44 \end{bmatrix}_{2\times1} = \begin{bmatrix} 42 - 77 \\ -28 + 66 \end{bmatrix}_{2\times1} = \begin{bmatrix} -35 \\ 38 \end{bmatrix}_{2\times1}$$

and $\bar{x}_1 = -35$ and $\bar{x}_2 = 38$.

11.28. The equilibrium conditions for two related markets (pork and beef) are given by

$$18P_b - P_p = 87$$
$$-2P_b + 36P_p = 98$$

Find the equilibrium price for each market.

$$\begin{bmatrix} 18 & -1 \\ -2 & 36 \end{bmatrix}\begin{bmatrix} P_b \\ P_p \end{bmatrix} = \begin{bmatrix} 87 \\ 98 \end{bmatrix}$$

where $|\mathbf{A}| = 18(36) - (-1)(-2) = 646$.

$$\mathbf{C} = \begin{bmatrix} 36 & 2 \\ 1 & 18 \end{bmatrix} \qquad \text{Adj } \mathbf{A} = \begin{bmatrix} 36 & 1 \\ 2 & 18 \end{bmatrix}$$

and

$$\mathbf{A}^{-1} = \frac{1}{646}\begin{bmatrix} 36 & 1 \\ 2 & 18 \end{bmatrix} = \begin{bmatrix} \frac{18}{323} & \frac{1}{646} \\ \frac{1}{323} & \frac{9}{323} \end{bmatrix}$$

Thus,

$$\mathbf{X} = \begin{bmatrix} \frac{18}{323} & \frac{1}{646} \\ \frac{1}{323} & \frac{9}{323} \end{bmatrix}\begin{bmatrix} 87 \\ 98 \end{bmatrix} = \begin{bmatrix} \frac{1615}{323} \\ \frac{969}{323} \end{bmatrix} = \begin{bmatrix} 5 \\ 3 \end{bmatrix}$$

and $\bar{P}_b = 5$ and $\bar{P}_p = 3$.

This is the same solution as that obtained by simultaneous equations in Problem 2.12. For practice try the inverse matrix solution for Problem 2.13.

11.29. The equilibrium condition for two substitute goods is given by

$$5P_1 - 2P_2 = 15$$
$$-P_1 + 8P_2 = 16$$

Find the equilibrium prices.

$$\begin{bmatrix} 5 & -2 \\ -1 & 8 \end{bmatrix}\begin{bmatrix} P_1 \\ P_2 \end{bmatrix} = \begin{bmatrix} 15 \\ 16 \end{bmatrix}$$

where $|\mathbf{A}| = 5(8) - (-1)(-2) = 38$.

$$\mathbf{C} = \begin{bmatrix} 8 & 1 \\ 2 & 5 \end{bmatrix} \qquad \text{Adj } \mathbf{A} = \begin{bmatrix} 8 & 2 \\ 1 & 5 \end{bmatrix}$$

and

$$\mathbf{A}^{-1} = \frac{1}{38}\begin{bmatrix} 8 & 2 \\ 1 & 5 \end{bmatrix} = \begin{bmatrix} \frac{4}{19} & \frac{1}{19} \\ \frac{1}{38} & \frac{5}{38} \end{bmatrix}$$

Thus,

$$\mathbf{X} = \begin{bmatrix} \frac{4}{19} & \frac{1}{19} \\ \frac{1}{38} & \frac{5}{38} \end{bmatrix}\begin{bmatrix} 15 \\ 16 \end{bmatrix} = \begin{bmatrix} \frac{60 + 16}{19} \\ \frac{15 + 80}{38} \end{bmatrix} = \begin{bmatrix} 4 \\ 2.5 \end{bmatrix}$$

and $\bar{P}_1 = 4$ and $\bar{P}_2 = 2.5$.

11.30. Given: the *IS* equation $0.3Y + 100i - 252 = 0$ and the *LM* equation $0.25Y - 200i - 176 = 0$. Find the equilibrium level of income and rate of interest.

The *IS* and *LM* equations can be reduced to the form

$$0.3Y + 100i = 252$$
$$0.25Y - 200i = 176$$

and then expressed in matrix form where

$$\mathbf{A} = \begin{bmatrix} 0.3 & 100 \\ 0.25 & -200 \end{bmatrix} \qquad \mathbf{X} = \begin{bmatrix} Y \\ i \end{bmatrix} \qquad \mathbf{B} = \begin{bmatrix} 252 \\ 176 \end{bmatrix}$$

Thus,
$$|\mathbf{A}| = 0.3(-200) - 100(0.25) = -85$$

$$\mathbf{C} = \begin{bmatrix} -200 & -0.25 \\ -100 & 0.3 \end{bmatrix}$$

$$\text{Adj}\,\mathbf{A} = \begin{bmatrix} -200 & -100 \\ -0.25 & 0.3 \end{bmatrix}$$

and
$$\mathbf{A}^{-1} = -\frac{1}{85}\begin{bmatrix} -200 & -100 \\ -0.25 & 0.3 \end{bmatrix} = \begin{bmatrix} \dfrac{40}{17} & \dfrac{20}{17} \\ \dfrac{0.05}{17} & -\dfrac{0.06}{17} \end{bmatrix}$$

Thus,
$$\mathbf{X} = \begin{bmatrix} \dfrac{40}{17} & \dfrac{20}{17} \\ \dfrac{0.05}{17} & -\dfrac{0.06}{17} \end{bmatrix}\begin{bmatrix} 252 \\ 176 \end{bmatrix} = \begin{bmatrix} \dfrac{10{,}080 + 3520}{17} \\ \dfrac{12.6 - 10.56}{17} \end{bmatrix} = \begin{bmatrix} 800 \\ 0.12 \end{bmatrix}$$

In equilibrium $\bar{Y} = 800$ and $\bar{i} = 0.12$ as found in Problem 2.23 where simultaneous equations were used. On your own, practice with Problem 2.24.

11.31. Use matrix inversion to solve for the unknowns in the system of linear equations given below.

$$2x_1 + 4x_2 - 3x_3 = 12$$
$$3x_1 - 5x_2 + 2x_3 = 13$$
$$-x_1 + 3x_2 + 2x_3 = 17$$

$$\begin{bmatrix} 2 & 4 & -3 \\ 3 & -5 & 2 \\ -1 & 3 & 2 \end{bmatrix}\begin{bmatrix} x_1 \\ x_2 \\ x_3 \end{bmatrix} = \begin{bmatrix} 12 \\ 13 \\ 17 \end{bmatrix}$$

where $|\mathbf{A}| = 2(-16) - 4(8) - 3(4) = -76$.

$$\mathbf{C} = \begin{bmatrix} \begin{vmatrix} -5 & 2 \\ 3 & 2 \end{vmatrix} & -\begin{vmatrix} 3 & 2 \\ -1 & 2 \end{vmatrix} & \begin{vmatrix} 3 & -5 \\ -1 & 3 \end{vmatrix} \\ -\begin{vmatrix} 4 & -3 \\ 3 & 2 \end{vmatrix} & \begin{vmatrix} 2 & -3 \\ -1 & 2 \end{vmatrix} & -\begin{vmatrix} 2 & 4 \\ -1 & 3 \end{vmatrix} \\ \begin{vmatrix} 4 & -3 \\ -5 & 2 \end{vmatrix} & -\begin{vmatrix} 2 & -3 \\ 3 & 2 \end{vmatrix} & \begin{vmatrix} 2 & 4 \\ 3 & -5 \end{vmatrix} \end{bmatrix} = \begin{bmatrix} -16 & -8 & 4 \\ -17 & 1 & -10 \\ -7 & -13 & -22 \end{bmatrix}$$

$$\text{Adj}\,\mathbf{A} = \begin{bmatrix} -16 & -17 & -7 \\ -8 & 1 & -13 \\ 4 & -10 & -22 \end{bmatrix}$$

$$\mathbf{A}^{-1} = -\frac{1}{76}\begin{bmatrix} -16 & -17 & -7 \\ -8 & 1 & -13 \\ 4 & -10 & -22 \end{bmatrix} = \begin{bmatrix} \frac{16}{76} & \frac{17}{76} & \frac{7}{76} \\ \frac{8}{76} & -\frac{1}{76} & \frac{13}{76} \\ -\frac{4}{76} & \frac{10}{76} & \frac{22}{76} \end{bmatrix}$$

where the common denominator 76 is deliberately kept to simplify later calculations.

Thus,
$$\mathbf{X} = \begin{bmatrix} \frac{16}{76} & \frac{17}{76} & \frac{7}{76} \\ \frac{8}{76} & -\frac{1}{76} & \frac{13}{76} \\ -\frac{4}{76} & \frac{10}{76} & \frac{22}{76} \end{bmatrix}\begin{bmatrix} 12 \\ 13 \\ 17 \end{bmatrix} = \begin{bmatrix} \dfrac{192 + 221 + 119}{76} \\ \dfrac{96 - 13 + 221}{76} \\ \dfrac{-48 + 130 + 374}{76} \end{bmatrix} = \begin{bmatrix} 7 \\ 4 \\ 6 \end{bmatrix} = \begin{bmatrix} \bar{x}_1 \\ \bar{x}_2 \\ \bar{x}_3 \end{bmatrix}$$

11.32. The equilibrium condition for three related markets is given by

$$11P_1 - P_2 - P_3 = 31$$
$$-P_1 + 6P_2 - 2P_3 = 26$$
$$-P_1 - 2P_2 + 7P_3 = 24$$

Find the equilibrium price for each market.

$$\begin{bmatrix} 11 & -1 & -1 \\ -1 & 6 & -2 \\ -1 & -2 & 7 \end{bmatrix} \begin{bmatrix} P_1 \\ P_2 \\ P_3 \end{bmatrix} = \begin{bmatrix} 31 \\ 26 \\ 24 \end{bmatrix}$$

where $|\mathbf{A}| = 11(38) + 1(-9) - 1(8) = 401$.

$$\mathbf{C} = \begin{bmatrix} \begin{vmatrix} 6 & -2 \\ -2 & 7 \end{vmatrix} & -\begin{vmatrix} -1 & -2 \\ -1 & 7 \end{vmatrix} & \begin{vmatrix} -1 & 6 \\ -1 & -2 \end{vmatrix} \\ -\begin{vmatrix} -1 & -1 \\ -2 & 7 \end{vmatrix} & \begin{vmatrix} 11 & -1 \\ -1 & 7 \end{vmatrix} & -\begin{vmatrix} 11 & -1 \\ -1 & -2 \end{vmatrix} \\ \begin{vmatrix} -1 & -1 \\ 6 & -2 \end{vmatrix} & -\begin{vmatrix} 11 & -1 \\ -1 & -2 \end{vmatrix} & \begin{vmatrix} 11 & -1 \\ -1 & 6 \end{vmatrix} \end{bmatrix} = \begin{bmatrix} 38 & 9 & 8 \\ 9 & 76 & 23 \\ 8 & 23 & 65 \end{bmatrix}$$

$$\text{Adj } \mathbf{A} = \begin{bmatrix} 38 & 9 & 8 \\ 9 & 76 & 23 \\ 8 & 23 & 65 \end{bmatrix}$$

$$\mathbf{A}^{-1} = \frac{1}{401} \begin{bmatrix} 38 & 9 & 8 \\ 9 & 76 & 23 \\ 8 & 23 & 65 \end{bmatrix} = \begin{bmatrix} \frac{38}{401} & \frac{9}{401} & \frac{8}{401} \\ \frac{9}{401} & \frac{76}{401} & \frac{23}{401} \\ \frac{8}{401} & \frac{23}{401} & \frac{65}{401} \end{bmatrix}$$

$$\mathbf{X} = \begin{bmatrix} \frac{38}{401} & \frac{9}{401} & \frac{8}{401} \\ \frac{9}{401} & \frac{76}{401} & \frac{23}{401} \\ \frac{8}{401} & \frac{23}{401} & \frac{65}{401} \end{bmatrix} \begin{bmatrix} 31 \\ 26 \\ 24 \end{bmatrix} = \begin{bmatrix} \dfrac{1178 + 234 + 192}{401} \\ \dfrac{279 + 1976 + 552}{401} \\ \dfrac{248 + 598 + 1560}{401} \end{bmatrix} = \begin{bmatrix} 4 \\ 7 \\ 6 \end{bmatrix} = \begin{bmatrix} \bar{P}_1 \\ \bar{P}_2 \\ \bar{P}_3 \end{bmatrix}$$

See Problem 2.16 for the same solution with simultaneous equations.

11.33. Given $Y = C + I_0$, where $C = C_0 + bY$. Use matrix inversion to find the equilibrium level of Y and C.

The given equations can first be rearranged so that the endogenous variables C and Y, together with their coefficients $-b$, are on the left-hand side of the equation and the exogenous variables C_0 and I_0 are on the right.

$$Y - C = I_0$$
$$-bY + C = C_0$$

Thus,

$$\begin{bmatrix} 1 & -1 \\ -b & 1 \end{bmatrix} \begin{bmatrix} Y \\ C \end{bmatrix} = \begin{bmatrix} I_0 \\ C_0 \end{bmatrix}$$

The determinant of the coefficient matrix is $|\mathbf{A}| = 1(1) + 1(-b) = 1 - b$. The cofactor matrix is

$$\mathbf{C} = \begin{bmatrix} 1 & b \\ 1 & 1 \end{bmatrix}$$

$$\text{Adj } \mathbf{A} = \begin{bmatrix} 1 & 1 \\ b & 1 \end{bmatrix}$$

and
$$\mathbf{A}^{-1} = \frac{1}{1-b} \begin{bmatrix} 1 & 1 \\ b & 1 \end{bmatrix}$$

Letting $X = \begin{bmatrix} Y \\ C \end{bmatrix}$,
$$X = \frac{1}{1-b} \begin{bmatrix} 1 & 1 \\ b & 1 \end{bmatrix} \begin{bmatrix} I_0 \\ C_0 \end{bmatrix} = \frac{1}{1-b} \begin{bmatrix} I_0 + C_0 \\ bI_0 + C_0 \end{bmatrix}$$

Thus,
$$\bar{Y} = \frac{1}{1-b}(I_0 + C_0) \qquad \bar{C} = \frac{1}{1-b}(C_0 + bI_0)$$

Example 3 in Chapter 2 was solved for the equilibrium level of income without matrices.

CRAMER'S RULE

11.34. Use Cramer's rule to solve for the unknowns in each of the following:

a) $2x_1 + 6x_2 = 22$
    $-x_1 + 5x_2 = 53$

From Cramer's rule,

$$\bar{x}_i = \frac{|\mathbf{A}_i|}{|\mathbf{A}|}$$

where $\mathbf{A}_i$ is a special matrix formed by replacing the column of coefficients of x_i with the column of constants. Thus, from the original data,

$$\begin{bmatrix} 2 & 6 \\ -1 & 5 \end{bmatrix} \begin{bmatrix} x_1 \\ x_2 \end{bmatrix} = \begin{bmatrix} 22 \\ 53 \end{bmatrix}$$

where $|\mathbf{A}| = 2(5) - 6(-1) = 16$.

Replacing the first column of the coefficient matrix with the column of constants,

$$\mathbf{A}_1 = \begin{bmatrix} 22 & 6 \\ 53 & 5 \end{bmatrix}$$

where $|\mathbf{A}_1| = 22(5) - 6(53) = -208$. Thus,

$$\bar{x}_1 = \frac{|\mathbf{A}_1|}{|\mathbf{A}|} = -\frac{208}{16} = -13$$

Replacing the second column of the original coefficient matrix with the column of constants,

$$\mathbf{A}_2 = \begin{bmatrix} 2 & 22 \\ -1 & 53 \end{bmatrix}$$

where $|\mathbf{A}_2| = 2(53) - 22(-1) = 128$. Thus,

$$\bar{x}_2 = \frac{|\mathbf{A}_2|}{|\mathbf{A}|} = \frac{128}{16} = 8$$

b) $7p_1 + 2p_2 = 60$
    $p_1 + 8p_2 = 78$

$$\mathbf{A} = \begin{bmatrix} 7 & 2 \\ 1 & 8 \end{bmatrix}$$

where $|\mathbf{A}| = 7(8) - 2(1) = 54$.

$$\mathbf{A}_1 = \begin{bmatrix} 60 & 2 \\ 78 & 8 \end{bmatrix}$$

where $|\mathbf{A}_1| = 60(8) - 2(78) = 324$.

$$\mathbf{A}_2 = \begin{bmatrix} 7 & 60 \\ 1 & 78 \end{bmatrix}$$

where $|\mathbf{A}_2| = 7(78) - 60(1) = 486$.

$$\bar{p}_1 = \frac{|\mathbf{A}_1|}{|\mathbf{A}|} = \frac{324}{54} = 6 \quad \text{and} \quad \bar{p}_2 = \frac{|\mathbf{A}_2|}{|\mathbf{A}|} = \frac{486}{54} = 9$$

c) $18P_b - \quad P_p = 87$
    $-2P_b + 36P_p = 98$

$$\mathbf{A} = \begin{bmatrix} 18 & -1 \\ -2 & 36 \end{bmatrix}$$

where $|\mathbf{A}| = 18(36) - (-1)(-2) = 646$.

$$\mathbf{A}_1 = \begin{bmatrix} 87 & -1 \\ 98 & 36 \end{bmatrix}$$

where $|\mathbf{A}_1| = 87(36) + 1(98) = 3230$.

$$\mathbf{A}_2 = \begin{bmatrix} 18 & 87 \\ -2 & 98 \end{bmatrix}$$

where $|\mathbf{A}_2| = 18(98) - 87(-2) = 1938$.

$$\bar{P}_b = \frac{|\mathbf{A}_1|}{|\mathbf{A}|} = \frac{3230}{646} = 5 \quad \text{and} \quad \bar{P}_p = \frac{|\mathbf{A}_2|}{|\mathbf{A}|} = \frac{1938}{646} = 3$$

Compare the work involved in this method of solution with the work involved in Problems 2.12 and 11.28 where the same problem is treated first with simultaneous equations and then with matrix inversion.

11.35. Redo Problem 11.34 for each of the following:

a) $0.4Y + 150i = 209$
    $0.1Y - 250i = \quad 35$

$$\mathbf{A} = \begin{bmatrix} 0.4 & 150 \\ 0.1 & -250 \end{bmatrix}$$

where $|\mathbf{A}| = 0.4(-250) - 150(0.1) = -115$.

$$\mathbf{A}_1 = \begin{bmatrix} 209 & 150 \\ 35 & -250 \end{bmatrix}$$

where $|\mathbf{A}_1| = 209(-250) - 150(35) = -57{,}500$.

$$\mathbf{A}_2 = \begin{bmatrix} 0.4 & 209 \\ 0.1 & 35 \end{bmatrix}$$

where $|\mathbf{A}_2| = 0.4(35) - 209(0.1) = -6.9$.

$$\bar{Y} = \frac{|\mathbf{A}_1|}{|\mathbf{A}|} = \frac{-57{,}500}{-115} = 500 \quad \text{and} \quad \bar{i} = \frac{|\mathbf{A}_2|}{|\mathbf{A}|} = \frac{-6.9}{-115} = 0.06$$

Compare this method of solution with Problem 2.24.

b) $5x_1 - 2x_2 + 3x_3 = 16$
    $2x_1 + 3x_2 - 5x_3 = 2$
    $4x_1 - 5x_2 + 6x_3 = 7$

$$|\mathbf{A}| = \begin{bmatrix} 5 & -2 & 3 \\ 2 & 3 & -5 \\ 4 & -5 & 6 \end{bmatrix} = 5(18 - 25) + 2(12 + 20) + 3(-10 - 12) = -37$$

$$|\mathbf{A}_1| = \begin{bmatrix} 16 & -2 & 3 \\ 2 & 3 & -5 \\ 7 & -5 & 6 \end{bmatrix} = 16(18 - 25) + 2(12 + 35) + 3(-10 - 21) = -111$$

$$|\mathbf{A}_2| = \begin{bmatrix} 5 & 16 & 3 \\ 2 & 2 & -5 \\ 4 & 7 & 6 \end{bmatrix} = 5(12 + 35) - 16(12 + 20) + 3(14 - 8) = -259$$

$$|\mathbf{A}_3| = \begin{bmatrix} 5 & -2 & 16 \\ 2 & 3 & 2 \\ 4 & -5 & 7 \end{bmatrix} = 5(21 + 10) + 2(14 - 8) + 16(-10 - 12) = -185$$

$$\bar{x}_1 = \frac{|\mathbf{A}_1|}{|\mathbf{A}|} = \frac{-111}{-37} = 3 \qquad \bar{x}_2 = \frac{|\mathbf{A}_2|}{|\mathbf{A}|} = \frac{-259}{-37} = 7 \qquad \bar{x}_3 = \frac{|\mathbf{A}_3|}{|\mathbf{A}|} = \frac{-185}{-37} = 5$$

c) $2x_1 + 4x_2 - x_3 = 52$
    $-x_1 + 5x_2 + 3x_3 = 72$
    $3x_1 - 7x_2 + 2x_3 = 10$

$$|\mathbf{A}| = \begin{vmatrix} 2 & 4 & -1 \\ -1 & 5 & 3 \\ 3 & -7 & 2 \end{vmatrix} = 2(31) - 4(-11) - 1(-8) = 114$$

$$|\mathbf{A}_1| = \begin{vmatrix} 52 & 4 & -1 \\ 72 & 5 & 3 \\ 10 & -7 & 2 \end{vmatrix} = 52(31) - 4(114) - 1(-554) = 1710$$

$$|\mathbf{A}_2| = \begin{vmatrix} 2 & 52 & -1 \\ -1 & 72 & 3 \\ 3 & 10 & 2 \end{vmatrix} = 2(114) - 52(-11) - 1(-226) = 1026$$

$$|\mathbf{A}_3| = \begin{vmatrix} 2 & 4 & 52 \\ -1 & 5 & 72 \\ 3 & -7 & 10 \end{vmatrix} = 2(554) - 4(-226) + 52(-8) = 1596$$

$$\bar{x}_1 = \frac{|\mathbf{A}_1|}{|\mathbf{A}|} = \frac{1710}{114} = 15 \qquad \bar{x}_2 = \frac{|\mathbf{A}_2|}{|\mathbf{A}|} = \frac{1026}{114} = 9 \qquad \bar{x}_3 = \frac{|\mathbf{A}_3|}{|\mathbf{A}|} = \frac{1596}{114} = 14$$

d) $11p_1 - p_2 - p_3 = 31$
    $-p_1 + 6p_2 - 2p_3 = 26$
    $-p_1 - 2p_2 + 7p_3 = 24$

$$|\mathbf{A}| = \begin{vmatrix} 11 & -1 & -1 \\ -1 & 6 & -2 \\ -1 & -2 & 7 \end{vmatrix} = 11(38) + 1(-9) - 1(8) = 401$$

$$|\mathbf{A}_1| = \begin{vmatrix} 31 & -1 & -1 \\ 26 & 6 & -2 \\ 24 & -2 & 7 \end{vmatrix} = 31(38) + 1(230) - 1(-196) = 1604$$

$$|\mathbf{A}_2| = \begin{vmatrix} 11 & 31 & -1 \\ -1 & 26 & -2 \\ -1 & 24 & 7 \end{vmatrix} = 11(230) - 31(-9) - 1(2) = 2807$$

$$|\mathbf{A}_3| = \begin{vmatrix} 11 & -1 & 31 \\ -1 & 6 & 26 \\ -1 & -2 & 24 \end{vmatrix} = 11(196) + 1(2) + 31(8) = 2406$$

Thus, $\bar{p}_1 = \dfrac{|\mathbf{A}_1|}{|\mathbf{A}|} = \dfrac{1604}{401} = 4$ $\bar{p}_2 = \dfrac{|\mathbf{A}_2|}{|\mathbf{A}|} = \dfrac{2807}{401} = 7$ $\bar{p}_3 = \dfrac{|\mathbf{A}_3|}{|\mathbf{A}|} = \dfrac{2406}{401} = 6$

Compare the work involved in this type of solution with the work involved in Problems 2.16 and 11.32.

11.36. Use Cramer's rule to solve for x and y, given the first-order conditions for constrained optimization from Example 7 of Chapter 6:

$$\frac{\partial TC}{\partial x} = 16x - y - \lambda = 0$$

$$\frac{\partial TC}{\partial y} = 24y - x - \lambda = 0$$

$$\frac{\partial TC}{\partial \lambda} = 42 - x - y = 0$$

Rearrange the equations,

$$\begin{array}{rrrcr} 16x - & y - & \lambda = & & 0 \\ -x + & 24y - & \lambda = & & 0 \\ -x - & y & & = & -42 \end{array}$$

and set them in matrix form.

$$\begin{bmatrix} 16 & -1 & -1 \\ -1 & 24 & -1 \\ -1 & -1 & 0 \end{bmatrix} \begin{bmatrix} x \\ y \\ \lambda \end{bmatrix} = \begin{bmatrix} 0 \\ 0 \\ -42 \end{bmatrix}$$

Expanding along the third column,

$$|\mathbf{A}| = (-1)(1 + 24) - (-1)(-16 - 1) + 0 = -42$$

$$\mathbf{A}_1 = \begin{bmatrix} 0 & -1 & -1 \\ 0 & 24 & -1 \\ -42 & -1 & 0 \end{bmatrix}$$

Expanding along the first column, $|\mathbf{A}_1| = -42(1 + 24) = -1050$.

$$\mathbf{A}_2 = \begin{bmatrix} 16 & 0 & -1 \\ -1 & 0 & -1 \\ -1 & -42 & 0 \end{bmatrix}$$

Expanding along the second column, $|\mathbf{A}_2| = -(-42)(-16 - 1) = -714$.

$$\mathbf{A}_3 = \begin{bmatrix} 16 & -1 & 0 \\ -1 & 24 & 0 \\ -1 & -1 & -42 \end{bmatrix}$$

Expanding along the third column, $|\mathbf{A}_3| = -42(384 - 1) = -16{,}086$.

Thus,
$$\bar{x} = \frac{|\mathbf{A}_1|}{|\mathbf{A}|} = \frac{-1050}{-42} = 25 \qquad \bar{y} = \frac{|\mathbf{A}_2|}{|\mathbf{A}|} = \frac{-714}{-42} = 17$$

and
$$\bar{\lambda} = \frac{|\mathbf{A}_3|}{|\mathbf{A}|} = \frac{-16{,}086}{-42} = 383$$

11.37. Use Cramer's rule to find the critical values of Q_1 and Q_2, given the first-order conditions for constrained utility maximization in Problem 6.37: $Q_2 - 10\lambda = 0$, $Q_1 - 2\lambda = 0$, and $240 - 10Q_1 - 2Q_2 = 0$.

$$\begin{bmatrix} 0 & 1 & -10 \\ 1 & 0 & -2 \\ -10 & -2 & 0 \end{bmatrix} \begin{bmatrix} Q_1 \\ Q_2 \\ \lambda \end{bmatrix} = \begin{bmatrix} 0 \\ 0 \\ -240 \end{bmatrix}$$

Thus,
$$|\mathbf{A}| = -(1)(-20) + (-10)(-2) = 40.$$

$$\mathbf{A}_1 = \begin{bmatrix} 0 & 1 & -10 \\ 0 & 0 & -2 \\ -240 & -2 & 0 \end{bmatrix} = -240(-2) = 480$$

$$\mathbf{A}_2 = \begin{bmatrix} 0 & 0 & -10 \\ 1 & 0 & -2 \\ -10 & -240 & 0 \end{bmatrix} = -(-240)(10) = 2400$$

$$\mathbf{A}_3 = \begin{bmatrix} 0 & 1 & 0 \\ 1 & 0 & 0 \\ -10 & -2 & -240 \end{bmatrix} = -240(-1) = 240$$

Thus,
$$\bar{Q}_1 = \frac{|\mathbf{A}_1|}{|\mathbf{A}|} = \frac{480}{40} = 12 \qquad \bar{Q}_2 = \frac{|\mathbf{A}_2|}{|\mathbf{A}|} = \frac{2400}{40} = 60$$

and
$$\bar{\lambda} = \frac{|\mathbf{A}_3|}{|\mathbf{A}|} = \frac{240}{40} = 6$$

11.38. Given

$$ax_1 + bx_2 = g \qquad\qquad\qquad (11.9)$$
$$cx_1 + dx_2 = h \qquad\qquad\qquad (11.10)$$

Prove Cramer's rule by showing

$$\bar{x}_1 = \frac{\begin{vmatrix} g & b \\ h & d \end{vmatrix}}{\begin{vmatrix} a & b \\ c & d \end{vmatrix}} = \frac{|\mathbf{A}_1|}{|\mathbf{A}|} \qquad \bar{x}_2 = \frac{\begin{vmatrix} a & g \\ c & h \end{vmatrix}}{\begin{vmatrix} a & b \\ c & d \end{vmatrix}} = \frac{|\mathbf{A}_2|}{|\mathbf{A}|}$$

Dividing (11.9) by b,

$$\frac{a}{b}x_1 + x_2 = \frac{g}{b} \qquad\qquad\qquad (11.11)$$

Multiplying (*11.11*) by d and subtracting (*11.10*),

$$\frac{ad}{b} x_1 + dx_2 = \frac{dg}{b}$$

$$-cx_1 - dx_2 = -h$$

$$\overline{\left(\frac{ad - cb}{b}\right) x_1 \qquad = \frac{dg - hb}{b}}$$

$$\bar{x}_1 = \frac{dg - hb}{ad - cb} = \frac{\begin{vmatrix} g & b \\ h & d \end{vmatrix}}{\begin{vmatrix} a & b \\ c & d \end{vmatrix}} = \frac{|\mathbf{A}_1|}{|\mathbf{A}|}$$

Similarly, dividing (*11.9*) by a, $$x_1 + \frac{b}{a} x_2 = \frac{g}{a}$$ (*11.12*)

Multiplying (*11.12*) by $-c$ and adding to (*11.10*),

$$-cx_1 - \frac{bc}{a} x_2 = -\frac{cg}{a}$$

$$cx_1 + dx_2 = h$$

$$\overline{\left(\frac{ad - bc}{a}\right) x_2 = \frac{ah - cg}{a}}$$

$$\bar{x}_2 = \frac{ah - cg}{ad - bc} = \frac{\begin{vmatrix} a & g \\ c & h \end{vmatrix}}{\begin{vmatrix} a & b \\ c & d \end{vmatrix}} = \frac{|\mathbf{A}_2|}{|\mathbf{A}|} \qquad \text{Q.E.D.}$$

CHAPTER 12

Special Determinants and Matrices and Their Use in Economics

12.1 THE JACOBIAN

Section 11.1 showed how to test for linear dependence through the use of a simple determinant. In contrast, a *Jacobian determinant* permits testing for functional dependence, both linear *and* nonlinear. A Jacobian determinant $|\mathbf{J}|$ is composed of all the first-order partial derivatives of a system of equations, arranged in ordered sequence. Given

$$y_1 = f_1(x_1, x_2, x_3)$$
$$y_2 = f_2(x_1, x_2, x_3)$$
$$y_3 = f_3(x_1, x_2, x_3)$$

$$|\mathbf{J}| = \left| \frac{\partial y_1, \partial y_2, \partial y_3}{\partial x_1, \partial x_2, \partial x_3} \right| = \begin{vmatrix} \dfrac{\partial y_1}{\partial x_1} & \dfrac{\partial y_1}{\partial x_2} & \dfrac{\partial y_1}{\partial x_3} \\[2mm] \dfrac{\partial y_2}{\partial x_1} & \dfrac{\partial y_2}{\partial x_2} & \dfrac{\partial y_2}{\partial x_3} \\[2mm] \dfrac{\partial y_3}{\partial x_1} & \dfrac{\partial y_3}{\partial x_2} & \dfrac{\partial y_3}{\partial x_3} \end{vmatrix}$$

Notice that the elements of each row are the partial derivatives of one function y_i with respect to each of the independent variables x_1, x_2, x_3, and the elements of each column are the partial derivatives

254

of each of the functions y_1, y_2, y_3 with respect to one of the independent variables x_j. If $|\mathbf{J}| = 0$, the equations are functionally dependent; if $|\mathbf{J}| \neq 0$, the equations are functionally independent. See Example 1 and Problems 12.1 to 12.4.

EXAMPLE 1.   Use of the Jacobian to test for functional dependence is demonstrated below, given

$$y_1 = 5x_1 + 3x_2$$
$$y_2 = 25x_1^2 + 30x_1 x_2 + 9x_2^2$$

First, take the first-order partials,

$$\frac{\partial y_1}{\partial x_1} = 5 \qquad \frac{\partial y_1}{\partial x_2} = 3 \qquad \frac{\partial y_2}{\partial x_1} = 50x_1 + 30x_2 \qquad \frac{\partial y_2}{\partial x_2} = 30x_1 + 18x_2$$

Then set up the Jacobian,

$$|\mathbf{J}| = \begin{vmatrix} 5 & 3 \\ 50x_1 + 30x_2 & 30x_1 + 18x_2 \end{vmatrix}$$

and evaluate,

$$|\mathbf{J}| = 5(30x_1 + 18x_2) - 3(50x_1 + 30x_2) = 0$$

Since $|\mathbf{J}| = 0$, there is functional dependence between the equations. In this, the simplest of cases, $(5x_1 + 3x_2)^2 = 25x_1^2 + 30x_1 x_2 + 9x_2^2$.

12.2   THE HESSIAN

Given that the first-order conditions $z_x = z_y = 0$ are met, a sufficient condition for a multivariable function $z = f(x, y)$ to be at an optimum is

1)
$$z_{xx}, z_{yy} > 0 \qquad \text{for a minimum}$$
$$z_{xx}, z_{yy} < 0 \qquad \text{for a maximum}$$
2)
$$z_{xx} z_{yy} > (z_{xy})^2$$

See Section 5.4. A convenient test for this second-order condition is the Hessian. A *Hessian* $|\mathbf{H}|$ is a determinant composed of all the second-order partial derivatives, with the second-order direct partials on the principal diagonal and the second-order cross partials off the principal diagonal. Thus,

$$|\mathbf{H}| = \begin{vmatrix} z_{xx} & z_{xy} \\ z_{yx} & z_{yy} \end{vmatrix}$$

where $z_{xy} = z_{yx}$. If the first element on the principal diagonal, the *first principal minor*, $|H_1| = z_{xx}$ is positive and the *second principal minor*

$$|H_2| = \begin{vmatrix} z_{xx} & z_{xy} \\ z_{xy} & z_{yy} \end{vmatrix} = z_{xx} z_{yy} - (z_{xy})^2 > 0$$

the second-order conditions for a minimum are met. When $|H_1| > 0$ and $|H_2| > 0$, the Hessian $|\mathbf{H}|$ is called *positive definite*. A positive definite Hessian fulfills the second-order conditions for a minimum.

If the first principal minor $|H_1| = z_{xx} < 0$ and the second principal minor

$$|H_2| = \begin{vmatrix} z_{xx} & z_{xy} \\ z_{xy} & z_{yy} \end{vmatrix} > 0$$

the second-order conditions for a maximum are met. When $|H_1| < 0$, $|H_2| > 0$, the Hessian $|\mathbf{H}|$ is *negative definite*. A negative definite Hessian fulfills the second-order conditions for a maximum. See Example 2 and Problems 12.10 to 12.13.

EXAMPLE 2. In Problem 5.10(a) it was found that

$$z = 3x^2 - xy + 2y^2 - 4x - 7y + 12$$

is optimized at $x_0 = 1$ and $y_0 = 2$. The second partials were $z_{xx} = 6$, $z_{yy} = 4$, and $z_{xy} = -1$. Using the Hessian to test the second-order conditions,

$$|\mathbf{H}| = \begin{vmatrix} z_{xx} & z_{xy} \\ z_{yx} & z_{yy} \end{vmatrix} = \begin{vmatrix} 6 & -1 \\ -1 & 4 \end{vmatrix}$$

Taking the principal minors, $|H_1| = 6 > 0$ and

$$|H_2| = \begin{vmatrix} 6 & -1 \\ -1 & 4 \end{vmatrix} = 6(4) - (-1)(-1) = 23 > 0$$

With $|H_1| > 0$ and $|H_2| > 0$, the Hessian $|\mathbf{H}|$ is positive definite, and z is minimized at the critical values.

12.3   THE DISCRIMINANT

Determinants may be used to test for positive or negative definiteness of any quadratic form. The determinant of a quadratic form is called a *discriminant* $|\mathbf{D}|$. Given the quadratic form

$$z = ax^2 + bxy + cy^2$$

the discriminant is formed by placing the coefficients of the squared terms on the principal diagonal and dividing the coefficients of the nonsquared term equally between the off-diagonal positions. Thus,

$$|\mathbf{D}| = \begin{vmatrix} a & \dfrac{b}{2} \\ \dfrac{b}{2} & c \end{vmatrix}$$

Then evaluate the principal minors as in the Hessian test, where

$$|D_1| = a \quad \text{and} \quad |D_2| = \begin{vmatrix} a & \dfrac{b}{2} \\ \dfrac{b}{2} & c \end{vmatrix} = ac - \dfrac{b^2}{4}$$

If $|D_1|, |D_2| > 0$, $|\mathbf{D}|$ is positive definite and z is positive for all nonzero values of x and y. If $|D_1| < 0$ and $|D_2| > 0$, z is negative definite and z is negative for all nonzero values of x and y. If $|D_2| \not> 0$, z is not sign definite and z may assume both positive and negative values. See Example 3 and Problems 12.5 to 12.7.

EXAMPLE 3. To test for sign definiteness, given the quadratic form

$$z = 2x^2 + 5xy + 8y^2$$

form the discriminant as explained in Section 12.3.

$$|\mathbf{D}| = \begin{vmatrix} 2 & 2.5 \\ 2.5 & 8 \end{vmatrix}$$

Then evaluate the principal minors as in the Hessian test.

$$|D_1| = 2 > 0 \quad |D_2| = \begin{vmatrix} 2 & 2.5 \\ 2.5 & 8 \end{vmatrix} = 16 - 6.25 = 9.75 > 0$$

Thus, z is positive definite, meaning that it will be greater than zero for all nonzero values of x and y.

12.4 HIGHER-ORDER HESSIANS

Given $y = f(x_1, x_2, x_3)$, the third-order Hessian is

$$|\mathbf{H}| = \begin{vmatrix} y_{11} & y_{12} & y_{13} \\ y_{21} & y_{22} & y_{23} \\ y_{31} & y_{32} & y_{33} \end{vmatrix}$$

where the elements are the various second-order partial derivatives of y:

$$y_{11} = \frac{\partial^2 y}{\partial x_1^2} \qquad y_{12} = \frac{\partial^2 y}{\partial x_2\, \partial x_1} \qquad y_{23} = \frac{\partial^2 y}{\partial x_3\, \partial x_2} \qquad \text{etc.}$$

Conditions for a relative minimum or maximum depend on the signs of the first, second, and third principal minors, respectively. If $|H_1| = y_{11} > 0$,

$$|H_2| = \begin{vmatrix} y_{11} & y_{12} \\ y_{21} & y_{22} \end{vmatrix} > 0 \qquad \text{and} \qquad |H_3| = |\mathbf{H}| > 0$$

where $|H_3|$ is the *third principal minor*, $|\mathbf{H}|$ is positive definite and fulfills the second-order conditions for a minimum. If $|H_1| = y_{11} < 0$,

$$|H_2| = \begin{vmatrix} y_{11} & y_{12} \\ y_{21} & y_{22} \end{vmatrix} > 0 \qquad \text{and} \qquad |H_3| = |\mathbf{H}| < 0$$

$|\mathbf{H}|$ is negative definite and will fulfill the second-order conditions for a maximum. Higher-order Hessians follow in analogous fashion. If all the principal minors of $|\mathbf{H}|$ are positive, $|\mathbf{H}|$ is positive definite and the second-order conditions for a relative minimum are met. If all the principal minors of $|\mathbf{H}|$ alternate in sign between negative and positive, $|\mathbf{H}|$ is negative definite and the second-order conditions for a relative maximum are met. See Example 4 and Problems 12.8, 12.9, and 12.14 to 12.18.

EXAMPLE 4. The function

$$y = -5x_1^2 + 10x_1 + x_1 x_3 - 2x_2^2 + 4x_2 + 2x_2 x_3 - 4x_3^2$$

is optimized as follows, using the Hessian to test the second-order conditions.

The first-order conditions are

$$\frac{\partial y}{\partial x_1} = y_1 = -10x_1 + 10 + x_3 = 0$$

$$\frac{\partial y}{\partial x_2} = y_2 = -4x_2 + 2x_3 + 4 = 0$$

$$\frac{\partial y}{\partial x_3} = y_3 = x_1 + 2x_2 - 8x_3 = 0$$

which can be expressed in matrix form as

$$\begin{bmatrix} -10 & 0 & 1 \\ 0 & -4 & 2 \\ 1 & 2 & -8 \end{bmatrix} \begin{bmatrix} x_1 \\ x_2 \\ x_3 \end{bmatrix} = \begin{bmatrix} -10 \\ -4 \\ 0 \end{bmatrix} \tag{12.1}$$

Using Cramer's rule (see Section 11.9) and taking the different determinants, $|\mathbf{A}| = -10(28) + 1(4) = -276 \neq 0$. Since $|\mathbf{A}|$ in this case is the Jacobian and does not equal zero, the three equations are functionally independent.

$$|\mathbf{A}_1| = -10(28) + 1(-8) = -288 \qquad |\mathbf{A}_2| = -10(32) - (-10)(-2) + 1(4) = -336$$

$$|\mathbf{A}_3| = -10(8) - 10(4) = -120$$

Thus, $\bar{x}_1 = \dfrac{|\mathbf{A}_1|}{|\mathbf{A}|} = \dfrac{-288}{-276} \cong 1.04$ $\bar{x}_2 = \dfrac{|\mathbf{A}_2|}{|\mathbf{A}|} = \dfrac{-336}{-276} \cong 1.22$ $\bar{x}_3 = \dfrac{|\mathbf{A}_3|}{|\mathbf{A}|} = \dfrac{-120}{-276} \cong 0.43$

taking the second partial derivatives from the first-order conditions to prepare the Hessian,

$$
\begin{array}{lll}
y_{11} = -10 & y_{12} = 0 & y_{13} = 1 \\
y_{21} = 0 & y_{22} = -4 & y_{23} = 2 \\
y_{31} = 1 & y_{32} = 2 & y_{33} = -8
\end{array}
$$

Thus, $|\mathbf{H}| = \begin{vmatrix} -10 & 0 & 1 \\ 0 & -4 & 2 \\ 1 & 2 & -8 \end{vmatrix}$

which has the same elements as the coefficient matrix in (12.1) since the first-order partials are all linear. Finally, applying the Hessian test, by checking the signs of the first, second, and third principal minors, respectively,

$$
|H_1| = -10 < 0 \qquad |H_2| = \begin{vmatrix} -10 & 0 \\ 0 & -4 \end{vmatrix} = 40 > 0 \qquad |H_3| = |\mathbf{H}| = |\mathbf{A}| = -276 < 0
$$

Since the principal minors alternate correctly in sign, the Hessian is negative definite and the function is maximized at $\bar{x}_1 = 1.04$, $\bar{x}_2 = 1.22$, and $\bar{x}_3 = 0.43$.

12.5 THE BORDERED HESSIAN FOR CONSTRAINED OPTIMIZATION

To optimize a function $f(x, y)$ subject to a constraint $g(x, y)$, Section 5.5 showed that a new function could be formed $F(x, y, \lambda) = f(x, y) + \lambda[k - g(x, y)]$, where the first-order conditions are $F_x = F_y = F_\lambda = 0$.

The second-order conditions can now be expressed in terms of a *bordered Hessian* $|\bar{\mathbf{H}}|$ in either of two ways:

$$
|\bar{\mathbf{H}}| = \begin{vmatrix} F_{xx} & F_{xy} & g_x \\ F_{yx} & F_{yy} & g_y \\ g_x & g_y & 0 \end{vmatrix} \qquad \text{or} \qquad \begin{vmatrix} 0 & g_x & g_y \\ g_x & F_{xx} & F_{xy} \\ g_y & F_{yx} & F_{yy} \end{vmatrix}
$$

which is simply the plain Hessian

$$
\begin{vmatrix} F_{xx} & F_{xy} \\ F_{yx} & F_{yy} \end{vmatrix}
$$

bordered by the first derivatives of the constraint with zero on the principal diagonal. The order of a *bordered principal minor* is determined by the order of the principal minor being bordered. Hence $|\bar{\mathbf{H}}|$ above represents a second bordered principal minor $|\bar{H}_2|$, because the principal minor being bordered is 2×2.

For a function in n variables $f(x_1, x_2, ..., x_n)$, subject to $g(x_1, x_2, ..., x_n)$,

$$
|\bar{\mathbf{H}}| = \begin{vmatrix} F_{11} & F_{12} & \cdots & F_{1n} & g_1 \\ F_{21} & F_{22} & \cdots & F_{2n} & g_2 \\ \hline F_{n1} & F_{n2} & \cdots & F_{nn} & g_n \\ g_1 & g_2 & \cdots & g_n & 0 \end{vmatrix} \qquad \text{or} \qquad \begin{vmatrix} 0 & g_1 & g_2 & \cdots & g_n \\ g_1 & F_{11} & F_{12} & \cdots & F_{1n} \\ g_2 & F_{21} & F_{22} & \cdots & F_{2n} \\ \hline g_n & F_{n1} & F_{n2} & \cdots & F_{nn} \end{vmatrix}
$$

where $|\bar{\mathbf{H}}| = |\bar{H}_n|$, because of the $n \times n$ principal minor being bordered.

If all the principal minors are negative, i.e., if $|\bar{H}_2|, |\bar{H}_3|, ..., |\bar{H}_n| < 0$, the bordered Hessian is *positive* definite, and a positive definite Hessian always satisfies the sufficient condition for a relative minimum.

If the principal minors alternate consistently in sign from positive to negative, i.e., if $|\bar{H}_2| > 0$,

$|\bar{H}_3| < 0, |\bar{H}_4| > 0$, etc., the bordered Hessian is *negative* definite, and a negative definite Hessian always meets the sufficient condition for a relative maximum. Further tests beyond the scope of the present book are needed if the criteria are not met, since the given criteria represent sufficient conditions, and not necessary conditions. See Examples 5 and 6 and Problems 12.19 to 12.27. For a 4×4 bordered Hessian, see Problem 12.28.

EXAMPLE 5. Refer to Example 9 in Chapter 5. The bordered Hessian can be used to check the second-order conditions of the optimized function and to determine if Z is maximized or minimized, as demonstrated below.

From Equations (5.8) and (5.9), $Z_{xx} = 8, Z_{yy} = 12, Z_{xy} = Z_{yx} = 3$. From the constraint, $x + y = 56, g_x = 1$, and $g_y = 1$. Thus,

$$|\bar{\mathbf{H}}| = \begin{vmatrix} 8 & 3 & 1 \\ 3 & 12 & 1 \\ 1 & 1 & 0 \end{vmatrix}$$

Starting with the second principal minor $|\bar{H}_2|$,

$$|\bar{H}_2| = |\bar{\mathbf{H}}| = 8(-1) - 3(-1) + 1(3 - 12) = -14$$

With $|\bar{H}_2| < 0, |\bar{\mathbf{H}}|$ is positive definite, which means that Z is at a minimum. See Problems 12.19 to 12.22.

EXAMPLE 6. The bordered Hessian is applied below to test the second-order condition of the generalized Cobb-Douglas production function maximized in Example 10 of Chapter 6.

From Equations (6.15) and (6.16), $Q_{KK} = -0.24K^{-1.6}L^{0.5}, Q_{LL} = -0.25K^{0.4}L^{-1.5}, Q_{KL} = Q_{LK} = 0.2K^{-0.6}L^{-0.5}$; and from the constraint, $3K + 4L = 108, g_K = 3, g_L = 4$,

$$|\bar{\mathbf{H}}| = \begin{vmatrix} -0.24K^{-1.6}L^{0.5} & 0.2K^{-0.6}L^{-0.5} & 3 \\ 0.2K^{-0.6}L^{-0.5} & -0.25K^{0.4}L^{-1.5} & 4 \\ 3 & 4 & 0 \end{vmatrix}$$

Starting with $|\bar{H}_2|$ and expanding along the third row,

$$|\bar{H}_2| = 3(0.8K^{-0.6}L^{-0.5} + 0.75K^{0.4}L^{-1.5}) - 4(-0.96K^{-1.6}L^{0.5} - 0.6K^{-0.6}L^{-0.5})$$

$$= 2.25K^{0.4}L^{-1.5} + 4.8K^{-0.6}L^{-0.5} + 3.84K^{-1.6}L^{0.5} = \frac{2.25K^{0.4}}{L^{1.5}} + \frac{4.8}{K^{0.6}L^{0.5}} + \frac{3.84L^{0.5}}{K^{1.6}} > 0$$

With $|\bar{H}_2| > 0, |\bar{\mathbf{H}}|$ is negative definite and Q is maximized. See Problems 12.23 to 12.28.

12.6 INPUT-OUTPUT ANALYSIS

In a modern economy where the production of one good requires the input of many other goods as *intermediate goods* in the production process (steel requires coal, iron ore, electricity, etc.), total demand x for product i will be the summation of all intermediate demand for the product plus the *final demand b* for the product arising from consumers, investors, the government, and exporters, as ultimate users. If a_{ij} is a *technical coefficient* expressing the value of input i required to produce one dollar's worth of product j, the total demand for product i can be expressed as

$$x_i = a_{i1}x_1 + a_{i2}x_2 + \cdots + a_{in}x_n + b_i$$

for $i = 1, 2, \ldots, n$. In matrix form this can be expressed as

$$\mathbf{X} = \mathbf{A}\mathbf{X} + \mathbf{B} \tag{12.2}$$

where
$$\mathbf{X} = \begin{bmatrix} x_1 \\ x_2 \\ \vdots \\ x_n \end{bmatrix} \quad \mathbf{A} = \begin{bmatrix} a_{11} & a_{12} & \cdots & a_{1n} \\ a_{21} & a_{22} & \cdots & a_{2n} \\ \hline a_{n1} & a_{n2} & \cdots & a_{nn} \end{bmatrix} \quad \mathbf{B} = \begin{bmatrix} b_1 \\ b_2 \\ \vdots \\ b_n \end{bmatrix}$$

and $\mathbf{A}$ is called the *matrix of technical coefficients*. To find the level of total output (intermediate and final) needed to satisfy final demand, we can solve for $\mathbf{X}$ in terms of the matrix of technical coefficients and the column vector of final demand, both of which are given. From (*12.2*),

$$\mathbf{X} - \mathbf{AX} = \mathbf{B}$$
$$(\mathbf{I} - \mathbf{A})\mathbf{X} = \mathbf{B}$$
$$\mathbf{X} = (\mathbf{I} - \mathbf{A})^{-1}\mathbf{B} \qquad (12.3)$$

Thus, for a three-sector economy

$$\begin{bmatrix} x_1 \\ x_2 \\ x_3 \end{bmatrix} = \begin{bmatrix} 1 - a_{11} & -a_{12} & -a_{13} \\ -a_{21} & 1 - a_{22} & -a_{23} \\ -a_{31} & -a_{32} & 1 - a_{33} \end{bmatrix}^{-1} \begin{bmatrix} b_1 \\ b_2 \\ b_3 \end{bmatrix}$$

where the $\mathbf{I} - \mathbf{A}$ matrix is called the *Leontief matrix*. In a complete input-output table, labor and capital would also be included as inputs, constituting value added by the firm. The vertical summation of elements along column j in such a model would equal 1: the input cost of producing one unit or one dollar's worth of the commodity, as seen in Problem 12.39. See Example 7 and Problems 12.29 to 12.39.

EXAMPLE 7.   Determine the total demand x for industries 1, 2, and 3, given the matrix of technical coefficients $\mathbf{A}$ and the final demand vector $\mathbf{B}$.

$$\mathbf{A} = \begin{bmatrix} 0.3 & 0.4 & 0.1 \\ 0.5 & 0.2 & 0.6 \\ 0.1 & 0.3 & 0.1 \end{bmatrix} \qquad \mathbf{B} = \begin{bmatrix} 20 \\ 10 \\ 30 \end{bmatrix}$$

From (*12.3*), $\mathbf{X} = (\mathbf{I} - \mathbf{A})^{-1}\mathbf{B}$, where

$$\mathbf{I} - \mathbf{A} = \begin{bmatrix} 1 & 0 & 0 \\ 0 & 1 & 0 \\ 0 & 0 & 1 \end{bmatrix} - \begin{bmatrix} 0.3 & 0.4 & 0.1 \\ 0.5 & 0.2 & 0.6 \\ 0.1 & 0.3 & 0.1 \end{bmatrix} = \begin{bmatrix} 0.7 & -0.4 & -0.1 \\ -0.5 & 0.8 & -0.6 \\ -0.1 & -0.3 & 0.9 \end{bmatrix}$$

Taking the inverse,

$$(\mathbf{I} - \mathbf{A})^{-1} = \frac{1}{0.151} \begin{bmatrix} 0.54 & 0.39 & 0.32 \\ 0.51 & 0.62 & 0.47 \\ 0.23 & 0.25 & 0.36 \end{bmatrix}$$

and substituting in (*12.3*),

$$\mathbf{X} = \frac{1}{0.151} \begin{bmatrix} 0.54 & 0.39 & 0.32 \\ 0.51 & 0.62 & 0.47 \\ 0.23 & 0.25 & 0.36 \end{bmatrix} \begin{bmatrix} 20 \\ 10 \\ 30 \end{bmatrix} = \frac{1}{0.151} \begin{bmatrix} 24.3 \\ 30.5 \\ 17.9 \end{bmatrix} = \begin{bmatrix} 160.93 \\ 201.99 \\ 118.54 \end{bmatrix} = \begin{bmatrix} x_1 \\ x_2 \\ x_3 \end{bmatrix}$$

12.7 CHARACTERISTIC ROOTS AND VECTORS (EIGENVALUES, EIGENVECTORS)

To this point, the sign definiteness of a Hessian and a quadratic form has been tested by using the principal minors. Sign definiteness can also be tested by using the characteristic roots of a matrix. Given a square matrix $\mathbf{A}$, if it is possible to find a vector $\mathbf{V} \neq 0$ and a scalar c such that

$$\mathbf{AV} = c\mathbf{V} \qquad (12.4)$$

the scalar c is called the *characteristic root*, *latent root*, or *eigenvalue*; and the vector is called the *characteristic vector*, *latent vector*, or *eigenvector*. Equation (*12.4*) can also be expressed

$$\mathbf{AV} = c\mathbf{IV}$$

which can be rearranged so that

$$\mathbf{AV} - c\mathbf{IV} = 0$$
$$(\mathbf{A} - c\mathbf{I})\mathbf{V} = 0 \tag{12.5}$$

where $\mathbf{A} - c\mathbf{I}$ is called the *characteristic matrix* of $\mathbf{A}$. Since by assumption $\mathbf{V} \neq 0$, the characteristic matrix $\mathbf{A} - c\mathbf{I}$ must be singular (see Problem 10.49) and thus its determinant must vanish. If $\mathbf{A} = 3 \times 3$ matrix, then

$$|\mathbf{A} - c\mathbf{I}| = \begin{vmatrix} a_{11} - c & a_{12} & a_{13} \\ a_{21} & a_{22} - c & a_{23} \\ a_{31} & a_{32} & a_{33} - c \end{vmatrix} = 0$$

With $|\mathbf{A} - c\mathbf{I}| = 0$ in (12.5), there will be an infinite number of solutions for $\mathbf{V}$. To force a unique solution, the solution may be *normalized* by requiring of the elements v_i of $\mathbf{V}$ that $\Sigma v_i^2 = 1$, as shown in Example 9.

If

1) All characteristic roots (c) are positive, $\mathbf{A}$ is positive definite.
2) All c's are negative, $\mathbf{A}$ is negative definite.
3) All c's are nonnegative and at least one $c = 0$, $\mathbf{A}$ is positive semidefinite.
4) All c's are nonpositive and at least one $c = 0$, $\mathbf{A}$ is negative semidefinite.
5) Some c's are positive and others negative, $\mathbf{A}$ is sign indefinite.

See Examples 8 and 9 and Problems 12.40 to 12.45, and Section 9.3.

EXAMPLE 8.   Given

$$\mathbf{A} = \begin{bmatrix} -6 & 3 \\ 3 & -6 \end{bmatrix}$$

To find the characteristic roots of $\mathbf{A}$, the determinant of the characteristic matrix $\mathbf{A} - c\mathbf{I}$ must equal zero. Thus,

$$|\mathbf{A} - c\mathbf{I}| = \begin{vmatrix} -6 - c & 3 \\ 3 & -6 - c \end{vmatrix} = 0 \tag{12.6}$$

$$(-6 - c)(-6 - c) - (3)(3) = 0$$
$$c^2 + 12c + 27 = 0 \qquad (c + 9)(c + 3) = 0$$
$$c_1 = -9 \qquad c_2 = -3$$

Testing for sign definiteness, since both characteristic roots are negative, $\mathbf{A}$ is negative definite. Note (1) that $c_1 + c_2$ must equal the sum of the elements on the principal diagonal of $\mathbf{A}$ and (2) $c_1 c_2$ must equal the determinant of the original matrix $|\mathbf{A}|$.

EXAMPLE 9.   Continuing with Example 8, the first root $c_1 = -9$ is now used to find the characteristic vector. Substituting $c = -9$ in (12.6),

$$\begin{bmatrix} -6 - (-9) & 3 \\ 3 & -6 - (-9) \end{bmatrix} \begin{bmatrix} v_1 \\ v_2 \end{bmatrix} = 0$$

$$\begin{bmatrix} 3 & 3 \\ 3 & 3 \end{bmatrix} \begin{bmatrix} v_1 \\ v_2 \end{bmatrix} = 0 \tag{12.7}$$

Since the coefficient matrix is linearly dependent, (12.7) is capable of an infinite number of solutions. The product of the matrices gives two equations which are identical.

$$3v_1 + 3v_2 = 0$$

Solving for v_2 in terms of v_1,

$$v_2 = -v_1 \tag{12.8}$$

Then, normalizing the solution in (12.8) so that

$$v_1^2 + v_2^2 = 1 \qquad (12.9)$$

$v_2 = -v_1$ is substituted in (12.9), getting

$$v_1^2 + (-v_1)^2 = 1$$

Thus, $2v_1^2 = 1$, $v_1^2 = \frac{1}{2}$. Then taking the positive square root, $v_1 = \sqrt{\frac{1}{2}} = \sqrt{0.5}$. From (12.8), $v_2 = -v_1$. Thus, $v_2 = -\sqrt{0.5}$, and the first characteristic vector is

$$\mathbf{V}_1 = \begin{bmatrix} \sqrt{0.5} \\ -\sqrt{0.5} \end{bmatrix}$$

When the second characteristic root $c_2 = -3$ is used,

$$\begin{bmatrix} -6-(-3) & 3 \\ 3 & -6-(-3) \end{bmatrix}\begin{bmatrix} v_1 \\ v_2 \end{bmatrix} = \begin{bmatrix} -3 & 3 \\ 3 & -3 \end{bmatrix}\begin{bmatrix} v_1 \\ v_2 \end{bmatrix} = 0$$

Multiplying the 2×2 matrix by the column vector,

$$-3v_1 + 3v_2 = 0$$
$$3v_1 - 3v_2 = 0$$

Thus, $v_1 = v_2$. Normalizing,

$$v_1^2 + v_2^2 = 1$$
$$(v_2)^2 + v_2^2 = 1$$
$$2v_2^2 = 1$$
$$v_2 = \sqrt{0.5} \qquad v_1 = \sqrt{0.5}$$

Thus,

$$\mathbf{V}_2 = \begin{bmatrix} \sqrt{0.5} \\ \sqrt{0.5} \end{bmatrix}$$

Solved Problems

THE JACOBIAN

12.1. Use the Jacobian to test for functional dependence in the following system of equations:

$$y_1 = 6x_1 + 4x_2$$
$$y_2 = 7x_1 + 9x_2$$

Taking the first-order partials to set up the Jacobian $|\mathbf{J}|$,

$$\frac{\partial y_1}{\partial x_1} = 6 \qquad \frac{\partial y_1}{\partial x_2} = 4 \qquad \frac{\partial y_2}{\partial x_1} = 7 \qquad \frac{\partial y_2}{\partial x_2} = 9$$

Thus,

$$|\mathbf{J}| = \begin{vmatrix} 6 & 4 \\ 7 & 9 \end{vmatrix} = 6(9) - 7(4) = 26$$

Since $|\mathbf{J}| \neq 0$, there is no functional dependence. Notice that in a system of linear equations the Jacobian $|\mathbf{J}|$ equals the determinant $|\mathbf{A}|$ of the coefficient matrix, and all their elements are identical. See Section 11.1, where the determinant test for nonsingularity of a matrix is nothing more than an application of the Jacobian to a system of linear equations.

12.2. Redo Problem 12.1, given

$$y_1 = 3x_1 - 4x_2$$
$$y_2 = 9x_1^2 - 24x_1x_2 + 16x_2^2$$

The first-order partials are

$$\frac{\partial y_1}{\partial x_1} = 3 \qquad \frac{\partial y_1}{\partial x_2} = -4 \qquad \frac{\partial y_2}{\partial x_1} = 18x_1 - 24x_2 \qquad \frac{\partial y_2}{\partial x_2} = -24x_1 + 32x_2$$

Thus, $\qquad |\mathbf{J}| = \begin{vmatrix} 3 & -4 \\ 18x_1 - 24x_2 & -24x_1 + 32x_2 \end{vmatrix} = 3(-24x_1 + 32x_2) + 4(18x_1 - 24x_2) = 0$

There is functional dependence: $(3x_1 - 4x_2)^2 = 9x_1^2 - 24x_1x_2 + 16x_2^2$.

12.3. Redo Problem 12.1, given

$$y_1 = x_1^2 - 3x_2 + 5$$
$$y_2 = x_1^4 - 6x_1^2x_2 + 9x_2^2$$

$$\frac{\partial y_1}{\partial x_1} = 2x_1 \qquad \frac{\partial y_1}{\partial x_2} = -3 \qquad \frac{\partial y_2}{\partial x_1} = 4x_1^3 - 12x_1x_2 \qquad \frac{\partial y_2}{\partial x_2} = -6x_1^2 + 18x_2$$

$$|\mathbf{J}| = \begin{vmatrix} 2x_1 & -3 \\ 4x_1^3 - 12x_1x_2 & -6x_1^2 + 18x_2 \end{vmatrix} = 2x_1(-6x_1^2 + 18x_2) + 3(4x_1^3 - 12x_1x_2) = 0$$

There is functional dependence: $y_2 = (y_1 - 5)^2$, where

$$y_1 - 5 = x_1^2 - 3x_2 + 5 - 5 = x_1^2 - 3x_2$$

and $\qquad (x_1^2 - 3x_2)^2 = x_1^4 - 6x_1^2x_2 + 9x_2^2$

12.4. Test for functional dependence in each of the following by means of the Jacobian:

a) $y_1 = 4x_1 - x_2$
$y_2 = 16x_1^2 + 8x_1x_2 + x_2^2$

$$|\mathbf{J}| = \begin{vmatrix} 4 & -1 \\ 32x_1 + 8x_2 & 8x_1 + 2x_2 \end{vmatrix} = 4(8x_1 + 2x_2) + 1(32x_1 + 8x_2) = 64x_1 + 16x_2 \neq 0$$

The equations are functionally independent.

b) $y_1 = 1.5x_1^2 + 12x_1x_2 + 24x_2^2$
$y_2 = 2x_1 + 8x_2$

$$|\mathbf{J}| = \begin{vmatrix} 3x_1 + 12x_2 & 12x_1 + 48x_2 \\ 2 & 8 \end{vmatrix} = 8(3x_1 + 12x_2) - 2(12x_1 + 48x_2) = 0$$

There is functional dependence between the equations.

c) $y_1 = 4x_1^2 + 3x_2 + 9$
$y_2 = 16x_1^4 + 24x_1^2x_2 + 9x_2^2 + 12$

$$|\mathbf{J}| = \begin{vmatrix} 8x_1 & 3 \\ 64x_1^3 + 48x_1x_2 & 24x_1^2 + 18x_2 \end{vmatrix} = 8x_1(24x_1^2 + 18x_2) - 3(64x_1^3 + 48x_1x_2) = 0$$

The equations are functionally dependent.

DISCRIMINANTS AND SIGN DEFINITENESS OF QUADRATIC FUNCTIONS

12.5. Use discriminants to determine whether each of the following quadratic functions is positive or negative definite:

a) $y = -3x_1^2 + 4x_1x_2 - 4x_2^2$

Since the coefficients of the squared terms are placed on the principal diagonal and the coefficient of the nonsquared term x_1x_2 is divided evenly between the a_{12} and a_{21} positions, in this case,

$$|\mathbf{D}| = \begin{vmatrix} -3 & 2 \\ 2 & -4 \end{vmatrix}$$

where $|D_1| = -3 < 0$ $|D_2| = \begin{vmatrix} -3 & 2 \\ 2 & -4 \end{vmatrix} = (-3)(-4) - (2)(2) = 8 > 0$

With $|D_1| < 0$ and $|D_2| > 0$, y is negative definite and y will be negative for all nonzero values of x_1 and x_2.

b) $y = 5x_1^2 - 2x_1x_2 + 7x_2^2$

The discriminant is $|\mathbf{D}| = \begin{vmatrix} 5 & -1 \\ -1 & 7 \end{vmatrix}$

where $|D_1| = 5 > 0$ and $|D_2| = |\mathbf{D}| = 5(7) - (-1)(-1) = 34 > 0$. With $|D_1| > 0$ and $|D_2| > 0$, y is positive definite and y will be positive for all nonzero values of x_1 and x_2.

12.6. Redo Problem 12.5 for $y = 5x_1^2 - 6x_1x_2 + 3x_2^2 - 2x_2x_3 + 8x_3^2 - 3x_1x_3$.

For a quadratic form in three variables, the coefficients of the squared terms continue to go on the principal diagonal, while the coefficient of x_1x_3 is divided evenly between the a_{13} and a_{31} positions, the coefficient of x_2x_3 is divided between the a_{23} and a_{32} positions, etc. Thus,

$$|\mathbf{D}| = \begin{vmatrix} 5 & -3 & -1.5 \\ -3 & 3 & -1 \\ -1.5 & -1 & 8 \end{vmatrix}$$

where $|D_1| = 5 > 0$ $|D_2| = \begin{vmatrix} 5 & -3 \\ -3 & 3 \end{vmatrix} = 6 > 0$

and $|D_3| = |\mathbf{D}| = 5(23) + 3(-25.5) - 1.5(7.5) = 27.25 > 0$. Therefore, y is positive definite.

12.7. Use discriminants to determine the sign definiteness of the following functions:

a) $y = -2x_1^2 + 4x_1x_2 - 5x_2^2 + 2x_2x_3 - 3x_3^2 + 2x_1x_3$

$$|\mathbf{D}| = \begin{vmatrix} -2 & 2 & 1 \\ 2 & -5 & 1 \\ 1 & 1 & -3 \end{vmatrix}$$

where $|D_1| = -2 < 0$ $|D_2| = \begin{vmatrix} -2 & 2 \\ 2 & -5 \end{vmatrix} = 6 > 0$

and $|D_3| = |\mathbf{D}| = -2(14) - 2(-7) + 1(7) = -7 < 0$. So y is negative definite.

b) $y = -7x_1^2 - 2x_2^2 + 2x_2x_3 - 4x_3^2 - 6x_1x_3$

$$|\mathbf{D}| = \begin{vmatrix} -7 & 0 & -3 \\ 0 & -2 & 1 \\ -3 & 1 & -4 \end{vmatrix}$$

where

$$|D_1| = -7 \qquad |D_2| = \begin{vmatrix} -7 & 0 \\ 0 & -2 \end{vmatrix} = 14$$

and $|D_3| = |\mathbf{D}| = -7(7) - 3(-6) = -31$. And y is negative definite.

THE HESSIAN IN OPTIMIZATION PROBLEMS

12.8. Optimize the following function, using (*a*) Cramer's rule for the first-order condition and (*b*) the Hessian for the second-order condition:

$$y = 3x_1^2 - 5x_1 - x_1x_2 + 6x_2^2 - 4x_2 + 2x_2x_3 + 4x_3^2 + 2x_3 - 3x_1x_3$$

a) The first-order conditions are

$$\begin{aligned} y_1 &= 6x_1 - 5 - x_2 - 3x_3 = 0 \\ y_2 &= -x_1 + 12x_2 - 4 + 2x_3 = 0 \\ y_3 &= 2x_2 + 8x_3 + 2 - 3x_1 = 0 \end{aligned} \qquad (12.10)$$

which in matrix form is

$$\begin{bmatrix} 6 & -1 & -3 \\ -1 & 12 & 2 \\ -3 & 2 & 8 \end{bmatrix} \begin{bmatrix} x_1 \\ x_2 \\ x_3 \end{bmatrix} = \begin{bmatrix} 5 \\ 4 \\ -2 \end{bmatrix}$$

Using Cramer's rule, $|\mathbf{A}| = 6(92) + 1(-2) - 3(34) = 448$. Since $|\mathbf{A}|$ also equals $|\mathbf{J}|$, the equations are functionally independent.

$$\begin{aligned} |\mathbf{A}_1| &= 5(92) + 1(36) - 3(32) = 400 \\ |\mathbf{A}_2| &= 6(36) - 5(-2) - 3(14) = 184 \\ |\mathbf{A}_3| &= 6(-32) + 1(14) + 5(34) = -8 \end{aligned}$$

Thus,

$$\bar{x}_1 = \frac{400}{448} \approx 0.89 \qquad \bar{x}_2 = \frac{184}{448} \approx 0.41 \qquad \bar{x}_3 = \frac{-8}{448} \approx -0.02$$

b) Testing the second-order condition by taking the second-order partials of (*12.10*) to form the Hessian,

$$\begin{aligned} y_{11} &= 6 & y_{12} &= -1 & y_{13} &= -3 \\ y_{21} &= -1 & y_{22} &= 12 & y_{23} &= 2 \\ y_{31} &= -3 & y_{32} &= 2 & y_{33} &= 8 \end{aligned}$$

Thus,

$$|\mathbf{H}| = \begin{vmatrix} 6 & -1 & -3 \\ -1 & 12 & 2 \\ -3 & 2 & 8 \end{vmatrix}$$

where

$$|H_1| = 6 > 0 \qquad |H_2| = \begin{vmatrix} 6 & -1 \\ -1 & 12 \end{vmatrix} = 71 > 0$$

and $|H_3| = |\mathbf{H}| = |\mathbf{A}| = 448 > 0$. With $|\mathbf{H}|$ positive definite, y is minimized at the critical values.

12.9. Redo Problem 12.8, given $y = -5x_1^2 + 10x_1 + x_1x_3 - 2x_2^2 + 4x_2 + 2x_2x_3 - 4x_3^2$.

a)

$$\begin{aligned} y_1 &= -10x_1 + 10 + x_3 = 0 \\ y_2 &= -4x_2 + 4 + 2x_3 = 0 \\ y_3 &= x_1 + 2x_2 - 8x_3 = 0 \end{aligned} \qquad (12.11)$$

In matrix form,

$$\begin{bmatrix} -10 & 0 & 1 \\ 0 & -4 & 2 \\ 1 & 2 & -8 \end{bmatrix} \begin{bmatrix} x_1 \\ x_2 \\ x_3 \end{bmatrix} = \begin{bmatrix} -10 \\ -4 \\ 0 \end{bmatrix}$$

Using Cramer's rule,
$$|\mathbf{A}| = -10(28) + 1(4) = -276$$
$$|\mathbf{A}_1| = -10(28) + 1(-8) = -288$$
$$|\mathbf{A}_2| = -10(32) + 10(-2) + 1(4) = -336$$
$$|\mathbf{A}_3| = -10(8) - 10(4) = -120$$

Thus,
$$\bar{x}_1 = \frac{-288}{-276} \approx 1.04 \qquad \bar{x}_2 = \frac{-336}{-276} \approx 1.22 \qquad \bar{x}_3 = \frac{-120}{-276} \approx 0.43$$

b) Taking the second partials of (12.11) and forming the Hessian,

$$|\mathbf{H}| = \begin{vmatrix} -10 & 0 & 1 \\ 0 & -4 & 2 \\ 1 & 2 & -8 \end{vmatrix}$$

where $|H_1| = -10 < 0$, $|H_2| = 40 > 0$, and $|H_3| = |\mathbf{A}| = -276 < 0$. Thus, $|\mathbf{H}|$ is negative definite, and y is maximized.

12.10. A firm produces two goods in pure competition and has the following total revenue and total cost functions:

$$\text{TR} = 15Q_1 + 18Q_2 \qquad \text{TC} = 2Q_1^2 + 2Q_1 Q_2 + 3Q_2^2$$

The two goods are *technically related in production*, since the marginal cost of one is dependent on the level of output of the other (for example, $\partial \text{TC}/\partial Q_1 = 4Q_1 + 2Q_2$). Maximize profits for the firm, using (a) Cramer's rule for the first-order condition and (b) the Hessian for the second-order condition.

a)
$$\Pi = \text{TR} - \text{TC} = 15Q_1 + 18Q_2 - 2Q_1^2 - 2Q_1 Q_2 - 3Q_2^2$$

The first-order conditions are

$$\Pi_1 = 15 - 4Q_1 - 2Q_2 = 0$$
$$\Pi_2 = 18 - 2Q_1 - 6Q_2 = 0$$

In matrix form,

$$\begin{bmatrix} -4 & -2 \\ -2 & -6 \end{bmatrix} \begin{bmatrix} Q_1 \\ Q_2 \end{bmatrix} = \begin{bmatrix} -15 \\ -18 \end{bmatrix}$$

Solving by Cramer's rule,

$$|\mathbf{A}| = 24 - 4 = 20 \qquad |\mathbf{A}_1| = 90 - 36 = 54 \qquad |\mathbf{A}_2| = 72 - 30 = 42$$

Thus,
$$\bar{Q}_1 = \frac{54}{20} = 2.7 \qquad \bar{Q}_2 = \frac{42}{20} = 2.1$$

b) Using the Hessian to test for the second-order condition,

$$|\mathbf{H}| = \begin{vmatrix} -4 & -2 \\ -2 & -6 \end{vmatrix}$$

where $|H_1| = -4$ and $|H_2| = 20$. With $|\mathbf{H}|$ negative definite, Π is maximized.

12.11. Using the techniques of Problem 12.10, maximize profits for the competitive firm whose goods are not technically related in production. The firm's total revenue and total cost functions are

$$\text{TR} = 7Q_1 + 9Q_2 \qquad \text{TC} = Q_1^2 + 2Q_1 + 5Q_2 + 2Q_2^2$$

a)
$$\Pi = 7Q_1 + 9Q_2 - Q_1^2 - 2Q_1 - 5Q_2 - 2Q_2^2$$
$$\Pi_1 = 7 - 2Q_1 - 2 = 0 \qquad \bar{Q}_1 = 2.5$$
$$\Pi_2 = 9 - 5 - 4Q_2 = 0 \qquad \bar{Q}_2 = 1$$

b)
$$|\mathbf{H}| = \begin{vmatrix} -2 & 0 \\ 0 & -4 \end{vmatrix}$$

where $|H_1| = -2$ and $|H_2| = 8$. So $|\mathbf{H}|$ is negative definite, and Π is maximized.

12.12. Maximize profits for a monopolistic firm producing two related goods, i.e.,
$$P_1 = f(Q_1, Q_2)$$
when the goods are substitutes and the demand and total cost functions are
$$P_1 = 80 - 5Q_1 - 2Q_2 \qquad P_2 = 50 - Q_1 - 3Q_2 \qquad TC = 3Q_1^2 + Q_1Q_2 + 2Q_2^2$$
Use (a) Cramer's rule and (b) the Hessian, as in Problem 12.10.

a) $\Pi = TR - TC$, where $TR = P_1Q_1 + P_2Q_2$.

$$\Pi = (80 - 5Q_1 - 2Q_2)Q_1 + (50 - Q_1 - 3Q_2)Q_2 - (3Q_1^2 + Q_1Q_2 + 2Q_2^2)$$
$$= 80Q_1 + 50Q_2 - 4Q_1Q_2 - 8Q_1^2 - 5Q_2^2$$
$$\Pi_1 = 80 - 4Q_2 - 16Q_1 = 0 \qquad \Pi_2 = 50 - 4Q_1 - 10Q_2 = 0$$

In matrix form,
$$\begin{bmatrix} -16 & -4 \\ -4 & -10 \end{bmatrix} \begin{bmatrix} Q_1 \\ Q_2 \end{bmatrix} = \begin{bmatrix} -80 \\ -50 \end{bmatrix}$$

$$|\mathbf{A}| = 160 - 16 = 144 \qquad |\mathbf{A}_1| = 800 - 200 = 600 \qquad |\mathbf{A}_2| = 800 - 320 = 480$$

and
$$\bar{Q}_1 = \frac{600}{144} \approx 4.17 \qquad \bar{Q}_2 = \frac{480}{144} \approx 3.33$$

b)
$$|\mathbf{H}| = \begin{vmatrix} -16 & -4 \\ -4 & -10 \end{vmatrix}$$

where $|H_1| = -16$ and $|H_2| = 144$. So Π is maximized.

12.13. Maximize profits for a producer of two substitute goods, given
$$P_1 = 130 - 4Q_1 - Q_2 \qquad P_2 = 160 - 2Q_1 - 5Q_2 \qquad TC = 2Q_1^2 + 2Q_1Q_2 + 4Q_2^2$$
Use (a) Cramer's rule for the first-order condition and (b) the Hessian for the second-order condition.

a)
$$\Pi = (130 - 4Q_1 - Q_2)Q_1 + (160 - 2Q_1 - 5Q_2)Q_2 - (2Q_1^2 + 2Q_1Q_2 + 4Q_2^2)$$
$$= 130Q_1 + 160Q_2 - 5Q_1Q_2 - 6Q_1^2 - 9Q_2^2$$
$$\Pi_1 = 130 - 5Q_2 - 12Q_1 = 0 \qquad \Pi_2 = 160 - 5Q_1 - 18Q_2 = 0$$

Thus,
$$\begin{bmatrix} -12 & -5 \\ -5 & -18 \end{bmatrix} \begin{bmatrix} Q_1 \\ Q_2 \end{bmatrix} = \begin{bmatrix} -130 \\ -160 \end{bmatrix}$$
$$|\mathbf{A}| = 191$$

$$|\mathbf{A}_1| = 1540 \qquad \bar{Q}_1 = \frac{1540}{191} \approx 8.06$$

$$|\mathbf{A}_2| = 1270 \qquad \bar{Q}_2 = \frac{1270}{191} \approx 6.65$$

b)
$$|\mathbf{H}| = \begin{vmatrix} -12 & -5 \\ -5 & -18 \end{vmatrix}$$

$|H_1| = -12$ and $|H_2| = 191$. So Π is maximized.

12.14. Redo Problem 12.13 for a monopolistic firm producing three related goods, when the demand functions and the cost function are

$$P_1 = 180 - 3Q_1 - Q_2 - 2Q_3 \qquad P_2 = 200 - Q_1 - 4Q_2 \qquad P_3 = 150 - Q_2 - 3Q_3$$
$$TC = Q_1^2 + Q_1 Q_2 + Q_2^2 + Q_2 Q_3 + Q_3^2$$

a)

$$\Pi = (180 - 3Q_1 - Q_2 - 2Q_3)Q_1 + (200 - Q_1 - 4Q_2)Q_2 + (150 - Q_2 - 3Q_3)Q_3$$
$$- (Q_1^2 + Q_1 Q_2 + Q_2^2 + Q_2 Q_3 + Q_3^2)$$
$$= 180Q_1 + 200Q_2 + 150Q_3 - 3Q_1 Q_2 - 2Q_2 Q_3 - 2Q_1 Q_3 - 4Q_1^2 - 5Q_2^2 - 4Q_3^2$$
$$\Pi_1 = 180 - 3Q_2 - 2Q_3 - 8Q_1 = 0 \qquad \Pi_2 = 200 - 3Q_1 - 2Q_3 - 10Q_2 = 0$$
$$\Pi_3 = 150 - 2Q_2 - 2Q_1 - 8Q_3 = 0$$

In matrix form,

$$\begin{bmatrix} -8 & -3 & -2 \\ -3 & -10 & -2 \\ -2 & -2 & -8 \end{bmatrix} \begin{bmatrix} Q_1 \\ Q_2 \\ Q_3 \end{bmatrix} = \begin{bmatrix} -180 \\ -200 \\ -150 \end{bmatrix}$$

$$|\mathbf{A}| = -8(76) + 3(20) - 2(-14) = -520$$
$$|\mathbf{A}_1| = -180(76) + 3(1300) - 2(-1100) = -7580$$
$$|\mathbf{A}_2| = -8(1300) + 180(20) - 2(50) = -6900$$
$$|\mathbf{A}_3| = -8(1100) + 3(50) - 180(-14) = -6130$$

Thus, $\bar{Q}_1 = \dfrac{-7580}{-520} \approx 14.58 \qquad \bar{Q}_2 = \dfrac{-6900}{-520} \approx 13.27 \qquad \bar{Q}_3 = \dfrac{-6130}{-520} \approx 11.79$

b)

$$|\mathbf{H}| = \begin{vmatrix} -8 & -3 & -2 \\ -3 & -10 & -2 \\ -2 & -2 & -8 \end{vmatrix}$$

where $|H_1| = -8$, $|H_2| = 71$, and $|H_3| = |\mathbf{H}| = |\mathbf{A}| = -520$. And Π is maximized.

12.15. Maximize profits as in Problem 12.14, given

$$P_1 = 70 - 2Q_1 - Q_2 - Q_3 \qquad P_2 = 120 - Q_1 - 4Q_2 - 2Q_3 \qquad P_3 = 90 - Q_1 - Q_2 - 3Q_3$$
$$TC = Q_1^2 + Q_1 Q_2 + 2Q_2^2 + 2Q_2 Q_3 + Q_3^2 + Q_1 Q_3$$

a)

$$\Pi = 70Q_1 + 120Q_2 + 90Q_3 - 3Q_1 Q_2 - 5Q_2 Q_3 - 3Q_1 Q_3 - 3Q_1^2 - 6Q_2^2 - 4Q_3^2$$
$$\Pi_1 = 70 - 3Q_2 - 3Q_3 - 6Q_1 = 0 \qquad \Pi_2 = 120 - 3Q_1 - 5Q_3 - 12Q_2 = 0$$
$$\Pi_3 = 90 - 3Q_1 - 5Q_2 - 8Q_3 = 0$$

Thus,

$$\begin{bmatrix} -6 & -3 & -3 \\ -3 & -12 & -5 \\ -3 & -5 & -8 \end{bmatrix} \begin{bmatrix} Q_1 \\ Q_2 \\ Q_3 \end{bmatrix} = \begin{bmatrix} -70 \\ -120 \\ -90 \end{bmatrix}$$

$$|\mathbf{A}| = -336$$

$$|\mathbf{A}_1| = -2000 \qquad \bar{Q}_1 = \dfrac{-2000}{-336} \approx 5.95$$

$$|\mathbf{A}_2| = -2160 \qquad \bar{Q}_2 = \dfrac{-2160}{-336} \approx 6.43$$

$$|\mathbf{A}_3| = -1680 \qquad \bar{Q}_3 = \dfrac{-1680}{-336} = 5$$

b)
$$|\mathbf{H}| = \begin{vmatrix} -6 & -3 & -3 \\ -3 & -12 & -5 \\ -3 & -5 & -8 \end{vmatrix}$$

where $|H_1| = -6$, $|H_2| = 63$, and $|H_3| = |\mathbf{H}| = |\mathbf{A}| = -336$. Π is maximized.

12.16. Given that $Q = F(P)$, maximize profits by (a) finding the inverse function $P = f(Q)$, (b) using Cramer's rule for the first-order condition, and (c) using the Hessian for the second-order condition. The demand functions and total cost function are

$$Q_1 = 100 - 3P_1 + 2P_2 \qquad Q_2 = 75 + 0.5P_1 - P_2 \qquad TC = Q_1^2 + 2Q_1 Q_2 + Q_2^2$$

where Q_1 and Q_2 are substitute goods, as indicated by the opposite signs for P_1 and P_2 in each equation (i.e., an increase in P_2 will increase demand for Q_1 and an increase in P_1 will increase demand for Q_2).

a) Since the markets are interrelated, the inverse functions must be found simultaneously. Rearranging the demand functions to get $P = f(Q)$, in order ultimately to maximize Π as a function of Q alone,

$$-3P_1 + 2P_2 = Q_1 - 100$$
$$0.5P_1 - P_2 = Q_2 - 75$$

In matrix form,

$$\begin{bmatrix} -3 & 2 \\ 0.5 & -1 \end{bmatrix} \begin{bmatrix} P_1 \\ P_2 \end{bmatrix} = \begin{bmatrix} Q_1 - 100 \\ Q_2 - 75 \end{bmatrix}$$

Using Cramer's rule,

$$|\mathbf{A}| = 2$$

$$|\mathbf{A}_1| = \begin{vmatrix} Q_1 - 100 & 2 \\ Q_2 - 75 & -1 \end{vmatrix} = -Q_1 + 100 - 2Q_2 + 150 = 250 - Q_1 - 2Q_2$$

$$P_1 = \frac{|\mathbf{A}_1|}{|\mathbf{A}|} = \frac{250 - Q_1 - 2Q_2}{2} = 125 - 0.5Q_1 - Q_2$$

$$|\mathbf{A}_2| = \begin{vmatrix} -3 & Q_1 - 100 \\ 0.5 & Q_2 - 75 \end{vmatrix} = -3Q_2 + 225 - 0.5Q_1 + 50 = 275 - 0.5Q_1 - 3Q_2$$

$$P_2 = \frac{|\mathbf{A}_2|}{|\mathbf{A}|} = \frac{275 - 0.5Q_1 - 3Q_2}{2} = 137.5 - 0.25Q_1 - 1.5Q_2$$

b)
$$\Pi = (125 - 0.5Q_1 - Q_2)Q_1 + (137.5 - 0.25Q_1 - 1.5Q_2)Q_2 - (Q_1^2 + 2Q_1 Q_2 + Q_2^2)$$
$$= 125Q_1 + 137.5Q_2 - 3.25Q_1 Q_2 - 1.5Q_1^2 - 2.5Q_2^2$$
$$\Pi_1 = 125 - 3.25Q_2 - 3Q_1 = 0 \qquad \Pi_2 = 137.5 - 3.25Q_1 - 5Q_2 = 0$$

Thus,
$$\begin{bmatrix} -3 & -3.25 \\ -3.25 & -5 \end{bmatrix} \begin{bmatrix} Q_1 \\ Q_2 \end{bmatrix} = \begin{bmatrix} -125 \\ -137.5 \end{bmatrix}$$

$$|\mathbf{A}| = 4.4375$$

$$|\mathbf{A}_1| = 178.125 \qquad \bar{Q}_1 = \frac{178.125}{4.4375} \approx 40.14$$

$$|\mathbf{A}_2| = 6.25 \qquad \bar{Q}_2 = \frac{6.25}{4.4375} \approx 1.4$$

c)
$$|\mathbf{H}| = \begin{vmatrix} -3 & -3.25 \\ -3.25 & -5 \end{vmatrix}$$

$|H_1| = -3$, $|H_2| = |\mathbf{H}| = |\mathbf{A}| = 4.4375$, and Π is maximized.

12.17. Redo Problem 12.16 by maximizing profits for

$$Q_1 = 90 - 6P_1 - 2P_2 \qquad Q_2 = 80 - 2P_1 - 4P_2 \qquad TC = 2Q_1^2 + 3Q_1Q_2 + 2Q_2^2$$

where Q_1 and Q_2 are complements, as indicated by the same sign for P_1 and P_2 in each equation.

a) Converting the demand functions to functions of Q,

$$\begin{bmatrix} -6 & -2 \\ -2 & -4 \end{bmatrix}\begin{bmatrix} P_1 \\ P_2 \end{bmatrix} = \begin{bmatrix} Q_1 - 90 \\ Q_2 - 80 \end{bmatrix}$$

$$|\mathbf{A}| = 20$$

$$|\mathbf{A}_1| = \begin{vmatrix} Q_1 - 90 & -2 \\ Q_2 - 80 & -4 \end{vmatrix} = -4Q_1 + 360 + 2Q_2 - 160 = 200 - 4Q_1 + 2Q_2$$

$$P_1 = \frac{200 - 4Q_1 + 2Q_2}{20} = 10 - 0.2Q_1 + 0.1Q_2$$

$$|\mathbf{A}_2| = \begin{vmatrix} -6 & Q_1 - 90 \\ -2 & Q_2 - 80 \end{vmatrix} = -6Q_2 + 480 + 2Q_1 - 180 = 300 - 6Q_2 + 2Q_1$$

$$P_2 = \frac{300 - 6Q_2 + 2Q_1}{20} = 15 - 0.3Q_2 + 0.1Q_1$$

b)
$$\Pi = (10 - 0.2Q_1 + 0.1Q_2)Q_1 + (15 - 0.3Q_2 + 0.1Q_1)Q_2 - (2Q_1^2 + 3Q_1Q_2 + 2Q_2^2)$$
$$= 10Q_1 + 15Q_2 - 2.8Q_1Q_2 - 2.2Q_1^2 - 2.3Q_2^2$$
$$\Pi_1 = 10 - 2.8Q_2 - 4.4Q_1 = 0 \qquad \Pi_2 = 15 - 2.8Q_1 - 4.6Q_2 = 0$$

Thus,
$$\begin{bmatrix} -4.4 & -2.8 \\ -2.8 & -4.6 \end{bmatrix}\begin{bmatrix} Q_1 \\ Q_2 \end{bmatrix} = \begin{bmatrix} -10 \\ -15 \end{bmatrix}$$

$$|\mathbf{A}| = 12.4$$

$$|\mathbf{A}_1| = 4 \qquad \bar{Q}_1 = \frac{4}{12.4} \approx 0.32$$

$$|\mathbf{A}_2| = 38 \qquad \bar{Q}_2 = \frac{38}{12.4} \approx 3.06$$

c)
$$|\mathbf{H}| = \begin{vmatrix} -4.4 & -2.8 \\ -2.8 & -4.6 \end{vmatrix}$$

$|H_1| = -4.4$, $|H_2| = |\mathbf{A}| = 12.4$, and Π is maximized.

12.18. Redo Problem 12.16, given

$$Q_1 = 150 - 3P_1 + P_2 + P_3 \qquad Q_2 = 180 + P_1 - 4P_2 + 2P_3 \qquad Q_3 = 200 + 2P_1 + P_2 - 5P_3$$
$$TC = Q_1^2 + Q_1Q_2 + 2Q_2^2 + Q_2Q_3 + Q_3^2 + Q_1Q_3$$

a) Finding the inverses of the demand functions

$$\begin{bmatrix} -3 & 1 & 1 \\ 1 & -4 & 2 \\ 2 & 1 & -5 \end{bmatrix}\begin{bmatrix} P_1 \\ P_2 \\ P_3 \end{bmatrix} = \begin{bmatrix} Q_1 - 150 \\ Q_2 - 180 \\ Q_3 - 200 \end{bmatrix}$$

$$|\mathbf{A}| = -36$$
$$|\mathbf{A}_1| = (Q_1 - 150)(20 - 2) - 1(-5Q_2 + 900 - 2Q_3 + 400) + 1(Q_2 - 180 + 4Q_3 - 800)$$
$$= -4980 + 18Q_1 + 6Q_2 + 6Q_3$$
$$P_1 = \frac{-4980 + 18Q_1 + 6Q_2 + 6Q_3}{-36} = 138.33 - 0.5Q_1 - 0.17Q_2 - 0.17Q_3$$

$$|\mathbf{A}_2| = -3(-5Q_2 + 900 - 2Q_3 + 400) - (Q_1 - 150)(-5 - 4) + 1(Q_3 - 200 - 2Q_2 + 360)$$
$$= -5090 + 9Q_1 + 13Q_2 + 7Q_3$$
$$P_2 = \frac{-5090 + 9Q_1 + 13Q_2 + 7Q_3}{-36} = 141.39 - 0.25Q_1 - 0.36Q_2 - 0.19Q_3$$
$$|\mathbf{A}_3| = -3(-4Q_3 + 800 - Q_2 + 180) - 1(Q_3 - 200 - 2Q_2 + 360) + (Q_1 - 150)(1 + 8)$$
$$= -4450 + 9Q_1 + 5Q_2 + 11Q_3$$
$$P_3 = \frac{-4450 + 9Q_1 + 5Q_2 + 11Q_3}{-36} = 123.61 - 0.25Q_1 - 0.14Q_2 - 0.31Q_3$$

b)
$$\Pi = P_1Q_1 + P_2Q_2 + P_3Q_3 - \text{TC}$$
$$= 138.33Q_1 + 141.39Q_2 + 123.61Q_3 - 1.42Q_1Q_2$$
$$- 1.33Q_2Q_3 - 1.42Q_1Q_3 - 1.5Q_1^2 - 2.36Q_2^2 - 1.31Q_3^2$$
$$\Pi_1 = 138.33 - 1.42Q_2 - 1.42Q_3 - 3Q_1 = 0$$
$$\Pi_2 = 141.39 - 1.42Q_1 - 1.33Q_3 - 4.72Q_2 = 0$$
$$\Pi_3 = 123.61 - 1.33Q_2 - 1.42Q_1 - 2.62Q_3 = 0$$

Thus,
$$\begin{bmatrix} -3 & -1.42 & -1.42 \\ -1.42 & -4.72 & -1.33 \\ -1.42 & -1.33 & -2.62 \end{bmatrix} \begin{bmatrix} Q_1 \\ Q_2 \\ Q_3 \end{bmatrix} = \begin{bmatrix} -138.33 \\ -141.39 \\ -123.61 \end{bmatrix}$$

$$|\mathbf{A}| = -22.37$$

$$|\mathbf{A}_1| = -612.27 \qquad \bar{Q}_1 = \frac{-612.27}{-22.37} \approx 27.37$$

$$|\mathbf{A}_2| = -329.14 \qquad \bar{Q}_2 = \frac{-329.14}{-22.37} \approx 14.71$$

$$|\mathbf{A}_3| = -556.64 \qquad \bar{Q}_3 = \frac{-556.64}{-22.37} \approx 24.88$$

c)
$$|H_1| = \begin{vmatrix} -3 & -1.42 & -1.42 \\ -1.42 & -4.72 & -1.33 \\ -1.42 & -1.33 & -2.62 \end{vmatrix}$$

$|H_1| = -3$, $|H_2| = 12.14$, and $|H_3| = |\mathbf{A}| = -22.37$. And Π is maximized.

THE BORDERED HESSIAN IN CONSTRAINED OPTIMIZATION

12.19. Maximize utility $u = 2xy$ subject to a budget constraint equal to $3x + 4y = 90$ by (a) finding the critical values $\bar{x}$, $\bar{y}$, and $\bar{\lambda}$ and (b) using the bordered Hessian $|\bar{\mathbf{H}}|$ to test the second-order condition.

a) The Lagrangian function is
$$U = 2xy + \lambda(90 - 3x - 4y)$$

The first-order conditions are
$$U_x = 2y - 3\lambda = 0 \qquad U_y = 2x - 4\lambda = 0 \qquad U_\lambda = 90 - 3x - 4y = 0$$

In matrix form,
$$\begin{bmatrix} 0 & 2 & -3 \\ 2 & 0 & -4 \\ -3 & -4 & 0 \end{bmatrix} \begin{bmatrix} x \\ y \\ \lambda \end{bmatrix} = \begin{bmatrix} 0 \\ 0 \\ -90 \end{bmatrix} \tag{12.12}$$

Solving by Cramer's rule, $|\mathbf{A}| = 48$, $|\mathbf{A}_1| = 720$, $|\mathbf{A}_2| = 540$, and $|\mathbf{A}_3| = 360$. Thus, $\bar{x} = 15$, $\bar{y} = 11.25$, and $\bar{\lambda} = 7.5$.

b) Taking the second partials of U with respect to x and y and the first partials of the constraint with respect to x and y to form the bordered Hessian,

$$U_{xx} = 0 \qquad U_{yy} = 0 \qquad U_{xy} = 2 = U_{yx} \qquad c_x = 3 \qquad c_y = 4$$

From Section 12.5,

$$|\bar{\mathbf{H}}| = \begin{vmatrix} 0 & 2 & 3 \\ 2 & 0 & 4 \\ 3 & 4 & 0 \end{vmatrix} \qquad \text{or} \qquad |\bar{\mathbf{H}}| = \begin{vmatrix} 0 & 3 & 4 \\ 3 & 0 & 2 \\ 4 & 2 & 0 \end{vmatrix}$$

$$|\bar{H}_2| = |\bar{\mathbf{H}}| = -2(-12) + 3(8) = 48 > 0 \qquad |\bar{H}_2| = |\bar{\mathbf{H}}| = -3(-8) + 4(6) = 48 > 0$$

The bordered Hessian can be set up in either of the above forms without affecting the value of the principal minor. With $|\bar{\mathbf{H}}| = |\mathbf{A}| > 0$, from the rules of Section 12.5 $|\bar{\mathbf{H}}|$ is negative definite, and U is maximized.

12.20. Maximize utility $u = xy + x$ subject to the budget constraint $6x + 2y = 110$, by using the techniques of Problem 12.19.

a)
$$U = xy + x + \lambda(110 - 6x - 2y)$$
$$U_x = y + 1 - 6\lambda = 0 \qquad U_y = x - 2\lambda = 0 \qquad U_\lambda = 110 - 6x - 2y = 0$$

In matrix form,

$$\begin{bmatrix} 0 & 1 & -6 \\ 1 & 0 & -2 \\ -6 & -2 & 0 \end{bmatrix} \begin{bmatrix} x \\ y \\ \lambda \end{bmatrix} = \begin{bmatrix} -1 \\ 0 \\ -110 \end{bmatrix}$$

Solving by Cramer's rule, $\bar{x} = 9\frac{1}{3}$, $\bar{y} = 27$, and $\bar{\lambda} = 4\frac{2}{3}$.

b) Since $U_{xx} = 0$, $U_{yy} = 0$, $U_{xy} = 1 = U_{yx}$, $c_x = 6$, and $c_y = 2$,

$$|\bar{\mathbf{H}}| = \begin{vmatrix} 0 & 1 & 6 \\ 1 & 0 & 2 \\ 6 & 2 & 0 \end{vmatrix} \qquad |\bar{H}_2| = |\bar{\mathbf{H}}| = 24$$

With $|\bar{H}_2| > 0$, $|\bar{\mathbf{H}}|$ is negative definite, and U is maximized.

12.21. Minimize a firm's total costs $c = 45x^2 + 90xy + 90y^2$ when the firm has to meet a production quota g equal to $2x + 3y = 60$ by (a) finding the critical values and (b) using the bordered Hessian to test the second-order conditions.

a)
$$C = 45x^2 + 90xy + 90y^2 + \lambda(60 - 2x - 3y)$$
$$C_x = 90x + 90y - 2\lambda = 0 \qquad C_y = 90x + 180y - 3\lambda = 0$$
$$C_\lambda = 60 - 2x - 3y = 0$$

In matrix form,

$$\begin{bmatrix} 90 & 90 & -2 \\ 90 & 180 & -3 \\ -2 & -3 & 0 \end{bmatrix} \begin{bmatrix} x \\ y \\ \lambda \end{bmatrix} = \begin{bmatrix} 0 \\ 0 \\ -60 \end{bmatrix}$$

Solving by Cramer's rule, $\bar{x} = 12$, $\bar{y} = 12$, and $\bar{\lambda} = 1080$.

b) Since $C_{xx} = 90$, $C_{yy} = 180$, $C_{xy} = 90 = C_{yx}$, $g_x = 2$, and $g_y = 3$,

$$|\bar{\mathbf{H}}| = \begin{vmatrix} 90 & 90 & 2 \\ 90 & 180 & 3 \\ 2 & 3 & 0 \end{vmatrix}$$

$|\bar{H}_2| = -450$. With $|\bar{H}_2| < 0$, $|\bar{\mathbf{H}}|$ is positive definite and C is minimized.

12.22. Minimize a firm's costs $c = 3x^2 + 5xy + 6y^2$ when the firm must meet a production quota of $5x + 7y = 732$, using the techniques of Problem 12.21.

a)
$$C = 3x^2 + 5xy + 6y^2 + \lambda(732 - 5x - 7y)$$
$$C_x = 6x + 5y - 5\lambda = 0 \qquad C_y = 5x + 12y - 7\lambda = 0$$
$$C_\lambda = 732 - 5x - 7y = 0$$

Solving simultaneously, $\qquad \bar{x} = 75 \qquad \bar{y} = 51 \qquad \bar{\lambda} = 141$

b) With $C_{xx} = 6$, $C_{yy} = 12$, $C_{xy} = 5 = C_{yx}$, $g_x = 5$, and $g_y = 7$,

$$|\bar{\mathbf{H}}| = \begin{vmatrix} 6 & 5 & 5 \\ 5 & 12 & 7 \\ 5 & 7 & 0 \end{vmatrix}$$

$|\bar{H}_2| = 5(35 - 60) - 7(42 - 25) = -244$. Thus, $|\bar{\mathbf{H}}|$ is positive definite, and C is minimized.

12.23. Redo Problem 12.21 by maximizing utility $u = x^{0.5}y^{0.3}$ subject to the budget constraint $10x + 3y = 140$.

a)
$$U = x^{0.5}y^{0.3} + \lambda(140 - 10x - 3y)$$
$$U_x = 0.5x^{-0.5}y^{0.3} - 10\lambda = 0 \qquad U_y = 0.3x^{0.5}y^{-0.7} - 3\lambda = 0$$
$$U_\lambda = 140 - 10x - 3y = 0$$

Solving simultaneously, as shown in Example 10 of Chapter 6,

$$\bar{x} = 8.75 \qquad \bar{y} = 17.5 \qquad \text{and} \qquad \bar{\lambda} = 0.04$$

b) With $U_{xx} = -0.25x^{-1.5}y^{0.3}$, $U_{yy} = -0.21x^{0.5}y^{-1.7}$, $U_{xy} = U_{yx} = 0.15x^{-0.5}y^{-0.7}$, $g_x = 10$, and $g_y = 3$,

$$|\bar{\mathbf{H}}| = \begin{vmatrix} -0.25x^{-1.5}y^{0.3} & 0.15x^{-0.5}y^{-0.7} & 10 \\ 0.15x^{-0.5}y^{-0.7} & -0.21x^{0.5}y^{-1.7} & 3 \\ 10 & 3 & 0 \end{vmatrix}$$

Expanding along the third column,

$$|\bar{H}_2| = 10(0.45x^{-0.5}y^{-0.7} + 2.1x^{0.5}y^{-1.7}) - 3(-0.75x^{-1.5}y^{0.3} - 1.5x^{-0.5}y^{-0.7})$$
$$= 21x^{0.5}y^{-1.7} + 9x^{-0.5}y^{-0.7} + 2.25x^{-1.5}y^{0.3} > 0$$

since x and $y > 0$, and a positive number x raised to a negative power $-n$ equals $1/x^n$, which is also positive. With $|\bar{H}_2| > 0$, $|\bar{\mathbf{H}}|$ is negative definite, and U is maximized.

12.24. Maximize utility $u = x^{0.25}y^{0.4}$ subject to the budget constraint $2x + 8y = 104$, as in Problem 12.23.

a)
$$U = x^{0.25}y^{0.4} + \lambda(104 - 2x - 8y)$$
$$U_x = 0.25x^{-0.75}y^{0.4} - 2\lambda = 0 \qquad U_y = 0.4x^{0.25}y^{-0.6} - 8\lambda = 0$$
$$U_\lambda = 104 - 2x - 8y = 0$$

Solving simultaneously, $\bar{x} = 20$, $\bar{y} = 8$, and $\bar{\lambda} = 0.03$.

b)
$$|\bar{\mathbf{H}}| = \begin{vmatrix} -0.1875x^{-1.75}y^{0.4} & 0.1x^{-0.75}y^{-0.6} & 2 \\ 0.1x^{-0.75}y^{-0.6} & -0.24x^{0.25}y^{-1.6} & 8 \\ 2 & 8 & 0 \end{vmatrix}$$

Expanding along the third row,

$$|\bar{H}_2| = 2(0.8x^{-0.75}y^{-0.6} + 0.48x^{0.25}y^{-1.6}) - 8(-1.5x^{-1.75}y^{0.4} - 0.2x^{-0.75}y^{-0.6})$$
$$= 0.96x^{0.25}y^{-1.6} + 3.2x^{-0.75}y^{-0.6} + 12x^{-1.75}y^{0.4} > 0$$

Thus, $|\bar{\mathbf{H}}|$ is negative definite, and U is maximized.

12.25. Minimize costs $c = 3x + 4y$ subject to the constraint $2xy = 337.5$, using the techniques of Problem 12.21(a) and (b). (c) Discuss the relationship between this solution and that for Problem 12.19.

a)

$$C = 3x + 4y + \lambda(337.5 - 2xy)$$

$$C_x = 3 - 2\lambda y = 0 \qquad \lambda = \frac{1.5}{y} \tag{12.13}$$

$$C_y = 4 - 2\lambda x = 0 \qquad \lambda = \frac{2}{x} \tag{12.14}$$

$$C_\lambda = 337.5 - 2xy = 0 \tag{12.15}$$

Equate λ's in (12.13) and (12.14).

$$\frac{1.5}{y} = \frac{2}{x} \qquad y = 0.75x$$

Substitute in (12.15).

$$337.5 = 2x(0.75x) = 1.5x^2$$
$$x^2 = 225 \qquad \bar{x} = 15$$

Thus, $\bar{y} = 11.25$ and $\bar{\lambda} = 0.133$.

b)   With $C_{xx} = 0$, $C_{yy} = 0$, and $C_{xy} = C_{yx} = -2\lambda$ and from the constraint $2xy = 337.5$, $g_x = 2y$, and $g_y = 2x$,

$$|\bar{H}| = \begin{vmatrix} 0 & -2\lambda & 2y \\ -2\lambda & 0 & 2x \\ 2y & 2x & 0 \end{vmatrix}$$

$|\bar{H}_2| = -(-2\lambda)(-4xy) + 2y(-4x\lambda) = -16\lambda xy$. With $\bar{\lambda}$, $\bar{x}$, $\bar{y} > 0$, $|\bar{H}_2| < 0$. Hence $|\bar{H}|$ is positive definite, and C is minimized.

c)   This problem and Problem 12.19 are the same, except that the objective functions and constraints are reversed. In Problem 12.19, the objective function $u = 2xy$ was maximized subject to the constraint $3x + 4y = 90$; in this problem the objective function $c = 3x + 4y$ was minimized subject to the constraint $2xy = 337.5$. Therefore, one may maximize utility subject to a budget constraint *or* minimize the cost of achieving a given level of utility.

12.26. Minimize the cost of 434 units of production for a firm when $Q = 10K^{0.7}L^{0.1}$ and $P_K = 28$, $P_L = 10$ by (a) finding the critical values and (b) using the bordered Hessian. (c) Check the answer with that of Problem 6.41(b).

a)   The objective function is $c = 28K + 10L$, and the constraint is $10K^{0.7}L^{0.1} = 434$. Thus,

$$C = 28K + 10L + \lambda(434 - 10K^{0.7}L^{0.1})$$
$$C_K = 28 - 7\lambda K^{-0.3}L^{0.1} = 0 \tag{12.16}$$
$$C_L = 10 - \lambda K^{0.7}L^{-0.9} = 0 \tag{12.17}$$
$$C_\lambda = 434 - 10K^{0.7}L^{0.1} = 0 \tag{12.18}$$

Rearranging and dividing (12.16) by (12.17) to eliminate λ,

$$\frac{28}{10} = \frac{7\lambda K^{-0.3}L^{0.1}}{\lambda K^{0.7}L^{-0.9}}$$

$$2.8 = \frac{7L}{K} \qquad K = 2.5L$$

Substituting in (12.18) and using a calculator,

$$434 = 10(2.5)^{0.7} L^{0.7} L^{0.1} \qquad 434 = 19 L^{0.8}$$
$$\bar{L} = (22.8)^{1/0.8} = (22.8)^{1.25} \approx 50$$

Thus, $\bar{K} = 125$ and $\bar{\lambda} = 11.5$.

b) With $C_{KK} = 2.1\lambda K^{-1.3} L^{0.1}$, $C_{LL} = 0.9\lambda K^{0.7} L^{-1.9}$, and $C_{KL} = -0.7\lambda K^{-0.3} L^{-0.9} = C_{LK}$ and from the constraint $g_K = 7K^{-0.3} L^{0.1}$ and $g_L = K^{0.7} L^{-0.9}$,

$$|\bar{\mathbf{H}}| = \begin{vmatrix} 2.1\lambda K^{-1.3} L^{0.1} & -0.7\lambda K^{-0.3} L^{-0.9} & 7K^{-0.3} L^{0.1} \\ -0.7\lambda K^{-0.3} L^{-0.9} & 0.9\lambda K^{0.7} L^{-1.9} & K^{0.7} L^{-0.9} \\ 7K^{-0.3} L^{0.1} & K^{0.7} L^{-0.9} & 0 \end{vmatrix}$$

Expanding along the third row,

$$|\bar{H}_2| = 7K^{-0.3} L^{0.1}(-0.7\lambda K^{0.4} L^{-1.8} - 6.3\lambda K^{0.4} L^{-1.8}) - K^{0.7} L^{-0.9}(2.1\lambda K^{-0.6} L^{-0.8} + 4.9\lambda K^{-0.6} L^{-0.8})$$
$$= -49\lambda K^{0.1} L^{-1.7} - 7\lambda K^{0.1} L^{-1.7} = -56\lambda K^{0.1} L^{-1.7}$$

With $K, L, \lambda > 0$, $|\bar{H}_2| < 0$; $|\bar{\mathbf{H}}|$ is positive definite, and C is minimized.

c) The answers are identical with those in Problem 6.41(b), but note the difference in the work involved when the linear function is selected as the objective function and not the constraint. See also the bordered Hessian for Problem 6.41(b), which is calculated in Problem 12.27(c).

12.27. Use the bordered Hessian to check the second-order conditions for (a) Example 7 of Chapter 6, (b) Problem 6.41(a), and (c) Problem 6.41(b).

a)
$$|\bar{\mathbf{H}}| = \begin{vmatrix} 16 & -1 & 1 \\ -1 & 24 & 1 \\ 1 & 1 & 0 \end{vmatrix}$$

$|\bar{H}_2| = 1(-1 - 24) - 1(16 + 1) = -42$. With $|\bar{H}_2| < 0$, $|\bar{\mathbf{H}}|$ is positive definite and C is minimized.

b)
$$|\bar{\mathbf{H}}| = \begin{vmatrix} -0.21 K^{-1.7} L^{0.5} & 0.15 K^{-0.7} L^{-0.5} & 6 \\ 0.15 K^{-0.7} L^{-0.5} & -0.25 K^{0.3} L^{-1.5} & 2 \\ 6 & 2 & 0 \end{vmatrix}$$

$$|\bar{H}_2| = 6(0.30 K^{-0.7} L^{-0.5} + 1.5 K^{0.3} L^{-1.5}) - 2(-0.42 K^{-1.7} L^{0.5} - 0.9 K^{-0.7} L^{-0.5})$$
$$= 9 K^{0.3} L^{-1.5} + 3.6 K^{-0.7} L^{-0.5} + 0.84 K^{-1.7} L^{0.5} > 0$$

With $|\bar{H}_2| > 0$, $|\bar{\mathbf{H}}|$ is negative definite, and Q is maximized.

c)
$$|\bar{\mathbf{H}}| = \begin{vmatrix} -2.1 K^{-1.3} L^{0.1} & 0.7 K^{-0.3} L^{-0.9} & 28 \\ 0.7 K^{-0.3} L^{-0.9} & -0.9 K^{0.7} L^{-1.9} & 10 \\ 28 & 10 & 0 \end{vmatrix}$$

$$|\bar{H}_2| = 28(7 K^{-0.3} L^{-0.9} + 25.2 K^{0.7} L^{-1.9}) - 10(-21 K^{-1.3} L^{0.1} - 19.6 K^{-0.3} L^{-0.9})$$
$$= 705.6 K^{0.7} L^{-1.9} + 392 K^{-0.3} L^{-0.9} + 210 K^{-1.3} L^{0.1} > 0$$

With $|\bar{H}_2| > 0$, $|\bar{\mathbf{H}}|$ is negative definite, and Q is maximized.

12.28. Use the bordered Hessian to check the second-order conditions in Problem 5.12(c), where $4xyz^2$ was optimized subject to the constraint $x + y + z = 56$; the first-order conditions were $F_x = 4yz^2 - \lambda = 0$, $F_y = 4xz^2 - \lambda = 0$, and $F_z = 8xyz - \lambda = 0$; and the critical values were $\bar{x} = 14$, $\bar{y} = 14$, and $\bar{z} = 28$.

Take the second partial derivatives of F and the first partials of the constraint, and set up the bordered Hessian, as follows:

$$|\bar{\mathbf{H}}| = \begin{vmatrix} 0 & g_x & g_y & g_z \\ g_x & \Gamma_{xx} & \Gamma_{xy} & \Gamma_{xz} \\ g_y & F_{yx} & F_{yy} & F_{yz} \\ g_z & F_{zx} & F_{zy} & F_{zz} \end{vmatrix} = \begin{vmatrix} 0 & 1 & 1 & 1 \\ 1 & 0 & 4z^2 & 8yz \\ 1 & 4z^2 & 0 & 8xz \\ 1 & 8yz & 8xz & 8xy \end{vmatrix}$$

Start with $|\bar{H}_2|$, the 3×3 submatrix in the upper left-hand corner.

$$|\bar{H}_2| = 0 - 1(-4z^2) + 1(4z^2) = 8z^2 > 0$$

Next evaluate $|\bar{H}_3|$, which here equals $|\bar{\mathbf{H}}|$. Expanding along the first row,

$$|\bar{H}_3| = 0 - 1\begin{vmatrix} 1 & 4z^2 & 8yz \\ 1 & 0 & 8xz \\ 1 & 8xz & 8xy \end{vmatrix} + 1\begin{vmatrix} 1 & 0 & 8yz \\ 1 & 4z^2 & 8xz \\ 1 & 8yz & 8xy \end{vmatrix} - 1\begin{vmatrix} 1 & 0 & 4z^2 \\ 1 & 4z^2 & 0 \\ 1 & 8yz & 8xz \end{vmatrix}$$

$$\begin{aligned} |\bar{H}_3| = {}&-1[1(0 - 8xz \cdot 8xz) - 4z^2(8xy - 8xz) + 8yz(8yz - 0)] \\ &+ 1[1(4z^2 \cdot 8xy - 8yz \cdot 8xz) - 0 + 8yz(8yz - 4z^2)] \\ &- 1[1(4z^2 \cdot 8xz - 0) - 0 + 4z^2(8yz - 4z^2)] \\ |\bar{H}_3| = {}&-1(-64x^2z^2 - 32xyz^2 + 32xz^3 + 64xyz^2) + 1(32xyz^2 - 64xyz^2 + 64y^2z^2 - 32yz^3) \\ &-1(32xz^3 + 32yz^3 - 16z^4) \\ |\bar{H}_3| = {}&16z^4 - 64xz^3 - 64yz^3 - 64xyz^2 + 64x^2z^2 + 64y^2z^2 \end{aligned}$$

Evaluated at $\bar{x} = 14$, $\bar{y} = 14$, $\bar{z} = 28$,

$$|\bar{H}_3| = -19{,}668{,}992 < 0$$

With $|\bar{H}_2| > 0$ and $|\bar{H}_3| < 0$, $|\bar{\mathbf{H}}|$ is negative definite and the function is maximized.

INPUT-OUTPUT ANALYSIS

12.29. Determine the total demand for industries 1, 2, and 3, given the matrix of technical coefficients $\mathbf{A}$ and the final demand vector $\mathbf{B}$ below.

$$\mathbf{A} = \begin{matrix} & \text{Output industry} \\ & \begin{matrix} 1 & 2 & 3 \end{matrix} \\ \begin{bmatrix} 0.2 & 0.3 & 0.2 \\ 0.4 & 0.1 & 0.3 \\ 0.3 & 0.5 & 0.2 \end{bmatrix} & \begin{matrix} 1 \\ 2 \\ 3 \end{matrix} \end{matrix} \begin{matrix} \\ \text{Input} \\ \text{industry} \end{matrix} \qquad \mathbf{B} = \begin{bmatrix} 150 \\ 200 \\ 210 \end{bmatrix}$$

From (12.3), the total demand vector is $\mathbf{X} = (\mathbf{I} - \mathbf{A})^{-1}\mathbf{B}$, where

$$\mathbf{I} - \mathbf{A} = \begin{bmatrix} 0.8 & -0.3 & -0.2 \\ -0.4 & 0.9 & -0.3 \\ -0.3 & -0.5 & 0.8 \end{bmatrix}$$

Taking the inverse of $\mathbf{I} - \mathbf{A}$,

$$(\mathbf{I} - \mathbf{A})^{-1} = \frac{1}{0.239}\begin{bmatrix} 0.57 & 0.34 & 0.27 \\ 0.41 & 0.58 & 0.32 \\ 0.47 & 0.49 & 0.60 \end{bmatrix}$$

Substituting in $\mathbf{X} = (\mathbf{I} - \mathbf{A})^{-1}\mathbf{B}$,

$$\mathbf{X} = \frac{1}{0.239}\begin{bmatrix} 0.57 & 0.34 & 0.27 \\ 0.41 & 0.58 & 0.32 \\ 0.47 & 0.49 & 0.60 \end{bmatrix}\begin{bmatrix} 150 \\ 200 \\ 210 \end{bmatrix} = \frac{1}{0.239}\begin{bmatrix} 210.2 \\ 244.7 \\ 294.5 \end{bmatrix} = \begin{bmatrix} 879.50 \\ 1023.85 \\ 1232.22 \end{bmatrix} = \begin{bmatrix} x_1 \\ x_2 \\ x_3 \end{bmatrix}$$

12.30. Determine the new level of total demand X_2 for Problem 12.29 if final demand increases by 40 in industry 1, 20 in industry 2, and 25 in industry 3.

$$\Delta X = (I - A)^{-1} \Delta B$$

$$\Delta X = \frac{1}{0.239} \begin{bmatrix} 0.57 & 0.34 & 0.27 \\ 0.41 & 0.58 & 0.32 \\ 0.47 & 0.49 & 0.60 \end{bmatrix} \begin{bmatrix} 40 \\ 20 \\ 25 \end{bmatrix} = \frac{1}{0.239} \begin{bmatrix} 36.35 \\ 36.00 \\ 43.60 \end{bmatrix} = \begin{bmatrix} 152.09 \\ 150.63 \\ 182.43 \end{bmatrix}$$

$$X_2 = X_1 + \Delta X = \begin{bmatrix} 879.50 \\ 1023.85 \\ 1232.22 \end{bmatrix} + \begin{bmatrix} 152.09 \\ 150.63 \\ 182.43 \end{bmatrix} = \begin{bmatrix} 1031.59 \\ 1174.48 \\ 1414.65 \end{bmatrix}$$

12.31. Determine the total demand for industries 1, 2, and 3, given the matrix of technical coefficients **A** and the final demand vector **B** below.

$$\begin{array}{c} \text{Output industry} \\ \begin{array}{ccc} 1 & 2 & 3 \end{array} \end{array}$$

$$A = \begin{bmatrix} 0.4 & 0.3 & 0.1 \\ 0.2 & 0.2 & 0.3 \\ 0.2 & 0.4 & 0.2 \end{bmatrix} \begin{array}{l} 1 \\ 2 \\ 3 \end{array} \begin{array}{l} \text{Input} \\ \text{industry} \end{array} \qquad B = \begin{bmatrix} 140 \\ 220 \\ 180 \end{bmatrix}$$

$$X = (I - A)^{-1} B$$

where

$$I - A = \begin{bmatrix} 0.6 & -0.3 & -0.1 \\ -0.2 & 0.8 & -0.3 \\ -0.2 & -0.4 & 0.8 \end{bmatrix}$$

and the inverse

$$(I - A)^{-1} = \frac{1}{0.222} \begin{bmatrix} 0.52 & 0.28 & 0.17 \\ 0.22 & 0.46 & 0.20 \\ 0.24 & 0.30 & 0.42 \end{bmatrix}$$

Thus,

$$X = \frac{1}{0.222} \begin{bmatrix} 0.52 & 0.28 & 0.17 \\ 0.22 & 0.46 & 0.20 \\ 0.24 & 0.30 & 0.42 \end{bmatrix} \begin{bmatrix} 140 \\ 220 \\ 180 \end{bmatrix} = \begin{bmatrix} 743.24 \\ 756.76 \\ 789.19 \end{bmatrix} = \begin{bmatrix} x_1 \\ x_2 \\ x_3 \end{bmatrix}$$

12.32. Determine the new total demand X_2 if final demand increases by 30 for industry 1 and decreases by 15 and 35 for industries 2 and 3, respectively, in Problem 12.31.

$$\Delta X = (I - A)^{-1} \Delta B$$

$$\Delta X = \frac{1}{0.222} \begin{bmatrix} 0.52 & 0.28 & 0.17 \\ 0.22 & 0.46 & 0.20 \\ 0.24 & 0.30 & 0.42 \end{bmatrix} \begin{bmatrix} 30 \\ -15 \\ -35 \end{bmatrix} = \frac{1}{0.222} \begin{bmatrix} 5.45 \\ -7.30 \\ -12.00 \end{bmatrix} = \begin{bmatrix} 24.55 \\ -32.88 \\ -54.05 \end{bmatrix}$$

$$X_2 = X_1 + \Delta X = \begin{bmatrix} 743.24 \\ 756.76 \\ 789.19 \end{bmatrix} + \begin{bmatrix} 24.55 \\ -32.88 \\ -54.05 \end{bmatrix} = \begin{bmatrix} 767.79 \\ 723.88 \\ 735.14 \end{bmatrix}$$

12.33. Given the interindustry transaction demand table in millions of dollars below, find the matrix of technical coefficients.

Sector of Origin	Sector of Destination				Final Demand	Total Demand
	Steel	Coal	Iron	Auto		
Steel	80	20	110	230	160	600
Coal	200	50	90	120	140	600
Iron	220	110	30	40	0	400
Auto	60	140	160	240	400	1000
Value added	40	280	10	370		
Gross production	600	600	400	1000		

The technical coefficient a_{ij} expresses the number of units or dollars of input i required to produce one unit or one dollar of product j. Thus a_{11} = the percentage of steel in one dollar of steel, a_{21} = the percentage of coal in one dollar of steel, a_{31} = the percentage of iron in one dollar of steel, and a_{41} = the percentage of autos in one dollar of steel. To find the technical coefficients, simply divide every element in each column by the value of gross production at the bottom of the column, omitting value added. Thus,

$$A = \begin{bmatrix} \frac{80}{600} & \frac{20}{600} & \frac{110}{400} & \frac{230}{1000} \\ \frac{200}{600} & \frac{50}{600} & \frac{90}{400} & \frac{120}{1000} \\ \frac{220}{600} & \frac{110}{600} & \frac{30}{400} & \frac{40}{1000} \\ \frac{60}{600} & \frac{140}{600} & \frac{160}{400} & \frac{240}{1000} \end{bmatrix} = \begin{bmatrix} 0.133 & 0.033 & 0.275 & 0.23 \\ 0.333 & 0.083 & 0.225 & 0.12 \\ 0.367 & 0.183 & 0.075 & 0.04 \\ 0.10 & 0.233 & 0.40 & 0.24 \end{bmatrix}$$

12.34. Check the matrix of technical coefficients **A** in Problem 12.33.

To check matrix **A**, multiply it by the column vector of total demand **X**. The product should equal the intermediate demand which is total demand **X** − final demand **B**. Allow for slight errors due to rounding.

$$AX = \begin{bmatrix} 0.133 & 0.033 & 0.275 & 0.23 \\ 0.333 & 0.083 & 0.225 & 0.12 \\ 0.367 & 0.183 & 0.075 & 0.04 \\ 0.10 & 0.233 & 0.40 & 0.24 \end{bmatrix} \begin{bmatrix} 600 \\ 600 \\ 400 \\ 1000 \end{bmatrix} = \begin{bmatrix} 439.6 \\ 459.6 \\ 400 \\ 599.8 \end{bmatrix}$$

$$X - B = \begin{bmatrix} 600 \\ 600 \\ 400 \\ 1000 \end{bmatrix} - \begin{bmatrix} 160 \\ 140 \\ 0 \\ 400 \end{bmatrix} = \begin{bmatrix} 440 \\ 460 \\ 400 \\ 600 \end{bmatrix}$$

12.35. Given the interindustry transaction demand table below, (a) find the matrix of technical coefficients and (b) check your answer.

Sector of Origin	Sector of Destination			Final Demand	Total Demand
	1	2	3		
1	20	60	10	50	140
2	50	10	80	10	150
3	40	30	20	40	130
Value added	30	50	20		
Gross production	140	150	130		

a)
$$\mathbf{A} = \begin{bmatrix} \frac{20}{140} & \frac{60}{150} & \frac{10}{130} \\ \frac{50}{140} & \frac{10}{150} & \frac{80}{130} \\ \frac{40}{140} & \frac{30}{150} & \frac{20}{130} \end{bmatrix} = \begin{bmatrix} 0.143 & 0.4 & 0.077 \\ 0.357 & 0.067 & 0.615 \\ 0.286 & 0.2 & 0.154 \end{bmatrix}$$

b)
$$\mathbf{AX} = \begin{bmatrix} 0.143 & 0.4 & 0.077 \\ 0.357 & 0.067 & 0.615 \\ 0.286 & 0.2 & 0.154 \end{bmatrix} \begin{bmatrix} 140 \\ 150 \\ 130 \end{bmatrix} = \begin{bmatrix} 90 \\ 140 \\ 90 \end{bmatrix}$$

$$\mathbf{X} - \mathbf{B} = \begin{bmatrix} 140 \\ 150 \\ 130 \end{bmatrix} - \begin{bmatrix} 50 \\ 10 \\ 40 \end{bmatrix} = \begin{bmatrix} 90 \\ 140 \\ 90 \end{bmatrix}$$

12.36. Find the new level of total demand in Problem 12.35 if in year 2 final demand is 70 in industry 1, 25 in industry 2, and 50 in industry 3.

$$\mathbf{X} = (\mathbf{I} - \mathbf{A})^{-1}\mathbf{B}$$

where $\mathbf{I} - \mathbf{A} = \begin{bmatrix} 0.857 & -0.4 & -0.077 \\ -0.357 & 0.933 & -0.615 \\ -0.286 & -0.2 & 0.846 \end{bmatrix}$ and $(\mathbf{I} - \mathbf{A})^{-1} = \dfrac{1}{0.354}\begin{bmatrix} 0.666 & 0.354 & 0.318 \\ 0.478 & 0.703 & 0.555 \\ 0.338 & 0.286 & 0.657 \end{bmatrix}$

Thus,
$$\mathbf{X} = \frac{1}{0.354}\begin{bmatrix} 0.666 & 0.354 & 0.318 \\ 0.478 & 0.703 & 0.555 \\ 0.338 & 0.286 & 0.657 \end{bmatrix}\begin{bmatrix} 70 \\ 25 \\ 50 \end{bmatrix} = \frac{1}{0.354}\begin{bmatrix} 71.37 \\ 78.79 \\ 63.66 \end{bmatrix} = \begin{bmatrix} 201.61 \\ 222.57 \\ 179.83 \end{bmatrix}$$

12.37. Having found the inverse of $\mathbf{I} - \mathbf{A}$, use it to check the accuracy of the matrix of coefficients derived in Problem 12.35; i.e., check to see if $(\mathbf{I} - \mathbf{A})^{-1}\mathbf{B} = \mathbf{X}$.

$$(\mathbf{I} - \mathbf{A})^{-1}\mathbf{B} = \frac{1}{0.354}\begin{bmatrix} 0.666 & 0.354 & 0.318 \\ 0.478 & 0.703 & 0.555 \\ 0.338 & 0.286 & 0.657 \end{bmatrix}\begin{bmatrix} 50 \\ 10 \\ 40 \end{bmatrix} = \frac{1}{0.354}\begin{bmatrix} 49.56 \\ 53.13 \\ 46.04 \end{bmatrix} = \begin{bmatrix} 140 \\ 150 \\ 130 \end{bmatrix}$$

12.38. Assume in Problem 12.35 that value added is composed entirely of the primary input labor. How much labor would be necessary to obtain the final demand (a) in Problem 12.35 and (b) in Problem 12.36? (c) If the amount of labor available in the economy is 100, is the output mix feasible?

a)   To get the technical coefficient of labor a_{Lj} in Problem 12.35, simply divide the value added in each column by the gross production. Thus, $a_{L1} = \frac{30}{140} = 0.214$, $a_{L2} = \frac{50}{150} = 0.333$, and $a_{L3} = \frac{20}{130} = 0.154$. The amount of labor needed to meet the final demand will then equal the row of technical coefficients for labor times the column vector of total demand, since labor must also be used to produce the intermediate products. Thus,

$$L_1 = [0.214 \quad 0.333 \quad 0.154]\begin{bmatrix} 140 \\ 150 \\ 130 \end{bmatrix} = 99.93$$

b)
$$L_2 = [0.214 \quad 0.333 \quad 0.154]\begin{bmatrix} 201.61 \\ 222.57 \\ 179.83 \end{bmatrix} = 144.95$$

c)   Final demand in Problem 12.35 is feasable since $99.93 < 100$. Final demand in Problem 12.36 is not feasible since society does not have sufficient labor resources to produce it.

12.39. Check the accuracy of the technical coefficients found in Problem 12.38.

Having found the technical coefficients of labor for Problem 12.35, where value added was due totally to labor inputs, the accuracy of the technical coefficients can be easily checked. Since each dollar of output must be completely accounted for in terms of inputs, simply add each column of technical coefficients to be sure it equals 1.

	1	2	3
1	0.143	0.4	0.077
2	0.357	0.067	0.615
3	0.286	0.2	0.154
Value added (labor)	0.214	0.333	0.154
	1.000	1.000	1.000

EIGENVALUES, EIGENVECTORS

12.40. Use eigenvalues (characteristic roots, latent roots) to determine sign definiteness for

$$\mathbf{A} = \begin{bmatrix} 10 & 3 \\ 3 & 4 \end{bmatrix}$$

To find the characteristic roots of $\mathbf{A}$, the determinant of the characteristic matrix $\mathbf{A} - c\mathbf{I}$ must equal zero. Thus,

$$|\mathbf{A} - c\mathbf{I}| = \begin{vmatrix} 10 - c & 3 \\ 3 & 4 - c \end{vmatrix} = 0$$

$$40 + c^2 - 14c - 9 = 0 \qquad c^2 - 14c + 31 = 0$$

Using the quadratic formula,

$$c = \frac{14 \pm \sqrt{196 - 4(31)}}{2} = \frac{14 \pm 8.485}{2}$$

$$c_1 = 11.2425 \qquad c_2 = 2.7575$$

With both characteristic roots positive, $\mathbf{A}$ is positive definite.

12.41. Redo Problem 12.40, given

$$\mathbf{A} = \begin{bmatrix} -4 & -2 \\ -2 & -6 \end{bmatrix}$$

$$|\mathbf{A} - c\mathbf{I}| = \begin{vmatrix} -4 - c & -2 \\ -2 & -6 - c \end{vmatrix} = 0$$

$$24 + c^2 + 10c - 4 = 0 \qquad c^2 + 10c + 20 = 0$$

$$c = \frac{-10 \pm \sqrt{100 - 4(20)}}{2} = \frac{-10 \pm 4.4721}{2}$$

$$c_1 = \frac{-5.5279}{2} = -2.764 \qquad c_2 = \frac{-14.4721}{2} = -7.236$$

With both characteristic roots negative, $\mathbf{A}$ is negative definite.

12.42. Redo Problem 12.40, given

$$\mathbf{A} = \begin{bmatrix} 6 & 2 \\ 2 & 2 \end{bmatrix}$$

$$|\mathbf{A} - c\mathbf{I}| = \begin{vmatrix} 6 - c & 2 \\ 2 & 2 - c \end{vmatrix} = 0$$

$$12 + c^2 - 8c - 4 = 0 \qquad c^2 - 8c + 8 = 0$$

$$c = \frac{8 \pm \sqrt{64 - 4(8)}}{2} = \frac{8 \pm 5.66}{2}$$

$$c_1 = 1.17 \qquad c_2 = 6.83$$

$\mathbf{A}$ is positive definite.

12.43. Redo Problem 12.40, given

$$\mathbf{A} = \begin{bmatrix} 4 & 6 & 3 \\ 0 & 2 & 5 \\ 0 & 1 & 3 \end{bmatrix}$$

$$|\mathbf{A} - c\mathbf{I}| = \begin{vmatrix} 4 - c & 6 & 3 \\ 0 & 2 - c & 5 \\ 0 & 1 & 3 - c \end{vmatrix} = 0$$

Expanding along the first column,

$$|\mathbf{A} - c\mathbf{I}| = (4 - c)[(2 - c)(3 - c) - 5] = 0 \tag{12.19}$$

$$-c^3 + 9c^2 - 21c + 4 = 0 \tag{12.20}$$

To solve (12.20), we may use a standard formula for finding cube roots or note that (12.20) will equal zero if in (12.19)

$$4 - c = 0 \qquad \text{or} \qquad (2 - c)(3 - c) - 5 = 0$$

Thus, the characteristic roots are

$$4 - c = 0 \qquad (2 - c)(3 - c) - 5 = 0$$

$$c_1 = 4 \qquad c^2 - 5c + 1 = 0$$

$$c = \frac{5 \pm \sqrt{25 - 4}}{2} = \frac{5 \pm 4.58}{2}$$

$$c_2 = 4.79 \qquad c_3 = 0.21$$

With all three characteristic roots positive, $\mathbf{A}$ is positive definite.

12.44. Redo Problem 12.40, given

$$\mathbf{A} = \begin{bmatrix} 6 & 1 & 0 \\ 13 & 4 & 0 \\ 5 & 1 & 9 \end{bmatrix}$$

$$|\mathbf{A} - c\mathbf{I}| = \begin{vmatrix} 6 - c & 1 & 0 \\ 13 & 4 - c & 0 \\ 5 & 1 & 9 - c \end{vmatrix} = 0$$

Expanding along the third column,

$$|\mathbf{A} - c\mathbf{I}| = (9 - c)[(6 - c)(4 - c) - 13] = 0 \tag{12.21}$$

$$-c^3 + 19c^2 - 101c + 99 = 0 \tag{12.22}$$

which will equal zero if in (12.21)

$$9 - c = 0 \quad \text{or} \quad (6 - c)(4 - c) - 13 = 0$$

Thus,
$$c_1 = 9 \qquad c^2 - 10c + 11 = 0$$

$$c = \frac{10 \pm \sqrt{100 - 4(11)}}{2} = \frac{10 \pm 7.48}{2}$$

$$c_2 = 8.74 \qquad c_3 = 1.26$$

With all latent roots positive, **A** is positive definite.

12.45. Redo Problem 12.40, given

$$\mathbf{A} = \begin{bmatrix} -5 & 1 & 2 \\ 0 & -2 & 0 \\ 4 & 2 & -3 \end{bmatrix}$$

$$|\mathbf{A} - c\mathbf{I}| = \begin{vmatrix} -5 - c & 1 & 2 \\ 0 & -2 - c & 0 \\ 4 & 2 & -3 - c \end{vmatrix} = 0$$

Expanding along the second row,

$$|\mathbf{A} - c\mathbf{I}| = (-2 - c)[(-5 - c)(-3 - c) - 8] = 0$$

Thus,
$$-2 - c = 0 \quad \text{or} \quad (-5 - c)(-3 - c) - 8 = 0$$
$$c_1 = -2 \qquad c^2 + 8c + 7 = 0$$
$$(c + 7)(c + 1) = 0$$
$$c_2 = -7 \qquad c_3 = -1$$

With all latent roots negative, **A** is negative definite.

12.46. Given
$$\mathbf{A} = \begin{bmatrix} 6 & 6 \\ 6 & -3 \end{bmatrix}$$

Find (a) the characteristic roots and (b) the characteristic vectors.

a)
$$|\mathbf{A} - c\mathbf{I}| = \begin{vmatrix} 6 - c & 6 \\ 6 & -3 - c \end{vmatrix} = 0$$

$$-18 + c^2 - 3c - 36 = 0$$
$$c^2 - 3c - 54 = 0$$
$$(c - 9)(c + 6) = 0$$
$$c_1 = 9 \qquad c_2 = -6$$

With one root positive and the other negative, **A** is sign indefinite.

b)   Using $c_1 = 9$ for the first characteristic vector $\mathbf{V}_1$,

$$\begin{bmatrix} 6 - 9 & 6 \\ 6 & -3 - 9 \end{bmatrix} \begin{bmatrix} v_1 \\ v_2 \end{bmatrix} = \begin{bmatrix} -3 & 6 \\ 6 & -12 \end{bmatrix} \begin{bmatrix} v_1 \\ v_2 \end{bmatrix} = 0$$

$$v_1 = 2v_2$$

Normalizing, as in Example 9,

$$(2v_2)^2 + v_2^2 = 1$$
$$5v_2^2 = 1$$
$$v_2 = \sqrt{0.2} \qquad v_1 = 2v_2 = 2\sqrt{0.2}$$

Thus,

$$\mathbf{V}_1 = \begin{bmatrix} 2\sqrt{0.2} \\ \sqrt{0.2} \end{bmatrix}$$

Using $c_2 = -6$ for the second characteristic vector,

$$\begin{bmatrix} 6-(-6) & 6 \\ 6 & -3-(-6) \end{bmatrix}\begin{bmatrix} v_1 \\ v_2 \end{bmatrix} = \begin{bmatrix} 12 & 6 \\ 6 & 3 \end{bmatrix}\begin{bmatrix} v_1 \\ v_2 \end{bmatrix} = 0$$

$$v_2 = -2v_1$$

Normalizing,

$$v_1^2 + (-2v_1)^2 = 1$$
$$5v_1^2 = 1$$
$$v_1 = \sqrt{0.2} \qquad v_2 = -2v_1 = -2\sqrt{0.2}$$

Thus,

$$\mathbf{V}_2 = \begin{bmatrix} \sqrt{0.2} \\ -2\sqrt{0.2} \end{bmatrix}$$

12.47. Redo Problem 12.46, given

$$\mathbf{A} = \begin{bmatrix} 6 & 3 \\ 3 & -2 \end{bmatrix}$$

a)

$$|\mathbf{A} - c\mathbf{I}| = \begin{vmatrix} 6-c & 3 \\ 3 & -2-c \end{vmatrix} = 0$$

$$c^2 - 4c - 21 = 0$$
$$c_1 = 7 \qquad c_2 = -3$$

With $c_1 > 0$ and $c_2 < 0$, $\mathbf{A}$ is sign indefinite.

b)   Using $c_1 = 7$ to form the first characteristic vector,

$$\begin{bmatrix} 6-7 & 3 \\ 3 & -2-7 \end{bmatrix}\begin{bmatrix} v_1 \\ v_2 \end{bmatrix} = \begin{bmatrix} -1 & 3 \\ 3 & -9 \end{bmatrix}\begin{bmatrix} v_1 \\ v_2 \end{bmatrix} = 0$$

$$v_1 = 3v_2$$

Normalizing,

$$(3v_2)^2 + v_2^2 = 1$$
$$9v_2^2 + v_2^2 = 1$$
$$10v_2^2 = 1$$
$$v_2 = \sqrt{0.1} \quad \text{and} \quad v_1 = 3v_2 = 3\sqrt{0.1}$$

Thus,

$$\mathbf{V}_1 = \begin{bmatrix} 3\sqrt{0.1} \\ \sqrt{0.1} \end{bmatrix}$$

Using $c_2 = -3$,

$$\begin{bmatrix} 6-(-3) & 3 \\ 3 & -2-(-3) \end{bmatrix}\begin{bmatrix} v_1 \\ v_2 \end{bmatrix} = \begin{bmatrix} 9 & 3 \\ 3 & 1 \end{bmatrix}\begin{bmatrix} v_1 \\ v_2 \end{bmatrix} = 0$$

$$v_2 = -3v_1$$

Normalizing,

$$v_1^2 + (-3v_1)^2 = 1$$
$$10v_1^2 = 1$$
$$v_1 = \sqrt{0.1} \quad \text{and} \quad v_2 = -3v_1 = -3\sqrt{0.1}$$

Thus,

$$\mathbf{V}_2 = \begin{bmatrix} \sqrt{0.1} \\ -3\sqrt{0.1} \end{bmatrix}$$

CHAPTER 13

Comparative Statics and Concave Programming

13.1 INTRODUCTION TO COMPARATIVE STATICS

Comparative-static analysis, more commonly known as *comparative statics*, compares the different equilibrium values of the endogenous variables resulting from changes in the values of the exogenous variables and parameters in the model. Comparative statics allows economists to estimate such things as the responsiveness of consumer demand to a projected excise tax, tariff, or subsidy; the effect on national income of a change in investment, government spending, or the interest rate; and the likely price of a commodity given some change in weather conditions, price of inputs, or availability of transportation. Comparative statics essentially involves finding the appropriate derivative, as we saw earlier in Section 6.2.

13.2 COMPARATIVE STATICS WITH ONE ENDOGENOUS VARIABLE

Comparative statics can be used both with specific and general functions. Example 1 provides a specific function illustration; Example 2 demonstrates the method with a general function. In the case of specific functions the prerequisite derivatives can be derived from either explicit or implicit functions. In the case of general functions, implicit functions must be used. Whenever there is more than one independent variable (Problem 13.2), partial derivatives are the appropriate derivatives and are found in a similar fashion (Problem 13.3).

EXAMPLE 1. Assume the demand Q_D and supply Q_S of a commodity are given by specific functions, here expressed in terms of parameters.

$$Q_D = m - nP + kY \qquad m, n, k > 0$$
$$Q_S = a + bP \qquad a, b > 0$$

284

where P = price and Y = consumers' income. The equilibrium condition is

$$Q_D = Q_S$$

Substituting from above and solving for the equilibrium price level P^*, we have

$$m - nP + kY = a + bP \tag{13.1}$$

$$m - a + kY = (b + n)P$$

$$P^* = \frac{m - a + kY}{b + n} \tag{13.2}$$

Using comparative statics we can now determine how the equilibrium level of the endogenous variable P^* will change for a change in the single exogenous variable (Y) or any of the five parameters (a, b, m, n, k), should the latter be of interest. Comparative-static analysis simply involves taking the desired derivative and determining its sign. To gauge the responsiveness of the equilibrium price to changes in income, we have from the explicit function (13.2),

$$\frac{dP^*}{dY} = \frac{k}{b + n} > 0 \tag{13.3}$$

This means that an increase in consumers' income in this model will lead to an increase in the equilibrium price of the good. If the values of the parameters are known, as in Problem 13.1, the specific size of the price increase can also be estimated.

Comparative statics can be applied equally well to implicit functions. By moving everything to the left in (13.1), so that $Q_D - Q_S = 0$, or excess demand equals zero, we can derive the implicit function F for the equilibrium condition:

$$F = m - nP + kY - a - bP = 0 \tag{13.4}$$

Then the implicit function rule (Section 5.10) can be employed to find the desired comparative-static derivative. Assuming $F_p \neq 0$,

$$\frac{dP^*}{dY} = -\frac{F_Y}{F_P}$$

where from (13.4), $F_Y = k$ and $F_P = -(n + b)$. Substituting and simplifying, we have

$$\frac{dP^*}{dY} = -\frac{k}{-(n + b)} = \frac{k}{b + n} > 0$$

Comparative statics can also be used to estimate the effect on P^* of a change in any of the parameters (m, n, k, a, b), but since these merely represent intercepts and slopes in demand and supply analysis, such as we have above, they generally have little practical relevance for economics. In other instances, however, such as income determination models (Problem 13.3), the parameters will frequently have economic significance and may warrant comparative-static derivatives of their own.

EXAMPLE 2. Now assume a general model in which the supply and demand of a commodity are given solely by general functions:

$$\text{Demand} = D(P, Y) \qquad D_p < 0, D_Y > 0,$$

$$\text{Supply} = S(P) \qquad S_P > 0$$

The equilibrium price level P^* can be found where demand equals supply:

$$D(P, Y) = S(P)$$

or equivalently where excess demand equals zero,

$$D(P, Y) - S(P) = 0 \tag{13.5}$$

With general functions, only implicit forms such as (*13.5*) are helpful in finding the comparative-static derivatives. Assuming $F_P \neq 0$,

$$\frac{dP^*}{dY} = -\frac{F_Y}{F_P}$$

where from (*13.5*), $F_Y = D_Y$ and $F_P = D_P - S_P$. Substituting,

$$\frac{dP^*}{dY} = -\frac{D_Y}{D_P - S_P}$$

From theory, we always expect $S_P > 0$. If the good is a normal good, then $D_Y > 0$ and $D_P < 0$. Substituting above, we have in the case of a normal good

$$\frac{dP^*}{dY} = -\frac{(+)}{(-) - (+)} > 0$$

If the good is an inferior good but not a Giffen good, then $D_Y \leq 0$ and $D_P < 0$ so that $dP^*/dY \leq 0$; and if the good is a Giffen good, then $D_Y < 0$ and $D_P > 0$ and the sign of the derivative will be indeterminate and depend on the sign of the denominator. See Problems 13.1 to 13.7.

13.3 COMPARATIVE STATICS WITH MORE THAN ONE ENDOGENOUS VARIABLE

In a model with more than one endogenous variable, comparative statics requires that there be a unique equilibrium condition for each of the endogenous variables. A system of n endogenous variables must have n equilibrium conditions. Measuring the effect of a particular exogenous variable on any or all of the endogenous variables involves taking the total derivative of each of the equilibrium conditions with respect to the particular exogenous variable and solving simultaneously for each of the desired partial derivatives. If the functions have continuous derivatives and the Jacobian consisting of the partial derivatives of all the functions with respect to the endogenous variables does not equal zero, then from the implicit function theorem the optimal values of the endogenous variables can be expressed as functions of the exogenous variables, as outlined in Problem 13.8, and the comparative-static derivatives can be estimated with the help of Cramer's rule, as demonstrated in Example 3 below. The method is also illustrated in terms of a typical economic problem in Example 4.

EXAMPLE 3. For simplicity of exposition, assume a model with only two endogenous variables and two exogenous variables, expressed in terms of implicit general functions in which the endogenous variables are listed first, followed by the exogenous variables, with a semicolon separating the former from the latter. The model can be easily expanded to any number of endogenous variables (n) and any number of exogenous variables (m), where n need not equal m.

$$F^1(y_1, y_2; x_1, x_2) = 0$$
$$F^2(y_1, y_2; x_1, x_2) = 0$$

To find the comparative-static partial derivatives of the system with respect to one of the independent variables, say x_1, we take the total derivative (Section 5.9) of both functions with respect to x_1.

$$\frac{\partial F^1}{\partial y_1} \cdot \frac{\partial y_1}{\partial x_1} + \frac{\partial F^1}{\partial y_2} \cdot \frac{\partial y_2}{\partial x_1} + \frac{\partial F^1}{\partial x_1} = 0$$

$$\frac{\partial F^2}{\partial y_1} \cdot \frac{\partial y_1}{\partial x_1} + \frac{\partial F^2}{\partial y_2} \cdot \frac{\partial y_2}{\partial x_1} + \frac{\partial F^2}{\partial x_1} = 0$$

When evaluated at the equilibrium point, which is also frequently indicated by a bar, all the partial derivatives will have fixed values and so can be expressed in matrix notation. Moving the partial derivatives of both functions with respect to x_1 to the right, we have

$$
\begin{bmatrix} \dfrac{\partial F^1}{\partial y_1} & \dfrac{\partial F^1}{\partial y_2} \\[2ex] \dfrac{\partial F^2}{\partial y_1} & \dfrac{\partial F^2}{\partial y_2} \end{bmatrix} \begin{bmatrix} \dfrac{\partial \bar{y}_1}{\partial x_1} \\[2ex] \dfrac{\partial \bar{y}_2}{\partial x_1} \end{bmatrix} = \begin{bmatrix} -\dfrac{\partial F^1}{\partial x_1} \\[2ex] -\dfrac{\partial F^2}{\partial x_1} \end{bmatrix}
$$

$$\mathbf{JX} = \mathbf{B}$$

If both functions have continuous first and second derivatives and the Jacobian $|\mathbf{J}|$, consisting of all the first-order partial derivatives of both functions (F^i) with respect to both *endogenous* variables (y_j), does not equal zero,

$$
|\mathbf{J}| = \frac{\partial F^1}{\partial y_1} \cdot \frac{\partial F^2}{\partial y_2} - \frac{\partial F^2}{\partial y_1} \cdot \frac{\partial F^1}{\partial y_2} \neq 0,
$$

then by making use of the implicit function theorem we can express the optimal values of the endogenous values as implicit functions of the exogenous variables and solve for the desired comparative-static derivatives in $\mathbf{X}$ using Cramer's rule. Specifically, assuming $|\mathbf{J}| \neq 0$,

$$
\frac{\partial \bar{y}_i}{\partial x_1} = \frac{|\mathbf{J}_i|}{|\mathbf{J}|} \tag{13.6}
$$

Thus, to solve for the first derivative, $\partial \bar{y}_1 / \partial x_1$, we form a new matrix $|\mathbf{J}_1|$ by replacing the first column of $\mathbf{J}$ with the column vector $\mathbf{B}$ and then substitute it above in (13.6).

$$
\frac{\partial \bar{y}_1}{\partial x_1} = \frac{|\mathbf{J}_1|}{|\mathbf{J}|} = \frac{\begin{vmatrix} -\dfrac{\partial F^1}{\partial x_1} & \dfrac{\partial F^1}{\partial y_2} \\[2ex] -\dfrac{\partial F^2}{\partial x_1} & \dfrac{\partial F^2}{\partial y_2} \end{vmatrix}}{\begin{vmatrix} \dfrac{\partial F^1}{\partial y_1} & \dfrac{\partial F^1}{\partial y_2} \\[2ex] \dfrac{\partial F^2}{\partial y_1} & \dfrac{\partial F^2}{\partial y_2} \end{vmatrix}} = \frac{-\left(\dfrac{\partial F^1}{\partial x_1} \cdot \dfrac{\partial F^2}{\partial y_2} - \dfrac{\partial F^2}{\partial x_1} \cdot \dfrac{\partial F^1}{\partial y_2} \right)}{\dfrac{\partial F^1}{\partial y_1} \cdot \dfrac{\partial F^2}{\partial y_2} - \dfrac{\partial F^2}{\partial y_1} \cdot \dfrac{\partial F^1}{\partial y_2}}
$$

Similarly,

$$
\frac{\partial \bar{y}_2}{\partial x_1} = \frac{|\mathbf{J}_2|}{|\mathbf{J}|} = \frac{\begin{vmatrix} \dfrac{\partial F^1}{\partial y_1} & -\dfrac{\partial F^1}{\partial x_1} \\[2ex] \dfrac{\partial F^2}{\partial y_1} & -\dfrac{\partial F^2}{\partial x_1} \end{vmatrix}}{\begin{vmatrix} \dfrac{\partial F^1}{\partial y_1} & \dfrac{\partial F^1}{\partial y_2} \\[2ex] \dfrac{\partial F^2}{\partial y_1} & \dfrac{\partial F^2}{\partial y_2} \end{vmatrix}} = \frac{-\left(\dfrac{\partial F^1}{\partial y_1} \cdot \dfrac{\partial F^2}{\partial x_1} - \dfrac{\partial F^2}{\partial y_1} \cdot \dfrac{\partial F^1}{\partial x_1} \right)}{\dfrac{\partial F^1}{\partial y_1} \cdot \dfrac{\partial F^2}{\partial y_2} - \dfrac{\partial F^2}{\partial y_1} \cdot \dfrac{\partial F^1}{\partial y_2}}
$$

The partials with respect to x_2 are found in like fashion, after starting with the total derivatives of both functions with respect to x_2. See Problem 13.9.

EXAMPLE 4.   Assume that equilibrium in the goods and services market (IS curve) and the money market (LM curve) are given, respectively, by

$$
F^1(Y, i; C_0, M_0, P) = Y - C_0 - C(Y, i) = 0 \qquad 0 < C_Y < 1, C_i < 0 \tag{13.7}
$$

$$
F^2(Y, i; C_0, M_0, P) = L(Y, i) - M_0/P = 0 \qquad L_Y > 0, L_i < 0 \tag{13.8}
$$

where $L(Y, i)$ = the demand for money, M_0 = the supply of money, C_0 = autonomous consumption, and P = the price level, which makes M_0/P the supply of real rather than nominal money. For simplicity, we will hold P

constant. The effect on the equilibrium levels of Y and i of a change in C_0 using comparative statics is demonstrated below.

a) Take the total derivative of the equilibrium conditions, (13.7) and (13.8), with respect to the desired exogenous variable, here C_0.

$$\frac{\partial Y}{\partial C_0} - 1 - \left(C_Y \cdot \frac{\partial Y}{\partial C_0} \right) - \left(C_i \cdot \frac{\partial i}{\partial C_0} \right) = 0$$

$$\left(L_Y \cdot \frac{\partial Y}{\partial C_0} \right) + \left(L_i \cdot \frac{\partial i}{\partial C_0} \right) = 0$$

b) Rearranging and setting in matrix form,

$$\begin{bmatrix} 1 - C_Y & -C_i \\ L_Y & L_i \end{bmatrix} \begin{bmatrix} \dfrac{\partial \bar{Y}}{\partial C_0} \\ \dfrac{\partial \bar{i}}{\partial C_0} \end{bmatrix} = \begin{bmatrix} 1 \\ 0 \end{bmatrix}$$

$$\mathbf{JX} = \mathbf{B}$$

c) Then check to make sure the Jacobian $|\mathbf{J}| \neq 0$ so the implicit function theorem holds.

$$|\mathbf{J}| = (1 - C_Y)L_i + C_i L_Y$$

Applying the signs, $|\mathbf{J}| = (+)(-) + (-)(+) = (-) < 0$

Therefore, $|\mathbf{J}| \neq 0$.

d) Solve for the first derivative, $\partial \bar{Y}/\partial C_0$ by forming a new matrix $|\mathbf{J}_1|$ in which the first column of $\mathbf{J}$ is replaced with the column vector $\mathbf{B}$ and substituted in (13.6).

$$|\mathbf{J}_1| = \begin{vmatrix} 1 & -C_i \\ 0 & L_i \end{vmatrix} = L_i$$

Thus, $$\frac{\partial \bar{Y}}{\partial C_0} = \frac{|\mathbf{J}_1|}{|\mathbf{J}|} = \frac{L_i}{(1 - C_Y)L_i + C_i L_Y} = \frac{(-)}{(-)} > 0$$

An increase in autonomous consumption C_0 will lead to an increase in the equilibrium level of income.

e) Solve for the second derivative, $\partial \bar{i}/\partial C_0$ by forming $|\mathbf{J}_2|$ in which the second column of $\mathbf{J}$ is replaced with the column vector $\mathbf{B}$ and substituted in (13.6).

$$|\mathbf{J}_2| = \begin{vmatrix} 1 - C_Y & 1 \\ L_Y & 0 \end{vmatrix} = -L_Y$$

and $$\frac{\partial \bar{i}}{\partial C_0} = \frac{|\mathbf{J}_2|}{|\mathbf{J}|} = \frac{-L_Y}{(1 - C_Y)L_i + C_i L_Y} = \frac{-(+)}{(-)} > 0$$

An increase in C_0 will also lead to an increase in the equilibrium level of interest. The effect on $\bar{Y}$ and $\bar{i}$ of a change in M_0 is treated in Problem 13.10. See also Problems 13.8 to 13.18.

13.4   COMPARATIVE STATICS FOR OPTIMIZATION PROBLEMS

In addition to their general interest in the effects of changes in exogenous variables on the equilibrium values of their models, economists frequently also want to study the effects of changes in exogenous variables on the solution values of optimization problems. This is done by applying comparative-static techniques to the first-order conditions from which the initial optimal values are determined. Since first-order conditions involve first-order derivatives, comparative-static analysis of optimization problems, as we shall see, is intimately tied up with second-order derivatives and Hessian determinants. The methodology is set forth in Example 5.

EXAMPLE 5.   A price-taking firm has a strictly concave production function $Q(K, L)$. Given $P = $ output price, $r = $ rental rate of capital, and $w = $ wage, its profit function is

$$\pi = PQ(K, L) - rK - wL$$

If we take the derivatives, $\partial\pi/\partial K$ and $\partial\pi/\partial L$, for the first-order optimization conditions, and express them as implicit functions, we have

$$F^1(K, L; r, w, P) = PQ_K(\bar{K}, \bar{L}) - r = 0$$
$$F^2(K, L; r, w, P) = PQ_L(\bar{K}, \bar{L}) - w = 0$$

where the bars indicate that the first derivatives Q_K and Q_L are evaluated at the optimal values of the profit function. From these first-order conditions we can determine the effects of a change in the exogenous variables (r, w) on the optimal values of the endogenous variables $(\bar{K}, \bar{L})$ by use of comparative statics as follows:

a)   Take the total derivatives of the first-order conditions with respect to either of the exogenous variables and set them in the now familiar matrix form. Starting with the rental rate of capital r and noting that each of the first derivatives Q_K and Q_L is a function of both K and L, we have

$$\begin{bmatrix} \dfrac{\partial F^1}{\partial K} & \dfrac{\partial F^1}{\partial L} \\ \dfrac{\partial F^2}{\partial K} & \dfrac{\partial F^2}{\partial L} \end{bmatrix} \begin{bmatrix} \dfrac{\partial \bar{K}}{\partial r} \\ \dfrac{\partial \bar{L}}{\partial r} \end{bmatrix} = \begin{bmatrix} -\dfrac{\partial F^1}{\partial r} \\ -\dfrac{\partial F^2}{\partial r} \end{bmatrix}$$

or specifically,

$$\begin{bmatrix} PQ_{KK} & PQ_{KL} \\ PQ_{LK} & PQ_{LL} \end{bmatrix} \begin{bmatrix} \dfrac{\partial \bar{K}}{\partial r} \\ \dfrac{\partial \bar{L}}{\partial r} \end{bmatrix} = \begin{bmatrix} 1 \\ 0 \end{bmatrix}$$

$$\mathbf{JX} = \mathbf{B}$$

$$|\mathbf{J}| = P^2(Q_{KK}Q_{LL} - Q_{LK}Q_{KL})$$

Provided the second-order sufficient conditions, as expressed within the parentheses above, are met, $|\mathbf{J}| > 0$. Here we observe that when finding the comparative-static derivatives from the first derivatives of the first-order optimization conditions, $|\mathbf{J}| = |\mathbf{H}|$, the Hessian (Section 12.2). For optimization of a (2×2) system, we also recall $|\mathbf{H}| > 0$.

b)   Since $|\mathbf{J}| = |\mathbf{H}| \neq 0$, and assuming continuous first and second derivatives, the conditions of the implicit function theorem are met and we can use Cramer's rule to find the desired derivatives.

$$\frac{\partial \bar{K}}{\partial r} = \frac{|\mathbf{J}_1|}{|\mathbf{J}|} = \frac{\begin{vmatrix} 1 & PQ_{KL} \\ 0 & PQ_{LL} \end{vmatrix}}{|\mathbf{J}|} = \frac{PQ_{LL}}{P^2(Q_{KK}Q_{LL} - Q_{KL}Q_{LK})} < 0$$

where $\partial\bar{K}/\partial r < 0$ because we are assuming strictly concave production functions, which means $Q_{LL} < 0$, $Q_{KK} < 0$, and $Q_{KK}Q_{LL} > Q_{KL}Q_{LK}$ over the entire domain of the function. We also know from microtheory that a profit-maximizing firm will only produce where the marginal productivity of inputs (Q_L, Q_K) is declining. Hence at the optimal level of production, $Q_{LL} < 0$ and $Q_{KK} < 0$. Similarly, we can find

$$\frac{\partial \bar{L}}{\partial r} = \frac{|\mathbf{J}_2|}{|\mathbf{J}|} = \frac{\begin{vmatrix} PQ_{KK} & 1 \\ PQ_{LK} & 0 \end{vmatrix}}{|\mathbf{J}|} = \frac{-PQ_{LK}}{P^2(Q_{KK}Q_{LL} - Q_{KL}Q_{LK})}$$

To be able to sign this comparative-static derivative, we need to know the sign of the cross partial Q_{LK}, which is the effect of a change in capital on the marginal productivity of labor Q_L. If we assume it is positive, which is likely, an increase in the interest rate will cause a decrease in the use of labor due to the negative sign in the numerator. For the effects of a change in wage w on $\bar{K}, \bar{L}$, see Problem 13.19. See also Problems 13.20 to 13.24.

13.5 COMPARATIVE STATICS USED IN CONSTRAINED OPTIMIZATION

Comparative-static analysis can also be applied to constrained optimization problems. In constrained optimization, the Lagrangian multiplier is an endogenous variable and in comparative-static analysis it is evaluated at its optimal value $(\bar{\lambda})$. If the second-order sufficient condition is satisfied, the bordered Hessian $|\bar{\mathbf{H}}|$ may be positive or negative, depending on the type of optimization, but it will never equal zero. If $|\bar{\mathbf{H}}| \neq 0$, the Jacobian will not equal zero since $|\mathbf{J}| = |\bar{\mathbf{H}}|$, as seen in Example 6. When $|\mathbf{J}| \neq 0$ and assuming continuous first and second derivatives, we know from the implicit function theorem that the optimal values of the endogenous variables can be expressed as implicit functions of the exogenous variables and the desired comparative-static derivatives found by means of Cramer's rule. An illustration is provided in Example 6.

EXAMPLE 6. Assume a firm operating in perfectly competitive input and output markets wants to maximize its output $q(K, L)$ subject to a given budgetary constraint

$$rK + wL = B$$

The Lagrangian function to be maximized is

$$Q = q(K, L) + \lambda(B - rK - wL)$$

and the three first-order derivatives $(\partial Q/\partial K, \partial Q/\partial L, \partial Q/\partial \lambda)$ representing the first-order conditions can be expressed as the following implicit functions:

$$F^1(\bar{K}, \bar{L}, \bar{\lambda}; r, w, B) = Q_K(\bar{K}, \bar{L}) - r\bar{\lambda} = 0$$
$$F^2(\bar{K}, \bar{L}, \bar{\lambda}; r, w, B) = Q_L(\bar{K}, \bar{L}) - w\bar{\lambda} = 0$$
$$F^3(\bar{K}, \bar{L}, \bar{\lambda}; r, w, B) = B - r\bar{K} - w\bar{L} = 0$$

From these first-order conditions for constrained optimization, assuming continuous derivatives and satisfaction of the second-order sufficient condition, we can determine the effects of a change in any of the exogenous variables (r, w, B) on the optimal values of the three endogenous variables $(\bar{K}, \bar{L}, \bar{\lambda})$ with comparative-static analysis.

To find the effect of a change in the budget B on the optimal values of the endogenous variables, we take the total derivative of each of the three functions with respect to B.

$$\begin{bmatrix} \dfrac{\partial F^1}{\partial \bar{K}} & \dfrac{\partial F^1}{\partial \bar{L}} & \dfrac{\partial F^1}{\partial \bar{\lambda}} \\[2ex] \dfrac{\partial F^2}{\partial \bar{K}} & \dfrac{\partial F^2}{\partial \bar{L}} & \dfrac{\partial F^2}{\partial \bar{\lambda}} \\[2ex] \dfrac{\partial F^3}{\partial \bar{K}} & \dfrac{\partial F^3}{\partial \bar{L}} & \dfrac{\partial F^3}{\partial \bar{\lambda}} \end{bmatrix} \begin{bmatrix} \dfrac{\partial \bar{K}}{\partial B} \\[2ex] \dfrac{\partial \bar{L}}{\partial B} \\[2ex] \dfrac{\partial \bar{\lambda}}{\partial B} \end{bmatrix} = \begin{bmatrix} -\dfrac{\partial F^1}{\partial B} \\[2ex] -\dfrac{\partial F^2}{\partial B} \\[2ex] -\dfrac{\partial F^3}{\partial B} \end{bmatrix}$$

or specifically,

$$\begin{bmatrix} Q_{KK} & Q_{KL} & -r \\ Q_{LK} & Q_{LL} & -w \\ -r & -w & 0 \end{bmatrix} \begin{bmatrix} \dfrac{\partial \bar{K}}{\partial B} \\[2ex] \dfrac{\partial \bar{L}}{\partial B} \\[2ex] \dfrac{\partial \bar{\lambda}}{\partial B} \end{bmatrix} = \begin{bmatrix} 0 \\ 0 \\ -1 \end{bmatrix}$$

$$|\mathbf{J}| = Q_{KK}(-w^2) - Q_{KL}(-rw) - r(rQ_{LL} - wQ_{LK})$$

$$|\mathbf{J}| = -w^2 Q_{KK} + rw Q_{KL} - r^2 Q_{LL} + rw Q_{LK} > 0$$

since $|\mathbf{J}| = |\bar{\mathbf{H}}|$ (Section 12.5) and, if the second-order sufficient condition is met, $|\bar{\mathbf{H}}| > 0$ for constrained maximization. Since profit-maximizing firms in perfect competition operate only in the area of decreasing marginal productivity of inputs $(Q_{KK}, Q_{LL} < 0)$, the second-order condition will be fulfilled whenever K and L are complements $(Q_{KL}, Q_{LK} > 0)$ and will depend on the relative strength of the direct and cross partials when K and L are substitutes $(Q_{KL}, Q_{LK} < 0)$.

With $|\mathbf{J}| = |\bar{\mathbf{H}}| \neq 0$, and assuming continuous first- and second-order derivatives, we can now use Cramer's rule to find the desired derivatives.

1.
$$\frac{\partial \bar{K}}{\partial B} = \frac{|\mathbf{J}_1|}{|\mathbf{J}|} = \frac{\begin{bmatrix} 0 & Q_{KL} & -r \\ 0 & Q_{LL} & -w \\ -1 & -w & 0 \end{bmatrix}}{|\mathbf{J}|} = \frac{wQ_{KL} - rQ_{LL}}{|\mathbf{J}|}$$

$\partial \bar{K}/\partial B > 0$ when K and L are complements and indeterminate when K and L are substitutes.

2.
$$\frac{\partial \bar{L}}{\partial B} = \frac{|\mathbf{J}_2|}{|\mathbf{J}|} = \frac{\begin{bmatrix} Q_{KK} & 0 & -r \\ Q_{LK} & 0 & -w \\ -r & -1 & 0 \end{bmatrix}}{|\mathbf{J}|} = \frac{rQ_{LK} - wQ_{KK}}{|\mathbf{J}|}$$

$\partial \bar{L}/\partial B > 0$ when K and L are complements and indeterminate when K and L are substitutes.

3.
$$\frac{\partial \bar{\lambda}}{\partial B} = \frac{|\mathbf{J}_3|}{|\mathbf{J}|} = \frac{\begin{bmatrix} Q_{KK} & Q_{KL} & 0 \\ Q_{LK} & Q_{LL} & 0 \\ -r & -w & -1 \end{bmatrix}}{|\mathbf{J}|} = \frac{-1(Q_{KK}Q_{LL} - Q_{KL}Q_{LK})}{|\mathbf{J}|}$$

$\partial \bar{\lambda}/\partial B$ is indeterminate. See also Problems 13.25 to 13.29.

13.6 THE ENVELOPE THEOREM

The envelope theorem enables us to measure the effect of a change in any of the exogenous variables on the optimal value of the objective function by merely taking the derivative of the Lagrangian function with respect to the desired exogenous variable and evaluating the derivative at the values of the optimal solution. The rationale is set forth in Example 7 and an illustration is offered in Example 8. The envelope theorem also provides the rationale for our earlier description of the Lagrange multiplier as an approximation of the marginal effect on the optimized value of the objective function due to a small change in the constant of the constraint (Section 5.6). One important implication of the envelope theorem for subsequent work in concave programming is that if $\bar{\lambda} = 0$ at the point at which the function is optimized, the constraint must be nonbinding. Conversely, if the constraint is nonbinding, $\bar{\lambda} = 0$.

EXAMPLE 7.   Assume one wishes to maximize the function

$$z(x, y; a, b)$$

subject to

$$f(x, y; a, b)$$

The Lagrangian function is

$$Z(x, y, \lambda; a, b) = z(x, y; a, b) + \lambda f(x, y; a, b)$$

and the first-order conditions are

$$Z_x = z_x(\bar{x}, \bar{y}; a, b) + \bar{\lambda} f_x(\bar{x}, \bar{y}; a, b) = 0$$
$$Z_y = z_y(\bar{x}, \bar{y}; a, b) + \bar{\lambda} f_y(\bar{x}, \bar{y}; a, b) = 0$$
$$Z_\lambda = f(\bar{x}, \bar{y}; a, b) = 0$$

If we assume all the functions have continuous first- and second-order derivatives and if

$$|\mathbf{J}| = \begin{vmatrix} z_{xx} + \bar{\lambda} f_{xx} & z_{xy} + \bar{\lambda} f_{xy} & f_x \\ z_{yx} + \bar{\lambda} f_{yx} & z_{yy} + \bar{\lambda} f_{yy} & f_y \\ f_x & f_y & 0 \end{vmatrix} \neq 0$$

we know from the implicit function theorem that we can express the optimal values of the endogenous variables as functions of the exogenous variables.

$$Z(\bar{x}, \bar{y}, \bar{\lambda}; a, b) = z[\bar{x}(a,b), \bar{y}(a,b); a, b] + \bar{\lambda}(a,b)f[\bar{x}(a,b), \bar{y}(a,b); a, b]$$

The objective function, when evaluated at the values of the optimal solution, is known as the *indirect objective function*.

$$\bar{z}(a,b) \equiv z[\bar{x}(a,b), \bar{y}(a,b); a, b] \equiv z(a,b)$$

The envelope theorem states that the partial derivative of the indirect objective function with respect to any one of the exogenous variables, say b, equals the partial derivative of the Lagrangian function with respect to the same exogenous variable. To prove the envelope theorem, then, we need to show

$$\frac{\partial \bar{z}}{\partial b} = \frac{\partial Z}{\partial b}$$

Making use of the chain rule to take the derivative of the indirect objective function, and recalling that it is always evaluated at the optimal solution, we have

$$\frac{\partial \bar{z}}{\partial b} = z_x \frac{\partial \bar{x}}{\partial b} + z_y \frac{\partial \bar{y}}{\partial b} + z_b$$

Then substituting $z_x = -\bar{\lambda}f_x$, $z_y = -\bar{\lambda}f_y$ from the first two first-order conditions,

$$\frac{\partial \bar{z}}{\partial b} = -\bar{\lambda}\left(f_x \frac{\partial \bar{x}}{\partial b} + f_y \frac{\partial \bar{y}}{\partial b} \right) + z_b \qquad (13.9)$$

Also from the third first-order condition, we know

$$f[\bar{x}(a,b), \bar{y}(a,b), a, b] \equiv 0$$

Taking the derivative with respect to b and rearranging,

$$f_x \frac{\partial \bar{x}}{\partial b} + f_y \frac{\partial \bar{y}}{\partial b} = -f_b$$

Then substituting $-f_b$ in (13.9) and rearranging,

$$\frac{\partial \bar{z}}{\partial b} = (z_b + \bar{\lambda}f_b) = \frac{\partial Z}{\partial b} \qquad \text{Q.E.D.}$$

The derivative of the Lagrangian function with respect to a specific exogenous variable, when evaluated at the optimal values of the problem, is a reliable measure of the effect of that exogenous variable on the *optimal* value of the objective function.

EXAMPLE 8. Assume a utility maximization problem subject to a budget constraint:

$$\text{maximize } u(x,y) \qquad \text{subject to} \qquad p_x x + p_y y = B$$

If there are continuous first and second derivatives and the Jacobian determinant consisting of the derivatives of the first-order conditions with respect to the endogenous variables does not equal zero (or vanish), then the Lagrangian function can be written

$$U(\bar{x}, \bar{y}, \bar{\lambda}; p_x, p_y, B) = u(\bar{x}, \bar{y}) + \bar{\lambda}(B - p_x \bar{x} - p_y \bar{y})$$

and the indirect objective function is

$$V(\bar{x}, \bar{y}; p_x, p_y, B) = u[\bar{x}(p_x, p_y, B), \bar{y}(p_x, p_y, B)]$$

Then using the envelope theorem to estimate the effect on the optimal value of the objective function of a change in any of the three exogenous variables, we have

a)
$$\frac{\partial V}{\partial p_x} = \frac{\partial U}{\partial p_x} = -\bar{\lambda}\bar{x}$$

b)
$$\frac{\partial V}{\partial p_y} = \frac{\partial U}{\partial p_y} = -\bar{\lambda}\bar{y}$$

c)
$$\frac{\partial V}{\partial B} = \frac{\partial U}{\partial B} = \bar{\lambda}$$

In (*c*), $\bar{\lambda}$ can be called the *marginal utility of money*, i.e., the extra utility the consumer would derive from a small change in his or her budget or income. Notice that with the budget constraint (*B*) appearing only in the constraint, the derivative comes more easily from the Lagrangian function. In (*a*) and (*b*), assuming positive utility for income from (*c*), a change in the price of the good will have a negative impact weighted by the quantity of the good consumed when utility is being maximized. In both cases, with prices appearing only in the constraint, the derivatives once again come more readily from the Lagrangian function. See also Problems 13.30 to 13.32.

13.7 CONCAVE PROGRAMMING AND INEQUALITY CONSTRAINTS

In the classical method for constrained optimization seen thus far, the constraints have always been strict equalities. Some economic problems call for weak inequality constraints, however, as when individuals want to maximize utility subject to spending *not more than x* dollars or business seeks to minimize costs subject to producing *no less than x* units of output. *Concave programming*, so called because the objective and constraint functions are all assumed to be concave, is a form of *nonlinear programming* designed to optimize functions subject to inequality constraints. Convex functions are by no means excluded, however, because the negative of a convex function is concave. Typically set up in the format of a maximization problem, concave programming can nevertheless also minimize a function by maximizing the negative of that function.

Given an optimization problem subject to an inequality constraint with the following differentiable concave objective and constraint functions,

$$\text{maximize } f(x_1, x_2) \qquad \text{subject to } g(x_1, x_2) \geq 0 \qquad x_1, x_2 \geq 0$$

and the corresponding Lagrangian function,

$$F(x_1, x_2, \lambda) = f(x_1, x_2) + \lambda g(x_1, x_2)$$

the first-order necessary and sufficient conditions for maximization, called the *Kuhn-Tucker* conditions, are

1. *a)*
$$\frac{\partial F}{\partial x_i} = f_i(\bar{x}_1, \bar{x}_2) + \bar{\lambda} g_i(\bar{x}_1, \bar{x}_2) \leq 0$$

   b)
$$x_i \geq 0$$

   c)
$$\bar{x}_i \frac{\partial F}{\partial x_i} = 0, \qquad i = 1, 2$$

2. *a)*
$$\frac{\partial F}{\partial \lambda} = g(\bar{x}_1, \bar{x}_2) \geq 0$$

   b)
$$\bar{\lambda} \geq 0$$

   c)
$$\bar{\lambda} \frac{\partial F}{\partial \lambda} = 0$$

where the conditions in (*c*) are called the *complementary-slackness conditions*, meaning that both $\bar{x}$ and $f'(\bar{x})$ cannot simultaneously both be nonzero. Since a linear function is concave and convex, though not strictly concave or strictly convex, a concave-programming problem consisting solely of linear functions that meet the Kuhn-Tucker conditions will always satisfy the necessary and sufficient conditions for a maximum.

Note: (1) condition 1(*a*) requires that the Lagrangian function be *maximized* with respect to x_1 and x_2, while condition 2(*a*) calls for the Lagrangian to be *minimized* with respect to λ. This means concave programming is designed to seek out a saddle point in the Lagrangian function in order to optimize the objective function subject to an inequality constraint.

(2) in the Kuhn-Tucker conditions the constraint is always expressed as greater than or equal to zero. This means that unlike equality constraints set equal to zero, where it makes no difference whether you subtract the constant from the variables in the constraint or the variables from the constant, the order of subtraction is important in concave programming (see Problem 13.33).

The rationale for the tripartite conditions (*a*)–(*c*) can be found in Example 9 and a demonstration of the basic maximization method is offered in Example 10. For minimization, see Problem 13.34. For multiple constraints, see Problem 13.39. For other applications, see Problems 13.33 to 13.42.

EXAMPLE 9.   Consider a single-variable function for which we seek a local maximum in the first quadrant where $x \geq 0$. Three scenarios are possible, each with slightly different conditions, as seen in Fig. 13-1.

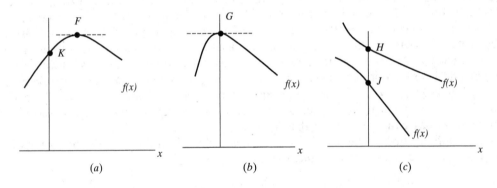

Fig. 13-1

a)   For the maximum at *F*, an interior solution,

$$f'(x) = 0 \qquad \text{and} \qquad x > 0$$

b)   For the maximum at *G*, a boundary solution,

$$f'(x) = 0 \qquad \text{and} \qquad x = 0$$

c)   For the maximum at *H* or *J*, both boundary solutions,

$$f'(x) < 0 \qquad \text{and} \qquad x = 0$$

All the possibilities for a maximum in the first quadrant can be summarized more succinctly, however, as

$$f'(x) \leq 0 \qquad x \geq 0 \qquad \text{and} \qquad xf'(x) = 0$$

which we recognize as part of the Kuhn-Tucker conditions. Note that the conditions automatically exclude a point like *K* in (*a*) which is not a maximum, because $f'(K) > 0$.

EXAMPLE 10.   A consumer wishing to maximize utility while spending *no more than* a predetermined budget faces the following concave-programming problem,

$$\text{maximize } u(x, y) \qquad \text{subject to } B - p_x x - p_y y \geq 0 \qquad x, y \geq 0$$

and Lagrangian function,

$$U = u(x,y) + \lambda(B - p_x x - p_y y)$$

Using the Kuhn-Tucker conditions, the Lagrangian function is first *maximized* with respect to the choice variables x and y and the related conditions checked.

1. *a)* 　　　　$\dfrac{\partial U}{\partial x} = u_x - \bar{\lambda}p_x \leq 0$　　　　$\dfrac{\partial U}{\partial y} = u_y - \bar{\lambda}p_y \leq 0$

　b)　　　　$\bar{x} \geq 0$　　　　　　　　　$\bar{y} \geq 0$

　c)　　　　$\bar{x}(u_x - \bar{\lambda}p_x) = 0$　　　$\bar{y}(u_y - \bar{\lambda}p_y) = 0$

Then the Lagrangian is *minimized* with respect to the constraint variable λ and the related conditions checked.

2. *a)* 　　　　$\dfrac{\partial U}{\partial \lambda} = B - p_x \bar{x} - p_y \bar{y} \geq 0$

　b)　　　　$\bar{\lambda} \geq 0$

　c)　　　　$\bar{\lambda}(B - p_x \bar{x} - p_y \bar{y}) = 0$

This leaves three categories of solutions that are nontrivial: (*a*) $\bar{x}, \bar{y} > 0$, (*b*) $\bar{x} = 0$, $\bar{y} > 0$, and (*c*) $\bar{x} > 0$, $\bar{y} = 0$. We deal with the first two below and leave the third to you as a private exercise.

a) First scenario. If $\bar{x}, \bar{y} > 0$, then from 1(*c*),

$$u_x - \bar{\lambda}p_x = 0 \qquad u_y - \bar{\lambda}p_y = 0$$

Therefore,　　　　$\bar{\lambda} = \dfrac{u_x}{p_x} \qquad \bar{\lambda} = \dfrac{u_y}{p_y}$　　　　　　　(*13.10*)

With $p_x, p_y > 0$, and assuming nonsatiation of the consumer, i.e., $u_x, u_y > 0$,

$$\bar{\lambda} > 0$$

If $\bar{\lambda} > 0$, then from 2(*c*),

$$B - p_x \bar{x} - p_y \bar{y} = 0$$

and the budget constraint holds as an exact equality, not a weak inequality. This means the optimal point, $\bar{x}, \bar{y}$, will lie somewhere on the budget line, and not below it.

By reconfiguring (*13.10*), we can also see

$$\frac{u_x}{u_y} = \frac{p_x}{p_y}$$

Since u_x/u_y = the slope of the indifference curve and p_x/p_y = the slope of the budget line, whenever both $\bar{x}, \bar{y} > 0$, the indifference curve will be tangent to the budget line at the point of optimization and we have an interior solution. With the budget constraint functioning as an exact equality, this first scenario, in which both $\bar{x}, \bar{y} > 0$, exactly parallels the classical constrained optimization problem, as seen in Fig. 13-2(*a*).

b) Second scenario. If $\bar{x} = 0$, $\bar{y} > 0$, then from 1(*c*),

$$u_x - \bar{\lambda}p_x \leq 0 \qquad u_y - \bar{\lambda}p_y = 0$$

and　　　　$\dfrac{u_x}{p_x} \leq \bar{\lambda} \qquad \bar{\lambda} = \dfrac{u_y}{p_y}$　　　　　　　(*13.11*)

Assuming $p_x, p_y, u_x, u_y > 0$, then $\bar{\lambda} > 0$. From 2(*c*), therefore, the budget constraint holds as an exact equality, not a weak inequality, even though only one variable is greater than zero and the other equals zero. This means that once again the optimal point, $\bar{x}, \bar{y}$, will lie on the budget line, and not below it.

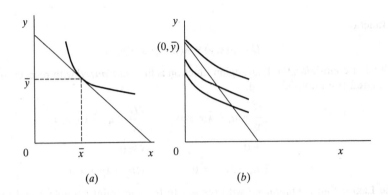

Fig. 13-2

Substituting $\bar{\lambda} = u_y/p_y$ from the equality on the right in (*13.11*) for $\bar{\lambda}$ in the inequality on the left in (*13.11*), we have

$$\frac{u_x}{p_x} \leq \frac{u_y}{p_y} \quad \text{or} \quad \frac{u_x}{u_y} \leq \frac{p_x}{p_y}$$

This means that along the budget line the indifference curves are everywhere flatter than the budget line, leading to a corner solution in the upper left, as seen in Fig. 13-2(*b*). At the corner solution, the slope of the highest indifference curve that just touches the budget line may be flatter than or equal to the slope of the budget line.

Solved Problems

COMPARATIVE STATICS WITH ONE EXOGENOUS VARIABLE

13.1. Given the model from Example 1,

$$Q_D = m - nP + kY \qquad m, n, k > 0$$
$$Q_S = a + bP \qquad\qquad\;\; a, b > 0$$

and now assuming we know the value of income and the parameters where

$$Y = 100, \, m = 60, \, n = 2, \, k = 0.1, \, a = 10, \text{ and } b = 0.5$$

(*a*) find the equilibrium price and quantity, (*b*) use comparative statics to estimate the effect on the equilibrium price P^* of a \$1 change in income, and (*c*) confirm the comparative statics results by reestimating the equilibrium price.

a)
$$Q_D = Q_S$$
$$60 - 2P + 0.1(100) = 10 + 0.5P$$
$$-2.5P = -60$$
$$P^* = 24 \qquad Q^* = 22$$

b) From (*13.3*),
$$\frac{dP^*}{dY} = \frac{k}{b+n} > 0$$

Substituting $k = 0.1$, $b = 0.5$, $n = 2$,

$$\frac{dP^*}{dY} = \frac{0.1}{0.5+2} = 0.04$$

A \$1 increase in income can be expected to cause a 4¢ increase in the equilibrium price of the commodity.

c)   Reestimating the equilibrium equation for $Y = 101$,

$$60 - 2P + 0.1(101) = 10 + 0.5P$$
$$-2.5P = -60.1$$
$$P^* = 24.04 \quad \text{Q.E.D.}$$

13.2.   Assume

$$Q_D = m - nP - cP_c + sP_s \qquad m, n, c, s > 0$$
$$Q_S = a + bP - iP_i \qquad a, b, i > 0$$

where P = price of the good, P_c = price of a complement, P_s = price of a substitute, and P_i = price of an input. (a) Find the equilibrium price P^*. (b) Find the comparative-static derivatives resulting from a change in P_c and P_i. (c) Find the implicit function for the equilibrium condition and (d) from it derive the comparative-static derivative for a change in P_s.

a)   The equilibrium condition is

$$m - nP - cP_c + sP_s = a + bP - iP_i \tag{13.12}$$

$$m - a - cP_c + sP_s + iP_i = (b + n)P$$

$$P^* = \frac{m - a - cP_c + sP_s + iP_i}{b + n}$$

b)   Whenever there is more than one independent variable, we have to work in terms of partial derivatives which will keep the other independent variables constant.

$$\frac{\partial P^*}{\partial P_c} = \frac{-c}{b + n} \qquad \frac{\partial P^*}{\partial P_i} = \frac{i}{b + n}$$

c)   From (13.12),

$$m - nP - cP_c + sP_s - a - bP + iP_i = 0$$

d)

$$\frac{\partial P^*}{\partial P_s} = -\frac{F_{P_s}}{F_P} = \frac{s}{b + n}$$

13.3.   Assume a two-sector income determination model where consumption depends on income and investment is autonomous, so that

$$C = bY, \ I = I_0, \ 0 < b < 1$$

and equilibrium occurs where $Y = C + I$. (a) Solve for the equilibrium level of income Y^* explicitly. (b) Use comparative statics to estimate the effect on Y^* of a change in autonomous investment I_0. (c) Find the same comparative-static derivative from the implicit function. (d) Evaluate the effect on Y^* of a change in the marginal propensity to consume b explicitly and (e) implicitly.

a)

$$Y = C + I$$

Substituting,

$$Y = bY + I_0 \tag{13.13}$$

$$Y - bY = I_0$$

$$Y^* = \frac{I_0}{1 - b} \tag{13.14}$$

b) The effect on Y^* of a change in autonomous investment I_0 is

$$\frac{dY^*}{dI_0} = \frac{1}{1-b} > 0$$

since $0 < b < 1$.

c) Moving everything to the left in (13.13), we obtain the implicit function,

$$Y - bY - I_0 = 0 \qquad\qquad (13.15)$$

From the implicit function rule, always under the usual assumptions,

$$\frac{dY^*}{dI_0} = -\frac{F_{I_0}}{F_Y}$$

where $F_{I_0} = -1$ and $F_Y = 1 - b$. So,

$$\frac{dY^*}{dI_0} = -\frac{-1}{1-b} = \frac{1}{1-b} > 0$$

d) If we treat the marginal propensity to consume b as an exogenous variable instead of as a parameter, we have a function of more than one independent variable and must find the partial derivative, which will hold I_0 constant. First applying the quotient rule on the explicit function in (13.14),

$$\frac{\partial Y^*}{\partial b} = \frac{I_0}{(1-b)^2}$$

Substituting from (13.14) where $Y^* = I_0/(1-b)$, we have

$$\frac{\partial Y^*}{\partial b} = \frac{Y^*}{(1-b)}$$

e) Next using the implicit function rule on (13.15) when evaluated at Y^*,

$$\frac{\partial Y^*}{\partial b} = -\frac{F_b}{F_Y} = \frac{Y^*}{1-b}$$

Though seemingly difficult at first, implicit functions are frequently faster and easier to work with in the long run. See Problems 13.5 to 13.6.

13.4. From (a) the explicit and (b) the implicit function, find the effect on the profit-maximizing level of output of a per-unit tax t placed on a monopoly with total revenue and total cost functions:

$$\text{TR} = mQ - nQ^2, \ \text{TC} = kQ \qquad m, n, k > 0$$

a) Profit π for the monopoly is

$$\pi = mQ - nQ^2 - kQ - tQ$$

$$\frac{d\pi}{dQ} = m - 2nQ - k - t = 0 \qquad\qquad (13.16)$$

$$Q^* = \frac{m - k - t}{2n} \qquad \text{profit-maximizing level of output}$$

Then from the explicit function above, the comparative-static derivative estimating the effect on Q^* of a change in the per-unit tax t is

$$\frac{dQ^*}{dt} = -\frac{1}{2n} < 0$$

b)   From (*13.16*), the implicit function for the optimum condition is

$$m - 2nQ^* - k - t = 0$$

By the implicit function rule,

$$\frac{dQ^*}{dt} = -\frac{F_t}{F_{Q^*}} = -\frac{1}{2n} < 0$$

13.5.   Assume a two-sector income determination model expressed in general functions:

$$C = C(Y), \; I = I_0$$

with equilibrium when $Y = C + I$. (*a*) Determine the implicit function for the equilibrium level of income Y^*. (*b*) Estimate the effect on Y^* of a change in autonomous investment I_0.

a)
$$Y - C(Y^*) - I_0 = 0$$

b)
$$\frac{dY^*}{dI_0} = -\frac{F_{I_0}}{F_{Y^*}} = \frac{1}{1 - C_{Y^*}}$$

13.6.   In the model of Problem 6.4 we found the equilibrium condition and explicit function for the equilibrium level of income $\bar{Y}$ derived from it were, respectively,

$$Y = C_0 + bY - bT_0 - btY + I_0 + G_0 \tag{13.17}$$

and
$$\bar{Y} = \frac{1}{1 - b + bt}(C_0 - bT_0 + I_0 + G_0)$$

We also found, after considerable simplification, the comparative-static derivative for the effect on the equilibrium level of income $\bar{Y}$ of a change in the tax rate t was

$$\frac{\partial \bar{Y}}{\partial t} = \frac{-b\bar{Y}}{1 - b + bt}$$

(*a*) Find the implicit function for the equilibrium condition in (*13.17*) and (*b*) from it derive the same derivative $\partial\bar{Y}/\partial t$ to convince yourself of the convenience of working with implicit functions.

a)   From (*13.17*), the implicit function for the equilibrium condition is

$$\bar{Y} - C_0 - b\bar{Y} + bT_0 + bt\bar{Y} - I_0 - G_0 = 0$$

b)   From the implicit function rule,

$$\frac{\partial \bar{Y}}{\partial t} = -\frac{F_t}{F_{\bar{Y}}} = \frac{-b\bar{Y}}{1 - b + bt}$$

13.7.   In Problems 6.7 and 6.8 we found the equilibrium condition and explicit function for the equilibrium level of income derived from it were, respectively,

$$Y = C_0 + b(Y - T_0 - tY) + I_0 + G_0 + X_0 - Z_0 - z(Y - T_0 - tY) \tag{13.18}$$

$$\bar{Y} = \frac{1}{1 - b + bt + z - zt}(C_0 - bT_0 + I_0 + G_0 + X_0 - Z_0 + zT_0)$$

We also found, after considerable simplification, the comparative-static derivative for the effect on $\bar{Y}$ of a change in the marginal propensity to import z was

$$\frac{\partial \bar{Y}}{\partial z} = \frac{-\bar{Y}_d}{1 - b + bt + z - zt}$$

(a) Find the implicit function for the equilibrium condition in (13.18) and (b) from it derive the same derivative $\partial \bar{Y}/\partial z$ to see once again the convenience of working with implicit functions.

a) From (13.18), the implicit function for the equilibrium condition is

$$\bar{Y} - C_0 - b(\bar{Y} - T_0 - t\bar{Y}) - I_0 - G_0 - X_0 + Z_0 + z(\bar{Y} - T_0 - t\bar{Y})$$

b) From the implicit function rule,

$$\frac{\partial \bar{Y}}{\partial z} = -\frac{F_z}{F_{\bar{Y}}} = -\frac{\bar{Y} - T_0 - t\bar{Y}}{1 - b + bt + z - zt} = \frac{-\bar{Y}_d}{1 - b + bt + z - zt}$$

COMPARATIVE STATICS WITH MORE THAN ONE ENDOGENOUS VARIABLE

13.8. Set forth the implicit function theorem.

Given the set of simultaneous equations,

$$f^1(y_1, y_2, \ldots, y_n; x_1, x_2, \ldots, x_m) = 0$$
$$f^2(y_1, y_2, \ldots, y_n; x_1, x_2, \ldots, x_m) = 0$$
$$\ldots\ldots\ldots\ldots\ldots\ldots\ldots\ldots\ldots\ldots\ldots\ldots\ldots$$
$$f^n(y_1, y_2, \ldots, y_n; x_1, x_2, \ldots, x_m) = 0$$

if all the above functions in f have continuous partial derivatives with respect to all the x and y variables, and if at a point $(y_{10}, y_{20}, \ldots, y_{n0}; x_{10}, x_{20}, \ldots, x_{m0})$, the Jacobian consisting of the partial derivatives of all the functions f_i with respect to all of the dependent variables y_i is nonzero, as indicated below:

$$|\mathbf{J}| \equiv \begin{vmatrix} \dfrac{\partial f^1}{\partial y_1} & \dfrac{\partial f^1}{\partial y_2} & \cdots & \dfrac{\partial f^1}{\partial y_n} \\ \dfrac{\partial f^2}{\partial y_1} & \dfrac{\partial f^2}{\partial y_2} & \cdots & \dfrac{\partial f^2}{\partial y_n} \\ \cdots & \cdots & \cdots & \cdots \\ \dfrac{\partial f^n}{\partial y_1} & \dfrac{\partial f^n}{\partial y_2} & \cdots & \dfrac{\partial f^n}{\partial y_n} \end{vmatrix} \neq 0$$

then there exists an m-dimensional neighborhood N in which the variables $y_1, y_2, \ldots, y_n$ are implicit functions of the variables $x_1, x_2, \ldots, x_m$ in the form

$$y_{10} = f^1(x_{10}, x_{20}, \ldots, x_{m0})$$
$$y_{20} = f^2(x_{10}, x_{20}, \ldots, x_{m0})$$
$$\ldots\ldots\ldots\ldots\ldots\ldots\ldots\ldots\ldots$$
$$y_{n0} = f^n(x_{10}, x_{20}, \ldots, x_{m0})$$

The implicit functions, $f^1, f^2, \ldots, f^n$, are continuous and have continuous partial derivatives with respect to all the independent variables. Derivation of the comparative-static derivatives is explained in Example 3.

13.9. Working with the model from Example 3 in which

$$F^1(y_1, y_2; x_1, x_2) = 0$$
$$F^2(y_1, y_2; x_1, x_2) = 0$$

find the comparative-static partial derivatives of the system with respect to x_2.

Taking the total derivative of both functions with respect to x_2,

$$\frac{\partial F^1}{\partial y_1} \cdot \frac{\partial y_1}{\partial x_2} + \frac{\partial F^1}{\partial y_2} \cdot \frac{\partial y_2}{\partial x_2} + \frac{\partial F^1}{\partial x_2} = 0$$

$$\frac{\partial F^2}{\partial y_1} \cdot \frac{\partial y_1}{\partial x_2} + \frac{\partial F^2}{\partial y_2} \cdot \frac{\partial y_2}{\partial x_2} + \frac{\partial F^2}{\partial x_2} = 0$$

Moving the partials of the functions with respect to x_2 to the right, we have

$$\begin{bmatrix} \dfrac{\partial F^1}{\partial y_1} & \dfrac{\partial F^1}{\partial y_2} \\[2mm] \dfrac{\partial F^2}{\partial y_1} & \dfrac{\partial F^2}{\partial y_2} \end{bmatrix} \begin{bmatrix} \dfrac{\partial y_1}{\partial x_2} \\[2mm] \dfrac{\partial y_2}{\partial x_2} \end{bmatrix} = \begin{bmatrix} -\dfrac{\partial F^1}{\partial x_2} \\[2mm] -\dfrac{\partial F^2}{\partial x_2} \end{bmatrix}$$

$$\mathbf{JX} = \mathbf{B}$$

where

$$|\mathbf{J}| = \frac{\partial F^1}{\partial y_1} \cdot \frac{\partial F^2}{\partial y_2} - \frac{\partial F^2}{\partial y_1} \cdot \frac{\partial F^1}{\partial y_2} \neq 0$$

To solve for the first derivative, $\partial y_1 / \partial x_2$, form a new matrix $|\mathbf{J}_1|$, replacing the first column of $\mathbf{J}$ with the column vector $\mathbf{B}$ and substitute in (13.6).

$$\frac{\partial y_1}{\partial x_2} = \frac{|\mathbf{J}_1|}{|\mathbf{J}|} = \frac{\begin{vmatrix} -\dfrac{\partial F^1}{\partial x_2} & \dfrac{\partial F^1}{\partial y_2} \\[2mm] -\dfrac{\partial F^2}{\partial x_2} & \dfrac{\partial F^2}{\partial y_2} \end{vmatrix}}{\begin{vmatrix} \dfrac{\partial F^1}{\partial y_1} & \dfrac{\partial F^1}{\partial y_2} \\[2mm] \dfrac{\partial F^2}{\partial y_1} & \dfrac{\partial F^2}{\partial y_2} \end{vmatrix}} = \frac{-\left(\dfrac{\partial F^1}{\partial x_2} \cdot \dfrac{\partial F^2}{\partial y_2} - \dfrac{\partial F^2}{\partial x_2} \cdot \dfrac{\partial F^1}{\partial y_2} \right)}{\dfrac{\partial F^1}{\partial y_1} \cdot \dfrac{\partial F^2}{\partial y_2} - \dfrac{\partial F^2}{\partial y_1} \cdot \dfrac{\partial F^1}{\partial y_2}}$$

Similarly,

$$\frac{\partial y_2}{\partial x_2} = \frac{|\mathbf{J}_2|}{|\mathbf{J}|} = \frac{\begin{vmatrix} \dfrac{\partial F^1}{\partial y_1} & -\dfrac{\partial F^1}{\partial x_2} \\[2mm] \dfrac{\partial F^2}{\partial y_1} & -\dfrac{\partial F^2}{\partial x_2} \end{vmatrix}}{\begin{vmatrix} \dfrac{\partial F^1}{\partial y_1} & \dfrac{\partial F^1}{\partial y_2} \\[2mm] \dfrac{\partial F^2}{\partial y_1} & \dfrac{\partial F^2}{\partial y_2} \end{vmatrix}} = \frac{-\left(\dfrac{\partial F^1}{\partial y_1} \cdot \dfrac{\partial F^2}{\partial x_2} - \dfrac{\partial F^2}{\partial y_1} \cdot \dfrac{\partial F^1}{\partial x_2} \right)}{\dfrac{\partial F^1}{\partial y_1} \cdot \dfrac{\partial F^2}{\partial y_2} - \dfrac{\partial F^2}{\partial y_1} \cdot \dfrac{\partial F^1}{\partial y_2}}$$

13.10. Assume the model from Example 4,

$$Y - C_0 - C(Y, i) = 0 \qquad 0 < C_Y < 1, C_i < 0$$
$$L(Y, i) - M_0/P = 0 \qquad L_Y > 0, L_i < 0$$

and use comparative-static analysis to find the effect on the equilibrium levels of Y and i of a change in the money supply M_0, recalling that P is constant.

Taking the total derivatives with respect to M_0,

$$\frac{\partial Y}{\partial M_0} - \left(C_Y \cdot \frac{\partial Y}{\partial M_0} \right) - \left(C_i \cdot \frac{\partial i}{\partial M_0} \right) = 0$$

$$\left(L_Y \cdot \frac{\partial Y}{\partial M_0} \right) + \left(L_i \cdot \frac{\partial i}{\partial M_0} \right) - \frac{1}{P} = 0$$

and setting them in matrix form,

$$\begin{bmatrix} 1 - C_Y & -C_i \\ L_Y & L_i \end{bmatrix} \begin{bmatrix} \dfrac{\partial \bar{Y}}{\partial M_0} \\[2mm] \dfrac{\partial \bar{i}}{\partial M_0} \end{bmatrix} = \begin{bmatrix} 0 \\ 1/P \end{bmatrix}$$

$$\mathbf{JX} = \mathbf{B}$$

where
$$|\mathbf{J}| = (1 - C_Y)L_i + C_i L_Y$$
$$|\mathbf{J}| = (+)(-) + (-)(+) = (-) < 0$$

We then solve for the first derivative, $\partial \bar{Y}/\partial M_0$ by forming the new matrix $|\mathbf{J}_1|$.

$$|\mathbf{J}_1| = \begin{vmatrix} 0 & -C_i \\ 1/P & L_i \end{vmatrix} = \frac{C_i}{P}$$

and
$$\frac{\partial \bar{Y}}{\partial M_0} = \frac{|\mathbf{J}_1|}{|\mathbf{J}|} = \frac{C_i}{P[(1 - C_Y)L_i + C_i L_Y]} = \frac{(-)}{(-)} > 0$$

An increase in the money supply M_0 will cause the equilibrium level of income to increase. For $\partial \bar{i}/\partial M_0$,

$$|\mathbf{J}_2| = \begin{vmatrix} 1 - C_Y & 0 \\ L_Y & 1/P \end{vmatrix} = \frac{1 - C_Y}{P}$$

and
$$\frac{\partial \bar{i}}{\partial M_0} = \frac{|\mathbf{J}_2|}{|\mathbf{J}|} = \frac{1 - C_Y}{P[(1 - C_Y)L_i + C_i L_Y]} = \frac{(+)}{(-)} < 0$$

An increase in M_0 will cause the equilibrium interest rate to fall.

13.11. Working with a system of three simultaneous equations similar to Problem 6.4,

$$Y = C + I_0 + G_0 \qquad C = C_0 + b(Y - T) \qquad T = T_0 + tY$$

(a) Express the system of equations as both general and specific implicit functions. (b) Use the simultaneous equations approach to comparative statics to find the Jacobian for both the general functions and the specific functions. (c) Express in matrix form the total derivatives of both the general and the specific functions with respect to G_0. Then find and sign (d) $\partial \bar{Y}/\partial G_0$, (e) $\partial \bar{C}/\partial G_0$, and (f) $\partial \bar{T}/\partial G_0$.

a)
$$F^1(Y, C, T; C_0, I_0, G_0, T_0, b, t) = Y - C - I_0 - G_0 = 0$$
$$F^2(Y, C, T; C_0, I_0, G_0, T_0, b, t) = C - C_0 - b(Y - T) = 0$$
$$F^3(Y, C, T; C_0, I_0, G_0, T_0, b, t) = T - T_0 - tY = 0$$

b)   The Jacobian consists of the partial derivatives of all the equations with respect to all the endogenous or dependent variables.

$$|\mathbf{J}| = \begin{vmatrix} \dfrac{\partial F^1}{\partial Y} & \dfrac{\partial F^1}{\partial C} & \dfrac{\partial F^1}{\partial T} \\[2mm] \dfrac{\partial F^2}{\partial Y} & \dfrac{\partial F^2}{\partial C} & \dfrac{\partial F^2}{\partial T} \\[2mm] \dfrac{\partial F^3}{\partial Y} & \dfrac{\partial F^3}{\partial C} & \dfrac{\partial F^3}{\partial T} \end{vmatrix} = \begin{vmatrix} 1 & -1 & 0 \\ -b & 1 & b \\ -t & 0 & 1 \end{vmatrix}$$

Expanding along the first row,

$$|\mathbf{J}| = 1(1) - (-1)(-b + bt) = 1 - b + bt \neq 0$$

c) Taking the total derivatives of the general functions with respect to G_0, setting them in the now familiar matrix form, and using bars over the endogenous variables to indicate that they are to be evaluated at the point of equilibrium in the model, we have

$$\begin{bmatrix} \dfrac{\partial F^1}{\partial Y} & \dfrac{\partial F^1}{\partial C} & \dfrac{\partial F^1}{\partial T} \\[2ex] \dfrac{\partial F^2}{\partial Y} & \dfrac{\partial F^2}{\partial C} & \dfrac{\partial F^2}{\partial T} \\[2ex] \dfrac{\partial F^3}{\partial Y} & \dfrac{\partial F^3}{\partial C} & \dfrac{\partial F^3}{\partial T} \end{bmatrix} \begin{bmatrix} \dfrac{\partial \bar{Y}}{\partial G_0} \\[2ex] \dfrac{\partial \bar{C}}{\partial G_0} \\[2ex] \dfrac{\partial \bar{T}}{\partial G_0} \end{bmatrix} = \begin{bmatrix} -\dfrac{\partial F^1}{\partial G_0} \\[2ex] -\dfrac{\partial F^2}{\partial G_0} \\[2ex] -\dfrac{\partial F^3}{\partial G_0} \end{bmatrix}$$

The same derivatives in terms of the specific functions are

$$\begin{bmatrix} 1 & -1 & 0 \\ -b & 1 & b \\ -t & 0 & 1 \end{bmatrix} \begin{bmatrix} \dfrac{\partial \bar{Y}}{\partial G_0} \\[2ex] \dfrac{\partial \bar{C}}{\partial G_0} \\[2ex] \dfrac{\partial \bar{T}}{\partial G_0} \end{bmatrix} = \begin{bmatrix} 1 \\ 0 \\ 0 \end{bmatrix} \qquad (13.19)$$

$$\mathbf{JX = B}$$

where the signs in the **B** matrix change as the matrix is moved to the right of the equal sign.

d) To find $\partial \bar{Y}/\partial G_0$, the a_{11} element in the **X** matrix in (13.19), we create a new matrix $|\mathbf{J}_1|$ by substituting **B** in the first column of **J** and using (13.6).

$$\frac{\partial \bar{Y}}{\partial G_0} = \frac{|\mathbf{J}_1|}{|\mathbf{J}|} = \frac{\begin{bmatrix} 1 & -1 & 0 \\ 0 & 1 & b \\ 0 & 0 & 1 \end{bmatrix}}{|\mathbf{J}|} = \frac{1}{1 - b + bt} > 0$$

e) To find $\partial \bar{C}/\partial G_0$, the a_{21} element in the **X** matrix in (13.19), we create $|\mathbf{J}_2|$ by substituting **B** in the second column of **J** and using (13.6).

$$\frac{\partial \bar{C}}{\partial G_0} = \frac{|\mathbf{J}_2|}{|\mathbf{J}|} = \frac{\begin{bmatrix} 1 & 1 & 0 \\ -b & 0 & b \\ -t & 0 & 1 \end{bmatrix}}{|\mathbf{J}|} = \frac{b(1 - t)}{1 - b + bt} > 0$$

f) To find $\partial \bar{T}/\partial G_0$, the a_{31} element in the **X** matrix in (13.19), we create $|\mathbf{J}_3|$ by substituting **B** in the third column of **J** and using (13.6).

$$\frac{\partial \bar{T}}{\partial G_0} = \frac{|\mathbf{J}_3|}{|\mathbf{J}|} = \frac{\begin{bmatrix} 1 & -1 & 1 \\ -b & 1 & 0 \\ -t & 0 & 0 \end{bmatrix}}{|\mathbf{J}|} = \frac{1}{1 - b + bt} > 0$$

13.12. Using the previous model in Problem 13.11, where the original Jacobian $|\mathbf{J}|$ will remain the same, (a) express in matrix form the total derivatives of both the general and the specific functions with respect to T_0. Then find and sign (b) $\partial \bar{Y}/\partial T_0$, (c) $\partial \bar{C}/\partial T_0$, and (d) $\partial \bar{T}/\partial T_0$.

a)   In matrix form, the total derivatives of the general functions with respect to T_0 are

$$
\begin{bmatrix}
\dfrac{\partial F^1}{\partial Y} & \dfrac{\partial F^1}{\partial C} & \dfrac{\partial F^1}{\partial T} \\[2mm]
\dfrac{\partial F^2}{\partial Y} & \dfrac{\partial F^2}{\partial C} & \dfrac{\partial F^2}{\partial T} \\[2mm]
\dfrac{\partial F^3}{\partial Y} & \dfrac{\partial F^3}{\partial C} & \dfrac{\partial F^3}{\partial T}
\end{bmatrix}
\begin{bmatrix}
\dfrac{\partial \bar Y}{\partial T_0} \\[2mm]
\dfrac{\partial \bar C}{\partial T_0} \\[2mm]
\dfrac{\partial \bar T}{\partial T_0}
\end{bmatrix}
=
\begin{bmatrix}
-\dfrac{\partial F^1}{\partial T_0} \\[2mm]
-\dfrac{\partial F^2}{\partial T_0} \\[2mm]
-\dfrac{\partial F^3}{\partial T_0}
\end{bmatrix}
$$

The same derivatives in terms of the specific function are

$$
\begin{bmatrix}
1 & -1 & 0 \\
-b & 1 & b \\
-t & 0 & 1
\end{bmatrix}
\begin{bmatrix}
\dfrac{\partial \bar Y}{\partial T_0} \\[2mm]
\dfrac{\partial \bar C}{\partial T_0} \\[2mm]
\dfrac{\partial \bar T}{\partial T_0}
\end{bmatrix}
=
\begin{bmatrix}
0 \\ 0 \\ 1
\end{bmatrix}
$$

b)   To find $\partial \bar Y/\partial T_0$, the a_{11} element in the $\mathbf{X}$ matrix, create $|\mathbf{J}_1|$ by substituting $\mathbf{B}$ in the first column of $\mathbf{J}$ and using (13.6).

$$
\frac{\partial \bar Y}{\partial T_0} = \frac{|\mathbf{J}_1|}{|\mathbf{J}|} = \frac{\begin{bmatrix} 0 & -1 & 0 \\ 0 & 1 & b \\ 1 & 0 & 1 \end{bmatrix}}{|\mathbf{J}|} = \frac{-b}{1-b+bt} < 0
$$

c)   For $\partial \bar C/\partial T_0$, the a_{21} element in the $\mathbf{X}$ matrix, create $|\mathbf{J}_2|$ by substituting $\mathbf{B}$ in the second column of $\mathbf{J}$.

$$
\frac{\partial \bar C}{\partial T_0} = \frac{|\mathbf{J}_2|}{|\mathbf{J}|} = \frac{\begin{bmatrix} 1 & 0 & 0 \\ -b & 0 & b \\ -t & 1 & 1 \end{bmatrix}}{|\mathbf{J}|} = \frac{-b}{1-b+bt} < 0
$$

d)   For $\partial \bar T/\partial T_0$, substitute $\mathbf{B}$ in the third column of $\mathbf{J}$ and use (13.6).

$$
\frac{\partial \bar T}{\partial T_0} = \frac{|\mathbf{J}_3|}{|\mathbf{J}|} = \frac{\begin{bmatrix} 1 & -1 & 0 \\ -b & 1 & 0 \\ -t & 0 & 1 \end{bmatrix}}{|\mathbf{J}|} = \frac{1-b}{1-b+bt} > 0
$$

13.13. Retaining the same model from Problem 13.11, (a) express in matrix form the total derivatives of both the general and the specific functions with respect to the tax rate t. Then find and sign (b) $\partial \bar Y/\partial t$, (c) $\partial \bar C/\partial t$, and (d) $\partial \bar T/\partial t$.

a)   The total derivatives of the general functions with respect to t are

$$
\begin{bmatrix}
\dfrac{\partial F^1}{\partial Y} & \dfrac{\partial F^1}{\partial C} & \dfrac{\partial F^1}{\partial T} \\[2mm]
\dfrac{\partial F^2}{\partial Y} & \dfrac{\partial F^2}{\partial C} & \dfrac{\partial F^2}{\partial T} \\[2mm]
\dfrac{\partial F^3}{\partial Y} & \dfrac{\partial F^3}{\partial C} & \dfrac{\partial F^3}{\partial T}
\end{bmatrix}
\begin{bmatrix}
\dfrac{\partial \bar Y}{\partial t} \\[2mm]
\dfrac{\partial \bar C}{\partial t} \\[2mm]
\dfrac{\partial \bar T}{\partial t}
\end{bmatrix}
=
\begin{bmatrix}
-\dfrac{\partial F^1}{\partial t} \\[2mm]
-\dfrac{\partial F^2}{\partial t} \\[2mm]
-\dfrac{\partial F^3}{\partial t}
\end{bmatrix}
$$

The same derivatives in terms of the specific functions are

$$
\begin{bmatrix} 1 & -1 & 0 \\ -b & 1 & b \\ -t & 0 & 1 \end{bmatrix}
\begin{bmatrix} \dfrac{\partial \bar{Y}}{\partial t} \\[2mm] \dfrac{\partial \bar{C}}{\partial t} \\[2mm] \dfrac{\partial \bar{T}}{\partial t} \end{bmatrix}
= \begin{bmatrix} 0 \\ 0 \\ \bar{Y} \end{bmatrix}
$$

b)   For $\partial \bar{Y}/\partial t$,

$$
\frac{\partial \bar{Y}}{\partial t} = \frac{|\mathbf{J}_1|}{|\mathbf{J}|} = \frac{\begin{bmatrix} 0 & -1 & 0 \\ 0 & 1 & b \\ \bar{Y} & 0 & 1 \end{bmatrix}}{|\mathbf{J}|} = \frac{-b\bar{Y}}{1-b+bt} < 0
$$

c)   For $\partial \bar{C}/\partial t$,

$$
\frac{\partial \bar{C}}{\partial t} = \frac{|\mathbf{J}_2|}{|\mathbf{J}|} = \frac{\begin{bmatrix} 1 & 0 & 0 \\ -b & 0 & b \\ -t & \bar{Y} & 1 \end{bmatrix}}{|\mathbf{J}|} = \frac{-b\bar{Y}}{1-b+bt} < 0
$$

d)   For $\partial \bar{T}/\partial t$,

$$
\frac{\partial \bar{T}}{\partial t} = \frac{|\mathbf{J}_3|}{|\mathbf{J}|} = \frac{\begin{bmatrix} 1 & -1 & 0 \\ -b & 1 & 0 \\ -t & 0 & \bar{Y} \end{bmatrix}}{|\mathbf{J}|} = \frac{(1-b)\bar{Y}}{1-b+bt} > 0
$$

13.14. Given the income determination model

$$
Y = C + I_0 + G_0 + X_0 - Z \qquad C = C_0 + bY \qquad Z = Z_0 + zY
$$

where X = exports, Z = imports, and a zero subscript indicates an exogenously fixed variable, (a) express the system of equations as both general and specific implicit functions. (b) Express in matrix form the total derivatives of both the general and the specific functions with respect to exports X_0. Then find and sign (c), $\partial \bar{Y}/\partial X_0$, (d) $\partial \bar{C}/\partial X_0$, and (e) $\partial \bar{Z}/\partial X_0$.

a) $$F^1(Y, C, Z; C_0, I_0, G_0, X_0, Z_0, b, z) = Y - C - I_0 - G_0 - X_0 + Z = 0$$
$$F^2(Y, C, Z; C_0, I_0 \, G_0, X_0, Z_0, b, z) = C - C_0 - bY = 0$$
$$F^3(Y, C, Z; C_0, I_0, G_0, X_0, Z_0, b, z) = Z - Z_0 - zY = 0$$

b)
$$
\begin{bmatrix} \dfrac{\partial F^1}{\partial Y} & \dfrac{\partial F^1}{\partial C} & \dfrac{\partial F^1}{\partial Z} \\[3mm] \dfrac{\partial F^2}{\partial Y} & \dfrac{\partial F^2}{\partial C} & \dfrac{\partial F^2}{\partial Z} \\[3mm] \dfrac{\partial F^3}{\partial Y} & \dfrac{\partial F^3}{\partial C} & \dfrac{\partial F^3}{\partial Z} \end{bmatrix}
\begin{bmatrix} \dfrac{\partial \bar{Y}}{\partial X_0} \\[3mm] \dfrac{\partial \bar{C}}{\partial X_0} \\[3mm] \dfrac{\partial \bar{Z}}{\partial X_0} \end{bmatrix}
= \begin{bmatrix} -\dfrac{\partial F^1}{\partial X_0} \\[3mm] -\dfrac{\partial F^2}{\partial X_0} \\[3mm] -\dfrac{\partial F^3}{\partial X_0} \end{bmatrix}
$$

$$
\begin{bmatrix} 1 & -1 & 1 \\ -b & 1 & 0 \\ -Z & 0 & 1 \end{bmatrix}
\begin{bmatrix} \dfrac{\partial \bar{Y}}{\partial X_0} \\[2mm] \dfrac{\partial \bar{C}}{\partial X_0} \\[2mm] \dfrac{\partial \bar{Z}}{\partial X_0} \end{bmatrix}
= \begin{bmatrix} 1 \\ 0 \\ 0 \end{bmatrix}
$$

$$
|\mathbf{J}| = 1 - b + z > 0
$$

c) For $\partial \bar{Y}/\partial X_0$,

$$\frac{\partial \bar{Y}}{\partial X_0} = \frac{|\mathbf{J}_1|}{|\mathbf{J}|} = \frac{\begin{bmatrix} 1 & -1 & 1 \\ 0 & 1 & 0 \\ 0 & 0 & 1 \end{bmatrix}}{|\mathbf{J}|} = \frac{1}{1-b+z} > 0$$

d) For $\partial \bar{C}/\partial X_0$,

$$\frac{\partial \bar{C}}{\partial X_0} = \frac{|\mathbf{J}_2|}{|\mathbf{J}|} = \frac{\begin{bmatrix} 1 & 1 & 1 \\ -b & 0 & 0 \\ -z & 0 & 1 \end{bmatrix}}{|\mathbf{J}|} = \frac{b}{1-b+z} > 0$$

e) For $\partial \bar{Z}/\partial X_0$,

$$\frac{\partial \bar{Z}}{\partial X_0} = \frac{|\mathbf{J}_3|}{|\mathbf{J}|} = \frac{\begin{bmatrix} 1 & -1 & 1 \\ -b & 1 & 0 \\ -z & 0 & 0 \end{bmatrix}}{|\mathbf{J}|} = \frac{z}{1-b+z} > 0$$

13.15. Using the model from Problem 13.14, (*a*) express in matrix form the total derivatives of the specific functions with respect to the marginal propensity to consume *b*. Then find and sign (*b*) $\partial \bar{Y}/\partial b$, (*c*) $\partial \bar{C}/\partial b$, and (*d*) $\partial \bar{Z}/\partial b$.

a)
$$\begin{bmatrix} 1 & -1 & 1 \\ -b & 1 & 0 \\ -z & 0 & 1 \end{bmatrix} \begin{bmatrix} \dfrac{\partial \bar{Y}}{\partial b} \\[2mm] \dfrac{\partial \bar{C}}{\partial b} \\[2mm] \dfrac{\partial \bar{Z}}{\partial b} \end{bmatrix} = \begin{bmatrix} 0 \\ \bar{Y} \\ 0 \end{bmatrix}$$

b)
$$\frac{\partial \bar{Y}}{\partial b} = \frac{|\mathbf{J}_1|}{|\mathbf{J}|} = \frac{\begin{bmatrix} 0 & -1 & 1 \\ \bar{Y} & 1 & 0 \\ 0 & 0 & 1 \end{bmatrix}}{|\mathbf{J}|} = \frac{\bar{Y}}{1-b+z} > 0$$

c)
$$\frac{\partial \bar{C}}{\partial b} = \frac{|\mathbf{J}_2|}{|\mathbf{J}|} = \frac{\begin{bmatrix} 1 & 0 & 1 \\ -b & \bar{Y} & 0 \\ -z & 0 & 1 \end{bmatrix}}{|\mathbf{J}|} = \frac{(1+z)\bar{Y}}{1-b+z} > 0$$

d)
$$\frac{\partial \bar{Z}}{\partial b} = \frac{|\mathbf{J}_3|}{|\mathbf{J}|} = \frac{\begin{bmatrix} 1 & -1 & 0 \\ -b & 1 & \bar{Y} \\ -z & 0 & 0 \end{bmatrix}}{|\mathbf{J}|} = \frac{z\bar{Y}}{1-b+z} > 0$$

13.16. Continuing with the model from Problem 13.14, (*a*) express in matrix form the total derivatives of the specific functions with respect to the marginal propensity to import *z*. Then find and sign (*b*) $\partial \bar{Y}/\partial z$, (*c*) $\partial \bar{C}/\partial z$, and (*d*) $\partial \bar{Z}/\partial z$.

a)
$$\begin{bmatrix} 1 & -1 & 1 \\ -b & 1 & 0 \\ -z & 0 & 1 \end{bmatrix} \begin{bmatrix} \dfrac{\partial \bar{Y}}{\partial z} \\[2mm] \dfrac{\partial \bar{C}}{\partial z} \\[2mm] \dfrac{\partial \bar{Z}}{\partial z} \end{bmatrix} = \begin{bmatrix} 0 \\ 0 \\ \bar{Y} \end{bmatrix}$$

b)   For $\partial \bar{Y}/\partial z$,

$$\frac{\partial \bar{Y}}{\partial z} = \frac{|\mathbf{J}_1|}{|\mathbf{J}|} = \frac{\begin{bmatrix} 0 & -1 & 1 \\ 0 & 1 & 0 \\ \bar{Y} & 0 & 1 \end{bmatrix}}{|\mathbf{J}|} = \frac{-\bar{Y}}{1 - b + z} < 0$$

c)   For $\partial \bar{C}/\partial z$,

$$\frac{\partial \bar{C}}{\partial z} = \frac{|\mathbf{J}_2|}{|\mathbf{J}|} = \frac{\begin{bmatrix} 1 & 0 & 1 \\ -b & 0 & 0 \\ -z & \bar{Y} & 1 \end{bmatrix}}{|\mathbf{J}|} = \frac{-b\bar{Y}}{1 - b + z} < 0$$

d)   For $\partial \bar{Z}/\partial z$,

$$\frac{\partial \bar{Z}}{\partial z} = \frac{|\mathbf{J}_3|}{|\mathbf{J}|} = \frac{\begin{bmatrix} 1 & -1 & 0 \\ -b & 1 & 0 \\ -z & 0 & \bar{Y} \end{bmatrix}}{|\mathbf{J}|} = \frac{(1 - b)\bar{Y}}{1 - b + z} > 0$$

13.17.   Having introduced the foreign trade market to the goods market in our national income model, let us now combine them with the money market. Assume

The goods market: $I = I(i)$ $(I_i < 0)$
                         $S = S(Y, i)$ $(0 < S_Y < 1; S_i > 0)$

The foreign trade market: $Z = Z(Y, i)$ $0 < Z_Y < 1; Z_i < 0$
                            $X = X_0$

and the money market: $M_D = L(Y, i)$ $L_Y > 0, L_i < 0$
                         $M_S = M_0$

where Z = imports, S = savings, X_0 = autonomous exports, M_D = demand for money, M_S = supply of money, and all the other symbols are familiar. (*a*) Express the equilibrium conditions for the combined goods market and the money market. (*b*) Express the equilibrium conditions as general and specific implicit functions. (*c*) Express in matrix form the total derivatives of these functions with respect to M_0. Find and sign (*d*) the Jacobian, (*e*) $\partial \bar{Y}/\partial M_0$, and (*f*) $\partial \bar{i}/\partial M_0$.

a)   The combined goods market is in equilibrium when injections equal leakages:

$$I(i) + X_0 = S(Y, i) + Z(Y, i)$$

The money market is in equilibrium when the demand for money equals the money supply:

$$L(Y, i) = M_0$$

b)

$$F^1(Y, i; M_0, X_0) = I(i) + X_0 - S(Y, i) - Z(Y, i)$$
$$F^2(Y, i; M_0, X_0) = L(Y, i) - M_0$$

c)

$$\begin{bmatrix} \dfrac{\partial F^1}{\partial Y} & \dfrac{\partial F^1}{\partial i} \\[2ex] \dfrac{\partial F^2}{\partial Y} & \dfrac{\partial F^2}{\partial i} \end{bmatrix} \begin{bmatrix} \dfrac{\partial \bar{Y}}{\partial M_0} \\[2ex] \dfrac{\partial \bar{i}}{\partial M_0} \end{bmatrix} = \begin{bmatrix} -\dfrac{\partial F^1}{\partial M_0} \\[2ex] -\dfrac{\partial F^2}{\partial M_0} \end{bmatrix}$$

$$\begin{bmatrix} -S_Y - Z_Y & I_i - S_i - Z_i \\ L_Y & L_i \end{bmatrix} \begin{bmatrix} \dfrac{\partial \bar{Y}}{\partial M_0} \\[2ex] \dfrac{\partial \bar{i}}{\partial M_0} \end{bmatrix} = \begin{bmatrix} 0 \\ 1 \end{bmatrix}$$

$$\mathbf{JX} = \mathbf{B}$$

d)
$$|\mathbf{J}| = L_i(-S_Y - Z_Y) - L_Y(I_i - S_i - Z_i) > 0$$

e) Using Cramer's rule for $\partial \bar{Y} / \partial M_0$,

$$\frac{\partial \bar{Y}}{\partial M_0} = \frac{|\mathbf{J}_1|}{|\mathbf{J}|} = \frac{\begin{vmatrix} 0 & I_i - S_i - Z_i \\ 1 & L_i \end{vmatrix}}{|\mathbf{J}|} = \frac{-(I_i - S_i - Z_i)}{L_i(-S_Y - Z_Y) - L_Y(I_i - S_i - Z_i)} > 0$$

f) For $\partial \bar{i} / \partial M_0$,

$$\frac{\partial \bar{i}}{\partial M_0} = \frac{|\mathbf{J}_2|}{|\mathbf{J}|} = \frac{\begin{vmatrix} -S_Y - Z_Y & 0 \\ L_Y & 1 \end{vmatrix}}{|\mathbf{J}|} = \frac{-(S_Y + Z_Y)}{L_i(-S_Y - Z_Y) - L_Y(I_i - S_i - Z_i)} < 0$$

13.18. Using the model in Problem 13.17, (*a*) express in matrix form the total derivatives of the specific functions with respect to exports X_0. Then find and sign (*b*) $\partial \bar{Y}/\partial X_0$ and (*c*) $\partial \bar{i}/\partial X_0$.

a)
$$\begin{bmatrix} -S_Y - Z_Y & I_i - S_i - Z_i \\ L_Y & L_i \end{bmatrix} \begin{bmatrix} \dfrac{\partial \bar{Y}}{\partial X_0} \\[2mm] \dfrac{\partial \bar{i}}{\partial X_0} \end{bmatrix} = \begin{bmatrix} -1 \\ 0 \end{bmatrix}$$

b) For $\partial \bar{Y} / \partial X_0$,

$$\frac{\partial \bar{Y}}{\partial X_0} = \frac{|\mathbf{J}_1|}{|\mathbf{J}|} = \frac{\begin{vmatrix} -1 & I_i - S_i - Z_i \\ 0 & L_i \end{vmatrix}}{|\mathbf{J}|} = \frac{-L_i}{L_i(-S_Y - Z_Y) - L_Y(I_i - S_i - Z_i)} > 0$$

c) For $\partial \bar{i} / \partial X_0$,

$$\frac{\partial \bar{i}}{\partial X_0} = \frac{|\mathbf{J}_2|}{|\mathbf{J}|} = \frac{\begin{vmatrix} -S_Y - Z_Y & -1 \\ L_Y & 0 \end{vmatrix}}{|\mathbf{J}|} = \frac{L_Y}{L_i(-S_Y - Z_Y) - L_Y(I_i - S_i - Z_i)} > 0$$

COMPARATIVE STATIC ANALYSIS IN OPTIMIZATION PROBLEMS

13.19. Returning to the model in Example 5, where the first-order conditions were

$$F^1(K, L; r, w, P) = PQ_K(\bar{K}, \bar{L}) - r = 0$$
$$F^2(K, L; r, w, P) = PQ_L(\bar{K}, \bar{L}) - w = 0$$

(*a*) express the total derivatives of the functions with respect to the wage w in matrix form. Then find and sign (*b*) $\partial \bar{K}/\partial w$ and (*c*) $\partial \bar{L}/\partial w$.

a)
$$\begin{bmatrix} PQ_{KK} & PQ_{KL} \\ PQ_{LK} & PQ_{LL} \end{bmatrix} \begin{bmatrix} \dfrac{\partial \bar{K}}{\partial w} \\[2mm] \dfrac{\partial \bar{L}}{\partial w} \end{bmatrix} = \begin{bmatrix} 0 \\ 1 \end{bmatrix}$$

$$|\mathbf{J}| = P^2(Q_{KK}Q_{LL} - Q_{LK}Q_{KL}) > 0$$

b) For $\partial \bar{K} / \partial w$,

$$\frac{\partial \bar{K}}{\partial w} = \frac{|\mathbf{J}_1|}{|\mathbf{J}|} = \frac{\begin{vmatrix} 0 & PQ_{KL} \\ 1 & PQ_{LL} \end{vmatrix}}{|\mathbf{J}|} = \frac{-PQ_{KL}}{P^2(Q_{KK}Q_{LL} - Q_{KL}Q_{LK})}$$

Without specific knowledge of the sign of the cross partial Q_{KL}, it is impossible to sign the derivative. Assuming the marginal productivity of capital will increase for an increase in labor, $Q_{KL} > 0$ and

$\partial \bar{K}/\partial w < 0$, meaning that the optimal level of capital will likely fall in response to an increase in wage.

c)   For $\partial \bar{L}/\partial w$,

$$\frac{\partial \bar{L}}{\partial w} = \frac{|\mathbf{J}_2|}{|\mathbf{J}|} = \frac{\begin{vmatrix} PQ_{KK} & 0 \\ PQ_{LK} & 1 \end{vmatrix}}{|\mathbf{J}|} = \frac{PQ_{KK}}{P^2(Q_{KK}Q_{LL} - Q_{KL}Q_{LK})} < 0$$

The optimal level of labor will decrease for an increase in wage since $Q_{KK} < 0$.

13.20. Staying with the model in Example 5, where the first-order conditions were

$$F^1(K, L; r, w, P) = PQ_K(\bar{K}, \bar{L}) - r = 0$$
$$F^2(K, L; r, w, P) = PQ_L(\bar{K}, \bar{L}) - w = 0$$

(a) express the total derivatives of the functions with respect to the commodity price P in matrix form. Then find and sign (b) $\partial \bar{K}/\partial P$ and (c) $\partial \bar{L}/\partial P$.

a)

$$\begin{bmatrix} PQ_{KK} & PQ_{KL} \\ PQ_{LK} & PQ_{LL} \end{bmatrix} \begin{bmatrix} \dfrac{\partial \bar{K}}{\partial P} \\[2ex] \dfrac{\partial \bar{L}}{\partial P} \end{bmatrix} = \begin{bmatrix} -Q_K \\ -Q_L \end{bmatrix}$$

b)   For $\partial \bar{K}/\partial P$,

$$\frac{\partial \bar{K}}{\partial P} = \frac{|\mathbf{J}_1|}{|\mathbf{J}|} = \frac{\begin{vmatrix} -Q_K & PQ_{KL} \\ -Q_L & PQ_{LL} \end{vmatrix}}{|\mathbf{J}|} = \frac{P(-Q_K Q_{LL} + Q_L Q_{KL})}{P^2(Q_{KK}Q_{LL} - Q_{KL}Q_{LK})} = \frac{(Q_L Q_{KL} - Q_K Q_{LL})}{P(Q_{KK}Q_{LL} - Q_{KL}Q_{LK})}$$

Since $Q_K = MP_K > 0$, $Q_L = MP_L > 0$, $Q_{LL} < 0$ for maximization, and $|\mathbf{J}| > 0$ in the denominator, the sign depends completely on the cross partial Q_{KL}. If K and L are complements, so that an increased use of one input will lead to an increase in the MP of the other input, $Q_{KL} > 0$ and the comparative-static derivative $\partial \bar{K}/\partial P > 0$. If $Q_{KL} < 0$, the sign of $\partial \bar{K}/\partial P$ is indeterminate.

c)   For $\partial \bar{L}/\partial P$,

$$\frac{\partial \bar{L}}{\partial P} = \frac{|\mathbf{J}_2|}{|\mathbf{J}|} = \frac{\begin{vmatrix} PQ_{KK} & -Q_K \\ PQ_{LK} & -Q_L \end{vmatrix}}{|\mathbf{J}|} = \frac{P(-Q_L Q_{KK} + Q_K Q_{LK})}{P^2(Q_{KK}Q_{LL} - Q_{KL}Q_{LK})} = \frac{(Q_K Q_{LK} - Q_L Q_{KK})}{P(Q_{KK}Q_{LL} - Q_{KL}Q_{LK})}$$

and the sign will depend on the cross partial Q_{LK}, exactly as in (b).

13.21. Assume now a firm seeks to optimize the discounted value of its profit function,

$$\pi = P_0 Q(X, Y) e^{-rt} - P_x X - P_y Y$$

where the first derivatives of the first-order conditions, $\partial \pi/\partial X$ and $\partial \pi/\partial Y$, when expressed as implicit functions, are

$$F^1(X, Y; P_0, P_x, P_y, r, t) = P_0 Q_x(\bar{X}, \bar{Y}) e^{-rt} - P_x = 0$$
$$F^2(X, Y; P_0, P_x, P_y, r, t) = P_0 Q_y(\bar{X}, \bar{Y}) e^{-rt} - P_y = 0$$

(a) Express the total derivatives of the functions with respect to P_0 in matrix form, recalling that r and t are constants. Then find and sign (b) the Jacobian, (c) $\partial \bar{X}/\partial P_0$ and (d) $\partial \bar{Y}/\partial P_0$.

a)

$$\begin{bmatrix} P_0 Q_{xx} e^{-rt} & P_0 Q_{xy} e^{-rt} \\ P_0 Q_{yx} e^{-rt} & P_0 Q_{yy} e^{-rt} \end{bmatrix} \begin{bmatrix} \dfrac{\partial \bar{X}}{\partial P_0} \\[2ex] \dfrac{\partial \bar{Y}}{\partial P_0} \end{bmatrix} = \begin{bmatrix} -Q_x e^{-rt} \\ -Q_y e^{-rt} \end{bmatrix}$$

b)
$$|\mathbf{J}| = P_0^2 e^{-2rt}(Q_{xx}Q_{yy} - Q_{yx}Q_{xy}) > 0$$

With $P_0^2 e^{-2rt} > 0$ and $(Q_{xx}Q_{yy} - Q_{yx}Q_{xy}) > 0$ from the second-order sufficient condition, $|\mathbf{J}| > 0$.

c) For $\partial \bar{X}/\partial P_0$,

$$\frac{\partial \bar{X}}{\partial P_0} = \frac{|\mathbf{J}_1|}{|\mathbf{J}|} = \frac{\begin{vmatrix} -Q_x e^{-rt} & P_0 Q_{xy} e^{-rt} \\ -Q_y e^{-rt} & P_0 Q_{yy} e^{-rt} \end{vmatrix}}{|\mathbf{J}|} = \frac{P_0 e^{-2rt}(Q_y Q_{xy} - Q_x Q_{yy})}{P_0^2 e^{-2rt}(Q_{xx}Q_{yy} - Q_{yx}Q_{xy})} = \frac{(Q_y Q_{xy} - Q_x Q_{yy})}{P_0(Q_{xx}Q_{yy} - Q_{yx}Q_{xy})}$$

As in Problem 13.20(b), the sign will depend on the cross partial Q_{xy}.

d) For $\partial \bar{Y}/\partial P_0$,

$$\frac{\partial \bar{Y}}{\partial P_0} = \frac{|\mathbf{J}_2|}{|\mathbf{J}|} = \frac{\begin{vmatrix} P_0 Q_{xx} e^{-rt} & -Q_x e^{-rt} \\ P_0 Q_{yx} e^{-rt} & -Q_y e^{-rt} \end{vmatrix}}{|\mathbf{J}|} = \frac{P_0 e^{-2rt}(Q_x Q_{yx} - Q_y Q_{xx})}{P_0^2 e^{-2rt}(Q_{xx}Q_{yy} - Q_{yx}Q_{xy})} = \frac{(Q_x Q_{yx} - Q_y Q_{xx})}{P_0(Q_{xx}Q_{yy} - Q_{yx}Q_{xy})}$$

13.22. Using the same model in Problem 13.21, (a) express the total derivatives of the functions with respect to time t in matrix form. Then find and sign (b) $\partial \bar{X}/\partial t$ and (c) $\partial \bar{Y}/\partial t$.

a)
$$\begin{bmatrix} P_0 Q_{xx} e^{-rt} & P_0 Q_{xy} e^{-rt} \\ P_0 Q_{yx} e^{-rt} & P_0 Q_{yy} e^{-rt} \end{bmatrix} \begin{bmatrix} \dfrac{\partial \bar{X}}{\partial t} \\ \dfrac{\partial \bar{Y}}{\partial t} \end{bmatrix} = \begin{bmatrix} rP_0 Q_x e^{-rt} \\ rP_0 Q_y e^{-rt} \end{bmatrix}$$

b)
$$\frac{\partial \bar{X}}{\partial t} = \frac{|\mathbf{J}_1|}{|\mathbf{J}|} = \frac{\begin{vmatrix} rP_0 Q_x e^{-rt} & P_0 Q_{xy} e^{-rt} \\ rP_0 Q_y e^{-rt} & P_0 Q_{yy} e^{-rt} \end{vmatrix}}{|\mathbf{J}|} = \frac{rP_0^2 e^{-2rt}(Q_x Q_{yy} - Q_y Q_{xy})}{P_0^2 e^{-2rt}(Q_{xx}Q_{yy} - Q_{yx}Q_{xy})} = \frac{r(Q_x Q_{yy} - Q_y Q_{xy})}{(Q_{xx}Q_{yy} - Q_{yx}Q_{xy})}$$

c)
$$\frac{\partial \bar{Y}}{\partial t} = \frac{|\mathbf{J}_2|}{|\mathbf{J}|} = \frac{\begin{vmatrix} P_0 Q_{xx} e^{-rt} & rP_0 Q_x e^{-rt} \\ P_0 Q_{yx} e^{-rt} & rP_0 Q_y e^{-rt} \end{vmatrix}}{|\mathbf{J}|} = \frac{rP_0^2 e^{-2rt}(Q_y Q_{xx} - Q_x Q_{yx})}{P_0^2 e^{-2rt}(Q_{xx}Q_{yy} - Q_{yx}Q_{xy})} = \frac{r(Q_y Q_{xx} - Q_x Q_{yx})}{(Q_{xx}Q_{yy} - Q_{yx}Q_{xy})}$$

In both cases, if the cross partials are positive, the comparative-static derivatives will be negative and if the cross partials are negative, the comparative-static derivatives will be indeterminate.

13.23. Assume the production function in Example 5 is specified as a Cobb-Douglas function with decreasing returns to scale (Section 6.9), so that the competitive firm's profit function is

$$\pi = PAK^\alpha L^\beta - rK - wL$$

and the first derivatives $\partial \pi/\partial K$ and $\partial \pi/\partial L$ from the first-order conditions are

$$F^1(K, L; r, w, P, A, \alpha, \beta) = \alpha PAK^{\alpha-1}L^\beta - r = 0$$
$$F^2(K, L; r, w, P, A, \alpha, \beta) = \beta PAK^\alpha L^{\beta-1} - w = 0$$

(a) Express the total derivatives of the functions with respect to the wage w in matrix form. Then find and sign (b) the Jacobian, (c) $\partial \bar{K}/\partial w$, and (d) $\partial \bar{L}/\partial w$.

a)
$$\begin{bmatrix} \alpha(\alpha-1)PAK^{\alpha-2}L^\beta & \alpha\beta PAK^{\alpha-1}L^{\beta-1} \\ \alpha\beta PAK^{\alpha-1}L^{\beta-1} & \beta(\beta-1)PAK^\alpha L^{\beta-2} \end{bmatrix} \begin{bmatrix} \dfrac{\partial \bar{K}}{\partial w} \\ \dfrac{\partial \bar{L}}{\partial w} \end{bmatrix} = \begin{bmatrix} 0 \\ 1 \end{bmatrix}$$

b)
$$|\mathbf{J}| = \alpha(\alpha-1)PAK^{\alpha-2}L^\beta \cdot \beta(\beta-1)PAK^\alpha L^{\beta-2} - (\alpha\beta PAK^{\alpha-1}L^{\beta-1})^2$$
$$|\mathbf{J}| = \alpha\beta(1 - \alpha - \beta)P^2 A^2 K^{2\alpha-2}L^{2\beta-2} > 0$$

since $|\mathbf{J}| = |\mathbf{H}|$ and $|\mathbf{H}| = |\mathbf{H}_2| > 0$ in unconstrained optimization problems. This condition implies that a profit-maximizing firm in perfect competition operates under decreasing returns to scale since $|\mathbf{J}| > 0$ requires $(\alpha + \beta) < 1$.

c) For $\partial \bar{K} / \partial w$,

$$
\frac{\partial \bar{K}}{\partial w} = \frac{|\mathbf{J}_1|}{|\mathbf{J}|} = \frac{\begin{vmatrix} 0 & \alpha\beta PAK^{\alpha-1}L^{\beta-1} \\ 1 & \beta(\beta-1)PAK^{\alpha}L^{\beta-2} \end{vmatrix}}{|\mathbf{J}|} = \frac{-\alpha\beta PAK^{\alpha-1}L^{\beta-1}}{|\mathbf{J}|} < 0
$$

since the numerator, independent of the negative sign, is unambiguously positive and the denominator is positive, the comparative-static derivative $\partial \bar{K} / \partial w$ is unquestionably negative. An increase in the wage will decrease the demand for capital. Through further simplification, if desired, we can also see

$$
\frac{\partial \bar{K}}{\partial w} = \frac{-\alpha\beta PAK^{\alpha-1}L^{\beta-1}}{\alpha\beta(1-\alpha-\beta)P^2 A^2 K^{2\alpha-2}L^{2\beta-2}} = \frac{-KL}{(1-\alpha-\beta)\text{TR}} < 0
$$

d) For $\partial \bar{L} / \partial w$,

$$
\frac{\partial \bar{L}}{\partial w} = \frac{|\mathbf{J}_2|}{|\mathbf{J}|} = \frac{\begin{vmatrix} \alpha(\alpha-1)PAK^{\alpha-2}L^{\beta} & 0 \\ \alpha\beta PAK^{\alpha-1}L^{\beta-1} & 1 \end{vmatrix}}{|\mathbf{J}|} = \frac{\alpha(\alpha-1)PAK^{\alpha-2}L^{\beta}}{|\mathbf{J}|} < 0
$$

since $\alpha < 1$, making $(\alpha - 1) < 0$. An increase in the wage will lead to a reduction in the optimal level of labor used. Through further simplification, we can also see

$$
\frac{\partial \bar{L}}{\partial w} = \frac{\alpha(\alpha-1)PAK^{\alpha-2}L^{\beta}}{\alpha\beta(1-\alpha-\beta)P^2 A^2 K^{2\alpha-2}L^{2\beta-2}} = \frac{-(1-\alpha)L^2}{(1-\alpha-\beta)\text{TR}} < 0
$$

13.24. Working with the same model in Problem 13.23, (*a*) express the total derivatives of the functions with respect to output price P in matrix form. Then find and sign (*b*) $\partial \bar{K} / \partial P$ and (*c*) $\partial \bar{L} / \partial P$.

a)

$$
\begin{bmatrix} \alpha(\alpha-1)PAK^{\alpha-2}L^{\beta} & \alpha\beta PAK^{\alpha-1}L^{\beta-1} \\ \alpha\beta PAK^{\alpha-1}L^{\beta-1} & \beta(\beta-1)PAK^{\alpha}L^{\beta-2} \end{bmatrix} \begin{bmatrix} \dfrac{\partial \bar{K}}{\partial P} \\ \dfrac{\partial \bar{L}}{\partial P} \end{bmatrix} = \begin{bmatrix} -\alpha AK^{\alpha-1}L^{\beta} \\ -\beta AK^{\alpha}L^{\beta-1} \end{bmatrix}
$$

b) For $\partial \bar{K} / \partial P$,

$$
\frac{\partial \bar{K}}{\partial P} = \frac{|\mathbf{J}_1|}{|\mathbf{J}|} = \frac{\begin{vmatrix} -\alpha AK^{\alpha-1}L^{\beta} & \alpha\beta PAK^{\alpha-1}L^{\beta-1} \\ -\beta AK^{\alpha}L^{\beta-1} & \beta(\beta-1)PAK^{\alpha}L^{\beta-2} \end{vmatrix}}{|\mathbf{J}|} = \frac{\alpha\beta PA^2 K^{2\alpha-1}L^{2\beta-2}}{\alpha\beta(1-\alpha-\beta)P^2 A^2 K^{2\alpha-2}L^{2\beta-2}}
$$

$$
\frac{\partial \bar{K}}{\partial P} = \frac{K}{(1-\alpha-\beta)P} > 0
$$

c) For $\partial \bar{L} / \partial P$,

$$
\frac{\partial \bar{L}}{\partial P} = \frac{|\mathbf{J}_2|}{|\mathbf{J}|} = \frac{\begin{vmatrix} \alpha(\alpha-1)PAK^{\alpha-1}L^{\beta} & -\alpha AK^{\alpha-1}L^{\beta} \\ \alpha\beta PAK^{\alpha-1}L^{\beta-1} & -\beta AK^{\alpha}L^{\beta-1} \end{vmatrix}}{|\mathbf{J}|} = \frac{\alpha\beta PA^2 K^{2\alpha-2}L^{2\beta-1}}{\alpha\beta(1-\alpha-\beta)P^2 A^2 K^{2\alpha-2}L^{2\beta-2}}
$$

$$
\frac{\partial \bar{L}}{\partial P} = \frac{L}{(1-\alpha-\beta)P} > 0
$$

COMPARATIVE STATICS IN CONSTRAINED OPTIMIZATION

13.25. A consumer wants to maximize utility $u(a,b)$ subject to the constraint $p_a a + p_b b = Y$, a constant. Given the Lagrangian function

$$U = u(a,b) + \lambda(Y - p_a a - p_b b)$$

and assuming the second-order sufficient condition is met so that $|\bar{\mathbf{H}}| = |\mathbf{J}| \neq 0$, the endogenous variables in the first-order conditions can be expressed as implicit functions of the exogenous variables, such that

$$
\begin{aligned}
F^1(a,b,\lambda; p_a, p_b, Y) &= U_a - \lambda p_a = 0 \\
F^2(a,b,\lambda; p_a, p_b, Y) &= U_b - \lambda p_b = 0 \\
F^3(a,b,\lambda; p_a, p_b, Y) &= Y - p_a a - p_b b = 0
\end{aligned}
\qquad (13.20)
$$

(a) Express the total derivatives of the functions with respect to p_a in matrix form and (b) find $\partial \bar{a}/\partial p_a$.

a)
$$
\begin{bmatrix}
U_{aa} & U_{ab} & -p_a \\
U_{ba} & U_{bb} & -p_b \\
-p_a & -p_b & 0
\end{bmatrix}
\begin{bmatrix}
\dfrac{\partial \bar{a}}{\partial p_a} \\[2mm]
\dfrac{\partial \bar{b}}{\partial p_a} \\[2mm]
\dfrac{\partial \bar{\lambda}}{\partial p_a}
\end{bmatrix}
=
\begin{bmatrix}
\bar{\lambda} \\ 0 \\ \bar{a}
\end{bmatrix}
$$

$$|\mathbf{J}| = -p_a^2 U_{bb} + p_a p_b U_{ab} + p_a p_b U_{ba} - p_b^2 U_{aa} > 0$$

since $|\mathbf{J}| = |\bar{\mathbf{H}}| > 0$ from the second-order sufficient condition for constrained maximization. But theory leaves unspecified the signs of the individual second partials.

b) For $\partial \bar{a}/\partial p_a$,

$$
\frac{\partial \bar{a}}{\partial p_a} = \frac{|\mathbf{J}_1|}{|\mathbf{J}|} = \frac{
\begin{vmatrix}
\bar{\lambda} & U_{ab} & -p_a \\
0 & U_{bb} & -p_b \\
\bar{a} & -p_b & 0
\end{vmatrix}
}{|\mathbf{J}|}
= \frac{\bar{a}(p_a U_{bb} - p_b U_{ab}) - \bar{\lambda} p_b^2}{|\mathbf{J}|}
\qquad (13.21)
$$

where the sign is indeterminate because the signs of the second partials are unknown.

13.26. Working with the same model in Problem 13.25, (a) express the total derivatives of the functions with respect to p_b in matrix form and (b) find $\partial \bar{b}/\partial p_b$.

a)
$$
\begin{bmatrix}
U_{aa} & U_{ab} & -p_a \\
U_{ba} & U_{bb} & -p_b \\
-p_a & -p_b & 0
\end{bmatrix}
\begin{bmatrix}
\dfrac{\partial \bar{a}}{\partial p_b} \\[2mm]
\dfrac{\partial \bar{b}}{\partial p_b} \\[2mm]
\dfrac{\partial \bar{\lambda}}{\partial p_b}
\end{bmatrix}
=
\begin{bmatrix}
0 \\ \bar{\lambda} \\ \bar{b}
\end{bmatrix}
$$

b) For $\partial \bar{b}/\partial p_b$,

$$
\frac{\partial \bar{b}}{\partial p_b} = \frac{|\mathbf{J}_2|}{|\mathbf{J}|} = \frac{
\begin{vmatrix}
U_{aa} & 0 & -p_a \\
U_{ba} & \bar{\lambda} & -p_b \\
-p_a & \bar{b} & 0
\end{vmatrix}
}{|\mathbf{J}|}
= \frac{\bar{b}(p_b U_{aa} - p_a U_{ba}) - \bar{\lambda} p_a^2}{|\mathbf{J}|}
\qquad (13.22)
$$

which is also indeterminate.

13.27. Continuing with the same model in Problem 13.25, (a) express the total derivatives of the functions with respect to Y in matrix form and find (b) $\partial \bar{a}/\partial Y$ and (c) $\partial \bar{b}/\partial Y$.

a)
$$
\begin{bmatrix}
U_{aa} & U_{ab} & -p_a \\
U_{ba} & U_{bb} & -p_b \\
-p_a & -p_b & 0
\end{bmatrix}
\begin{bmatrix}
\dfrac{\partial \bar{a}}{\partial Y} \\[2mm]
\dfrac{\partial \bar{b}}{\partial Y} \\[2mm]
\dfrac{\partial \bar{\lambda}}{\partial Y}
\end{bmatrix}
=
\begin{bmatrix}
0 \\ 0 \\ -1
\end{bmatrix}
$$

b) For $\partial \bar{a}/\partial Y$,

$$
\frac{\partial \bar{a}}{\partial Y} = \frac{|\mathbf{J}_1|}{|\mathbf{J}|} = \frac{\begin{vmatrix} 0 & U_{ab} & -p_a \\ 0 & U_{bb} & -p_b \\ -1 & -p_b & 0 \end{vmatrix}}{|\mathbf{J}|} = \frac{-(p_a U_{bb} - p_b U_{ab})}{|\mathbf{J}|} \tag{13.23}
$$

which cannot be signed from the mathematics but can be signed from the economics. If a is a normal good, $\partial \bar{a}/\partial Y > 0$; if a is a weakly inferior good, $\partial \bar{a}/\partial Y = 0$; and if a is a strictly inferior good, $\partial \bar{a}/\partial Y < 0$.

c) For $\partial \bar{b}/\partial Y$,

$$
\frac{\partial \bar{b}}{\partial Y} = \frac{|\mathbf{J}_2|}{|\mathbf{J}|} = \frac{\begin{vmatrix} U_{aa} & 0 & -p_a \\ U_{ba} & 0 & -p_b \\ -p_a & -1 & 0 \end{vmatrix}}{|\mathbf{J}|} = \frac{-(p_b U_{aa} - p_a U_{ba})}{|\mathbf{J}|} \tag{13.24}
$$

which can also be signed according to the nature of the good, as in (b) above.

13.28. Derive the Slutsky equation for the effect of a change in p_a on the optimal quantity of the good demanded $\bar{a}$ and determine the sign of the comparative-static derivative $\partial \bar{a}/\partial p_a$ from the information gained in Problems 13.25 to 13.27.

From (13.21), with slight rearrangement,

$$
\frac{\partial \bar{a}}{\partial p_a} = -\frac{\bar{\lambda} p_b^2}{|\mathbf{J}|} + \frac{\bar{a}(p_a U_{bb} - p_b U_{ab})}{|\mathbf{J}|} \tag{13.25}
$$

But from (13.23),

$$
\frac{\partial \bar{a}}{\partial Y} = \frac{-(p_a U_{bb} - P_b U_{ab})}{|\mathbf{J}|}
$$

Substituting $-\partial \bar{a}/\partial Y$ in (13.25), we get the Slutsky equation for $\bar{a}$, where the first term on the right is the *substitution effect* and the second term is the *income effect*.

$$
\frac{\partial \bar{a}}{\partial p_a} = -\frac{\bar{\lambda} p_b^2}{|\mathbf{J}|} - \bar{a}\left(\frac{\partial \bar{a}}{\partial Y} \right)
$$

$$
\underbrace{\qquad\qquad}_{\text{Substitution effect}} \qquad \underbrace{\qquad\qquad}_{\text{Income effect}}
$$

Since $|\mathbf{J}| = |\bar{\mathbf{H}}| > 0$ for constrained maximization and from (13.20),

$$
\bar{\lambda} = \frac{U_a}{p_a} = \frac{MU_a}{p_a} > 0
$$

the substitution effect in the first term is unambiguously negative. The income effect in the second term will depend on the nature of the good. For a normal good, $\partial \bar{a}/\partial Y > 0$ and the income effect above will be negative, making $\partial \bar{a}/\partial p_a < 0$. For a weakly inferior good, $\partial \bar{a}/\partial Y = 0$, and $\partial \bar{a}/\partial p_a < 0$. For a strictly inferior good, $\partial \bar{a}/\partial Y < 0$ and the sign of $\partial \bar{a}/\partial p_a$ will depend on the relative magnitude of the different effects. If the income effect overwhelms the substitution effect, as in the case of a *Giffen* good, $\partial \bar{a}/\partial p_a > 0$ and the demand curve will be positively sloped.

13.29. Derive the Slutsky equation for the effect of a change in p_b on the optimal quantity of the good demanded $\bar{b}$ and determine the sign of the comparative-static derivative $\partial \bar{b} / \partial p_b$.

From (13.22),
$$\frac{\partial \bar{b}}{\partial p_b} = -\frac{\bar{\lambda} p_a^2}{|\mathbf{J}|} + \frac{\bar{b}(p_b U_{aa} - p_a U_{ba})}{|\mathbf{J}|}$$

But from (13.24),
$$\frac{\partial \bar{b}}{\partial Y} = \frac{-(p_b U_{aa} - p_a U_{ba})}{|\mathbf{J}|}$$

Substituting above,
$$\frac{\partial \bar{b}}{\partial p_b} = -\frac{\bar{\lambda} p_a^2}{|\mathbf{J}|} - \bar{b}\left(\frac{\partial \bar{b}}{\partial Y}\right)$$

where the substitution effect in the first term on the right is unquestionably negative and the income effect in the second term will depend on the nature of the good, as in Problem 13.28.

THE ENVELOPE THEOREM

13.30. A firm in perfect competition with the production function $Q = f(K, L)$ and a production limit of Q_0 seeks to maximize profit

$$\pi = PQ - rK - wL$$

Assuming conditions are satisfied for the implicit function theorem, the Lagrangian function and the indirect objective function can be written, respectively,

$$\Pi(K, L, Q, \lambda; r, w, P, Q_0) = PQ(r, w, P, Q_0) - rK(r, w, P, Q_0) - wL(r, w, P, Q_0)$$
$$+ \lambda[Q_0 - f(K, L)]$$
$$\bar{\pi}(\bar{K}, \bar{L}, \bar{Q}; r, w, P, Q_0) = P\bar{Q}(r, w, P, Q_0) - r\bar{K}(r, w, P, Q_0) - w\bar{L}(r, w, P, Q_0)$$

Use the envelope theorem to find and comment on the changes in the indirect objective function signified by (a) $\partial \bar{\pi} / \partial r$, (b) $\partial \bar{\pi} / \partial w$, (c) $\partial \bar{\pi} / \partial P$, (d) $\partial \bar{\pi} / \partial Q_0$.

a)
$$\frac{\partial \bar{\pi}}{\partial r} = \frac{\partial \Pi}{\partial r} = -\bar{K}(r, w, P, Q_0)$$

b)
$$\frac{\partial \bar{\pi}}{\partial w} = \frac{\partial \Pi}{\partial w} = -\bar{L}(r, w, P, Q_0)$$

Differentiating the profit function with respect to input prices gives the firm's demand for inputs. Notice here where input prices (r, w) appear in the objective function and not in the constraint, the desired derivatives can be readily found from either function.

c)
$$\frac{\partial \bar{\pi}}{\partial P} = \frac{\partial \Pi}{\partial P} = \bar{Q}(r, w, P, Q_0)$$

Differentiating the profit function with respect to output prices gives the firm's supply function. Since output price (P) appears only in the objective function, the derivative can once again easily be found from either function.

d)
$$\frac{\partial \bar{\pi}}{\partial Q_0} = \frac{\partial \Pi}{\partial Q_0} = \bar{\lambda}$$

Differentiating the profit function with respect to an output constraint gives the marginal value of relaxing that constraint, i.e., the extra profit the firm could earn if it could increase output by one unit. Notice here where the output limit (Q_0) appears only in the constraint, the derivative can be derived more quickly and readily from the Lagrangian function.

13.31. A consumer wants to minimize the cost of attaining a specific level of utility:

$$c = p_x x + p_y y \qquad \text{subject to} \qquad u(x, y) = U_0$$

If the implicit function theorem conditions are fulfilled, the Lagrangian and indirect object functions are

$$C(x, y, \lambda; p_x, p_y, U_0) = p_x x(p_x, p_y, U_0) + p_y y(p_x, p_y, U_0) + \lambda[U_0 - u(x, y)]$$
$$\bar{c}(\bar{x}, \bar{y}; p_x, p_y, U_0) = p_x \bar{x}(p_x, p_y, U_0) + p_y \bar{y}(p_x, p_y, U_0)$$

Use the envelope theorem to find and comment on the changes in the indirect objective function signified by (a) $\partial \bar{c} / \partial p_x$, (b) $\partial \bar{c} / \partial p_y$, (c) $\partial \bar{c} / \partial U_0$.

a)
$$\frac{\partial \bar{c}}{\partial p_x} = \frac{\partial C}{\partial p_x} = \bar{x}(p_x, p_y, U_0)$$

b)
$$\frac{\partial \bar{c}}{\partial p_y} = \frac{\partial C}{\partial p_y} = \bar{y}(p_x, p_y, U_0)$$

In both cases, a change in the price of a good has a positive effect on the cost that is weighted by the amount of the good consumed. Since prices appear only in the objective function and not the constraint, the desired derivatives can be easily taken from either function.

c)
$$\frac{\partial \bar{c}}{\partial U_0} = \frac{\partial C}{\partial U_0} = \bar{\lambda}$$

Here $\bar{\lambda}$ measures the marginal cost of changing the given level of utility. Since the utility limit U_0 appears only in the constraint, the derivative is more easily found from the Lagrangian function.

13.32. Assume the model in Example 7 is a function with only a single exogenous variable a. Show that at the optimal solution the total derivative of the Lagrangian function with respect to a is equal to the partial derivative of the same Lagrangian function with respect to a.

The new Lagrangian function and first-order conditions are

$$Z(x, y, \lambda; a) = z[x(a), y(a); a] + \lambda(a) f[x(a), y(a); a]$$
$$Z_x = z_x(\bar{x}, \bar{y}; a) + \bar{\lambda} f_x(\bar{x}, \bar{y}; a) = 0$$
$$Z_y = z_y(\bar{x}, \bar{y}; a) + \bar{\lambda} f_y(\bar{x}, \bar{y}; a) = 0$$
$$Z_\lambda = f(\bar{x}, \bar{y}; a) = 0$$

Taking the total derivative of the original Lagrangian function with respect to a,

$$\frac{dZ}{da} = (z_x + \bar{\lambda} f_x) \frac{d\bar{x}}{da} + (z_y + \bar{\lambda} f_y) \frac{d\bar{y}}{da} + f \frac{d\bar{\lambda}}{da} + (z_a + \bar{\lambda} f_a)$$

But from the first-order conditions,

$$z_x + \bar{\lambda} f_x = 0, \qquad z_y + \bar{\lambda} f_y = 0, \qquad \text{and} \qquad f = 0$$

so the first three terms cancel and the total derivative of the function with respect to the exogenous variable a ends up equal to the partial derivative of the function with respect to the exogenous variable a:

$$\frac{dZ}{da} = (z_a + \bar{\lambda} f_a) = \frac{\partial Z}{\partial a}$$

This suggests we can find the total effect of a change in a single exogenous variable on the optimal value of the Lagrangian function by simply taking the partial derivative of the Lagrangian function with respect to that exogenous variable.

CONCAVE PROGRAMMING

13.33. Given the typical format for constrained optimization of general functions below,

$$(a) \text{ maximize } f(x, y) \text{ subject to } g(x, y) \leq B$$
$$(b) \text{ minimize } f(x, y) \text{ subject to } g(x, y) \geq B$$

express them in suitable form for concave programming and write out the Lagrangians.

a) For less than or equal to constraints in maximization problems, subtract the variables in the constraint from the constant of the constraint.

$$\text{Maximize } f(x, y) \quad \text{subject to} \quad B - g(x, y) \geq 0 \quad \bar{x}, \bar{y} \geq 0$$
$$\text{Max } F = f(x, y) + \lambda[B - g(x, y)]$$

b) For minimization, multiply the objective function by -1 to make it negative and then maximize the negative of the original function. For the corresponding greater than or equal to constraints in minimization problems, subtract the constant of the constraint from the variables in the constraint.

$$\text{Maximize } -f(x, y) \quad \text{subject to} \quad g(x, y) - B \geq 0 \quad \bar{x}, \bar{y} \geq 0$$
$$\text{Max } F = -f(x, y) + \lambda[g(x, y) - B]$$

13.34. Assume a firm with the production function $Q(K, L)$ and operating in a purely competitive market for inputs wishes to minimize cost while producing *no less than* a specific amount of output, given by

$$\text{minimize } c = rK + wL \quad \text{subject to} \quad Q(K, L) \geq Q_0$$

(a) Express the problem in concave-programming format, (b) write out the Lagrangian function, and (c) solve the problem.

a) Multiplying the objective function by -1 to make it negative and maximizing it,

$$\text{Maximize } -rK - wL \quad \text{subject to} \quad Q(K, L) - Q_0 \geq 0 \quad K, L \geq 0$$

b) $$\text{Maximize } C = -rK - wL + \lambda[Q(K, L) - Q_0]$$

c) Applying the Kuhn-Tucker conditions, we first *maximize* the Lagrangian function with respect to the choice variables K and L and check the related conditions.

1. a) $$\frac{\partial C}{\partial K} = -r + \bar{\lambda}Q_K \leq 0 \qquad \frac{\partial C}{\partial L} = -w + \bar{\lambda}Q_L \leq 0$$

   b) $$\bar{K} \geq 0 \qquad\qquad\qquad \bar{L} \geq 0$$

   c) $$\bar{K}(-r + \bar{\lambda}Q_K) = 0 \qquad \bar{L}(-w + \bar{\lambda}Q_L) = 0$$

We then *minimize* the Lagrangian with respect to the constraint variable λ and check the related conditions.

2. a) $$\frac{\partial C}{\partial \lambda} = Q(\bar{K}, \bar{L}) - Q_0 \geq 0$$

   b) $$\bar{\lambda} \geq 0$$

   c) $$\bar{\lambda}[Q(\bar{K}, \bar{L}) - Q_0] = 0$$

Assuming production depends on both inputs, $\bar{K}, \bar{L} > 0$, the two expressions within the parentheses in 1(c) have to be equalities.

$$-r + \bar{\lambda}Q_K = 0 \qquad -w + \bar{\lambda}Q_L = 0$$

Rearranging, we see

$$\bar{\lambda} = \frac{r}{Q_K} = \frac{w}{Q_L} > 0 \qquad\qquad\qquad (13.26)$$

since input prices $r, w > 0$ and marginal productivities $Q_K, Q_L > 0$. With $\bar{\lambda} > 0$, from 2(c) the budget constraint binds as an equality, with

$$Q(\bar{K}, \bar{L}) = Q_0$$

Rearranging (13.26), we have an interior solution with the isoquant tangent to the isocost line,

$$\frac{Q_L}{Q_K} = \frac{w}{r}$$

13.35. Maximize profits

$$\pi = 64x - 2x^2 + 96y - 4y^2 - 13$$

subject to the production constraint

$$x + y \leq 20$$

We first set up the Lagrangian function

$$\Pi = 64x - 2x^2 + 96y - 4y^2 - 13 + \lambda(20 - x - y)$$

and set down the Kuhn-Tucker conditions.

1. a) $\quad \Pi_x = 64 - 4\bar{x} - \bar{\lambda} \leq 0 \qquad \Pi_y = 96 - 8\bar{y} - \bar{\lambda} \leq 0$
   b) $\quad \bar{x} \geq 0 \qquad\qquad\qquad \bar{y} \geq 0$
   c) $\quad \bar{x}(64 - 4\bar{x} - \bar{\lambda}) = 0 \qquad \bar{y}(96 - 8\bar{y} - \bar{\lambda}) = 0$

2. a) $\quad \Pi_\lambda = 20 - \bar{x} - \bar{y} \geq 0$
   b) $\quad \bar{\lambda} \geq 0$
   c) $\quad \bar{\lambda}(20 - \bar{x} - \bar{y}) = 0$

We then test the Kuhn-Tucker conditions methodically.

1. Check the possibility that $\bar{\lambda} = 0$ or $\bar{\lambda} > 0$.
   If $\bar{\lambda} = 0$, then from 1(a),

$$64 - 4\bar{x} \leq 0 \qquad 96 - 8\bar{y} \leq 0$$

Therefore, $\qquad\qquad\qquad 4\bar{x} \geq 64 \qquad\qquad 8\bar{y} \geq 96$

$$\bar{x} \geq 16 \qquad\qquad \bar{y} \geq 12$$

But this violates the initial constraint since $\bar{x} + \bar{y} = 28 > 20$. Hence $\bar{\lambda} \neq 0$ and from 2(b) we conclude $\bar{\lambda} > 0$.

2. If $\bar{\lambda} > 0$, from 2(c), the constraint holds as an equality, with

$$20 - \bar{x} - \bar{y} = 0$$

3. Next check to see if either of the choice variables $\bar{x}$ or $\bar{y}$ can equal zero.

   a) If $\bar{x} = 0$, $\bar{y} = 20$ and the second condition in 1(c) is violated.

$$20[96 - 8(20) - (\bar{\lambda} > 0)] \neq 0$$

   b) If $\bar{y} = 0$, $\bar{x} = 20$ and the first condition in 1(c) is violated.

$$20[64 - 4(20) - (\bar{\lambda} > 0)] \neq 0$$

So neither choice variable can equal zero and from 1(b),

$$\text{if } \bar{x} \neq 0, \bar{x} > 0 \qquad \text{and} \qquad \text{if } \bar{y} \neq 0, \bar{y} > 0$$

4. If $\bar{x}$, $\bar{y}$, $\bar{\lambda} > 0$, then from 1(c) and 2(c), the following equalities hold,

$$64 - 4\bar{x} - \bar{\lambda} = 0$$
$$96 - 8\bar{y} - \bar{\lambda} = 0$$
$$20 - \bar{x} - \bar{y} = 0$$

Setting them down in matrix form,

$$\begin{bmatrix} -4 & 0 & -1 \\ 0 & -8 & -1 \\ -1 & -1 & 0 \end{bmatrix} \begin{bmatrix} \bar{x} \\ \bar{y} \\ \bar{\lambda} \end{bmatrix} = \begin{bmatrix} -64 \\ -96 \\ -20 \end{bmatrix}$$

and solving by Cramer's rule where $|\mathbf{A}| = 12$, $|\mathbf{A}_1| = 128$, $|\mathbf{A}_2| = 112$, and $|\mathbf{A}_3| = 256$, we get the solution:

$$\bar{x} = 10.67, \quad \bar{y} = 9.33, \quad \text{and} \quad \bar{\lambda} = 21.33$$

which we know to be optimal because none of the Kuhn-Tucker conditions is violated. With $\bar{\lambda} = 21.33$, a unit increase in the constant of the production constraint will cause profits to increase by approximately 21.33.

13.36. Maximize the profit function in Problem 13.35,

$$\pi = 64x - 2x^2 + 96y - 4y^2 - 13$$

subject to the new production constraint

$$x + y \leq 36$$

The Lagrangian function and the Kuhn-Tucker conditions are

$$\Pi = 64x - 2x^2 + 96y - 4y^2 - 13 + \lambda(36 - x - y)$$

1. $a)$ $\qquad \Pi_x = 64 - 4\bar{x} - \bar{\lambda} \leq 0 \qquad \Pi_y = 96 - 8\bar{y} - \bar{\lambda} \leq 0$

   $b)$ $\qquad \bar{x} \geq 0 \qquad\qquad\qquad\qquad \bar{y} \geq 0$

   $c)$ $\qquad \bar{x}(64 - 4\bar{x} - \bar{\lambda}) = 0 \qquad \bar{y}(96 - 8\bar{y} - \bar{\lambda}) = 0$

2. $a)$ $\qquad\qquad \Pi_\lambda = 36 - \bar{x} - \bar{y} \geq 0$

   $b)$ $\qquad\qquad \bar{\lambda} \geq 0$

   $c)$ $\qquad\qquad \bar{\lambda}(36 - \bar{x} - \bar{y}) = 0$

We then check the Kuhn-Tucker conditions systematically.

1. Test the possibility of $\bar{\lambda} = 0$ or $\bar{\lambda} > 0$.
    If $\bar{\lambda} = 0$, then from 1(a),

$$64 - 4\bar{x} \leq 0 \qquad 96 - 8\bar{y} \leq 0$$

Therefore, $\qquad\qquad\qquad\qquad \bar{x} \geq 16 \qquad \bar{y} \geq 12$

Since $\bar{x} + \bar{y} = 28 < 36$, no condition is violated. Therefore, it is possible that $\bar{\lambda} = 0$ or $\bar{\lambda} > 0$.

2. Now check to see if either of the choice variables $\bar{x}$ or $\bar{y}$ can equal zero.

   $a)$ If $\bar{x} = 0$, $\bar{y} = 36$, and the second condition in 1(c) is violated.

$$36[96 - 8(36) - (\bar{\lambda} \geq 0)] \neq 0$$

   $b)$ If $\bar{y} = 0$, $\bar{x} = 36$, and the first condition in 1(c) is violated.

$$36[64 - 4(36) - (\bar{\lambda} \geq 0)] \neq 0$$

Therefore, neither choice variable can equal zero and from 1(b),

$$\bar{x} > 0 \quad \text{and} \quad \bar{y} > 0$$

3.   Check the solutions when (a) $\bar{\lambda} > 0$ and (b) $\bar{\lambda} = 0$.

a)   If $\bar{\lambda}, \bar{x}, \bar{y} > 0$, then from the Kuhn-Tucker conditions listed under (c),

$$64 - 4\bar{x} - \bar{\lambda} = 0$$
$$96 - 8\bar{y} - \bar{\lambda} = 0$$
$$36 - x - \bar{y} = 0$$

In matrix form,

$$\begin{bmatrix} -4 & 0 & -1 \\ 0 & -8 & -1 \\ -1 & -1 & 0 \end{bmatrix} \begin{bmatrix} \bar{x} \\ \bar{y} \\ \bar{\lambda} \end{bmatrix} = \begin{bmatrix} -64 \\ -96 \\ -36 \end{bmatrix}$$

Using Cramer's rule where $|\mathbf{A}| = 12$, $|\mathbf{A}_1| = 256$, $|\mathbf{A}_2| = 176$, and $|\mathbf{A}_3| = -256$, we get

$$\bar{x} = 21.33, \quad \bar{y} = 14.67, \quad \text{and} \quad \bar{\lambda} = -21.33$$

which cannot be optimal because $\bar{\lambda} < 0$ in violation of 2(b) of the Kuhn-Tucker conditions. With $\bar{\lambda} < 0$, from 2(c), the constraint is a strict equality and decreasing the level of output will increase the level of profit.

b)   If $\bar{\lambda} = 0$ and $\bar{x}, \bar{y} > 0$, then from 1(c),

$$64 - 4\bar{x} = 0, \quad \bar{x} = 16$$
$$96 - 8\bar{y} = 0, \quad \bar{y} = 12$$

This gives us the optimal solution, $\bar{x} = 16$, $\bar{y} = 12$, and $\bar{\lambda} = 0$, which we know is optimal because it violates none of the Kuhn-Tucker conditions. With $\bar{\lambda} = 0$, the constraint is nonbinding as we see from the optimal solution $\bar{x} + \bar{y} = 28 < 36$.

13.37.   Minimize cost

$$c = 5x^2 - 80x + y^2 - 32y$$

subject to

$$x + y \geq 30$$

Multiplying the objective function by -1 and setting up the Lagrangian, we have

$$\text{Max } C = -5x^2 + 80x - y^2 + 32y + \lambda(x + y - 30)$$

where the Kuhn-Tucker conditions are,

1.   a) $C_x = -10\bar{x} + 80 + \bar{\lambda} \leq 0$ $C_y = -2\bar{y} + 32 + \bar{\lambda} \leq 0$

     b) $\bar{x} \geq 0$ $\bar{y} \geq 0$

     c) $\bar{x}(-10\bar{x} + 80 + \bar{\lambda}) = 0$ $\bar{y}(-2\bar{y} + 32 + \bar{\lambda}) = 0$

2.   a) $C_\lambda = \bar{x} + \bar{y} - 30 \geq 0$

     b) $\bar{\lambda} \geq 0$

     c) $\bar{\lambda}(\bar{x} + \bar{y} - 30) = 0$

1.   Check the possibility of $\bar{\lambda} = 0$.
     If $\bar{\lambda} = 0$, then from 1(a),

$$-10\bar{x} + 80 \leq 0 \qquad -2\bar{y} + 32 \leq 0$$

Therefore, $\bar{x} \geq 8$ $\bar{y} \geq 16$

But this is a violation because $\bar{x} + \bar{y} \geq 24$ fails to satisfy the initial constraint $\bar{x} + \bar{y} \geq 30$. So $\bar{\lambda} > 0$.

2.   Check to see if $\bar{x}$ or $\bar{y}$ can equal zero.
     From 1(a), if $\bar{x} = 0$, $\bar{\lambda} \leq -80$, and if $\bar{y} = 0$, $\bar{\lambda} \leq -32$, both of which violate the nonnegativity constraint on variables. So $\bar{x}, \bar{y} > 0$.

3. Now check the Kuhn-Tucker conditions when $\bar{x}, \bar{y} > 0$ and $\bar{\lambda} > 0$. If $\bar{\lambda} > 0$ and $\bar{x}, \bar{y} > 0$, all the first partials are strict equalities and we have

$$\begin{bmatrix} -10 & 0 & 1 \\ 0 & -2 & 1 \\ 1 & 1 & 0 \end{bmatrix} \begin{bmatrix} \bar{x} \\ \bar{y} \\ \bar{\lambda} \end{bmatrix} = \begin{bmatrix} -80 \\ -32 \\ 30 \end{bmatrix}$$

where $|\mathbf{A}| = 12$, $|\mathbf{A}_1| = 109$, $|\mathbf{A}_2| = 252$, and $|\mathbf{A}_3| = 440$, giving the optimal solution, which violates none of the Kuhn-Tucker conditions:

$$\bar{x} = 9, \ \bar{y} = 21, \ \bar{\lambda} = 36.67$$

13.38. Minimize the same function as above,

$$5x^2 - 80x + y^2 - 32y$$

subject to a new constraint,

$$x + y \geq 20$$

The Lagrangian function and Kuhn-Tucker conditions are,

$$\text{Max } C = -5x^2 + 80x - y^2 + 32y + \lambda(x + y - 20)$$

1. *a)* $C_x = -10\bar{x} + 80 + \bar{\lambda} \leq 0$ $C_y = -2\bar{y} + 32 + \bar{\lambda} \leq 0$

     b) $\bar{x} \geq 0$ $\bar{y} \geq 0$

     c) $\bar{x}(-10\bar{x} + 80 + \bar{\lambda}) = 0$ $\bar{y}(-2\bar{y} + 32 + \bar{\lambda}) = 0$

2. *a)* $C_\lambda = \bar{x} + \bar{y} - 20 \geq 0$

     b) $\bar{\lambda} \geq 0$

     c) $\bar{\lambda}(\bar{x} + \bar{y} - 20) = 0$

1. Check the possibility of $\bar{\lambda} = 0$.
    If $\bar{\lambda} = 0$, then from 1(*a*),

$$-10\bar{x} + 80 \leq 0 \qquad\qquad -2\bar{y} + 32 \leq 0$$

Therefore, $\bar{x} \geq 8$ $\bar{y} \geq 16$

This violates no constraint because $\bar{x} + \bar{y} \geq 24$ satisfies $\bar{x} + \bar{y} \geq 20$. So $\bar{\lambda} = 0$ or $\bar{\lambda} > 0$.

2. By the same reasoning as in step 2 above, we can show $\bar{x}, \bar{y} > 0$.

3. So we are left with two possibilities, depending on whether $\bar{\lambda} = 0$, or $\bar{\lambda} > 0$.

    a) If $\bar{\lambda} > 0$ and $\bar{x}, \bar{y} > 0$, all the derivatives are strict equalities and we have

$$\begin{bmatrix} -10 & 0 & 1 \\ 0 & -2 & 1 \\ 1 & 1 & 0 \end{bmatrix} \begin{bmatrix} \bar{x} \\ \bar{y} \\ \bar{\lambda} \end{bmatrix} = \begin{bmatrix} -80 \\ -32 \\ 20 \end{bmatrix}$$

With $|\mathbf{A}| = 12$, $|\mathbf{A}_1| = 88$, $|\mathbf{A}_2| = 152$, and $|\mathbf{A}_3| = -80$, the solution is

$$\bar{x} = 7.33, \ \bar{y} = 12.67, \ \bar{\lambda} = -6.67$$

Since $\bar{\lambda}$ is negative, condition 2(*b*) is violated and the solution is nonoptimal. The solution suggests we can reduce the cost by increasing output.

    b) If $\bar{\lambda} = 0$ and $\bar{x}, \bar{y} > 0$, from 1(*c*),

$$-10\bar{x} + 80 = 0 \qquad\qquad -2\bar{y} + 32 = 0$$

$$\bar{x} = 8 \qquad\qquad\qquad \bar{y} = 16$$

This satisfies the new constraint $x + y \geq 20$ and violates none of the Kuhn-Tucker conditions, so the optimal solution is

$$\bar{x} = 8, \ \bar{y} = 16, \ \bar{\lambda} = 0$$

With $\bar{\lambda} = 0$, the constraint is nonbinding. The firm can minimize costs while exceeding its production quota.

13.39. Maximize the utility function,

$$u = xy$$

subject to the following budget and dietary constraints,

$$3x + 4y \leq 144 \qquad \text{budget constraint}$$
$$5x + 2y \leq 120 \qquad \text{dietary constraint}$$

The Lagrangian function and Kuhn-Tucker conditions are

$$\text{Max } U = xy + \lambda_1(144 - 3x - 4y) + \lambda_2(120 - 5x - 2y)$$

1. *a)* $U_x = \bar{y} - 3\bar{\lambda}_1 - 5\bar{\lambda}_2 \leq 0$ $U_y = \bar{x} - 4\bar{\lambda}_1 - 2\bar{\lambda}_2 \leq 0$

   b) $\bar{x} \geq 0$ $\bar{y} \geq 0$

   c) $\bar{x}(\bar{y} - 3\bar{\lambda}_1 - 5\bar{\lambda}_2) = 0$ $\bar{y}(\bar{x} - 4\bar{\lambda}_1 - 2\bar{\lambda}_2) = 0$

2. *a)* $U_{\lambda_1} = 144 - 3\bar{x} - 4\bar{y} \geq 0$ $U_{\lambda_2} = 120 - 5\bar{x} - 2\bar{y} \geq 0$

   b) $\bar{\lambda}_1 \geq 0$ $\bar{\lambda}_2 \geq 0$

   c) $\bar{\lambda}_1(144 - 3\bar{x} - 4\bar{y}) = 0$ $\bar{\lambda}_2(120 - 5\bar{x} - 2\bar{y}) = 0$

Given the nature of the objective function, we assume neither of the choice variables, $\bar{x}, \bar{y}$, can equal zero. Otherwise the utility function, $u = xy = 0$. If $\bar{x}, \bar{y} > 0$, from 1(*c*),

$$\bar{y} - 3\lambda_1 - 5\lambda_2 = 0$$
$$\bar{x} - 4\lambda_1 - 2\lambda_2 = 0$$

Noting that the $MU_x = u_x = \bar{y}$, $MU_y = u_y = \bar{x}$, and assuming MU's > 0, both λ's cannot equal zero, so at least one of the constraints must be binding. This leaves us with three possibilities: (*a*) $\bar{\lambda}_1 > 0$, $\bar{\lambda}_2 = 0$, (*b*) $\bar{\lambda}_1 = 0$, $\bar{\lambda}_2 > 0$, and (*c*) $\bar{\lambda}_1 > 0$, $\bar{\lambda}_2 > 0$. We examine each in turn.

a) $\bar{\lambda}_1 > 0$, $\bar{\lambda}_2 = 0$, $\bar{x}, \bar{y} > 0$. From the 1(*c*) and 2(*c*) conditions we have four equalities, three of which still call for solution.

$$\bar{y} - 3\bar{\lambda}_1 - 5\bar{\lambda}_2 = 0$$
$$\bar{x} - 4\bar{\lambda}_1 - 2\bar{\lambda}_2 = 0$$
$$144 - 3\bar{x} - 4\bar{y} = 0$$
$$\bar{\lambda}_2 = 0$$

Putting the latter in matrix form,

$$\begin{bmatrix} 0 & 1 & -3 \\ 1 & 0 & -4 \\ -3 & -4 & 0 \end{bmatrix} \begin{bmatrix} \bar{x} \\ \bar{y} \\ \bar{\lambda}_1 \end{bmatrix} = \begin{bmatrix} 0 \\ 0 \\ -144 \end{bmatrix}$$

and solving by Cramer's rule, where $|\mathbf{A}| = 24$, $|\mathbf{A}_1| = 576$, $|\mathbf{A}_2| = 432$, and $|\mathbf{A}_3| = 144$, we have

$$\bar{x} = 24, \ \bar{y} = 18, \ \bar{\lambda}_1 = 6$$

But checking the answer for internal consistency, we find

$$U_{\lambda_2} = 120 - 5(24) - 2(18) = -36 < 0$$

This is clearly a violation of the Kuhn-Tucker conditions which require $\partial U / \partial \lambda_i \geq 0$ and hence the solution is not optimal.

b) $\bar{\lambda}_1 = 0$, $\bar{\lambda}_2 > 0$, $\bar{x}, \bar{y} > 0$. From $1(c)$ and $2(c)$, we have four equalities, three of which must still be solved.

$$\bar{y} - 3\bar{\lambda}_1 - 5\bar{\lambda}_2 = 0$$
$$\bar{x} - 4\bar{\lambda}_1 - 2\bar{\lambda}_2 = 0$$
$$\bar{\lambda}_1 = 0$$
$$120 - 5\bar{x} - 2\bar{y} = 0$$

Setting the latter in matrix form and solving by Cramer's rule,

$$\begin{bmatrix} 0 & 1 & -5 \\ 1 & 0 & -2 \\ -5 & -2 & 0 \end{bmatrix} \begin{bmatrix} \bar{x} \\ \bar{y} \\ \bar{\lambda}_2 \end{bmatrix} = \begin{bmatrix} 0 \\ 0 \\ -120 \end{bmatrix}$$

we have $|\mathbf{A}| = 20$, $|\mathbf{A}_1| = 240$, $|\mathbf{A}_2| = 600$, and $|\mathbf{A}_3| = 120$, and

$$\bar{x} = 12, \ \bar{y} = 30, \ \bar{\lambda}_2 = 6$$

But checking for internal consistency once again, we find

$$U_{\lambda_1} = 144 - 3(12) - 4(30) = -12 < 0$$

This violates the Kuhn-Tucker conditions and so the solution cannot be optimal.

c) $\bar{\lambda}_1 > 0$, $\bar{\lambda}_2 > 0$, $\bar{x}, \bar{y} > 0$. From $1(c)$ and $2(c)$, all four derivatives are strict equalities which we set down immediately in matrix form.

$$\begin{bmatrix} 0 & 1 & -3 & -5 \\ 1 & 0 & -4 & -2 \\ -3 & -4 & 0 & 0 \\ -5 & -2 & 0 & 0 \end{bmatrix} \begin{bmatrix} \bar{x} \\ \bar{y} \\ \bar{\lambda}_1 \\ \bar{\lambda}_2 \end{bmatrix} = \begin{bmatrix} 0 \\ 0 \\ -144 \\ -120 \end{bmatrix}$$

From Cramer's rule, where $|\mathbf{A}| = 196$, $|\mathbf{A}_1| = 2688$, $|\mathbf{A}_2| = 5040$, $|\mathbf{A}_3| = 240$, and $|\mathbf{A}_4| = 864$, we find the optimal solution, which violates none of the conditions:

$$\bar{x} = 13.71, \ \bar{y} = 25.71, \ \bar{\lambda}_1 = 1.22, \ \bar{\lambda}_2 = 4.41$$

13.40. Confirm the results of Problem 13.39 with (a) a graph and (b) explain what the graph illustrates.

a) See Fig. 13-3.

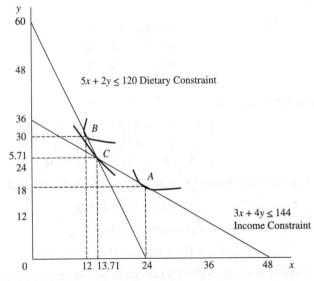

Fig. 13-3

b) At point A, $\bar{x} = 24$, $\bar{y} = 18$, the income constraint is exactly fulfilled, but the dietary constraint is violated. Hence A cannot be optimal. At point B, $\bar{x} = 12$, $\bar{y} = 30$, the dietary constraint is exactly fulfilled, but the income constraint is violated. Hence B cannot be optimal. At point C, $\bar{x} = 13.71$, $\bar{y} = 25.71$, and both the income constraint and the dietary constraint are binding with no Kuhn-Tucker condition violated. Hence C is optimal and note that it occurs exactly at the intersection of the two constraints.

13.41. Following a fair-rate-of-return policy, regulators of natural monopolies, such as utility companies, restrict profits to a certain fixed proportion of the capital employed. The policy, however, leads to a distortion of inputs on the part of regulated industries that has been termed the Averch-Johnson effect. Assuming a firm wishes to maximize profits

$$\pi(K, L) = E(K, L) - rK - wL$$

subject to the fair-rate-of-return constraint,

$$E(K, L) - wL \leq mK$$

where E = earnings and $E_K, E_L > 0$, $E_{KK}, E_{LL} < 0$, and $E_{KK}E_{LL} > E_{KL}E_{LK}$, r = cost of capital, w = wage, m = maximum rate of return on capital, and $m - r > 0$, use concave programming to demonstrate the possibility of a distorting effect predicted by Averch-Johnson.

The Lagrangian function and the Kuhn-Tucker conditions are

$$\Pi(K, L) = E(K, L) - rK - wL + \lambda[mK - E(K, L) + wL]$$

1. *a)* $\Pi_K = E_K - r + \bar{\lambda}m - \bar{\lambda}E_K \leq 0$ $\qquad$ $\Pi_L = E_L - w - \bar{\lambda}E_L + \bar{\lambda}w \leq 0$

    b) $\bar{K} \geq 0$ $\qquad\qquad\qquad\qquad\qquad\qquad$ $\bar{L} \geq 0$

    c) $\bar{K}(E_K - r + \bar{\lambda}m - \bar{\lambda}E_K) = 0$ $\qquad$ $\bar{L}(E_L - w - \bar{\lambda}E_L + \bar{\lambda}w) = 0$

2. *a)* $\qquad\qquad\qquad\qquad\qquad$ $\Pi_\lambda = m\bar{K} + w\bar{L} - E(\bar{K}, \bar{L}) \geq 0$

    b) $\qquad\qquad\qquad\qquad\qquad$ $\bar{\lambda} \geq 0$

    c) $\qquad\qquad\qquad\qquad\qquad$ $\bar{\lambda}[m\bar{K} + w\bar{L} - E(\bar{K}, \bar{L})] = 0$

Making the common sense assumption that $\bar{K}, \bar{L} \geq 0$, adding and subtracting $\bar{\lambda}r$ within the parentheses of the equation on the left in 1(c), and rearranging we have,

$$(1 - \bar{\lambda})(E_K - r) + \bar{\lambda}(m - r) = 0 \tag{13.27}$$

$$(1 - \bar{\lambda})(E_L - w) = 0 \tag{13.28}$$

$$m\bar{K} + w\bar{L} - E(\bar{K}, \bar{L}) \geq 0$$

$$\bar{\lambda}[m\bar{K} + w\bar{L} - E(\bar{K}, \bar{L})] = 0$$

If $\bar{\lambda} = 1$ in (13.27), $m - r = 0$ and this would contradict the assumption that the maximum allowable rate of return is greater than the cost of capital $m > r$. So $\bar{\lambda} \neq 1$. But if $\bar{\lambda} \neq 1$, from (13.27),

$$E_K = r - \frac{\bar{\lambda}(m - r)}{(1 - \bar{\lambda})} \tag{13.29}$$

and from (13.28),

$$E_L = w \tag{13.30}$$

Dividing (13.29) by (13.30),

$$\frac{E_K}{E_L} = \frac{r - \left[\dfrac{\bar{\lambda}(m - r)}{(1 - \bar{\lambda})}\right]}{w} = \frac{r}{w} - \frac{\bar{\lambda}(m - r)}{(1 - \bar{\lambda})w} \tag{13.31}$$

If $\bar{\lambda} = 0$, the constraint is not binding and we have the *unregulated* optimum,

$$\frac{E_K}{E_L} = \frac{r}{w}$$

where $E_K = MRP_K$ and $E_L = MRP_L$. Dividing numerator and denominator on the left by the common output price, the above expression is equivalent to the familiar,

$$\frac{MP_K}{MP_L} = \frac{r}{w}$$

But if $\bar{\lambda} \neq 0$ and the constraint is binding, regulation interferes with the economically optimal solution in which the ratio of marginal products exactly equals the ratio of respective prices. Thus if $\bar{\lambda} \neq 0$, there will be a distorting effect on the profit-maximizing level of output.

13.42. (*a*) Specify the direction of the distortion the Averch-Johnson effect predicts and (*b*) demonstrate the conditions necessary to verify it in terms of the previous model.

a)   The Averch-Johnson effect predicts the distortion will lead to a higher K/L ratio than in an unregulated market. If more than the optimal amount of capital is used, the marginal productivity of capital will be diminished and the result predicted by the Averch-Johnson effect will be

$$\frac{MP_K}{MP_L} < \frac{r}{w}$$

b)   In terms of (*13.31*), the bias towards greater than optimal capital intensity will be true whenever

$$\frac{\bar{\lambda}(m - r)}{(1 - \bar{\lambda})w} > 0 \tag{13.32}$$

Since we know $r > 0$, $w > 0$, and by the assumption from common practice, $m - r > 0$, (*13.32*) will be positive whenever $\bar{\lambda} < 1$. To determine the sign of $\bar{\lambda}$, we revert to comparative-static techniques.

Having determined that $\bar{x}, \bar{y}, \bar{\lambda} \neq 0$, we know that all three partial derivatives in the Kuhn-Tucker conditions must hold as equalities:

$$(1 - \bar{\lambda})E_K - r + \bar{\lambda}m = 0$$
$$(1 - \bar{\lambda})(E_L - w) = 0$$
$$m\bar{K} + w\bar{L} - E(\bar{K}, \bar{L}) = 0$$

These equations are the same as the first-order conditions we would obtain if we had maximized the original function subject to the equality constraint:

$$\text{Max } \pi(K, L) = E(K, L) - rK - wL$$

subject to $$E(K, L) - wL = mK$$

The second-order conditions for maximization require that the Bordered Hessian be positive:

$$|\bar{\mathbf{H}}| = \begin{vmatrix} (1 - \bar{\lambda})E_{KK} & (1 - \bar{\lambda})E_{KL} & m - E_K \\ (1 - \bar{\lambda})E_{LK} & (1 - \bar{\lambda})E_{LL} & w - E_L \\ m - E_K & w - E_L & 0 \end{vmatrix} > 0$$

Expanding along the third row,

$$|\mathbf{H}| = (m - E_K)[(w - E_L)(1 - \bar{\lambda})E_{KL} - (m - E_K)(1 - \bar{\lambda})E_{LL}]$$
$$- (w - E_L)[(w - E_L)(1 - \bar{\lambda})E_{KK} - (m - E_K)(1 - \bar{\lambda})E_{LK}]$$

Since $w = E_L$ at the optimum, all the $(w - E_L)$ terms $= 0$, leaving

$$|\bar{\mathbf{H}}| = -(m - E_K)^2(1 - \bar{\lambda})E_{LL}$$

For $|\mathbf{H}| > 0$,

$$(1 - \bar{\lambda})E_{LL} < 0$$

Since $E_{LL} < 0$ from our earlier assumption of strict concavity, it follows that

$$\bar{\lambda} < 1$$

With $\bar{\lambda} < 1$, $(13.32) > 0$ and from (13.31),

$$\frac{MP_K}{MP_L} < \frac{r}{w} \qquad \text{Q.E.D.}$$

CHAPTER 14

Integral Calculus: The Indefinite Integral

14.1 INTEGRATION

Chapters 3 to 6 were devoted to differential calculus, which measures the rate of change of functions. Differentiation, we learned, is the process of finding the derivative $F'(x)$ of a function $F(x)$. Frequently in economics, however, we know the rate of change of a function $F'(x)$ and want to find the original function. Reversing the process of differentiation and finding the original function from the derivative is called *integration*, or *antidifferentiation*. The original function $F(x)$ is called the *integral*, or *antiderivative*, of $F'(x)$.

EXAMPLE 1.   Letting $f(x) = F'(x)$ for simplicity, the antiderivative of $f(x)$ is expressed mathematically as

$$\int f(x)\,dx = F(x) + c$$

Here the left-hand side of the equation is read, "the *indefinite integral* of f of x with respect to x." The symbol $\int$ is an *integral sign*, $f(x)$ is the *integrand*, and c is the *constant of integration*, which is explained in Example 3.

14.2 RULES OF INTEGRATION

The following rules of integration are obtained by reversing the corresponding rules of differentiation. Their accuracy is easily checked, since the derivative of the integral must equal the integrand. Each rule is illustrated in Example 2 and Problems 14.1 to 14.6.

Rule 1.   The integral of a constant k is

$$\int k\,dx = kx + c$$

Rule 2. The integral of 1, written simply as dx, not $1\,dx$, is

$$\int dx = x + c$$

Rule 3. The integral of a power function x^n, where $n \neq -1$, is given by the *power rule*:

$$\int x^n \, dx = \frac{1}{n+1} x^{n+1} + c \qquad n \neq -1$$

Rule 4. The integral of x^{-1} (or $1/x$) is

$$\int x^{-1} \, dx = \ln x + c \qquad x > 0$$

The condition $x > 0$ is added because only positive numbers have logarithms. For negative numbers,

$$\int x^{-1} \, dx = \ln |x| + c \qquad x \neq 0$$

Rule 5. The integral of an exponential function is

$$\int a^{kx} \, dx = \frac{a^{kx}}{k \ln a} + c$$

Rule 6. The integral of a natural exponential function is

$$\int e^{kx} \, dx = \frac{e^{kx}}{k} + c \qquad \text{since} \qquad \ln e = 1$$

Rule 7. The integral of a constant times a function equals the constant times the integral of the function.

$$\int k f(x) \, dx = k \int f(x) \, dx$$

Rule 8. The integral of the sum or difference of two or more functions equals the sum or difference of their integrals.

$$\int [f(x) + g(x)] \, dx = \int f(x) \, dx + \int g(x) \, dx$$

Rule 9. The integral of the negative of a function equals the negative of the integral of that function.

$$\int -f(x) \, dx = - \int f(x) \, dx$$

EXAMPLE 2. The rules of integration are illustrated below. Check each answer on your own by making sure that the derivative of the integral equals the integrand.

i)
$$\int 3 \, dx = 3x + c$$
(Rule 1)

ii)
$$\int x^2 \, dx = \frac{1}{2+1} x^{2+1} + c = \frac{1}{3} x^3 + c$$
(Rule 3)

iii)

$$\int 5x^4\, dx = 5 \int x^4\, dx \qquad\qquad \text{(Rule 7)}$$

$$= 5\left(\frac{1}{5}x^5 + c_1\right) \qquad\qquad \text{(Rule 3)}$$

$$= x^5 + c$$

where c_1 and c are arbitrary constants and $5c_1 = c$. Since c is an arbitrary constant, it can be ignored in the preliminary calculation and included only in the final solution.

iv)

$$\int (3x^3 - x + 1)\, dx = 3 \int x^3\, dx - \int x\, dx + \int dx \qquad\qquad \text{(Rules 7, 8, and 9)}$$

$$= 3(\tfrac{1}{4}x^4) - \tfrac{1}{2}x^2 + x + c \qquad\qquad \text{(Rules 2 and 3)}$$

$$= \tfrac{3}{4}x^4 - \tfrac{1}{2}x^2 + x + c$$

v)

$$\int 3x^{-1}\, dx = 3 \int x^{-1}\, dx \qquad\qquad \text{(Rule 7)}$$

$$= 3 \ln|x| + c \qquad\qquad \text{(Rule 4)}$$

vi)

$$\int 2^{3x}\, dx = \frac{2^{3x}}{3 \ln 2} + c \qquad\qquad \text{(Rule 5)}$$

vii)

$$\int 9e^{-3x}\, dx = \frac{9e^{-3x}}{-3} + c \qquad\qquad \text{(Rule 6)}$$

$$= -3e^{-3x} + c$$

EXAMPLE 3. Functions which differ by only a constant have the same derivative. The function $F(x) = 2x + k$ has the same derivative, $F'(x) = f(x) = 2$, for any infinite number of possible values for k. If the process is reversed, it is clear that $\int 2\, dx$ must be the antiderivative or indefinite integral for an infinite number of functions differing from each other by only a constant. The constant of integration c thus represents the value of any constant which was part of the primitive function but precluded from the derivative by the rules of differentiation.

The graph of an indefinite integral $\int f(x)\, dx = F(x) + c$, where c is unspecified, is a family of curves parallel in the sense that the slope of the tangent to any of them at x is $f(x)$. Specifying c specifies the curve; changing c shifts the curve. This is illustrated in Fig. 14-1 for the indefinite integral $\int 2\, dx = 2x + c$ where $c = -7, -3, 1,$ and 5, respectively. If $c = 0$, the curve begins at the origin.

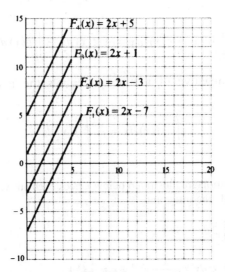

Fig. 14-1

14.3 INITIAL CONDITIONS AND BOUNDARY CONDITIONS

In many problems an *initial condition* ($y = y_0$ when $x = 0$) or a *boundary condition* ($y = y_0$ when $x = x_0$) is given which uniquely determines the constant of integration. By permitting a unique determination of c, the initial or boundary condition singles out a specific curve from the family of curves illustrated in Example 3 and Problems 14.3 to 14.5.

EXAMPLE 4. Given the boundary condition $y = 11$ when $x = 3$, the integral $y = \int 2\,dx$ is evaluated as follows:

$$y = \int 2\,dx = 2x + c$$

Substituting $y = 11$ when $x = 3$,

$$11 = 2(3) + c \qquad c = 5$$

Therefore, $y = 2x + 5$. Note that even though c is specified, $\int 2\,dx$ remains an *indefinite* integral because x is unspecified. Thus, the integral $2x + 5$ can assume an infinite number of possible values.

14.4 INTEGRATION BY SUBSTITUTION

Integration of a product or quotient of two differentiable functions of x, such as

$$\int 12x^2(x^3 + 2)\,dx$$

cannot be done directly by using the simple rules above. However, if the integrand can be expressed as a *constant* multiple of another function u and its derivative du/dx, integration by substitution is possible. By expressing the integrand $f(x)$ as a function of u and its derivative du/dx and integrating with respect to x,

$$\int f(x)\,dx = \int \left(u \frac{du}{dx} \right) dx$$

$$\int f(x)\,dx = \int u\,du = F(u) + c$$

The substitution method reverses the operation of the chain rule and the generalized power function rule in differential calculus. See Examples 5 and 6 and Problems 14.7 to 14.18.

EXAMPLE 5. The substitution method is used below to determine the indefinite integral

$$\int 12x^2(x^3 + 2)\,dx$$

1. Be sure that the integrand can be converted to a product of another function u and its derivative du/dx times a *constant* multiple. (*a*) Let u equal the function in which the independent variable is raised to the higher power in terms of absolute value; here let $u = x^3 + 2$. (*b*) Take the derivative of u; $du/dx = 3x^2$. (*c*) Solve algebraically for dx; $dx = du/3x^2$. (*d*) Then substitute u for $x^3 + 2$ and $du/3x^2$ for dx in the original integrand:

$$\int 12x^2(x^3 + 2)\,dx = \int 12x^2 \cdot u \cdot \frac{du}{3x^2} = \int 4u\,du = 4\int u\,du$$

    where 4 is a *constant* multiple of u.

2. Integrate with respect to u, using Rule 3 and ignoring c in the first step of the calculation.

$$4\int u\,du = 4(\tfrac{1}{2}u^2) = 2u^2 + c$$

3. Convert back to the terms of the original problem by substituting $x^3 + 2$ for u.

$$\int 12x^2(x^3 + 2)\,dx = 2u^2 + c = 2(x^3 + 2)^2 + c$$

4. Check the answer by differentiating with the generalized power function rule or chain rule.

$$\frac{d}{dx}[2(x^3 + 2)^2 + c] = 4(x^3 + 2)(3x^2) = 12x^2(x^3 + 2)$$

See also Problems 14.7 to 14.18.

EXAMPLE 6. Determine the integral $\int 4x(x + 1)^3\,dx$.

Let $u = x + 1$. Then $du/dx = 1$ and $dx = du/1 = du$. Substitute $u = x + 1$ and $dx = du$ in the original integrand.

$$\int 4x(x + 1)^3\,dx = \int 4xu^3\,du = 4\int xu^3\,du$$

Since x is a *variable* multiple which cannot be factored out, the original integrand cannot be transformed to a *constant* multiple of $u\,du/dx$. Hence the substitution method is ineffectual. Integration by parts (Section 14.5) may be helpful.

14.5 INTEGRATION BY PARTS

If an integrand is a product or quotient of differentiable functions of x and cannot be expressed as a constant multiple of $u\,du/dx$, integration by parts is frequently useful. The method is derived by reversing the process of differentiating a product. From the product rule in Section 3.7.5,

$$\frac{d}{dx}[f(x)\,g(x)] = f(x)\,g'(x) + g(x)\,f'(x)$$

Taking the integral of the derivative gives

$$f(x)\,g(x) = \int f(x)\,g'(x)\,dx + \int g(x)\,f'(x)\,dx$$

Then solving algebraically for the first integral on the right-hand side,

$$\int f(x)\,g'(x)\,dx = f(x)\,g(x) - \int g(x)\,f'(x)\,dx \qquad (14.1)$$

See Examples 7 and 8 and Problems 14.19 to 14.24.

For more complicated functions, *integration tables* are generally used. Integration tables provide formulas for the integrals of as many as 500 different functions, and they can be found in mathematical handbooks.

EXAMPLE 7. Integration by parts is used below to determine

$$\int 4x(x + 1)^3\,dx$$

1. Separate the integrand into two parts amenable to the formula in *(14.1)*. As a general rule, consider first the simpler function for $f(x)$ and the more complicated function for $g'(x)$. By letting $f(x) = 4x$ and $g'(x) = (x + 1)^3$, then $f'(x) = 4$ and $g(x) = \int (x + 1)^3\,dx$, which can be integrated by using the simple power rule (Rule 3):

$$g(x) = \int (x + 1)^3\,dx = \tfrac{1}{4}(x + 1)^4 + c_1$$

2. Substitute the values for $f(x)$, $f'(x)$, and $g(x)$ in (14.1); and note that $g'(x)$ is not used in the formula.

$$\int 4x(x+1)^3\, dx = f(x) \cdot g(x) - \int [g(x) \cdot f'(x)]\, dx$$

$$= 4x[\tfrac{1}{4}(x+1)^4 + c_1] - \int [\tfrac{1}{4}(x+1)^4 + c_1](4)\, dx$$

$$= x(x+1)^4 + 4c_1 x - \int [(x+1)^4 + 4c_1]\, dx$$

3. Use Rule 3 to compute the final integral and substitute.

$$\int 4x(x+1)^3\, dx = x(x+1)^4 + 4c_1 x - \tfrac{1}{5}(x+1)^5 - 4c_1 x + c$$

$$= x(x+1)^4 - \tfrac{1}{5}(x+1)^5 + c$$

Note that the c_1 term does not appear in the final solution. Since this is common to integration by parts, c_1 will henceforth be assumed equal to 0 and not formally included in future problem solving.

4. Check the answer by letting $y(x) = x(x+1)^4 - \tfrac{1}{5}(x+1)^5 + c$ and using the product and generalized power function rules.

$$y'(x) = [x \cdot 4(x+1)^3 + (x+1)^4 \cdot 1] - (x+1)^4 = 4x(x+1)^3$$

EXAMPLE 8. The integral $\int 2xe^x\, dx$ is determined as follows:
Let $f(x) = 2x$ and $g'(x) = e^x$; then $f'(x) = 2$, and by Rule 6, $g(x) = \int e^x\, dx = e^x$. Substitute in ($14.1$).

$$\int 2xe^x\, dx = f(x) \cdot g(x) - \int g(x) \cdot f'(x)\, dx$$

$$= 2x \cdot e^x - \int e^x \cdot 2\, dx = 2xe^x - 2\int e^x\, dx$$

Apply Rule 6 again and remember the constant of integration.

$$\int 2xe^x\, dx = 2xe^x - 2e^x + c$$

Then let $y(x) = 2xe^x - 2e^x + c$ and check the answer.

$$y'(x) = 2x \cdot e^x + e^x \cdot 2 - 2e^x = 2xe^x$$

14.6 ECONOMIC APPLICATIONS

Net investment I is defined as the rate of change in capital stock formation K over time t. If the process of capital formation is continuous over time, $I(t) = dK(t)/dt = K'(t)$. From the rate of investment, the level of capital stock can be estimated. Capital stock is the integral with respect to time of net investment:

$$K_t = \int I(t)\, dt = K(t) + c = K(t) + K_0$$

where c = the initial capital stock K_0.

Similarly, the integral can be used to estimate total cost from marginal cost. Since marginal cost is the change in total cost from an incremental change in output, $MC = dTC/dQ$, and only variable costs change with the level of output

$$TC = \int MC\, dQ = VC + c = VC + FC$$

since c = the fixed or initial cost FC. Economic analysis which traces the time path of variables or attempts to determine whether variables will converge toward equilibrium over time is called *dynamics*. For similar applications, see Example 9 and Problems 14.25 to 14.35.

EXAMPLE 9. The rate of net investment is given by $I(t) = 140t^{3/4}$, and the initial stock of capital at $t = 0$ is 150. Determining the function for capital K, the time path $K(t)$,

$$K = \int 140t^{3/4}\, dt = 140 \int t^{3/4}\, dt$$

By the power rule,

$$K = 140(\tfrac{4}{7}t^{7/4}) + c = 80t^{7/4} + c$$

But $c = K_0 = 150$. Therefore, $K = 80t^{7/4} + 150$.

Solved Problems

INDEFINITE INTEGRALS

14.1. Determine the following integrals. Check the answers on your own by making sure that the derivative of the integral equals the integrand.

a) $\int 3.5\, dx$

$$\int 3.5\, dx = 3.5x + c \tag{Rule 1}$$

b) $\int -\tfrac{1}{2}\, dx$

$$\int -\tfrac{1}{2}\, dx = -\int \tfrac{1}{2}\, dx = -\tfrac{1}{2}x + c \tag{Rules 1 and 9}$$

c) $\int dx$

$$\int dx = x + c \tag{Rule 2}$$

d) $\int x^5\, dx$

$$\int x^5\, dx = \tfrac{1}{6}x^6 + c \tag{Rule 3}$$

e) $\int 4x^3\, dx$

$$\int 4x^3\, dx = 4\int x^3\, dx \tag{Rule 7}$$
$$= 4(\tfrac{1}{4}x^4) + c = x^4 + c \tag{Rule 3}$$

f) $\int x^{2/3}\, dx$

$$\int x^{2/3}\, dx = \tfrac{3}{5}x^{5/3} + c \tag{Rule 3}$$

g) $\int x^{-1/5}\, dx$

$$\int x^{-1/5}\, dx = \tfrac{5}{4}x^{4/5} + c \tag{Rule 3}$$

h) $\displaystyle\int 4x^{-2}\, dx$

$$\int 4x^{-2}\, dx = -4x^{-1} + c = -\frac{4}{x} + c \qquad\qquad \text{(Rule 3)}$$

i) $\displaystyle\int x^{-5/2}\, dx$

$$\int x^{-5/2}\, dx = -\frac{2}{3}x^{-3/2} + c = \frac{-2}{3\sqrt{x^3}} + c \qquad\qquad \text{(Rule 3)}$$

14.2. Redo Problem 14.1 for each of the following:

a) $\displaystyle\int \frac{dx}{x}$

$$\int \frac{dx}{x} = \int \frac{1}{x}\, dx = \ln|x| + c \qquad\qquad \text{(Rule 4)}$$

b) $\displaystyle\int 5x^{-1}\, dx$

$$\int 5x^{-1}\, dx = 5\ln|x| + c \qquad\qquad \text{(Rules 7 and 4)}$$

c) $\displaystyle\int \frac{1}{3x}\, dx$

$$\int \frac{1}{3x}\, dx = \frac{1}{3}\int \frac{1}{x}\, dx = \frac{1}{3}\ln|x| + c \qquad\qquad \text{(Rules 7 and 4)}$$

d) $\displaystyle\int \sqrt{x}\, dx$

$$\int \sqrt{x}\, dx = \int x^{1/2}\, dx = \tfrac{2}{3}x^{3/2} + c \qquad\qquad \text{(Rule 3)}$$

e) $\displaystyle\int \frac{dx}{x^4}$

$$\int \frac{dx}{x^4} = \int x^{-4}\, dx = -\frac{1}{3}x^{-3} + c \qquad\qquad \text{(Rule 3)}$$

f) $\displaystyle\int \frac{dx}{\sqrt[3]{x}}$

$$\int \frac{dx}{\sqrt[3]{x}} = \int x^{-1/3}\, dx = \frac{3}{2}x^{2/3} + c \qquad\qquad \text{(Rule 3)}$$

g) $\displaystyle\int (5x^3 + 2x^2 + 3x)\, dx$

$$\int (5x^3 + 2x^2 + 3x)\, dx = 5\int x^3\, dx + 2\int x^2\, dx + 3\int x\, dx \qquad\qquad \text{(Rules 7 and 8)}$$

$$= 5(\tfrac{1}{4}x^4) + 2(\tfrac{1}{3}x^3) + 3(\tfrac{1}{2}x^2) + c \qquad\qquad \text{(Rule 3)}$$

$$= \tfrac{5}{4}x^4 + \tfrac{2}{3}x^3 + \tfrac{3}{2}x^2 + c$$

h) $\displaystyle\int (2x^6 - 3x^4)\,dx$

$$\int (2x^6 - 3x^4)\,dx = \tfrac{2}{7}x^7 - \tfrac{3}{5}x^5 + c \qquad\qquad \text{(Rules 3, 7, 8, and 9)}$$

14.3. Find the integral for $y = \int (x^{1/2} + 3x^{-1/2})\,dx$, given the initial condition $y = 0$ when $x = 0$.

$$y = \int (x^{1/2} + 3x^{-1/2})\,dx = \tfrac{2}{3}x^{3/2} + 6x^{1/2} + c$$

Substituting the initial condition $y = 0$ when $x = 0$ above, $c = 0$. Hence, $y = \tfrac{2}{3}x^{3/2} + 6x^{1/2}$.

14.4. Find the integral for $y = \int (2x^5 - 3x^{-1/4})\,dx$, given the initial condition $y = 6$ when $x = 0$.

$$y = \int (2x^5 - 3x^{-1/4})\,dx = \tfrac{1}{3}x^6 - 4x^{3/4} + c$$

Substituting $y = 6$ and $x = 0$, $c = 6$. Thus, $y = \tfrac{1}{3}x^6 - 4x^{3/4} + 6$.

14.5. Find the integral for $y = \int (10x^4 - 3)\,dx$, given the boundary condition $y = 21$ when $x = 1$.

$$y = \int (10x^4 - 3)\,dx = 2x^5 - 3x + c$$

Substituting $y = 21$ and $x = 1$, $21 = 2(1)^5 - 3(1) + c \qquad c = 22$
$$y = 2x^5 - 3x + 22$$

14.6. Redo Problem 14.1 for each of the following:

a) $\displaystyle\int 2^{4x}\,dx$

$$\int 2^{4x}\,dx = \frac{2^{4x}}{4\ln 2} + c \qquad \text{(Rule 5)}$$

b) $\displaystyle\int 8^x\,dx$

$$\int 8^x\,dx = \frac{8^x}{\ln 8} + c$$

c) $\displaystyle\int e^{5x}\,dx$

$$\int e^{5x}\,dx = \frac{e^{5x}}{5} + c \qquad \text{(Rule 6)}$$
$$= \tfrac{1}{5}e^{5x} + c$$

d) $\displaystyle\int 16e^{-4x}\,dx$

$$\int 16e^{-4x}\,dx = \frac{16e^{-4x}}{-4} + c = -4e^{-4x} + c$$

e) $\displaystyle\int (6e^{3x} - 8e^{-2x})\,dx$

$$\int (6e^{3x} - 8e^{-2x})\,dx = \frac{6e^{3x}}{3} - \frac{8e^{-2x}}{-2} + c = 2e^{3x} + 4e^{-2x} + c$$

INTEGRATION BY SUBSTITUTION

14.7. Determine the following integral, using the substitution method. Check the answer on your own. Given $\int 10x(x^2 + 3)^4\,dx$.

Let $u = x^2 + 3$. Then $du/dx = 2x$ and $dx = du/2x$. Substituting in the original integrand to reduce it to a function of $u \, du/dx$,

$$\int 10x(x^2 + 3)^4 \, dx = \int 10xu^4 \frac{du}{2x} = 5 \int u^4 \, du$$

Integrating by the power rule, $5 \int u^4 \, du = 5(\tfrac{1}{5}u^5) = u^5 + c$

Substituting $u = x^2 + 3$, $\int 10x(x^2 + 3)^4 \, dx = u^5 + c = (x^2 + 3)^5 + c$

14.8. Redo Problem 14.7, given $\int x^4 (2x^5 - 5)^4 \, dx$.

Let $u = 2x^5 - 5$, $du/dx = 10x^4$, and $dx = du/10x^4$. Substituting in the original integrand,

$$\int x^4 (2x^5 - 5)^4 \, dx = \int x^4 u^4 \frac{du}{10x^4} = \frac{1}{10} \int u^4 \, du$$

Integrating, $\dfrac{1}{10} \int u^4 \, du = \dfrac{1}{10}\left(\dfrac{1}{5}u^5\right) = \dfrac{1}{50} u^5 + c$

Substituting, $\displaystyle\int x^4 (2x^5 - 5)^4 \, dx = \dfrac{1}{50} u^5 + c = \dfrac{1}{50}(2x^5 - 5)^5 + c$

14.9. Redo Problem 14.7, given $\int (x - 9)^{7/4} \, dx$.

Let $u = x - 9$. Then $du/dx = 1$ and $dx = du$. Substituting,

$$\int (x - 9)^{7/4} \, dx = \int u^{7/4} \, du$$

Integrating, $\displaystyle\int u^{7/4} \, du = \dfrac{4}{11} u^{11/4} + c$

Substituting, $\displaystyle\int (x - 9)^{7/4} \, dx = \dfrac{4}{11}(x - 9)^{11/4} + c$

Whenever $du/dx = 1$, the power rule can be used immediately for integration by substitution.

14.10. Redo Problem 14.7, given $\int (6x - 11)^{-5} \, dx$.

Let $u = 6x - 11$. Then $du/dx = 6$ and $dx = du/6$. Substituting,

$$\int (6x - 11)^{-5} \, dx = \int u^{-5} \frac{du}{6} = \frac{1}{6} \int u^{-5} \, du$$

Integrating, $\dfrac{1}{6} \int u^{-5} \, du = \dfrac{1}{6}\left(\dfrac{1}{-4} u^{-4}\right) = -\dfrac{1}{24} u^{-4} + c$

Substituting, $\displaystyle\int (6x - 11)^{-5} \, dx = -\tfrac{1}{24}(6x - 11)^{-4} + c$

Notice that here $du/dx = 6 \neq 1$, and the power rule cannot be used directly.

14.11. Redo Problem 14.7, given

$$\int \frac{x^2}{(4x^3 + 7)^2} \, dx$$

$$\int \frac{x^2}{(4x^3 + 7)^2} \, dx = \int x^2 (4x^3 + 7)^{-2} \, dx$$

Let $u = 4x^3 + 7$, $du/dx = 12x^2$, and $dx = du/12x^2$. Substituting,

$$\int x^2 u^{-2} \frac{du}{12x^2} = \frac{1}{12} \int u^{-2} \, du$$

Integrating,

$$\frac{1}{12} \int u^{-2} \, du = -\frac{1}{12} u^{-1} + c$$

Substituting,

$$\int \frac{x^2}{(4x^3 + 7)^2} \, dx = -\frac{1}{12(4x^3 + 7)} + c$$

14.12. Redo Problem 14.7, given

$$\int \frac{6x^2 + 4x + 10}{(x^3 + x^2 + 5x)^3} \, dx$$

Let $u = x^3 + x^2 + 5x$. Then $du/dx = 3x^2 + 2x + 5$ and $dx = du/(3x^2 + 2x + 5)$. Substituting,

$$\int (6x^2 + 4x + 10) u^{-3} \frac{du}{3x^2 + 2x + 5} = 2 \int u^{-3} \, du$$

Integrating,

$$2 \int u^{-3} \, du = -u^{-2} + c$$

Substituting,

$$\int \frac{6x^2 + 4x + 10}{(x^3 + x^2 + 5x)^3} \, dx = -\frac{1}{(x^3 + x^2 + 5x)^2} + c$$

14.13. Redo Problem 14.7, given

$$\int \frac{dx}{9x - 5}$$

$$\int \frac{dx}{9x - 5} = \int (9x - 5)^{-1} \, dx$$

Let $u = 9x - 5$, $du/dx = 9$, and $dx = du/9$. Substituting,

$$\int u^{-1} \frac{du}{9} = \frac{1}{9} \int u^{-1} \, du$$

Integrating with Rule 4, $\frac{1}{9} \int u^{-1} \, du = \frac{1}{9} \ln |u| + c$. Since u may be $\gtrless 0$, and only positive numbers have logs, always use the absolute value of u. See Rule 4. Substituting,

$$\int \frac{dx}{9x - 5} = \frac{1}{9} \ln |9x - 5| + c$$

14.14. Redo Problem 14.7, given

$$\int \frac{3x^2 + 2}{4x^3 + 8x} \, dx$$

Let $u = 4x^3 + 8x$, $du/dx = 12x^2 + 8$, and $dx = du/(12x^2 + 8)$. Substituting,

$$\int (3x^2 + 2) u^{-1} \frac{du}{12x^2 + 8} = \frac{1}{4} \int u^{-1} \, du$$

Integrating,

$$\frac{1}{4} \int u^{-1} \, du = \frac{1}{4} \ln |u| + c$$

Substituting,
$$\int \frac{3x^2 + 2}{4x^3 + 8x} dx = \frac{1}{4} \ln |4x^3 + 8x| + c$$

14.15. Use the substitution method to find the integral for $\int x^3 e^{x^4} dx$. Check your answer.

Let $u = x^4$. Then $du/dx = 4x^3$ and $dx = du/4x^3$. Substituting, and noting that u is now an exponent,

$$\int x^3 e^u \frac{du}{4x^3} = \frac{1}{4} \int e^u \, du$$

Integrating with Rule 6,
$$\frac{1}{4} \int e^u \, du = \frac{1}{4} e^u + c$$

Substituting,
$$\int x^3 e^{x^4} dx = \frac{1}{4} e^{x^4} + c$$

14.16. Redo Problem 14.15, given $\int 24x e^{3x^2} dx$.

Let $u = 3x^2$, $du/dx = 6x$, and $dx = du/6x$. Substituting,

$$\int 24x e^u \frac{du}{6x} = 4 \int e^u \, du$$

Integrating,
$$4 \int e^u \, du = 4e^u + c$$

Substituting,
$$\int 24x e^{3x^2} dx = 4e^{3x^2} + c$$

14.17. Redo Problem 14.15, given $\int 14 e^{2x+7} dx$.

Let $u = 2x + 7$; then $du/dx = 2$ and $dx = du/2$. Substituting,

$$\int 14 e^u \frac{du}{2} = 7 \int e^u \, du = 7e^u + c$$

Substituting,
$$\int 14 e^{2x+7} dx = 7e^{2x+7} + c$$

14.18. Redo Problem 14.15, given $\int 5x e^{5x^2+3} dx$.

Let $u = 5x^2 + 3$, $du/dx = 10x$, and $dx = du/10x$. Substituting,

$$\int 5x e^u \frac{du}{10x} = \frac{1}{2} \int e^u \, du$$

Integrating,
$$\frac{1}{2} \int e^u \, du = \frac{1}{2} e^u + c$$

Substituting,
$$\int 5x e^{5x^2+3} dx = \frac{1}{2} e^{5x^2+3} + c$$

INTEGRATION BY PARTS

14.19. Use integration by parts to evaluate the following integral. Keep in the habit of checking your answers. Given $\int 15x(x + 4)^{3/2} dx$.

Let $f(x) = 15x$, then $f'(x) = 15$. Let $g'(x) = (x+4)^{3/2}$, then $g(x) = \int (x+4)^{3/2}\, dx = \frac{2}{5}(x+4)^{5/2}$. Substituting in (14.1),

$$\int 15x(x+4)^{3/2}\, dx = f(x)g(x) - \int g(x)f'(x)\, dx$$

$$= 15x[\tfrac{2}{5}(x+4)^{5/2}] - \int \tfrac{2}{5}(x+4)^{5/2}\, 15\, dx = 6x(x+4)^{5/2} - 6\int (x+4)^{5/2}\, dx$$

Evaluating the remaining integral,

$$\int 15x(x+4)^{3/2}\, dx = 6x(x+4)^{5/2} - \tfrac{12}{7}(x+4)^{7/2} + c$$

14.20. Redo Problem 14.19, given

$$\int \frac{2x}{(x-8)^3}\, dx$$

Let $f(x) = 2x$, $f'(x) = 2$, and $g'(x) = (x-8)^{-3}$; then $g(x) = \int (x-8)^{-3}\, dx = -\frac{1}{2}(x-8)^{-2}$. Substituting in (14.1),

$$\int \frac{2x}{(x-8)^3}\, dx = 2x\left[-\frac{1}{2}(x-8)^{-2}\right] - \int -\frac{1}{2}(x-8)^{-2}\ 2\, dx = -x(x-8)^{-2} + \int (x-8)^{-2}\, dx$$

Integrating for the last time,

$$\int \frac{2x}{(x-8)^3}\, dx = -x(x-8)^{-2} - (x-8)^{-1} + c = \frac{-x}{(x-8)^2} - \frac{1}{x-8} + c$$

14.21. Redo Problem 14.19, given

$$\int \frac{5x}{(x-1)^2}\, dx$$

Let $f(x) = 5x$, $f'(x) = 5$, and $g'(x) = (x-1)^{-2}$; then $g(x) = \int (x-1)^{-2}\, dx = -(x-1)^{-1}$. Substituting in (14.1),

$$\int \frac{5x}{(x-1)^2}\, dx = 5x[-(x-1)^{-1}] - \int -(x-1)^{-1}5\, dx = -5x(x-1)^{-1} + 5\int (x-1)^{-1}\, dx$$

Integrating again,

$$\int \frac{5x}{(x-1)^2}\, dx = -5x(x-1)^{-1} + 5\ln|x-1| + c = \frac{-5x}{x-1} + 5\ln|x-1| + c$$

14.22. Redo Problem 14.19, given $\int 6xe^{x+7}\, dx$.

Let $f(x) = 6x$, $f'(x) = 6$, $g'(x) = e^{x+7}$, and $g(x) = \int e^{x+7}\, dx = e^{x+7}$. Using (14.1),

$$\int 6xe^{x+7}\, dx = 6xe^{x+7} - \int e^{x+7}6\, dx = 6xe^{x+7} - 6\int e^{x+7}\, dx$$

Integrating again,

$$\int 6xe^{x+7}\, dx = 6xe^{x+7} - 6e^{x+7} + c$$

14.23. Use integration by parts to evaluate $\int 16xe^{-(x+9)} dx$.

Let $f(x) = 16x$, $f'(x) = 16$, $g'(x) = e^{-(x+9)}$, and $g(x) = \int e^{-(x+9)} dx = -e^{-(x+9)}$. Using (14.1),

$$\int 16xe^{-(x+9)} dx = -16xe^{-(x+9)} - \int -e^{-(x+9)} 16 dx = -16xe^{-(x+9)} + 16 \int e^{-(x+9)} dx$$

Integrating once more,

$$\int 16xe^{-(x+9)} dx = -16xe^{-(x+9)} - 16e^{-(x+9)} + c$$

14.24. Redo Problem 14.23, given $\int x^2 e^{2x} dx$.

Let $f(x) = x^2$, $f'(x) = 2x$, $g'(x) = e^{2x}$, and $g(x) = \int e^{2x} dx = \frac{1}{2}e^{2x}$. Substituting in (14.1).

$$\int x^2 e^{2x} dx = x^2(\tfrac{1}{2}e^{2x}) - \int \tfrac{1}{2}e^{2x}(2x) dx = \tfrac{1}{2}x^2 e^{2x} - \int xe^{2x} dx \qquad (14.2)$$

Using parts again for the remaining integral, $f(x) = x$, $f'(x) = 1$, $g'(x) = e^{2x}$, and $g(x) = \int e^{2x} dx = \frac{1}{2}e^{2x}$. Using (14.1),

$$\int xe^{2x} dx = x(\tfrac{1}{2}e^{2x}) - \int \tfrac{1}{2}e^{2x} dx = \tfrac{1}{2}xe^{2x} - \tfrac{1}{2}(\tfrac{1}{2}e^{2x})$$

Finally, substituting in (14.2),

$$\int x^2 e^{2x} dx = \tfrac{1}{2}x^2 e^{2x} - \tfrac{1}{2}xe^{2x} + \tfrac{1}{4}e^{2x} + c$$

ECONOMIC APPLICATIONS

14.25. The rate of net investment is $I = 40t^{3/5}$, and capital stock at $t = 0$ is 75. Find the capital function K.

$$K = \int I dt = \int 40t^{3/5} dt = 40(\tfrac{5}{8}t^{8/5}) + c = 25t^{8/5} + c$$

Substituting $t = 0$ and $K = 75$,

$$75 = 0 + c \qquad c = 75$$

Thus, $K = 25t^{8/5} + 75$.

14.26. The rate of net investment is $I = 60t^{1/3}$, and capital stock at $t = 1$ is 85. Find K.

$$K = \int 60t^{1/3} dt = 45t^{4/3} + c$$

At $t = 1$ and $K = 85$,

$$85 = 45(1) + c \qquad c = 40$$

Thus, $K = 45t^{4/3} + 40$.

14.27. Marginal cost is given by $MC = dTC/dQ = 25 + 30Q - 9Q^2$. Fixed cost is 55. Find the (*a*) total cost, (*b*) average cost, and (*c*) variable cost functions.

a)
$$TC = \int MC \, dQ = \int (25 + 30Q - 9Q^2) \, dQ = 25Q + 15Q^2 - 3Q^3 + c$$

With $FC = 55$, at $Q = 0$, $TC = FC = 55$. Thus, $c = FC = 55$ and $TC = 25Q + 15Q^2 - 3Q^3 + 55$.

b)
$$AC = \frac{TC}{Q} = 25 + 15Q - 3Q^2 + \frac{55}{Q}$$

c)
$$VC = TC - FC = 25Q + 15Q^2 - 3Q^3$$

14.28. Given $MC = dTC/dQ = 32 + 18Q - 12Q^2$, $FC = 43$. Find the *(a)* TC, *(b)* AC, and *(c)* VC functions.

a)
$$TC = \int MC\,dQ = \int (32 + 18Q - 12Q^2)\,dQ = 32Q + 9Q^2 - 4Q^3 + c$$

At $Q = 0$, $TC = FC = 43$, $TC = 32Q + 9Q^2 - 4Q^3 + 43$.

b)
$$AC = \frac{TC}{Q} = 32 + 9Q - 4Q^2 + \frac{43}{Q}$$

c)
$$VC = TC - FC = 32Q + 9Q^2 - 4Q^3$$

14.29. Marginal revenue is given by $MR = dTR/dQ = 60 - 2Q - 2Q^2$. Find *(a)* the TR function and *(b)* the demand function $P = f(Q)$.

a)
$$TR = \int MR\,dQ = \int (60 - 2Q - 2Q^2)\,dQ = 60Q - Q^2 - \tfrac{2}{3}Q^3 + c$$

At $Q = 0$, $TR = 0$. Therefore $c = 0$. Thus, $TR = 60Q - Q^2 - \tfrac{2}{3}Q^3$.

b) $TR = PQ$. Therefore, $P = TR/Q$, which is the same as saying that the demand function and the average revenue function are identical. Thus, $P = AR = TR/Q = 60 - Q - \tfrac{2}{3}Q^2$.

14.30. Find *(a)* the total revenue function and *(b)* the demand function, given
$$MR = 84 - 4Q - Q^2$$

a)
$$TR = \int MR\,dQ = \int (84 - 4Q - Q^2)\,dQ = 84Q - 2Q^2 - \tfrac{1}{3}Q^3 + c$$

At $Q = 0$, $TR = 0$. Therefore $c = 0$. Thus, $TR = 84Q - 2Q^2 - \tfrac{1}{3}Q^3$.

b)
$$P = AR = \frac{TR}{Q} = 84 - 2Q - \frac{1}{3}Q^2$$

14.31. With $C = f(Y)$, the marginal propensity to consume is given by $MPC = dC/dY = f'(Y)$. If the $MPC = 0.8$ and consumption is 40 when income is zero, find the consumption function.
$$C = \int f'(Y)\,dY = \int 0.8\,dY = 0.8Y + c$$

At $Y = 0$, $C = 40$. Thus, $c = 40$ and $C = 0.8Y + 40$.

14.32. Given $dC/dY = 0.6 + 0.1/\sqrt[3]{Y} = MPC$ and $C = 45$ when $Y = 0$. Find the consumption function.
$$C = \int \left(0.6 + \frac{0.1}{\sqrt[3]{Y}}\right) dY = \int (0.6 + 0.1Y^{-1/3})\,dY = 0.6Y + 0.15Y^{2/3} + c$$

At $Y = 0$, $C = 45$. Thus, $C = 0.6Y + 0.15Y^{2/3} + 45$.

14.33. The marginal propensity to save is given by $dS/dY = 0.5 - 0.2Y^{-1/2}$. There is dissaving of 3.5 when income is 25, that is, $S = -3.5$ when $Y = 25$. Find the savings function.

$$S = \int (0.5 - 0.2Y^{-1/2})\, dY = 0.5Y - 0.4Y^{1/2} + c$$

At $Y = 25$, $S = -3.5$.

$$-3.5 = 0.5(25) - 0.4(\sqrt{25}) + c \qquad c = -14$$

Thus, $S = 0.5Y - 0.4Y^{1/2} - 14$.

14.34. Given MC $= dTC/dQ = 12e^{0.5Q}$ and FC $= 36$. Find the total cost.

$$TC = \int 12e^{0.5Q}\, dQ = 12\frac{1}{0.5}e^{0.5Q} + c = 24e^{0.5Q} + c$$

With FC $= 36$, TC $= 36$ when $Q = 0$. Substituting, $36 = 24e^{0.5(0)} + c$. Since $e^0 = 1$, $36 = 24 + c$, and $c = 12$. Thus, TC $= 24e^{0.5Q} + 12$. Notice that c does not always equal FC.

14.35. Given MC $= 16e^{0.4Q}$ and FC $= 100$. Find TC.

$$TC = \int 16e^{0.4Q}\, dQ = 16\left(\frac{1}{0.4}\right)e^{0.4Q} + c = 40e^{0.4Q} + c$$

At $Q = 0$, TC $= 100$.

$$100 = 40e^0 + c \qquad c = 60$$

Thus, TC $= 40e^{0.4Q} + 60$.

CHAPTER 15

Integral Calculus: The Definite Integral

15.1 AREA UNDER A CURVE

There is no geometric formula for the area under an irregularly shaped curve, such as $y = f(x)$ between $x = a$ and $x = b$ in Fig. 15-1(a). If the interval $[a, b]$ is divided into n subintervals $[x_1, x_2]$, $[x_2, x_3]$, etc., and rectangles are constructed such that the height of each is equal to the smallest value of the function in the subinterval, as in Fig. 15-1(b), then the sum of the areas of the rectangles $\sum_{i=1}^{n} f(x_i) \Delta x_i$, called a *Riemann sum*, will approximate, but underestimate, the actual area under the curve. The smaller the subintervals (the smaller the Δx_i), the more rectangles are created and the closer the combined area of the rectangles $\sum_{i=1}^{n} f(x_i) \Delta x_i$ approaches the actual area under the curve. If

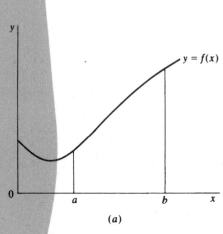

Fig. 15-1

342

the number of subintervals is increased so that $n \rightarrow \infty$, each subinterval becomes infinitesimal ($\Delta x_i = dx_i = dx$) and the area A under the curve can be expressed mathematically as

$$A = \lim_{n \to \infty} \sum_{i=1}^{n} f(x_i)\, \Delta x_i$$

15.2 THE DEFINITE INTEGRAL

The area under a graph of a continuous function such as that in Fig. 15-1 from a to b ($a < b$) can be expressed more succinctly as the *definite integral* of $f(x)$ over the interval a to b. Put mathematically,

$$\int_a^b f(x)\, dx = \lim_{n \to \infty} \sum_{i=1}^{n} f(x_i)\, \Delta x_i$$

Here the left-hand side is read, "the integral from a to b of f of $x\, dx$." Here a is called the *lower limit* of integration and b the *upper limit* of integration. Unlike the indefinite integral which is a set of functions containing all the antiderivatives of $f(x)$, as explained in Example 3 of Chapter 14, the definite integral is a real number which can be evaluated by using the fundamental theorem of calculus (Section 15.3).

15.3 THE FUNDAMENTAL THEOREM OF CALCULUS

The *fundamental theorem of calculus* states that the numerical value of the definite integral of a continuous function $f(x)$ over the interval from a to b is given by the indefinite integral $F(x) + c$ evaluated at the upper limit of integration b, minus the same indefinite integral $F(x) + c$ evaluated at the lower limit of integration a. Since c is common to both, the constant of integration is eliminated in subtraction. Expressed mathematically,

$$\int_a^b f(x)\, dx = F(x)\Big|_a^b = F(b) - F(a)$$

where the symbol $|_a^b$, $]_a^b$, or $[\cdots]_a^b$ indicates that b and a are to be substituted successively for x. See Examples 1 and 2 and Problems 15.1 to 15.10.

EXAMPLE 1.   The definite integrals given below

$$(1) \quad \int_1^4 10x\, dx \qquad (2) \quad \int_1^3 (4x^3 + 6x)\, dx$$

are evaluated as follows:

1)
$$\int_1^4 10x\, dx = 5x^2 \Big|_1^4 = 5(4)^2 - 5(1)^2 = 75$$

2)
$$\int_1^3 (4x^3 + 6x)\, dx = [x^4 + 3x^2]_1^3 = [(3)^4 + 3(3)^2] - [(1)^4 + 3(1)^2] = 108 - 4 = 104$$

EXAMPLE 2.   The definite integral is used below to determine the area under the curve in Fig. 15-2 over the interval 0 to 20 as follows:

$$A = \int_0^{20} \tfrac{1}{2}x\, dx = \tfrac{1}{4}x^2 \Big|_0^{20} = \tfrac{1}{4}(20)^2 - \tfrac{1}{4}(0)^2 = 100$$

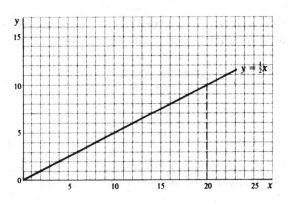

Fig. 15-2

The answer can be checked by using the geometric formula $A = \frac{1}{2}xy$:

$$A = \frac{1}{2}xy = \frac{1}{2}(20)(10) = 100$$

15.4 PROPERTIES OF DEFINITE INTEGRALS

1. Reversing the order of the limits changes the sign of the definite integral.

$$\int_a^b f(x)\,dx = -\int_b^a f(x)\,dx \tag{15.1}$$

2. If the upper limit of integration equals the lower limit of integration, the value of the definite integral is zero.

$$\int_a^a f(x)\,dx = F(a) - F(a) = 0 \tag{15.2}$$

3. The definite integral can be expressed as the sum of component subintegrals.

$$\int_a^c f(x)\,dx = \int_a^b f(x)\,dx + \int_b^c f(x)\,dx \qquad a \le b \le c \tag{15.3}$$

4. The sum or difference of two definite integrals with identical limits of integration is equal to the definite integral of the sum or difference of the two functions.

$$\int_a^b f(x)\,dx \pm \int_a^b g(x)\,dx = \int_a^b [f(x) \pm g(x)]\,dx \tag{15.4}$$

5. The definite integral of a constant times a function is equal to the constant times the definite integral of the function.

$$\int_a^b kf(x)\,dx = k\int_a^b f(x)\,dx \tag{15.5}$$

See Example 3 and Problems 15.11 to 15.14.

EXAMPLE 3. To illustrate a sampling of the properties presented above, the following definite integrals are evaluated:

1. $\displaystyle\int_1^3 2x^3\,dx = -\int_3^1 2x^3\,dx$

$$\int_1^3 2x^3\,dx = \tfrac{1}{2}x^4\Big|_1^3 = \tfrac{1}{2}(3)^4 - \tfrac{1}{2}(1)^4 = 40$$

Checking this answer,

$$\int_3^1 2x^3\,dx = \tfrac{1}{2}x^4\Big|_3^1 = \tfrac{1}{2}(1)^4 - \tfrac{1}{2}(3)^4 = -40$$

2. $\displaystyle\int_5^5 (2x+3)\,dx = 0$

Checking this answer,

$$\int_5^5 (2x+3)\,dx = [x^2 + 3x]_5^5 = [(5)^2 + 3(5)] - [(5)^2 + 3(5)] = 0$$

3. $\displaystyle\int_0^4 6x\,dx = \int_0^3 6x\,dx + \int_3^4 6x\,dx$

$$\int_0^4 6x\,dx = 3x^2\Big|_0^4 = 3(4)^2 - 3(0)^2 = 48$$

$$\int_0^3 6x\,dx = 3x^2\Big|_0^3 = 3(3)^2 - 3(0)^2 = 27$$

$$\int_3^4 6x\,dx = 3x^2\Big|_3^4 = 3(4)^2 - 3(3)^2 = 21$$

Checking this answer, $48 = 27 + 21$

15.5 AREA BETWEEN CURVES

The area of a region between two or more curves can be evaluated by applying the properties of definite integrals outlined above. The procedure is demonstrated in Example 4 and treated in Problems 15.15 to 15.18.

EXAMPLE 4. Using the properties of integrals, the area of the region between two functions such as $y_1 = 3x^2 - 6x + 8$ and $y_2 = -2x^2 + 4x + 1$ from $x = 0$ to $x = 2$ is found in the following way:

a) Draw a rough sketch of the graph of the functions and shade in the desired area as in Fig. 15-3.

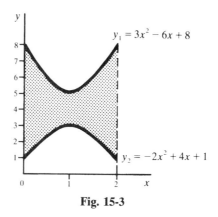

Fig. 15-3

b)   Note the relationship between the curves. Since y_1 lies above y_2, the desired region is simply the area under y_1 minus the area under y_2 between $x = 0$ and $x = 2$. Hence,

$$A = \int_0^2 (3x^2 - 6x + 8)\, dx - \int_0^2 (-2x^2 + 4x + 1)\, dx$$

From (*15.4*),

$$A = \int_0^2 [(3x^2 - 6x + 8) - (-2x^2 + 4x + 1)]\, dx$$

$$= \int_0^2 (5x^2 - 10x + 7)\, dx$$

$$= (\tfrac{5}{3}x^3 - 5x^2 + 7x)\big|_0^2 = 7\tfrac{1}{3} - 0 = 7\tfrac{1}{3}$$

15.6   IMPROPER INTEGRALS

The area under some curves that extend infinitely far along the x axis, as in Fig. 15-4(*a*), may be estimated with the help of improper integrals. A definite integral with infinity for either an upper or lower limit of integration is called an *improper integral*.

$$\int_a^\infty f(x)\, dx \qquad \text{and} \qquad \int_{-\infty}^b f(x)\, dx$$

are improper integrals because ∞ is not a number and cannot be substituted for x in $F(x)$. They can, however, be defined as the limits of other integrals, as shown below.

$$\int_a^\infty f(x)\, dx = \lim_{b \to \infty} \int_a^b f(x)\, dx \qquad \text{and} \qquad \int_{-\infty}^b f(x)\, dx = \lim_{a \to -\infty} \int_a^b f(x)\, dx$$

If the limit in either case exists, the improper integral is said to *converge*. The integral has a definite value, and the area under the curve can be evaluated. If the limit does not exist, the improper integral *diverges* and is meaningless. See Example 5 and Problems 15.19 to 15.25.

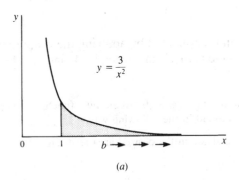
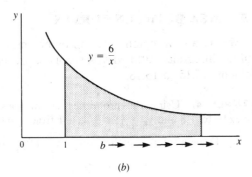

(*a*)   (*b*)

Fig. 15-4

EXAMPLE 5.   The improper integrals given below

$$(a) \quad \int_1^\infty \frac{3}{x^2}\, dx \qquad (b) \quad \int_1^\infty \frac{6}{x}\, dx$$

are sketched in Fig. 15-4(*a*) and (*b*) and evaluated as follows:

a)

$$\int_1^\infty \frac{3}{x^2}\, dx = \lim_{b \to \infty} \int_1^b \frac{3}{x^2}\, dx = \lim_{b \to \infty} \left[\frac{-3}{x} \right]_1^b$$

$$= \lim_{b \to \infty} \left[\frac{-3}{b} - \frac{(-3)}{1} \right] = \lim_{b \to \infty} \left(\frac{-3}{b} + 3 \right) = 3$$

because as $b \to \infty$, $-3/b \to 0$. Hence the improper integral is convergent and the area under the curve in Fig. 15-4(a) equals 3.

b)
$$\int_1^\infty \frac{6}{x}\,dx = \lim_{b \to \infty} \int_1^b \frac{6}{x}\,dx$$
$$= \lim_{b \to \infty} [6 \ln |x|]_1^b = \lim_{b \to \infty} [6 \ln |b| - 6 \ln |1|]$$
$$= \lim_{b \to \infty} [6 \ln |b|] \qquad \text{since} \qquad \ln |1| = 0$$

As $b \to \infty$, $6 \ln |b| \to \infty$. The improper integral diverges and has no definite value. The area under the curve in Fig. 15-4(b) cannot be computed even though the graph is deceptively similar to the one in (a).

15.7   L'HÔPITAL'S RULE

If the limit of a function $f(x) = g(x)/h(x)$ as $x \to a$ cannot be evaluated, such as (1) when both numerator and denominator approach zero, giving rise to the indeterminate form 0/0, or (2) when both numerator and denominator approach infinity, giving rise to the indeterminate form ∞/∞, *L'Hôpital's rule* can often be helpful. L'Hôpital's rule states:

$$\lim_{x \to a} \frac{g(x)}{h(x)} = \lim_{x \to a} \frac{g'(x)}{h'(x)} \tag{15.6}$$

It is illustrated in Example 6 and Problem 15.26.

EXAMPLE 6.   The limits of the functions given below are found as follows, using L'Hôpital's rule. Note that numerator and denominator are differentiated separately, not as a quotient.

$$(a) \quad \lim_{x \to 4} \frac{x - 4}{16 - x^2} \qquad (b) \quad \lim_{x \to \infty} \frac{6x - 2}{7x + 4}$$

a)   As $x \to 4$, $x - 4$ and $16 - x^2 \to 0$. Using (*15.6*), therefore, and differentiating numerator and denominator separately,

$$\lim_{x \to 4} \frac{x - 4}{16 - x^2} = \lim_{x \to 4} \frac{1}{-2x} = -\frac{1}{8}$$

b)   As $x \to \infty$, both $6x - 2$ and $7x + 4 \to \infty$. Using (*15.6*),

$$\lim_{x \to \infty} \frac{6x - 2}{7x + 4} = \lim_{x \to \infty} \frac{6}{7} = \frac{6}{7}$$

15.8   CONSUMERS' AND PRODUCERS' SURPLUS

A demand function $P_1 = f_1(Q)$, as in Fig. 15-5(a), represents the different prices consumers are willing to pay for different quantities of a good. If equilibrium in the market is at (Q_0, P_0), then the consumers who would be willing to pay more than P_0 benefit. Total benefit to consumers is represented by the shaded area and is called *consumers' surplus*. Mathematically,

$$\text{Consumers' surplus} = \int_0^{Q_0} f_1(Q)\,dQ - Q_0 P_0 \tag{15.7}$$

A supply function $P_2 = f_2(Q)$, as in Fig. 15-5(b), represents the prices at which different quantities of a good will be supplied. If market equilibrium occurs at (Q_0, P_0), the producers who would supply at a lower price than P_0 benefit. Total gain to producers is called *producers' surplus* and is designated by the shaded area. Mathematically,

$$\text{Producers' surplus} = Q_0 P_0 - \int_0^{Q_0} f_2(Q)\,dQ \tag{15.8}$$

See Example 7 and Problems 15.27 to 15.31.

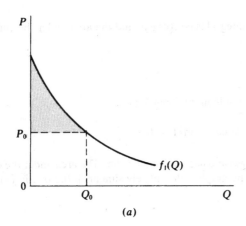
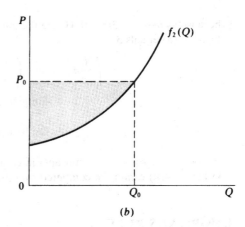

(a)   (b)

Fig. 15-5

EXAMPLE 7.   Given the demand function $P = 42 - 5Q - Q^2$. Assuming that the equilibrium price is 6, the consumers' surplus is evaluated as follows:

At $P_0 = 6$,

$$42 - 5Q - Q^2 = 6$$
$$36 - 5Q - Q^2 = 0$$
$$(Q + 9)(-Q + 4) = 0$$

So $Q_0 = 4$, because $Q = -9$ is not feasible. Substituting in (15.7),

$$\text{Consumers' surplus} = \int_0^4 (42 - 5Q - Q^2)\, dQ - (4)(6)$$
$$= [42Q - 2.5Q^2 - \tfrac{1}{3}Q^3]_0^4 - 24$$
$$= (168 - 40 - 21\tfrac{1}{3}) - 0 - 24 = 82\tfrac{2}{3}$$

15.9   THE DEFINITE INTEGRAL AND PROBABILITY

The probability P that an event will occur can be measured by the corresponding area under a probability density function. A *probability density* or *frequency function* is a continuous function $f(x)$ such that:

1. $f(x) \geq 0$. Probability cannot be negative.
2. $\int_{-\infty}^{\infty} f(x)\, dx = 1$. The probability of the event occurring over the entire range of x is 1.
3. $P(a < x < b) = \int_a^b f(x)\, dx$. The probability of the value of x falling within the interval $[a, b]$ is the value of the definite integral from a to b.

See Example 8 and Problems 15.32 and 15.33.

EXAMPLE 8.   The time in minutes between cars passing on a highway is given by the frequency function $f(t) = 2e^{-2t}$ for $t \geq 0$. The probability of a car passing in 0.25 minute is calculated as follows:

$$P = \int_0^{0.25} 2e^{-2t}\, dt = -e^{-2t}\Big|_0^{0.25} = -e^{-0.5} - (-e^0) = -0.606531 + 1 = 0.393469$$

Solved Problems

DEFINITE INTEGRALS

15.1. Evaluate the following definite integrals:

a) $\displaystyle\int_0^6 5x\,dx$

$$\int_0^6 5x\,dx = 2.5x^2\,\bigg|_0^6 = 2.5(6)^2 - 2.5(0)^2 = 90$$

b) $\displaystyle\int_1^{10} 3x^2\,dx$

$$\int_1^{10} 3x^2\,dx = x^3\,\bigg|_1^{10} = (10)^3 - (1)^3 = 999$$

c) $\displaystyle\int_1^{64} x^{-2/3}\,dx$

$$\int_1^{64} x^{-2/3}\,dx = 3x^{1/3}\,\bigg|_1^{64} = 3\sqrt[3]{64} - 3\sqrt[3]{1} = 9$$

d) $\displaystyle\int_1^3 (x^3 + x + 6)\,dx$

$$\int_1^3 (x^3 + x + 6)\,dx = (\tfrac{1}{4}x^4 + \tfrac{1}{2}x^2 + 6x)\,\bigg|_1^3 = \tfrac{1}{4}(3)^4 + \tfrac{1}{2}(3)^2 + 6(3) - [\tfrac{1}{4}(1)^4 + \tfrac{1}{2}(1)^2 + 6(1)] = 36$$

e) $\displaystyle\int_1^4 (x^{-1/2} + 3x^{1/2})\,dx$

$$\int_1^4 (x^{-1/2} + 3x^{1/2})\,dx = (2x^{1/2} + 2x^{3/2})\,\bigg|_1^4 = 2\sqrt{4} + 2\sqrt{4^3} - (2\sqrt{1} + 2\sqrt{1^3}) = 16$$

f) $\displaystyle\int_0^3 4e^{2x}\,dx$

$$\int_0^3 4e^{2x}\,dx = 2e^{2x}\,\bigg|_0^3 = 2(e^{2(3)} - e^{2(0)})$$
$$= 2(403.4 - 1) = 804.8$$

g) $\displaystyle\int_0^{10} 2e^{-2x}\,dx$

$$\int_0^{10} 2e^{-2x}\,dx = -e^{-2x}\,\bigg|_0^{10} = -e^{-2(10)} - (-e^{-2(0)}) = -e^{-20} + e^0 = 1$$

SUBSTITUTION METHOD

15.2. Use the substitution method to integrate the following definite integral:

$$\int_0^3 8x(2x^2 + 3)\,dx$$

Let $u = 2x^2 + 3$. Then $du/dx = 4x$ and $dx = du/4x$. Ignore the limits of integration for the moment, and treat the integral as an indefinite integral. Substituting in the original integrand,

$$\int 8x(2x^2 + 3)\, dx = \int 8xu\, \frac{du}{4r} = 2\int u\, du$$

Integrating with respect to u,

$$2\int u\, du = 2\left(\frac{u^2}{2}\right) + c = u^2 + c \qquad (15.9)$$

Finally, by substituting $u = 2x^2 + 3$ in (15.9) and recalling that c will drop out in the integration, the definite integral can be written in terms of x, incorporating the original limits:

$$\int_0^3 8x(2x^2 + 3)\, dx = (2x^2 + 3)^2 \bigg|_0^3 = [2(3)^2 + 3]^2 - [2(0)^2 + 3]^2 = 441 - 9 = 432$$

Because in the original substitution $u \neq x$ but $2x^2 + 3$, the limits of integration in terms of x will differ from the limits of integration in terms of u. The limits can be expressed in terms of u, if so desired. Since we have set $u = 2x^2 + 3$ and x ranges from 0 to 3, the limits in terms of u are $u = 2(3)^2 + 3 = 21$ and $u = 2(0)^2 + 3 = 3$. Using these limits with the integral expressed in terms of u, as in (15.9),

$$2\int_3^{21} u\, du = u^2 \bigg|_3^{21} = 441 - 9 = 432$$

15.3.   Redo Problem 15.2, given $\int_1^2 x^2(x^3 - 5)^2\, dx$.

Let $u = x^3 - 5$, $du/dx = 3x^2$, and $dx = du/3x^2$. Substituting independently of the limits,

$$\int x^2(x^3 - 5)^2\, dx = \int x^2 u^2\, \frac{du}{3x^2} = \frac{1}{3}\int u^2\, du$$

Integrating with respect to u and ignoring the constant,

$$\frac{1}{3}\int u^2\, du = \frac{1}{3}(\tfrac{1}{3}u^3) = \tfrac{1}{9}u^3$$

Substituting $u = x^3 - 5$ and incorporating the limits for x,

$$\int_1^2 x^2(x^3 - 5)^2\, dx = [\tfrac{1}{9}(x^3 - 5)^3]_1^2$$
$$= \tfrac{1}{9}[(2)^3 - 5]^3 - \tfrac{1}{9}[(1)^3 - 5]^3 = \tfrac{1}{9}(27) - \tfrac{1}{9}(-64) = 10.11$$

Since $u = x^3 - 5$ and the limits for x are $x = 1$ and $x = 2$, by substitution the limits for u are $u = (1)^3 - 5 = -4$ and $u = (2)^3 - 5 = 3$. Incorporating these limits for the integral with respect to u,

$$\frac{1}{3}\int_{-4}^3 u^2\, du = [\tfrac{1}{9}u^3]_{-4}^3 = \tfrac{1}{9}(3)^3 - \tfrac{1}{9}(-4)^3 = 10.11$$

15.4.   Redo Problem 15.2, given

$$\int_0^2 \frac{3x^2}{(x^3 + 1)^2}\, dx$$

Let $u = x^3 + 1$. Then $du/dx = 3x^2$ and $dx = du/3x^2$. Substituting,

$$\int \frac{3x^2}{(x^3 + 1)^2}\, dx = \int 3x^2 u^{-2}\, \frac{du}{3x^2} = \int u^{-2}\, du$$

Integrating with respect to u and ignoring the constant,

$$\int u^{-2}\,du = -u^{-1}$$

Substituting $u = x^3 + 1$ with the original limits,

$$\int_0^2 \frac{3x^2}{(x^3+1)^2}\,dx = -(x^3+1)^{-1}\bigg|_0^2 = \frac{-1}{2^3+1} - \frac{-1}{0^3+1} = -\frac{1}{9} + 1 = \frac{8}{9}$$

With $u = x^3 + 1$, and the limits of x ranging from 0 to 2, the limits of u are $u = (0)^3 + 1 = 1$ and $u = (2)^3 + 1 = 9$. Thus,

$$\int_1^9 u^{-2}\,du = -u^{-1}\bigg|_1^9 = \left(-\tfrac{1}{9}\right) - \left(-\tfrac{1}{1}\right) = \tfrac{8}{9}$$

15.5. Integrate the following definite integral by means of the substitution method:

$$\int_0^3 \frac{6x}{x^2+1}\,dx$$

Let $u = x^2 + 1$, $du/dx = 2x$, and $dx = du/2x$. Substituting,

$$\int \frac{6x}{x^2+1}\,dx = \int 6xu^{-1}\frac{du}{2x} = 3\int u^{-1}\,du$$

Integrating with respect to u,

$$3\int u^{-1}\,du = 3\ln|u|$$

Substituting $u = x^2 + 1$,

$$\int_0^3 \frac{6x}{x^2+1}\,dx = 3\ln|x^2+1|\,\bigg|_0^3$$

$$= 3\ln|3^2+1| - 3\ln|0^2+1| = 3\ln 10 - 3\ln 1$$

Since $\ln 1 = 0$, $= 3\ln 10 = 6.9078$

The limits of u are $u = (0)^2 + 1 = 1$ and $u = (3)^2 + 1 = 10$. Integrating with respect to u,

$$3\int_1^{10} u^{-1}\,du = 3\ln|u|\,\bigg|_1^{10} = 3\ln 10 - 3\ln 1 = 3\ln 10 = 6.9078$$

15.6. Redo Problem 15.5, given $\int_1^2 4xe^{x^2+2}\,dx$.

Let $u = x^2 + 2$. Then $du/dx = 2x$ and $dx = du/2x$. Substituting,

$$\int 4xe^{x^2+2}\,dx = \int 4xe^u \frac{du}{2x} = 2\int e^u\,du$$

Integrating with respect to u and ignoring the constant,

$$2\int e^u\,du = 2e^u$$

Substituting $u = x^2 + 2$,

$$\int_1^2 4xe^{x^2+2}\,dx = 2e^{x^2+2}\,\bigg|_1^2 = 2(e^{(2)^2+2} - e^{(1)^2+2}) = 2(e^6 - e^3)$$

$$= 2(403.43 - 20.09) = 766.68$$

With $u = x^2 + 2$, the limits of u are $u = (1)^2 + 2 = 3$ and $u = (2)^2 + 2 = 6$.

$$2 \int_3^6 e^u \, du = 2e^u \Big|_3^6 = 2(e^6 - e^3) = 766.68$$

15.7. Redo Problem 15.5, given $\int_0^1 3x^2 e^{2x^3+1} \, dx$.

Let $u = 2x^3 + 1$, $du/dx = 6x^2$, and $dx = du/6x^2$. Substituting,

$$\int 3x^2 e^{2x^3+1} \, dx = \int 3x^2 e^u \frac{du}{6x^2} = \frac{1}{2} \int e^u \, du$$

Integrating with respect to u,

$$\tfrac{1}{2} \int e^u \, du = \tfrac{1}{2} e^u$$

Substituting $u = 2x^3 + 1$,

$$\int_0^1 3x^2 e^{2x^3+1} \, dx = \tfrac{1}{2} e^{2x^3+1} \Big|_0^1 = \tfrac{1}{2}(e^3 - e^1) = \tfrac{1}{2}(20.086 - 2.718) = 8.684$$

With $u = 2x^3 + 1$, the limits of u are $u = 2(0)^3 + 1 = 1$ and $u = 2(1)^3 + 1 = 3$. Thus

$$\tfrac{1}{2} \int_1^3 e^u \, du = \tfrac{1}{2} e^u \Big|_1^3 = \tfrac{1}{2}(e^3 - e^1) = 8.68$$

INTEGRATION BY PARTS

15.8. Integrate the following definite integral, using the method of integration by parts:

$$\int_2^5 \frac{3x}{(x+1)^2} \, dx$$

Let $f(x) = 3x$; then $f'(x) = 3$. Let $g'(x) = (x+1)^{-2}$; then $g(x) = \int (x+1)^{-2} \, dx = -(x+1)^{-1}$. Substituting in (14.1),

$$\int \frac{3x}{(x+1)^2} \, dx = 3x[-(x+1)^{-1}] - \int -(x+1)^{-1} 3 \, dx$$
$$= -3x(x+1)^{-1} + 3 \int (x+1)^{-1} \, dx$$

Integrating and ignoring the constant,

$$\int \frac{3x}{(x+1)^2} \, dx = -3x(x+1)^{-1} + 3 \ln|x+1|$$

Applying the limits,

$$\int_2^5 \frac{3x}{(x+1)^2} \, dx = [-3x(x+1)^{-1} + 3\ln|x+1|]_2^5$$
$$= \left[-\frac{3(5)}{5+1} + 3\ln|5+1| \right] - \left[-\frac{3(2)}{2+1} + 3\ln|2+1| \right]$$
$$= -\tfrac{5}{2} + 3\ln 6 + 2 - 3\ln 3$$
$$= 3(\ln 6 - \ln 3) - \tfrac{1}{2} = 3(1.7918 - 1.0986) - 0.5 = 1.5796$$

15.9. Redo Problem 15.8, given

$$\int_1^3 \frac{4x}{(x+2)^3}\, dx$$

Let $f(x) = 4x$, $f'(x) = 4$, $g'(x) = (x+2)^{-3}$, and $g(x) = \int (x+2)^{-3}\, dx = -\frac{1}{2}(x+2)^{-2}$. Substituting in (14.1),

$$\int \frac{4x}{(x+2)^3}\, dx = 4x\left[-\frac{1}{2}(x+2)^{-2}\right] - \int -\frac{1}{2}(x+2)^{-2} 4\, dx$$

$$= -2x(x+2)^{-2} + 2\int (x+2)^{-2}\, dx$$

Integrating,

$$\int \frac{4x}{(x+2)^3}\, dx = -2x(x+2)^{-2} - 2(x+2)^{-1}$$

Applying the limits,

$$\int_1^3 \frac{4x}{(x+2)^3}\, dx = [-2x(x+2)^{-2} - 2(x+2)^{-1}]_1^3$$

$$= [-2(3)(3+2)^{-2} - 2(3+2)^{-1}] - [-2(1)(1+2)^{-2} - 2(1+2)^{-1}]$$

$$= -\tfrac{6}{25} - \tfrac{2}{5} + \tfrac{2}{9} + \tfrac{2}{3} = \tfrac{56}{225}$$

15.10. Redo Problem 15.8, given $\int_1^3 5xe^{x+2}\, dx$.

Let $f(x) = 5x$, $f'(x) = 5$, $g'(x) = e^{x+2}$, and $g(x) = \int e^{x+2}\, dx = e^{x+2}$. Applying (14.1),

$$\int 5xe^{x+2}\, dx = 5xe^{x+2} - \int e^{x+2} 5\, dx = 5xe^{x+2} - 5\int e^{x+2}\, dx$$

Integrating,

$$\int 5xe^{x+2}\, dx = 5xe^{x+2} - 5e^{x+2}$$

Applying the limits,

$$\int_1^3 5xe^{x+2}\, dx = [5xe^{x+2} - 5e^{x+2}]_1^3 = (15e^5 - 5e^5) - (5e^3 - 5e^3) = 10e^5 = 10(148.4) = 1484$$

PROPERTIES OF DEFINITE INTEGRALS

15.11. Show $\int_{-4}^4 (8x^3 + 9x^2)\, dx = \int_{-4}^0 (8x^3 + 9x^2)\, dx + \int_0^4 (8x^3 + 9x^2)\, dx$.

$$\int_{-4}^4 (8x^3 + 9x^2)\, dx = 2x^4 + 3x^3 \Big|_{-4}^4 = 704 - 320 = 384$$

$$\int_{-4}^0 (8x^3 + 9x^2)\, dx = 2x^4 + 3x^3 \Big|_{-4}^0 = 0 - 320 = -320$$

$$\int_0^4 (8x^3 + 9x^2)\, dx = 2x^4 + 3x^3 \Big|_0^4 = 704 - 0 = 704$$

Checking this answer, $-320 + 704 = 384$

15.12. Show $\int_0^{16} (x^{-1/2} + 3x)\,dx = \int_0^4 (x^{-1/2} + 3x)\,dx + \int_4^9 (x^{-1/2} + 3x)\,dx + \int_9^{16}(x^{-1/2} + 3x)\,dx.$

$$\int_0^{16} (x^{-1/2} + 3x)\,dx = 2x^{1/2} + 1.5x^2 \Big|_0^{16} = 392 - 0 = 392$$

$$\int_0^4 (x^{-1/2} + 3x)\,dx = 2x^{1/2} + 1.5x^2 \Big|_0^4 = 28 - 0 = 28$$

$$\int_4^9 (x^{-1/2} + 3x)\,dx = 2x^{1/2} + 1.5x^2 \Big|_4^9 = 127.5 - 28 = 99.5$$

$$\int_9^{16} (x^{-1/2} + 3x)\,dx = 2x^{1/2} + 1.5x^2 \Big|_9^{16} = 392 - 127.5 = 264.5$$

Checking this answer, $28 + 99.5 + 264.5 = 392$

15.13. Show

$$\int_0^3 \frac{6x}{x^2 + 1}\,dx = \int_0^1 \frac{6x}{x^2 + 1}\,dx + \int_1^2 \frac{6x}{x^2 + 1}\,dx + \int_2^3 \frac{6x}{x^2 + 1}\,dx$$

From Problem 15.5,

$$\int_0^3 \frac{6x}{x^2 + 1}\,dx = 3\ln|x^2 + 1| \Big|_0^3 = 3\ln 10$$

$$\int_0^1 \frac{6x}{x^2 + 1}\,dx = 3\ln|x^2 + 1| \Big|_0^1 = 3\ln 2 - 0 = 3\ln 2$$

$$\int_1^2 \frac{6x}{x^2 + 1}\,dx = 3\ln|x^2 + 1| \Big|_1^2 = 3\ln 5 - 3\ln 2$$

$$\int_2^3 \frac{6x}{x^2 + 1}\,dx = 3\ln|x^2 + 1| \Big|_2^3 = 3\ln 10 - 3\ln 5$$

Checking this answer, $3\ln 2 + 3\ln 5 - 3\ln 2 + 3\ln 10 - 3\ln 5 = 3\ln 10$

15.14. Show $\int_1^3 5xe^{x+2}\,dx = \int_1^2 5xe^{x+2}\,dx + \int_2^3 5xe^{x+2}\,dx.$

From Problem 15.10,

$$\int_1^3 5xe^{x+2}\,dx = [5xe^{x+2} - 5e^{x+2}]_1^3 = 10e^5$$

$$\int_1^2 5xe^{x+2}\,dx = [5xe^{x+2} - 5e^{x+2}]_1^2 = (10e^4 - 5e^4) - (5e^3 - 5e^3) = 5e^4$$

$$\int_2^3 5xe^{x+2}\,dx = [5xe^{x+2} - 5e^{x+2}]_2^3 = (15e^5 - 5e^5) - (10e^4 - 5e^4) = 10e^5 - 5e^4$$

Checking this answer, $5e^4 + 10e^5 - 5e^4 = 10e^5$

AREA BETWEEN CURVES

15.15. (*a*) Draw the graphs of the following functions, and (*b*) evaluate the area between the curves over the stated interval:

$$y_1 = 7 - x \quad \text{and} \quad y_2 = 4x - x^2 \quad \text{from } x = 1 \text{ to } x = 4$$

a)   See Fig. 15-6.
b)   From Fig. 15-6, the desired region is the area under the curve specified by $y_1 = 7 - x$ from $x = 1$ to

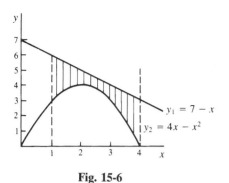

Fig. 15-6

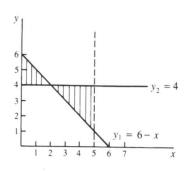

Fig. 15-7

$x = 4$ minus the area under the curve specified by $y_2 = 4x - x^2$ from $x = 1$ to $x = 4$. Using the properties of definite integrals,

$$A = \int_1^4 (7 - x)\,dx - \int_1^4 (4x - x^2)\,dx = \int_1^4 (x^2 - 5x + 7)\,dx$$

$$= [\tfrac{1}{3}x^3 - 2.5x^2 + 7x]_1^4$$

$$= [\tfrac{1}{3}(4)^3 - 2.5(4)^2 + 7(4)] - [\tfrac{1}{3}(1)^3 - 2.5(1)^2 + 7(1)] = 4.5$$

15.16. Redo Problem 15.15, given

$$y_1 = 6 - x \qquad \text{and} \qquad y_2 = 4 \qquad \text{from } x = 0 \text{ to } x = 5$$

Notice the shift in the relative positions of the curves at the point of intersection.

a) See Fig. 15-7.

b) From Fig. 15-7, the desired area is the area between $y_1 = 6 - x$ and $y_2 = 4$ from $x = 0$ to $x = 2$ plus the area between $y_2 = 4$ and $y_1 = 6 - x$ from $x = 2$ to $x = 5$. Mathematically,

$$A = \int_0^2 [(6 - x) - 4]\,dx + \int_2^5 [4 - (6 - x)]\,dx$$

$$= \int_0^2 (2 - x)\,dx + \int_2^5 (x - 2)\,dx$$

$$= [2x - \tfrac{1}{2}x^2]_0^2 + [\tfrac{1}{2}x^2 - 2x]_2^5 = 2 - 0 + 2.5 - (-2) = 6.5$$

15.17. Redo Problem 15.15, given

$$y_1 = x^2 - 4x + 8 \qquad \text{and} \qquad y_2 = 2x \qquad \text{from } x = 0 \text{ to } x = 3$$

a) See Fig. 15-8.

b)

$$A = \int_0^2 [(x^2 - 4x + 8) - 2x]\,dx + \int_2^3 [2x - (x^2 - 4x + 8)]\,dx$$

$$= \int_0^2 (x^2 - 6x + 8)\,dx + \int_2^3 (-x^2 + 6x - 8)\,dx$$

$$= [\tfrac{1}{3}x^3 - 3x^2 + 8x]_0^2 + [-\tfrac{1}{3}x^3 + 3x^2 - 8x]_2^3 = 7\tfrac{1}{3}$$

15.18. Redo Problem 15.15, given

$$y_1 = x^2 - 4x + 12 \qquad \text{and} \qquad y_2 = x^2 \qquad \text{from } x = 0 \text{ to } x = 4$$

a) See Fig. 15-9.

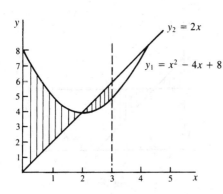

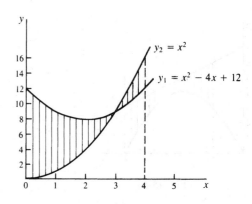

Fig. 15-8   Fig. 15-9

b)
$$A = \int_0^3 \left[(x^2 - 4x + 12) - x^2\right] dx + \int_3^4 \left[x^2 - (x^2 - 4x + 12)\right] dx$$

$$= \int_0^3 (12 - 4x)\, dx + \int_3^4 (4x - 12)\, dx$$

$$= \left[12x - 2x^2\right]_0^3 + \left[2x^2 - 12x\right]_3^4 = 20$$

IMPROPER INTEGRALS AND L'HÔPITAL'S RULE

15.19. (*a*) Specify why the integral given below is improper and (*b*) test for convergence. Evaluate where possible.

$$\int_1^\infty \frac{2x}{(x^2 + 1)^2}\, dx$$

a) This is an example of an improper integral because the upper limit of integration is infinite.

b)
$$\int_1^\infty \frac{2x}{(x^2 + 1)^2}\, dx = \lim_{b \to \infty} \int_1^b \frac{2x}{(x^2 + 1)^2}\, dx$$

Let $u = x^2 + 1$, $du/dx = 2x$, and $dx = du/2x$. Substituting,

$$\int \frac{2x}{(x^2 + 1)^2}\, dx = \int 2xu^{-2} \frac{du}{2x} = \int u^{-2}\, du$$

Integrating with respect to u and ignoring the constant,

$$\int u^{-2}\, du = -u^{-1}$$

Substituting $u = x^2 + 1$ and incorporating the limits of x,

$$\int_1^\infty \frac{2x}{(x^2 + 1)^2}\, dx = \lim_{b \to \infty} \int_1^b \frac{2x}{(x^2 + 1)^2}\, dx = -(x^2 + 1)^{-1} \Big|_1^b$$

$$= \frac{-1}{b^2 + 1} + \frac{1}{(1)^2 + 1} = \frac{1}{2} - \frac{1}{b^2 + 1}$$

As $b \to \infty$, $1/(b^2 + 1) \to 0$. The integral converges and has a value of $\frac{1}{2}$.

15.20. Redo Problem 15.19, given

$$\int_{1}^{\infty} \frac{dx}{x+7}$$

a)   This is an improper integral because one of its limits of integration is infinite.

b)
$$\int_{1}^{\infty} \frac{dx}{x+7} = \lim_{b \to \infty} \int_{1}^{b} \frac{dx}{x+7} = \ln|x+7| \Big|_{1}^{b}$$
$$= \ln|b+7| - \ln|1+7|$$

As $b \to \infty$, $\ln|b+7| \to \infty$. The integral diverges and is meaningless.

15.21. Redo Problem 15.19, given $\int_{-\infty}^{0} e^{3x}\, dx$.

a)   The lower limit is infinite.

b)
$$\int_{-\infty}^{0} e^{3x}\, dx = \lim_{a \to -\infty} \int_{a}^{0} e^{3x}\, dx = \tfrac{1}{3}e^{3x} \Big|_{a}^{0}$$
$$= \tfrac{1}{3}e^{3(0)} - \tfrac{1}{3}e^{3a} = \tfrac{1}{3} - \tfrac{1}{3}e^{3a}$$

As $a \to -\infty$, $\tfrac{1}{3}e^{3a} \to 0$. The integral converges and has a value of $\tfrac{1}{3}$.

15.22. (a) Specify why the integral given below is improper and (b) test for convergence. Evaluate where possible:

$$\int_{-\infty}^{0} (5-x)^{-2}\, dx$$

a)   The lower limit is infinite.

b)
$$\int_{-\infty}^{0} (5-x)^{-2}\, dx = \lim_{a \to -\infty} \int_{a}^{0} (5-x)^{-2}\, dx$$

Let $u = 5 - x$, $du/dx = -1$, and $dx = -du$. Substituting,

$$\int (5-x)^{-2}\, dx = \int u^{-2}(-du) = -\int u^{-2}\, du$$

Integrating with respect to u,

$$-\int u^{-2}\, du = u^{-1}$$

Substituting $u = 5 - x$ and incorporating the limits of x,

$$\int_{-\infty}^{0} (5-x)^{-2}\, dx = \lim_{a \to -\infty} \int_{a}^{0} (5-x)^{-2}\, dx = (5-x)^{-1} \Big|_{a}^{0}$$
$$= \frac{1}{5-0} - \frac{1}{5-a} = \frac{1}{5} - \frac{1}{5-a}$$

As $a \to -\infty$, $1/(5-a) \to 0$. The integral converges and equals $\tfrac{1}{5}$.

15.23. Redo Problem 15.22, given $\int_{-\infty}^{0} 2xe^{x}\, dx$.

a)   The lower limit is infinite.

b)
$$\int_{-\infty}^{0} 2xe^{x}\, dx = \lim_{a \to -\infty} \int_{a}^{0} 2xe^{x}\, dx$$

Using integration by parts, let $f(x) = 2x$, $f'(x) = 2$, $g'(x) = e^x$, and $g(x) = \int e^x \, dx = e^x$. Substituting in (14.1),

$$\int 2xe^x \, dx = 2xe^x - \int e^x 2 \, dx$$

Integrating once again,

$$\int 2xe^x \, dx = 2xe^x - 2e^x$$

Incorporating the limits,

$$\int_{-\infty}^{0} 2xe^x \, dx = \lim_{a \to -\infty} \int_{a}^{0} 2xe^x \, dx = (2xe^x - 2e^x)\Big|_{a}^{0}$$
$$= [2(0)e^0 - 2e^0] - (2ae^a - 2e^a)$$
$$= -2 - 2ae^a + 2e^a \quad \text{since} \quad e^0 = 1$$

As $a \to -\infty$, $e^a \to 0$. Therefore the integral converges and has a value of -2.

15.24. Redo Problem 15.22, given

$$\int_{0}^{6} \frac{dx}{x - 6}$$

a) This is also an improper integral because, as x approaches 6 from the left $(x \to 6^-)$, the integrand $\to -\infty$.

b)
$$\int_{0}^{6} \frac{dx}{x - 6} = \lim_{b \to 6} \int_{0}^{b} \frac{dx}{x - 6} = \ln|x - 6| \Big|_{0}^{b}$$
$$= \ln|b - 6| - \ln|0 - 6|$$

As $b \to 6^-$, $|b - 6| \to 0$ and $\ln 0$ is undefined. Therefore, the integral diverges and is meaningless.

15.25. Redo Problem 15.22, given $\int_{0}^{8} (8 - x)^{-1/2} \, dx$.

a) As $x \to 8^-$, the integrand approaches infinity.

b)
$$\int_{0}^{8} (8 - x)^{-1/2} \, dx = \lim_{b \to 8} \int_{0}^{b} (8 - x)^{-1/2} \, dx = -2(8 - x)^{1/2} \Big|_{0}^{b}$$
$$= (-2\sqrt{8 - b}) - (-2\sqrt{8 - 0}) = 2\sqrt{8} - 2\sqrt{8 - b}$$

As $b \to 8^-$, $-2\sqrt{8 - b} \to 0$. The integral converges and has a value of $2\sqrt{8} = 4\sqrt{2}$.

15.26. Use L'Hôpital's rule to evaluate the following limits:

a) $\lim_{x \to \infty} \dfrac{5x - 9}{e^x}$

As $x \to \infty$, both $5x - 9$ and e^x tend to ∞, giving rise to the indeterminate form ∞/∞. Using (15.6), therefore, and differentiating numerator and denominator separately,

$$\lim_{x \to \infty} \frac{5x - 9}{e^x} = \lim_{x \to \infty} \frac{5}{e^x} = \frac{5}{\infty} = 0$$

b) $\lim_{x \to \infty} \dfrac{1 - e^{1/x}}{1/x}$

As $x \to \infty$, $1 - e^{1/x}$ and $1/x \to 0$. Using (15.6), therefore, and recalling that $1/x = x^{-1}$,

$$\lim_{x \to \infty} \frac{1 - e^{1/x}}{1/x} = \lim_{x \to \infty} \frac{-(-1/x^2)e^{1/x}}{-1/x^2}$$

Simplifying algebraically,

$$\lim_{x \to \infty} \frac{1 - e^{1/x}}{1/x} = \lim_{x \to \infty} \left(-e^{1/x}\right) = -e^0 = -1$$

c) $\lim\limits_{x \to \infty} \dfrac{\ln 2x}{e^{5x}}$

As $x \to \infty$, $\ln 2x$ and $e^{5x} \to \infty$. Again using (15.6),

$$\lim_{x \to \infty} \frac{\ln 2x}{e^{5x}} = \lim_{x \to \infty} \frac{1/x}{5e^{5x}} = \frac{0}{\infty} = 0 \qquad \text{since } \frac{0}{\infty} \text{ is not an indeterminate form.}$$

d) $\lim\limits_{x \to \infty} \dfrac{6x^3 - 7}{3x^2 + 9}$

$$\lim_{x \to \infty} \frac{6x^3 - 7}{3x^2 + 9} = \lim_{x \to \infty} \frac{18x^2}{6x} = \lim_{x \to \infty} 3x = \infty$$

e) $\lim\limits_{x \to \infty} \dfrac{3x^2 - 7x}{4x^2 - 21}$

$$\lim_{x \to \infty} \frac{3x^2 - 7x}{4x^2 - 21} = \lim_{x \to \infty} \frac{6x - 7}{8x}$$

Whenever application of L'Hôpital's rule gives rise to a new quotient whose limit is also an indeterminate form, L'Hôpital's rule must be applied again. Thus,

$$\lim_{x \to \infty} \frac{6x - 7}{8x} = \lim_{x \to \infty} \frac{6}{8} = \frac{3}{4} \qquad \text{See Problem 3.4(c).}$$

f) $\lim\limits_{x \to \infty} \dfrac{8x^3 - 5x^2 + 13x}{2x^3 + 7x^2 - 18x}$

Using L'Hôpital's rule repeatedly,

$$\lim_{x \to \infty} \frac{8x^3 - 5x^2 + 13x}{2x^3 + 7x^2 - 18x} = \lim_{x \to \infty} \frac{24x^2 - 10x + 13}{6x^2 + 14x - 18} = \lim_{x \to \infty} \frac{48x - 10}{12x + 14}$$

$$= \lim_{x \to \infty} \tfrac{48}{12} = 4$$

CONSUMERS' AND PRODUCERS' SURPLUS

15.27. Given the demand function $P = 45 - 0.5Q$, find the consumers' surplus CS when $P_0 = 32.5$ and $Q_0 = 25$.

Using (15.7),

$$CS = \int_0^{25} (45 - 0.5Q)\, dQ - (32.5)(25) = [45Q - 0.25Q^2]_0^{25} - 812.5$$

$$= [45(25) - 0.25(25)^2] - 0 - 812.5 = 156.25$$

15.28. Given the supply function $P = (Q+3)^2$, find the producers' surplus PS at $P_0 = 81$ and $Q_0 = 6$.

From (*15.8*),

$$PS = (81)(6) - \int_0^6 (Q+3)^2 \, dQ = 486 - [\tfrac{1}{3}(Q+3)^3]_0^6$$

$$= 486 - [\tfrac{1}{3}(6-3)^3 - \tfrac{1}{3}(0+3)^3] = 252$$

15.29. Given the demand function $P_d = 25 - Q^2$ and the supply function $P_s = 2Q + 1$. Assuming pure competition, find (*a*) the consumers' surplus and (*b*) the producers' surplus.

For market equilibrium, $s = d$. Thus,

$$2Q + 1 = 25 - Q^2 \qquad Q^2 + 2Q - 24 = 0$$
$$(Q+6)(Q-4) = 0 \qquad Q_0 = 4 \qquad P_0 = 9$$

since Q_0 cannot equal -6.

a)
$$CS = \int_0^4 (25 - Q^2) \, dQ - (9)(4) = [25Q - \tfrac{1}{3}Q^3]_0^4 - 36$$

$$= [25(4) - \tfrac{1}{3}(4)^3] - 0 - 36 = 42.67$$

b)
$$PS = (9)(4) - \int_0^4 (2Q + 1) \, dQ$$

$$= 36 - [Q^2 + Q]_0^4 = 16$$

15.30. Given the demand function $P_d = 113 - Q^2$ and the supply function $P_s = (Q+1)^2$ under pure competition, find (*a*) CS and (*b*) PS.

Multiplying the supply function out and equating supply and demand,

$$Q^2 + 2Q + 1 = 113 - Q^2 \qquad 2(Q^2 + Q - 56) = 0$$
$$(Q+8)(Q-7) = 0 \qquad Q_0 = 7 \qquad P_0 = 64$$

a)
$$CS = \int_0^7 (113 - Q^2) \, dQ - (64)(7) = [113Q - \tfrac{1}{3}Q^3]_0^7 - 448 = 228.67$$

b)
$$PS = (64)(7) - \int_0^7 (Q+1)^2 \, dQ = 448 - [\tfrac{1}{3}(Q+1)^3]_0^7 = 448 - (170.67 - 0.33) = 277.67$$

15.31. Under a monopoly, the quantity sold and market price are determined by the demand function. If the demand function for a profit-maximizing monopolist is $P = 274 - Q^2$ and $MC = 4 + 3Q$, find the consumers' surplus.

Given $P = 274 - Q^2$,

$$TR = PQ = (274 - Q^2)Q = 274Q - Q^3$$

and

$$MR = \frac{dTR}{dQ} = 274 - 3Q^2$$

The monopolist maximizes profit at $MR = MC$. Thus,

$$274 - 3Q^2 = 4 + 3Q \qquad 3(Q^2 + Q - 90) = 0$$
$$(Q+10)(Q-9) = 0 \qquad Q_0 = 9 \qquad P_0 = 193$$

and

$$CS = \int_0^9 (274 - Q^2) \, dQ - (193)(9) = [274Q - \tfrac{1}{3}Q^3]_0^9 - 1737 = 486$$

FREQUENCY FUNCTIONS AND PROBABILITY

15.32. The probability in minutes of being waited on in a large chain restaurant is given by the frequency function $f(t) = \frac{4}{81}t^3$ for $0 \le t \le 3$. What is the probability of being waited on between 1 and 2 minutes?

$$P = \int_1^2 \frac{4}{81}t^3\,dt = \frac{1}{81}t^4\,\Big|_1^2 = \frac{1}{81}(16) - \frac{1}{81}(1) = 0.1852$$

15.33. The proportion of assignments completed within a given day is described by the probability density function $f(x) = 12(x^2 - x^3)$ for $0 \le x \le 1$. What is the probability that (a) 50 percent or less of the assignments will be completed within the day and (b) 50 percent or more will be completed?

a)
$$P_a = \int_0^{0.5} 12(x^2 - x^3)\,dx = 12\left[\frac{x^3}{3} - \frac{x^4}{4}\right]_0^{0.5}$$
$$= 12\left[\left(\frac{0.125}{3} - \frac{0.0625}{4}\right) - 0\right] = 0.3125$$

b)
$$P_b = \int_{0.5}^1 12(x^2 - x^3)\,dx = 12\left[\frac{x^3}{3} - \frac{x^4}{4}\right]_{0.5}^1$$
$$= 12\left[\left(\frac{1}{3} - \frac{1}{4}\right) - \left(\frac{0.125}{3} - \frac{0.0625}{4}\right)\right] = 0.6875$$

As expected, $P_a + P_b = 0.3125 + 0.6875 = 1$.

OTHER ECONOMIC APPLICATIONS

15.34. Given $I(t) = 9t^{1/2}$, find the level of capital formation in (a) 8 years and (b) for the fifth through the eighth years (interval [4, 8]).

a)
$$K = \int_0^8 9t^{1/2}\,dt = 6t^{3/2}\,\Big|_0^8 = 6(8)^{3/2} - 0 = 96\sqrt{2} = 135.76$$

b)
$$K = \int_4^8 9t^{1/2}\,dt = 6t^{3/2}\,\Big|_4^8 = 6(8)^{3/2} - 6(4)^{3/2} = 135.76 - 48 = 87.76$$

CHAPTER 16

First-Order Differential Equations

16.1 DEFINITIONS AND CONCEPTS

A *differential equation* is an equation which expresses an explicit or implicit relationship between a function $y = f(t)$ and one or more of its derivatives or differentials. Examples of differential equations include

$$\frac{dy}{dt} = 5t + 9 \qquad y' = 12y \qquad \text{and} \qquad y'' - 2y' + 19 = 0$$

Equations involving a single independent variable, such as those above, are called *ordinary differential equations*. The *solution* or *integral* of a differential equation is any equation, without derivative or differential, that is defined over an interval and satisfies the differential equation for all the values of the independent variable(s) in the interval. See Example 1.

The *order* of a differential equation is the order of the highest derivative in the equation. The *degree* of a differential equation is the highest power to which the derivative of highest order is raised. See Example 2 and Problem 16.1.

EXAMPLE 1. To solve the differential equation $y''(t) = 7$ for all the functions $y(t)$ which satisfy the equation, simply integrate both sides of the equation to find the integrals.

$$y'(t) = \int 7 \, dt = 7t + c_1$$

$$y(t) = \int (7t + c_1) \, dt = 3.5t^2 + c_1 t + c$$

This is called a *general solution* which indicates that when c is unspecified, a differential equation has an infinite number of possible solutions. If c can be specified, the differential equation has a *particular* or *definite solution* which alone of all possible solutions is relevant.

EXAMPLE 2.   The order and degree of differential equations are shown below.

1. $\dfrac{dy}{dt} = 2x + 6$ first-order, first-degree

2. $\left(\dfrac{dy}{dt}\right)^4 - 5t^5 = 0$ first-order, fourth-degree

3. $\dfrac{d^2y}{dt^2} + \left(\dfrac{dy}{dt}\right)^3 + x^2 = 0$ second-order, first-degree

4. $\left(\dfrac{d^2y}{dt^2}\right)^7 + \left(\dfrac{d^3y}{dt^3}\right)^5 = 75y$ third-order, fifth-degree

16.2   GENERAL FORMULA FOR FIRST-ORDER LINEAR DIFFERENTIAL EQUATIONS

For a first-order *linear* differential equation, dy/dt and y must be of the first degree, and no product $y(dy/dt)$ may occur. For such an equation

$$\frac{dy}{dt} + vy = z$$

where v and z may be constants or functions of time, the formula for a *general solution* is

$$y(t) = e^{-\int v\,dt}\left(A + \int z e^{\int v\,dt}\,dt\right) \tag{16.1}$$

where A is an arbitrary constant. A solution is composed of two parts: $e^{-\int v\,dt}A$ is called the *complementary function*, and $e^{-\int v\,dt}\int z e^{\int v\,dt}\,dt$ is called the *particular integral*. The particular integral y_p equals the *intertemporal equilibrium level* of $y(t)$; the complementary function y_c represents the *deviation from the equilibrium*. For $y(t)$ to be *dynamically stable*, y_c must approach zero as t approaches infinity (that is, k in e^{kt} must be negative). The solution of a differential equation can always be checked by differentiation. See Examples 3 to 5, Problems 16.2 to 16.12, and Problem 20.33.

EXAMPLE 3.   The general solution for the differential equation $dy/dt + 4y = 12$ is calculated as follows. Since $v = 4$ and $z = 12$, substituting in (*16.1*) gives

$$y(t) = e^{-\int 4\,dt}\left(A + \int 12 e^{\int 4\,dt}\,dt\right)$$

From Section 14.2, $\int 4\,dt = 4t + c$. When (*16.1*) is used, c is always ignored and subsumed under A. Thus,

$$y(t) = e^{-4t}\left(A + \int 12 e^{4t}\,dt\right) \tag{16.2}$$

Integrating the remaining integral gives $\int 12 e^{4t}\,dt = 3e^{4t} + c$. Ignoring the constant again and substituting in (*16.2*),

$$y(t) = e^{-4t}(A + 3e^{4t}) = Ae^{-4t} + 3 \tag{16.3}$$

since $e^{-4t}e^{4t} = e^0 = 1$. As $t \to \infty$, $y_c = Ae^{-4t} \to 0$ and $y(t)$ approaches $y_p = 3$, the intertemporal equilibrium level. $y(t)$ is dynamically stable.

To check this answer, which is a general solution because A has not been specified, start by taking the derivative of (*16.3*).

$$\frac{dy}{dt} = -4Ae^{-4t}$$

From the original problem,

$$\frac{dy}{dt} + 4y = 12 \qquad \frac{dy}{dt} = 12 - 4y$$

Substituting $y = Ae^{-4t} + 3$ from (16.3),

$$\frac{dy}{dt} = 12 - 4(Ae^{-4t} + 3) = -4Ae^{-4t}$$

EXAMPLE 4. Given $dy/dt + 3t^2 y = t^2$ where $v = 3t^2$ and $z = t^2$. To find the general solution, first substitute in (16.1),

$$y(t) = e^{-\int 3t^2\, dt}\left(A + \int t^2 e^{\int 3t^2\, dt}\, dt\right) \tag{16.4}$$

Integrating the exponents, $\int 3t^2\, dt = t^3$. Substituting in (16.4),

$$y(t) = e^{-t^3}\left(A + \int t^2 e^{t^3}\, dt\right) \tag{16.5}$$

Integrating the remaining integral in (16.5) calls for the substitution method. Letting $u = t^3$, $du/dt = 3t^2$, and $dt = du/3t^2$,

$$\int t^2 e^{t^3}\, dt = \int t^2 e^u \frac{du}{3t^2} = \frac{1}{3}\int e^u\, du = \frac{1}{3}e^u = \frac{1}{3}e^{t^3}$$

Finally, substituting in (16.5),

$$y(t) = e^{-t^3}(A + \tfrac{1}{3}e^{t^3}) = Ae^{-t^3} + \tfrac{1}{3} \tag{16.6}$$

As $t \to \infty$, $y_c = Ae^{-t^3} \to 0$ and $y(t)$ approaches $\frac{1}{3}$. The equilibrium is dynamically stable.

Differentiating (16.6) to check the general solution, $dy/dt = -3t^2 Ae^{-t^3}$. From the original problem,

$$\frac{dy}{dt} + 3t^2 y = t^2 \qquad \frac{dy}{dt} = t^2 - 3t^2 y$$

Substituting y from (16.6),

$$\frac{dy}{dt} = t^2 - 3t^2\left(Ae^{-t^3} + \frac{1}{3}\right) = -3t^2 Ae^{-t^3}$$

EXAMPLE 5. Suppose that $y(0) = 1$ in Example 4. The definite solution is calculated as follows: From (16.6), $y = Ae^{-t^3} + \frac{1}{3}$. At $t = 0$, $y(0) = 1$. Hence, $1 = A + \frac{1}{3}$ since $e^0 = 1$, and $A = \frac{2}{3}$. Substituting $A = \frac{2}{3}$ in (16.6), the definite solution is $y = \frac{2}{3}e^{-t^3} + \frac{1}{3}$.

16.3 EXACT DIFFERENTIAL EQUATIONS AND PARTIAL INTEGRATION

Given a function of more than one independent variable, such as $F(y,t)$ where $M = \partial F/\partial y$ and $N = \partial F/\partial t$, the total differential is written

$$dF(y,t) = M\, dy + N\, dt \tag{16.7}$$

Since F is a function of more than one independent variable, M and N are partial derivatives and Equation (16.7) is called a *partial differential equation*. If the differential is set equal to zero, so that $M\, dy + N\, dt = 0$, it is called an *exact differential equation* because the left side exactly equals the differential of the primitive function $F(y,t)$. For an exact differential equation, $\partial M/\partial t$ must equal $\partial N/\partial y$, that is, $\partial^2 F/(\partial t\, \partial y) = \partial^2 F/(\partial y\, \partial t)$. For proof of this proposition, see Problem 16.49.

Solution of an exact differential equation calls for successive integration with respect to one independent variable at a time while holding constant the other independent variable(s). The procedure, called *partial integration*, reverses the process of partial differentiation. See Example 6 and Problems 16.13 to 16.17.

EXAMPLE 6.   Solve the exact nonlinear differential equation

$$(6yt + 9y^2)\, dy + (3y^2 + 8t)\, dt = 0 \qquad\qquad (16.8)$$

1. Test to see if it is an exact differential equation. Here $M = 6yt + 9y^2$ and $N = 3y^2 + 8t$. Thus, $\partial M/\partial t = 6y$ and $\partial N/\partial y = 6y$. If $\partial M/\partial t \neq \partial N/\partial y$, it is not an exact differential equation.
2. Since $M = \partial F/\partial y$ is a partial derivative, integrate M partially with respect to y by treating t as a constant, and add a new function $Z(t)$ for any additive terms of t which would have been eliminated by the original differentiation with respect to y. Note that ∂y replaces dy in partial integration.

$$F(y,t) = \int (6yt + 9y^2)\, \partial y + Z(t) = 3y^2 t + 3y^3 + Z(t) \qquad\qquad (16.9)$$

This gives the original function except for the unknown additive terms of t, $Z(t)$.
3. Differentiate (*16.9*) with respect to t to find $\partial F/\partial t$ (earlier called N). Thus,

$$\frac{\partial F}{\partial t} = 3y^2 + Z'(t) \qquad\qquad (16.10)$$

Since $\partial F/\partial t = N$ and $N = 3y^2 + 8t$ from (*16.8*), substitute $\partial F/\partial t = 3y^2 + 8t$ in (*16.10*).

$$3y^2 + 8t = 3y^2 + Z'(t) \qquad Z'(t) = 8t$$

4. Next integrate $Z'(t)$ with respect to t to find the missing t terms.

$$Z(t) = \int Z'(t)\, dt = \int 8t\, dt = 4t^2 \qquad\qquad (16.11)$$

5. Substitute (*16.11*) in (*16.9*), and add a constant of integration.

$$F(y,t) = 3y^2 t + 3y^3 + 4t^2 + c$$

This is easily checked by differentiation.

16.4   INTEGRATING FACTORS

Not all differential equations are exact. However, some can be made exact by means of an *integrating factor*. This is a multiplier which permits the equation to be integrated. See Example 7 and Problems 16.18 to 16.22.

EXAMPLE 7.   Testing the nonlinear differential equation $5yt\, dy + (5y^2 + 8t)\, dt = 0$ reveals that it is not exact. With $M = 5yt$ and $N = 5y^2 + 8t$, $\partial M/\partial t = 5y \neq \partial N/\partial y = 10y$. Multiplying by an integrating factor of t, however, makes it exact: $5yt^2\, dy + (5y^2 t + 8t^2)\, dt = 0$. Now $\partial M/\partial t = 10yt = \partial N/\partial y$, and the equation can be solved by the procedure outlined above. See Problem 16.22.

To check the answer to a problem in which an integrating factor was used, take the total differential of the answer and then divide by the integrating factor.

16.5   RULES FOR THE INTEGRATING FACTOR

Two rules will help to find the integrating factor for a nonlinear first-order differential equation, if such a factor exists. Assuming $\partial M/\partial t \neq \partial N/\partial y$,

Rule 1.   If $\dfrac{1}{N}\left(\dfrac{\partial M}{\partial t} - \dfrac{\partial N}{\partial y}\right) = f(y)$ alone, then $e^{\int f(y)\, dy}$ is an integrating factor.

Rule 2.   If $\dfrac{1}{M}\left(\dfrac{\partial N}{\partial y}-\dfrac{\partial M}{\partial t}\right)=g(t)$ alone, then $e^{\int g(t)\,dt}$ is an integrating factor.

See Example 8 and Problems 16.23 to 16.28.

EXAMPLE 8.   To illustrate the rules above, find the integrating factor given in Example 7, where

$$5yt\,dy+(5y^2+8t)\,dt=0 \qquad M=5yt \qquad N=5y^2+8t \qquad \frac{\partial M}{\partial t}=5y\neq\frac{\partial N}{\partial y}=10y$$

Applying Rule 1,

$$\frac{1}{5y^2+8t}(5y-10y)=\frac{-5y}{5y^2+8t}$$

which is not a function of y alone and will not supply an integrating factor for the equation. Applying Rule 2,

$$\frac{1}{5yt}(10y-5y)=\frac{5y}{5yt}=\frac{1}{t}$$

which is a function of t alone. The integrating factor, therefore, is $e^{\int (1/t)\,dt}=e^{\ln t}=t$.

16.6   SEPARATION OF VARIABLES

Solution of nonlinear first-order first-degree differential equations is complex. (A first-order first-degree differential equation is one in which the highest derivative is the first derivative dy/dt and that derivative is raised to a power of 1. It is *nonlinear* if it contains a product of y and dy/dt, or y raised to a power other than 1.) If the equation is exact or can be rendered exact by an integrating factor, the procedure outlined in Example 6 can be used. If, however, the equation can be written in the form of *separated variables* such that $R(y)\,dy+S(t)\,dt=0$, where R and S, respectively, are functions of y and t alone, the equation can be solved simply by ordinary integration. The procedure is illustrated in Examples 9 and 10 and Problems 16.29 to 16.37.

EXAMPLE 9.   The following calculations illustrate the separation of variables procedure to solve the nonlinear differential equation

$$\frac{dy}{dt}=y^2 t \tag{16.12}$$

First, separating the variables by rearranging terms,

$$\frac{dy}{y^2}=t\,dt$$

where $R=1/y^2$ and $S=t$. Then integrating both sides,

$$\int y^{-2}\,dy=\int t\,dt$$

$$-y^{-1}+c_1=\frac{t^2}{2}+c_2$$

$$-\frac{1}{y}=\frac{t^2+2c_2-2c_1}{2}$$

Letting $c=2c_2-2c_1$,

$$y=\frac{-2}{t^2+c} \tag{16.13}$$

Since the constant of integration is arbitrary until it is evaluated to obtain a particular solution, it will be treated generally and not specifically in the initial steps of the solution. e^c and $\ln c$ can also be used to express the constant.

This solution can be checked as follows: Taking the derivatives of $y = -2(t^2 + c)^{-1}$ by the generalized power function rule,

$$\frac{dy}{dt} = (-1)(-2)(t^2 + c)^{-2}(2t) = \frac{4t}{(t^2 + c)^2}$$

From (16.12), $dy/dt = y^2 t$. Substituting into (16.12) from (16.13),

$$\frac{dy}{dt} = \left(\frac{-2}{t^2 + c}\right)^2 t = \frac{4t}{(t^2 + c)^2}$$

EXAMPLE 10.   Given the nonlinear differential equation

$$t^2 \, dy + y^3 \, dt = 0 \tag{16.14}$$

where $M \neq f(y)$ and $N \neq f(t)$. But multiplying (16.14) by $1/(t^2 y^3)$ to separate the variables gives

$$\frac{1}{y^3} dy + \frac{1}{t^2} dt = 0 \tag{16.14a}$$

Integrating the separated variables,

$$\int y^{-3} \, dy + \int t^{-2} \, dt = -\tfrac{1}{2} y^{-2} - t^{-1} + c$$

and

$$F(y, t) = -\tfrac{1}{2} y^{-2} - t^{-1} + c$$

$$= -\frac{1}{2y^2} - \frac{1}{t} + c$$

For complicated functions, the answer is frequently left in this form. It can be checked by differentiating and comparing with (16.14a), which can be reduced to (16.14) through multiplication by $y^3 t^2$. For other forms in which an answer can be expressed, see Problems 16.20 to 16.22 and 16.29 to 16.35.

16.7 ECONOMIC APPLICATIONS

Differential equations serve many functions in economics. They are used to determine the conditions for dynamic stability in microeconomic models of market equilibria and to trace the time path of growth under various conditions in macroeconomics. Given the growth rate of a function, differential equations enable the economist to find the function whose growth is described; from point elasticity, they enable the economist to estimate the demand function (see Example 11 and Problems 16.38 to 16.47). In Section 14.6 they were used to estimate capital functions from investment functions and total cost and total revenue functions from marginal cost and marginal revenue functions.

EXAMPLE 11.   Given the demand function $Q_d = c + bP$ and the supply function $Q_s = g + hP$, the equilibrium price is

$$\bar{P} = \frac{c - g}{h - b} \tag{16.15}$$

Assume that the rate of change of price in the market dP/dt is a positive linear function of *excess demand* $Q_d - Q_s$ such that

$$\frac{dP}{dt} = m(Q_d - Q_s) \qquad m = \text{a constant} > 0 \tag{16.16}$$

The conditions for dynamic price stability in the market [i.e., under what conditions $P(t)$ will converge to $\bar{P}$ as $t \to \infty$] can be calculated as shown below.

Substituting the given parameters for Q_d and Q_s in (16.16),

$$\frac{dP}{dt} = m[(c + bP) - (g + hP)] = m(c + bP - g - hP)$$

Rearranging to fit the general format of Section 16.2, $dP/dt + m(h - b)P = m(c - g)$. Letting $v = m(h - b)$ and $z = m(c - g)$, and using (16.1),

$$P(t) = e^{-\int v\, dt}\left(A + \int z e^{\int v\, dt}\, dt\right) = e^{-vt}\left(A + \int z e^{vt}\, dt\right)$$

$$= e^{-vt}\left(A + \frac{z e^{vt}}{v}\right) = A e^{-vt} + \frac{z}{v} \tag{16.17}$$

At $t = 0$, $P(0) = A + z/v$ and $A = P(0) - z/v$.

Substituting in (16.17),

$$P(t) = \left[P(0) - \frac{z}{v}\right]e^{-vt} + \frac{z}{v}$$

Finally, replacing $v = m(h - b)$ and $z = m(c - g)$,

$$P(t) = \left[P(0) - \frac{c - g}{h - b}\right]e^{-m(h-b)t} + \frac{c - g}{h - b}$$

and making use of (16.15), the time path is

$$P(t) = [P(0) - \bar{P}]e^{-m(h-b)t} + \bar{P} \tag{16.18}$$

Since $P(0)$, $\bar{P}$, $m > 0$, the first term on the right-hand side will converge toward zero as $t \to \infty$, and thus $P(t)$ will converge toward $\bar{P}$ only if $h - b > 0$. For normal cases where demand is negatively sloped ($b < 0$) and supply is positively sloped ($h > 0$), the dynamic stability condition is assured. Markets with positively sloped demand functions or negatively sloped supply functions will also be dynamically stable as long as $h > b$.

16.8 PHASE DIAGRAMS FOR DIFFERENTIAL EQUATIONS

Many nonlinear differential equations cannot be solved explicitly as functions of time. *Phase diagrams*, however, offer qualitative information about the stability of equations that is helpful in determining whether the equations will converge to an intertemporal (steady-state) equilibrium or not. A phase diagram of a differential equation depicts the derivative which we now express as $\dot{y}$ for simplicity of notation as a function of y. The steady-state solution is easily identified on a phase diagram as any point at which the graph crosses the horizontal axis, because at that point $\dot{y} = 0$ and the function is not changing. For some equations there may be more than one intersection and hence more than one solution.

Diagrammatically, the stability of the steady-state solution(s) is indicated by the *arrows of motion*. The arrows of motion will point to the right (indicating y is increasing) any time the graph of $\dot{y}$ is above the horizontal axis (indicating $\dot{y} > 0$) and to the left (indicating y is decreasing) any time the graph of $\dot{y}$ is below the horizontal axis (indicating $\dot{y} < 0$). If the arrows of motion point towards a steady-state solution, the solution is stable; if the arrows of motion point away from a steady-state solution, the solution is unstable.

Mathematically, the slope of the phase diagram as it passes through a steady-state equilibrium point tells us if the equilibrium point is stable or not. When evaluated at a steady-state equilibrium point,

$$\text{if } \frac{d\dot{y}}{dy} < 0, \qquad \text{the equilibrium is stable}$$

$$\text{if } \frac{d\dot{y}}{dy} > 0, \qquad \text{the point is unstable}$$

Phase diagrams are illustrated in Example 12. The derivative test for stability is used in Example 13. See also Problems 16.48 to 16.50.

EXAMPLE 12.   Given the nonlinear differential equation

$$\dot{y} = 8y - 2y^2$$

a phase diagram can be constructed and employed in six easy steps.

1.   The intertemporal or steady-state solution(s), where there is no pressure for change, is found by setting $\dot{y} = 0$ and solving algebraically.

$$\dot{y} = 8y - 2y^2 = 0$$
$$2y(4 - y) = 0$$
$$\bar{y} = 0 \qquad \bar{y} = 4 \qquad \text{steady-state solutions}$$

The phase diagram will pass through the horizontal axis at $y = 0$, $y = 4$.

2.   Since the function passes through the horizontal axis twice, it has one turning point. We next determine whether that point is a maximum or minimum.

$$\frac{d\dot{y}}{dy} = 8 - 4y = 0 \qquad y = 2 \text{ is a critical value}$$

$$\frac{d^2\dot{y}}{dy^2} = -4 < 0 \qquad \text{concave, relative maximum}$$

3.   A rough, but accurate, sketch of the phase diagram can then easily be drawn. See Fig. 16-1.

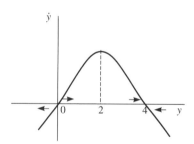

Fig. 16-1

4.   The arrows of motion complete the graph. As explained above, where the graph lies above the horizontal axis, $\dot{y} > 0$ and the arrows must point to the right; where the graph lies below the horizontal axis, $\dot{y} < 0$ and the arrows must point to the left.
5.   The stability of the steady-state equilibrium points can now be read from the graph. Since the arrows of motion point away from the first intertemporal equilibrium $\bar{y}_1 = 0$, $\bar{y}_1$ is an unstable equilibrium. With the arrows of motion pointing toward the second intertemporal equilibrium $\bar{y}_2 = 4$, $\bar{y}_2$ is a stable equilibrium.
6.   The slope of the phase diagram at the steady-state solutions can also be used to test stability independently of the arrows of motion. Since the slope of the phase diagram is positive at $\bar{y}_1 = 0$, we can conclude $\bar{y}_1$ is an unstable equilibrium. Since the slope of the phase diagram is negative at $\bar{y}_2 = 4$, we know $\bar{y}_2$ must be stable.

EXAMPLE 13.   Without even resorting to a phase diagram, we can use the simple first-derivative evaluated at the intertemporal equilibrium level(s) to determine the stability of differential equations. Given $\dot{y} = 8y - 2y^2$,

$$\frac{d\dot{y}}{dy} = 8 - 4y$$

Evaluated at the steady-state levels, $\bar{y}_1 = 0$ and $\bar{y}_2 = 4$,

$$\frac{d\dot{y}}{dy}(0) = 8 - 4(0) = 8 > 0 \qquad \frac{d\dot{y}}{dy}(4) = 8 - 4(4) = -8 < 0$$

$$\bar{y}_1 = 0 \text{ is unstable} \qquad \bar{y}_2 = 4 \text{ is stable}$$

Solved Problems

ORDER AND DEGREE

16.1. Specify the order and degree of the following differential equations:

a) $\dfrac{d^2y}{dx^2} + \left(\dfrac{dy}{dx}\right)^3 = 12x$

b) $\dfrac{dy}{dx} = 3x^2$

c) $\left(\dfrac{d^3y}{dx^3}\right)^4 + \left(\dfrac{d^2y}{dx^2}\right)^6 = 4 - y$

d) $\left(\dfrac{d^2y}{dx^2}\right)^3 + \dfrac{d^4y}{dx^4} - 75y = 0$

e) $\dfrac{d^3y}{dx^3} + x^2y\left(\dfrac{d^2y}{dx^2}\right) - 4y^4 = 0$

(*a*) Second order, first degree; (*b*) first order, first degree; (*c*) third order, fourth degree; (*d*) fourth order, first degree; (*e*) third order, first degree.

FIRST-ORDER FIRST-DEGREE LINEAR DIFFERENTIAL EQUATIONS

16.2. (*a*) Use the formula for a general solution to solve the following equation. (*b*) Check your answer.

$$\frac{dy}{dt} + 5y = 0 \tag{16.19}$$

a)   Here $v = 5$ and $z = 0$. Substituting in (*16.1*),

$$y(t) = e^{-\int 5\,dt}\left(A + \int 0 e^{\int 5\,dt}\,dt\right)$$

Integrating the exponents, $\int 5\,dt = 5t + c$, where c can be ignored because it is subsumed under A. Thus, $y(t) = e^{-5t}(A + \int 0\,dt)$. And $\int 0\,dt = k$, a constant, which can also be subsumed under A. Hence,

$$y(t) = e^{-5t}A = Ae^{-5t} \tag{16.20}$$

b)   Taking the derivative of (*16.20*), $dy/dt = -5Ae^{-5t}$. From (*16.19*), $dy/dt = -5y$. Substituting y from (*16.20*),

$$\frac{dy}{dt} = -5(Ae^{-5t}) = -5Ae^{-5t}$$

16.3. Redo Problem 16.2, given

$$\frac{dy}{dt} = 3y \qquad y(0) = 2 \tag{16.21}$$

a)   Rearranging to obtain the general format,

$$\frac{dy}{dt} - 3y = 0$$

Here $v = -3$ and $z = 0$. Substituting in *(16.1)*,

$$y(t) = e^{-\int -3\,dt}\left(A + \int 0e^{\int -3\,dt}\,dt\right)$$

Substituting $\int -3\,dt = -3t$, $y(t) = e^{3t}(A + \int 0\,dt) = Ae^{3t}$. At $t = 0$, $y = 2$. Thus, $2 = Ae^{3(0)}$, $A = 2$. Substituting,

$$y(t) = 2e^{3t} \tag{16.22}$$

b)   Taking the derivative of *(16.22)*, $dy/dt = 6e^{3t}$. From *(16.21)*, $dy/dt = 3y$. Substituting y from *(16.22)*, $dy/dt = 3(2e^{3t}) = 6e^{3t}$.

16.4.   Redo Problem 16.2, given

$$\frac{dy}{dt} = 15 \tag{16.23}$$

a)   Here $v = 0$ and $z = 15$. Thus,

$$y(t) = e^{-\int 0\,dt}\left(A + \int 15e^{\int 0\,dt}\,dt\right)$$

where $\int 0\,dt = k$, a constant. Substituting and recalling that e^k is also a constant,

$$y(t) = e^{-k}\left(A + \int 15e^k\,dt\right)$$
$$= e^{-k}(A + 15te^k) = Ae^{-k} + 15t = 15t + A \tag{16.24}$$

where A is an arbitrary constant equal to Ae^{-k} or simply c. Whenever the derivative is equal to a constant, simply integrate as in Example 1.

b)   Taking the derivative of *(16.24)*, $dy/dt = 15$. From *(16.23)*, $dy/dt = 15$.

16.5.   Redo Problem 16.2, given

$$\frac{dy}{dt} - 6y = 18 \tag{16.25}$$

a)   Here $v = -6$, $z = 18$, and $\int -6\,dt = -6t$. Substituting in *(16.1)*,

$$y(t) = e^{6t}\left(A + \int 18e^{-6t}\,dt\right)$$

where $\int 18e^{-6t}\,dt = -3e^{-6t}$. Thus,

$$y(t) = e^{6t}(A - 3e^{-6t}) = Ae^{6t} - 3 \tag{16.26}$$

b)   Taking the derivative of *(16.26)*, $dy/dt = 6Ae^{6t}$. From *(16.25)*, $dy/dt = 18 + 6y$. Substituting y from *(16.26)*, $dy/dt = 18 + 6(Ae^{6t} - 3) = 6Ae^{6t}$.

16.6.   Redo Problem 16.2, given

$$\frac{dy}{dt} + 4y = -20 \qquad y(0) = 10 \tag{16.27}$$

a) Here $v = 4$, $z = -20$, and $\int 4\,dt = 4t$. Thus,

$$y(t) = e^{-4t}\left(A + \int -20e^{4t}\,dt\right)$$

where $\int -20e^{4t}\,dt = -5e^{4t}$. Substituting, $y(t) = e^{-4t}(A - 5e^{4t}) = Ae^{-4t} - 5$. At $t = 0$, $y = 10$. Thus, $10 = Ae^{-4(0)} - 5$, and $A = 15$. Substituting,

$$y(t) = 15e^{-4t} - 5 = 5(3e^{-4t} - 1) \tag{16.28}$$

b) The derivative of *(16.28)* is $dy/dt = -60e^{-4t}$. From *(16.27)*, $dy/dt = -20 - 4y$. Substituting from *(16.28)* for y, $dy/dt = -20 - 4(15e^{-4t} - 5) = -60e^{-4t}$.

16.7. Redo Problem 16.2, given

$$\frac{dy}{dt} + 4ty = 6t \tag{16.29}$$

a) $v = 4t$, $z = 6t$, and $\int 4t\,dt = 2t^2$. Thus,

$$y(t) = e^{-2t^2}\left(A + \int 6te^{2t^2}\,dt\right) \tag{16.30}$$

Using the substitution method for the remaining integral, let $u = 2t^2$, $du/dt = 4t$, and $dt = du/4t$. Thus,

$$\int 6te^{2t^2}\,dt = \int 6te^{u}\,\frac{du}{4t} = 1.5\int e^{u}\,du = 1.5e^{2t^2}$$

Substituting back in *(16.30)*,

$$y(t) = e^{-2t^2}(A + 1.5e^{2t^2}) = Ae^{-2t^2} + 1.5 \tag{16.31}$$

b) The derivative of *(16.31)* is $dy/dt = -4tAe^{-2t^2}$. From *(16.29)*, $dy/dt = 6t - 4ty$. Substituting from *(16.31)*, $dy/dt = 6t - 4t(Ae^{-2t^2} + 1.5) = -4tAe^{-2t^2}$.

16.8. *(a)* Solve the equation below using the formula for a general solution. *(b)* Check your answer.

$$2\frac{dy}{dt} - 2t^2 y = 9t^2 \qquad y(0) = -2.5 \tag{16.32}$$

a) Dividing through by 2, $dy/dt - t^2 y = 4.5t^2$. Thus, $v = -t^2$, $z = 4.5t^2$, and $\int -t^2\,dt = -\frac{1}{3}t^3$. Substituting,

$$y(t) = e^{(1/3)t^3}\left(A + \int 4.5t^2 e^{-(1/3)t^3}\,dt\right) \tag{16.33}$$

Let $u = -\frac{1}{3}t^3$, $du/dt = -t^2$, and $dt = -du/t^2$. Thus,

$$\int 4.5t^2 e^{-(1/3)t^3}\,dt = \int 4.5t^2 e^{u}\,\frac{du}{-t^2} = -4.5\int e^{u}\,du = -4.5e^{-(1/3)t^3}$$

Substituting in *(16.33)*,

$$y(t) = e^{(1/3)t^3}(A - 4.5e^{-(1/3)t^3}) = Ae^{(1/3)t^3} - 4.5$$

At $t = 0$, $-2.5 = A - 4.5$; $A = 2$. Thus,

$$y(t) = 2e^{(1/3)t^3} - 4.5 \tag{16.34}$$

b) Taking the derivative of *(16.34)*, $dy/dt = 2t^2 e^{(1/3)t^3}$. From *(16.32)*, $dy/dt = 4.5t^2 + t^2 y$. Substituting from *(16.34)*, $dy/dt = 4.5t^2 + t^2(2e^{(1/3)t^3} - 4.5) = 2t^2 e^{(1/3)t^3}$.

16.9. Redo Problem 16.8, given

$$\frac{dy}{dt} - 2ty = e^{t^2} \tag{16.35}$$

a) $v = -2t$, $z = e^{t^2}$, and $\int -2t\,dt = -t^2$. Thus,

$$y(t) = e^{t^2}\left(A + \int e^{t^2}e^{-t^2}\,dt\right) = e^{t^2}\left(A + \int e^0\,dt\right)$$

where $e^0 = 1$ and $\int 1\,dt = t$. Substituting back,

$$y(t) = e^{t^2}(A + t) \tag{16.36}$$

b)   The derivative of (16.36), by the product rule, is $dy/dt = 2te^{t^2}(A+t) + e^{t^2}(1) = 2tAe^{t^2} + 2t^2e^{t^2} + e^{t^2}$. From (16.35), $dy/dt = e^{t^2} + 2ty$. Substituting from (16.36),

$$\frac{dy}{dt} = e^{t^2} + 2t[e^{t^2}(A+t)] = e^{t^2} + 2tAe^{t^2} + 2t^2e^{t^2}$$

16.10. Redo Problem 16.8, given

$$\frac{dy}{dt} + 3y = 6t \qquad y(0) = \frac{1}{3} \tag{16.37}$$

a) $v = 3$, $z = 6t$, and $\int 3\,dt = 3t$. Then,

$$y(t) = e^{-3t}\left(A + \int 6te^{3t}\,dt\right) \tag{16.38}$$

Using integration by parts for the remaining integral, let $f(t) = 6t$, then $f'(t) = 6$; let $g'(t) = e^{3t}$, then $g(t) = \int e^{3t}\,dt = \frac{1}{3}e^{3t}$. Substituting in (14.1),

$$\int 6te^{3t}\,dt = 6t(\tfrac{1}{3}e^{3t}) - \int \tfrac{1}{3}e^{3t}6\,dt$$

$$= 2te^{3t} - 2\int e^{3t}\,dt = 2te^{3t} - \tfrac{2}{3}e^{3t}$$

Substituting back in (16.38),

$$y(t) = e^{-3t}(A + 2te^{3t} - \tfrac{2}{3}e^{3t}) = Ae^{-3t} + 2t - \tfrac{2}{3}$$

At $t = 0$, $\frac{1}{3} = Ae^{-3(0)} + 2(0) - \frac{2}{3}$; $A = 1$. Thus,

$$y(t) = e^{-3t} + 2t - \tfrac{2}{3} \tag{16.39}$$

b)   Taking the derivative of (16.39), $dy/dt = -3e^{-3t} + 2$. From (16.37), $dy/dt = 6t - 3y$. Substituting (16.39) directly above, $dy/dt = 6t - 3(e^{-3t} + 2t - \frac{2}{3}) = -3e^{-3t} + 2$.

16.11. Redo Problem 16.8, given

$$\frac{dy}{dt} - \frac{y}{t} = 0 \qquad y(3) = 12 \tag{16.40}$$

a) $v = -1/t$, $z = 0$, and $\int -(1/t)\,dt = -\ln t$. Thus,

$$y(t) = e^{\ln t}\left(A + \int 0\,dt\right) = At$$

since $e^{\ln t} = t$. At $t = 3$, $12 = A(3)$; $A = 4$. Thus,

$$y(t) = 4t \tag{16.41}$$

b) The derivative of (*16.41*) is $dy/dt = 4$. From (*16.40*), $dy/dt = y/t$. Substituting from (*16.41*), $dy/dt = 4t/t = 4$.

16.12. Redo Problem 16.8, given

$$\frac{dy}{dt} = -y \qquad y(3) = 20 \qquad\qquad (16.42)$$

a) With rearranging, $dy/dt + y = 0$. Therefore, $v = 1$, $z = 0$, and $\int 1\,dt = t$. Thus,

$$y(t) = e^{-t}\left(A + \int 0\,dt\right) = Ae^{-t}$$

At $t = 3$, $20 = Ae^{-3}$; $20 = A(0.05)$, so $A = 400$. Thus,

$$y(t) = 400e^{-t} \qquad\qquad (16.43)$$

b) Taking the derivative of (*16.43*), $dy/dt = -400e^{-t}$. From (*16.42*), $dy/dt = -y$. Substituting from (*16.43*), $dy/dt = -(400e^{-t}) = -400e^{-t}$.

EXACT DIFFERENTIAL EQUATIONS AND PARTIAL INTEGRATION

16.13. Solve the following exact differential equation. Check the answer on your own.

$$(4y + 8t^2)\,dy + (16yt - 3)\,dt = 0$$

As outlined in Example 6,

1. Check to see if it is an exact differential equation. Letting $M = 4y + 8t^2$ and $N = 16yt - 3$, $\partial M/\partial t = 16t = \partial N/\partial y$.
2. Integrate M partially with respect to y and add $Z(t)$ to get $F(y, t)$.

$$F(y,t) = \int (4y + 8t^2)\,\partial y + Z(t) = 2y^2 + 8t^2 y + Z(t) \qquad\qquad (16.44)$$

3. Differentiate $F(y, t)$ partially with respect to t and equate with N above.

$$\frac{\partial F}{\partial t} = 16ty + Z'(t)$$

But $\partial F/\partial t = N = 16yt - 3$, so

$$16ty + Z'(t) = 16yt - 3 \qquad Z'(t) = -3$$

4. Integrate $Z'(t)$ with respect to t to get $Z(t)$.

$$Z(t) = \int Z'(t)\,dt = \int -3\,dt = -3t \qquad\qquad (16.45)$$

5. Substitute (*16.45*) in (*16.44*) and add a constant of integration.

$$F(y,t) = 2y^2 + 8t^2 y - 3t + c$$

16.14. Redo Problem 16.13, given $(12y + 7t + 6)\,dy + (7y + 4t - 9)\,dt = 0$.

1. $\partial M/\partial t = 7 = \partial N/\partial y$.

2. $$F(y,t) = \int (12y + 7t + 6)\,\partial y + Z(t) = 6y^2 + 7yt + 6y + Z(t)$$

3. $\partial F/\partial t = 7y + Z'(t)$. But $\partial F/\partial t = N = 7y + 4t - 9$, so

$$7y + Z'(t) = 7y + 4t - 9 \qquad Z'(t) = 4t - 9$$

4.
$$Z(t) = \int (4t - 9)\, dt = 2t^2 - 9t$$

5.
$$F(y, t) = 6y^2 + 7yt + 6y + 2t^2 - 9t + c$$

16.15. Redo Problem 16.13, given $(12y^2t^2 + 10y)\, dy + (8y^3t)\, dt = 0$.

1. $\partial M/\partial t = 24y^2t = \partial N/\partial y$.

2.
$$F(y, t) = \int (12y^2t^2 + 10y)\, \partial y + Z(t) = 4y^3t^2 + 5y^2 + Z(t)$$

3. $\partial F/\partial t = 8y^3t + Z'(t)$. But $N = 8y^3t$, so
$$8y^3t = 8y^3t + Z'(t) \qquad Z'(t) = 0$$

4. $Z(t) = \int 0\, dt = k$, which will be subsumed under c.

5.
$$F(y, t) = 4y^3t^2 + 5y^2 + c$$

16.16. Redo Problem 16.13, given $8tyy' = -(3t^2 + 4y^2)$.

By rearranging,
$$8ty\, dy = -(3t^2 + 4y^2)\, dt \qquad 8ty\, dy + (3t^2 + 4y^2)\, dt = 0$$

1. $\partial M/\partial t = 8y = \partial N/\partial y$.

2.
$$F(y, t) = \int 8ty\, \partial y + Z(t) = 4ty^2 + Z(t)$$

3. $\partial F/\partial t = 4y^2 + Z'(t)$. But $\partial F/\partial t = N = 3t^2 + 4y^2$, so
$$4y^2 + Z'(t) = 3t^2 + 4y^2 \qquad Z'(t) = 3t^2$$

4.
$$Z(t) = \int 3t^2\, dt = t^3$$

5.
$$F(y, t) = t^3 + 4ty^2 + c$$

16.17. Redo Problem 16.13, given $60ty^2y' = -(12t^3 + 20y^3)$.

By rearranging, $\qquad 60ty^2\, dy + (12t^3 + 20y^3)\, dt = 0$

1. $\partial M/\partial t = 60y^2 = \partial N/\partial y$.

2.
$$F(y, t) = \int 60ty^2\, \partial y + Z(t) = 20ty^3 + Z(t)$$

3. $\partial F/\partial t = 20y^3 + Z'(t)$. But $\partial F/\partial t = N = 12t^3 + 20y^3$, so
$$20y^3 + Z'(t) = 12t^3 + 20y^3 \qquad Z'(t) = 12t^3$$

4.
$$Z(t) = \int 12t^3\, dt = 3t^4$$

5.
$$F(y, t) = 3t^4 + 20ty^3 + c$$

INTEGRATING FACTORS

16.18. Use the integrating factors provided in parentheses to solve the following differential equation. Check the answer on your own (remember to divide by the integrating factor after taking the total differential of the answer).

$$6t\, dy + 12y\, dt = 0 \qquad (t)$$

1. $\partial M/\partial t = 6 \neq \partial N/\partial y = 12$. But multiplying by the integrating factor t,

$$6t^2 \, dy + 12yt \, dt = 0$$

where $\partial M/\partial t = 12t = \partial N/\partial y$. Continuing with the new function,

2. $$F(y,t) = \int 6t^2 \, \partial y + Z(t) = 6t^2 y + Z(t)$$

3. $\partial F/\partial t = 12ty + Z'(t)$. But $\partial F/\partial t = N = 12ty$, so $Z'(t) = 0$.

4. $Z(t) = \int 0 \, dt = k$, which will be subsumed under the c below.

5. $$F(y,t) = 6t^2 y + c$$

16.19. Redo Problem 16.18, given

$$t^2 \, dy + 3yt \, dt = 0 \qquad (t)$$

1. $\partial M/\partial t = 2t \neq \partial N/\partial y = 3t$. But multiplying by t,

$$t^3 \, dy + 3yt^2 \, dt = 0$$

where $\partial M/\partial t = 3t^2 = \partial N/\partial y$.

2. $$F(y,t) = \int t^3 \, \partial y + Z(t) = t^3 y + Z(t)$$

3. $\partial F/\partial t = 3t^2 y + Z'(t)$. But $\partial F/\partial t = N = 3t^2 y$, so $Z'(t) = 0$ and $F(y,t) = t^3 y + c$.

16.20. Redo Problem 16.18, given

$$\frac{dy}{dt} = \frac{y}{t} \qquad \left(\frac{1}{ty}\right)$$

Rearranging, $t \, dy = y \, dt$ $t \, dy - y \, dt = 0$

1. $\partial M/\partial t = 1 \neq \partial N/\partial y = -1$. Multiplying by $1/(ty)$,

$$\frac{dy}{y} - \frac{dt}{t} = 0$$

where $\partial M/\partial t = 0 = \partial N/\partial y$, since neither function contains the variable with respect to which it is being partially differentiated.

2. $$F(y,t) = \int \frac{1}{y} \, \partial y + Z(t) = \ln y + Z(t)$$

3. $\partial F/\partial t = Z'(t)$. But $\partial F/\partial t = N = -1/t$, so $Z'(t) = -1/t$.

4. $$Z(t) = \int -\frac{1}{t} \, dt = -\ln t$$

5. $F(y,t) = \ln y - \ln t + c$ which can be expressed in different ways. Since c is an arbitrary constant, we can write $\ln y - \ln t = c$. Making use of the laws of logs (Section 7.3), $\ln y - \ln t = \ln (y/t)$. Thus, $\ln (y/t) = c$. Finally, expressing each side of the equation as exponents of e, and recalling that $e^{\ln x} = x$,

$$e^{\ln (y/t)} = e^c$$

$$\frac{y}{t} = e^c \qquad \text{or} \qquad y = te^c$$

For other treatments of c, see Problems 16.29 to 16.37.

16.21. Redo Problem 16.18, given

$$4t \, dy + (16y - t^2) \, dt = 0 \qquad (t^3)$$

1. $\partial M/\partial t = 4 \neq \partial N/\partial y = 16$. Multiplying by t^3, $4t^4 \, dy + (16t^3 y - t^5) \, dt = 0$ where $\partial M/\partial t = 16t^3 = \partial N/\partial y$.

2.
$$F(y, t) = \int 4t^4 \, \partial y + Z(t) = 4t^4 y + Z(t)$$

3. $\partial F/\partial t = 16t^3 y + Z'(t)$. But $\partial F/\partial t = N = 16t^3 y - t^5$, so

$$16t^3 y + Z'(t) = 16t^3 y - t^5 \qquad Z'(t) = -t^5$$

4.
$$Z(t) = \int -t^5 \, dt = -\tfrac{1}{6}t^6$$

5.
$$F(y, t) = 4t^4 y - \tfrac{1}{6}t^6 + c = 24t^4 y - t^6 + c$$

or
$$24t^4 y - t^6 = c$$

16.22. Redo Problem 16.18, given

$$5yt \, dy + (5y^2 + 8t) \, dt = 0 \qquad (t)$$

1. $\partial M/\partial t = 5y \neq \partial N/\partial y = 10y$. Multiplying by t, as in Example 7, $5yt^2 \, dy + (5y^2 t + 8t^2) \, dt = 0$ where $\partial M/\partial t = 10yt = \partial N/\partial y$.

2.
$$F(y, t) = \int 5yt^2 \, \partial y + Z(t) = 2.5y^2 t^2 + Z(t)$$

3. $\partial F/\partial t = 5y^2 t + Z'(t)$. But $\partial F/\partial t = N = 5y^2 t + 8t^2$, so

$$5y^2 t + Z'(t) = 5y^2 t + 8t^2 \qquad Z'(t) = 8t^2$$

4.
$$Z(t) = \int 8t^2 \, dt = \tfrac{8}{3}t^3$$

5.
$$F(y, t) = 2.5y^2 t^2 + \tfrac{8}{3}t^3 + c = 7.5y^2 t^2 + 8t^3 + c$$

FINDING THE INTEGRATING FACTOR

16.23. (*a*) Find the integrating factor for the differential equation given below, and (*b*) solve the equation, using the five steps from Example 6.

$$(7y + 4t^2) \, dy + 4ty \, dt = 0 \tag{16.46}$$

a) $\partial M/\partial t = 8t \neq \partial N/\partial y = 4t$. Applying Rule 1 from Section 16.5, since $M = 7y + 4t^2$ and $N = 4ty$,

$$\frac{1}{4ty}(8t - 4t) = \frac{4t}{4ty} = \frac{1}{y} = f(y) \qquad \text{alone}$$

Thus the integrating factor is

$$e^{\int (1/y) \, dy} = e^{\ln y} = y$$

b) Multiplying (*16.46*) by the integrating factor y, $(7y^2 + 4yt^2) \, dy + 4ty^2 \, dt = 0$.

1. $\partial M/\partial t = 8yt = \partial N/\partial y$. Thus,

2.
$$F(y, t) = \int (7y^2 + 4yt^2) \, \partial y + Z(t) = \tfrac{7}{3}y^3 + 2y^2 t^2 + Z(t)$$

3.
$$\frac{\partial F}{\partial t} = 4y^2 t + Z'(t)$$

4. $\partial F/\partial t = N = 4y^2 t$, so $Z'(t) = 0$ and $Z(t)$ is a constant. Thus,

5. $F(y,t) = \frac{7}{3}y^3 + 2y^2 t^2 + c = 7y^3 + 6y^2 t^2 + c$

16.24. Redo Problem 16.23, given

$$y^3 t\, dy + \tfrac{1}{2}y^4\, dt = 0 \qquad\qquad (16.47)$$

a) $\partial M/\partial t = y^3 \neq \partial N/\partial y = 2y^3$. Applying Rule 1,

$$\frac{1}{\frac{1}{2}y^4}(y^3 - 2y^3) = \frac{2}{y^4}(-y^3) = -\frac{2}{y} = f(y) \qquad \text{alone}$$

Thus, $e^{\int -2y^{-1}\, dy} = e^{-2\ln y} = e^{\ln y^{-2}} = y^{-2}$

b) Multiplying (16.47) by y^{-2}, $yt\, dy + \tfrac{1}{2}y^2\, dt = 0$.

    1. $\partial M/\partial t = y = \partial N/\partial y$. Thus,

    2. $F(y,t) = \displaystyle\int yt\, \partial y + Z(t) = \tfrac{1}{2}y^2 t + Z(t)$

    3. $\dfrac{\partial F}{\partial t} = \dfrac{1}{2}y^2 + Z'(t)$

    4. $\partial F/\partial t = N = \tfrac{1}{2}y^2$, so $Z'(t) = 0$, and $Z(t)$ is a constant. Thus,

    5. $F(y,t) = \tfrac{1}{2}y^2 t + c$

16.25. Redo Problem 16.23, given

$$4t\, dy + (16y - t^2)\, dt = 0 \qquad\qquad (16.48)$$

a) $M = 4t$, $N = 16y - t^2$, and $\partial M/\partial t = 4 \neq \partial N/\partial y = 16$. Applying Rule 1,

$$\frac{1}{16y - t^2}(4 - 16) = \frac{-12}{16y - t^2} \neq f(y) \qquad \text{alone}$$

Applying Rule 2,

$$\frac{1}{4t}(16 - 4) = \frac{3}{t} = g(t) \qquad \text{alone}$$

Thus, $e^{\int 3t^{-1}\, dt} = e^{3\ln t} = e^{\ln t^3} = t^3$

b) Multiplying (16.48) by t^3, $4t^4\, dy + (16yt^3 - t^5)\, dt = 0$ which was solved in Problem 16.21.

16.26. Redo Problem 16.23, given

$$t^2\, dy + 3yt\, dt = 0 \qquad\qquad (16.49)$$

a) Here $M = t^2$, $N = 3yt$, and $\partial M/\partial t = 2t \neq \partial N/\partial y = 3t$. Applying Rule 1,

$$\frac{1}{3yt}(2t - 3t) = \frac{-t}{3yt} = \frac{-1}{3y} = f(y) \qquad \text{alone}$$

Thus, $e^{\int (-1/3y)\, dy} = e^{-(1/3)\ln y} = e^{\ln y^{-1/3}} = y^{-1/3}$

Consequently, $y^{-1/3}$ is an integrating factor for the equation, although in Problem 16.19 t was given as an integrating factor. Let us check $y^{-1/3}$ first.

b) Multiplying (16.49) by $y^{-1/3}$, $t^2 y^{-1/3}\, dy + 3ty^{2/3}\, dt = 0$.

    1. $\partial M/\partial t = 2ty^{-1/3} = \partial N/\partial y$. Thus,

2.
$$F(y,t) = \int t^2 y^{-1/3}\, \partial y + Z(t) = 1.5t^2 y^{2/3} + Z(t)$$

3.
$$\frac{\partial F}{\partial t} = 3ty^{2/3} + Z'(t)$$

4. $\partial F/\partial t = N = 3ty^{2/3}$, so $Z'(t) = 0$ and $Z(t)$ is a constant. Hence,

5.
$$F_1(y,t) = 1.5t^2 y^{2/3} + c \qquad\qquad (16.50)$$

Here F_1 is used to distinguish this function from the function F_2 below.

16.27. Test to see if t is a possible integrating factor in Problem 16.26.

Applying Rule 2 to the original equation,

$$\frac{1}{t^2}(3t - 2t) = \frac{t}{t^2} = \frac{1}{t}$$

Thus,
$$e^{\int (1/t)\,dt} = e^{\ln t} = t$$

Hence t is also a possible integrating factor, as demonstrated in Problem 16.19, where the solution was $F_2(y,t) = t^3 y + c$. This differs from (16.50) but is equally correct, as you can check on your own.

16.28. Redo Problem 16.23, given

$$(y - t)\, dy - dt = 0 \qquad\qquad (16.51)$$

a) $M = y - t$, $N = -1$, and $\partial M/\partial t = -1 \neq \partial N/\partial y = 0$. Applying Rule 1,

$$\frac{1}{-1}(-1 - 0) = 1 = f(y) \qquad \text{alone}$$

Thus,
$$e^{\int 1\, dy} = e^y$$

b)   Multiplying (16.51) by e^y,

$$(y - t)e^y\, dy - e^y\, dt = 0 \qquad\qquad (16.52)$$

1. $\partial M/\partial t = -e^y = \partial N/\partial y$. Thus.

2.
$$F(y,t) = \int (y - t)e^y\, \partial y + Z(t) \qquad\qquad (16.53)$$

which requires integration by parts. Let

$$f(y) = y - t \qquad f'(y) = 1 \qquad g'(y) = e^y \qquad g(y) = \int e^y\, dy = e^y$$

Substituting in (14.1),

$$\int (y - t)e^y\, \partial y = (y - t)e^y - \int e^y 1\, dy = (y - t)e^y - e^y$$

Substituting in (16.53), $F(y,t) = (y - t)e^y - e^y + Z(t)$.

3.
$$\frac{\partial F}{\partial t} = -e^y + Z'(t)$$

4. $\partial F/\partial t = N = -e^y$ in (16.52), so $Z'(t) = 0$ and $Z(t)$ is a constant. Thus,

5.
$$F(y,t) = (y - t)e^y - e^y + c \qquad \text{or} \qquad (y - 1)e^y - te^y + c$$

SEPARATION OF VARIABLES

16.29. Solve the following differential equation, using the procedure for separating variables described in Section 16.6.

$$\frac{dy}{dt} = \frac{-5t}{y}$$

Separating the variables,

$$y\,dy = -5t\,dt \qquad y\,dy + 5t\,dt = 0$$

Integrating each term separately,

$$\frac{y^2}{2} + \frac{5t^2}{2} = c_1$$

$$y^2 + 5t^2 = 2c_1$$

Letting $c = 2c_1$, $\qquad\qquad\qquad y^2 + 5t^2 = c$

16.30. Redo Problem 16.29, given

a) $\dfrac{dy}{dt} = \dfrac{t^5}{y^4}$

b) $\quad t^2\,dy - y^2\,dt = 0$

$$y^4\,dy - t^5\,dt = 0 \qquad\qquad\qquad\qquad \frac{dy}{y^2} - \frac{dt}{t^2} = 0$$

Integrating, $\quad\dfrac{y^5}{5} - \dfrac{t^6}{6} = c_1$ $\qquad$ Integrating, $\quad -\dfrac{1}{y} + \dfrac{1}{t} = c$

$$6y^5 - 5t^6 = 30c_1 \qquad\qquad\qquad\qquad\qquad y - t = cty$$

Letting $c = 30c_1$, $\quad 6y^5 - 5t^6 = c$

16.31. Redo Problem 16.29, given $t\,dy + y\,dt = 0$.

$$\frac{dy}{y} + \frac{dt}{t} = 0$$

Integrating, $\qquad\qquad \ln y + \ln t = \ln c \quad$ (an arbitrary constant)

By the rule of logs, $\qquad\qquad \ln yt = \ln c \qquad yt = c$

16.32. Use separation of variables to solve the following differential equation.

$$\frac{dy}{dt} = -y$$

Separating the variables,

$$\frac{dy}{y} = -dt$$

Integrating both sides and using $\ln c$ for the constant of integration,

$$\ln y = -t + \ln c$$

Then playing with the constant of integration for ultimate simplicity of the solution,

$$\ln y - \ln c = -t$$

$$\ln \frac{y}{c} = -t$$

Setting both sides as exponents of e,

$$\frac{y}{c} = e^{-t}$$

$$y = ce^{-t}$$

16.33. Redo Problem 16.32, given

$$\frac{dy}{dt} = b - ay$$

Separating the variables and then multiplying both sides by -1,

$$\frac{dy}{b - ay} = dt$$

$$\frac{dy}{ay - b} = -dt$$

Integrating both sides and being creative once again with the constant of integration,

$$\frac{1}{a}\ln(ay - b) = -t + \frac{1}{a}\ln c$$

Multiplying both sides by a and rearranging,

$$\ln(ay - b) = -at + \ln c$$

$$\ln\left(\frac{ay - b}{c}\right) = -at$$

$$\frac{ay - b}{c} = e^{-at}$$

$$ay - b = ce^{-at}$$

$$y = Ce^{-at} + \frac{b}{a} \qquad \text{where } C = \frac{c}{a}$$

16.34. Redo Problem 16.32, given $(t + 5)\, dy - (y + 9)\, dt = 0$.

$$\frac{dy}{y + 9} - \frac{dt}{t + 5} = 0$$

Integrating, $$\ln(y + 9) - \ln(t + 5) = \ln c$$

By the rule of logs, $$\ln\frac{y + 9}{t + 5} = \ln c$$

$$\frac{y + 9}{t + 5} = c \qquad \text{or} \qquad y + 9 = c(t + 5)$$

16.35. Using the procedure for separating variables, solve the differential equation $dy = 3t^2 y\, dt$.

$$\frac{dy}{y} - 3t^2\, dt = 0$$

Integrating, $$\ln y - t^3 = \ln c$$

Expressing each side of the equation as an exponent of e,

$$e^{\ln y - t^3} = e^{\ln c}$$
$$e^{\ln y} e^{-t^3} = e^{\ln c}$$
$$y e^{-t^3} = c$$
$$y = c e^{t^3}$$

16.36. Redo Problem 16.35, given $y^2(t^3 + 1)\, dy + t^2(y^3 - 5)\, dt = 0$.

$$\frac{y^2}{y^3 - 5}\, dy + \frac{t^2}{t^3 + 1}\, dt = 0$$

Integrating by substitution,

$$\tfrac{1}{3}\ln(y^3 - 5) + \tfrac{1}{3}\ln(t^3 + 1) = \ln c$$
$$\ln[(y^3 - 5)(t^3 + 1)] = \ln c \qquad (y^3 - 5)(t^3 + 1) = c$$

16.37. Redo Problem 16.35, given

$$3\, dy + \frac{t}{t^2 - 1}\, dt = 0$$

Integrating, $$3y + \tfrac{1}{2}\ln(t^2 - 1) = c$$

Setting the left-hand side as an exponent of e and ignoring c, because, as an arbitrary constant, it can be expressed equally well as c or e^c,

$$e^{3y + (1/2)\ln(t^2 - 1)} = c$$
$$e^{3y} e^{\ln(t^2 - 1)^{1/2}} = c \qquad e^{3y}(t^2 - 1)^{1/2} = c$$

USE OF DIFFERENTIAL EQUATIONS IN ECONOMICS

16.38. Find the demand function $Q = f(P)$ if point elasticity ϵ is -1 for all $P > 0$.

$$\epsilon = \frac{dQ}{dP}\frac{P}{Q} = -1 \qquad \frac{dQ}{dP} = -\frac{Q}{P}$$

Separating the variables,

$$\frac{dQ}{Q} + \frac{dP}{P} = 0$$

Integrating, $\ln Q + \ln P = \ln c$

$$QP = c \qquad Q = \frac{c}{P}$$

16.39. Find the demand function $Q = f(P)$ if $\epsilon = -k$, a constant.

$$\epsilon = \frac{dQ}{dP}\frac{P}{Q} = -k \qquad \frac{dQ}{dP} = -\frac{kQ}{P}$$

Separating the variables,

$$\frac{dQ}{Q} + \frac{k}{P}\, dP = 0$$
$$\ln Q + k \ln P = c$$
$$QP^k = c \qquad Q = cP^{-k}$$

16.40. Find the demand function $Q = f(P)$ if $\epsilon = -(5P + 2P^2)/Q$ and $Q = 500$ when $P = 10$.

$$\epsilon = \frac{dQ}{dP} \frac{P}{Q} = \frac{-(5P + 2P^2)}{Q}$$

$$\frac{dQ}{dP} = \frac{-(5P + 2P^2)}{Q} \frac{Q}{P} = -(5 + 2P)$$

Separating the variables,

$$dQ + (5 + 2P)\, dP = 0$$

Integrating, $Q + 5P + P^2 = c$ $Q = -P^2 - 5P + c$

At $P = 10$ and $Q = 500$,

$$500 = -100 - 50 + c \qquad c = 650$$

Thus, $Q = 650 - 5P - P^2$.

16.41. Derive the formula $P = P(0)\, e^{it}$ for the total value of an initial sum of money $P(0)$ set out for t years at interest rate i, when i is compounded continuously.

If i is compounded continuously,

$$\frac{dP}{dt} = iP$$

Separating the variables,

$$\frac{dP}{P} - i\, dt = 0$$

Integrating, $\ln P - it = c$

Setting the left-hand side as an exponent of e,

$$e^{\ln P - it} = c$$

$$\mathrm{P}e^{-it} = c \qquad P = ce^{it}$$

At $t = 0$, $P = P(0)$. Thus $P(0) = ce^0$, $c = P(0)$, and $P = P(0)\, e^{it}$.

16.42. Determine the stability condition for a two-sector income determination model in which $\hat{C}$, $\hat{I}$, $\hat{Y}$ are deviations of consumption, investment, and income, respectively, from their equilibrium values C_e, I_e, Y_e. That is, $\hat{C} = C(t) - C_e$, etc., where $\hat{C}$ is read "C hat." Income changes at a rate proportional to excess demand $C + I - Y$, and

$$\hat{C}(t) = g\hat{Y}(t) \qquad \hat{I}(t) = b\hat{Y}(t) \qquad \frac{d\hat{Y}(t)}{dt} = a(\hat{C} + \hat{I} - \hat{Y}) \qquad 0 < a, b, g < 1$$

Substituting the first two equations in the third,

$$\frac{d\hat{Y}}{dt} = a(g + b - 1)\,\hat{Y}$$

Separating the variables and then integrating,

$$\frac{d\hat{Y}}{\hat{Y}} = a(g + b - 1)\, dt$$

$$\ln \hat{Y} = a(g + b - 1)t + c$$

$$e^{\ln \hat{Y}} = e^{a(g+b-1)t+c}$$

Letting the constant $e^c = c$,

$$\hat{Y} = ce^{a(g+b-1)t}$$

At $t = 0$, $\hat{Y} = Y(0) - Y_e = c$. Substituting above, $\hat{Y} = [Y(0) - Y_e]e^{a(g+b-1)t}$. Since $\hat{Y} = Y(t) - Y_e$, $Y(t) = Y_e + \hat{Y}$. Thus,

$$Y(t) = Y_e + [Y(0) - Y_e]e^{a(g+b-1)t}$$

As $t \to \infty$, $Y(t) \to Y_e$ only if $g + b < 1$. The sum of the marginal propensity to consume g and the marginal propensity to invest b must be less than 1.

16.43. In Example 11 we found $P(t) = [P(0) - \bar{P}]e^{-m(h-b)t} + \bar{P}$. (a) Explain the time path if (1) the initial price $P(0) = \bar{P}$, (2) $P(0) > \bar{P}$, and (3) $P(0) < \bar{P}$. (b) Graph your findings.

a) 1) If the initial price equals the equilibrium price, $P(0) = \bar{P}$, the first term on the right disappears and $P(t) = \bar{P}$. The time path is a horizontal line, and adjustment is immediate. See Fig. 16-2.
    2) If $P(0) > \bar{P}$, the first term on the right is positive. Thus $P(t) > \bar{P}$ and $P(t)$ approaches $\bar{P}$ from above as $t \to \infty$ and the first term on the right $\to 0$.
    3) If $P(0) < \bar{P}$, the first term on the right is negative. And $P(t) < \bar{P}$ and approaches it from below as $t \to \infty$ and the first term $\to 0$.

b) See Fig. 16-2.

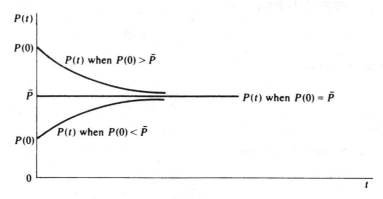

Fig. 16-2

16.44. A change in the rate of investment will affect both aggregate demand and the productive capability of an economy. The Domar model seeks to find the time path along which an economy can grow while maintaining full utilization of its productive capacity. If the marginal propensity to save s and the marginal capital-output ratio k are constant, find the investment function needed for the desired growth.

The change in aggregate demand is equal to the change in investment times the multiplier $1/s$,

$$\frac{dY}{dt} = \frac{1}{s}\frac{dI}{dt} \tag{16.54}$$

The change in productive capacity is equal to the change in the capital stock times the reciprocal of the marginal capital-output ratio,

$$\frac{dQ}{dt} = \frac{1}{k}\frac{dK}{dt} = \frac{1}{k}I \qquad \text{since} \qquad \frac{dK}{dt} = I \tag{16.55}$$

Equating (16.54) and (16.55) for fully utilized capacity,

$$\frac{1}{s}\frac{dI}{dt} = \frac{1}{k}I \qquad \frac{1}{s}dI = \frac{1}{k}I\,dt$$

Separating the variables,

$$\frac{dI}{I} - \frac{s}{k} dt = 0$$

Integrating,

$$\ln I - \frac{s}{k} t = c$$

$$I e^{-(s/k)t} = c \qquad I = c e^{(s/k)t}$$

At $t = 0$, $I(0) = c$, and $I = I(0) e^{(s/k)t}$.

Investment must grow at a constant rate determined by s/k: the savings rate divided by the capital-output ratio.

16.45. The Solow model examines equilibrium growth paths with full employment of both capital and labor. Based on the assumptions that

1. Output is a linearly homogeneous function of capital and labor exhibiting constant returns to scale,

$$Y = f(K, L) \tag{16.56}$$

2. A constant proportion s of output is saved and invested,

$$\frac{dK}{dt} \equiv \dot{K} = sY \tag{16.57}$$

3. The supply of labor is growing at a constant rate r,

$$L = L_0 e^{rt} \tag{16.58}$$

derive the differential equation in terms of the single variable K/L, which serves as the basis of the model.

Substituting Y from (16.56) in (16.57),

$$\frac{dK}{dt} = sf(K, L) \tag{16.59}$$

Substituting L from (16.58) in (16.59),

$$\frac{dK}{dt} = sf(K, L_0 e^{rt}) \tag{16.60}$$

This is the time path capital formation (dK/dt) must follow for full employment of a growing labor force. Preparing to convert to a function of K/L, let $z = K/L$, then $K = zL$. Making use of (16.58),

$$K = zL_0 e^{rt} \tag{16.61}$$

Taking the derivative of (16.61) and using the product rule since z is a function of t,

$$\frac{dK}{dt} = z(rL_0 e^{rt}) + L_0 e^{rt}\frac{dz}{dt} = \left(zr + \frac{dz}{dt} \right) L_0 e^{rt} \tag{16.62}$$

Equating (16.60) and (16.62),

$$sf(K, L_0 e^{rt}) = \left(zr + \frac{dz}{dt} \right) L_0 e^{rt} \tag{16.63}$$

Since the left-hand side of (16.63) is a linearly homogeneous production function, we may divide both inputs by $L_0 e^{rt}$ and multiply the function itself by $L_0 e^{rt}$ without changing its value. Thus,

$$sf(K, L_0 e^{rt}) = sL_0 e^{rt} f\left(\frac{K}{L_0 e^{rt}}, 1 \right) \tag{16.64}$$

Substituting (16.64) in (16.63) and dividing both sides by $L_0 e^{rt}$,

$$sf\left(\frac{K}{L_0 e^{rt}}, 1\right) = zr + \frac{dz}{dt} \tag{16.65}$$

Finally, substituting z for $K/L_0 e^{rt}$ and subtracting zr from both sides,

$$\frac{dz}{dt} = sf(z, 1) - zr \tag{16.66}$$

which is a differential equation in terms of the single variable z and two parameters r and s, where $z = K/L$, r = the rate of growth of the labor force, and s = the savings rate.

16.46. Assume that the demand for money is for transaction purposes only. Thus,

$$M_d = kP(t)Q \tag{16.67}$$

where k is constant, P is the price level, and Q is real output. Assume $M_s = M_d$ and is exogenously determined by monetary authorities. If inflation or the rate of change of prices is proportional to excess demand for goods in society and, from Walras' law, an excess demand for goods is the same thing as an excess supply of money, so that

$$\frac{dP(t)}{dt} = b(M_s - M_d) \tag{16.68}$$

find the stability conditions, when real output Q is constant.

Substituting (16.67) in (16.68),

$$\frac{dP(t)}{dt} = bM_s - bkP(t)Q \tag{16.69}$$

If we let

$$\hat{P} = P(t) - P_e \tag{16.70}$$

where $\hat{P}$ is the deviation of prices from the equilibrium price level P_e, then taking the derivative of (16.70),

$$\frac{d\hat{P}}{dt} = \frac{dP(t)}{dt} - \frac{dP_e}{dt}$$

But in equilibrium $dP_e/dt = 0$. Hence,

$$\frac{d\hat{P}}{dt} = \frac{dP(t)}{dt} \tag{16.71}$$

Substituting in (16.69),

$$\frac{d\hat{P}}{dt} = bM_s - bkP(t)Q \tag{16.72}$$

In equilibrium, $M_s = M_d = kP_eQ$. Hence $M_s - kP_eQ = 0$ and $b(M_s - kP_eQ) = 0$. Subtracting this from (16.72),

$$\frac{d\hat{P}}{dt} = bM_s - bkP(t)Q - bM_s + bkP_eQ = -bkQ[P(t) - P_e] = -bkQ\hat{P} \tag{16.73}$$

which is a differential equation. Separating the variables,

$$\frac{d\hat{P}}{\hat{P}} = -bkQ\, dt$$

Integrating, $\ln \hat{P} = -bkQt + c$, $\hat{P} = Ae^{-bkQt}$, where $e^c = A$.

Since $b, k, Q > 0$, $\hat{P} \to 0$ as $t \to \infty$, and the system is stable. To find the time path $P(t)$ from $\hat{P}$, see the conclusion of Problem 16.42, where $Y(t)$ was derived from $\hat{Y}$.

16.47. If the expectation of inflation is a positive function of the present rate of inflation

$$\left[\frac{dP(t)}{dt}\right]_E = h \frac{dP(t)}{dt} \qquad (16.74)$$

and the expectation of inflation reduces people's desire to hold money, so that

$$M_d = kP(t)Q - g\left[\frac{dP(t)}{dt}\right]_E \qquad (16.75)$$

check the stability conditions, assuming that the rate of inflation is proportional to the excess supply of money as in (16.68).

Substituting (16.74) in (16.75),

$$M_d = kP(t)Q - gh\frac{dP(t)}{dt} \qquad (16.76)$$

Substituting (16.76) in (16.68),

$$\frac{dP(t)}{dt} = bM_s - b\left[kP(t)Q - gh\frac{dP(t)}{dt}\right]$$

By a process similar to the steps involving (16.70) to (16.73),

$$\frac{d\hat{P}}{dt} = bM_s - bkP(t)Q + bgh\frac{dP(t)}{dt} - bM_s + bkP_eQ = -bkQ\hat{P} + bgh\frac{dP(t)}{dt} \qquad (16.77)$$

Substituting (16.71) for $dP(t)/dt$ in (16.77),

$$\frac{d\hat{P}}{dt} = -bkQ\hat{P} + bgh\frac{d\hat{P}}{dt} = \frac{-bkQ\hat{P}}{1 - bgh}$$

Separating the variables,

$$\frac{d\hat{P}}{\hat{P}} = \frac{-bkQ}{1 - bgh}dt$$

Integrating, $\ln \hat{P} = -bkQt/(1 - bgh)$

$$\hat{P} = Ae^{-bkQt/(1-bgh)}$$

Since $b, k, Q > 0$, $\hat{P} \to 0$ as $t \to \infty$, if $bgh < 1$. Hence even if h is greater than 1, meaning people expect inflation to accelerate, the economy need not be unstable, as long as b and g are sufficiently small.

PHASE DIAGRAMS FOR DIFFERENTIAL EQUATIONS

16.48. (a) Construct a phase diagram for the following nonlinear differential equation and test the dynamic stability using (b) the arrows of motion, (c) the slope of the phase line, and (d) the derivative test.

$$\dot{y} = 3y^2 - 18y$$

a)   By setting $\dot{y} = 0$, we find the intertemporal equilibrium solution(s) where the phase diagram crosses the horizontal axis.

$$3y(y - 6) = 0$$
$$\bar{y}_1 = 0 \qquad \bar{y}_2 = 6$$

We then find the critical value and whether it represents a maximum or minimum.

$$\frac{d\dot{y}}{dy} = 6y - 18 = 0 \qquad y = 3 \qquad \text{critical value}$$

$$\frac{d^2\dot{y}}{dy^2} = 6 > 0 \qquad \text{relative minimum}$$

Armed with this information, we can then draw a rough but accurate sketch of the graph, as in Fig. 16-3.

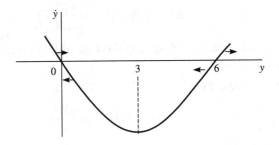

The phase diagram for $\dot{y} = 3y^2 - 18y$

Fig. 16-3

b) Above the horizontal axis, where $\dot{y} > 0$, the arrows of motion point to the right; below the horizontal axis, $\dot{y} < 0$ and the arrows of motion point to the left. Since the arrows of motion point towards $\bar{y}_1 = 0$ and away from $\bar{y}_2 = 6$, $\bar{y}_1$ is stable and $\bar{y}_2$ is unstable.

c) With the slope of the phase diagram negative as it passes through $\bar{y}_1 = 0$, we know $\bar{y}_1$ must be stable. With a positive slope at $\bar{y}_2 = 6$, $\bar{y}_2$ must be unstable.

d) Taking the derivative of the equation, independently of the graph, and evaluating it at the critical values, we see

$$\frac{d\dot{y}}{dy} = 6y - 18$$

$$\frac{d\dot{y}}{dy}(0) = 6(0) - 18 = -18 < 0 \qquad \bar{y}_1 = 0 \text{ is stable}$$

$$\frac{d\dot{y}}{dy}(6) = 6(6) - 18 = 18 > 0 \qquad \bar{y}_2 = 6 \text{ is unstable}$$

16.49. Repeat the exercise in Problem 16.48 for

$$\dot{y} = -y^2 + 6y - 5$$

a) Setting $\dot{y} = 0$,

$$(y - 1)(-y + 5) = 0$$
$$\bar{y}_1 = 1 \qquad \bar{y}_2 = 5$$

Optimizing,

$$\frac{d\dot{y}}{dy} = -2y + 6 = 0$$

$$y = 3 \qquad \text{critical value}$$

$$\frac{d^2\dot{y}}{dy^2} = -2 < 0 \qquad \text{relative maximum}$$

Then sketching the graph, as in Fig. 16-4.

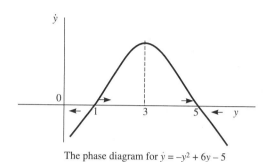

The phase diagram for $\dot{y} = -y^2 + 6y - 5$

Fig. 16-4

b)   Since the arrows of motions point away from $\bar{y}_1 = 1$ and towards $\bar{y}_2 = 5$, $\bar{y}_1$ is unstable while $\bar{y}_2$ is a stable intertemporal equilibrium.

c)   The positive slope at $\bar{y}_1 = 1$ and the negative slope at $\bar{y}_2 = 5$ indicate that $\bar{y}_1$ is an unstable equilibrium and $\bar{y}_2$ is a stable equilibrium.

d)
$$\frac{d\dot{y}}{dy} = -2y + 6$$

$$\frac{d\dot{y}}{dy}(1) = -2(1) + 6 = 4 > 0 \qquad \bar{y}_1 = 1 \text{ is unstable}$$

$$\frac{d\dot{y}}{dy}(5) = -2(5) + 6 = -4 < 0 \qquad \bar{y}_2 = 5 \text{ is stable}$$

16.50. Repeat Problem 16.49, given
$$\dot{y} = y^2 - 10y + 16$$

a)
$$(y - 2)(y - 8) = 0$$

$$\bar{y}_1 = 2 \qquad \bar{y}_2 = 8$$

$$\frac{d\dot{y}}{dy} = 2y - 10 = 0 \qquad y = 5 \qquad \text{critical value}$$

$$\frac{d^2\dot{y}}{dy^2} = 2 > 0 \qquad \text{relative minimum}$$

Then sketching the graph, as in Fig. 16.5.

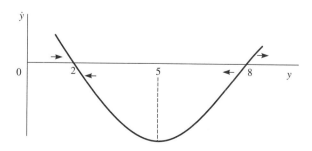

The phase diagram for $\dot{y} = y^2 - 10y + 16$

Fig. 16-5

b) Since the arrows of motion point towards $\bar{y}_1 = 2$ and away from $\bar{y}_2 = 8$, $\bar{y}_1$ is stable while $\bar{y}_2$ is an unstable intertemporal equilibrium.

c) The negative slope at $\bar{y}_1 = 2$ and the positive slope at $\bar{y}_2 = 8$ indicate that $\bar{y}_1$ is a stable equilibrium and $\bar{y}_2$ is an unstable equilibrium.

d)

$$\frac{d\dot{y}}{dy} = 2y - 10$$

$$\frac{d\dot{y}}{dy}(2) = 2(2) - 10 = -6 < 0 \qquad \bar{y}_1 = 2 \text{ is stable}$$

$$\frac{d\dot{y}}{dy}(8) = 2(8) - 10 = 6 > 0 \qquad \bar{y}_2 = 8 \text{ is unstable}$$

CHAPTER 17

First-Order Difference Equations

17.1 DEFINITIONS AND CONCEPTS

A *difference equation* expresses a relationship between a dependent variable and a lagged independent variable (or variables) which changes at discrete intervals of time, for example, $I_t = f(Y_{t-1})$, where I and Y are measured at the end of each year. The *order* of a difference equation is determined by the greatest number of periods lagged. A *first-order* difference equation expresses a time lag of one period; a *second-order*, two periods; etc. The change in y as t changes from t to $t + 1$ is called the *first difference of y*. It is written

$$\frac{\Delta y}{\Delta t} = \Delta y_t = y_{t+1} - y_t \tag{17.1}$$

where Δ is an operator replacing d/dt that is used to measure continuous change in differential equations. The *solution* of a difference equation defines y for every value of t and does not contain a difference expression. See Examples 1 and 2.

EXAMPLE 1. Each of the following is a difference equation of the order indicated.

$$
\begin{array}{ll}
I_t = a(Y_{t-1} - Y_{t-2}) & \text{order 2} \\
Q_s = a + bP_{t-1} & \text{order 1} \\
y_{t+3} - 9y_{t+2} + 2y_{t+1} + 6y_t = 8 & \text{order 3} \\
\Delta y_t = 5y_t & \text{order 1}
\end{array}
$$

Substituting from (*17.1*) for Δy_t above,

$$y_{t+1} - y_t = 5y_t \qquad y_{t+1} = 6y_t \qquad \text{order 1}$$

EXAMPLE 2. Given that the initial value of y is y_0, in the difference equation

$$y_{t+1} = by_t \tag{17.2}$$

a solution is found as follows. By successive substitutions of $t = 0, 1, 2, 3$, etc. in (17.2),

$$y_1 = by_0 \qquad\qquad y_3 = by_2 = b(b^2 y_0) = b^3 y_0$$
$$y_2 = by_1 = b(by_0) = b^2 y_0 \qquad y_4 = by_3 = b(b^3 y_0) = b^4 y_0$$

Thus, for any period t,

$$y_t = b^t y_0$$

This method is called the *iterative method*. Since y_0 is a constant, notice the crucial role b plays in determining values for y as t changes.

17.2 GENERAL FORMULA FOR FIRST-ORDER LINEAR DIFFERENCE EQUATIONS

Given a first-order difference equation which is *linear* (i.e., all the variables are raised to the first power and there are no cross products),

$$y_t = by_{t-1} + a \tag{17.3}$$

where b and a are constants, the general formula for a *definite solution* is

$$y_t = \left(y_0 - \frac{a}{1-b}\right)b^t + \frac{a}{1-b} \qquad \text{when} \quad b \neq 1 \tag{17.4}$$

$$y_t = y_0 + at \qquad\qquad\qquad \text{when} \quad b = 1 \tag{17.4a}$$

If no initial condition is given, an arbitrary constant A is used for $y_0 - a/(1-b)$ in (17.4) and for y_0 in ($17.4a$). This is called a *general solution*. See Example 3 and Problems 17.1 to 17.13.

EXAMPLE 3.   Consider the difference equation $y_t = -7y_{t-1} + 16$ and $y_0 = 5$. In the equation, $b = -7$ and $a = 16$. Since $b \neq 1$, it is solved by using (17.4), as follows:

$$y_t = \left(5 - \frac{16}{1+7}\right)(-7)^t + \frac{16}{1+7} = 3(-7)^t + 2 \tag{17.5}$$

To check the answer, substitute $t = 0$ and $t = 1$ in (17.5).

$$y_0 = 3(-7)^0 + 2 = 5 \qquad \text{since} \quad (-7)^0 = 1$$
$$y_1 = 3(-7)^1 + 2 = -19$$

Substituting $y_1 = -19$ for y_t and $y_0 = 5$ for y_{t-1} in the original equation,

$$-19 = -7(5) + 16 = -35 + 16$$

17.3 STABILITY CONDITIONS

Equation (17.4) can be expressed in the general form

$$y_t = Ab^t + c \tag{17.6}$$

where $A = y_0 - a/(1-b)$ and $c = a/(1-b)$. Here Ab^t is called the *complementary function* and c is the *particular solution*. The particular solution expresses the *intertemporal equilibrium level of y*; the complementary function represents the *deviations from that equilibrium*. Equation (17.6) will be dynamically stable, therefore, only if the complementary function $Ab^t \to 0$, as $t \to \infty$. All depends on the base b. Assuming $A = 1$ and $c = 0$ for the moment, the exponential expression b^t will generate seven different time paths depending on the value of b, as illustrated in Example 4. As seen there, if $|b| > 1$, the time path will explode and move farther and farther away from equilibrium; if $|b| < 1$, the time path will be damped and move toward equilibrium. If $b < 0$, the time path will oscillate between positive and negative values; if $b > 0$, the time path will be nonoscillating. If $A \neq 1$, the value of the

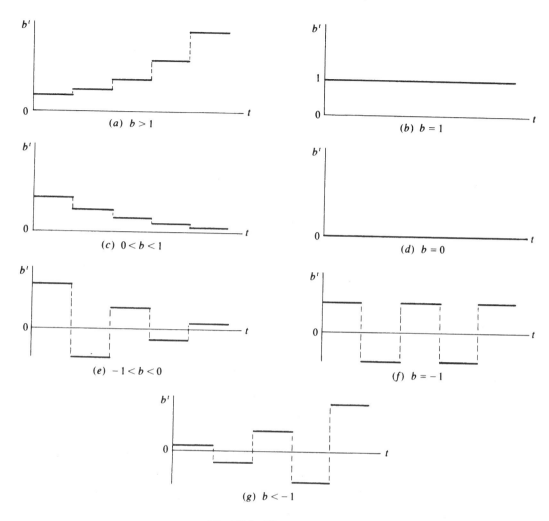

Fig. 17-1   Time path of b^t

multiplicative constant will scale up or down the magnitude of b^t, but will not change the basic pattern of movement. If $A = -1$, a mirror image of the time path of b^t with respect to the horizontal axis will be produced. If $c \neq 0$, the vertical intercept of the graph is affected, and the graph shifts up or down accordingly. See Examples 4 and 5 and Problems 17.1 to 17.13.

EXAMPLE 4.   In the equation $y_t = b^t$, b can range from $-\infty$ to ∞. Seven different time paths can be generated, each of which is explained below and graphed in Fig. 17-1.

1. If $b > 1$, b^t increases at an increasing rate as t increases, thus moving farther and farther away from the horizontal axis. This is illustrated in Fig. 17-1(a), which is a step function representing changes at discrete intervals of time, not a continuous function. Assume $b = 3$. Then as t goes from 0 to 4, $b^t = 1$, 3, 9, 27, 81.

2. If $b = 1$, $b^t = 1$ for all values of t. This is represented by a horizontal line in Fig. 17-1(b).

3. If $0 < b < 1$, then b is a positive fraction and b^t decreases as t increases, drawing closer and closer to the horizontal axis, but always remaining positive, as illustrated in Fig. 17-1(c). Assume $b = \frac{1}{3}$. Then as t goes from 0 to 4, $b^t = 1, \frac{1}{3}, \frac{1}{9}, \frac{1}{27}, \frac{1}{81}$.

4. If $b = 0$, then $b^t = 0$ for all values of t. See Fig. 17-1(d).

5. If $-1 < b < 0$, then b is a negative fraction; b^t will alternate in sign and draw closer and closer to the horizontal axis as t increases. See Fig. 17-1(e). Assume $b = -\frac{1}{3}$. Then as t goes from 0 to 4, $b^t = 1, -\frac{1}{3}, \frac{1}{9}, -\frac{1}{27}, \frac{1}{81}$.

6. If $b = -1$, then b^t oscillates between $+1$ and -1. See Fig. 17-1(f).
7. If $b < -1$, then b^t will oscillate and move farther and farther away from the horizontal axis, as illustrated in Fig. 17-1(g). Assume $b = -3$. Then $b^t = 1, -3, 9, -27, 81$, as t goes from 0 to 4.

In short, if $\quad\quad$ $|b| > 1$ $\quad$ the time path explodes
$\quad\quad\quad\quad\quad\quad$ $|b| < 1$ $\quad$ the time path converges
$\quad\quad\quad\quad\quad\quad\quad$ $b > 0$ $\quad$ the time path is nonoscillating
$\quad\quad\quad\quad\quad\quad\quad$ $b < 0$ $\quad$ the time path oscillates

EXAMPLE 5. In the equation $y_t = 6(-\frac{1}{4})^t + 6$, since $b = -\frac{1}{4} < 0$, the time path oscillates. Since $|b| < 1$, the time path converges.

When $y_t = 5(6)^t + 9$ and $b = 6 > 0$, there is no oscillation. With $|b| > 1$, the time path explodes.

17.4 LAGGED INCOME DETERMINATION MODEL

In the simple income determination model of Section 2.3 there were no lags. Now assume that consumption is a function of the previous period's income, so that

$$C_t = C_0 + cY_{t-1} \quad\quad Y_t = C_t + I_t$$

where $I_t = I_0$. Thus, $Y_t = C_0 + cY_{t-1} + I_0$. Rearranging terms to conform with (17.3),

$$Y_t = cY_{t-1} + C_0 + I_0 \quad\quad\quad\quad (17.7)$$

where $b = c$ and $a = C_0 + I_0$. Substituting these values in (17.4), since the marginal propensity to consume c cannot equal 1, and assuming $Y_t = Y_0$ at $t = 0$,

$$Y_t = \left(Y_0 - \frac{C_0 + I_0}{1 - c}\right)(c)^t + \frac{C_0 + I_0}{1 - c} \qu\quad\quad (17.8)$$

The stability of the time path thus depends on c. Since $0 < \text{MPC} < 1$, $|c| < 1$ and the time path will converge. Since $c > 0$, there will be no oscillations. The equilibrium is stable, and as $t \to \infty$, $Y_t \to (C_0 + I_0)/(1 - c)$, which is the intertemporal equilibrium level of income. See Example 6 and Problems 17.14 to 17.20.

EXAMPLE 6. Given $Y_t = C_t + I_t$, $C_t = 200 + 0.9Y_{t-1}$, $I_t = 100$, and $Y_0 = 4500$. Solving for Y_t,

$$Y_t = 200 + 0.9Y_{t-1} + 100 = 0.9Y_{t-1} + 300 \quad\quad\quad\quad (17.9)$$

Using (17.4),

$$Y_t = \left(4500 - \frac{300}{1 - 0.9}\right)(0.9)^t + \frac{300}{1 - 0.9} = 1500(0.9)^t + 3000 \qu\quad\quad (17.10)$$

With $|0.9| < 1$, the time path converges; with $0.9 > 0$, there is no oscillation. Thus, Y_t is dynamically stable. As $t \to \infty$, the first term on the right-hand side goes to zero, and Y_t approaches the intertemporal equilibrium level of income: $300/(1 - 0.9) = 3000$.

To check this answer, let $t = 0$ and $t = 1$ in (17.10). Thus,

$$Y_0 = 1500(0.9)^0 + 3000 = 4500$$
$$Y_1 = 1500(0.9)^1 + 3000 = 4350$$

Substituting $Y_1 = 4350$ for Y_t and $Y_0 = 4500$ for Y_{t-1} in (17.9),

$$4350 - 0.9(4500) = 300$$
$$4350 - 4050 = 300$$

17.5 THE COBWEB MODEL

For many products, such as agricultural commodities, which are planted a year before marketing, current supply depends on last year's prices. This poses interesting stability questions. If

$$Q_{dt} = c + bP_t \quad \text{and} \quad Q_{st} = g + hP_{t-1}$$

in equilibrium,

$$c + bP_t = g + hP_{t-1} \tag{17.11}$$

$$bP_t = hP_{t-1} + g - c \tag{17.12}$$

Dividing (17.12) by b to conform to (17.3),

$$P_t = \frac{h}{b}P_{t-1} + \frac{g-c}{b}$$

Since $b < 0$ and $h > 0$ under normal demand and supply conditions, $h/b \neq 1$. Using (17.4),

$$P_t = \left[P_0 - \frac{(g-c)/b}{1 - h/b}\right]\left(\frac{h}{b}\right)^t + \frac{(g-c)/b}{1 - h/b}$$

$$= \left(P_0 - \frac{g-c}{b-h}\right)\left(\frac{h}{b}\right)^t + \frac{g-c}{b-h} \tag{17.13}$$

When the model is in equilibrium, $P_t = P_{t-1}$. Substituting P_e for P_t and P_{t-1} in (17.11)

$$P_e = \frac{g-c}{b-h} \tag{17.13a}$$

Substituting in (17.13),

$$P_t = (P_0 - P_e)\left(\frac{h}{b}\right)^t + P_e$$

With an ordinary negative demand function and positive supply function, $b < 0$ and $h > 0$. Therefore, $h/b < 0$ and the time path will oscillate.

If $|h| > |b|$, $|h/b| > 1$, and the time path P_t explodes.
If $|h| = |b|$, $h/b = -1$, and the time oscillates uniformly.
If $|h| < |b|$, $|h/b| < 1$, and the time path converges, and P_t approaches P_e.

In short, when $Q = f(P)$ in supply-and-demand analysis, as is common in mathematics, the supply curve must be flatter than the demand curve for stability. See Example 7 and Problems 17.21 to 17.25. But if $P = f(Q)$, as is typical in economics, the reverse is true. The demand curve must be flatter, or more elastic, than the supply curve if the model is to be stable.

EXAMPLE 7. Given $Q_{dt} = 86 - 0.8P_t$ and $Q_{st} = -10 + 0.2P_{t-1}$, the market price P_t for any time period and the equilibrium price P_e can be found as follows. Equating demand and supply,

$$86 - 0.8P_t = -10 + 0.2P_{t-1} \qquad -0.8P_t = 0.2P_{t-1} - 96$$

Dividing through by -0.8 to conform to (17.3), $P_t = -0.25P_{t-1} + 120$. Using (17.4),

$$P_t = \left(P_0 - \frac{120}{1 + 0.25}\right)(-0.25)^t + \frac{120}{1 + 0.25} = (P_0 - 96)(-0.25)^t + 96$$

which can be checked by substituting the appropriate values in (17.13). From (17.13a), $P_e = (-10 - 86)/(-0.8 - 0.2) = (-96)/(-1) = 96$.

With the base $b = -0.25$, which is negative and less than 1, the time path oscillates and converges. The equilibrium is stable, and P_t will converge to $P_e = 96$ as $t \to \infty$.

17.6   THE HARROD MODEL

The Harrod model attempts to explain the dynamics of growth in the economy. It assumes

$$S_t = sY_t$$

where s is a constant equal to both the MPS and APS. It also assumes the *acceleration principle*, i.e., investment is proportional to the rate of change of national income over time.

$$I_t = a(Y_t - Y_{t-1})$$

where a is a constant equal to both the marginal and average capital-output ratios. In equilibrium, $I_t = S_t$. Therefore,

$$a(Y_t - Y_{t-1}) = sY_t \qquad (a - s)Y_t = aY_{t-1}$$

Dividing through by $a - s$ to conform to (17.3), $Y_t = [a/(a - s)]Y_{t-1}$. Using (17.4) since $a/(a - s) \neq 1$,

$$Y_t = (Y_0 - 0)\left(\frac{a}{a - s}\right)^t + 0 = \left(\frac{a}{a - s}\right)^t Y_0 \qquad (17.14)$$

The stability of the time path thus depends on $a/(a - s)$. Since $a =$ the capital-output ratio, which is normally larger than 1, and since $s =$ MPS which is larger than 0 and less than 1, the base $a/(a - s)$ will be larger than 0 and usually larger than 1. Therefore, Y_t is explosive but nonoscillating. Income will expand indefinitely, which means it has no bounds. See Examples 8 and 9 and Problems 17.26 and 17.27. For other economic applications, see Problems 17.28 to 17.30.

EXAMPLE 8.   The *warranted rate of growth* (i.e., the path the economy must follow to have equilibrium between saving and investment each year) can be found as follows in the Harrod model.

From (17.14) Y_t increases indefinitely. Income in one period is $a/(a - s)$ times the income of the previous period.

$$Y_1 = \left(\frac{a}{a - s}\right) Y_0 \qquad (17.15)$$

The rate of growth G between the periods is defined as

$$G = \frac{Y_1 - Y_0}{Y_0}$$

Substituting from (17.15),

$$G = \frac{[a/(a - s)]Y_0 - Y_0}{Y_0} = \frac{[a/(a - s) - 1]Y_0}{Y_0}$$

$$= \frac{a}{a - s} - 1 = \frac{a}{a - s} - \frac{a - s}{a - s} = \frac{s}{a - s}$$

The warranted rate of growth, therefore, is

$$G_w = \frac{s}{a - s} \qquad (17.16)$$

EXAMPLE 9.   Assume that the marginal propensity to save in the Harrod model above is 0.12 and the capital-output ratio is 2.12. To find Y_t from (17.14),

$$Y_t = \left(\frac{2.12}{2.12 - 0.12}\right)^t Y_0 = (1.06)^t Y_0$$

The warranted rate of growth, from (17.16), is

$$G_w = \frac{0.12}{2.12 - 0.12} = \frac{0.12}{2} = 0.06$$

17.7 PHASE DIAGRAMS FOR DIFFERENCE EQUATIONS

While linear difference equations can be solved explicitly, nonlinear difference equations in general cannot. Important information about stability conditions, however, can once again be gleaned from phase diagrams. A phase diagram of a difference equation depicts y_t as a function of y_{t-1}. If we restrict the diagram to the first quadrant for economic reasons so all the variables will be nonnegative, a 45° line from the origin will capture all the possible steady-state equilibrium points where $y_t = y_{t-1}$. Consequently, any point at which the phase diagram intersects the 45° line will indicate an intertemporal equilibrium solution. The stability of a solution can be tested diagrammatically (Example 10) and mathematically (Example 11). Mathematically, the test depends on the following criteria for the first derivative of the phase line when it is evaluated at a steady-state point.

1. If $\left| \dfrac{dy_t}{dy_{t-1}}(\bar{y}) \right| < 1$, $\bar{y}$ is locally stable. If $\left| \dfrac{dy_t}{dy_{t-1}}(\bar{y}) \right| \geq 1$, $\bar{y}$ is locally unstable.

2. If $\dfrac{dy_t}{dy_{t-1}}(\bar{y}) \geq 0$, no oscillation. If $\dfrac{dy_t}{dy_{t-1}}(\bar{y}) < 0$, oscillation.

EXAMPLE 10. Given a nonlinear difference equation, such as

$$y_t = y_{t-1}^{0.5} = \sqrt{y_{t-1}}$$

we can construct a phase diagram in a few easy steps.

1. Find the steady-state solution(s), where $y_t = y_{t-1}$, by setting both y_t and $y_{t-1} = \bar{y}$ and solving algebraically for $\bar{y}$.

$$\bar{y} = \bar{y}^{0.5}$$
$$\bar{y}^{0.5} - \bar{y} = 0$$
$$\bar{y}\left(\frac{\bar{y}^{0.5}}{\bar{y}} - 1\right) = \bar{y}(\bar{y}^{-0.5} - 1) = 0$$
$$\bar{y}_1 = 0 \qquad \bar{y}_2 = 1 \qquad \text{steady-state solutions}$$

The phase diagram must intersect the 45° line at $\bar{y}_1 = 0$ and $\bar{y}_2 = 1$.

2. Take the first-derivative to see if the slope is positive or negative.

$$\frac{dy_t}{dy_{t-1}} = 0.5y_{t-1}^{-0.5} = \frac{0.5}{\sqrt{y_{t-1}}} > 0$$

Assuming $y_t, y_{t-1} > 0$, the phase diagram must be positively sloped.

3. Take the second derivative to see if the phase line is concave or convex.

$$\frac{d^2y_t}{dy_{t-1}^2} = -0.25y_{t-1}^{-1.5} = -0.25y_{t-1}^{-3/2} = \frac{-0.25}{\sqrt{y_{t-1}^3}} < 0 \qquad \text{concave}$$

4. Draw a rough sketch of the graph, as in Fig. 17-2.

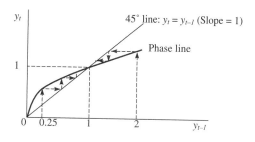

Fig. 17-2

5. To analyze the phase diagram for stability conditions, we pick an arbitrary value for y_{t-1}, say $y_{t-1} = 0.25$, and in a repeated sequence of steps (shown as dotted arrows) move up vertically to the phase diagram, then across horizontally to the 45° line, to see if the process converges to an equilibrium point or diverges from an equilibrium point. Here, starting from a value $0 < y_{t-1} < 1$, the process converges towards $\bar{y}_2 = 1$ and diverges from $\bar{y}_1 = 0$. We conclude, therefore, that starting from a value $0 < y_{t-1} < 1$, $\bar{y}_2 = 1$ is a locally stable equilibrium and $\bar{y}_1 = 0$ is locally unstable.

We next pick a value $y_{t-1} > 1$, here $y_{t-1} = 2$, and repeat the process. From the pattern of dotted arrows that emerges, we conclude that starting from a value $y_{t-1} > 1$, $\bar{y}_2 = 1$ is also locally stable when approached from a value > 1.

EXAMPLE 11. Here we confirm the results of the phase diagram test with the simple calculus test described above. Assuming once again, $y_t = y_{t-1}^{0.5}$,

$$\frac{dy_t}{dy_{t-1}} = 0.5 y_{t-1}^{-0.5}$$

Evaluated in absolute value at $\bar{y}_2 = 1$,

$$\left| 0.5(1)^{-0.5} \right| = \left| \frac{0.5}{\sqrt{1}} \right| = 0.5 < 1 \qquad \text{locally stable.}$$

Evaluated simply as $\bar{y}_2 = 1$,

$$0.5(1)^{-0.5} = \frac{0.5}{\sqrt{1}} = 0.5 > 0 \qquad \text{no oscillation.}$$

Evaluated at $\bar{y}_1 = 0$, the derivative is undefined but approaches infinity as $y_{t-1} \to 0$. Therefore, $\bar{y}_1 = 0$ is locally unstable. See Problems 17.31 to 17.33.

Solved Problems

USE OF GENERAL FORMULA FOR FIRST-ORDER LINEAR DIFFERENCE EQUATIONS

17.1. (a) Solve the difference equation given below; (b) check your answer, using $t = 0$ and $t = 1$; and (c) comment on the nature of the time path.

$$y_t = 6 y_{t-1}$$

a) Here $b = 6$ and $a = 0$. Using (17.4) for all cases in which $b \neq 1$,

$$y_t = (y_0 - 0)(6)^t + 0 = y_0(6)^t = A(6)^t \qquad (17.17)$$

where A, as a more generally used unspecified constant, replaces y_0.

b) Estimating (17.17) at $t = 0$ and $t = 1$,

$$y_0 = A(6)^0 = A \qquad y_1 = A(6) = 6A$$

Substituting $y_0 = A$ for y_{t-1} and $y_1 = 6A$ for y_t in the original problems, $6A = 6(A)$.

c) With the base $b = 6$ in (17.17) positive and greater than 1, that is, $b > 0$ and $|b| > 1$, the time path is nonoscillating and explosive.

17.2. Redo Problem 17.1 for $y_t = \frac{1}{8} y_{t-1}$.

a) Using (17.4),

$$y_t = (y_0 - 0)(\tfrac{1}{8})^t + 0 = y_0(\tfrac{1}{8})^t = A(\tfrac{1}{8})^t$$

b)   At $t = 0$, $y_0 = A(\frac{1}{8})^0 = A$. At $t = 1$, $y_1 = A(\frac{1}{8}) = \frac{1}{8}A$. Substituting $y_0 = A$ for y_{t-1} and $y_1 = \frac{1}{8}A$ for y_t in the original equation, $\frac{1}{8}A = \frac{1}{8}(A)$.

c)   With $b = \frac{1}{8}$, $b > 0$ and $|b| < 1$. The time path is nonoscillating and converging.

17.3.   Redo Problem 17.1, given $y_t = -\frac{1}{4}y_{t-1} + 60$ and $y_0 = 8$.

a)
$$y_t = \left(8 - \frac{60}{1 + \frac{1}{4}}\right)\left(-\frac{1}{4}\right)^t + \frac{60}{1 + \frac{1}{4}} = -40\left(-\frac{1}{4}\right)^t + 48$$

b)   At $t = 0$, $y_0 = -40(-\frac{1}{4})^0 + 48 = 8$. At $t = 1$, $y_1 = -40(-\frac{1}{4}) + 48 = 58$. Substituting in the original equation, $58 = -\frac{1}{4}(8) + 60 = 58$.

c)   With $b = -\frac{1}{4}$, $b < 0$ and $|b| < 1$. The time path oscillates and converges.

17.4.   Redo Problem 17.1, given $x_t + 3x_{t-1} + 8 = 0$ and $x_0 = 16$.

a)   Rearranging to conform with (17.3),

$$x_t = -3x_{t-1} - 8$$

Thus, $b = -3$ and $a = -8$. Substituting in (17.4),

$$x_t = \left(16 + \frac{8}{1 + 3}\right)(-3)^t - \frac{8}{1 + 3} = 18(-3)^t - 2$$

b)   At $t = 0$, $x_0 = 18(-3)^0 - 2 = 16$. At $t = 1$, $x_1 = 18(-3) - 2 = -56$. Substituting in the original, $-56 + 3(16) + 8 = 0$.

c)   With $b = -3$, $b < 0$ and $|b| > 1$. The time path oscillates and explodes.

17.5.   Redo Problem 17.1, given $y_t - y_{t-1} = 17$.

a)   Rearranging, $y_t = y_{t-1} + 17$. Here $b = 1$. Using (17.4a), therefore, $y_t = y_0 + 17t = A + 17t$.

b)   At $t = 0$, $y_0 = A$. At $t = 1$, $y_1 = A + 17$. Substituting in the original, $A + 17 - A = 17$.

c)   Here $b = 1$. Thus $b > 0$ and y_t will not oscillate. But with $|b| = 1$, $1 \not< |b| \not< 1$. This presents a special case. With $a \neq 0$, unless $y_0 = A = 0$, the time path is *divergent* because the complementary function A does not approach 0 as $t \to \infty$. Thus, y_t approaches $A + at$, and not the particular solution, at, itself. For $b = 1$ and $a = 0$, see Problem 17.17.

17.6.   Redo Problem 17.1, given $g_t = g_{t-1} - 25$ and $g_0 = 40$.

a)   Using (17.4a), $g_t = 40 - 25t$.

b)   At $t = 0$, $g_0 = 40$. At $t = 1$, $g_t = 15$. Substituting in the original, $15 = 40 - 25$.

c)   With $b = 1$, $a \neq 0$ and $A = g_0 \neq 0$. The time path is nonoscillatory and divergent.

17.7.   Redo Problem 17.1, given $2y_t = y_{t-1} - 18$.

a)   Dividing through by 2 to conform to (17.3) and then using (17.4),

$$y_t = \frac{1}{2}y_{t-1} - 9 = \left(y_0 + \frac{9}{1 - \frac{1}{2}}\right)\left(\frac{1}{2}\right)^t - \frac{9}{1 - \frac{1}{2}} = A\left(\frac{1}{2}\right)^t - 18$$

where A is an arbitrary constant for $y_0 + 18$.

b)   At $t = 0$, $y_0 = A - 18$. At $t = 1$, $y_1 = \frac{1}{2}A - 18$. Substituting in the original, $2(\frac{1}{2}A - 18) = A - 18 - 18$; $A - 36 = A - 36$.

c)   With $b = \frac{1}{2}$, $b > 0$ and $|b| < 1$. So y_t is nonoscillating and convergent.

17.8. (*a*) Solve the following difference equation; (*b*) check the answer, using $t = 0$ and $t = 1$; and (*c*) comment on the nature of the time path.

$$5y_t + 2y_{t-1} - 140 = 0 \qquad y_0 = 30$$

a)   Dividing by 5, rearranging terms, and using (*17.4*),

$$y_t = -0.4y_{t-1} + 28 = \left(30 - \frac{28}{1 + 0.4}\right)(-0.4)^t + \frac{28}{1 + 0.4} = 10(-0.4)^t + 20$$

b)   At $t = 0$, $y_0 = 30$. At $t = 1$, $y_1 = 16$. Substituting in the original, $5(16) + 2(30) - 140 = 0$.

c)   With $b = -0.4$, $b < 0$ and $|b| < 1$. So y_t oscillates and converges.

17.9. Redo Problem 17.8, given $x_{t+1} = 4x_t - 36$.

a)   Shifting the time periods back one period to conform with (*17.3*), $x_t = 4x_{t-1} - 36$. Using (*17.4*) and allowing A to replace $x_0 - a/(1 - b)$ as in Problem 17.7.

$$x_t = A(4)^t - \frac{36}{1 - 4} = A(4)^t + 12$$

b)   At $t = 0$, $x_0 = A + 12$. At $t = 1$, $x_1 = 4A + 12$. Substituting $x_1 = 4A + 12$ for x_{t+1} and $x_0 = A + 12$ for x_t in the original equation, $4A + 12 = 4(A + 12) - 36$; $4A + 12 = 4A + 12$.

c)   With $b = 4$, $b > 0$ and $|b| > 1$. So x_t does not oscillate but it explodes.

17.10. Redo Problem 17.8, given $y_{t+5} + 2y_{t+4} + 57 = 0$ and $y_0 = 11$.

a)   Moving the time periods back 5 periods, rearranging terms, and using (*17.4*),

$$y_t = -2y_{t-1} - 57 = \left(11 + \frac{57}{1 + 2}\right)(-2)^t - \frac{57}{1 + 2} = 30(-2)^t - 19$$

b)   At $t = 0$, $y_0 = 11$. At $t = 1$, $y_1 = -79$. Substituting y_1 for y_{t+5} and y_0 for y_{t+4} in the original equation, $-79 + 2(11) + 57 = 0$.

c)   With $b = -2$, $b < 0$ and $|b| > 1$. y_t oscillates and explodes.

17.11. Redo Problem 17.8, given $8y_{t-2} - 2y_{t-3} = 120$ and $y_0 = 28$.

a)   Divide through by 8, shift the time periods ahead by 2, and rearrange terms.

$$y_t = \frac{1}{4}y_{t-1} + 15 = \left(28 - \frac{15}{1 - \frac{1}{4}}\right)\left(\frac{1}{4}\right)^t + \frac{15}{1 - \frac{1}{4}} = 8\left(\frac{1}{4}\right)^t + 20$$

b)   At $t = 0$, $y_0 = 28$. At $t = 1$, $y_1 = 22$. Substitute y_1 for y_{t-2} and y_0 for y_{t-3}.

$$8(22) - 2(28) = 120 \qquad 120 = 120$$

c)   With $b = \frac{1}{4}$, $b > 0$ and $|b| < 1$. So y_t is nonoscillating and convergent.

17.12. Redo Problem 17.8, given $\Delta g_t = 14$.

a)   Substituting (*17.1*) for Δg_t,

$$g_{t+1} - g_t = 14 \qquad\qquad\qquad (17.18)$$

Set the time periods back 1 and rearrange terms.

$$g_t = g_{t-1} + 14$$

Using (*17.4a*), $g_t = g_0 + 14t = A + 14t$.

b) At $t = 0$, $g_0 = A$. At $t = 1$, $g_1 = A + 14$. Substituting g_1 for g_{t+1} and g_0 for g_t in (17.18), $A + 14 - A = 14$.

c) With $b = 1$, g_t is nonoscillatory. If $A \neq 0$, g_t is divergent.

17.13. Redo Problem 17.8, given $\Delta y_t = y_t + 13$ and $y_0 = 45$.

a) Substituting from (17.1), moving the time periods back 1, and rearranging terms,

$$y_t = 2y_{t-1} + 13 \qquad (17.19)$$

Using (17.4), $y_t = \left(45 - \dfrac{13}{1-2}\right)(2)^t + \dfrac{13}{1-2} = 58(2)^t - 13$

b) At $t = 0$, $y_0 = 45$. At $t = 1$, $y_1 = 103$. Substituting in (17.19), $103 = 2(45) + 13$; $103 = 103$.

c) With $b = 2$, $b > 0$ and $|b| > 1$. Thus y_t is nonoscillatory and explosive.

LAGGED INCOME DETERMINATION MODELS

17.14. Given the data below, (a) find the time path of national income Y_t; (b) check your answer, using $t = 0$ and $t = 1$; and (c) comment on the stability of the time path.

$$C_t = 90 + 0.8Y_{t-1} \qquad I_t = 50 \qquad Y_0 = 1200$$

a) In equilibrium, $Y_t = C_t + I_t$. Thus,

$$Y_t = 90 + 0.8Y_{t-1} + 50 = 0.8Y_{t-1} + 140 \qquad (17.20)$$

Using (17.4), $Y_t = \left(1200 - \dfrac{140}{1-0.8}\right)(0.8)^t + \dfrac{140}{1-0.8} = 500(0.8)^t + 700$

b) $Y_0 = 1200$; $Y_1 = 1100$. Substituting in (17.20),

$$1100 = 0.8(1200) + 140 \qquad\qquad 1100 = 1100$$

c) With $b = 0.8$, $b > 0$ and $|b| < 1$. The time path Y_t is nonoscillating and convergent. Y_t converges to the equilibrium level of income 700.

17.15. Redo Problem 17.14, given $C_t = 200 + 0.75Y_{t-1}$, $I_t = 50 + 0.15Y_{t-1}$, and $Y_0 = 3000$.

a) $Y_t = 200 + 0.75Y_{t-1} + 50 + 0.15Y_{t-1} = 0.9Y_{t-1} + 250$

Using (17.4), $Y_t = \left(3000 - \dfrac{250}{1-0.9}\right)(0.9)^t + \dfrac{250}{1-0.9} = 500(0.9)^t + 2500$

b) $Y_0 = 3000$; $Y_1 = 2950$. Substituting above, $2950 = 0.9(3000) + 250$; $2950 = 2950$.

c) With $b = 0.9$, the time path Y_t is nonoscillatory and converges toward 2500.

17.16. Redo Problem 17.14, given $C_t = 300 + 0.87Y_{t-1}$, $I_t = 150 + 0.13Y_{t-1}$, and $Y_0 = 6000$.

a) $Y_t = 300 + 0.87Y_{t-1} + 150 + 0.13Y_{t-1} = Y_{t-1} + 450 \qquad (17.21)$

Using (17.4a), $Y_t = 6000 + 450t$.

b) $Y_0 = 6000$; $Y_1 = 6450$. Substituting in (17.21) above, $6450 = 6000 + 450$.

c) With $b = 1$ and $A \neq 0$, the time path Y_t is nonoscillatory but divergent. See Problem 17.5.

17.17. Redo Problem 17.14, given $C_t = 0.92Y_{t-1}$, $I_t = 0.08Y_{t-1}$, and $Y_0 = 4000$.

a) $$Y_t = 0.92Y_{t-1} + 0.08Y_{t-1} = Y_{t-1}$$

Using $(17.4a)$, $Y_t = 4000 + 0 = 4000$.

b) $Y_0 = 4000 = Y_1$.

c)   When $b = 1$ and $a = 0$, Y_t is a stationary path.

17.18. Redo Problem 17.14, given $C_t = 400 + 0.6Y_t + 0.35Y_{t-1}$, $I_t = 240 + 0.15Y_{t-1}$, and $Y_0 = 7000$.

a) $Y_t = 400 + 0.6Y_t + 0.35Y_{t-1} + 240 + 0.15Y_{t-1}$ $0.4Y_t = 0.5Y_{t-1} + 640$

Divide through by 0.4 and then use (17.4).

$$Y_t = 1.25Y_{t-1} + 1600 = \left(7000 - \frac{1600}{1 - 1.25}\right)(1.25)^t + \frac{1600}{1 - 1.25} = 13{,}400(1.25)^t - 6400$$

b) $Y_0 = 7000$; $Y_1 = 10{,}350$. Substituting in the initial equation,

$$10{,}350 = 400 + 0.6(10{,}350) + 0.35(7000) + 240 + 0.15(7000) = 10{,}350$$

c)   With $b = 1.25$, the time path Y_t is nonoscillatory and explosive.

17.19. Redo Problem 17.14, given $C_t = 300 + 0.5Y_t + 0.4Y_{t-1}$, $I_t = 200 + 0.2Y_{t-1}$, and $Y_0 = 6500$.

a) $Y_t = 300 + 0.5Y_t + 0.4Y_{t-1} + 200 + 0.2Y_{t-1}$ $0.5Y_t = 0.6Y_{t-1} + 500$

Dividing through by 0.5 and then using (17.4),

$$Y_t = 1.2Y_{t-1} + 1000 = \left(6500 - \frac{1000}{1 - 1.2}\right)(1.2)^t + \frac{1000}{1 - 1.2} = 11{,}500(1.2)^t - 5000$$

b) $Y_0 = 6500$; $Y_1 = 8800$. Substituting in the initial equation,

$$8800 = 300 + 0.5(8800) + 0.4(6500) + 200 + 0.2(6500) = 8800$$

c)   With $b = 1.2$, Y_t is nonoscillatory and explosive.

17.20. Redo Problem 17.14, given $C_t = 200 + 0.5Y_t$, $I_t = 3(Y_t - Y_{t-1})$, and $Y_0 = 10{,}000$.

a) $Y_t = 200 + 0.5Y_t + 3(Y_t - Y_{t-1})$ $-2.5Y_t = -3Y_{t-1} + 200$

Dividing through by -2.5 and then using (17.4),

$$Y_t = 1.2Y_{t-1} - 80 = \left(10{,}000 - \frac{80}{1 - 1.2}\right)(1.2)^t - \frac{80}{1 - 1.2} = 9600(1.2)^t + 400$$

b) $Y_0 = 10{,}000$; $Y_1 = 11{,}920$. Substituting in the initial equation,

$$11{,}920 = 200 + 0.5(11{,}920) + 3(11{,}920 - 10{,}000) = 11{,}920$$

c)   With $b = 1.2$, the time path Y_t explodes but does not oscillate.

THE COBWEB MODEL

17.21. For the data given below, determine *(a)* the market price P_t in any time period, *(b)* the equilibrium price P_e, and *(c)* the stability of the time path.

$$Q_{dt} = 180 - 0.75P_t \qquad Q_{st} = -30 + 0.3P_{t-1} \qquad P_0 = 220$$

a)   Equating demand and supply,

$$180 - 0.75P_t = -30 + 0.3P_{t-1} \tag{17.22}$$
$$-0.75P_t = 0.3P_{t-1} - 210$$

Dividing through by -0.75 and using (17.4),

$$P_t = -0.4P_{t-1} + 280 = \left(220 - \frac{280}{1+0.4}\right)(-0.4)^t + \frac{280}{1+0.4} = 20(-0.4)^t + 200 \tag{17.23}$$

b)   If the market is in equilibrium, $P_t = P_{t-1}$. Substituting P_e for P_t and P_{t-1} in (17.22),

$$180 - 0.75P_e = -30 + 0.3P_e \qquad P_e = 200$$

which is the second term on the right-hand side of (17.23).

c)   With $b = -0.4$, the time path P_t will oscillate and converge.

17.22.   Check the answer to Problem 17.21(a), using $t = 0$ and $t = 1$.

From (17.23), $P_0 = 20(-0.4)^0 + 200 = 220$ and $P_1 = 20(-0.4) + 200 = 192$. Substituting P_1 for P_t and P_0 for P_{t-1} in (17.22),

$$180 - 0.75(192) = -30 + 0.3(220)$$
$$36 = 36$$

17.23.   Redo Problem 17.21, given $Q_{dt} = 160 - 0.8P_t$, $Q_{st} = -20 + 0.4P_{t-1}$, and $P_0 = 153$.

a)
$$160 - 0.8P_t = -20 + 0.4P_{t-1} \tag{17.24}$$
$$-0.8P_t = 0.4P_{t-1} - 180$$

Dividing through by -0.8 and using (17.4),

$$P_t = -0.5P_{t-1} + 225 = \left(153 - \frac{225}{1+0.5}\right)(-0.5)^t + \frac{225}{1+0.5} = 3(-0.5)^t + 150 \tag{17.25}$$

b)   As shown in Problem 17.21(b), $P_e = 150$. See also Section 17.5.

c)   With $b = -0.5$, P_t oscillates and converges toward 150.

17.24.   Check the answer to Problem 17.23(a), using $t = 0$ and $t = 1$.

From (17.25), $P_0 = 3(-0.5)^0 + 150 = 153$ and $P_1 = 3(-0.5) + 150 = 148.5$. Substituting in (17.24),

$$160 - 0.8(148.5) = -20 + 0.4(153)$$
$$41.2 = 41.2$$

17.25.   Redo Problem 17.21, given $Q_{dt} = 220 - 0.4P_t$, $Q_{st} = -30 + 0.6P_{t-1}$, and $P_0 = 254$.

a)
$$220 - 0.4P_t = -30 + 0.6P_{t-1}$$
$$-0.4P_t = 0.6P_{t-1} - 250$$

Dividing through by -0.4 and then using (17.4),

$$P_t = -1.5P_{t-1} + 625 = \left(254 - \frac{625}{1+1.5}\right)(-1.5)^t + \frac{625}{1+1.5} = 4(-1.5)^t + 250$$

b)
$$P_e = 250$$

c)   With $b = -1.5$, P_t oscillates and explodes.

THE HARROD GROWTH MODEL

17.26. For the following data, find (*a*) the level of income Y_t for any period and (*b*) the warranted rate of growth.

$$I_t = 2.66(Y_t - Y_{t-1}) \qquad S_t = 0.16Y_t \qquad Y_0 = 9000$$

a)   In equilibrium,

$$2.66(Y_t - Y_{t-1}) = 0.16Y_t \qquad 2.5Y_t = 2.66Y_{t-1}$$

Dividing through by 2.5 and then using (*17.4*),

$$Y_t = 1.064Y_{t-1} = (9000 - 0)(1.064)^t + 0 = 9000(1.064)^t$$

b)   From (*17.16*), $G_w = 0.16/(2.66 - 0.16) = 0.064$.

17.27. Redo Problem 17.26, given $I_t = 4.2(Y_t - Y_{t-1})$, $S_t = 0.2Y_t$, and $Y_0 = 5600$.

a)

$$4.2(Y_t - Y_{t-1}) = 0.2Y_t$$
$$4Y_t = 4.2Y_{t-1}$$
$$Y_t = 1.05Y_{t-1}$$

Using (*17.4*), $Y_t = 5600(1.05)^t$.

b)

$$G_w = \frac{0.2}{4.2 - 0.2} = 0.05$$

OTHER ECONOMIC APPLICATIONS

17.28. Derive the formula for the value P_t of an initial amount of money P_0 deposited at i interest for t years when compounded annually.

When interest is compounded annually,

$$P_{t+1} = P_t + iP_t = (1 + i)P_t$$

Moving the time periods back one to conform with (*17.3*),

$$P_t = (1 + i)P_{t-1}$$

Using (*17.4*) since $i \neq 0$, $\qquad P_t = (P_0 + 0)(1 + i)^t + 0 = P_0(1 + i)^t$

17.29. Assume that $Q_{dt} = c + zP_t$, $Q_{st} = g + hP_t$, and

$$P_{t+1} = P_t - a(Q_{st} - Q_{dt}) \tag{17.26}$$

i.e., price is no longer determined by a market-clearing mechanism but by the level of inventory $Q_{st} - Q_{dt}$. Assume, too, that $a > 0$ since a buildup in inventory ($Q_{st} > Q_{dt}$) will tend to reduce price and a depletion of inventory ($Q_{st} < Q_{dt}$) will cause prices to rise. (*a*) Find the price P_t for any period and (*b*) comment on the stability conditions of the time path.

a)   Substituting Q_{st} and Q_{dt} in (*17.26*),

$$P_{t+1} = P_t - a(g + hP_t - c - zP_t)$$
$$= [1 - a(h - z)]P_t - a(g - c) = [1 + a(z - h)]P_t - a(g - c)$$

Shifting the time periods back 1 to conform to (*17.3*) and using (*17.4*),

$$P_t = \left\{ P_0 + \frac{a(g - c)}{1 - [1 + a(z - h)]} \right\} [1 + a(z - h)]^t - \frac{a(g - c)}{1 - [1 + a(z - h)]}$$

$$= \left(P_0 - \frac{g - c}{z - h} \right) [1 + a(z - h)]^t + \frac{g - c}{z - h} \tag{17.27}$$

Substituting as in (17.13a),

$$P_t = (P_0 - P_e)[1 + a(z - h)]^t + P_e \tag{17.28}$$

b)   The stability of the time path depends on $b = 1 + a(z - h)$. Since $a > 0$ and under normal conditions $z < 0$ and $h > 0$, $a(z - h) < 0$. Thus,

If $0 < |a(z - h)| < 1$,	$0 < b < 1$;	P_t converges and is nonoscillatory.
If $a(z - h) = -1$,	$b = 0$;	P_t remains in equilibrium ($P_t = P_0$).
If $-2 < a(z - h) < -1$,	$-1 < b < 0$;	P_t converges with oscillation.
If $a(z - h) = -2$,	$b = -1$;	uniform oscillation takes place.
If $a(z - h) < -2$,	$b < -1$;	P_t oscillates and explodes.

17.30. Given the following data, (a) find the price P_t for any time period; (b) check the answer, using $t = 0$ and $t = 1$; and (c) comment on the stability conditions.

$$Q_{dt} = 120 - 0.5P_t \qquad Q_{st} = -30 + 0.3P_t \qquad P_{t+1} = P_t - 0.2(Q_{st} - Q_{dt}) \qquad P_0 = 200$$

a)   Substituting, $P_{t+1} = P_t - 0.2(-30 + 0.3P_t - 120 + 0.5P_t) = 0.84P_t + 30$

Shifting time periods back 1 and using (17.4),

$$P_t = 0.84P_{t-1} + 30 = \left(200 - \frac{30}{1 - 0.84}\right)(0.84)^t + \frac{30}{1 - 0.84} = 12.5(0.84)^t + 187.5$$

b) $P_0 = 200$; $P_1 = 198$. Substituting in the first equation of the solution, $198 = 200 - 0.2[-30 + 0.3(200) - 120 + 0.5(200)] = 198$.

c)   With $b = 0.84$, P_t converges without oscillation toward 187.5.

PHASE DIAGRAMS FOR DIFFERENCE EQUATIONS

17.31. (a) Construct a phase diagram for the nonlinear difference equation below, (b) use it to test for dynamic stability, and (c) confirm your results with the derivative test.

$$y_t = y_{t-1}^3$$

a)   Setting $y_t = y_{t-1} = \bar{y}$ for the intertemporal equilibrium solution,

$$\bar{y} = \bar{y}^3$$
$$\bar{y}(1 - \bar{y}^2) = 0$$

$\bar{y}_1 = 0 \qquad \bar{y}_2 = 1 \qquad$ intertemporal equilibrium levels

The phase diagram will intersect the 45° line at $y = 0$ and $y = 1$. We next take the first and second derivative to determine the slope and concavity of the phase line.

$$\frac{dy_t}{dy_{t-1}} = 3y_{t-1}^2 > 0 \qquad \text{positive slope}$$

$$\frac{d^2 y_t}{dy_{t-1}^2} = 6y_{t-1} > 0 \qquad \text{convex}$$

With the above information we then draw a rough sketch of the graph, as in Fig. 17-3.

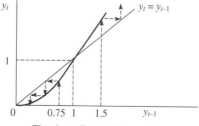

The phase diagram for $y_t = y_{t-1}^3$

Fig. 17-3

$b)$ Starting from a value of $y_t = 0.75$ and following the series of moves indicated by the dotted arrows we see the function converges to $\bar{y}_1 = 0$ and diverges from $\bar{y}_2 = 1$. From a starting point of $y_t = 1.5$, the function also diverges from $\bar{y}_2 = 1$. We conclude, therefore, that $\bar{y}_2 = 1$ is an unstable equilibrium when approached from either side and $\bar{y}_1 = 0$ is a stable equilibrium when approached from a positive value.

$c)$ Independently of the phase diagram, we can also test the stability conditions by taking the first derivative of the equation and evaluating it at the steady-state solutions.

$$\frac{dy_t}{dy_{t-1}} = 3y_{t-1}^2$$

$$\left| \frac{dy_t}{dy_{t-1}}(0) \right| = 3(0) = 0 < 1 \qquad \text{locally stable}$$

$$\left| \frac{dy_t}{dy_{t-1}}(1) \right| = |3| > 1 \qquad \text{locally unstable}$$

With the derivative positive at $\bar{y}_1 = 0$ and $\bar{y}_2 = 1$, there is no oscillation.

17.32. Repeat the steps in Problem 17.31 for the nonlinear difference equation,

$$y_t = y_{t-1}^{-0.25} = \frac{1}{\sqrt[4]{y_{t-1}}}$$

$a)$ Setting $y_t = y_{t-1} = \bar{y}$, and substituting above,

$$\bar{y} = \bar{y}^{-0.25}$$

$$\bar{y} - \bar{y}^{-0.25} = 0$$

$$\bar{y}(1 - \bar{y}^{-1.25}) = 0$$

$$\bar{y} = 0 \qquad \bar{y} = 1$$

Since $\bar{y} = 0$ is undefined at $\bar{y}^{-1.25}$, there is only one intertemporal equilibrium, $\bar{y} = 1$. Taking the derivatives,

$$\frac{dy_t}{dy_{t-1}} = -0.25y_{t-1}^{-1.25} = \frac{-0.25}{y_{t-1}^{1.25}} < 0 \qquad \text{negative slope}$$

$$\frac{d^2y_t}{dy_{t-1}^2} = 0.3125y_{t-1}^{-2.25} > 0 \qquad \text{convex}$$

We can then sketch the graph, as in Fig. 17-4.

$b)$ Starting at a value less than 1, say $y_{t-1} = 0.75$, the function oscillates between values larger and smaller than one but converges to $\bar{y} = 1$. If we start at a value larger than 1, we also get the same results.

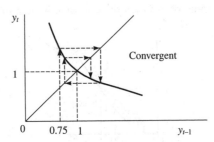

The phase diagram for $y_t = y_{t-1}^{-0.25}$

Fig. 17-4

c) Working simply with the derivative and evaluating it in absolute value at the steady-state equilibrium solution, we have

$$\frac{dy_t}{dy_{t-1}} = -0.25y_{t-1}^{-1.25}$$

$$\left|\frac{dy_t}{dy_{t-1}}(1)\right| = |-0.25(1)| < 1 \qquad \text{locally stable}$$

$$\frac{dy_t}{dy_{t-1}}(1) = -0.25(1) < 0 \qquad \text{oscillation}$$

17.33. Redo Problem 17.31 for the equation

$$y_t = y_{t-1}^{-1.5}$$

a) The intertemporal equilibrium solution is again $\bar{y} = 1$.

$$\frac{dy_t}{dy_{t-1}} = -1.5y_{t-1}^{-2.5} < 0 \qquad \text{negative slope}$$

$$\frac{d^2y_t}{dy_{t-1}^2} = 3.75y_{t-1}^{-3.5} > 0 \qquad \text{convex}$$

See Fig. 17-5.

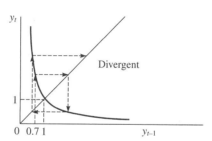

The phase diagram for $y_t = y_{t-1}^{-1.5}$

Fig. 17-5

b) Starting at a value less than 1, say $y_{t-1} = 0.7$, the function oscillates between values larger and smaller than one and ultimately diverges from $\bar{y} = 1$.

c) Working solely with the first derivative of the function,

$$\left|\frac{dy_t}{dy_{t-1}}(1)\right| = |-1.5| > 1 \qquad \text{locally unstable}$$

$$\frac{dy_t}{dy_{t-1}}(1) = -1.5 < 0 \qquad \text{oscillation}$$

CHAPTER 18

Second-Order Differential Equations and Difference Equations

18.1 SECOND-ORDER DIFFERENTIAL EQUATIONS

Second-order differential equations require separate solutions for the complementary function y_c and the particular integral y_p. The general solution is the sum of the two: $y(t) = y_c + y_p$. Given the second-order linear differential equation

$$y''(t) + b_1 y'(t) + b_2 y(t) = a \qquad (18.1)$$

where b_1, b_2, and a are constants, the particular integral will be

$$y_p = \frac{a}{b_2} \qquad b_2 \neq 0 \qquad (18.2)$$

$$y_p = \frac{a}{b_1} t \qquad b_2 = 0 \qquad b_1 \neq 0 \qquad (18.2a)$$

$$y_p = \frac{a}{2} t^2 \qquad b_1 = b_2 = 0 \qquad (18.2b)$$

The complementary function is

$$y_c = y_1 + y_2 \qquad (18.3)$$

408

where

$$y_1 = A_1 e^{r_1 t} \qquad (18.3a)$$

$$y_2 = A_2 e^{r_2 t} \qquad (18.3b)$$

and

$$r_1, r_2 = \frac{-b_1 \pm \sqrt{b_1^2 - 4b_2}}{2} \qquad (18.4)$$

Here, A_1 and A_2 are arbitrary constants, and $b_1^2 \neq 4b_2$. r_1 and r_2 are referred to as *characteristic roots*, and (18.4) is the solution to the *characteristic* or *auxiliary equation*: $r^2 + b_1 r + b_2 = 0$. See Examples 1 to 4 and Problems 18.1 to 18.11, 20.9, 20.10, 20.13, 20.14 and 20.16 to 20.20.

EXAMPLE 1.   The particular integral for each of the following equations

$$(1) \quad y''(t) - 5y'(t) + 4y(t) = 2 \qquad (2) \quad y''(t) + 3y'(t) = 12 \qquad (3) \quad y''(t) = 16$$

is found as shown below.

For (1), using (18.2), $\qquad\qquad\qquad y_p = \frac{2}{4} = \frac{1}{2} \qquad\qquad\qquad\qquad (18.5)$

For (2), using (18.2a), $\qquad\qquad\quad y_p = \frac{12}{3}t = 4t \qquad\qquad\qquad\qquad (18.5a)$

For (3), using (18.2b), $\qquad\qquad\quad y_p = \frac{16}{2}t^2 = 8t^2 \qquad\qquad\qquad\quad (18.5b)$

EXAMPLE 2.   The complementary functions for equations (1) and (2) in Example 1 are calculated below. Equation (3) will be treated in Example 9.

For (1), from (18.4),

$$r_1, r_2 = \frac{+5 \pm \sqrt{(-5)^2 - 4(4)}}{2} = \frac{5 \pm 3}{2} = 1, 4$$

Substituting in (18.3a) and (18.3b), and finally in (18.3),

$$y_c = A_1 e^t + A_2 e^{4t} \qquad (18.6)$$

For (2), $\qquad\qquad r_1, r_2 = \dfrac{-3 \pm \sqrt{(3)^2 - 4(0)}}{2} = \dfrac{-3 \pm 3}{2} = 0, -3$

Thus, $\qquad\qquad\qquad y_c = A_1 e^0 + A_2 e^{-3t} = A_1 + A_2 e^{-3t} \qquad (18.6a)$

EXAMPLE 3.   The general solution of a differential equation is composed of the complementary function and the particular integral (Section 16.2), that is, $y(t) = y_c + y_p$. As applied to the equations in Example 1,

For (1), from (18.6) and (18.5), $\qquad y(t) = A_1 e^t + A_2 e^{4t} + \frac{1}{2} \qquad\qquad\qquad (18.7)$

For (2), from (18.6a) and (18.5a), $\qquad y(t) = A_1 + A_2 e^{-3t} + 4t \qquad\qquad\qquad (18.7a)$

EXAMPLE 4.   The definite solution for (1) in Example 3 is calculated below. Assume $y(0) = 5\frac{1}{2}$ and $y'(0) = 11$.

From (18.7),

$$y(t) = A_1 e^t + A_2 e^{4t} + \frac{1}{2} \qquad (18.8)$$

Thus, $\qquad\qquad\qquad\qquad\qquad y'(t) = A_1 e^t + 4A_2 e^{4t} \qquad\qquad\qquad\qquad (18.8a)$

Evaluating (18.8) and (18.8a) at $t = 0$, and setting $y(0) = 5\frac{1}{2}$ and $y'(0) = 11$ from the initial conditions,

$$y(0) = A_1 e^0 + A_2 e^{4(0)} + \tfrac{1}{2} = 5\tfrac{1}{2} \qquad \text{thus} \qquad A_1 + A_2 = 5$$

$$y'(0) = A_1 e^0 + 4A_2 e^{4(0)} = 11 \qquad \text{thus} \qquad A_1 + 4A_2 = 11$$

Solving simultaneously, $A_1 = 3$ and $A_2 = 2$. Substituting in (18.7),

$$y(t) = 3e^t + 2e^{4t} + \tfrac{1}{2} \qquad (18.9)$$

To check this solution, from (18.9),

$$y(t) = 3e^t + 2e^{4t} + \tfrac{1}{2}$$

Thus, $$y'(t) = 3e^t + 8e^{4t} \qquad y''(t) = 3e^t + 32e^{4t}$$

Substituting in the original equation [(1) in Example 1],

$$(3e^t + 32e^{4t}) - 5(3e^t + 8e^{4t}) + 4(3e^t + 2e^{4t} + \tfrac{1}{2}) = 2$$

18.2 SECOND-ORDER DIFFERENCE EQUATIONS

The general solution of a second-order difference equation is composed of a complementary function and a particular solution: $y(t) = y_c + y_p$. Given the second-order linear difference equation

$$y_t + b_1 y_{t-1} + b_2 y_{t-2} = a \tag{18.10}$$

where b_1, b_2, and a are constants, the particular solution is

$$y_p = \frac{a}{1 + b_1 + b_2} \qquad b_1 + b_2 \neq -1 \tag{18.11}$$

$$y_p = \frac{a}{2 + b_1} t \qquad b_1 + b_2 = -1 \qquad b_1 \neq -2 \tag{18.11a}$$

$$y_p = \frac{a}{2} t^2 \qquad b_1 + b_2 = -1 \qquad b_1 = -2 \tag{18.11b}$$

The complementary function is

$$y_c = A_1 r_1^t + A_2 r_2^t \tag{18.12}$$

where A_1 and A_2 are arbitrary constants and the characteristic roots r_1 and r_2 are found by using (18.4), assuming $b_1^2 \neq 4b_2$. See Examples 5 to 8 and Problems 18.12 to 18.20.

EXAMPLE 5. The particular solution for each of the following equations:

1) $y_t - 10y_{t-1} + 16y_{t-2} = 14$ 2) $y_t - 6y_{t-1} + 5y_{t-2} = 12$ 3) $y_t - 2y_{t-1} + y_{t-2} = 8$

is found as shown below.

For (1), using (18.11), $$y_p = \frac{14}{1 - 10 + 16} = 2 \tag{18.13}$$

For (2), using $(18.11a)$, $$y_p = \frac{12}{2 - 6} t = -3t \tag{18.13a}$$

For (3), using $(18.11b)$, $$y_p = \tfrac{8}{2} t^2 = 4t^2 \tag{18.13b}$$

EXAMPLE 6. From Example 5, the complementary functions for (1) and (2) are calculated below. For (3), see Example 9.

For (1), using (18.4) and then substituting in (18.12),

$$r_1, r_2 = \frac{10 \pm \sqrt{100 - 4(16)}}{2} = \frac{10 \pm 6}{2} = 2, 8$$

Thus, $$y_c = A_1(2)^t + A_2(8)^t \tag{18.14}$$

For (2), $$r_1, r_2 = \frac{6 \pm \sqrt{36 - 4(5)}}{2} = \frac{6 \pm 4}{2} = 1, 5$$

Thus, $$y_c = A_1(1)^t + A_2(5)^t = A_1 + A_2(5)^t \tag{18.14a}$$

EXAMPLE 7.   The general solutions for (1) and (2) from Example 5 are calculated below.
For (1), $y(t) = y_c + y_p$. From (18.14) and (18.13),

$$y(t) = A_1(2)^t + A_2(8)^t + 2 \tag{18.15}$$

For (2), from (18.14a) and (18.13a),

$$y(t) = A_1 + A_2(5)^t - 3t \tag{18.15a}$$

EXAMPLE 8.   Given $y(0) = 10$ and $y(1) = 36$, the definite solution for (1) in Example 7 is calculated as follows:
Letting $t = 0$ and $t = 1$ successively in (18.15),

$$y(0) = A_1(2)^0 + A_2(8)^0 + 2 = A_1 + A_2 + 2 \qquad y(1) = A_1(2) + A_2(8) + 2 = 2A_1 + 8A_2 + 2$$

Setting $y(0) = 10$ and $y(1) = 36$ from the initial conditions,

$$\begin{aligned} A_1 + \ A_2 + 2 &= 10 \\ 2A_1 + 8A_2 + 2 &= 36 \end{aligned}$$

Solving simultaneously, $A_1 = 5$ and $A_2 = 3$. Finally, substituting in (18.15),

$$y(t) = 5(2)^t + 3(8)^t + 2 \tag{18.16}$$

This answer is checked by evaluating (18.16) at $t = 0$, $t = 1$, and $t = 2$,

$$y(0) = 5 + 3 + 2 = 10 \qquad y(1) = 10 + 24 + 2 = 36 \qquad y(2) = 20 + 192 + 2 = 214$$

Substituting $y(2)$ for y_t, $y(1)$ for y_{t-1}, and $y(0)$ for y_{t-2} in $y_t - 10y_{t-1} + 16y_{t-2} = 14$ of Equation (1) in Example 5,
$214 - 10(36) + 16(10) = 14$.

18.3   CHARACTERISTIC ROOTS

A characteristic equation can have three different types of roots.

1. *Distinct real roots.* If $b_1^2 > 4b_2$, the square root in (18.4) will be a real number, and r_1 and r_2 will
    be distinct real numbers as in (18.6) and (18.6a).
2. *Repeated real roots.* If $b_1^2 = 4b_2$, the square root in (18.4) will vanish, and r_1 and r_2 will equal
    the same real number. In the case of repeated real roots, the formulas for y_c in (18.3) and
    (18.12) must be changed to

$$y_c = A_1 e^{rt} + A_2 t e^{rt} \tag{18.17}$$

$$y_c = A_1 r^t + A_2 t r^t \tag{18.18}$$

3. *Complex roots.* If $b_1^2 < 4b_2$, (18.4) contains the square root of a negative number, which is
    called an *imaginary number*. In this case r_1 and r_2 are complex numbers. A *complex number*
    contains a real part and an imaginary part; for example, $(12 + i)$ where $i = \sqrt{-1}$.

[As a simple test to check your answers when using (18.4), assuming the coefficient of the $y''(t)$ term is 1,
$r_1 + r_2$ must equal $-b_1$; $r_1 \times r_2$ must equal b_2.]

EXAMPLE 9.   The complementary function for Equation (3) in Example 1, where $y''(t) = 16$, is found as follows:
From (18.4),

$$r_1, r_2 = \frac{0 \pm \sqrt{0 - 4(0)}}{2} = 0$$

Using (18.17) since $r_1 = r_2 = 0$, which is a case of repeated real roots, $y_c = A_1 e^0 + A_2 t e^0 = A_1 + A_2 t$.
In Equation (3) of Example 5, $y_t - 2y_{t-1} + y_{t-2} = 8$. Solving for the complementary function, from (18.4),

$$r_1, r_2 = \frac{2 \pm \sqrt{4 - 4(1)}}{2} = \frac{2 \pm 0}{2} = 1$$

Using (18.18) because $r_1 = r_2 = 1$, $y_c = A_1(1)^t + A_2 t(1)^t = A_1 + A_2 t$.

18.4   CONJUGATE COMPLEX NUMBERS

If $b_1^2 < 4b_2$ in (18.4), factoring out $\sqrt{-1}$ gives

$$r_1, r_2 = \frac{-b_1 \pm \sqrt{-1}\sqrt{4b_2 - b_1^2}}{2} = \frac{-b_1 \pm i\sqrt{4b_2 - b_1^2}}{2}$$

Put more succinctly, $\qquad\qquad\qquad r_1, r_2 = g \pm hi$

where $\qquad\qquad g = -\tfrac{1}{2}b_1 \qquad \text{and} \qquad h = \tfrac{1}{2}\sqrt{4b_2 - b_1^2}$ $\qquad\qquad$ (18.19)

$g \pm hi$ are called *conjugate* complex numbers because they always appear together. Substituting (18.19) in (18.3) and (18.12) to find y_c for cases of complex roots,

$$y_c = A_1 e^{(g+hi)t} + A_2 e^{(g-hi)t} = e^{gt}(A_1 e^{hit} + A_2 e^{-hit}) \qquad (18.20)$$

$$y_c = A_1(g + hi)^t + A_2(g - hi)^t \qquad (18.21)$$

See Example 10 and Problems 18.28 to 18.35, 20.11 and 20.12.

EXAMPLE 10. The complementary function for $y''(t) + 2y'(t) + 5y(t) = 18$ is calculated as shown below. Using (18.19) since $b_1^2 < 4b_2$,

$$g = -\tfrac{1}{2}(2) = -1 \qquad h = \tfrac{1}{2}\sqrt{4(5) - (2)^2} = \tfrac{1}{2}(4) = 2$$

Thus, $r_1, r_2 = -1 \pm 2i$. Substituting in (18.20), $y_c = e^{-t}(A_1 e^{2it} + A_2 e^{-2it})$.

18.5   TRIGONOMETRIC FUNCTIONS

Trigonometric functions are often used in connection with complex numbers. Given the angle θ in Fig. 18-1, which is at the center of a circle of radius k and measured counterclockwise, the trigonometric functions of θ are

$$\text{sine (sin) } \theta = \frac{h}{k} \qquad\qquad \text{cosine (cos) } \theta = \frac{g}{k}$$

$$\text{tangent (tan) } \theta = \frac{h}{g} \qquad\qquad \text{cotangent (cot) } \theta = \frac{g}{h}$$

$$\text{secant (sec) } \theta = \frac{k}{g} \qquad\qquad \text{cosecant (csc) } \theta = \frac{k}{h}$$

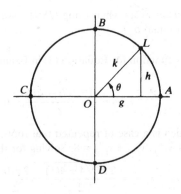

Fig. 18-1

The signs of the trigonometric functions in each of the four quadrants are

$$\begin{array}{c|c} + & + \\ \hline - & - \end{array} \qquad \begin{array}{c|c} - & + \\ \hline - & + \end{array} \qquad \begin{array}{c|c} - & + \\ \hline + & - \end{array}$$

sin, csc   cos, sec   tan, cot

The angle θ is frequently measured in *radians*. Since there are 2π radians in a circle, $1° = \pi/180$ radian. Thus $360° = 2\pi$ radians, $180° = \pi$ radians, $90° = \pi/2$ radians, and $45° = \pi/4$ radians.

EXAMPLE 11.   If the radius OL in Fig. 18-1 starts at A and moves counterclockwise 360°, $\sin \theta = h/k$ goes from 0 at A, to 1 at B, to 0 at C, to -1 at D, and back to 0 at A. Cosine $\theta = g/k$ goes from 1 at A, to 0 at B, to -1 at C, to 0 at D, and back to 1 at A. This is summarized in Table 18-1 and graphed in Fig. 18-2. Notice that both functions are *periodic* with a *period* of 2π (i.e., they repeat themselves every 360° or 2π radians). Both have an *amplitude* of fluctuation of 1 and differ only in *phase* or location of their peaks.

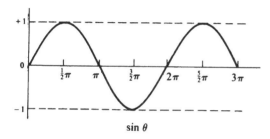
sin θ

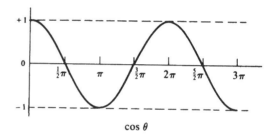
cos θ

Fig. 18-2

Table 18-1

Degrees	0	90	180	270	360
Radians	0	$\dfrac{\pi}{2}$	π	$\dfrac{3}{2}\pi$	2π
$\sin \theta$	0	1	0	-1	0
$\cos \theta$	1	0	-1	0	1

18.6   DERIVATIVES OF TRIGONOMETRIC FUNCTIONS

Given that u is a differentiable function of x,

1) $\dfrac{d}{dx}(\sin u) = \cos u \dfrac{du}{dx}$

4) $\dfrac{d}{dx}(\cot u) = -\csc^2 u \dfrac{du}{dx}$

2) $\dfrac{d}{dx}(\cos u) = -\sin u \dfrac{du}{dx}$

5) $\dfrac{d}{dx}(\sec u) = \sec u \tan u \dfrac{du}{dx}$

3) $\dfrac{d}{dx}(\tan u) = \sec^2 u \dfrac{du}{dx}$

6) $\dfrac{d}{dx}(\csc u) = -\csc u \cot u \dfrac{du}{dx}$

See Example 12 and Problems 18.21 to 18.27.

EXAMPLE 12.   The derivatives for the trigonometric functions

1) $y = \sin(3x^2 + 6)$ 2) $y = 4\cos 2x$ 3) $y = (1 + \tan x)^2$

are calculated as follows:

1) $\dfrac{dy}{dx} = 6x\cos(3x^2 + 6)$ 2) $\dfrac{dy}{dx} = -8\sin 2x$ 3) $\dfrac{dy}{dx} = 2(1 + \tan x)(\sec^2 x) = 2\sec^2 x(1 + \tan x)$

18.7   TRANSFORMATION OF IMAGINARY AND COMPLEX NUMBERS

Three rules are helpful in transforming imaginary and complex numbers to trigonometric functions.

1. g and h in Fig. 18-1, which are *Cartesian coordinates*, can be expressed in terms of θ and k, which are called *polar coordinates*, by the simple formula

$$g = k\cos\theta \qquad h = k\sin\theta \qquad k > 0$$

Thus for the conjugate complex number $(g \pm hi)$,

$$g \pm hi = k\cos\theta \pm ik\sin\theta = k(\cos\theta \pm i\sin\theta) \tag{18.22}$$

2. By what are called *Euler relations*,

$$e^{\pm i\theta} = \cos\theta \pm i\sin\theta \tag{18.23}$$

Thus, by substituting (18.23) in (18.22) we can also express $(g \pm hi)$ as

$$g \pm hi = ke^{\pm i\theta} \tag{18.23a}$$

3. From $(18.23a)$, raising a conjugate complex number to the nth power means

$$(g \pm hi)^n = (ke^{\pm i\theta})^n = k^n e^{\pm in\theta} \tag{18.24}$$

Or, by making use of (18.23) and noting that $n\theta$ replaces θ, we have *De Moivre's theorem*:

$$(g \pm hi)^n = k^n(\cos n\theta \pm i\sin n\theta) \tag{18.25}$$

See Examples 13 to 15 and Problems 18.28 to 18.35, 20.11 and 20.12.

EXAMPLE 13.   The value of the imaginary exponential function $e^{2i\pi}$ is found as follows. Using (18.23), where $\theta = 2\pi$,

$$e^{2i\pi} = \cos 2\pi + i\sin 2\pi$$

From Table 18-1, $\cos 2\pi = 1$ and $\sin 2\pi = 0$. Thus, $e^{2i\pi} = 1 + i(0) = 1$.

EXAMPLE 14.   The imaginary exponential expressions in (18.20) and (18.21) are transformed to trigonometric functions as shown below.

From (18.20), $y_c = e^{gt}(A_1 e^{hit} + A_2 e^{-hit})$. Using (18.23) where $\theta = ht$,

$$\begin{aligned}
y_c &= e^{gt}[A_1(\cos ht + i\sin ht) + A_2(\cos ht - i\sin ht)] \\
&= e^{gt}[(A_1 + A_2)\cos ht + (A_1 - A_2)i\sin ht] \\
&= e^{gt}(B_1\cos ht + B_2\sin ht) \tag{18.26}
\end{aligned}$$

where $B_1 = A_1 + A_2$ and $B_2 = (A_1 - A_2)i$.

From (18.21), $y_c = A_1(g + hi)^t + A_2(g - hi)^t$. Using (18.25) and substituting t for n,

$$\begin{aligned}
y_c &= A_1 k^t(\cos t\theta + i\sin t\theta) + A_2 k^t(\cos t\theta - i\sin t\theta) \\
&= k^t[(A_1 + A_2)\cos t\theta + (A_1 - A_2)i\sin t\theta] \\
&= k^t(B_1\cos t\theta + B_2\sin t\theta) \tag{18.27}
\end{aligned}$$

where $B_1 = A_1 + A_2$ and $B_2 = (A_1 - A_2)i$.

EXAMPLE 15.   The time paths of (*18.26*) and (*18.27*) are evaluated as follows: Examining each term in (*18.26*),

1.   Here $B_1 \cos ht$ is a cosine function of t, as in Fig. 18-2, with period $2\pi/h$ instead of 2π and amplitude of the multiplicative constant B_1 instead of 1.
2.   Likewise $B_2 \sin ht$ is a sine function of t with period $2\pi/h$ and amplitude of B_2.
3.   With the first two terms constantly fluctuating, stability depends on e^{gt}.

> If $g > 0$, e^{gt} gets increasingly larger as t increases. This increases the amplitude and leads to explosive fluctuations of y_c, precluding convergence.
>
> If $g = 0$, $e^{gt} = 1$ and y_c displays uniform fluctuations determined by the sine and cosine functions. This also precludes convergence.
>
> If $g < 0$, e^{gt} approaches zero as t increases. This diminishes the amplitude, produces damped fluctuations, and leads to convergence. See Fig. 18-3.

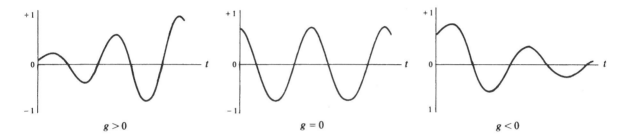

Fig. 18-3   Time path of $y_c(t)$.

Since (*18.27*) concerns a difference equation in which t can only change at discrete intervals, y_c is a step function rather than a continuous function (see Fig. 17-1). Like (18.26), it will fluctuate, and stability will depend on k^t. If $|k| < 1$, y_c will converge. See Problems 18.8 to 18.11, 18.18 to 18.20, and 18.31 to 18.35.

18.8   STABILITY CONDITIONS

For a second-order linear differential equation with distinct or repeated real roots, both roots must be negative for convergence. If one of the roots is positive, the exponential term with the positive root approaches infinity as t approaches infinity, thereby precluding convergence. See Problems 18.8 to 18.11. In the case of complex roots, g in e^{gt} of (18.26) must be negative, as illustrated in Example 15.

For a second-order linear difference equation with distinct or repeated real roots, the root with the largest absolute value is called the *dominant root* because it dominates the time path. For convergence, the absolute value of the dominant root must be less than 1. See Problems 18.18 to 18.20. In the case of complex roots, the absolute value of k in (*18.27*) must be less than 1, as explained in Example 15. For economic applications, see Problems 18.36 and 18.37.

Solved Problems

SECOND-ORDER LINEAR DIFFERENTIAL EQUATIONS

Distinct Real Roots

18.1. For the following equation, find (*a*) the particular integral y_p, (*b*) the complementary function y_c, and (*c*) the general solution $y(t)$.

$$y''(t) + 9y'(t) + 14y(t) = 7$$

a) Using (*18.2*), $y_p = \frac{7}{14} = \frac{1}{2}$.

b) Using (*18.4*),

$$r_1, r_2 = \frac{-9 \pm \sqrt{81 - 4(14)}}{2} = \frac{-9 \pm 5}{2} = -2, -7$$

Substituting in (*18.3*), $y_c = A_1 e^{-2t} + A_2 e^{-7t}$.

c)
$$y(t) = y_c + y_p = A_1 e^{-2t} + A_2 e^{-7t} + \tfrac{1}{2} \qquad\qquad (18.28)$$

18.2. Redo Problem 18.1, given $y''(t) - 12y'(t) + 20y(t) = -100$.

a) From (*18.2*), $y_p = -\frac{100}{20} = -5$.

b) From (*18.4*),

$$r_1, r_2 = \frac{12 \pm \sqrt{144 - 4(20)}}{2} = \frac{12 \pm 8}{2} = 2, 10$$

Thus, $y_c = A_1 e^{2t} + A_2 e^{10t}$.

c)
$$y(t) = y_c + y_p = A_1 e^{2t} + A_2 e^{10t} - 5 \qquad\qquad (18.29)$$

18.3. Redo Problem 18.1, given $y''(t) - 4y'(t) - 5y(t) = 35$.

a) From (*18.2*), $y_p = \dfrac{35}{-5} = -7$.

b) From (*18.4*)

$$r_1, r_2 = \frac{4 \pm \sqrt{16 - 4(-5)}}{2} = \frac{4 \pm 6}{2} = 5, -1$$

Thus, $y_c = A_1 e^{5t} + A_2 e^{-t}$.

c)
$$y(t) = A_1 e^{5t} + A_2 e^{-t} - 7 \qquad\qquad (18.30)$$

18.4. Redo Problem 18.1, given $y''(t) + 7y'(t) = 28$.

a) Using (*18.2a*), $y_p = \frac{28}{7}t = 4t$.

b) From (*18.4*),

$$r_1, r_2 = \frac{-7 \pm \sqrt{49 - 4(0)}}{2} = \frac{-7 \pm 7}{2} = 0, -7$$

Thus, $y_c = A_1 e^{(0)t} + A_2 e^{-7t} = A_1 + A_2 e^{-7t}$

c)
$$y(t) = A_1 + A_2 e^{-7t} + 4t \qquad\qquad (18.31)$$

18.5. Redo Problem 18.1, given $y''(t) - \frac{1}{2}y'(t) = 13$.

a) From (*18.2a*), $y_p = \frac{13}{-\frac{1}{2}}t = -26t$.

b)
$$r_1, r_2 = \frac{\frac{1}{2} \pm \sqrt{\frac{1}{4} - 4(0)}}{2} = \frac{\frac{1}{2} \pm \frac{1}{2}}{2} = 0, \frac{1}{2}$$

Thus, $y_c = A_1 + A_2 e^{(1/2)t}$

c)
$$y(t) = A_1 + A_2 e^{(1/2)t} - 26t \tag{18.32}$$

Repeated Real Roots

18.6. Find (*a*) the particular integral y_p, (*b*) the complementary function y_c, and (*c*) the general solution $y(t)$, given $y''(t) - 12y'(t) + 36y(t) = 108$.

a)
$$y_p = \frac{108}{36} = 3$$

b)
$$r_1, r_2 = \frac{12 \pm \sqrt{144 - 4(36)}}{2} = \frac{12 \pm 0}{2} = 6$$

Using (*18.17*) since $r_1 = r_2 = 6$, $y_c = A_1 e^{6t} + A_2 t e^{6t}$.

c)
$$y(t) = A_1 e^{6t} + A_2 t e^{6t} + 3 \tag{18.33}$$

18.7. Redo Problem 18.6, given $y''(t) + y'(t) + \frac{1}{4}y(t) = 9$.

a)
$$y_p = \frac{9}{\frac{1}{4}} = 36$$

b)
$$r_1, r_2 = \frac{-1 \pm \sqrt{1 - 4(\frac{1}{4})}}{2} = \frac{-1 \pm 0}{2} = -\frac{1}{2}$$

Using (*18.17*) since $r_1 = r_2 = -\frac{1}{2}$, $y_c = A_1 e^{-(1/2)t} + A_2 t e^{-(1/2)t}$.

c)
$$y(t) = A_1 e^{-(1/2)t} + A_2 t e^{-(1/2)t} + 36 \tag{18.34}$$

DEFINITE SOLUTIONS AND STABILITY CONDITIONS

18.8. Find (*a*) the definite solution for the following equation, (*b*) check your answer, and (*c*) comment on the dynamic stability of the time path, given $y''(t) + 9y'(t) + 14y(t) = 7$, $y(0) = -2\frac{1}{2}$, and $y'(0) = 31$.

a) From (*18.28*),
$$y(t) = A_1 e^{-2t} + A_2 e^{-7t} + \frac{1}{2} \tag{18.35}$$

Thus,
$$y'(t) = -2A_1 e^{-2t} - 7A_2 e^{-7t} \tag{18.35a}$$

Evaluating (*18.35*) and (*18.35a*) at $t = 0$,
$$y(0) = A_1 + A_2 + \frac{1}{2} \qquad y'(0) = -2A_1 - 7A_2$$

Setting $y(0) = -2\frac{1}{2}$ and $y'(0) = 31$ from the initial conditions,
$$A_1 + A_2 + \frac{1}{2} = -2\frac{1}{2}$$
$$-2A_1 - 7A_2 = 31$$

Solving simultaneously, $A_1 = 2$ and $A_2 = -5$, which when substituted in (*18.35*) gives
$$y(t) = 2e^{-2t} - 5e^{-7t} + \frac{1}{2} \tag{18.36}$$

b) From (18.36), $y(t) = 2e^{-2t} - 5e^{-7t} + \frac{1}{2}$. Thus,

$$y'(t) = -4e^{-2t} + 35e^{-7t} \qquad y''(t) = 8e^{-2t} - 245e^{-7t}$$

Substituting these values in the original problem, where $y'' + 9y'(t) + 14y(t) = 7$,

$$8e^{-2t} - 245e^{-7t} + 9(-4e^{-2t} + 35e^{-7t}) + 14(2e^{-2t} - 5e^{-7t} + \frac{1}{2}) = 7$$

c) With both characteristic roots negative, (18.36) will approach $\frac{1}{2}$ as $t \to \infty$. Therefore $y(t)$ is convergent. Any time both characteristic roots are negative, the time path will converge.

18.9. Redo Problem 18.8, given $y''(t) - 4y'(t) - 5y(t) = 35$, $y(0) = 5$, and $y'(0) = 6$.

a) From (18.30),

$$y(t) = A_1 e^{5t} + A_2 e^{-t} - 7 \tag{18.37}$$

Thus,

$$y'(t) = 5A_1 e^{5t} - A_2 e^{-t} \tag{18.37a}$$

Evaluating (18.37) and (18.37a) at $t = 0$ and setting them equal to the initial conditions where $y(0) = 5$ and $y'(0) = 6$,

$$y(0) = A_1 + A_2 - 7 = 5 \qquad \text{thus} \qquad A_1 + A_2 = 12$$
$$y'(0) = 5A_1 - A_2 = 6$$

Solving simultaneously, $A_1 = 3$ and $A_2 = 9$, which when substituted in (18.37) gives

$$y(t) = 3e^{5t} + 9e^{-t} - 7 \tag{18.38}$$

b) From (18.38), $y(t) = 3e^{5t} + 9e^{-t} - 7$. Thus, $y'(t) = 15e^{5t} - 9e^{-t}$ and $y''(t) = 75e^{5t} + 9e^{-t}$.
Substituting these values in the original problem, where $y''(t) - 4y'(t) - 5y(t) = 35$,

$$75e^{5t} + 9e^{-t} - 4(15e^{5t} - 9e^{-t}) - 5(3e^{5t} + 9e^{-t} - 7) = 35$$

c) With one characteristic root positive and the other negative, the time path is divergent. The positive root dominates the negative root independently of their relative absolute values because as $t \to \infty$, the positive root $\to \infty$ and the negative root $\to 0$.

18.10. Redo Problem 18.8, given $y''(t) - \frac{1}{2}y'(t) = 13$, $y(0) = 17$, and $y'(0) = -19\frac{1}{2}$.

a) From (18.32),

$$y(t) = A_1 + A_2 e^{(1/2)t} - 26t \tag{18.39}$$

Thus,

$$y'(t) = \frac{1}{2}A_2 e^{(1/2)t} - 26 \tag{18.39a}$$

Evaluating (18.39) and (18.39a) at $t = 0$ and setting them equal to the initial conditions,

$$y(0) = A_1 + A_2 = 17$$
$$y'(0) = \frac{1}{2}A_2 - 26 = -19\frac{1}{2} \qquad A_2 = 13$$

With $A_2 = 13$, $A_1 = 4$. Substituting in (18.39) and rearranging terms,

$$y(t) = 13e^{(1/2)t} - 26t + 4 \tag{18.40}$$

b) From (18.40), $y(t) = 13e^{(1/2)t} - 26t + 4$. Thus,

$$y'(t) = 6.5e^{(1/2)t} - 26 \qquad y''(t) = 3.25e^{(1/2)t}$$

Substituting these values in the original equation,

$$3.25e^{(1/2)t} - \frac{1}{2}(6.5e^{(1/2)t} - 26) = 13$$

c) With both characteristic roots positive, the time path will diverge.

18.11. Redo Problem 18.8, given $y''(t) + y'(t) + \frac{1}{4}y(t) = 9$, $y(0) = 30$, and $y'(0) = 15$.

a) From (18.34),

$$y(t) = A_1 e^{-(1/2)t} + A_2 t e^{-(1/2)t} + 36 \qquad (18.41)$$

Using the product rule for the derivative of the second term,

$$y'(t) = -\tfrac{1}{2}A_1 e^{-(1/2)t} - \tfrac{1}{2}A_2 t e^{-(1/2)t} + A_2 e^{-(1/2)t} \qquad (18.41a)$$

Evaluating (18.41) and (18.41a) at $t = 0$ and equating to the initial conditions,

$$y(0) = A_1 + 36 = 30 \qquad A_1 = -6$$
$$y'(0) = -\tfrac{1}{2}A_1 + A_2 = 15$$

With $A_1 = -6$, $A_2 = 12$. Substituting in (18.41),

$$y(t) = 12t e^{-(1/2)t} - 6e^{-(1/2)t} + 36 \qquad (18.42)$$

b) From (18.42), $y(t) = 12t e^{-(1/2)t} - 6e^{-(1/2)t} + 36$. By the product rule,

$$y'(t) = -6t e^{-(1/2)t} + 12 e^{-(1/2)t} + 3e^{-(1/2)t}$$
$$y''(t) = 3t e^{-(1/2)t} - 6e^{-(1/2)t} - 6e^{-(1/2)t} - 1.5 e^{-(1/2)t} = 3t e^{-(1/2)t} - 13.5 e^{-(1/2)t}$$

Substituting in the original equation,

$$(3t e^{-(1/2)t} - 13.5 e^{-(1/2)t}) + (-6t e^{-(1/2)t} + 15 e^{-(1/2)t}) + \tfrac{1}{4}(12t e^{-(1/2)t} - 6e^{-(1/2)t} + 36) = 9$$

c) With the repeated characteristic roots negative, the time path will converge since te^{rt} follows basically the same time path as e^{rt}.

SECOND-ORDER LINEAR DIFFERENCE EQUATIONS

Distinct Real Roots

18.12. Find (a) the particular solution, (b) the complementary function, and (c) the general solution, given $y_t + 7y_{t-1} + 6y_{t-2} = 42$.

a) From (18.11),

$$y_p = \frac{42}{1 + 7 + 6} = 3$$

b) From (18.4),

$$r_1, r_2 = \frac{-7 \pm \sqrt{49 - 4(6)}}{2} = \frac{-7 \pm 5}{2} = -1, -6$$

From (18.12),

$$y_c = A_1(-1)^t + A_2(-6)^t$$

c)

$$y(t) = y_c + y_p = A_1(-1)^t + A_2(-6)^t + 3 \qquad (18.43)$$

18.13. Redo Problem 18.12, given $y_t + 12y_{t-1} + 11y_{t-2} = 6$.

a) From (18.11),

$$y_p = \frac{6}{1 + 12 + 11} = \frac{1}{4}$$

b)

$$r_1, r_2 = \frac{-12 \pm \sqrt{144 - 4(11)}}{2} = \frac{-12 \pm 10}{2} = -1, -11$$

Thus,

$$y_c = A_1(-1)^t + A_2(-11)^t$$

c)

$$y(t) = A_1(-1)^t + A_2(-11)^t + \tfrac{1}{4} \qquad (18.44)$$

18.14. Redo Problem 18.12, given $y_{t+2} - 11y_{t+1} + 10y_t = 27$.

a) Shifting the time periods back 2 to conform with (*18.10*), $y_t - 11y_{t-1} + 10y_{t-2} = 27$. Then from (*18.11a*),

$$y_p = \frac{27}{2 - 11}t = -3t$$

b)

$$r_1, r_2 = \frac{11 \pm \sqrt{121 - 4(10)}}{2} = \frac{11 \pm 9}{2} = 1,\ 10$$

$$y_c = A_1 + A_2(10)^t$$

c)

$$y(t) = A_1 + A_2(10)^t - 3t \qquad (18.45)$$

18.15. Redo Problem 18.12, given $y_t + 7y_{t-1} - 8y_{t-2} = 45$.

a) From (*18.11a*),

$$y_p = \frac{45}{2 + 7}t = 5t$$

b)

$$r_1, r_2 = \frac{-7 \pm \sqrt{49 - 4(-8)}}{2} = \frac{-7 \pm 9}{2} = 1,\ -8$$

$$y_c = A_1 + A_2(-8)^t$$

c)

$$y(t) = A_1 + A_2(-8)^t + 5t \qquad (18.46)$$

Repeated Real Roots

18.16. Redo Problem 18.12, given $y_t - 10y_{t-1} + 25y_{t-2} = 8$.

a)

$$y_p = \frac{8}{1 - 10 + 25} = \frac{1}{2}$$

b)

$$r_1, r_2 = \frac{10 \pm \sqrt{100 - 4(25)}}{2} = \frac{10 \pm 0}{2} = 5$$

Using (*18.18*) because $r_1 = r_2 = 5$, $y_c = A_1(5)^t + A_2 t(5)^t$.

c)

$$y(t) = A_1(5)^t + A_2 t(5)^t + \tfrac{1}{2} \qquad (18.47)$$

18.17. Redo Problem 18.12, given $y_t + 14y_{t-1} + 49y_{t-2} = 128$.

a)

$$y_p = \frac{128}{1 + 14 + 49} = 2$$

b)

$$r_1, r_2 = \frac{-14 \pm \sqrt{196 - 4(49)}}{2} = \frac{-14 \pm 0}{2} = -7$$

From (*18.18*), $y_c = A_1(-7)^t + A_2 t(-7)^t$.

c)

$$y(t) = A_1(-7)^t + A_2 t(-7)^t + 2 \qquad (18.48)$$

DEFINITE SOLUTIONS AND STABILITY CONDITIONS

18.18. (*a*) Find the definite solution, (*b*) check the answer, and (*c*) comment on dynamic stability, given $y_t + 7y_{t-1} + 6y_{t-2} = 42$, $y(0) = 16$, and $y(1) = -35$.

a) From (*18.43*), $\qquad\qquad y(t) = A_1(-1)^t + A_2(-6)^t + 3 \qquad (18.49)$

Letting $t = 0$ and $t = 1$ successively in (*18.49*) and making use of the initial conditions,

$$y(0) = A_1 + A_2 + 3 = 16 \qquad y(1) = -A_1 - 6A_2 + 3 = -35$$

Solving simultaneously, $A_1 = 8$ and $A_2 = 5$. Substituting in (18.49),

$$y(t) = 8(-1)^t + 5(-6)^t + 3 \qquad\qquad (18.50)$$

b) Evaluating (18.50) at $t = 0$, $t = 1$, and $t = 2$ to check this answer,

$$y(0) = 8 + 5 + 3 = 16 \qquad y(1) = -8 - 30 + 3 = -35 \qquad y(2) = 8 + 180 + 3 = 191$$

Substituting in the initial equation with

$$y(2) = y_t \qquad y(1) = y_{t-1} \qquad y(0) = y_{t-2} \qquad 191 + 7(-35) + 6(16) = 42$$

c) The characteristic roots are -1 and -6. The characteristic root with the largest absolute value is called the *dominant root* because it dominates the time path. For convergence, the absolute value of the dominant root must be less than 1. Since $|-6| > 1$, the time path is divergent.

18.19. (a) Find the definite solution and (b) comment on dynamic stability, given

$$y_{t+2} - 11y_{t+1} + 10y_t = 27 \qquad y(0) = 2 \qquad y(1) = 53$$

a) From (18.45), $\qquad\qquad y(t) = A_1 + A_2(10)^t - 3t \qquad\qquad (18.51)$

Letting $t = 0$ and $t = 1$, and using the initial conditions,

$$y(0) = A_1 + A_2 = 2 \qquad y(1) = A_1 + 10A_2 - 3 = 53$$

Solving simultaneously, $A_1 = -4$ and $A_2 = 6$. Substituting in (18.51),

$$y(t) = 6(10)^t - 3t - 4$$

b) The time path is divergent because the dominant root 10 is greater than 1.

18.20. Redo Problem 18.19, given $y_t - 10y_{t-1} + 25y_{t-2} = 8$, $y(0) = 1$, and $y(1) = 5$.

a) From (18.47),

$$y(t) = A_1(5)^t + A_2 t(5)^t + \tfrac{1}{2} \qquad\qquad (18.52)$$

Letting $t = 0$ and $t = 1$, and using the initial conditions,

$$y(0) = A_1 + \tfrac{1}{2} = 1 \qquad A_1 = \tfrac{1}{2}$$
$$y(1) = 5A_1 + 5A_2 + \tfrac{1}{2} = 5$$

With $A_1 = \tfrac{1}{2}$, $A_2 = \tfrac{2}{5}$. Substituting in (18.52),

$$y(t) = \tfrac{1}{2}(5)^t + \tfrac{2}{5}t(5)^t + \tfrac{1}{2} \qquad\qquad (18.53)$$

b) Convergence in the case of repeated real roots likewise depends on $|r| < 1$ since the effect of r^t dominates the effect of t in the second term $A_2 t r^t$. Here with $r = 5 > 1$, the time path is divergent.

DERIVATIVES OF TRIGONOMETRIC FUNCTIONS

18.21. Find the first-order derivative for the following trigonometric functions. Note that they are also called *circular functions* or *sinusoidal functions*.

a) $y = \sin 7x$

$$\frac{dy}{dx} = 7\cos 7x$$

b) $y = \cos(5x + 2)$

$$\frac{dy}{dx} = -5\sin(5x + 2)$$

c) $y = \tan 11x$

$$\frac{dy}{dx} = 11\sec^2 11x$$

d) $y = \csc(8x + 3)$

$$\frac{dy}{dx} = -8[\csc(8x + 3)\cot(8x + 3)]$$

e) $y = \sin(3 - x^2)$ f) $y = \sin(5 - x)^2$

$\dfrac{dy}{dx} = -2x \cos(3 - x^2)$ $\dfrac{dy}{dx} = -2(5 - x) \cos(5 - x)^2$ (chain rule)

18.22. Redo Problem 18.21, given $y = x^2 \tan x$.

By the product rule, $\dfrac{dy}{dx} = x^2 (\sec^2 x) + (\tan x)(2x) = x^2 \sec^2 x + 2x \tan x$

18.23. Redo Problem 18.21, given $y = x^3 \sin x$.

$\dfrac{dy}{dx} = x^3 (\cos x) + (\sin x)(3x^2) = x^3 \cos x + 3x^2 \sin x$

18.24. Redo Problem 18.21, given $y = (1 + \cos x)^2$.

By the chain rule, $\dfrac{dy}{dx} = 2(1 + \cos x)(-\sin x) = (-2 \sin x)(1 + \cos x)$

18.25. Redo Problem 18.21, given $y = (\sin x + \cos x)^2$.

$\dfrac{dy}{dx} = 2(\sin x + \cos x)(\cos x - \sin x) = 2(\cos^2 x - \sin^2 x)$

18.26. Redo Problem 18.21, given $y = \sin^2 5x$, where $\sin^2 5x = (\sin 5x)^2$.

By the chain rule, $\dfrac{dy}{dx} = 2 \sin 5x \cos 5x(5) = 10 \sin 5x \cos 5x$

18.27. Redo Problem 18.21, given $y = \csc^2 12x$.

$\dfrac{dy}{dx} = (2 \csc 12x)[-\csc 12x \cot 12x(12)] = -24 \csc^2 12x \cot 12x$

COMPLEX ROOTS IN SECOND-ORDER DIFFERENTIAL EQUATIONS

18.28. Find (*a*) the particular integral, (*b*) the complementary function, and (*c*) the general solution, given the second-order linear differential equation $y''(t) + 2y'(t) + 10y(t) = 80$.

a) From (*18.2*), $y_p = \frac{80}{10} = 8$

b) Using (*18.19*) since $b_1^2 < 4b_2$, that is, $(2)^2 < 4(10)$,

$$g = -\tfrac{1}{2}(2) = -1 \qquad h = \tfrac{1}{2}\sqrt{4(10) - (2)^2} = 3$$

Thus, $r_1, r_2 = -1 \pm 3i$. Substituting g and h in (*18.26*),

$$y_c = e^{-t}(B_1 \cos 3t + B_2 \sin 3t)$$

c) $y(t) = y_c + y_p = e^{-t}(B_1 \cos 3t + B_2 \sin 3t) + 8$ (*18.54*)

18.29. Redo Problem 18.28, given $y''(t) - 6y'(t) + 25y(t) = 150$.

a) $y_p = \frac{150}{25} = 6$

b) From (*18.19*), $g = -\tfrac{1}{2}(-6) = 3$ and $h = \tfrac{1}{2}\sqrt{4(25) - (-6)^2} = 4$. Substituting in (*18.26*),

$$y_c = e^{3t}(B_1 \cos 4t + B_2 \sin 4t)$$

c) $y(t) = e^{3t}(B^1 \cos 4t + B_2 \sin 4t) + 6$ (*18.55*)

18.30. Redo Problem 18.28, given $y''(t) + 4y'(t) + 40y(t) = 10$.

    a)
$$y_p = \frac{10}{40} = \frac{1}{4}$$

    b)   From (18.19), $g = -2$ and $h = \frac{1}{2}\sqrt{160 - 16} = 6$. Thus, $y_c = e^{-2t}(B_1 \cos 6t + B_2 \sin 6t)$.

    c)
$$y(t) = e^{-2t}(B_1 \cos 6t + B_2 \sin 6t) + \tfrac{1}{4} \qquad\qquad (18.56)$$

18.31. (*a*) Find the definite solution for the following data. (*b*) Comment on the dynamic stability.

$$y''(t) + 2y'(t) + 10y(t) = 80 \qquad y(0) = 10 \qquad y'(0) = 13$$

    a)   From (18.54),
$$y(t) = e^{-t}(B_1 \cos 3t + B_2 \sin 3t) + 8 \qquad\qquad (18.57)$$

        By the product rule,

$$y'(t) = e^{-t}(-3B_1 \sin 3t + 3B_2 \cos 3t) + (B_1 \cos 3t + B_2 \sin 3t)(-e^{-t})$$
$$= e^{-t}(3B_2 \cos 3t - 3B_1 \sin 3t) - e^{-t}(B_1 \cos 3t + B_2 \sin 3t) \qquad (18.57a)$$

        Evaluating (18.57) and (18.57a) at $t = 0$ and equating them to the initial conditions,

$$y(0) = e^0(B_1 \cos 0 + B_2 \sin 0) + 8 = 10$$

        From Table 18-1, $\cos 0 = 1$ and $\sin 0 = 0$. Thus,

$$y(0) = B_1 + 0 + 8 = 10 \qquad B_1 = 2$$

        Similarly, $y'(0) = e^0(3B_2 \cos 0 - 3B_1 \sin 0) - e^0(B_1 \cos 0 + B_2 \sin 0) = 13$.

$$y'(0) = 3B_2 - B_1 = 13$$

        Since $B_1 = 2$ from above, $B_2 = 5$. Finally, substituting in (18.57),

$$y(t) = e^{-t}(2 \cos 3t + 5 \sin 3t) + 8$$

    b)   With $g = -1$, the time path converges, as it does in Fig. 18-3 (see Example 15).

18.32. Redo Problem 18.31, given $y''(t) - 6y'(t) + 25y(t) = 150$, $y(0) = 13$, and $y'(0) = 25$.

    a)   From (18.55),
$$y(t) = e^{3t}(B_1 \cos 4t + B_2 \sin 4t) + 6 \qquad\qquad (18.58)$$

        Thus,
$$y'(t) = e^{3t}(-4B_1 \sin 4t + 4B_2 \cos 4t) + 3e^{3t}(B_1 \cos 4t + B_2 \sin 4t) \qquad (18.58a)$$

        Evaluating (18.58) and (18.58a) at $t = 0$ and equating them to the initial conditions,

$$y(0) = e^0(B_1 \cos 0 + B_2 \sin 0) + 6 = 13$$
$$y(0) = B_1 + 0 + 6 = 13 \qquad B_1 = 7$$

        and $y'(0) = e^0(-4B_1 \sin 0 + 4B_2 \cos 0) + 3e^0(B_1 \cos 0 + B_2 \sin 0)$.

$$y'(0) = 4B_2 + 3B_1 = 25 \qquad B_2 = 1$$

        Substituting in (18.58), $y(t) = e^{3t}(7 \cos 4t + \sin 4t) + 6$.

    b)   With $g = 3$, the time path is divergent.

18.33. Redo Problem 18.31, given $y''(t) + 4y'(t) + 40y(t) = 10$, $y(0) = \frac{1}{2}$, and $y'(0) = 2\frac{1}{2}$.

    a)   From (18.56),
$$y(t) = e^{-2t} = (B_1 \cos 6t + B_2 \sin 6t) + \tfrac{1}{4} \qquad\qquad (18.59)$$

        Thus,
$$y'(t) = e^{-2t}(-6B_1 \sin 6t + 6B_2 \cos 6t) - 2e^{-2t}(B_1 \cos 6t + B_2 \sin 6t) \qquad (18.59a)$$

Evaluating (18.59) and $(18.59a)$ at $t = 0$ and equating them to the initial conditions,

$$y(0) = B_1 + \tfrac{1}{4} = \tfrac{1}{2} \qquad\qquad B_1 = \tfrac{1}{4}$$
$$y'(0) = 6B_2 - 2B_1 = 2\tfrac{1}{2} \qquad\qquad B_2 = \tfrac{1}{2}$$

Thus, $y(t) = e^{-2t}(\tfrac{1}{4}\cos 6t + \tfrac{1}{2}\sin 6t) + \tfrac{1}{4} = \tfrac{1}{4}e^{-2t}(\cos 6t + 2\sin 6t) + \tfrac{1}{4}$.

b)   With $g = -2$, the time path is convergent.

COMPLEX ROOTS IN SECOND-ORDER DIFFERENCE EQUATIONS

18.34. Find (a) the particular solution, (b) the complementary function, (c) the general solution, and (d) the definite solution. (e) Comment on the dynamic stability of the following second-order linear difference equation:

$$y_t + 4y_{t-2} = 15 \qquad y(0) = 12 \qquad y(1) = 11$$

a)   From (18.11),
$$y_p = \frac{15}{1 + 0 + 4} = 3$$

b)   From (18.19), $g = -\tfrac{1}{2}(0) = 0$ and $h = \tfrac{1}{2}\sqrt{4(4) - 0} = 2$. For second-order difference equations we now need k and θ. Applying the Pythagorean theorem to Fig. 18-1,
$$k^2 = g^2 + h^2 \qquad k = \sqrt{g^2 + h^2}$$

Substituting with the parameters of (18.19) for greater generality,

$$k = \sqrt{\frac{b_1^2 + 4b_2 - b_1^2}{4}} = \sqrt{b_2} \qquad\qquad (18.60)$$

Thus, $k = \sqrt{4} = 2$. From the definitions of Section 18.5,

$$\sin\theta = \frac{h}{k} \qquad\qquad \cos\theta = \frac{g}{k} \qquad\qquad (18.61)$$

Substituting the values from the present problem,

$$\sin\theta = \tfrac{2}{2} = 1 \qquad\qquad \cos\theta = \tfrac{0}{2} = 0$$

From Table 18-1, the angle with $\sin\theta = 1$ and $\cos\theta = 0$ is $\pi/2$. Thus, $\theta = \pi/2$. Substituting in (18.27),

$$y_c = 2^t\left[B_1\cos\left(\frac{\pi}{2}t\right) + B_2\sin\left(\frac{\pi}{2}t\right)\right]$$

c)
$$y(t) = 2^t\left[B_1\cos\left(\frac{\pi}{2}t\right) + B_2\sin\left(\frac{\pi}{2}t\right)\right] + 3 \qquad\qquad (18.62)$$

d)   Using Table 18-1 to evaluate (18.62) at $t = 0$ and $t = 1$ from the initial conditions,

$$y(0) = (B_1 + 0) + 3 = 12 \qquad\qquad B_1 = 9$$
$$y(1) = 2(0 + B_2) + 3 = 11 \qquad\qquad B_2 = 4$$

Thus,
$$y(t) = 2^t\left(9\cos\frac{\pi}{2}t + 4\sin\frac{\pi}{2}t\right) + 3$$

e)   With $k = 2$, the time path is divergent, as explained in Example 15.

18.35. Redo Problem 18.34, given $y_t + 2y_{t-2} = 24$, $y(0) = 11$, and $y(1) = 18$.

a)   From (18.11),
$$y_p = \frac{24}{1 + 0 + 2} = 8$$

b) From *(18.60)*, $k = \sqrt{b_2} = \sqrt{2}$. From *(18.19)*

$$g = -\tfrac{1}{2}(0) = 0 \qquad h = \tfrac{1}{2}\sqrt{4(2) - 0} = \sqrt{2}$$

From *(18.61)*, $\sin\theta = \dfrac{\sqrt{2}}{\sqrt{2}} = 1 \qquad \cos\theta = 0$

From Table 18-1, $\theta = \pi/2$. Substituting in *(18.27)*,

$$y_c = (\sqrt{2})^t\left[B_1\cos\left(\frac{\pi}{2}t\right) + B_2\sin\left(\frac{\pi}{2}t\right)\right]$$

c)

$$y(t) = (\sqrt{2})^t\left[B_1\cos\left(\frac{\pi}{2}t\right) + B_2\sin\left(\frac{\pi}{2}t\right)\right] + 8 \qquad (18.63)$$

d)
$$\begin{aligned} y(0) &= B_1 + 8 = 11 & B_1 &= 3 \\ y(1) &= \sqrt{2}B_2 + 8 = 18 & B_2 &= 7.07 \end{aligned}$$

Thus, $$y(t) = (\sqrt{2})^t\left(3\cos\frac{\pi}{2}t + 7.07\sin\frac{\pi}{2}t\right) + 8$$

e) With $k = \sqrt{2} > 1$, the time path is divergent.

ECONOMIC APPLICATIONS

18.36. In many markets supply and demand are influenced by current prices and price trends (i.e., whether prices are rising or falling and whether they are rising or falling at an increasing or decreasing rate). The economist, therefore, needs to know the current price $P(t)$, the first derivative $dP(t)/dt$, and the second derivative $d^2P(t)/dt^2$. Assume

$$Q_s = c_1 + w_1P + u_1P' + v_1P'' \qquad Q_d = c_2 + w_2P + u_2P' + v_2P'' \qquad (18.64)$$

Comment on the dynamic stability of the market if price clears the market at each point in time.

In equilibrium, $Q_s = Q_d$. Therefore,

$$c_1 + w_1P + u_1P' + v_1P'' = c_2 + w_2P + u_2P' + v_2P''$$
$$(v_1 - v_2)P'' + (u_1 - u_2)P' + (w_1 - w_2)P = -(c_1 - c_2)$$

Letting $v = v_1 - v_2$, $u = u_1 - u_2$, $w = w_1 - w_2$, $c = c_1 - c_2$, and dividing through by v to conform to *(18.1)*,

$$P'' + \frac{u}{v}P' + \frac{w}{v}P = -\frac{c}{v} \qquad (18.65)$$

Using *(18.2)* to find the particular integral, which will be the intertemporal equilibrium price $\bar{P}$,

$$\bar{P} = P_p = \frac{-c/v}{w/v} = -\frac{c}{w}$$

Since $c = c_1 - c_2$ and $w = w_1 - w_2$ where under ordinary supply conditions, $c_1 < 0$, $w_1 > 0$, and under ordinary demand conditions, $c_2 > 0$, $w_2 < 0$, $-c/w > 0$, as is necessary for $\bar{P}$. Using *(18.4)* to find the characteristic roots for the complementary function,

$$r_1, r_2 = \frac{-u/v \pm \sqrt{(u/v)^2 - 4w/v}}{2} \qquad (18.66)$$

which can assume three different types of solutions, depending on the specification of w, u, and v:

1. If $(u/v)^2 > 4w/v$, r_1 and r_2 will be *distinct real roots* solvable in terms of *(18.66)*; and $P(t) = A_1e^{r_1t} + A_2e^{r_2t} - c/w$.

2. If $(u/v)^2 = 4w/v$, r_1 and r_2 will be *repeated real roots*. Thus, (18.66) reduces to $-(u/v)/2$ or $-u/2v$. Then from (18.17), $P(t) = A_1 e^{-(u/2v)t} + A_2 t e^{-(u/2v)t} - c/w$.

3. If $(u/v)^2 < 4w/v$, r_1 and r_2 will be *complex roots* and from (18.26), $P(t) = e^{gt}(B_1 \cos ht + B_2 \sin ht) - c/w$, where from (18.19), $g = -u/(2v)$ and $h = \frac{1}{2}\sqrt{4w/v - (u/v)^2}$.

Specification of w, u, v depends on expectations. If people are bothered by inflationary psychology and expect prices to keep rising, u_2 in (18.64) will be positive; if they expect prices to ultimately fall and hold off buying because of that expectation, u_2 will be negative; and so forth.

18.37. In a model similar to Samuelson's interaction model between the multiplier and the accelerator, assume

$$Y_t = C_t + I_t + G_t \qquad (18.67)$$

$$C_t = C_0 + cY_{t-1} \qquad (18.68)$$

$$I_t = I_0 + w(C_t - C_{t-1}) \qquad (18.69)$$

where $0 < c < 1$, $w > 0$, and $G_t = G_0$. (*a*) Find the time path $Y(t)$ of national income and (*b*) comment on the stability conditions.

a)   Substituting (18.68) in (18.69),

$$I_t = I_0 + cw(Y_{t-1} - Y_{t-2}) \qquad (18.70)$$

Substituting $G_t = G_0$, (18.70), and (18.68) into (18.67), and then rearranging to conform with (18.10),

$$Y_t = C_0 + cY_{t-1} + I_0 + cw(Y_{t-1} - Y_{t-2}) + G_0$$

$$Y_t - c(1+w)Y_{t-1} + cwY_{t-2} = C_0 + I_0 + G_0 \qquad (18.71)$$

Using (18.11) for the particular solution,

$$Y_p = \frac{C_0 + I_0 + G_0}{1 - c(1+w) + cw} = \frac{C_0 + I_0 + G_0}{1 - c}$$

which is the intertemporal equilibrium level of income $\bar{Y}$. Using (18.4) to find the characteristic roots for the complementary function,

$$r_1, r_2 = \frac{c(1+w) \pm \sqrt{[-c(1+w)]^2 - 4cw}}{2} \qquad (18.72)$$

which can assume three different types of solutions depending on the values assigned to c and w:

1. If $c^2(1+w)^2 > 4cw$, or equivalently, if $c(1+w)^2 > 4w$, r_1 and r_2 will be *distinct real roots* solvable in terms of (18.72) and

$$Y(t) = A_1 r_1^t + A_2 r_2^t + \frac{C_0 + I_0 + G_0}{1 - c}$$

2. If $c(1+w)^2 = 4w$, r_1 and r_2 will be *repeated real roots*, and from (18.72) and (18.18),

$$Y(t) = A_1 \left[\frac{1}{2}c(1+w)\right]^t + A_2 t \left[\frac{1}{2}c(1+w)\right]^t + \frac{C_0 + I_0 + G_0}{1 - c}$$

3. If $c(1+w)^2 < 4w$, r_1 and r_2 will be *complex roots*; from (18.27),

$$Y(t) = k^t(B_1 \cos t\theta + B_2 \sin t\theta) + \frac{C_0 + I_0 + G_0}{1 - c}$$

where from (18.60), $k = \sqrt{cw}$, and from (18.61) θ must be such that

$$\sin \theta = \frac{h}{k} \qquad \cos \theta = \frac{g}{k}$$

where from (18.19), $g = \frac{1}{2}c(1+w)$ and $h = \frac{1}{2}\sqrt{4cw - c^2(1+w)^2}$.

b) For stability in the model under all possible initial conditions, the necessary and sufficient conditions are (1) $c < 1$ and (2) $cw < 1$. Since $c =$ MPC with respect to the previous year's income, c will be less than 1; for $cw < 1$, the product of the MPC and the marginal capital-output ratio must also be less than 1. If the characteristic roots are conjugate complex, the time path will oscillate.

CHAPTER 19

Simultaneous Differential and Difference Equations

19.1 MATRIX SOLUTION OF SIMULTANEOUS DIFFERENTIAL EQUATIONS, PART 1

Assume a system of n first-order, autonomous, linear differential equations in which no derivative is a function of another derivative and which we limit here to $n = 2$ for notational simplicity. *Autonomous* simply means all a_{ij} and b_i are constant.

$$\begin{aligned} \dot{y}_1 &= a_{11}y_1 + a_{12}y_2 + b_1 \\ \dot{y}_2 &= a_{21}y_1 + a_{22}y_2 + b_2 \end{aligned} \tag{19.1}$$

Expressed in matrix form,

$$\begin{bmatrix} \dot{y}_1 \\ \dot{y}_2 \end{bmatrix} = \begin{bmatrix} a_{11} & a_{12} \\ a_{21} & a_{22} \end{bmatrix} \begin{bmatrix} y_1 \\ y_2 \end{bmatrix} + \begin{bmatrix} b_1 \\ b_2 \end{bmatrix}$$

or

$$\dot{\mathbf{Y}} = \mathbf{A}\mathbf{Y} + \mathbf{B}$$

The complete solution to such a system will consist of n equations, each in turn composed of (1) a complementary solution y_c and (2) a particular solution y_p.

1. *a)* From our earlier work with single differential equations, we can expect the complementary solution of the system of equations, given distinct real roots, to take the general form,

$$y_c = \sum_{i=1}^{n} k_i \mathbf{C}_i e^{r_i t} = k_1 \mathbf{C}_1 e^{r_1 t} + k_2 \mathbf{C}_2 e^{r_2 t} \tag{19.2}$$

where k_i = a scalar or constant, $\mathbf{C}_i = (2 \times 1)$ column vector of constants called an *eigenvector*, and r_i = a scalar called the *characteristic root*. See Section 12.8.

428

b) As demonstrated in Problem 19.13, the characteristic roots, also called *eigenvalues*, can be found by solving the quadratic equation

$$r_i = \frac{\text{Tr}(\mathbf{A}) \pm \sqrt{[\text{Tr}(\mathbf{A})]^2 - 4|\mathbf{A}|}}{2} \qquad (19.3)$$

where $|\mathbf{A}|$ = determinant of $\mathbf{A}$, $\text{Tr}(\mathbf{A})$ = *trace* of $\mathbf{A}$, and $\text{Tr}(\mathbf{A})$ = Σ of all the elements on the principal diagonal of $\mathbf{A}$. Here where $\mathbf{A} = (2 \times 2)$,

$$\text{Tr}(\mathbf{A}) = a_{11} + a_{22}$$

c) As explained in Problem 19.12, the solution of the system of simultaneous equations requires that

$$(\mathbf{A} - r_i\mathbf{I})\mathbf{C}_i = 0 \qquad (19.4)$$

where $(\mathbf{A} - r_i\mathbf{I}) = \begin{bmatrix} a_{11} & a_{12} \\ a_{21} & a_{22} \end{bmatrix} - r_i \begin{bmatrix} 1 & 0 \\ 0 & 1 \end{bmatrix} = \begin{bmatrix} a_{11} - r_i & a_{12} \\ a_{21} & a_{22} - r_i \end{bmatrix}$

r_i is a scalar, and $\mathbf{I}$ is an identity matrix, here $\mathbf{I}_2$. Equation (19.4) is called the *eigenvalue problem*. The eigenvectors are found by solving (19.4) for $\mathbf{C}_i$. To preclude trivial, i.e., null-vector, solutions for $\mathbf{C}_i$, the matrix $(\mathbf{A} - r_i\mathbf{I})$ must be constrained to be singular.

2. The particular integral, y_p, is simply the intertemporal or steady-state solution. As demonstrated in Problem 19.14,

$$y_p = \bar{\mathbf{Y}} = -\mathbf{A}^{-1}\mathbf{B} \qquad (19.5)$$

where $\mathbf{A}^{-1}$ = the inverse of $\mathbf{A}$ and $\mathbf{B}$ = the column of constants.

The stability of the model depends on the characteristic roots.

> If all $r_i < 0$, the model is dynamically stable.
> If all $r_i > 0$, the model is dynamically unstable.

If the r_i are of different signs, the solution is at a saddle-point equilibrium and the model is unstable, except along the saddle path. See Section 19.5 and Examples 10 and 12.

EXAMPLE 1. Solve the following system of first-order, autonomous, linear differential equations,

$$\dot{y}_1 = 5y_1 - 0.5y_2 - 12 \qquad y_1(0) = 12$$
$$\dot{y}_2 = -2y_1 + 5y_2 - 24 \qquad y_2(0) = 4$$

1. Convert them to matrices for ease of computation.

$$\begin{bmatrix} \dot{y}_1 \\ \dot{y}_2 \end{bmatrix} = \begin{bmatrix} 5 & -0.5 \\ -2 & 5 \end{bmatrix} \begin{bmatrix} y_1 \\ y_2 \end{bmatrix} + \begin{bmatrix} -12 \\ -24 \end{bmatrix}$$

$$\dot{\mathbf{Y}} = \mathbf{A}\mathbf{Y} + \mathbf{B}$$

2. Then find the complementary functions. From (19.2), assuming distinct real roots,

$$y_c = k_1\mathbf{C}_1 e^{r_1 t} + k_2\mathbf{C}_2 e^{r_2 t}$$

But from (19.3), the characteristic roots are

$$r_1, r_2 = \frac{\text{Tr}(\mathbf{A}) \pm \sqrt{[\text{Tr}(\mathbf{A})]^2 - 4|\mathbf{A}|}}{2}$$

where $\text{Tr}(\mathbf{A}) = a_{11} + a_{22} = 5 + 5 = 10$

and $|\mathbf{A}| = 25 - 1 = 24$

Substituting,

$$r_1, r_2 = \frac{10 \pm \sqrt{(10)^2 - 4(24)}}{2} = \frac{10 \pm 2}{2}$$

$$r_1 = 4 \qquad r_2 = 6 \qquad \text{characteristic roots or eigenvalues}$$

3. We next find the eigenvectors. Using (19.4) and recalling $(\mathbf{A} - r_i\mathbf{I})$ is singular,

$$(\mathbf{A} - r_i\mathbf{I})\mathbf{C}_i = 0$$

where

$$(\mathbf{A} - r_i\mathbf{I})\mathbf{C}_i = \begin{bmatrix} a_{11} - r_i & a_{12} \\ a_{21} & a_{22} - r_i \end{bmatrix} \begin{bmatrix} c_1 \\ c_2 \end{bmatrix} = 0$$

a) Substituting first for $r_1 = 4$,

$$\begin{bmatrix} 5 - 4 & -0.5 \\ -2 & 5 - 4 \end{bmatrix} \begin{bmatrix} c_1 \\ c_2 \end{bmatrix} = \begin{bmatrix} 1 & -0.5 \\ -2 & 1 \end{bmatrix} \begin{bmatrix} c_1 \\ c_2 \end{bmatrix} = 0$$

Then, by simple multiplication of row by column, we have

$$\begin{aligned} c_1 - 0.5c_2 &= 0 & c_1 &= 0.5c_2 \\ -2c_1 + c_2 &= 0 & c_1 &= 0.5c_2 \end{aligned}$$

Since $(\mathbf{A} - r_i\mathbf{I})$ is constrained to be singular, there will always be linear dependence between the equations and we can work with either one. With linear dependence, there is also an infinite number of eigenvectors that will satisfy the equation. We can normalize the equation by choosing a vector whose length is unity, i.e., $c_1^2 + c_2^2 = 1$, which is called the *Euclidian distance condition*, or we can simply pick any arbitrary value for one element while maintaining the relationship between elements. Opting for the latter, let $c_1 = 1$.

$$\text{If } c_1 = 1, \text{ then } c_2 = \frac{1}{0.5} = 2$$

Thus, the eigenvector $\mathbf{C}_1$ corresponding to $r_1 = 4$ is

$$\mathbf{C}_1 = \begin{bmatrix} c_1 \\ c_2 \end{bmatrix} = \begin{bmatrix} 1 \\ 2 \end{bmatrix}$$

and the first elements of the complementary function of the general solution are

$$y_c^1 = k_1 \begin{bmatrix} 1 \\ 2 \end{bmatrix} e^{4t} = \begin{bmatrix} k_1 e^{4t} \\ 2k_1 e^{4t} \end{bmatrix}$$

b) Substituting next for $r_2 = 6$,

$$\begin{bmatrix} 5 - 6 & -0.5 \\ -2 & 5 - 6 \end{bmatrix} \begin{bmatrix} c_1 \\ c_2 \end{bmatrix} = \begin{bmatrix} -1 & -0.5 \\ -2 & -1 \end{bmatrix} \begin{bmatrix} c_1 \\ c_2 \end{bmatrix} = 0$$

Multiplying row by column,

$$\begin{aligned} -c_1 - 0.5c_2 &= 0 & c_1 &= -0.5c_2 \\ -2c_1 - c_2 &= 0 & c_1 &= -0.5c_2 \end{aligned}$$

$$\text{If } c_1 = 1, \text{ then } c_2 = \frac{1}{-0.5} = -2$$

Thus, the eigenvector $\mathbf{C}_2$ corresponding to $r_2 = 6$ is

$$\mathbf{C}_2 = \begin{bmatrix} c_1 \\ c_2 \end{bmatrix} = \begin{bmatrix} 1 \\ -2 \end{bmatrix}$$

and the second elements of the complementary function of the general solution are

$$y_c^2 = k_2 \begin{bmatrix} 1 \\ -2 \end{bmatrix} e^{6t} = \begin{bmatrix} k_2 e^{6t} \\ -2k_2 e^{6t} \end{bmatrix}$$

Putting them together for the complete complementary solution to the system,

$$y_1(t) = k_1 e^{4t} + k_2 e^{6t}$$
$$y_2(t) = 2k_1 e^{4t} - 2k_2 e^{6t}$$

4. Now we find the intertemporal or steady-state solutions for y_p. From (19.5),

$$y_p = \bar{Y} = -A^{-1}B$$

where $B = \begin{bmatrix} -12 \\ -24 \end{bmatrix}$, $A = \begin{bmatrix} 5 & -0.5 \\ -2 & 5 \end{bmatrix}$, $|A| = 25 - 1 = 24$,

the cofactor matrix is $C = \begin{bmatrix} 5 & 2 \\ 0.5 & 5 \end{bmatrix}$, the adjoint matrix is Adj. $A = C' = \begin{bmatrix} 5 & 0.5 \\ 2 & 5 \end{bmatrix}$,

and the inverse is $A^{-1} = \dfrac{1}{24}\begin{bmatrix} 5 & 0.5 \\ 2 & 5 \end{bmatrix}$

Substituting in (19.5),

$$\bar{Y} = -\frac{1}{24}\begin{bmatrix} 5 & 0.5 \\ 2 & 5 \end{bmatrix}\begin{bmatrix} -12 \\ -24 \end{bmatrix}$$

Multiplying row by column,

$$\bar{Y} = \begin{bmatrix} \bar{y}_1 \\ \bar{y}_2 \end{bmatrix} = \frac{1}{24}\begin{bmatrix} 72 \\ 144 \end{bmatrix} = \begin{bmatrix} 3 \\ 6 \end{bmatrix}$$

Thus the complete general solution, $y(t) = y_c + y_p$, is

$$y_1(t) = k_1 e^{4t} + k_2 e^{6t} + 3$$
$$y_2(t) = 2k_1 e^{4t} - 2k_2 e^{6t} + 6 \qquad\qquad (19.6)$$

With $r_1 = 4 > 0$ and $r_2 = 6 > 0$, the equilibrium is unstable. See also Problems 19.1 to 19.3.

EXAMPLE 2.   To find the definite solution for Example 1, we simply employ the initial conditions, $y_1(0) = 12$, $y_2(0) = 4$. When (19.6) is evaluated at $t = 0$, we have

$$y_1(0) = k_1 + k_2 + 3 = 12$$
$$y_2(0) = 2k_1 - 2k_2 + 6 = 4$$

Solved simultaneously, $k_1 = 4 \qquad k_2 = 5$

Substituting in (19.6), we have the definite solution,

$$y_1(t) = 4e^{4t} + 5e^{6t} + 3$$
$$y_2(t) = 8e^{4t} - 10e^{6t} + 6$$

which remains dynamically unstable because of the positive roots.

19.2 MATRIX SOLUTION OF SIMULTANEOUS DIFFERENTIAL EQUATIONS, PART 2

Assume a system of n first-order, autonomous, linear differential equations in which one or more derivatives is a function of another derivative, and which we limit here to $n = 2$ simply for notational simplicity,

$$a_{11}\dot{y}_1 + a_{12}\dot{y}_2 = a_{13}y_1 + a_{14}y_2 + b_1$$
$$a_{21}\dot{y}_1 + a_{22}\dot{y}_2 = a_{23}y_1 + a_{24}y_2 + b_2 \qquad\qquad (19.7)$$

In matrix form,

$$\begin{bmatrix} a_{11} & a_{12} \\ a_{21} & a_{22} \end{bmatrix} \begin{bmatrix} \dot{y}_1 \\ \dot{y}_2 \end{bmatrix} = \begin{bmatrix} a_{13} & a_{14} \\ a_{23} & a_{24} \end{bmatrix} \begin{bmatrix} y_1 \\ y_2 \end{bmatrix} + \begin{bmatrix} b_1 \\ b_2 \end{bmatrix}$$

$$\mathbf{A}_1 \dot{\mathbf{Y}} = \mathbf{A}_2 \mathbf{Y} + \mathbf{B}$$

The general solution $y(t)$ will consist of a complementary function y_c and a particular integral or solution y_p. As in previous examples, for distinct real roots we can expect the complementary function to assume the general form

$$y_c = k_1 \mathbf{C}_1 e^{r_1 t} + k_2 \mathbf{C}_2 e^{r_2 t} \tag{19.8}$$

As explained in Problems 19.15 to 19.16, the eigenvalue problem here is

$$(\mathbf{A}_2 - r_i \mathbf{A}_1) \mathbf{C}_i = 0 \tag{19.9}$$

where

$$(\mathbf{A}_2 - r_i \mathbf{A}_1) = \begin{bmatrix} a_{13} & a_{14} \\ a_{23} & a_{24} \end{bmatrix} - r_i \begin{bmatrix} a_{11} & a_{12} \\ a_{21} & a_{22} \end{bmatrix} = \begin{bmatrix} a_{13} - a_{11} r_i & a_{14} - a_{12} r_i \\ a_{23} - a_{21} r_i & a_{24} - a_{22} r_i \end{bmatrix}$$

The characteristic equation is

$$|\mathbf{A}_2 - r_i \mathbf{A}_1| = 0 \tag{19.10}$$

the particular integral is

$$\bar{\mathbf{Y}} = -\mathbf{A}_2^{-1} \mathbf{B} \tag{19.11}$$

and the stability conditions are the same as in Section 19.1.

EXAMPLE 3. Solve the following system of first-order, autonomous, nonlinear differential equations.

$$\dot{y}_1 = -3y_1 + 1.5y_2 - 2.5\dot{y}_2 + 2.4 \qquad y_1(0) = 14$$
$$\dot{y}_2 = 2y_1 - 5y_2 + 16 \qquad y_2(0) = 15.4$$

1. First rearrange the equations to conform with (19.7) and set them in matrix form,

$$\begin{bmatrix} 1 & 2.5 \\ 0 & 1 \end{bmatrix} \begin{bmatrix} \dot{y}_1 \\ \dot{y}_2 \end{bmatrix} = \begin{bmatrix} -3 & 1.5 \\ 2 & -5 \end{bmatrix} \begin{bmatrix} y_1 \\ y_2 \end{bmatrix} + \begin{bmatrix} 2.4 \\ 16 \end{bmatrix}$$

2. Assuming distinct real roots, find the complementary function.

$$y_c = k_1 \mathbf{C}_1 e^{r_1 t} + k_2 \mathbf{C}_2 e^{r_2 t}$$

   a) Start with the characteristic equation to find the characteristic roots. From (19.10),

$$|\mathbf{A}_2 - r_i \mathbf{A}_1| = 0$$

   b) Substituting and dropping the i subscript for simplicity,

$$|\mathbf{A}_2 - r\mathbf{A}_1| = \left| \begin{bmatrix} -3 & 1.5 \\ 2 & -5 \end{bmatrix} - r \begin{bmatrix} 1 & 2.5 \\ 0 & 1 \end{bmatrix} \right| = \begin{vmatrix} -3 - r & 1.5 - 2.5r \\ 2 & -5 - r \end{vmatrix} = 0$$

$$(-3 - r)(-5 - r) - 2(1.5 - 2.5r) = 0$$
$$r^2 + 13r + 12 = 0$$

$$r_1 = -1 \qquad r_2 = -12 \qquad \text{characteristic roots}$$

3. Find the eigenvectors, $\mathbf{C}_i$. From (19.9),

$$(\mathbf{A}_2 - r_i \mathbf{A}_1) \mathbf{C}_i = 0$$

where

$$(\mathbf{A}_2 - r_i \mathbf{A}_1) \mathbf{C}_i = \begin{bmatrix} -3 - r_i & 1.5 - 2.5r_i \\ 2 & -5 - r_i \end{bmatrix} \begin{bmatrix} c_1 \\ c_2 \end{bmatrix}$$

a)   Substituting first for $r_1 = -1$,

$$\begin{bmatrix} -3-(-1) & 1.5-2.5(-1) \\ 2 & -5-(-1) \end{bmatrix}\begin{bmatrix} c_1 \\ c_2 \end{bmatrix} = \begin{bmatrix} -2 & 4 \\ 2 & -4 \end{bmatrix}\begin{bmatrix} c_1 \\ c_2 \end{bmatrix} = 0$$

By simple matrix multiplication,

$$\begin{aligned} -2c_1 + 4c_2 &= 0 & c_1 &= 2c_2 \\ 2c_1 - 4c_2 &= 0 & c_1 &= 2c_2 \end{aligned}$$

If we let $c_1 = 2$, $c_2 = 1$. Thus,

$$\mathbf{C}_1 = \begin{bmatrix} c_1 \\ c_2 \end{bmatrix} = \begin{bmatrix} 2 \\ 1 \end{bmatrix}$$

and the first elements of the general complementary function for $r_1 = -1$ are

$$y_c^1 = k_1 \begin{bmatrix} 2 \\ 1 \end{bmatrix} e^{-t} = \begin{bmatrix} 2k_1 \\ k_1 \end{bmatrix} e^{-t}$$

b)   Now substituting for $r_2 = -12$,

$$\begin{bmatrix} -3-(-12) & 1.5-2.5(-12) \\ 2 & -5-(-12) \end{bmatrix}\begin{bmatrix} c_1 \\ c_2 \end{bmatrix} = \begin{bmatrix} 9 & 31.5 \\ 2 & 7 \end{bmatrix}\begin{bmatrix} c_1 \\ c_2 \end{bmatrix} = 0$$

Multiplying row by column,

$$\begin{aligned} 9c_1 + 31.5c_2 &= 0 & c_1 &= -3.5c_2 \\ 2c_1 + 7c_2 &= 0 & c_1 &= -3.5c_2 \end{aligned}$$

Letting $c_1 = -3.5$, $c_2 = 1$. So,

$$\mathbf{C}_2 = \begin{bmatrix} c_1 \\ c_2 \end{bmatrix} = \begin{bmatrix} -3.5 \\ 1 \end{bmatrix}$$

and the second elements of the complementary function for $r_2 = -12$ are

$$y_c^2 = k_2 \begin{bmatrix} -3.5 \\ 1 \end{bmatrix} e^{-12t} = \begin{bmatrix} -3.5k_2 \\ k_2 \end{bmatrix} e^{-12t}$$

Adding the two together, the complementary functions are

$$\begin{aligned} y_1(t) &= 2k_1 e^{-t} - 3.5k_2 e^{-12t} \\ y_2(t) &= k_1 e^{-t} + k_2 e^{-12t} \end{aligned} \tag{19.12}$$

4.   Find the particular integral y_p which is simply the intertemporal equilibrium $\bar{\mathbf{Y}}$. From (19.11),

$$\bar{\mathbf{Y}} = -\mathbf{A}_2^{-1}\mathbf{B}$$

where $\quad \mathbf{B} = \begin{bmatrix} 2.4 \\ 16 \end{bmatrix}, \quad \mathbf{A}_2 = \begin{bmatrix} -3 & 1.5 \\ 2 & -5 \end{bmatrix}, \quad |\mathbf{A}_2| = 15 - 3 = 12$

the cofactor matrix is $\mathbf{C} = \begin{bmatrix} -5 & -2 \\ -1.5 & -3 \end{bmatrix}$, the adjoint matrix is Adj. $\mathbf{A} = \mathbf{C}' = \begin{bmatrix} -5 & -1.5 \\ -2 & -3 \end{bmatrix}$

and the inverse is $\qquad \mathbf{A}_2^{-1} = \dfrac{1}{12}\begin{bmatrix} -5 & -1.5 \\ -2 & -3 \end{bmatrix}$

Substituting in (19.11),

$$\bar{\mathbf{Y}} = \begin{bmatrix} \bar{y}_1 \\ \bar{y}_2 \end{bmatrix} = -\frac{1}{12}\begin{bmatrix} -5 & -1.5 \\ -2 & -3 \end{bmatrix}\begin{bmatrix} 2.4 \\ 16 \end{bmatrix} = \begin{bmatrix} 3 \\ 4.4 \end{bmatrix}$$

5. By adding the particular integrals or steady-state solutions to the complementary function in (19.12), we derive the complete general solution.

$$y_1(t) = 2k_1 e^{-t} - 3.5k_2 e^{-12t} + 3$$
$$y_2(t) = k_1 e^{-t} + k_2 e^{-12t} + 4.4 \qquad (19.13)$$

With $r_1 = -1 < 0$, $r_2 = -12 < 0$, the system of equations is dynamically stable. See also Problems 19.4 to 19.6.

EXAMPLE 4. To find the definite solution for Example 3, we simply employ the initial conditions, $y_1(0) = 14$, $y_2(0) = 15.4$. When (19.13) is evaluated at $t = 0$, we have

$$y_1(0) = 2k_1 - 3.5k_2 + 3 = 14$$
$$y_2(0) = k_1 + k_2 + 4.4 = 15.4$$

Solved simultaneously, $k_1 = 9 \qquad k_2 = 2$

Substituting in (19.13) for the definite solution,

$$y_1(t) = 18e^{-t} - 7e^{-12t} + 3$$
$$y_2(t) = 9e^{-t} + 2e^{-12t} + 4.4$$

19.3 MATRIX SOLUTION OF SIMULTANEOUS DIFFERENCE EQUATIONS, PART 1

Assume a system of n linear first-order difference equations in which no difference is a function of another difference, the coefficients are constants, and we again set $n = 2$ for notational simplicity.

$$x_t = a_{11}x_{t-1} + a_{12}y_{t-1} + b_1$$
$$y_t = a_{21}x_{t-1} + a_{22}y_{t-1} + b_2$$

In matrix form,

$$\begin{bmatrix} x_t \\ y_t \end{bmatrix} = \begin{bmatrix} a_{11} & a_{12} \\ a_{21} & a_{22} \end{bmatrix} \begin{bmatrix} x_{t-1} \\ y_{t-1} \end{bmatrix} + \begin{bmatrix} b_1 \\ b_2 \end{bmatrix}$$

Letting $\mathbf{Y}_t = \begin{bmatrix} x_t \\ y_t \end{bmatrix}$, $\mathbf{Y}_{t-1} = \begin{bmatrix} x_{t-1} \\ y_{t-1} \end{bmatrix}$, $\mathbf{B} = \begin{bmatrix} b_1 \\ b_2 \end{bmatrix}$, and $\mathbf{A}$ = the coefficient matrix, we have

$$\mathbf{Y}_t = \mathbf{A}\mathbf{Y}_{t-1} + \mathbf{B} \qquad (19.14)$$

The complete solution will consist of n equations, each in turn composed of the complementary solution y_c and the particular solution y_p. Based on our earlier work with single difference equations and assuming distinct real roots, we can expect the complementary function will take the general form,

$$y_c = \sum_{i=1}^{n} k_i \mathbf{C}_i r_i^t = k_1 \mathbf{C}_1 r_1^t + k_2 \mathbf{C}_2 r_2^t \qquad (19.15)$$

As demonstrated in Problem 19.17, the eigenvalue problem breaks down to

$$(\mathbf{A} - r_i \mathbf{I})\mathbf{C}_i = 0 \qquad (19.16)$$

By similar steps to the demonstrations in Problems 19.13 to 19.14, it can be shown that the characteristic equation is

$$|\mathbf{A} - r_i \mathbf{I}| = 0$$

where the characteristic roots can be found with (19.3), and the particular solution is

$$y_p = (\mathbf{I} - \mathbf{A})^{-1}\mathbf{B} \qquad (19.17)$$

The stability conditions require that each of the n roots be less than 1 in absolute value for dynamic stability. If even one root is greater than 1 in absolute value, it will dominate the other(s) and the time path will be divergent.

EXAMPLE 5.   Solve the following system of first-order linear difference equations,

$$x_t = -4x_{t-1} + y_{t-1} + 12 \qquad x_0 = 16$$
$$y_t = 2x_{t-1} - 3y_{t-1} + 6 \qquad y_0 = 8$$

(19.18)

1. Set them in matrix form,

$$\begin{bmatrix} x_t \\ y_t \end{bmatrix} = \begin{bmatrix} -4 & 1 \\ 2 & -3 \end{bmatrix} \begin{bmatrix} x_{t-1} \\ y_{t-1} \end{bmatrix} + \begin{bmatrix} 12 \\ 6 \end{bmatrix}$$

$$\mathbf{Y}_t = \mathbf{A}\mathbf{Y}_{t-1} + \mathbf{B}$$

2. We next find the complementary functions. Assuming a case of distinct real roots,

$$y_c = k_1 \mathbf{C}_1 r_1^t + k_2 \mathbf{C}_2 r_2^t$$

and the characteristic roots are

$$r_1, r_2 = \frac{\text{Tr}(\mathbf{A}) \pm \sqrt{[\text{Tr}(\mathbf{A})]^2 - 4|\mathbf{A}|}}{2}$$

where

$$\text{Tr}(\mathbf{A}) = -4 - 3 = -7 \qquad \text{and} \qquad |\mathbf{A}| = 12 - 2 = 10$$

Substituting,

$$r_1, r_2 = \frac{-7 \pm \sqrt{(-7)^2 - 4(10)|}}{2} = \frac{-7 \pm 3}{2}$$

$$r_1 = -2 \qquad r_2 = -5 \qquad \text{characteristic roots or eigenvalues}$$

3. We then find the eigenvectors. Using (19.16) and recalling $(\mathbf{A} - r_i\mathbf{I})$ is singular,

$$(\mathbf{A} - r_i\mathbf{I})\mathbf{C}_i = \begin{bmatrix} a_{11} - r_i & a_{12} \\ a_{21} & a_{22} - r_i \end{bmatrix} \begin{bmatrix} c_1 \\ c_2 \end{bmatrix} = 0$$

a)   Substituting first for $r_1 = -2$,

$$\begin{bmatrix} -4 - (-2) & 1 \\ 2 & -3 - (-2) \end{bmatrix} \begin{bmatrix} c_1 \\ c_2 \end{bmatrix} = \begin{bmatrix} -2 & 1 \\ 2 & -1 \end{bmatrix} \begin{bmatrix} c_1 \\ c_2 \end{bmatrix} = 0$$

Then by multiplying row by column, we find

$$-2c_1 + c_2 = 0 \qquad c_2 = 2c_1$$
$$2c_1 - c_2 = 0 \qquad c_2 = 2c_1$$

Letting $c_1 = 1$, $c_2 = 2$

Hence the eigenvector corresponding to $r_1 = -2$ is

$$\mathbf{C}_1 = \begin{bmatrix} 1 \\ 2 \end{bmatrix}$$

and the a_{i1} elements of the complementary function are

$$k_1 \begin{bmatrix} 1 \\ 2 \end{bmatrix} (-2)^t = \begin{bmatrix} k_1(-2)^t \\ 2k_1(-2)^t \end{bmatrix}$$

b)   Then substituting for $r_2 = -5$,

$$\begin{bmatrix} -4 - (-5) & 1 \\ 2 & -3 - (-5) \end{bmatrix} \begin{bmatrix} c_1 \\ c_2 \end{bmatrix} = \begin{bmatrix} 1 & 1 \\ 2 & 2 \end{bmatrix} \begin{bmatrix} c_1 \\ c_2 \end{bmatrix} = 0$$

Multiplying row by column,

$$c_1 + c_2 = 0 \qquad\qquad c_2 = -c_1$$
$$2c_1 + 2c_2 = 0 \qquad\qquad c_2 = -c_1$$

$$\text{If } c_1 = 1, \ c_2 = -1$$

The eigenvector for $r_2 = -5$ is

$$\mathbf{C}_2 = \begin{bmatrix} 1 \\ -1 \end{bmatrix}$$

and the a_{i2} elements of the complementary function are

$$k_2 \begin{bmatrix} 1 \\ -1 \end{bmatrix} (-5)^t = \begin{bmatrix} k_2(-5)^t \\ -k_2(-5)^t \end{bmatrix}$$

Combining the two for the general complementary function, we have

$$x_c = k_1(-2)^t + k_2(-5)^t$$
$$y_c = 2k_1(-2)^t - k_2(-5)^t$$

4.   Now we find the particular solution for the steady-state solutions $\bar{x}, \bar{y}$. From (*19.17*),

$$y_p = (\mathbf{I} - \mathbf{A})^{-1}\mathbf{B}$$

where

$$(\mathbf{I} - \mathbf{A}) = \begin{bmatrix} 1 & 0 \\ 0 & 1 \end{bmatrix} - \begin{bmatrix} -4 & 1 \\ 2 & -3 \end{bmatrix} = \begin{bmatrix} 5 & -1 \\ -2 & 4 \end{bmatrix}$$

Following a series of steps similar to those in Example 3,

$$y_p = \begin{bmatrix} \bar{x} \\ \bar{y} \end{bmatrix} = \frac{1}{18} \begin{bmatrix} 4 & 1 \\ 2 & 5 \end{bmatrix} \begin{bmatrix} 12 \\ 6 \end{bmatrix} = \begin{bmatrix} 3 \\ 3 \end{bmatrix}$$

This makes the complete general solution,

$$x_t = k_1(-2)^t + k_2(-5)^t + 3$$
$$y_t = 2k_1(-2)^t - k_2(-5)^t + 3$$

With $|-2| > 1$ and $|-5| > 1$, the time path is divergent. See also Problems 19.7 to 19.8.

EXAMPLE 6.   For the specific solution, we need only employ the initial conditions. Given $x_0 = 16$, $y_0 = 8$, at $t = 0$, the general functions reduce to

$$k_1 + k_2 + 3 = 16$$
$$2k_1 - k_2 + 3 = 8$$

Solving simultaneously, $$k_1 = 6, \qquad k_2 = 7$$

By simple substitution, we then find the specific solution,

$$x_t = 6(-2)^t + 7(-5)^t + 3$$

$$y_t = 12(-2)^t - 7(-5)^t + 3$$

(*19.19*)

To check the answer, substitute $t = 1$ and $t = 0$ in (*19.19*).

$$x_1 = 6(-2)^1 + 7(-5)^1 + 3 = -44 \qquad x_0 = 6(-2)^0 + 7(-5)^0 + 3 = 16$$
$$y_1 = 12(-2)^1 - 7(-5)^1 + 3 = 14 \qquad y_0 = 12(-2)^0 - 7(-5)^0 + 3 = 8$$

Then go back to (*19.18*) and substitute x_1, y_1 for x_t and y_t, and $x_0, y_0 =$ for x_{t-1} and y_{t-1}.

$$-44 = -4(16) + 8 + 12 = -44$$
$$14 = 2(16) - 3(8) + 6 = 14$$

19.4   MATRIX SOLUTION OF SIMULTANEOUS DIFFERENCE EQUATIONS, PART 2

Assume a system of n linear first-order difference equations in which one or more differences is a function of another difference, the coefficients are constant, and we again set $n = 2$ for notational simplicity.

$$a_{11}x_t + a_{12}y_t = a_{13}x_{t-1} + a_{14}y_{t-1} + b_1$$
$$a_{21}x_t + a_{22}y_t = a_{23}x_{t-1} + a_{24}y_{t-1} + b_2 \tag{19.20}$$

or

$$\begin{bmatrix} a_{11} & a_{12} \\ a_{21} & a_{22} \end{bmatrix}\begin{bmatrix} x_t \\ y_t \end{bmatrix} = \begin{bmatrix} a_{13} & a_{14} \\ a_{23} & a_{24} \end{bmatrix}\begin{bmatrix} x_{t-1} \\ y_{t-1} \end{bmatrix} + \begin{bmatrix} b_1 \\ b_2 \end{bmatrix}$$

$$\mathbf{A}_1\mathbf{Y}_t = \mathbf{A}_2\mathbf{Y}_{t-1} + \mathbf{B}$$

From previous sections we can expect the general solution y_t will consist of a complementary function y_c and a particular solution y_p, where the complementary function for distinct real roots will take the general form

$$y_c = k_1\mathbf{C}_1(r_1)^t + k_2\mathbf{C}_2(r_2)^t$$

As demonstrated in Problems 19.18 to 19.19, the eigenvalue problem here reduces to

$$(\mathbf{A}_2 - r_i\mathbf{A}_1)\mathbf{C}_i = 0 \tag{19.21}$$

where

$$(\mathbf{A}_2 - r_i\mathbf{A}_1) = \begin{bmatrix} a_{13} & a_{14} \\ a_{23} & a_{24} \end{bmatrix} - r_i\begin{bmatrix} a_{11} & a_{12} \\ a_{21} & a_{22} \end{bmatrix} = \begin{bmatrix} a_{13} - a_{11}r_i & a_{14} - a_{12}r_i \\ a_{23} - a_{21}r_i & a_{24} - a_{22}r_i \end{bmatrix}$$

and the particular integral is

$$\bar{\mathbf{Y}} = (\mathbf{A}_1 - \mathbf{A}_2)^{-1}\mathbf{B} \tag{19.22}$$

The stability conditions remain the same as in Section 19.3.

EXAMPLE 7.   Solve the following system of linear first-order difference equations.

$$x_t = 4x_{t-1} - 2y_{t-1} + y_t - 10 \qquad x_0 = 20$$
$$y_t = 3x_{t-1} + 6y_{t-1} - 4 \qquad y_0 = 3 \tag{19.23}$$

1.   Rearrange to conform with (19.20) and set in matrix form.

$$\begin{bmatrix} 1 & -1 \\ 0 & 1 \end{bmatrix}\begin{bmatrix} x_t \\ y_t \end{bmatrix} = \begin{bmatrix} 4 & -2 \\ 3 & 6 \end{bmatrix}\begin{bmatrix} x_{t-1} \\ y_{t-1} \end{bmatrix} + \begin{bmatrix} -10 \\ -4 \end{bmatrix}$$

$$\mathbf{A}_1\mathbf{Y}_t = \mathbf{A}_2\mathbf{Y}_{t-1} + \mathbf{B}$$

2.   For the complementary function, begin with the characteristic equation derived from (19.21)

$$|\mathbf{A}_2 - r_i\mathbf{A}_1| = 0$$

and substitute the parameters of the problem,

$$|\mathbf{A}_2 - r_i\mathbf{A}_1| = \begin{vmatrix} 4 - r & -2 + r \\ 3 & 6 - r \end{vmatrix} = 0$$

$$(4 - r)(6 - r) - 3(-2 + r) = 0$$
$$r^2 - 13r + 30 = 0$$

$$r_1 = 3 \qquad r_2 = 10$$

3.   Find the nontrivial solutions for the eigenvectors $\mathbf{C}_i$. From (19.21),

$$(\mathbf{A}_2 - r_i\mathbf{A}_1)\mathbf{C}_i = 0$$

where
$$(\mathbf{A}_2 - r_i\mathbf{A}_1)\mathbf{C}_i = \begin{bmatrix} 4-r & -2+r \\ 3 & 6-r \end{bmatrix}\begin{bmatrix} c_1 \\ c_2 \end{bmatrix}$$

a) Substituting for $r_1 = 3$,
$$\begin{bmatrix} 4-3 & -2+3 \\ 3 & 6-3 \end{bmatrix}\begin{bmatrix} c_1 \\ c_2 \end{bmatrix} = \begin{bmatrix} 1 & 1 \\ 3 & 3 \end{bmatrix}\begin{bmatrix} c_1 \\ c_2 \end{bmatrix} = 0$$

Multiplying row by column,
$$c_1 + c_2 = 0 \qquad c_1 = -c_2$$
$$3c_1 + 3c_2 = 0 \qquad c_1 = -c_2$$

If $c_1 = 1$, $c_2 = -1$, the eigenvector is
$$\mathbf{C}_1 = \begin{bmatrix} 1 \\ -1 \end{bmatrix}$$

making the a_{i1} elements of the complementary function for $r_1 = 3$,
$$k_1\begin{bmatrix} 1 \\ -1 \end{bmatrix}(3)^t = \begin{bmatrix} k_1(3)^t \\ -k_1(3)^t \end{bmatrix}$$

b) For $r_2 = 10$,
$$\begin{bmatrix} 4-10 & -2+10 \\ 3 & 6-10 \end{bmatrix}\begin{bmatrix} c_1 \\ c_2 \end{bmatrix} = \begin{bmatrix} -6 & 8 \\ 3 & -4 \end{bmatrix}\begin{bmatrix} c_1 \\ c_2 \end{bmatrix} = 0$$

By matrix multiplication,
$$-6c_1 + 8c_2 = 0 \qquad c_2 = 0.75c_1$$
$$3c_1 - 4c_2 = 0 \qquad c_2 = 0.75c_1$$

Letting $c_1 = 1$, $c_2 = 0.75$, the corresponding eigenvector becomes
$$\mathbf{C}_1 = \begin{bmatrix} 1 \\ 0.75 \end{bmatrix}$$

and the a_{i2} elements of the complementary function for $r_2 = 10$ is
$$k_2\begin{bmatrix} 1 \\ 0.75 \end{bmatrix}(10)^t = \begin{bmatrix} k_2(10)^t \\ 0.75k_2(10)^t \end{bmatrix}$$

Combining the two, the complete complementary function becomes
$$x_c = k_1(3)^t + k_2(10)^t$$
$$y_c = -k_1(3)^t + 0.75k_2(10)^t \qquad\qquad (19.24)$$

4. Now find the particular or steady-state solution $y_p = \bar{\mathbf{Y}}$. From (19.22),
$$\bar{\mathbf{Y}} = (\mathbf{A}_1 - \mathbf{A}_2)^{-1}\mathbf{B}$$

where
$$\mathbf{A}_1 - \mathbf{A}_2 = \begin{bmatrix} 1 & -1 \\ 0 & 1 \end{bmatrix} - \begin{bmatrix} 4 & -2 \\ 3 & 6 \end{bmatrix} = \begin{bmatrix} -3 & 1 \\ -3 & -5 \end{bmatrix}$$

Hence
$$\bar{\mathbf{Y}} = \begin{bmatrix} \bar{x} \\ \bar{y} \end{bmatrix} = \frac{1}{18}\begin{bmatrix} -5 & -1 \\ 3 & -3 \end{bmatrix}\begin{bmatrix} -10 \\ -4 \end{bmatrix} = \begin{bmatrix} 3 \\ -1 \end{bmatrix}$$

5. The complete general solution, $y_0 = y_c + y_p$, then becomes
$$x_t = k_1(3)^t + k_2(10)^t + 3$$
$$y_t = -k_1(3)^t + 0.75k_2(10)^t - 1 \qquad\qquad (19.25)$$

Since $r_1 = 3$, $r_2 = 10 > |1|$, the system of equations is dynamically unstable. See also Problems 19.9 to 19.10.

EXAMPLE 8.   The specific solution is found with the help of the initial conditions. With $x_0 = 20$ and $y_0 = 3$, at $t = 0$, (19.24) reduces to

$$k_1 + k_2 + 3 = 20$$
$$-k_1 + 0.75k_2 - 1 = 3$$

Solving simultaneously, $$k_1 = 5, \qquad k_2 = 12$$

and by substituting into (19.25), we come to the definite solution,

$$x_t = 5(3)^t + 12(10)^t + 3$$
$$y_t = -5(3)^t + 9(10)^t - 1 \tag{19.26}$$

To check the answer, substitute $t = 1$ and $t = 0$ in (19.26).

$$x_1 = 5(3) + 12(10) + 3 = 138 \qquad x_0 = 5 + 12 + 3 = 20$$
$$y_1 = -5(3) + 9(10) - 1 = 74 \qquad y_0 = -5 + 9 - 1 = 3$$

Then substitute x_1, y_1 for x_t and y_t, and x_0, $y_0 =$ for x_{t-1} and y_{t-1} back in (19.23).

$$138 = 4(20) - 2(3) + 74 - 10 = 138$$
$$74 = 3(20) + 6(3) - 4 = 74$$

19.5   STABILITY AND PHASE DIAGRAMS FOR SIMULTANEOUS DIFFERENTIAL EQUATIONS

Given a system of linear autonomous differential equations, the intertemporal equilibrium level will be asymptotically stable, i.e., $y(t)$ will converge to $\bar{y}$ as $t \to \infty$, if and only if all the characteristic roots are negative. In the case of complex roots, the real part must be negative. If all the roots are positive, the system will be unstable. A saddle-point equilibrium, in which roots assume different signs, will generally be unstable. If, however, the initial conditions for y_1 and y_2 satisfy the condition

$$y_2 = \left(\frac{r_1 - a_{11}}{a_{12}} \right)(y_1 - \bar{y}_1) + \bar{y}_2$$

where $r_1 =$ the negative root, we have what is called a *saddle path*, and $y_1(t)$ and $y_2(t)$ will converge to their intertemporal equilibrium level (see Example 10).

A *phase diagram* for a system of two differential equations, linear or nonlinear, graphs y_2 on the vertical axis and y_1 on the horizontal axis. The y_1, y_2 plane is called the *phase plane*. Construction of a phase diagram is easiest explained in terms of an example.

EXAMPLE 9.   Given the system of linear autonomous differential equations,

$$\dot{y}_1 = -4y_1 + 16$$
$$\dot{y}_2 = -5y_2 + 15$$

a phase diagram is used below to test the stability of the model. Since neither variable is a function of the other variable in this simple model, each equation can be graphed separately.

1.   Find the intertemporal equilibrium level, $\bar{y}_1$, i.e., the locus of points at which $\dot{y}_1 = 0$.

$$\dot{y}_1 = -4y_1 + 16 = 0 \qquad \bar{y}_1 = 4$$

The graph of $\bar{y}_1 = 4$, a vertical line at $y_1 = 4$, is called the y_1 isocline. The y_1 isocline divides the phase plane into two regions called *isosectors*, one to the left of the y_1 isocline and one to the right.

2.   Find the intertemporal equilibrium level, $\bar{y}_2$, i.e., the locus of points at which $\dot{y}_2 = 0$.

$$\dot{y}_2 = -5y_2 + 15 = 0 \qquad \bar{y}_2 = 3$$

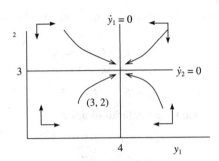

Fig. 19-1

The graph of $\bar{y}_2 = 3$ is a horizontal line at $y_2 = 3$, called the y_2 isocline. The y_2 isocline divides the phase plane into two isosectors, one above the y_2 isocline and the other below it. See Fig. 19-1.

The intersection of the isoclines demarcates the intertemporal equilibrium level,

$$(\bar{y}_1, \bar{y}_2) = (4, 3)$$

3. Determine the motion around the y_1 isocline, using arrows of horizontal motion.

    a) To the left of the y_1 isocline, $y_1 < 4$.   *b)* To the right of the y_1 isocline, $y_1 > 4$.

      By substituting these values successively in $\dot{y}_1 = -4y_1 + 16$, we see

    If $y_1 < 4$, $\dot{y}_1 > 0$, and there will be   If $y_1 > 4$, $\dot{y}_1 < 0$, and there will be
    motion to right   motion to left.

4. Determine the motion around the y_2 isocline, using arrows of vertical motion.

    a) Above the y_2 isocline, $y_2 > 3$.   *b)* Below the y_2 isocline, $y_2 < 3$.

      Substitution of these values successively in $\dot{y}_2 = -5y_2 + 15$, shows

    If $y_2 > 3$, $\dot{y}_2 < 0$, and the motion   If $y_2 < 3$, $\dot{y}_2 > 0$, and the motion
    will be downward.   will be upward.

The resulting arrows of motion in Fig. 19-1, all pointing to the intertemporal equilibrium, suggest the system of equations is convergent. Nevertheless, trajectory paths should be drawn because the arrows by themselves can be deceiving, as seen in Fig. 19-2. Starting from an arbitrary point, such as $(3, 2)$ in the southwest quadrant, or any point in any quadrant, we can see that the dynamics of the model will lead to the steady-state solution $(4, 3)$. Hence the time path converges to the steady-state solution, making that solution stable.

Since the equations are linear, the answer can be checked using the techniques of Chapter 16 or 19, getting

$$y_1(t) = k_1 e^{-4t} + 4$$
$$y_2(t) = k_2 e^{-5t} + 3$$

With both characteristic roots negative, the system must be stable.

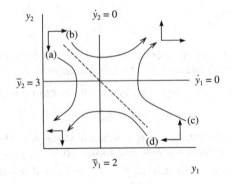

Fig. 19-2

EXAMPLE 10. A phase diagram is constructed in Fig. 19-2 and used below to test the dynamic stability of a saddle point equilibrium for the system of equations,

$$\dot{y}_1 = 2y_2 - 6$$
$$\dot{y}_2 = 8y_1 - 16$$

1. Find the y_1 isocline on which $\dot{y}_1 = 0$.

$$\dot{y}_1 = 2y_2 - 6 = 0 \qquad \bar{y}_2 = 3$$

Here the y_1 isocline is a horizontal line at $\bar{y}_2 = 3$.

2. Find the y_2 isocline on which $\dot{y}_2 = 0$.

$$\dot{y}_2 = 8y_1 - 16 \qquad \bar{y}_1 = 2$$

The y_2 isocline is a vertical line at $\bar{y}_1 = 2$. See Fig. 19-2.

3. Determine the motion around the y_1 isocline, using arrows of horizontal motion.

   a) Above the y_1 isocline, $y_2 > 3$.   b) Below the y_1 isocline, $y_2 < 3$.

   Substitution of these values successively in $\dot{y}_1 = 2y_2 - 6$, shows

   If $y_2 > 3$, $\dot{y}_1 > 0$, and the arrows of motion point to the right.   If $y_2 < 3$, $\dot{y}_1 < 0$, and the arrows of motion point to the left.

4. Determine the motion around the y_2 isocline, using arrows of vertical motion.

   a) To the left of the y_2 isocline, $y_1 < 2$.   b) To the right of the y_2 isocline, $y_1 > 2$.

   By substituting these values successively in $\dot{y}_2 = 8y_1 - 16$, we see

   If $y_1 < 2$, $\dot{y}_2 < 0$, and there will be motion downward.   If $y_1 > 2$, $\dot{y}_2 > 0$, and there will be motion upward.

   Despite appearances in Fig. 19-2, the system is unstable even in the northwest and southeast quadrants. As explained in Example 11, we can show by simply drawing trajectories that the time paths diverge in all four quadrants, whether we start at point a, b, c, or d.

EXAMPLE 11. The instability in the model in Fig. 19-2 is made evident by drawing a trajectory from any of the quadrants. We do two, one from a and one from b, and leave the other two for you as a practice exercise. In each case the path of the trajectory is best described in four steps.

1. Departure from point a.
   a) The trajectory moves in a southeasterly direction.
   b) But as the time path approaches the y_1 isocline where $\dot{y}_1 = 0$, the y_1 motion eastward slows down while the y_2 motion southward continues unabated.
   c) At the y_1 isocline, $\dot{y}_1 = 0$. Consequently, the trajectory must cross the y_1 isocline vertically.
   d) Below the y_1 isocline, the arrows of motion point in a southwesterly direction, taking the time path away from the equilibrium and hence indicating an unstable equilibrium.

2. Departure from point b.
   a) The trajectory once again moves in a southeasterly direction.
   b) But as the time path approaches the y_2 isocline where $\dot{y}_2 = 0$, the y_2 motion southward ebbs while the y_1 motion eastward continues unaffected.
   c) Since $\dot{y}_2 = 0$ at the y_2 isocline, the time path must cross the y_2 isocline horizontally.
   d) To the right of the y_2 isocline, the arrows of motion point in a northeasterly direction, taking the time path away from the equilibrium and belying the appearance of a stable equilibrium.

EXAMPLE 12. The dotted line in Fig. 19-2 is a saddle path. Only if the initial conditions fall on the saddle path will the steady-state equilibrium prove to be stable. The equation for the saddle path is

$$y_2 = \left(\frac{r_1 - a_{11}}{a_{12}}\right)(y_1 - \bar{y}_1) + \bar{y}_2$$

where we already know all but r_1, the negative root. From the original equations,

$$\mathbf{A} = \begin{vmatrix} a_{11} & a_{12} \\ a_{21} & a_{22} \end{vmatrix} = \begin{vmatrix} 0 & 2 \\ 8 & 0 \end{vmatrix} = -16$$

Any time $|\mathbf{A}| < 0$, we have a saddle-point equilibrium. Substituting in (19.3),

$$r_i = \frac{\text{Tr}(\mathbf{A}) \pm \sqrt{[\text{Tr}(\mathbf{A})]^2 - 4|\mathbf{A}|}}{2}$$

$$r_1, r_2 = \frac{0 \pm \sqrt{0 - 4(-16)}}{2} = -4, 4$$

Then substituting in the saddle-path equation above,

$$y_2 = \left(\frac{-4 - 0}{2}\right)(y_1 - 2) + 3$$

$$y_2 = 7 - 2y_1 \qquad \text{saddle path}$$

Note that the intertemporal equilibrium $(2,3)$ falls on the saddle path. Only if the initial conditions satisfy the saddle-path condition will the intertemporal equilibrium be stable.

Solved Problems

SIMULTANEOUS DIFFERENTIAL EQUATIONS

19.1. Solve the following system of first-order, autonomous, linear differential equations,

$$\dot{y}_1 = -8y_1 + 5y_2 + 4 \qquad y_1(0) = 7$$
$$\dot{y}_2 = 3.25y_1 - 4y_2 + 22 \qquad y_2(0) = 21.5$$

1. Putting them in matrix form for ease of computation,

$$\begin{bmatrix} \dot{y}_1 \\ \dot{y}_2 \end{bmatrix} = \begin{bmatrix} -8 & 5 \\ 3.25 & -4 \end{bmatrix} \begin{bmatrix} y_1 \\ y_2 \end{bmatrix} + \begin{bmatrix} 4 \\ 22 \end{bmatrix}$$

$$\dot{\mathbf{Y}} = \mathbf{AY} + \mathbf{B}$$

2. Then find the complementary functions. Assuming distinct real roots,

$$y_c = k_1 \mathbf{C}_1 e^{r_1 t} + k_2 \mathbf{C}_2 e^{r_2 t}$$

where

$$r_1, r_2 = \frac{\text{Tr}(\mathbf{A}) \pm \sqrt{[\text{Tr}(\mathbf{A})]^2 - 4|\mathbf{A}|}}{2}$$

$\text{Tr}(\mathbf{A}) = -12$, and $|\mathbf{A}| = 15.75$.

$$r_1, r_2 = \frac{-12 \pm \sqrt{(12)^2 - 4(15.75)}}{2} = \frac{-12 \pm 9}{2}$$

$$r_1 = -1.5 \qquad r_2 = -10.5$$

3. Next we find the eigenvectors $\mathbf{C}_i$ from

$$(\mathbf{A} - r_i \mathbf{I})\mathbf{C}_i = \begin{bmatrix} a_{11} - r_i & a_{12} \\ a_{21} & a_{22} - r_i \end{bmatrix} \begin{bmatrix} c_1 \\ c_2 \end{bmatrix} = 0$$

a) For $r_1 = -1.5$,

$$\begin{bmatrix} -8-(-1.5) & 5 \\ 3.25 & -4-(-1.5) \end{bmatrix}\begin{bmatrix} c_1 \\ c_2 \end{bmatrix} = \begin{bmatrix} -6.5 & 5 \\ 3.25 & -2.5 \end{bmatrix}\begin{bmatrix} c_1 \\ c_2 \end{bmatrix} = 0$$

By simple multiplication of row by column,

$$\begin{array}{ll} -6.5c_1 + 5c_2 = 0 & c_2 = 1.3c_1 \\ 3.25c_1 - 2.5c_2 = 0 & c_2 = 1.3c_1 \end{array}$$

If $c_1 = 1$, then $c_2 = 1.3$. Thus, the eigenvector $\mathbf{C}_1$ corresponding to $r_1 = -1.5$ is

$$\mathbf{C}_1 = \begin{bmatrix} c_1 \\ c_2 \end{bmatrix} = \begin{bmatrix} 1 \\ 1.3 \end{bmatrix}$$

and the first elements of the complementary function of the general solution are

$$y_c^1 = k_1\begin{bmatrix} 1 \\ 1.3 \end{bmatrix}e^{-1.5t} = \begin{bmatrix} k_1 e^{-1.5t} \\ 1.3k_1 e^{-1.5t} \end{bmatrix}$$

b) Substituting next for $r_2 = -10.5$,

$$\begin{bmatrix} -8-(-10.5) & 5 \\ 3.25 & -4-(-10.5) \end{bmatrix}\begin{bmatrix} c_1 \\ c_2 \end{bmatrix} = \begin{bmatrix} 2.5 & 5 \\ 3.25 & 6.5 \end{bmatrix}\begin{bmatrix} c_1 \\ c_2 \end{bmatrix} = 0$$

Then simply multiplying the first row by the column, since the final results will always be the same due to the singularity of the $(\mathbf{A} - r_i\mathbf{I})$ matrix

$$2.5c_1 + 5c_2 = 0 \qquad c_1 = -2c_2$$

If $c_2 = 1$, then $c_1 = -2$; the eigenvector $\mathbf{C}_2$ for $r_2 = -10.5$ is

$$\mathbf{C}_2 = \begin{bmatrix} c_1 \\ c_2 \end{bmatrix} = \begin{bmatrix} -2 \\ 1 \end{bmatrix}$$

and the second elements of the general complementary function are

$$y_c^2 = k_2\begin{bmatrix} -2 \\ 1 \end{bmatrix}e^{-10.5t} = \begin{bmatrix} -2k_2 e^{-10.5t} \\ k_2 e^{-10.5t} \end{bmatrix}$$

This makes the complete complementary solution,

$$y_1(t) = k_1 e^{-1.5t} - 2k_2 e^{-10.5t}$$
$$y_2(t) = 1.3k_1 e^{-1.5t} + k_2 e^{-10.5t}$$

4. Now find the intertemporal equilibrium solutions for y_p,

$$y_p = \bar{\mathbf{Y}} = -\mathbf{A}^{-1}\mathbf{B}$$

where $\mathbf{A} = \begin{bmatrix} -8 & 5 \\ 3.25 & -4 \end{bmatrix}$, $\mathbf{C} = \begin{bmatrix} -4 & -3.25 \\ -5 & -8 \end{bmatrix}$, Adj. $\mathbf{A} = \begin{bmatrix} -4 & -5 \\ -3.25 & -8 \end{bmatrix}$, $\mathbf{A}^{-1} = \dfrac{1}{15.75}\begin{bmatrix} -4 & -5 \\ -3.25 & -8 \end{bmatrix}$

Substituting above,

$$\bar{\mathbf{Y}} = \begin{bmatrix} \bar{y}_1 \\ \bar{y}_2 \end{bmatrix} = -\dfrac{1}{15.75}\begin{bmatrix} -4 & -5 \\ -3.25 & -8 \end{bmatrix}\begin{bmatrix} 4 \\ 22 \end{bmatrix} = \begin{bmatrix} 8 \\ 12 \end{bmatrix}$$

Thus the complete general solution, $y(t) = y_c + y_p$, is

$$y_1(t) = k_1 e^{-1.5t} - 2k_2 e^{-10.5t} + 8$$

$$y_2(t) = 1.3k_1 e^{-1.5t} + k_2 e^{-10.5t} + 12$$

(19.27)

5.   To find the definite solution, simply evaluate the above equations at $t = 0$ and use the initial conditions, $y_1(0) = 7$, $y_2(0) = 21.5$.

$$y_1(0) = k_1 - 2k_2 + 8 = 7$$
$$y_2(0) = 1.3k_1 + k_2 + 12 = 21.5$$

Solved simultaneously, $k_1 = 5$ $k_2 = 3$

Substituting in (19.27),

$$y_1(t) = 5e^{-1.5t} - 6e^{-10.5t} + 8$$
$$y_2(t) = 6.5e^{-1.5t} + 3e^{-10.5t} + 12$$

With $r_1 = -1.5 < 0$, $r_2 = -10.5 < 0$, the equilibrium is dynamically stable.

19.2.   Solve the following system of differential equations,

$$\dot{y}_1 = 2y_2 - 6 \qquad y_1(0) = 1$$
$$\dot{y}_2 = 8y_1 - 16 \qquad y_2(0) = 4$$

1.   In matrix form,

$$\begin{bmatrix} \dot{y}_1 \\ \dot{y}_2 \end{bmatrix} = \begin{bmatrix} 0 & 2 \\ 8 & 0 \end{bmatrix} \begin{bmatrix} y_1 \\ y_2 \end{bmatrix} + \begin{bmatrix} -6 \\ -16 \end{bmatrix}$$

$$\dot{\mathbf{Y}} = \mathbf{A}\mathbf{Y} + \mathbf{B}$$

2.   Finding the characteristic roots,

$$r_1, r_2 = \frac{\mathrm{Tr}(\mathbf{A}) \pm \sqrt{[\mathrm{Tr}(\mathbf{A})]^2 - 4|\mathbf{A}|}}{2}$$

$$r_1, r_2 = \frac{0 \pm \sqrt{(0)^2 - 4(-16)}}{2} = \frac{\pm 8}{2}$$

$$r_1 = -4 \qquad r_2 = 4$$

3.   Now determine the eigenvectors.

   a)   For $r_1 = -4$,

$$\begin{bmatrix} 0 - (-4) & 2 \\ 8 & 0 - (-4) \end{bmatrix} \begin{bmatrix} c_1 \\ c_2 \end{bmatrix} = \begin{bmatrix} 4 & 2 \\ 8 & 4 \end{bmatrix} \begin{bmatrix} c_1 \\ c_2 \end{bmatrix} = 0$$

$$4c_1 + 2c_2 = 0 \qquad c_2 = -2c_1$$

If $c_1 = 1$, then $c_2 = -2$, and the first elements of the complementary function are

$$y_c^1 = k_1 \begin{bmatrix} 1 \\ -2 \end{bmatrix} e^{-4t} = \begin{bmatrix} k_1 e^{-4t} \\ -2k_1 e^{-4t} \end{bmatrix}$$

   b)   For $r_1 = 4$,

$$\begin{bmatrix} 0 - 4 & 2 \\ 8 & 0 - 4 \end{bmatrix} \begin{bmatrix} c_1 \\ c_2 \end{bmatrix} = \begin{bmatrix} -4 & 2 \\ 8 & -4 \end{bmatrix} \begin{bmatrix} c_1 \\ c_2 \end{bmatrix} = 0$$

$$-4c_1 + 2c_2 = 0 \qquad c_2 = 2c_1$$

If $c_1 = 1$, then $c_2 = 2$, and the second elements of the complementary function are

$$y_c^2 = k_2 \begin{bmatrix} 1 \\ 2 \end{bmatrix} e^{4t} = \begin{bmatrix} k_2 e^{4t} \\ 2k_2 e^{4t} \end{bmatrix}$$

This means the general complementary functions are

$$y_1(t) = k_1 e^{-4t} + k_2 e^{4t}$$
$$y_2(t) = -2k_1 e^{-4t} + 2k_2 e^{4t}$$

4. For the steady-state solutions y_p,

$$y_p = \bar{\mathbf{Y}} = -\mathbf{A}^{-1}\mathbf{B}$$

$$\bar{\mathbf{Y}} = \begin{bmatrix} \bar{y}_1 \\ \bar{y}_2 \end{bmatrix} = -\frac{1}{-16}\begin{bmatrix} 0 & -2 \\ -8 & 0 \end{bmatrix}\begin{bmatrix} -6 \\ -16 \end{bmatrix} = \begin{bmatrix} 2 \\ 3 \end{bmatrix}$$

Thus the complete general solution, $y(t) = y_c + y_p$, is

$$y_1(t) = k_1 e^{-4t} + k_2 e^{4t} + 2$$
$$y_2(t) = -2k_1 e^{-4t} + 2k_2 e^{4t} + 3 \qquad\qquad (19.28)$$

5. We then find the definite solution from the initial conditions, $y_1(0) = 1$, $y_2(0) = 4$.

$$y_1(0) = k_1 + k_2 + 2 = 1$$
$$y_2(0) = -2k_1 + 2k_2 + 3 = 4$$

Solved simultaneously, $k_1 = -0.75$ $k_2 = -0.25$

Substituting in (19.28), we have the final solution.

$$y_1(t) = -0.75e^{-4t} - 0.25e^{4t} + 2$$
$$y_2(t) = 1.5e^{-4t} + 0.5k_2 e^{4t} + 3$$

With $r_1 = -4 < 0$ and $r_2 = 4 > 0$, we have a saddle-point solution. Saddle-point solutions are generally unstable unless the initial conditions fall on the saddle path:

$$y_2 = \left(\frac{r_1 - a_{11}}{a_{12}}\right)(y_1 - \bar{y}_1) + \bar{y}_2$$

Substituting, $$y_2 = \left(\frac{-4 - 0}{2}\right)(y_1 - 2) + 3$$
$$y_2 = 7 - 2y_1$$

This is the equation for the saddle path, which was graphed in Fig. 19-2 of Example 10. Substituting the initial conditoins, $y_1(0) = 1$, $y_2(0) = 4$, we see

$$4 \neq 7 - 2(1) = 5$$

Since the initial conditions do not fall on the saddle path, the system is unstable.

19.3. Solve the following system of equations.

$$\dot{y}_1 = 4y_1 + 7y_2 + 3 \qquad\qquad y_1(0) = 7$$
$$\dot{y}_2 = y_1 - 2y_2 + 4 \qquad\qquad y_2(0) = 10$$

1. Converting to matrices,

$$\begin{bmatrix} \dot{y}_1 \\ \dot{y}_2 \end{bmatrix} = \begin{bmatrix} 4 & 7 \\ 1 & -2 \end{bmatrix}\begin{bmatrix} y_1 \\ y_2 \end{bmatrix} + \begin{bmatrix} 31 \\ 4 \end{bmatrix}$$

$$\dot{\mathbf{Y}} = \mathbf{A}\mathbf{Y} + \mathbf{B}$$

2. The characteristic roots are

$$r_1, r_2 = \frac{2 \pm \sqrt{(2)^2 - 4(-15)}}{2} = \frac{2 \pm 8}{2}$$

$$r_1 = -3 \qquad r_2 = 5$$

3. The eigenvector for $r_1 = -3$,

$$\begin{bmatrix} 4-(-3) & 7 \\ 1 & -2-(-3) \end{bmatrix}\begin{bmatrix} c_1 \\ c_2 \end{bmatrix} = \begin{bmatrix} 7 & 7 \\ 1 & 1 \end{bmatrix}\begin{bmatrix} c_1 \\ c_2 \end{bmatrix} = 0$$

$$7c_1 + 7c_2 = 0 \qquad c_1 = -c_2$$

If $c_2 = 1$, then $c_1 = -1$, and

$$y_c^1 = k_1 \begin{bmatrix} -1 \\ 1 \end{bmatrix} e^{-3t} = \begin{bmatrix} -k_1 e^{-3t} \\ k_1 e^{-3t} \end{bmatrix}$$

For $r_2 = 5$,

$$\begin{bmatrix} 4-5 & 7 \\ 1 & -2-5 \end{bmatrix}\begin{bmatrix} c_1 \\ c_2 \end{bmatrix} = \begin{bmatrix} -1 & 7 \\ 1 & -7 \end{bmatrix}\begin{bmatrix} c_1 \\ c_2 \end{bmatrix} = 0$$

$$-c_1 + 7c_2 = 0 \qquad c_1 = 7c_2$$

If $c_2 = 1$, then $c_1 = 7$, and

$$y_c^2 = k_2 \begin{bmatrix} 7 \\ 1 \end{bmatrix} e^{5t} = \begin{bmatrix} 7k_2 e^{5t} \\ k_2 e^{5t} \end{bmatrix}$$

The general complementary functions are

$$y_1(t) = -k_1 e^{-3t} + 7k_2 e^{5t}$$
$$y_2(t) = k_1 e^{-3t} + k_2 e^{5t}$$

4. The steady-state solutions y_p are

$$y_p = \bar{Y} = -A^{-1}B$$

$$\bar{Y} = \begin{bmatrix} \bar{y}_1 \\ \bar{y}_2 \end{bmatrix} = -\frac{1}{-15}\begin{bmatrix} -2 & -7 \\ -1 & 4 \end{bmatrix}\begin{bmatrix} 31 \\ 4 \end{bmatrix} = \begin{bmatrix} -6 \\ -1 \end{bmatrix}$$

and the complete general solution is

$$y_1(t) = -k_1 e^{-3t} + 7k_2 e^{5t} - 6$$
$$y_2(t) = k_1 e^{-3t} + k_2 e^{5t} - 1 \tag{19.29}$$

5. The definite solution, given $y_1(0) = 7$, $y_2(0) = 10$, is

$$y_1(0) = -k_1 + 7k_2 - 6 = 7$$
$$y_2(0) = k_1 + k_2 - 1 = 10$$

$$k_1 = 8 \qquad k_2 = 3$$

Substituting in (19.29) for the final solution,

$$y_1(t) = -8e^{-3t} + 21e^{5t} - 6$$
$$y_2(t) = 8e^{-3t} + 3e^{5t} - 1$$

With $r_1 = -3 < 0$ and $r_2 = 5 > 0$, we again have a saddle-point solution which will be unstable unless the initial conditions fulfill the saddle-path equation:

$$y_2 = \left(\frac{r_1 - a_{11}}{a_{12}}\right)(y_1 - \bar{y}_1) + \bar{y}_2$$

Substituting,

$$y_2 = \left(\frac{-3-4}{7}\right)[y_1 - (-6)] + (-1)$$

$$y_2 = -7 - y_1$$

Employing the initial conditions, $y_1(0) = 7$, $y_2(0) = 10$,

$$10 \neq -7 - (7) = -14$$

Since the initial conditions do not satisfy the saddle-path equation, the system of equations is unstable.

19.4. Solve the following system of nonlinear, autonomous, first-order differential equations in which one or more derivative is a function of another derivative.

$$\dot{y}_1 = 4y_1 + y_2 + 6 \qquad\qquad y_1(0) = 9$$
$$\dot{y}_2 = 8y_1 + 5y_2 - \dot{y}_1 - 6 \qquad y_2(0) = 10$$

1. Rearranging the equations to conform with (19.7) and setting them in matrix form,

$$\begin{bmatrix} 1 & 0 \\ 1 & 1 \end{bmatrix}\begin{bmatrix} \dot{y}_1 \\ \dot{y}_2 \end{bmatrix} = \begin{bmatrix} 4 & 1 \\ 8 & 5 \end{bmatrix}\begin{bmatrix} y_1 \\ y_2 \end{bmatrix} + \begin{bmatrix} 6 \\ -6 \end{bmatrix}$$

$$\mathbf{A}_1\dot{\mathbf{Y}} = \mathbf{A}_2\mathbf{Y} + \mathbf{B}$$

2. Find the characteristic roots from the characteristic equation,

$$|\mathbf{A}_2 - r_i\mathbf{A}_1| = 0$$

where, dropping the i subscript for simplicity,

$$|\mathbf{A}_2 - r\mathbf{A}_1| = \left|\begin{bmatrix} 4 & 1 \\ 8 & 5 \end{bmatrix} - r\begin{bmatrix} 1 & 0 \\ 1 & 1 \end{bmatrix}\right| = \begin{vmatrix} 4-r & 1 \\ 8-r & 5-r \end{vmatrix} = 0$$

$$r^2 - 8r + 12 = 0$$

$$r_1 = 2 \qquad r_2 = 6$$

3. Find the eigenvectors $\mathbf{C}_i$ where

$$(\mathbf{A}_2 - r_i\mathbf{A}_1)\mathbf{C}_i = 0$$

and

$$(\mathbf{A}_2 - r_i\mathbf{A}_1)\mathbf{C}_i = \begin{bmatrix} 4-r_i & 1 \\ 8-r_i & 5-r_i \end{bmatrix}\begin{bmatrix} c_1 \\ c_2 \end{bmatrix}$$

Substituting for $r_1 = 2$,

$$\begin{bmatrix} 4-2 & 1 \\ 8-2 & 5-2 \end{bmatrix}\begin{bmatrix} c_1 \\ c_2 \end{bmatrix} = \begin{bmatrix} 2 & 1 \\ 6 & 3 \end{bmatrix}\begin{bmatrix} c_1 \\ c_2 \end{bmatrix} = 0$$

$$2c_1 + c_2 = 0 \qquad c_2 = -2c_1$$

If $c_1 = 1$, $c_2 = -2$, and

$$y_c^1 = k_1\begin{bmatrix} 1 \\ -2 \end{bmatrix}e^{2t} = \begin{bmatrix} k_1 \\ -2k_1 \end{bmatrix}e^{2t}$$

Now substituting for $r_2 = 6$,

$$\begin{bmatrix} 4-6 & 1 \\ 8-6 & 5-6 \end{bmatrix}\begin{bmatrix} c_1 \\ c_2 \end{bmatrix} = \begin{bmatrix} -2 & 1 \\ 2 & -1 \end{bmatrix}\begin{bmatrix} c_1 \\ c_2 \end{bmatrix} = 0$$

$$-2c_1 + c_2 = 0 \qquad c_2 = 2c_1$$

If $c_1 = 1$, $c_2 = 2$, and

$$y_c^2 = k_2\begin{bmatrix} 1 \\ 2 \end{bmatrix}e^{6t} = \begin{bmatrix} k_2 \\ 2k_2 \end{bmatrix}e^{6t}$$

Adding the two components of the complementary functions,

$$y_1(t) = k_1e^{2t} + k_2e^{6t}$$
$$y_2(t) = -2k_1e^{2t} + 2k_2e^{6t}$$

4. For the particular integral y_p,

$$\bar{Y} = -A_2^{-1}B$$

where $B = \begin{bmatrix} 6 \\ -6 \end{bmatrix}$, $A_2 = \begin{bmatrix} 4 & 1 \\ 8 & 5 \end{bmatrix}$, $|A_2| = 20 - 8 = 12$, and $A_2^{-1} = \dfrac{1}{12}\begin{bmatrix} 5 & -1 \\ -8 & 4 \end{bmatrix}$.

Substituting,

$$\bar{Y} = \begin{bmatrix} \bar{y}_1 \\ \bar{y}_2 \end{bmatrix} = -\frac{1}{12}\begin{bmatrix} 5 & -1 \\ -8 & 4 \end{bmatrix}\begin{bmatrix} 6 \\ -6 \end{bmatrix} = \begin{bmatrix} -3 \\ 6 \end{bmatrix}$$

Adding the particular integrals to the complementary functions,

$$y_1(t) = k_1 e^{2t} + k_2 e^{6t} - 3$$
$$y_2(t) = -2k_1 e^{2t} + 2k_2 e^{6t} + 6$$

(19.30)

5. For the definite solution, set $t = 0$ in (19.30) and use $y_1(0) = 9$, $y_2(0) = 10$.

$$y_1(0) = k_1 + k_2 - 3 = 9$$
$$y_2(0) = -2k_1 + 2k_2 + 6 = 10$$

$$k_1 = 5 \qquad k_2 = 7$$

Then substituting back in (19.30),

$$y_1(t) = 5e^{2t} + 7e^{6t} - 3$$
$$y_2(t) = -10e^{2t} + 14e^{6t} + 6$$

With $r_1 = 2 > 0$ and $r_2 = 6 > 0$, the system of equations will be dynamically unstable.

19.5. Solve the following system of differential equations.

$$\dot{y}_1 = -y_1 + 4y_2 - 0.5\dot{y}_2 - 1 \qquad y_1(0) = 4.5$$
$$\dot{y}_2 = 4y_1 - 2y_2 - 10 \qquad\qquad y_2(0) = 16$$

1. Rearranging and setting in matrix form,

$$\begin{bmatrix} 1 & 0.5 \\ 0 & 1 \end{bmatrix}\begin{bmatrix} \dot{y}_1 \\ \dot{y}_2 \end{bmatrix} = \begin{bmatrix} -1 & 4 \\ 4 & -2 \end{bmatrix}\begin{bmatrix} y_1 \\ y_2 \end{bmatrix} + \begin{bmatrix} -1 \\ -10 \end{bmatrix}$$

$$A_1\dot{Y} = A_2 Y + B$$

2. From the characteristic equation,

$$|A_2 - rA_1| = \begin{vmatrix} -1 - r & 4 - 0.5r \\ 4 & -2 - r \end{vmatrix} = 0$$

we find the characteristic roots,

$$r^2 + 5r - 14 = 0$$

$$r_1 = -7 \qquad r_2 = 2$$

3. We next find the eigenvectors C_i from the eigenvalue problem,

$$(A_2 - r_i A_1)C_i = \begin{bmatrix} -1 - r & 4 - 0.5r \\ 4 & -2 - r \end{bmatrix}\begin{bmatrix} c_1 \\ c_2 \end{bmatrix} = 0$$

Substituting for $r_1 = -7$,

$$\begin{bmatrix} -1 - (-7) & 4 - 0.5(-7) \\ 4 & -2 - (-7) \end{bmatrix}\begin{bmatrix} c_1 \\ c_2 \end{bmatrix} = \begin{bmatrix} 6 & 7.5 \\ 4 & 5 \end{bmatrix}\begin{bmatrix} c_1 \\ c_2 \end{bmatrix} = 0$$

$$6c_1 + 7.5c_2 = 0 \qquad c_1 = -1.25c_2$$

If $c_2 = 1$, $c_1 = -1.25$, and

$$y_c^1 = k_1 \begin{bmatrix} -1.25 \\ 1 \end{bmatrix} e^{-7t} = \begin{bmatrix} -1.25 k_1 e^{-7t} \\ k_1 e^{-7t} \end{bmatrix}$$

Substituting for $r_1 = 2$,

$$\begin{bmatrix} -1-2 & 4-0.5(2) \\ 4 & -2-2 \end{bmatrix} \begin{bmatrix} c_1 \\ c_2 \end{bmatrix} = \begin{bmatrix} -3 & 3 \\ 4 & -4 \end{bmatrix} \begin{bmatrix} c_1 \\ c_2 \end{bmatrix} = 0$$

$$-3c_1 + 3c_2 = 0 \qquad c_1 = c_2$$

If $c_2 = 1$, $c_1 = 1$, and

$$y_c^2 = k_2 \begin{bmatrix} 1 \\ 1 \end{bmatrix} e^{2t} = \begin{bmatrix} k_2 e^{2t} \\ k_2 e^{2t} \end{bmatrix}$$

This makes the complete general complementary functions,

$$y_1(t) = -1.25 k_1 e^{-7t} + k_2 e^{2t}$$
$$y_2(t) = k_1 e^{-7t} + k_2 e^{2t}$$

4. Finding the particular integral $\bar{\mathbf{Y}} = -\mathbf{A}_2^{-1} \mathbf{B}$,

where $\mathbf{A}_2 = \begin{bmatrix} -1 & 4 \\ 4 & -2 \end{bmatrix}$, $\mathbf{A}_2^{-1} = \dfrac{1}{-14} \begin{bmatrix} -2 & -4 \\ -4 & -1 \end{bmatrix}$, and

$$\bar{\mathbf{Y}} = \begin{bmatrix} \bar{y}_1 \\ \bar{y}_2 \end{bmatrix} = \frac{1}{14} \begin{bmatrix} -2 & -4 \\ -4 & -1 \end{bmatrix} \begin{bmatrix} -1 \\ -10 \end{bmatrix} = \begin{bmatrix} 3 \\ 1 \end{bmatrix}$$

By adding the particular integrals to the complementary functions, we get

$$y_1(t) = -1.25 k_1 e^{-7t} + k_2 e^{2t} + 3$$
$$y_2(t) = k_1 e^{-7t} + k_2 e^{2t} + 1$$

(19.31)

5. For the definite solution, we set $t = 0$ in (19.31) and use $y_1(0) = 4.5$, $y_2(0) = 16$.

$$y_1(0) = -1.25 k_1 + k_2 + 3 = 4.5$$
$$y_2(0) = k_1 + k_2 + 1 = 16$$
$$k_1 = 6 \qquad k_2 = 9$$

Finally, substituting in (19.31),

$$y_1(t) = -7.5 e^{-7t} + 9 e^{2t} + 3$$
$$y_2(t) = 6 e^{-7t} + 9 e^{2t} + 1$$

With characteristic roots of different signs, we have a saddle-point equilibrium which will be unstable unless the initial conditions happen to coincide with a point on the saddle path.

19.6. Solve the following.

$$\dot{y}_1 = -3y_1 - y_2 - 0.5\dot{y}_2 + 5 \qquad y_1(0) = 22.2$$
$$\dot{y}_2 = -2y_1 - 4y_2 - \dot{y}_1 + 10 \qquad y_2(0) = 3.9$$

1. In matrix form,

$$\begin{bmatrix} 1 & 0.5 \\ 1 & 1 \end{bmatrix} \begin{bmatrix} \dot{y}_1 \\ \dot{y}_2 \end{bmatrix} = \begin{bmatrix} -3 & -1 \\ -2 & -4 \end{bmatrix} \begin{bmatrix} y_1 \\ y_2 \end{bmatrix} + \begin{bmatrix} 5 \\ 10 \end{bmatrix}$$

$$\mathbf{A}_1 \dot{\mathbf{Y}} = \mathbf{A}_2 \mathbf{Y} + \mathbf{B}$$

2. The characteristic equation is

$$|\mathbf{A}_2 - r\mathbf{A}_1| = \begin{vmatrix} -3-r & -1-0.5r \\ -2-r & -4-r \end{vmatrix} = 0$$

$$0.5r^2 + 5r + 10 = 0$$

Multiplying by 2 and using the quadratic formula, the characteristic roots are

$$r_1 = -7.235 \qquad r_2 = -2.765$$

3. The eigenvector for $r_1 = -7.235$ is

$$\begin{bmatrix} -3-(-7.235) & -1-0.5(-7.235) \\ -2-(-7.235) & -4-(-7.235) \end{bmatrix}\begin{bmatrix} c_1 \\ c_2 \end{bmatrix} = \begin{bmatrix} 4.235 & 2.6175 \\ 5.235 & 3.235 \end{bmatrix}\begin{bmatrix} c_1 \\ c_2 \end{bmatrix} = 0$$

$$4.235c_1 + 2.6175c_2 = 0 \qquad c_2 \approx -1.62c_1$$

If $c_1 = 1$, $c_2 = -1.62$, and

$$y_c^1 = k_1\begin{bmatrix} 1 \\ -1.62 \end{bmatrix}e^{-7.235t} = \begin{bmatrix} k_1 e^{-7.235t} \\ -1.62k_1 e^{-7.235t} \end{bmatrix}$$

For $r_2 = -2.765$,

$$\begin{bmatrix} -3-(-2.765) & -1-0.5(-2.765) \\ -2-(-2.765) & -4-(-2.765) \end{bmatrix}\begin{bmatrix} c_1 \\ c_2 \end{bmatrix} = \begin{bmatrix} -0.235 & 0.3825 \\ 0.765 & -1.235 \end{bmatrix}\begin{bmatrix} c_1 \\ c_2 \end{bmatrix} = 0$$

$$-0.235c_1 + 0.3825c_2 = 0 \qquad c_1 \approx 1.62c_2$$

If $c_2 = 1$, $c_1 = 1.62$, and

$$y_c^2 = k_2\begin{bmatrix} 1.62 \\ 1 \end{bmatrix}e^{-2.765t} = \begin{bmatrix} 1.62k_2 e^{-2.765t} \\ k_2 e^{-2.765t} \end{bmatrix}$$

The complete complementary function, then, is

$$y_1(t) = k_1 e^{-7.235t} + 1.62k_2 e^{-2.765t}$$
$$y_2(t) = -1.62k_1 e^{-7.235t} + k_2 e^{-2.765t}$$

4. The particular integral $\bar{\mathbf{Y}} = -\mathbf{A}_2^{-1}\mathbf{B}$ is

$$\bar{\mathbf{Y}} = \begin{bmatrix} \bar{y}_1 \\ \bar{y}_2 \end{bmatrix} = -\frac{1}{10}\begin{bmatrix} -4 & 1 \\ 2 & -3 \end{bmatrix}\begin{bmatrix} 5 \\ 10 \end{bmatrix} = \begin{bmatrix} 1 \\ 2 \end{bmatrix}$$

and the general solution is

$$y_1(t) = k_1 e^{-7.235t} + 1.62k_2 e^{-2.765t} + 1$$
$$y_2(t) = -1.62k_1 e^{-7.235t} + k_2 e^{-2.765t} + 2$$

(19.32)

5. Using $y_1(0) = 22.2$, $y_2(0) = 3.9$ to solve for k_1 and k_2,

$$k_1 + 1.62k_2 + 1 = 22.2$$
$$-1.62k_1 + k_2 + 2 = 3.9$$
$$k_1 = 5 \qquad k_2 = 10$$

Substituting in (19.32) for the definite solution,

$$y_1(t) = 5e^{-7.235t} + 16.2e^{-2.765t} + 1$$
$$y_2(t) = -8.1e^{-7.235t} + 10e^{-2.765t} + 2$$

With both characteristic roots negative, the intertemporal equilibrium is stable.

SIMULTANEOUS DIFFERENCE EQUATIONS

19.7. Solve the following system of first-order linear difference equations in which no difference is a function of another difference.

$$x_t = 0.4x_{t-1} + 0.6y_{t-1} + 6 \qquad x_0 = 14$$
$$y_t = 0.1x_{t-1} + 0.3y_{t-1} + 5 \qquad y_0 = 23$$

1. Setting them in matrix form,

$$\begin{bmatrix} x_t \\ y_t \end{bmatrix} = \begin{bmatrix} 0.4 & 0.6 \\ 0.1 & 0.3 \end{bmatrix} \begin{bmatrix} x_{t-1} \\ y_{t-1} \end{bmatrix} + \begin{bmatrix} 6 \\ 5 \end{bmatrix}$$

$$\mathbf{Y}_t = \mathbf{A}\mathbf{Y}_{t-1} + \mathbf{B}$$

2. Using (*19.3*) on the characteristic equation $|\mathbf{A} - r_i\mathbf{I}| = 0$, find the characteristic roots.

$$r_1, r_2 = \frac{0.7 \pm \sqrt{(0.7)^2 - 4(0.06)}}{2} = \frac{0.7 \pm 0.5}{2}$$

$$r_1 = 0.6 \qquad r_2 = 0.1$$

3. The eigenvector for $r_1 = 0.6$ is

$$\begin{bmatrix} 0.4 - 0.6 & 0.6 \\ 0.1 & 0.3 - 0.6 \end{bmatrix} \begin{bmatrix} c_1 \\ c_2 \end{bmatrix} = \begin{bmatrix} -0.2 & 0.6 \\ 0.1 & -0.3 \end{bmatrix} \begin{bmatrix} c_1 \\ c_2 \end{bmatrix} = 0$$

$$-0.2c_1 + 0.6c_2 = 0 \qquad c_1 = 3c_2$$

If $c_2 = 1$, $c_1 = 3$, and we have

$$k_1 \begin{bmatrix} 3 \\ 1 \end{bmatrix} (0.6)^t = \begin{bmatrix} 3k_1(0.6)^t \\ k_1(0.6)^t \end{bmatrix}$$

For $r_2 = 0.1$,

$$\begin{bmatrix} 0.4 - 0.1 & 0.6 \\ 0.1 & 0.3 - 0.1 \end{bmatrix} \begin{bmatrix} c_1 \\ c_2 \end{bmatrix} = \begin{bmatrix} 0.3 & 0.6 \\ 0.1 & 0.2 \end{bmatrix} \begin{bmatrix} c_1 \\ c_2 \end{bmatrix} = 0$$

$$0.3c_1 + 0.6c_2 = 0 \qquad c_1 = -2c_2$$

If $c_2 = 1$, $c_1 = -2$, and

$$k_2 \begin{bmatrix} -2 \\ 1 \end{bmatrix} (0.1)^t = \begin{bmatrix} -2k_2(0.1)^t \\ k_2(0.1)^t \end{bmatrix}$$

Combining the two for the general complementary functions,

$$x_c = 3k_1(0.6)^t - 2k_2(0.1)^t$$
$$y_c = k_1(0.6)^t + k_2(0.1)^t$$

4. For the particular solution,

$$y_p = (\mathbf{I} - \mathbf{A})^{-1}\mathbf{B}$$

where

$$(\mathbf{I} - \mathbf{A}) = \begin{bmatrix} 1 & 0 \\ 0 & 1 \end{bmatrix} - \begin{bmatrix} 0.4 & 0.6 \\ 0.1 & 0.3 \end{bmatrix} = \begin{bmatrix} 0.6 & -0.6 \\ -0.1 & 0.7 \end{bmatrix}$$

and

$$y_p = \begin{bmatrix} \bar{x} \\ \bar{y} \end{bmatrix} = \frac{1}{0.36} \begin{bmatrix} 0.7 & 0.6 \\ 0.1 & 0.6 \end{bmatrix} \begin{bmatrix} 6 \\ 5 \end{bmatrix} = \begin{bmatrix} 20 \\ 10 \end{bmatrix}$$

This makes the complete general solution,

$$x_t = 3k_1(0.6)^t - 2k_2(0.1)^t + 20$$

$$y_t = k_1(0.6)^t + k_2(0.1)^t + 10 \tag{19.33}$$

5. Employing the initial conditions, $x_0 = 14$, $y_0 = 23$, (19.33) reduces to

$$3k_1 - 2k_2 + 20 = 14$$
$$k_1 + k_2 + 10 = 23$$

Solved simultaneously, $k_1 = 4$, $k_2 = 9$

Substituting in (19.33),

$$x_t = 12(0.6)^t - 18k_2(0.1)^t + 20$$
$$y_t = 4(0.6)^t + 9(0.1)^t + 10$$

With $|0.6| < 1$ and $|0.1| < 1$, the time path is convergent. With both roots positive, there will be no oscillation.

19.8. Solve the following system of first-order linear difference equations.

$$x_t = -0.6x_{t-1} + 0.1y_{t-1} + 9 \qquad x_0 = 7.02$$
$$y_t = 0.5x_{t-1} - 0.2y_{t-1} + 42 \qquad y_0 = 57.34$$

1. In matrix form,

$$\begin{bmatrix} x_t \\ y_t \end{bmatrix} = \begin{bmatrix} -0.6 & 0.1 \\ 0.5 & -0.2 \end{bmatrix} \begin{bmatrix} x_{t-1} \\ y_{t-1} \end{bmatrix} + \begin{bmatrix} 9 \\ 42 \end{bmatrix}$$

$$\mathbf{Y}_t = \mathbf{A}\mathbf{Y}_{t-1} + \mathbf{B}$$

2. The characteristic roots are

$$r_1, r_2 = \frac{-0.8 \pm \sqrt{(-0.8)^2 - 4(0.07)|}}{2} = \frac{-0.8 \pm 0.6}{2}$$

$$r_1 = -0.1 \qquad r_2 = -0.7$$

3. The eigenvector for $r_1 = -0.1$ is

$$\begin{bmatrix} -0.6 - (-0.1) & 0.1 \\ 0.5 & -0.2 - (-0.1) \end{bmatrix} \begin{bmatrix} c_1 \\ c_2 \end{bmatrix} = \begin{bmatrix} -0.5 & 0.1 \\ 0.5 & -0.1 \end{bmatrix} \begin{bmatrix} c_1 \\ c_2 \end{bmatrix} = 0$$

$$-0.5c_1 + 0.1c_2 = 0 \qquad c_2 = 5c_1$$

If $c_1 = 1$, $c_2 = 5$, and

$$k_1 \begin{bmatrix} 1 \\ 5 \end{bmatrix}(-0.1)^t = \begin{bmatrix} k_1(-0.1)^t \\ 5k_1(-0.1)^t \end{bmatrix}$$

For $r_2 = -0.7$,

$$\begin{bmatrix} -0.6 - (-0.7) & 0.1 \\ 0.5 & -0.2 - (-0.7) \end{bmatrix} \begin{bmatrix} c_1 \\ c_2 \end{bmatrix} = \begin{bmatrix} 0.1 & 0.1 \\ 0.5 & 0.5 \end{bmatrix} \begin{bmatrix} c_1 \\ c_2 \end{bmatrix} = 0$$

$$0.1c_1 + 0.1c_2 = 0 \qquad c_1 = -c_2$$

If $c_2 = 1$, $c_1 = -1$, and

$$k_2 \begin{bmatrix} -1 \\ 1 \end{bmatrix}(-0.7)^t = \begin{bmatrix} -k_2(-0.7)^t \\ k_2(-0.7)^t \end{bmatrix}$$

This makes the general complementary functions,

$$x_c = k_1(-0.1)^t - k_2(-0.7)^t$$
$$y_c = 5k_1(-0.1)^t + k_2(-0.7)^t$$

4. For the particular solution,

$$y_p = (\mathbf{I} - \mathbf{A})^{-1}\mathbf{B}$$

where

$$(\mathbf{I} - \mathbf{A}) = \begin{bmatrix} 1 & 0 \\ 0 & 1 \end{bmatrix} - \begin{bmatrix} -0.6 & 0.1 \\ 0.5 & -0.2 \end{bmatrix} = \begin{bmatrix} 1.6 & -0.1 \\ -0.5 & 1.2 \end{bmatrix}$$

and

$$y_p = \begin{bmatrix} \bar{x} \\ \bar{y} \end{bmatrix} = \frac{1}{1.87}\begin{bmatrix} 1.2 & 0.1 \\ 0.5 & 1.6 \end{bmatrix}\begin{bmatrix} 9 \\ 42 \end{bmatrix} = \begin{bmatrix} 8.02 \\ 38.34 \end{bmatrix}$$

This makes the complete general solution,

$$x_t = k_1(-0.1)^t - k_2(-0.7)^t + 8.02$$
$$y_t = 5k_1(-0.1)^t + k_2(-0.7)^t + 38.34$$

(19.34)

5. Using $x_0 = 7.02$, $y_0 = 57.34$, (19.34) reduces to

$$k_1 - k_2 + 8.02 = 7.02$$
$$5k_1 + k_2 + 38.34 = 57.34$$

Solved simultaneously, $\qquad k_1 = 3, \qquad k_2 = 4$

Substituting in (19.34),

$$x_t = 3(-0.1)^t - 4(-0.7)^t + 8.02$$
$$y_t = 15(-0.1)^t + 4(-0.7)^t + 38.34$$

With both characteristic roots in absolute value less than 1, the system of equations will approach a stable intertemporal equilibrium solution. With the roots negative, there will be oscillation.

19.9. Solve the following system of first-order linear difference equations in which one difference is a function of another difference.

$$x_t = -0.7x_{t-1} - 0.4y_{t-1} + 40 \qquad\qquad x_0 = 24$$
$$y_t = -0.575x_{t-1} - 0.5y_{t-1} - x_t + 6 \qquad y_0 = -32$$

1. Rearranging and setting in matrix form,

$$\begin{bmatrix} 1 & 0 \\ 1 & 1 \end{bmatrix}\begin{bmatrix} x_t \\ y_t \end{bmatrix} = \begin{bmatrix} -0.7 & -0.4 \\ -0.575 & -0.5 \end{bmatrix}\begin{bmatrix} x_{t-1} \\ y_{t-1} \end{bmatrix} + \begin{bmatrix} 40 \\ 6 \end{bmatrix}$$

$$\mathbf{A}_1\mathbf{Y}_t = \mathbf{A}_2\mathbf{Y}_{t-1} + \mathbf{B}$$

2. We then find the characteristic roots from the characteristic equation,

$$|\mathbf{A}_2 - r_i\mathbf{A}_1| = 0$$

$$\begin{bmatrix} -0.7 & -0.4 \\ -0.575 & -0.5 \end{bmatrix} - r_i\begin{bmatrix} 1 & 0 \\ 1 & 1 \end{bmatrix} = \begin{bmatrix} -0.7 - r_i & -0.4 \\ -0.575 - r_i & -0.5 - r_i \end{bmatrix} = 0$$

$$r^2 + 0.8r + 0.12 = 0$$

$$r_1 = -0.6 \qquad r_2 = -0.2$$

3. The eigenvector for $r_1 = -0.6$ is

$$\begin{bmatrix} -0.7 - (-0.6) & -0.4 \\ -0.575 - (-0.6) & -0.5 - (-0.6) \end{bmatrix}\begin{bmatrix} c_1 \\ c_2 \end{bmatrix} = \begin{bmatrix} -0.1 & -0.4 \\ 0.025 & 0.1 \end{bmatrix}\begin{bmatrix} c_1 \\ c_2 \end{bmatrix} = 0$$

$$-0.1c_1 - 0.4c_2 = 0 \qquad c_1 = -4c_2$$

If $c_2 = 1$, $c_1 = -4$ and the eigenvector is

$$k_1\begin{bmatrix} -4 \\ 1 \end{bmatrix}(-0.6)^t = \begin{bmatrix} -4k_1(-0.6)^t \\ k_1(-0.6)^t \end{bmatrix}$$

For $r_2 = -0.2$,

$$\begin{bmatrix} -0.7 - (-0.2) & -0.4 \\ -0.575 - (-0.2) & -0.5 - (-0.2) \end{bmatrix} \begin{bmatrix} c_1 \\ c_2 \end{bmatrix} = \begin{bmatrix} -0.5 & -0.4 \\ -0.375 & 0.3 \end{bmatrix} \begin{bmatrix} c_1 \\ c_2 \end{bmatrix} = 0$$

$$-0.5c_1 - 0.4c_2 = 0 \qquad c_2 = -1.25c_1$$

If $c_1 = 1$, $c_2 = -1.25$, and the eigenvector for $r_2 = -0.2$ is

$$k_2 \begin{bmatrix} 1 \\ -1.25 \end{bmatrix} (-0.2)^t = \begin{bmatrix} k_2(-0.2)^t \\ -1.25k_2(-0.2)^t \end{bmatrix}$$

Adding the two eigenvectors, the complementary functions are

$$x_c = -4k_1(-0.6)^t + k_2(-0.2)^t$$
$$y_c = k_1(-0.6)^t - 1.25k_2(-0.2)^t$$

4. For the particular solution $y_p = \bar{\mathbf{Y}}$,

$$\bar{\mathbf{Y}} = (\mathbf{A}_1 - \mathbf{A}_2)^{-1}\mathbf{B}$$

where

$$(\mathbf{A}_1 - \mathbf{A}_2) = \begin{bmatrix} 1 & 0 \\ 1 & 1 \end{bmatrix} - \begin{bmatrix} -0.7 & -0.4 \\ -0.575 & -0.5 \end{bmatrix} = \begin{bmatrix} 1.7 & 0.4 \\ 1.575 & 1.5 \end{bmatrix}$$

and

$$\bar{\mathbf{Y}} = \begin{bmatrix} \bar{x} \\ \bar{y} \end{bmatrix} = \frac{1}{1.92} \begin{bmatrix} 1.5 & -0.4 \\ -1.575 & 1.7 \end{bmatrix} \begin{bmatrix} 40 \\ 6 \end{bmatrix} = \begin{bmatrix} 30 \\ -27.5 \end{bmatrix}$$

Adding y_c and y_p, the complete general solution is

$$x_t = -4k_1(-0.6)^t + k_2(-0.2)^t + 30$$
$$y_t = k_1(-0.6)^t - 1.25k_2(-0.2)^t - 27.5$$

(19.35)

5. Finally, we apply the initial conditions, $x_0 = 24$ and $y_0 = -32$, to (19.35),

$$-4k_1 + k_2 + 30 = 24$$
$$k_1 - 1.25k_2 - 27.5 = -32$$
$$k_1 = 3 \qquad k_2 = 6$$

and substitute these values back in (19.35) for the definite solution.

$$x_t = -12(-0.6)^t + 6(-0.2)^t + 30$$
$$y_t = 3(-0.6)^t - 7.5(-0.2)^t - 27.5$$

With both characteristic roots less than 1 in absolute value, the solution is stable.

19.10. Solve the following system of first-order linear difference equations.

$$x_t = 0.6x_{t-1} + 0.85y_{t-1} - y_t + 15 \qquad x_0 = 27$$
$$y_t = 0.2x_{t-1} + 0.4y_{t-1} + 6 \qquad y_0 = 38$$

1. In matrix form,

$$\begin{bmatrix} 1 & 1 \\ 0 & 1 \end{bmatrix} \begin{bmatrix} x_t \\ y_t \end{bmatrix} = \begin{bmatrix} 0.6 & 0.85 \\ 0.2 & 0.4 \end{bmatrix} \begin{bmatrix} x_{t-1} \\ y_{t-1} \end{bmatrix} + \begin{bmatrix} 15 \\ 6 \end{bmatrix}$$

$$\mathbf{A}_1 \mathbf{Y}_t = \mathbf{A}_2 \mathbf{Y}_{t-1} + \mathbf{B}$$

2. For the characteristic roots, $|\mathbf{A}_2 - r_i\mathbf{A}_1| = 0$.

$$\begin{bmatrix} 0.6 & 0.85 \\ 0.2 & 0.4 \end{bmatrix} - r_i \begin{bmatrix} 1 & 1 \\ 0 & 1 \end{bmatrix} = \begin{bmatrix} 0.6 - r_i & 0.85 - r_i \\ 0.2 & 0.4 - r_i \end{bmatrix} = 0$$

$$r^2 - 0.8r + 0.07 = 0$$
$$r_1 = 0.7 \qquad r_2 = 0.1$$

3. For $r_1 = 0.7$,

$$\begin{bmatrix} 0.6 - 0.7 & 0.85 - 0.7 \\ 0.2 & 0.4 - 0.7 \end{bmatrix}\begin{bmatrix} c_1 \\ c_2 \end{bmatrix} = \begin{bmatrix} -0.1 & 0.15 \\ 0.2 & -0.3 \end{bmatrix}\begin{bmatrix} c_1 \\ c_2 \end{bmatrix} = 0$$

$$-0.1c_1 + 0.15c_2 = 0 \qquad c_1 = 1.5c_2$$

If $c_2 = 1$, $c_1 = 1.5$ and the eigenvector is

$$k_1\begin{bmatrix} 1.5 \\ 1 \end{bmatrix}(0.7)^t = \begin{bmatrix} 1.5k_1(0.7)^t \\ k_1(0.7)^t \end{bmatrix}$$

For $r_2 = 0.1$,

$$\begin{bmatrix} 0.6 - 0.1 & 0.85 - 0.1 \\ 0.2 & 0.4 - 0.1 \end{bmatrix}\begin{bmatrix} c_1 \\ c_2 \end{bmatrix} = \begin{bmatrix} 0.5 & 0.75 \\ 0.2 & 0.3 \end{bmatrix}\begin{bmatrix} c_1 \\ c_2 \end{bmatrix} = 0$$

$$0.5c_1 + 0.75c_2 = 0 \qquad c_1 = -1.5c_2$$

If $c_2 = 1$, $c_1 = -1.5$ and the eigenvector is

$$k_2\begin{bmatrix} -1.5 \\ 1 \end{bmatrix}(0.1)^t = \begin{bmatrix} -1.5k_2(0.1)^t \\ k_2(0.1)^t \end{bmatrix}$$

4. For the particular solution,

$$\bar{\mathbf{Y}} = (\mathbf{A}_1 - \mathbf{A}_2)^{-1}\mathbf{B}$$

Here

$$(\mathbf{A}_1 - \mathbf{A}_2) = \begin{bmatrix} 1 & 1 \\ 0 & 1 \end{bmatrix} - \begin{bmatrix} 0.6 & 0.85 \\ 0.2 & 0.4 \end{bmatrix} = \begin{bmatrix} 0.4 & 0.15 \\ -0.2 & 0.6 \end{bmatrix}$$

and

$$\bar{\mathbf{Y}} = \begin{bmatrix} \bar{x} \\ \bar{y} \end{bmatrix} = \frac{1}{0.27}\begin{bmatrix} 0.6 & -0.15 \\ 0.2 & 0.4 \end{bmatrix}\begin{bmatrix} 15 \\ 6 \end{bmatrix} = \begin{bmatrix} 30 \\ 20 \end{bmatrix}$$

Adding y_c and y_p,

$$x_t = 1.5k_1(0.7)^t - 1.5k_2(0.1)^t + 30$$

$$y_t = k_1(0.7)^t + k_2(0.1)^t + 20 \qquad\qquad (19.36)$$

5. For the definite solution, we apply $x_0 = 27$ and $y_0 = 38$ to (19.36),

$$1.5k_1 - 1.5k_2 + 30 = 27$$

$$k_1 + k_2 + 20 = 38$$

$$k_1 = 8 \qquad k_2 = 10$$

Substituting back in (19.36),

$$x_t = 12(0.7)^t - 15(0.1)^t + 30$$
$$y_t = 8(0.7)^t + 10(0.1)^t + 20$$

With both characteristic roots less than 1 in absolute value, the solution is stable.

PHASE DIAGRAMS FOR SIMULTANEOUS DIFFERENTIAL EQUATIONS

19.11. Use a phase diagram to test the stability of the system of equations,

$$\dot{y}_1 = 3y_1 - 18$$
$$\dot{y}_2 = -2y_2 + 16$$

1. Determine the steady-state solutions $\bar{y}_i$ where $\dot{y}_i = 0$ to find the isoclines.

$$\dot{y}_1 = 3y_1 - 18 = 0 \qquad\qquad \dot{y}_2 = -2y_2 + 16 = 0$$
$$\bar{y}_1 = 6 \quad \text{the } y_1 \text{ isocline} \qquad \bar{y}_2 = 8 \quad \text{the } y_2 \text{ isocline}$$

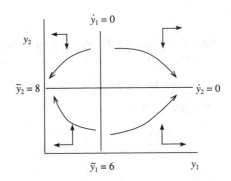

Fig. 19-3

As seen in Fig. 19-3, the intersection of the isoclines demarcates the intertemporal equilibrium level, $(\bar{y}_1, \bar{y}_2) = (6, 8)$.

2. Determine the motion around the y_1 isocline using arrows of horizontal motion.

   a) To the left of the y_1 isocline, $y_1 < 6$.   *b*) To the right of the y_1 isocline, $y_1 > 6$.

   Substituting these values successively in $\dot{y}_1 = 3y_1 - 18$, we see

   If $y_1 < 6$, $\dot{y}_1 < 0$, and there will be   If $y_1 > 6$, $\dot{y}_1 > 0$, and there will be
   motion to the left.   motion to the right.

3. Determine the motion around the y_2 isocline, using arrows of vertical motion.

   a) Above the y_2 isocline, $y_2 > 8$.   *b*) Below the y_2 isocline, $y_2 < 8$.

   Substitution of these values successively in $\dot{y}_2 = -2y_2 + 16$ shows

   If $y_2 > 8$, $\dot{y}_2 < 0$, and the motion   If $y_2 < 8$, $\dot{y}_2 > 0$, and the motion
   will be downward.   will be upward.

   The resulting arrows of motion in Fig. 19-3, all pointing away from the intertemporal equilibrium, suggest the system of equations is divergent. Drawing trajectory paths to be sure confirms that the system is indeed divergent.

Proofs and Demonstrations

19.12. Given

$$\begin{bmatrix} \dot{y}_1 \\ \dot{y}_2 \end{bmatrix} = \begin{bmatrix} a_{11} & a_{12} \\ a_{21} & a_{22} \end{bmatrix} \begin{bmatrix} y_1 \\ y_2 \end{bmatrix} + \begin{bmatrix} b_1 \\ b_2 \end{bmatrix}$$

or $$\dot{\mathbf{Y}} = \mathbf{AY} + \mathbf{B} \tag{19.37}$$

show in terms of Section 19.1 and Example 1 that

$$(\mathbf{A} - r_i \mathbf{I})\mathbf{C}_i = 0$$

Starting with the homogeneous form of the system of equations in which $\mathbf{B} = 0$, or a null vector, and assuming distinct real roots, we can expect the solution to be in the form

$$\mathbf{Y} = k_i \mathbf{C}_i e^{r_i t} \tag{19.38}$$

where k_i = a scalar, $\mathbf{C}_i = (2 \times 1)$ column vector of constants, and r_i = a scalar. Taking the derivative of (19.38) with respect to t, we have

$$\dot{\mathbf{Y}} = r_i k_i \mathbf{C}_i e^{r_i t} \tag{19.39}$$

Substituting (*19.38*) and (*19.39*) in the homogeneous form of (*19.37*) where $\mathbf{B} = 0$,

$$r_i k_i \mathbf{C}_i e^{r_i t} = \mathbf{A} k_i \mathbf{C}_i e^{r_i t}$$

Canceling the common k_i and $e^{r_i t}$ terms, we have

$$r_i \mathbf{C}_i = \mathbf{A} \mathbf{C}_i$$
$$\mathbf{A} \mathbf{C}_i - r_i \mathbf{C}_i = 0$$

Factoring out $\mathbf{C}_i$ and recalling that $\mathbf{A}$ is a (2×2) matrix while r_i is a scalar, we multiply r_i by a (2×2) identity matrix $\mathbf{I}_2$, or simply $\mathbf{I}$, to get

$$(\mathbf{A} - r_i \mathbf{I}) \mathbf{C}_i = 0 \qquad \text{Q.E.D.} \tag{19.40}$$

19.13. Continuing with the model in Problem 19.12,

show that
$$r_1, r_2 = \frac{\text{Tr}(\mathbf{A}) \pm \sqrt{[\text{Tr}(\mathbf{A})]^2 - 4|\mathbf{A}|}}{2}$$

If $(\mathbf{A} - r_i \mathbf{I})$ is nonsingular in (*19.40*), meaning it contains no linear dependence, then $\mathbf{C}_i$ must be a null column vector, making the solution trivial. To find a nontrivial solution, $(\mathbf{A} - r_i \mathbf{I})$ must be singular. A necessary condition for a nontrivial solution ($\mathbf{C}_i \neq 0$), then, is that the determinant

$$|\mathbf{A} - r_i \mathbf{I}| = 0 \tag{19.41}$$

where equation (*19.41*) is called the *characteristic equation* or *characteristic polynomial* for matrix $\mathbf{A}$. Dropping the subscript for simplicity and substituting from above, we have

$$\begin{vmatrix} a_{11} - r & a_{12} \\ a_{21} & a_{22} - r \end{vmatrix} = 0$$

$$a_{11} a_{22} - a_{11} r - a_{22} r + r^2 - a_{12} a_{21} = 0$$

Rearranging,
$$r^2 - (a_{11} + a_{22}) r + (a_{11} a_{22} - a_{12} a_{21}) = 0$$

Or, using matrix notion,

$$r^2 - \text{Tr}(\mathbf{A}) r + |\mathbf{A}| = 0$$

which is a quadratic equation that can be solved for r with the quadratic formula,

$$r_1, r_2 = \frac{\text{Tr}(\mathbf{A}) \pm \sqrt{[\text{Tr}(\mathbf{A})]^2 - 4|\mathbf{A}|}}{2} \qquad \text{Q.E.D.}$$

19.14. Continuing with the model in Problem 19.13, show that the particular integral or solution is

$$y_p = \bar{\mathbf{Y}} = -\mathbf{A}^{-1} \mathbf{B} \tag{19.42}$$

The particular integral is simply the intertemporal or steady-state solution $\bar{\mathbf{Y}}$. To find the steady-state solution, we simply set the column vector of derivatives equal to zero such that $\dot{\mathbf{Y}} = 0$. When $\dot{\mathbf{Y}} = 0$, there is no change and $\mathbf{Y} = \bar{\mathbf{Y}}$. Substituting in (*19.37*),

$$\dot{\mathbf{Y}} = \mathbf{A} \bar{\mathbf{Y}} + \mathbf{B} = 0$$
$$\mathbf{A} \bar{\mathbf{Y}} = -\mathbf{B}$$
$$\bar{\mathbf{Y}} = -\mathbf{A}^{-1} \mathbf{B} \qquad \text{Q.E.D.}$$

19.15. Given
$$\begin{bmatrix} a_{11} & a_{12} \\ a_{21} & a_{22} \end{bmatrix} \begin{bmatrix} \dot{y}_1 \\ \dot{y}_2 \end{bmatrix} = \begin{bmatrix} a_{13} & a_{14} \\ a_{23} & a_{24} \end{bmatrix} \begin{bmatrix} y_1 \\ y_2 \end{bmatrix} + \begin{bmatrix} b_1 \\ b_2 \end{bmatrix}$$

or
$$\mathbf{A}_1 \dot{\mathbf{Y}} = \mathbf{A}_2 \mathbf{Y} + \mathbf{B} \tag{19.43}$$

show in terms of Section 19.2 and Example 3 that to find the complementary function, one must solve the specific eigenvalue problem

$$(\mathbf{A}_2 - r_i \mathbf{A}_1)\mathbf{C}_i = 0$$

Starting with the homogeneous form of (19.43) in which $\mathbf{B} = 0$, or a null vector, and assuming distinct real roots, we can expect the solution and its derivative to take the forms

$$\mathbf{Y} = k_i \mathbf{C}_i e^{r_i t} \qquad \dot{\mathbf{Y}} = r_i k_i \mathbf{C}_i e^{r_i t} \tag{19.44}$$

Substituting from (19.44) into the homogeneous form of (19.43) where $\mathbf{B} = 0$,

$$\mathbf{A}_1 r_i k_i \mathbf{C}_i e^{r_i t} = \mathbf{A}_2 k_i \mathbf{C}_i e^{r_i t}$$

Canceling the common k_i and $e^{r_i t}$ terms, we have

$$\mathbf{A}_1 r_i \mathbf{C}_i = \mathbf{A}_2 \mathbf{C}_i$$
$$(\mathbf{A}_2 - r_i \mathbf{A}_1)\mathbf{C}_i = 0 \qquad \text{Q.E.D.}$$

19.16. In terms of the model in Problem 19.15, show that the particular integral is

$$y_p = \bar{\mathbf{Y}} = -\mathbf{A}_2^{-1}\mathbf{B} \tag{19.45}$$

The particular integral is the steady-state solution $\bar{\mathbf{Y}}$ when $\dot{\mathbf{Y}} = 0$. Substituting in (19.43),

$$\mathbf{A}_1 \dot{\mathbf{Y}} = \mathbf{A}_2 \bar{\mathbf{Y}} + \mathbf{B} = 0$$
$$\mathbf{A}_2 \bar{\mathbf{Y}} = -\mathbf{B}$$
$$\bar{\mathbf{Y}} = -\mathbf{A}_2^{-1}\mathbf{B} \qquad \text{Q.E.D.}$$

19.17. Given

$$\begin{bmatrix} x_t \\ y_t \end{bmatrix} = \begin{bmatrix} a_{11} & a_{12} \\ a_{21} & a_{22} \end{bmatrix}\begin{bmatrix} x_{t-1} \\ y_{t-1} \end{bmatrix} + \begin{bmatrix} b_1 \\ b_2 \end{bmatrix}$$

or

$$\mathbf{Y}_t = \mathbf{A}\mathbf{Y}_{t-1} + \mathbf{B} \tag{19.46}$$

show in terms of Section 19.3 and Example 5 that the eigenvalue problem for a system of simultaneous first-order linear difference equations when no difference is a function of another difference is

$$(\mathbf{A} - r_i \mathbf{I})\mathbf{C}_i = 0$$

Starting with the homogeneous form of the system of equations in which $\mathbf{B} = 0$, and assuming a case of distinct real roots, from what we know of individual difference equations, we can expect that

$$\mathbf{Y}_t = k_i \mathbf{C}_i (r_i)^t \qquad \text{and} \qquad \mathbf{Y}_{t-1} = k_i \mathbf{C}_i (r_i)^{t-1} \tag{19.47}$$

where k_i and r_i are scalars, and $\mathbf{C}_i = (2 \times 1)$ column vector of constants. Substituting in (19.46) when $\mathbf{B} = 0$, we have

$$k_i \mathbf{C}_i (r_i)^t = \mathbf{A} k_i \mathbf{C}_i (r_i)^{t-1}$$

Canceling the common k_i terms and rearranging,

$$\mathbf{A}\mathbf{C}_i (r_i)^{t-1} - \mathbf{C}_i (r_i)^t = 0$$

Evaluated at $t = 1$,

$$(\mathbf{A} - r_i \mathbf{I})\mathbf{C}_i = 0 \qquad \text{Q.E.D.}$$

19.18. Given

$$\begin{bmatrix} a_{11} & a_{12} \\ a_{21} & a_{22} \end{bmatrix}\begin{bmatrix} x_t \\ y_t \end{bmatrix} = \begin{bmatrix} a_{13} & a_{14} \\ a_{23} & a_{24} \end{bmatrix}\begin{bmatrix} x_{t-1} \\ y_{t-1} \end{bmatrix} + \begin{bmatrix} b_1 \\ b_2 \end{bmatrix}$$
$$\mathbf{A}_1 \mathbf{Y}_t = \mathbf{A}_2 \mathbf{Y}_{t-1} + \mathbf{B} \tag{19.48}$$

show in terms of Section 19.4 and Example 9 that for a system of simultaneous first-order linear difference equations when one or more differences is a function of another difference the eigenvalue problem is

$$(\mathbf{A}_2 - r_i \mathbf{A}_1)\mathbf{C}_i = 0$$

From earlier work, assuming distinct real roots, we can anticipate

$$\mathbf{Y}_t = k_i \mathbf{C}_i (r_i)^t \qquad \text{and} \qquad \mathbf{Y}_{t-1} = k_i \mathbf{C}_i (r_i)^{t-1} \qquad (19.49)$$

Substituting in the homogeneous form of (19.48) where $\mathbf{B} = 0$, we have

$$\mathbf{A}_1 k_i \mathbf{C}_i (r_i)^t = \mathbf{A}_2 k_i \mathbf{C}_i (r_i)^{t-1}$$
$$\mathbf{A}_2 \mathbf{C}_i (r_i)^{t-1} - \mathbf{A}_1 \mathbf{C}_i (r_i)^t = 0$$

Evaluated at $t = 1$,

$$(\mathbf{A}_2 - r_i \mathbf{A}_1)\mathbf{C}_i = 0 \qquad \text{Q.E.D.}$$

19.19. Remaining with the same model as in Problem 19.18, show that the particular solution is

$$y_p = \bar{\mathbf{Y}} = (\mathbf{A}_1 - \mathbf{A}_2)^{-1}\mathbf{B} \qquad (19.50)$$

For the particular or steady-state solution,

$$x_t = x_{t-1} = \bar{x} \qquad \text{and} \qquad y_t = y_{t-1} = \bar{y}$$

In matrix notation,

$$\mathbf{Y}_t = \mathbf{Y}_{t-1} = \bar{\mathbf{Y}}$$

Substituting in (19.48),

$$\mathbf{A}_1 \bar{\mathbf{Y}} = \mathbf{A}_2 \bar{\mathbf{Y}} + \mathbf{B}$$

Solving for $\bar{\mathbf{Y}}$,

$$\bar{\mathbf{Y}} = (\mathbf{A}_1 - \mathbf{A}_2)^{-1}\mathbf{B} \qquad \text{Q.E.D.}$$

CHAPTER 20

The Calculus of Variations

20.1 DYNAMIC OPTIMIZATION

In the *static* optimization problems studied in Chapters 4 and 5, we sought a *point* or *points* that would maximize or minimize a given function at a particular point or period of time. Given a function $y = y(x)$, the first-order condition for an optimal point x^* is simply $y'(x^*) = 0$. In *dynamic optimization* we seek a *curve* $x^*(t)$ which will maximize or minimize a given integral expression. The integral to be optimized typically defines the area under a curve F which is a function of the independent variable t, the function $x(t)$, and its derivative dx/dt. In brief, assuming a time period from $t_0 = 0$ to $t_1 = T$ and using $\dot{x}$ for the derivative dx/dt, we seek to maximize or minimize

$$\int_0^T F[t, x(t), \dot{x}(t)]\, dt \tag{20.1}$$

where F is assumed continuous for t, $x(t)$, and $\dot{x}(t)$ and to have continuous partial derivatives with respect to x and $\dot{x}$. An integral such as (20.1) which assumes a numerical value for each of the class of functions $x(t)$ is called a *functional*. A curve that maximizes or minimizes the value of a functional is called an *extremal*. Acceptable candidates for an extremal are the class of functions $x(t)$ which are continuously differentiable on the defined interval and which typically satisfy some fixed endpoint conditions. In our work with extremals, we start with the classical approach, called the *calculus of variations*, pioneered by Isaac Newton and James and John Bernoulli toward the end of the seventeenth century.

EXAMPLE 1. A firm wishing to maximize profits π from time $t_0 = 0$ to $t_1 = T$ finds that demand for its product depends on not only the price p of the product but also the rate of change of the price with respect to time dp/dt. By assuming that costs are fixed and that both p and dp/dt are functions of time, and employing $\dot{p}$ for dp/dt, the firm's objective can be expressed mathematically as

$$\max \int_0^T \pi[t, p(t), \dot{p}(t)]\, dt$$

A second firm has found that its total cost C depends on the level of production $x(t)$ and the rate of change of production $dx/dt = \dot{x}$, due to start-up and tapering-off costs. Assuming that the firm wishes to minimize costs and that x and $\dot{x}$ are functions of time, the firm's objective might be written

$$\min \int_{t_0}^{t_1} C[t, x(t), \dot{x}(t)]\, dt$$

subject to $x(t_0) = x_0$ and $x(t_1) = x_1$

These initial and terminal constraints are known as *endpoint conditions*.

20.2 DISTANCE BETWEEN TWO POINTS ON A PLANE

The length S of any nonlinear curve connecting two points on a plane, such as the curve connecting the points (t_0, x_0) and (t_1, x_1) in Fig. 20-1(a), can be approximated mathematically as follows. Subdivide the curve mentally into subintervals, as in Fig. 20-1(b), and recall from the Pythagorean theorem that the square of the length of the hypotenuse of a right triangle equals the sum of the squares of the lengths of the other two sides. Accordingly, the length of an individual subsegment ds is

$$(ds)^2 = (dt)^2 + (dx)^2$$

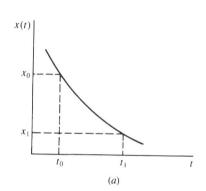

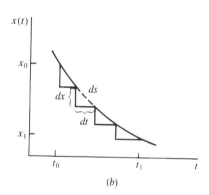

(a)   (b)

Fig. 20-1

Simplifying mathematically,

$$ds = \sqrt{(dt)^2 + (dx)^2}$$

Dividing, then multiplying, both sides by $\sqrt{(dt)^2}$, or simply dt,

$$\frac{ds}{dt} = \sqrt{1 + \left(\frac{dx}{dt}\right)^2}$$

$$ds = \sqrt{1 + \left(\frac{dx}{dt}\right)^2}\, dt$$

or, using the more compact symbol,

$$ds = \sqrt{1 + (\dot{x})^2}\, dt$$

from which the length of the total curve S from t_0 to t_1 can be estimated by simple integration to get

$$S = \int_{t_0}^{t_1} \sqrt{1 + (\dot{x})^2}\, dt$$

See Problems 20.1 to 20.3.

20.3 EULER'S EQUATION: THE NECESSARY CONDITION FOR DYNAMIC OPTIMIZATION

For a curve $X^* = x^*(t)$ connecting points (t_0, x_0) and (t_1, x_1) to be an extremal for (i.e., to optimize) a functional

$$\int_{t_0}^{t_1} F[t, x(t), \dot{x}(t)] \, dt$$

the necessary condition, called *Euler's equation*, is

$$\frac{\partial F}{\partial x} = \frac{d}{dt}\left(\frac{\partial F}{\partial \dot{x}}\right) \qquad (20.2a)$$

Although it is the equivalent of the first-order necessary conditions in static optimization, Euler's equation is actually a second-order differential equation which can perhaps be more easily understood in terms of slightly different notation. Using subscripts to denote partial derivatives and listing the arguments of the derivatives, which are themselves functions, we can express Euler's equation in $(20.2a)$ as

$$F_x(t, x, \dot{x}) = \frac{d}{dt}[F_{\dot{x}}(t, x, \dot{x})] \qquad (20.2b)$$

Then using the chain rule to take the derivative of $F_{\dot{x}}$ with respect to t and omitting the arguments for simplicity, we get

$$F_x = F_{\dot{x}t} + F_{\dot{x}x}(\dot{x}) + F_{\dot{x}\dot{x}}(\ddot{x}) \qquad (20.2c)$$

where $\ddot{x} = d^2x/dt^2$.

An explanation of why Euler's equation is a necessary condition for an extremal in dynamic optimization is offered in Example 2. See also Problems 20.26 to 20.33.

EXAMPLE 2. To show that Euler's equation in $(20.2a)$ is a necessary condition for an extremal, let $X^* = x^*(t)$ be the curve connecting points (t_0, x_0) and (t_1, x_1) in Fig. 20-2 which optimizes the functional (i.e., posits the optimizing function for)

$$\int_{t_0}^{t_1} F[t, x(t), \dot{x}(t)] \, dt \qquad (20.3)$$

Let $\hat{X} = x^*(t) + mh(t)$ be a neighboring curve joining these points, where m is an arbitrary constant and $h(t)$ is an arbitrary function. In order for the curve $\hat{X}$ to also pass through the points (t_0, x_0) and (t_1, x_1), that is, for $\hat{X}$ to also satisfy the endpoint conditions, it is necessary that

$$h(t_0) = 0 \qquad \text{and} \qquad h(t_1) = 0 \qquad (20.4)$$

By holding both $x^*(t)$ and $h(t)$ fixed, the value of the integral becomes a function of m alone and can be written

$$g(m) = \int_{t_0}^{t_1} F[t, x^*(t) + mh(t), \dot{x}^*(t) + m\dot{h}(t)] \, dt \qquad (20.5)$$

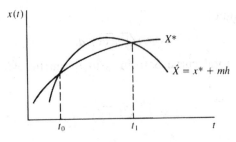

Fig. 20-2

Since $x^*(t)$ by definition optimizes the functional in (20.3), the function $g(m)$ in (20.5) can be optimized only when $m = 0$ and

$$\frac{dg}{dm}\bigg|_{m=0} = 0 \tag{20.6}$$

To differentiate under the integral sign in (20.5), we use *Leibnitz's rule* which states that given

$$g(m) = \int_{t_0}^{t_1} F(t, m)\, dt$$

where t_0 and t_1 are differentiable functions of m,

$$\frac{dg}{dm} = \int_{t_0}^{t_1} \frac{\partial F}{\partial m}\, dt + F(t_1, m)\frac{\partial t_1}{\partial m} - F(t_0, m)\frac{\partial t_0}{\partial m} \tag{20.7}$$

Since the boundaries of integration t_0 and t_1 are fixed in the present example, $\partial t_0/\partial m = \partial t_1/\partial m = 0$, and we have to consider only the first term in Leibnitz's rule. Applying the chain rule to (20.5) to find $\partial F/\partial m$, because F is a function of x and $\dot{x}$, which in turn are functions of m, and substituting in (20.7), we have

$$\frac{dg}{dm} = \int_{t_0}^{t_1} \left[\frac{\partial F}{\partial x}\frac{\partial(x^* + mh)}{\partial m} + \frac{\partial F}{\partial \dot{x}}\frac{\partial(\dot{x}^* + m\dot{h})}{\partial m} \right] dt$$

With $\partial(x^* + mh)/\partial m = h$ and $\partial(\dot{x}^* + m\dot{h})/\partial m = \dot{h}$, and using (20.6),

$$\frac{dg}{dm}\bigg|_{m=0} = \int_{t_0}^{t_1} \left[\frac{\partial F}{\partial x}h(t) + \frac{\partial F}{\partial \dot{x}}\dot{h}(t) \right] dt = 0 \tag{20.8}$$

Leaving the first term in the brackets in (20.8) untouched and integrating the second term by means of parts,

$$\frac{dg}{dm}\bigg|_{m=0} = \int_{t_0}^{t_1} \frac{\partial F}{\partial x}h(t)\, dt + \left[\frac{\partial F}{\partial \dot{x}}h(t) \right]_{t_0}^{t_1} - \int_{t_0}^{t_1} \frac{d}{dt}\left(\frac{\partial F}{\partial \dot{x}} \right)h(t)\, dt = 0$$

With $h(t_0) = h(t_1) = 0$ from (20.4), the second term above drops out. Combining the other two terms and rearranging,

$$\frac{dg}{dm}\bigg|_{m=0} = \int_{t_0}^{t_1} \left[\frac{\partial F}{\partial x} - \frac{d}{dt}\left(\frac{\partial F}{\partial \dot{x}} \right) \right] h(t)\, dt = 0 \tag{20.9}$$

Since $h(t)$ is an arbitrary function that need not equal zero, it follows that a necessary condition for an extremal is that the integrand within the brackets equal zero, namely,

$$\frac{\partial F}{\partial x} - \frac{d}{dt}\left(\frac{\partial F}{\partial \dot{x}} \right) = 0 \qquad \text{or} \qquad \frac{\partial F}{\partial x} = \frac{d}{dt}\left(\frac{\partial F}{\partial \dot{x}} \right)$$

which is Euler's equation. See also Problems 20.26 to 20.33.

20.4 FINDING CANDIDATES FOR EXTREMALS

Finding candidates for extremals to maximize or minimize a given integral subject to fixed endpoint conditions in dynamic optimization problems is facilitated by the following five steps:

1. Let the integrand equal F. Normally $F = F(t, x, \dot{x})$.
2. Take the partial derivatives of F with respect to x and $\dot{x}$ to find $\partial F/\partial x = F_x$ and $\partial F/\partial \dot{x} = F_{\dot{x}}$.
3. Substitute in Euler's equation from (20.2a) or (20.2b).
4. Take the derivative with respect to t of $F_{\dot{x}}$, recalling that the chain rule may be necessary because $F_{\dot{x}}$ can be a function of t, x, and $\dot{x}$, and x and $\dot{x}$ are functions of t.
5. If there are no derivative terms ($\dot{x}$ or $\ddot{x}$), solve immediately for x; if there are $\dot{x}$ or $\ddot{x}$ terms, integrate until all the derivatives are gone and then solve for x.

Illustrations of this technique are provided in Examples 3 and 4 and Problems 20.4 to 20.18.

EXAMPLE 3. Given
$$\int_0^T (6x^2 e^{3t} + 4t\dot{x})\, dt$$
the functional is optimized by using the procedure outlined in Section 20.4 and the notation from (20.2a), as follows:

1. Let
$$F = 6x^2 e^{3t} + 4t\dot{x}$$

2. Then
$$\frac{\partial F}{\partial x} = 12xe^{3t} \quad \text{and} \quad \frac{\partial F}{\partial \dot{x}} = 4t$$

3. Substituting in Euler's equation from (20.2a),
$$12xe^{3t} = \frac{d}{dt}(4t)$$

4. But $d(4t)/dt = 4$. Substituting above,
$$12xe^{3t} = 4$$

5. Solving for x directly since there are no $\dot{x}$ or $\ddot{x}$ terms, and expressing the solution as $x(t)$,
$$x(t) = \tfrac{1}{3}e^{-3t}$$

This satisfies the necessary condition for dynamic optimization, which only makes the solution a *candidate* for an extremal. The sufficiency conditions, which follow in Section 20.5, must also be applied.

EXAMPLE 4. The functional
$$\int_0^2 (4\dot{x}^2 + 12xt - 5t)\, dt$$
subject to
$$x(0) = 1 \qquad x(2) = 4$$
is optimized as above, but now with the notation from (20.2b).

1. Let
$$F = 4\dot{x}^2 + 12xt - 5t$$

2. Then
$$F_x = 12t \quad \text{and} \quad F_{\dot{x}} = 8\dot{x}$$

3. Substituting in Euler's equation from (20.2b),
$$12t = \frac{d}{dt}(8\dot{x})$$

4. Recalling that $\dot{x} = \dfrac{dx}{dt}$ and that $\dfrac{d}{dt}\left(\dfrac{dx}{dt}\right) = \dfrac{d^2x}{dt^2} = \ddot{x}$,
$$12t = 8\ddot{x}$$

5. Since an $\ddot{x}$ term remains, integrate both sides of the equation successively twice, using only one constant of integration term at each step.
$$\int 12t\, dt = \int 8\ddot{x}\, dt$$
$$6t^2 + c_1 = 8\dot{x}$$

Integrating again,
$$\int (6t^2 + c_1)\, dt = \int 8\dot{x}\, dt$$
$$2t^3 + c_1 t + c_2 = 8x$$

Solving for x,

$$x(t) = \frac{1}{4}t^3 + \frac{c_1}{8}t + \frac{c_2}{8}$$

Applying the boundary conditions,

$$x(0) = \frac{c_2}{8} = 1 \qquad c_2 = 8$$

$$x(2) = \frac{1}{4}(2)^3 + \frac{1}{8}(2)c_1 + 1 = 4 \qquad c_1 = 4$$

Substituting, $$x(t) = \frac{1}{4}t^3 + \frac{1}{2}t + 1$$

20.5 THE SUFFICIENCY CONDITIONS FOR THE CALCULUS OF VARIATIONS

Assuming the necessary conditions for an extremal are satisfied,

1. If the functional $F[t, x(t), \dot{x}(t)]$ is jointly concave in $x(t)$, $\dot{x}(t)$, then the necessary conditions are sufficient for a maximum.
2. If the functional $F[t, x(t), \dot{x}(t)]$ is jointly convex in $x(t)$, $\dot{x}(t)$, the necessary conditions are sufficient for a minimum.

Joint concavity and convexity are easily determined in terms of the sign definiteness of the quadratic form of the second derivatives of the functional. Given the discriminant,

$$|\mathbf{D}| = \begin{vmatrix} F_{xx} & F_{x\dot{x}} \\ F_{\dot{x}x} & F_{\dot{x}\dot{x}} \end{vmatrix}$$

1. a) If $|\mathbf{D}_1| = F_{xx} < 0$ and $|\mathbf{D}_2| = |\mathbf{D}| > 0$, $|\mathbf{D}|$ is negative definite and F is strictly concave, making the extremal a global maximum.
   b) If $|\mathbf{D}_1| = F_{xx} \leq 0$ and $|\mathbf{D}_2| = |\mathbf{D}| \geq 0$, when tested for all possible orderings of the variables, $|\mathbf{D}|$ is negative semidefinite and F is simply concave, which is sufficient for a relative maximum.
2. a) If $|\mathbf{D}_1| = F_{xx} > 0$ and $|\mathbf{D}_2| = |\mathbf{D}| > 0$, $|\mathbf{D}|$ is positive definite and F is strictly convex, making the extremal a global minimum.
   b) If $|\mathbf{D}_1| = F_{xx} \geq 0$ and $|\mathbf{D}_2| = |\mathbf{D}| \geq 0$, when tested for all possible orderings of the variables, $|\mathbf{D}|$ is positive semidefinite and F is simply convex, which is sufficient for a relative minimum. See Example 5 and Problems 20.4 to 20.18.

EXAMPLE 5. The sufficiency conditions are illustrated below for Example 3 where the functional was $F = 6x^2 e^{3t} + 4t\dot{x}$, $F_x = 12xe^{3t}$, and $F_{\dot{x}} = 4t$.

$$|\mathbf{D}^1| = \begin{vmatrix} F_{xx} & F_{x\dot{x}} \\ F_{\dot{x}x} & F_{\dot{x}\dot{x}} \end{vmatrix} = \begin{vmatrix} 12e^{3t} & 0 \\ 0 & 0 \end{vmatrix}$$

$$|\mathbf{D}_1^1| = 12e^{3t} > 0 \qquad |\mathbf{D}_2^1| = 0$$

$|\mathbf{D}^1|$ fails to meet the positive definite criteria for a global minimum, but may prove to be positive semidefinite for a relative minimum if the discriminant for the reversed order of variables is also positive semidefinite.

$$|\mathbf{D}^2| = \begin{vmatrix} F_{\dot{x}\dot{x}} & F_{\dot{x}x} \\ F_{x\dot{x}} & F_{xx} \end{vmatrix} = \begin{vmatrix} 0 & 0 \\ 0 & 12e^{3t} \end{vmatrix}$$

$$|\mathbf{D}_1^2| = 0, \qquad |\mathbf{D}_2^2| = 0$$

With $|\mathbf{D}^1| \geq 0$ and $|\mathbf{D}_2| \geq 0$ for both possible orderings of variables, $|\mathbf{D}|$ is positive semidefinite, which is sufficient to establish that the functional is at a relative minimum. The sufficiency conditions for Example 4 test out in a perfectly analogous fashion.

20.6 DYNAMIC OPTIMIZATION SUBJECT TO FUNCTIONAL CONSTRAINTS

To find an extremal that maximizes or minimizes a given integral

$$\int_0^T F[t, x(t), \dot{x}(t)] \, dt \qquad (20.10)$$

under a constraint that keeps the integral

$$\int_0^T G[t, x(t), \dot{x}(t)] \, dt = k \qquad (20.11)$$

where k is a constant, the Lagrangian multiplier method may be used. Multiply the constraint in (20.11) by λ, and add it to the objective function from (20.10) to form the Lagrangian function:

$$\int_0^T (F + \lambda G) \, dt \qquad (20.12)$$

The necessary, but not sufficient, condition to have an extremal for dynamic optimization is the Euler equation

$$\frac{\partial H}{\partial x} = \frac{d}{dt}\left(\frac{\partial H}{\partial \dot{x}}\right) \qquad \text{where} \quad H = F + \lambda G \qquad (20.13)$$

See Example 6 and Problem 20.25.

EXAMPLE 6.   Constrained optimization of functionals is commonly used in problems to determine a curve with a given perimeter that encloses the largest area. Such problems are called *isoperimetric problems* and are usually expressed in the functional notation of $y(x)$ rather than $x(t)$. Adjusting for this notation, to find the curve Y of given length k which encloses a maximum area A, where

$$A = \frac{1}{2}\int (x\dot{y} - y) \, dx$$

and the length of the curve is

$$\int_{x_0}^{x_1} \sqrt{1 + \dot{y}^2} \, dx = k$$

set up the Lagrangian function, as explained in Section 20.6.

$$\int_{x_0}^{x_1} [\tfrac{1}{2}(x\dot{y} - y) + \lambda\sqrt{1 + \dot{y}^2}] \, dx \qquad (20.14)$$

Letting H equal the integrand in (20.14), the Euler equation is

$$\frac{\partial H}{\partial y} = \frac{d}{dx}\left(\frac{\partial H}{\partial \dot{y}}\right)$$

where from (20.14),

$$\frac{\partial H}{\partial y} = -\frac{1}{2} \qquad \text{and} \qquad \frac{\partial H}{\partial \dot{y}} = \frac{1}{2}x + \frac{\lambda\dot{y}}{\sqrt{1 + \dot{y}^2}}$$

Substituting in Euler's equation,

$$-\frac{1}{2} = \frac{d}{dx}\left(\frac{1}{2}x + \frac{\lambda\dot{y}}{\sqrt{1 + \dot{y}^2}}\right)$$

$$-\frac{1}{2} = \frac{1}{2} + \frac{d}{dx}\left(\frac{\lambda\dot{y}}{\sqrt{1 + \dot{y}^2}}\right)$$

$$-1 = \frac{d}{dx}\left(\frac{\lambda\dot{y}}{\sqrt{1 + \dot{y}^2}}\right)$$

Integrating both sides directly and rearranging,

$$\frac{\lambda \dot{y}}{\sqrt{1 + \dot{y}^2}} = -(x - c_1)$$

Squaring both sides of the equation and solving algebraically for $\dot{y}$,

$$\lambda^2 \dot{y}^2 = (x - c_1)^2 (1 + \dot{y}^2)$$
$$\lambda^2 \dot{y}^2 - (x - c_1)^2 \dot{y}^2 = (x - c_1)^2$$
$$\dot{y}^2 = \frac{(x - c_1)^2}{\lambda^2 - (x - c_1)^2}$$
$$\dot{y} = \pm \frac{x - c_1}{\sqrt{\lambda^2 - (x - c_1)^2}}$$

Integrating both sides, using integration by substitution on the right, gives,

$$y - c_2 = \pm \sqrt{\lambda^2 - (x - c_1)^2}$$

which, by squaring both sides and rearranging, can be expressed as a circle

$$(x - c_1)^2 + (y - c_2)^2 = \lambda^2$$

where c_1, c_2, and λ are determined by x_0, x_1, and k.

20.7   VARIATIONAL NOTATION

A special symbol δ is used in the calculus of variations which has properties similar to the differential d in differential calculus.

Given a function $F[t, x(t), \dot{x}(t)]$ and considering t as constant, let

$$\Delta F = F[t, x(t) + mh(t), \dot{x}(t) + m\dot{h}(t)] - F[t, x(t), \dot{x}(t)] \qquad (20.15)$$

where m is an arbitrary constant, $h(t)$ is an arbitrary function as in Example 2, and the arguments are frequently omitted for succinctness. Using the *Taylor expansion* which approximates a function such as $x(t)$ by taking successive derivatives and summing them in ordered sequence to get

$$x(t) = x(t_0) + \dot{x}(t_0)(t - t_0) + \frac{\ddot{x}(t_0)(t - t_0)^2}{2!} + \cdots$$

we have

$$F(t, x + mh, \dot{x} + m\dot{h}) = F(t, x, \dot{x}) + \frac{\partial F}{\partial x} mh + \frac{\partial F}{\partial \dot{x}} m\dot{h} + \cdots \qquad (20.16)$$

Substituting (20.16) in (20.15) and subtracting as indicated,

$$\Delta F = \frac{\partial F}{\partial x} mh + \frac{\partial F}{\partial \dot{x}} m\dot{h} + \cdots \qquad (20.17)$$

where the sum of the first two terms in (20.17) is called the *variation* of F and is denoted by δF. Thus,

$$\delta F = \frac{\partial F}{\partial x} mh + \frac{\partial F}{\partial \dot{x}} m\dot{h} \qquad (20.18)$$

From (20.18) it is readily seen that if $F = x$, by substituting x for F, we have

$$\delta x = mh \qquad (20.19)$$

Similarly, if $F = \dot{x}$

$$\delta\dot{x} = m\dot{h} \qquad (20.20)$$

Hence an alternative expression for (20.18) is

$$\delta F = \frac{\partial F}{\partial x}\delta x + \frac{\partial F}{\partial \dot{x}}\delta\dot{x} \qquad (20.21)$$

and the necessary condition for finding an extremal in dynamic optimization can aslso be expressed as

$$\delta\int_{t_0}^{t_1} F[t, x(t), \dot{x}(t)]\,dt = 0$$

For proof, see Problems 20.34 and 20.35.

20.8 APPLICATIONS TO ECONOMICS

A firm wishes to minimize the present value at discount rate i of an order of N units to be delivered at time t_1. The firm's costs consist of production costs $a[\dot{x}(t)]^2$ and inventory costs $bx(t)$, where a and b are positive constants; $x(t)$ is the accumulated inventory by time t; the rate of change of inventory is the production rate $\dot{x}(t)$, where $\dot{x}(t) \geq 0$; and $a\dot{x}(t)$ is the per unit cost of production. Assuming $x(t_0) = 0$ and the firm wishes to achieve $x(t_1) = N$, in terms of the calculus of variations the firm must

$$\min\int_{t_0}^{t_1} e^{-it}(a\dot{x}^2 + bx)\,dt$$

subject to $x(t_0) = 0 \qquad x(t_1) = N$

To find a candidate for the extremal that will minimize the firm's cost, let

$$F[t, x(t), \dot{x}(t)] = e^{-it}(a\dot{x}^2 + bx)$$

then $F_x = be^{-it} \qquad$ and $\qquad F_{\dot{x}} = 2ae^{-it}\dot{x}$

Substituting in Euler's equation from (20.2b),

$$be^{-it} = \frac{d}{dt}(2ae^{-it}\dot{x})$$

Using the product rule and the chain rule to take the derivative on the right since $\dot{x}$ is a function of t, we have

$$be^{-it} = 2ae^{-it}(\ddot{x}) + \dot{x}(-i2ae^{-it})$$
$$= 2ae^{-it}\ddot{x} - 2aie^{-it}\dot{x}$$

Canceling the e^{-it} terms and rearranging to solve for $\ddot{x}$,

$$\ddot{x}(t) - i\dot{x}(t) = \frac{b}{2a} \qquad (20.22)$$

With $\ddot{x}(t) = i\dot{x}(t) + b/(2a)$ from (20.22) and $\dot{x}(t) \geq 0$ by assumption, $\ddot{x}(t)$ in (20.22) must be positive, indicating that the firm should maintain a strictly increasing rate of production over time.

Equation (20.22) is a second-order linear differential equation which can be solved with the method outlined in Section 18.1. Using Equation (18.2a) to find the particular integral, since in terms of (18.1) $b_1 = -i$, $b_2 = 0$, and $a = b/2a$, and adjusting the functional notation from $y = y(t)$ to $x = x(t)$, we have

$$x_p = \frac{b/2a}{-i}\,t = -\frac{b}{2ai}\,t$$

Using (18.4) to find r_1 and r_2 for the complementary function,

$$r_1, r_2 = \frac{-(-i) \pm \sqrt{(-i)^2 - 4(0)}}{2} = \frac{i \pm i}{2}$$

$$r_1 = i \qquad r_2 = 0$$

Substituting in (18.3) to find x_c and adding to x_p,

$$x(t) = A_1 e^{it} + A_2 - \frac{b}{2ai}\,t \tag{20.23}$$

Letting $t_0 = 0$ and $t_1 = T$, from the boundary conditions we have

$$x(0) = A_1 + A_2 = 0 \qquad A_2 = -A_1$$

$$x(T) = A_1 e^{iT} + (-A_1) - \frac{b}{2ai}\,T = N$$

Solving $x(T)$ for A_1,

$$A_1(e^{iT} - 1) = N + \frac{b}{2ai}\,T$$

$$A_1 = \frac{N + [b/(2ai)]\,T}{e^{iT} - 1} \tag{20.24}$$

Finally, substituting in (20.23), and recalling that $A_2 = -A_1$, we have as a candidate for an extremal:

$$x(t) = \left(\frac{N + [b/(2ai)]\,T}{e^{iT} - 1}\right) e^{it} - \left(\frac{N + [b/(2ai)]\,T}{e^{iT} - 1}\right) - \frac{b}{2ai}\,t$$

$$x(t) = \left(N + \frac{b}{2ai}\,T\right) \frac{e^{it} - 1}{e^{iT} - 1} - \frac{b}{2ai}\,t \qquad 0 \le t \le T$$

Then testing the sufficiency conditions, where $F_x = be^{-it}$ and $F_{\dot{x}} = 2ae^{-it}x$,

$$|\mathbf{D}^1| = \begin{vmatrix} F_{xx} & F_{x\dot{x}} \\ F_{\dot{x}x} & F_{\dot{x}\dot{x}} \end{vmatrix} = \begin{vmatrix} 0 & 0 \\ 0 & 2ae^{-it} \end{vmatrix}$$

$$|\mathbf{D}^1_1| = 0 \qquad |\mathbf{D}^1_2| = 0$$

$$|\mathbf{D}^2| = \begin{vmatrix} F_{\dot{x}\dot{x}} & F_{\dot{x}x} \\ F_{x\dot{x}} & F_{xx} \end{vmatrix} = \begin{vmatrix} 2ae^{-it} & 0 \\ 0 & 0 \end{vmatrix}$$

$$|\mathbf{D}^2_1| = 2ae^{-it} > 0 \qquad |\mathbf{D}^2_2| = 0$$

With the discriminant of the quadratic form of the second-order derivatives of the functional positive semidefinite when tested for both orderings of the variables, the sufficiency condition for a relative minimum is met. For further economic applications, see Problems 20.19 to 20.24.

Solved Problems

DISTANCE BETWEEN TWO POINTS ON A PLANE

20.1. Minimize the length of a curve S connecting the points (t_0, x_0) and (t_1, x_1) in Fig. 20-1 from Section 20.2, i.e.,

$$\min \int_{t_0}^{t_1} \sqrt{1 + \dot{x}^2}\, dt$$

subject to $\qquad\qquad\qquad x(t_0) = x_0 \qquad x(t_1) = x_1$

Using the procedure outlined in Section 20.4 to find a candidate for an extremal to minimize the functional,

1. Let $\qquad\qquad\qquad F = \sqrt{1 + \dot{x}^2} = (1 + \dot{x}^2)^{1/2}$

2. Take the partial derivatives F_x and $F_{\dot{x}}$, noting that there is no x term in F, only an $\dot{x}$ term, and that the chain rule or generalized power function rule is necessary for $F_{\dot{x}}$.

$$F_x = 0 \qquad F_{\dot{x}} = \frac{1}{2}(1 + \dot{x}^2)^{-1/2} \cdot 2\dot{x} = \frac{\dot{x}}{\sqrt{1 + \dot{x}^2}}$$

3. Substitute in Euler's equation.

$$0 = \frac{d}{dt}\left(\frac{\dot{x}}{\sqrt{1 + \dot{x}^2}}\right)$$

4. Since there are no variables on the left-hand side, integrate both sides immediately with respect to t. Integrating the derivative on the right-hand side will produce the original function. With $\int 0\, dt = c$, a constant, we have

$$c = \frac{\dot{x}}{\sqrt{1 + \dot{x}^2}}$$

Squaring both sides and rearranging to solve for $\dot{x}$,

$$c^2(1 + \dot{x}^2) = \dot{x}^2$$
$$c^2 = \dot{x}^2 - c^2\dot{x}^2 = (1 - c^2)\dot{x}^2$$
$$\dot{x} = \sqrt{\frac{c^2}{1 - c^2}} = k_1 \qquad \text{a constant}$$

5. With an $\dot{x}$ term remaining, integrate again to get

$$x(t) = k_1 t + k_2 \qquad\qquad\qquad\qquad (20.25)$$

6. With only one variable $\dot{x}$ in the functional, the sufficiency conditions of concavity or convexity can be determined solely by the sign of the second derivative. From $F_{\dot{x}} = \dot{x}(1 + \dot{x}^2)^{-1/2}$, we have by the product rule,

$$F_{\dot{x}\dot{x}} = (1 + \dot{x}^2)^{-1/2} - \dot{x}^2(1 + \dot{x}^2)^{-3/2}$$
$$F_{\dot{x}\dot{x}} = (1 + \dot{x}^2)^{-3/2}[(1 + \dot{x}^2) - \dot{x}^2]$$
$$F_{\dot{x}\dot{x}} = (1 + \dot{x}^2)^{-3/2} = \frac{1}{\sqrt{(1 + \dot{x}^2)^3}}$$

Since the square root of a distance can never be negative, $F_{\dot{x}\dot{x}} > 0$. The functional is convex and the sufficiency conditions for a minimum are satisfied.

Note the solution in (20.25) is linear, indicating that the shortest distance between two points is a straight line. The parameters k_1 (slope) and k_2 (vertical intercept) are uniquely determined by the boundary conditions, as is illustrated in Problem 20.2.

20.2. Minimize
$$\int_0^2 \sqrt{1 + \dot{x}^2} \, dt$$

subject to
$$x(0) = 3 \qquad x(2) = 8$$

From (20.25),
$$x(t) = k_1 t + k_2$$

Applying the boundary conditions,

$$x(0) = k_1(0) + k_2 = 3 \qquad k_2 = 3$$
$$x(2) = k_1(2) + 3 = 8 \qquad k_1 = 2.5$$

Substituting,

$$x(t) = 2.5t + 3$$

20.3. (*a*) Estimate the distance between the points (t_0, x_0) and (t_1, x_1) from Problem 20.2, using the functional; (*b*) draw a graph and check your answer geometrically.

a)   Given
$$\int_0^2 \sqrt{1 + \dot{x}^2} \, dt \qquad \text{and} \qquad x(t) = 2.5t + 3$$

by taking the derivative $\dot{x}(t) = 2.5$ and substituting, we have

$$\int_0^2 \sqrt{1 + (2.5)^2} \, dt = \int_0^2 \sqrt{7.25} \, dt = \sqrt{7.25}\, t \, \Big|_0^2$$
$$= 2.69258(2) - 2.69258(0) = 5.385$$

b)   Applying the Pythagorean theorem to Fig. 20-3,

$$x^2 = 5^2 + 2^2$$
$$x = \sqrt{29} = 5.385$$

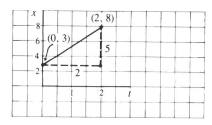

Fig. 20-3

FINDING CANDIDATES FOR EXTREMALS

20.4. Optimize

$$\int_{t_0}^{t_1} (2\dot{x}^2 - 42xt + 11t) \, dt$$

subject to
$$x(t_0) = x_0 \qquad x(t_1) = x_1$$

Using the now familiar six steps to find a candidate for an extremal,

1.
$$F = 2\dot{x}^2 - 42xt + 11t$$

2.
$$F_x = -42t \qquad F_{\dot{x}} = 4\dot{x}$$

3. Substituting in Euler's equation,

$$-42t = \frac{d}{dt}(4\dot{x})$$

$$-42t = 4\ddot{x}$$

4. Integrating both sides to eliminate the $\ddot{x}$ term and using only one constant of integration term at each step throughout,

$$-21t^2 + c_1 = 4\dot{x}$$

Integrating again to eliminate the $\dot{x}$ term,

$$-7t^3 + c_1 t + c_2 = 4x$$

5. Solving for x,

$$x(t) = -1.75t^3 + 0.25c_1 t + 0.25c_2$$

6. Testing the sufficiency conditions,

$$|\mathbf{D}^1| = \begin{vmatrix} F_{xx} & F_{x\dot{x}} \\ F_{\dot{x}x} & F_{\dot{x}\dot{x}} \end{vmatrix} = \begin{vmatrix} 0 & 0 \\ 0 & 4 \end{vmatrix}$$

$$|\mathbf{D}_1^1| = 0 \qquad |\mathbf{D}_2^1| = 0$$

$$|\mathbf{D}^2| = \begin{vmatrix} F_{\dot{x}\dot{x}} & F_{\dot{x}x} \\ F_{x\dot{x}} & F_{xx} \end{vmatrix} = \begin{vmatrix} 4 & 0 \\ 0 & 0 \end{vmatrix}$$

$$|\mathbf{D}_1^2| = 4 > 0 \qquad |\mathbf{D}_2^2| = 0$$

The discriminant of the second-order derivatives of the functional is positive semidefinite when tested for both orderings of the variables, which satisfies the sufficiency condition for a relative minimum.

20.5. Optimize

$$\int_{t_0}^{t_1} (\dot{x}^2 + 60t^3 x)\, dt$$

subject to $\qquad\qquad x(t_0) = x_0 \qquad x(t_1) = x_1$

1. $$F = \dot{x}^2 + 60t^3 x$$

2. $$F_x = 60t^3 \qquad F_{\dot{x}} = 2\dot{x}$$

3. Substituting in Euler's equation,

$$60t^3 = \frac{d}{dt}(2\dot{x})$$

$$60t^3 = 2\ddot{x}$$

4. Integrating both sides and combining constants of integration for each step,

$$15t^4 + c_1 = 2\dot{x}$$

Integrating again and solving for x,

$$3t^5 + c_1 t + c_2 = 2x$$

5. $$x(t) = 1.5t^5 + 0.5c_1 t + 0.5c_2$$

6. The sufficiency conditions, when tested as in step 6 of the previous problem, reveal the functional is at a relative minimum.

20.6. Optimize

$$\int_0^1 (13t - 3\dot{x}^2 + 36xt)\, dt$$

subject to $x(0) = 2 \qquad x(1) = 4$

1. $F = 13t - 3\dot{x}^2 + 36xt$

2. $F_x = 36t \qquad F_{\dot{x}} = -6\dot{x}$

3.
$$36t = \frac{d}{dt}(-6\dot{x})$$

$$36t = -6\ddot{x}$$

4.   Integrating both sides twice but using only one constant of integration each time,

$$18t^2 + c_1 = -6\dot{x}$$
$$6t^3 + c_1 t + c_2 = -6x$$

5. $x(t) = -t^3 - \dfrac{c_1}{6}t - \dfrac{c_2}{6}$

Applying the initial conditions,

$$x(0) = -\frac{c_2}{6} = 2 \qquad\qquad c_2 = -12$$

$$x(1) = -1 - \frac{c_1}{6} + 2 = 4 \qquad c_1 = -18$$

Then substituting above,

$$x(t) = -t^3 + 3t + 2$$

6.   Finally, testing the sufficiency conditions,

$$|\mathbf{D}^1| = \begin{vmatrix} F_{xx} & F_{x\dot{x}} \\ F_{\dot{x}x} & F_{\dot{x}\dot{x}} \end{vmatrix} = \begin{vmatrix} 0 & 0 \\ 0 & -6 \end{vmatrix}$$

$$|\mathbf{D}_1^1| = 0 \qquad |\mathbf{D}_2^1| = 0$$

$$|\mathbf{D}^2| = \begin{vmatrix} F_{\dot{x}\dot{x}} & F_{\dot{x}x} \\ F_{x\dot{x}} & F_{xx} \end{vmatrix} = \begin{vmatrix} -6 & 0 \\ 0 & 0 \end{vmatrix}$$

$$|\mathbf{D}_1^2| = -6 < 0 \qquad |\mathbf{D}_2^2| = 0$$

When tested for both orderings of the variables, the discriminant of the second-order derivatives is negative semidefinite. This fulfills the sufficiency conditions for a relative maximum.

20.7. Optimize

$$\int_{t_0}^{t_1} (3x^2 e^{5t} + 4t^3 \dot{x})\, dt$$

subject to $x(t_0) = x_0 \qquad x(t_1) = x_1$

1. $F = 3x^2 e^{5t} + 4t^3 \dot{x}$

2. $F_x = 6xe^{5t} \qquad F_{\dot{x}} = 4t^3$

3. Substituting in Euler's equation,

$$6xe^{5t} = \frac{d}{dt}(4t^3)$$

4.

$$6xe^{5t} = 12t^2$$

5. With no $\dot{x}$ terms left, simply solve for x algebraically.

$$x(t) = 2t^2 e^{-5t}$$

6. Checking the sufficiency conditions,

$$|\mathbf{D}^1| = \begin{vmatrix} F_{xx} & F_{x\dot{x}} \\ F_{\dot{x}x} & F_{\dot{x}\dot{x}} \end{vmatrix} = \begin{vmatrix} 6e^{5t} & 0 \\ 0 & 0 \end{vmatrix} \qquad |\mathbf{D}^2| = \begin{vmatrix} F_{\dot{x}\dot{x}} & F_{\dot{x}x} \\ F_{x\dot{x}} & F_{xx} \end{vmatrix} = \begin{vmatrix} 0 & 0 \\ 0 & 6e^{5t} \end{vmatrix}$$

$$|\mathbf{D}_1^1| = 6e^{5t} > 0 \qquad |\mathbf{D}_2^1| = 0 \qquad |\mathbf{D}_1^2| = 0 \qquad |\mathbf{D}_2^2| = 0$$

For both orderings of the variables, the discriminant of the second-order derivatives is positive semidefinite, fulfilling the sufficiency condition for a relative minimum.

20.8. Optimize

$$\int_{t_0}^{t_1} \frac{3\dot{x}^2}{8t^3} dt$$

subject to

$$x(t_0) = x_0 \qquad x(t_1) = x_1$$

1.

$$F = \frac{3\dot{x}^2}{8t^3}$$

2.

$$F_x = 0 \qquad F_{\dot{x}} = \frac{6\dot{x}}{8t^3} = \frac{3\dot{x}}{4t^3}$$

3. Substituting in Euler's equation,

$$0 = \frac{d}{dt}\left(\frac{3\dot{x}}{4t^3}\right)$$

4. With $F_x = 0$, integrate immediately.

$$c_1 = \frac{3\dot{x}}{4t^3}$$

$$3\dot{x} = 4c_1 t^3$$

With an $\dot{x}$ still remaining, integrate again.

$$3x = c_1 t^4 + c_2$$

5.

$$x(t) = \frac{c_1}{3}t^4 + \frac{c_2}{3}$$

$$= k_1 t^4 + k_2 \qquad \text{where} \quad k_1 = \frac{c_1}{3} \qquad k_2 = \frac{c_2}{3}$$

6. When tested, as above, the sufficiency conditions reveal a relative minimum.

20.9. Optimize

$$\int_{t_0}^{t_1} (5t^2 \dot{x} - 4x^{2e^{-0.7t}}) dt$$

subject to $\qquad\qquad\qquad x(t_0) = x_0 \qquad x(t_1) = x_1$

1. $\qquad\qquad\qquad\qquad F = 5t^2\dot{x} - 4x^{2e^{-0.7t}}$

2. $\qquad\qquad F_x = -8xe^{-0.7t} \qquad F_{\dot{x}} = 5t^2$

3. $\qquad\qquad\qquad -8xe^{-0.7t} = \dfrac{d}{dt}(5t^2) = 10t$

$\qquad\qquad\qquad\qquad -8xe^{-0.7t} = 10t$

4. With no derivative term remaining and no need to integrate, we solve directly for x.

5. $\qquad\qquad\qquad\qquad x(t) = -1.25te^{0.7t}$

6. Testing the sufficiency conditions,

$$|\mathbf{D}^1| = \begin{vmatrix} F_{xx} & F_{x\dot{x}} \\ F_{\dot{x}x} & F_{\dot{x}\dot{x}} \end{vmatrix} = \begin{vmatrix} -8e^{-0.7t} & 0 \\ 0 & 0 \end{vmatrix} \qquad |\mathbf{D}^2| = \begin{vmatrix} F_{\dot{x}\dot{x}} & F_{\dot{x}x} \\ F_{x\dot{x}} & F_{xx} \end{vmatrix} = \begin{vmatrix} 0 & 0 \\ 0 & -8e^{-0.7t} \end{vmatrix}$$

$$|\mathbf{D}_1^1| = -8e^{-0.7t} < 0 \qquad |\mathbf{D}_2^1| = 0 \qquad\qquad |\mathbf{D}_1^2| = 0 \qquad |\mathbf{D}_2^2| = 0$$

With both orderings of the variables, $|\mathbf{D}|$ is negative semidefinite, making F concave and fulfilling the sufficiency condition for a relative maximum.

20.10. Optimize

$$\int_{t_0}^{t_1} (7t^2 + 2\dot{x}^2 t)\, dt$$

subject to $\qquad\qquad\qquad x(t_0) = x_0 \qquad x(t_1) = x_1$

1. $\qquad\qquad\qquad\qquad F = 7t^2 + 2\dot{x}^2 t$

2. $\qquad\qquad\qquad F_x = 0 \qquad F_{\dot{x}} = 4\dot{x}t$

3. $\qquad\qquad\qquad\qquad 0 = \dfrac{d}{dt}(4\dot{x}t)$

4. Integrating immediately,

$$c_1 = 4\dot{x}t \qquad \dot{x} = \dfrac{c_1}{4t}$$

Integrating again,

$$x = \dfrac{c_1}{4}\ln t + c_2$$

5. $\qquad\qquad x(t) = k_1\ln t + k_2 \qquad \text{where} \quad k_1 = \dfrac{c_1}{4} \qquad k_2 = c_2$

6. The sufficiency conditions for a relative minimum are satisfied.

20.11. Optimize

$$\int_{t_0}^{t_1} (15x^2 - 132x + 19x\dot{x} + 12\dot{x}^2)\, dt$$

subject to $\qquad\qquad\qquad x(t_0) = x_0 \qquad x(t_1) = x_1$

1. $\qquad\qquad\qquad F = 15x^2 - 132x + 19x\dot{x} + 12\dot{x}^2$

2.
$$F_x = 30x - 132 + 19\dot{x} \qquad F_{\dot{x}} = 19x + 24\dot{x}$$

3.
$$30x - 132 + 19\dot{x} = \frac{d}{dt}(19x + 24\dot{x})$$

4.
$$30x - 132 + 19\dot{x} = 19\dot{x} + 24\ddot{x}$$

Solving algebraically,

$$\ddot{x} - 1.25x = -5.5 \qquad\qquad (20.26)$$

5. Equation (20.26) is a second-order differential equation which can be solved with the techniques of Section 18.1. Using Equation (18.2) to find the particular integral x_p, where $b_1 = 0$, $b_2 = -1.25$, and $a = -5.5$,

$$x_p = \frac{a}{b_2} = \frac{-5.5}{-1.25} = 4.4$$

Then using (18.4) to find the characteristic roots,

$$r_1, r_2 = \frac{0 \pm \sqrt{0 - 4(-1.25)}}{2} = \pm\sqrt{1.25}$$

and substituting in (18.3) to find the complementary function x_c,

$$x_c = A_1 e^{\sqrt{1.25}t} + A_2 e^{-\sqrt{1.25}t}$$

Finally, by adding x_c and x_p, we have
$$x(t) = A_1 e^{\sqrt{1.25}t} + A_2 e^{-\sqrt{1.25}t} + 4.4$$

6. Checking the sufficiency conditions,

$$|\mathbf{D}^1| = \begin{vmatrix} F_{xx} & F_{x\dot{x}} \\ F_{\dot{x}x} & F_{\dot{x}\dot{x}} \end{vmatrix} = \begin{vmatrix} 30 & 19 \\ 19 & 24 \end{vmatrix}$$

$$|\mathbf{D}_1^1| = 30 > 0 \qquad\qquad |\mathbf{D}_2^1| = 359 > 0$$

With $|\mathbf{D}_1| > 0$ and $|\mathbf{D}_2| > 0$, $|\mathbf{D}|$ is positive definite, which means F is strictly convex and we have a global minimum. There is no need to test in reverse order with $|\mathbf{D}^2|$.

20.12. Optimize

$$\int_0^1 (-16x^2 + 144x + 11x\dot{x} - 4\dot{x}^2)\,dt$$

subject to
$$x(0) = 8 \qquad x(1) = 8.6$$

1.
$$F = -16x^2 + 144x + 11x\dot{x} - 4\dot{x}^2$$

2.
$$F_x = -32x + 144 + 11\dot{x} \qquad F_{\dot{x}} = 11x - 8\dot{x}$$

3.
$$-32x + 144 + 11\dot{x} = \frac{d}{dt}(11x - 8\dot{x})$$

$$-32x + 144 + 11\dot{x} = 11\dot{x} - 8\ddot{x}$$

Simplifying and rearranging to conform with (18.1),

$$\ddot{x} - 4x = -18$$

4. From (18.2), the particular integral is

$$x_p = \frac{-18}{-4} = 4.5$$

From (18.4), the characteristic roots are

$$r_1, r_2 = \frac{0 \pm \sqrt{0 - 4(-4)}}{2}$$

$$r_1 = 2 \qquad r_2 = -2$$

Thus, $$x(t) = A_1 e^{2t} + A_2 e^{-2t} + 4.5$$

Applying the initial conditions, $x(0) = 8$, $x(1) = 8.6$,

$$x(0) = A_1 + A_2 + 4.5 = 8$$
$$x(1) = 7.3891 A_1 + 0.1353 A_2 + 4.5 = 8.6$$

Solving simultaneously, $$A_1 = 0.5, \ A_2 = 3$$

5. Substituting, $$x(t) = 0.5 e^{2t} + 3 e^{-2t} + 4.5$$

6. Applying the sufficiency conditions,

$$|\mathbf{D}^1| = \begin{vmatrix} F_{xx} & F_{x\dot{x}} \\ F_{\dot{x}x} & F_{\dot{x}\dot{x}} \end{vmatrix} = \begin{vmatrix} -32 & 11 \\ 11 & -8 \end{vmatrix}$$

$$|\mathbf{D}_1^1| = -32 < 0 \qquad |\mathbf{D}_2^1| = 135 > 0$$

Since $|\mathbf{D}_1| < 0$ and $|\mathbf{D}_2| > 0$, $|\mathbf{D}|$ is negative definite. F is strictly concave and there is a global maximum. Hence $|\mathbf{D}^2|$ need not be tested.

20.13. Optimize

$$\int_{t_0}^{t_1} (16x^2 + 9x\dot{x} + 8\dot{x}^2) \, dt$$

subject to $$x(t_0) = x_0 \qquad x(t_1) = x_1$$

1. $$F = 16x^2 + 9x\dot{x} + 8\dot{x}^2$$

2. $$F_x = 32x + 9\dot{x} \qquad F_{\dot{x}} = 9x + 16\dot{x}$$

3. $$32x + 9\dot{x} = \frac{d}{dt}(9x + 16\dot{x})$$

4. $$32x + 9\dot{x} = 9\dot{x} + 16\ddot{x}$$
$$\ddot{x} - 2x = 0$$

5. Using (18.2) where $b_1 = 0$, $b_2 = -2$, and $a = 0$ to find x_p,

$$x_p = \frac{a}{b_2} = \frac{0}{-2} = 0$$

Using (18.4) to find r_1 and r_2,

$$r_1, r_2 = \frac{\pm \sqrt{-4(-2)}}{2} = \pm \sqrt{2}$$

and substituting in (18.3) to find x_c,

$$x_c = A_1 e^{\sqrt{2}t} + A_2 e^{-\sqrt{2}t}$$

Then since $x_p = 0$,

$$x(t) = x_c = A_1 e^{\sqrt{2}t} + A_2 e^{-\sqrt{2}t}$$

6. Sufficiency conditions reveal a global minimum.

20.14. Optimize

$$\int_{t_0}^{t_1} (7\dot{x}^2 + 4x\dot{x} - 63x^2)\,dt$$

subject to $x(t_0) = x_0 \qquad x(t_1) = x_1$

1. $F = 7\dot{x}^2 + 4x\dot{x} - 63x^2$

2. $F_x = 4\dot{x} - 126x \qquad F_{\dot{x}} = 14\dot{x} + 4x$

3. $4\dot{x} - 126x = \dfrac{d}{dt}(14\dot{x} + 4x)$

4. $4\dot{x} - 126x = 14\ddot{x} + 4\dot{x}$

$$\ddot{x} + 9x = 0$$

where in terms of (18.1), $b_1 = 0$, $b_2 = 9$, and $a = 0$.

5. Using (18.2) for x_p, $x_p = \dfrac{0}{9} = 0$

Using (18.19) since $b_1^2 < 4b_2$, $r_1, r_2 = g \pm hi$

where $g = -\tfrac{1}{2}b_1 = -\tfrac{1}{2}(0) = 0 \qquad h = \tfrac{1}{2}\sqrt{4b_2 - b_1^2} = \tfrac{1}{2}\sqrt{36} = 3$

and $r_1, r_2 = 0 \pm 3i = \pm 3i$

Substituting in (18.26),

$$x_c = B_1 \cos 3t + B_2 \sin 3t$$

With $x_p = 0$,

$$x(t) = B_1 \cos 3t + B_2 \sin 3t$$

6. Sufficiency conditions indicate a global minimum.

20.15. Optimize

$$\int_{t_0}^{t_1} (5x^2 + 27x - 8x\dot{x} - \dot{x}^2)\,dt$$

subject to $x(t_0) = x_0 \qquad x(t_1) = x_1$

1. $F = 5x^2 + 27x - 8x\dot{x} - \dot{x}^2$

2. $F_x = 10x + 27 - 8\dot{x} \qquad F_{\dot{x}} = -8x - 2\dot{x}$

3. $10x + 27 - 8\dot{x} = \dfrac{d}{dt}(-8x - 2\dot{x})$

4. $10x + 27 - 8\dot{x} = -8\dot{x} - 2\ddot{x}$

$$\ddot{x} + 5x = -13.5$$

5. Using (18.2), $x_p = \dfrac{-13.5}{5} = -2.7$

Using (18.19), $g = -\tfrac{1}{2}(0) = 0 \qquad h = \tfrac{1}{2}\sqrt{4(5)} = \sqrt{5}$

$$r_1, r_2 = 0 \pm \sqrt{5}i = \pm \sqrt{5}i$$

Substituting in (18.26),

$$x_c = B_1 \cos \sqrt{5}t + B_2 \sin \sqrt{5}t$$

and $x(t) = B_1 \cos \sqrt{5}t + B_2 \sin \sqrt{5}t - 2.7$

6. For the second-order conditions,

$$|\mathbf{D}^1| = \begin{vmatrix} F_{xx} & F_{x\dot{x}} \\ F_{\dot{x}x} & F_{\dot{x}\dot{x}} \end{vmatrix} = \begin{vmatrix} 10 & -8 \\ -8 & -2 \end{vmatrix}$$

$$|\mathbf{D}_1^1| = 10 > 0 \qquad |\mathbf{D}_2^1| = -84 < 0$$

With $|\mathbf{D}_2| < 0$, $|\mathbf{D}|$ fails the test both for concavity and convexity. F is neither maximized nor minimized. It is at a saddle point.

20.16. Optimize

$$\int_{t_0}^{t_1} e^{0.12t}(5\dot{x}^2 - 18x)\, dt$$

subject to
$$x(t_0) = x_0 \qquad x(t_1) = x_1$$

1.
$$F = e^{0.12t}(5\dot{x}^2 - 18x)$$

2.
$$F_x = -18e^{0.12t} \qquad F_{\dot{x}} = 10\dot{x}e^{0.12t}$$

3.
$$-18e^{0.12t} = \frac{d}{dt}(10\dot{x}e^{0.12t})$$

4. Using the product rule,

$$-18e^{0.12t} = 10\dot{x}(0.12e^{0.12t}) + e^{0.12t}(10\ddot{x})$$

Canceling the $e^{0.12t}$ terms and rearranging algebraically,

$$\ddot{x} + 0.12\dot{x} = -1.8$$

5. Using (*18.2a*) and (*18.4*),

$$x_p = \left(\frac{-1.8}{0.12}\right)t = -15t$$

$$r_1, r_2 = \frac{-0.12 \pm \sqrt{(0.12^2 - 0)}}{2} = -0.12, 0$$

$$x_c = A_1 e^{-0.12t} + A_2$$

and
$$x(t) = A_1 e^{-0.12t} + A_2 - 15t$$

6. Checking the sufficiency conditions,

$$|\mathbf{D}^1| = \begin{vmatrix} F_{xx} & F_{x\dot{x}} \\ F_{\dot{x}x} & F_{\dot{x}\dot{x}} \end{vmatrix} = \begin{vmatrix} 0 & 0 \\ 0 & 10e^{0.12t} \end{vmatrix} \qquad\qquad |\mathbf{D}^2| = \begin{vmatrix} F_{\dot{x}\dot{x}} & F_{\dot{x}x} \\ F_{x\dot{x}} & F_{xx} \end{vmatrix} = \begin{vmatrix} 10e^{0.12t} & 0 \\ 0 & 0 \end{vmatrix}$$

$$|\mathbf{D}_1^1| = 0 \qquad |\mathbf{D}_2^1| = 0 \qquad\qquad\qquad\qquad |\mathbf{D}_1^2| = 10e^{0.12t} > 0 \qquad |\mathbf{D}_2^2| = 0$$

$|\mathbf{D}|$ is positive semidefinite, F is convex, and we have a relative minimum.

20.17. Optimize

$$\int_{t_0}^{t_1} e^{-0.05t}(4\dot{x}^2 + 15x)\, dt$$

subject to
$$x(t_0) = x_0 \qquad x(t_1) = x_1$$

1.
$$F = e^{-0.05t}(4\dot{x}^2 + 15x)$$

2.
$$F_x = 15e^{-0.05t} \qquad F_{\dot{x}} = 8\dot{x}e^{-0.05t}$$

3.
$$15e^{-0.05t} = \frac{d}{dt}(8\dot{x}e^{-0.05t})$$

4.
$$15e^{-0.05t} = 8\dot{x}(-0.05e^{-0.05t}) + e^{-0.05t}(8\ddot{x})$$

Canceling the $e^{-0.05t}$ terms and rearranging,

$$\ddot{x} - 0.05\dot{x} = 1.875$$

5. Using ($18.2a$) and (18.4),

$$x_p = \left(\frac{1.875}{-0.05}\right)t = -37.5t$$

$$r_1, r_2 = \frac{-(-0.05) \pm \sqrt{(-0.05)^2 - 0}}{2} = 0.05, 0$$

$$x_c = A_1 e^{0.05t} + A_2$$

and
$$x(t) = A_1 e^{0.05t} + A_2 - 37.5t$$

6. The sufficiency conditions indicate a relative minimum.

20.18. Find the curve connecting (t_0, x_0) and (t_1, x_1) which will generate the surface of minimal area when revolved around the t axis, as in Fig. 20.4. That is,

$$\text{Minimize} \quad 2\pi \int_{t_0}^{t_1} x(1 + \dot{x}^2)^{1/2}\, dt$$

subject to
$$x(t_0) = x_0 \qquad x(t_1) = x_1$$

1.
$$F = x(1 + \dot{x}^2)^{1/2}$$

2. Using the chain rule for $F_{\dot{x}}$,

$$F_x = (1 + \dot{x}^2)^{1/2} \qquad F_{\dot{x}} = x\dot{x}(1 + \dot{x}^2)^{-1/2}$$

3.
$$(1 + \dot{x}^2)^{1/2} = \frac{d}{dt}[x\dot{x}(1 + \dot{x}^2)^{-1/2}]$$

4. Using the product rule and the chain rule,

$$(1 + \dot{x}^2)^{1/2} = x\dot{x}[-\tfrac{1}{2}(1 + \dot{x}^2)^{-3/2} \cdot 2\dot{x}\ddot{x}] + (1 + \dot{x}^2)^{-1/2}(x\ddot{x} + \dot{x}\dot{x})$$
$$= -x\dot{x}^2\ddot{x}(1 + \dot{x}^2)^{-3/2} + (x\ddot{x} + \dot{x}^2)(1 + \dot{x}^2)^{-1/2}$$

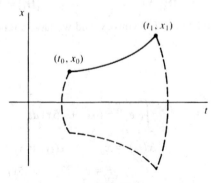

Fig. 20-4

Multiplying both sides by $(1 + \dot{x}^2)^{3/2}$,

$$(1 + \dot{x}^2)^2 = -x\dot{x}^2\ddot{x} + (x\ddot{x} + \dot{x}^2)(1 + \dot{x}^2)$$
$$1 + 2\dot{x}^2 + \dot{x}^4 = -x\dot{x}^2\ddot{x} + x\ddot{x} + \dot{x}^2 + x\dot{x}^2\ddot{x} + \dot{x}^4$$
$$\ddot{x} - \frac{\dot{x}^2}{x} = \frac{1}{x} \tag{20.27}$$

5.   Let $\dot{x} = u = dx/dt$, then

$$\ddot{x} = \dot{u} = \frac{du}{dt} = \frac{du}{dx} \cdot \frac{dx}{dt} = \frac{du}{dx} \cdot u = u\frac{du}{dx}$$

Substituting in (20.27),

$$u\frac{du}{dx} - \frac{u^2}{x} = \frac{1}{x}$$

Separating the variables and integrating,

$$u\frac{du}{dx} = \frac{1}{x}(1 + u^2)$$

$$\int \frac{u}{1 + u^2}\,du = \int \frac{1}{x}\,dx$$

$$\tfrac{1}{2}\ln(1 + u^2) + c_1 = \ln x$$

Solving for u,

$$e^{\ln\sqrt{1 + u^2} + c_1} = e^{\ln x}$$
$$c_2\sqrt{1 + u^2} = x$$

where $c_2 = e^{c_1}$. Squaring both sides and rearranging algebraically,

$$1 + u^2 = \frac{x^2}{c_2^2}$$

$$u = \frac{\sqrt{x^2 - c_2^2}}{c_2} = \frac{dx}{dt}$$

Separating variables again and integrating,

$$\int \frac{dx}{\sqrt{x^2 - c_2^2}} = \int \frac{dt}{c_2} \tag{20.28}$$

Using integral tables for the left-hand side,

$$\ln\left(x + \sqrt{x^2 - c_2^2}\right) = \frac{t + c_3}{c_2} \tag{20.29}$$

or by applying trigonometric substitution directly to (20.28),

$$\cosh^{-1}\frac{x}{c_2} = \frac{t + c_3}{c_2}$$

$$x(t) = c_2 \cosh\frac{t + c_3}{c_2} \tag{20.30}$$

The curve in (20.30) is called a *catenary* from the Latin word for chain because it depicts the shape a chain would assume if hung from points (t_0, x_0) and (t_1, x_1). The constants c_2 and c_3 can be found from either (20.29) or (20.30) by using the initial conditions $x(t_0) = x_0$ and $x(t_1) = x_1$.

ECONOMIC APPLICATIONS

20.19. The demand for a monopolist's product in terms of the number of units $x(t)$ she or he can sell depends on both the price $p(t)$ of the good and the rate of change of price $\dot{p}(t)$:

$$x(t) = ap(t) + b\dot{p}(t) + c \qquad (20.31)$$

Production costs $z(x)$ at rate of production x are

$$z(x) = mx^2 + nx + k \qquad (20.32)$$

Assuming $p(0) = p_0$ and the desired price at time T is $p(T) = p_1$, find the pricing policy to maximize profits over $0 \le t \le T$. That is,

$$\text{Maximize} \quad \int_0^T [p(t)x(t) - z(x)]\, dt$$

Substituting from (20.31) and (20.32),

$$\int_0^T [p(t)x(t) - z(x)]\, dt = \int_0^T [p(ap + b\dot{p} + c) - (mx^2 + nx + k)]\, dt$$

Substituting from (20.31) again,

$$
\begin{aligned}
\int_0^T [p(t)x(t) - z(x)]\, dt &= \int_0^T [ap^2 + bp\dot{p} + cp - m(ap + b\dot{p} + c)^2 - n(ap + b\dot{p} + c) - k]\, dt \\
&= \int_0^T (ap^2 + bp\dot{p} + cp - ma^2p^2 - mabp\dot{p} - macp - mabp\dot{p} - mb^2\dot{p}^2 \\
&\quad - mbc\dot{p} - macp - mbc\dot{p} - mc^2 - nap - nb\dot{p} - nc - k)\, dt \\
&= \int_0^T [a(1 - ma)p^2 + (c - 2mac - na)p + (b - 2mab)p\dot{p} \\
&\quad - (2mbc + nb)\dot{p} - mb^2\dot{p}^2 - mc^2 - nc - k]\, dt \qquad (20.33)
\end{aligned}
$$

Letting $F = $ the integrand in (20.33),

$$F_p = 2a(1 - ma)p + c - 2mac - na + (b - 2mab)\dot{p}$$

and

$$F_{\dot{p}} = (b - 2mab)p - 2mbc - nb - 2mb^2\dot{p}$$

Using the Euler equation,

$$2a(1 - ma)p + c - 2mac - na + (b - 2mab)\dot{p} = \frac{d}{dt}[(b - 2mab)p - 2mbc - nb - 2mb^2\dot{p}]$$

$$2a(1 - ma)p + c - 2mac - na + (b - 2mab)\dot{p} = (b - 2mab)\dot{p} - 2mb^2\ddot{p}$$

Rearranging algebraically,

$$2mb^2\ddot{p} + 2a(1 - ma)p = 2mac + na - c$$

$$\ddot{p} + \left[\frac{a(1 - ma)}{mb^2}\right]p = \frac{2mac + na - c}{2mb^2}$$

Using (18.2) and (18.4),

$$p_p = \frac{(2mac + na - c)/(2mb^2)}{a(1 - ma)(mb^2)} = \frac{2mac + na - c}{2a(1 - ma)}$$

$$r_1, r_2 = \frac{0 \pm \sqrt{0 - 4a(1 - ma)/(mb^2)}}{2} = \pm\sqrt{\frac{a(ma - 1)}{mb^2}}$$

Thus,

$$p(t) = A_1 \exp\left(\sqrt{\frac{a(ma-1)}{mb^2}}\right)t + A_2 \exp -\left(\sqrt{\frac{a(ma-1)}{mb^2}}\right)t + \frac{2mac + na - c}{2a(1-ma)}$$

If $a < 0$ and $m > 0$, as one would expect from economic theory, independently of the sign for b, $\sqrt{a(ma-1)/(mb^2)} > 0$ and the time path $p(t)$ will have distinct real roots of equal magnitude and opposite signs.

    Note: This is a classic case that has appeared in one form or another in the economic literature over the decades.

20.20. Maximize the stream of instantaneous utility $U(t)$ from the flow of consumption $C(t)$ where

$$\begin{array}{ccccc} C(t) & = & G[K(t)] & - & \dot{K}(t) \\ \downarrow & & \downarrow & & \downarrow \end{array} \qquad (20.34)$$

Flow of consumption = production − investment

and the endpoints are fixed at $K(0) = K_0$, $K(T) = K_T$. That is,

$$\text{Maximize} \quad \int_0^T U[C(t)]\, dt = \int_0^T U\{G[K(t)] - \dot{K}(t)\}\, dt$$

subject to $\qquad\qquad K(0) = K_0 \qquad K(T) = K_T$

Letting $F = U\{G[K(t)] - \dot{K}(t)\}$, $U' = dU/dC$, and $G' = dG/dK$,

$$F_K = U'[C(t)]G'[K(t)] \qquad F_{\dot{K}} = -U'[(C(t)]$$

Substituting in Euler's equation,

$$U'[C(t)]G'[K(t)] = \frac{d}{dt}\{-U'[C(t)]\}$$
$$= -U''[C(t)] \cdot \dot{C}$$

where, upon using the chain rule on (20.34),

$$\dot{C} = \frac{dC}{dt} = G'[K(t)] \cdot \dot{K} - \ddot{K}$$

Substituting above,

$$U'[C(t)]G'[K(t)] = -U''[C(t)] \cdot (G'[K(t)] \cdot \dot{K} - \ddot{K})$$

The Euler equation thus yields a second-order ordinary differential equation, the solution of which maximizes the given extremal.

20.21. Maximize the discounted stream of utility from consumption $C(t)$ over $0 \le t \le T$, that is,

$$\text{Maximize} \quad \int_0^T \{e^{-it} U[C(t)]\}\, dt \qquad (20.35)$$

where $C(t) = G[K(t)] - \dot{K}(t) - bK(t)$, $0 \le b \le 1$, $G[K(t)]$ is the rate of production, $\dot{K}(t) + bK(t)$ is investment, and b is a constant rate of capital depreciation.

    Substituting in (20.35), we seek to maximize

$$\int_0^T (e^{-it} U\{G[K(t)] - \dot{K}(t) - bK(t)\}\, dt$$

With $\qquad\qquad F = e^{-it} U\{G[K(t)] - \dot{K}(t) - bK(t)\}$

and letting $U' = dU/dC$ and $G' = dG/dK$, while retaining $\dot{K} = dK/dt$,

$$F_K = e^{-it}U'\{G[K(t)] - \dot{K}(t) - bK(t)\} \cdot \{G'[K(t)] - b\}$$
$$F_{\dot{K}} = -e^{-it}U'\{G[K(t)] - \dot{K}(t) - bK(t)\}$$

Substituting in Euler's equation and simplifying notation,

$$e^{-it}[U' \cdot (G' - b)] = \frac{d}{dt}(-e^{-it}U') \qquad (20.36)$$

Using the product and chain rules for the derivative on the right,

$$e^{-it}[U' \cdot (G' - b)] = -e^{-it}U'' \cdot (G'\dot{K} - \ddot{K} - b\dot{K}) + U'ie^{-it}$$

and canceling the e^{-it} terms,

$$U' \cdot (G' - b) = -U'' \cdot (G'\dot{K} - \ddot{K} - b\dot{K}) + U'i$$

With $U\{G[K(t)]\}$ unspecified, we cannot proceed further. Going back to (20.36) and using general notation for the derivative on the right, however, we see that

$$e^{-it}[U' \cdot (G' - b)] = -e^{-it}\frac{d}{dt}(U') + U'(ie^{-it})$$

Canceling the e^{-it} terms and rearranging,

$$\frac{d}{dt}(U') = U'i - U' \cdot (G' - b)$$

$$\frac{d(U')/dt}{U'} = i + b - G'$$

where the term on the left, the rate of change of the marginal utility, equals the discount rate plus the depreciation rate minus the marginal product of capital. In brief, if we consider the term on the left as capital gains, the optimal time path suggests that if capital gains are greater than the discount rate plus the depreciation rate minus the marginal product of capital, then more capital and hence more consumption should be forthcoming. If it is less, capital accumulation and consumption should be scaled back.

20.22. Maximize the discounted stream of utility from consumption $C(t)$ over $0 \le t < T$, that is,

$$\text{Maximize} \quad \int_0^T e^{-it}U[C(t)]\,dt \qquad (20.37)$$

given (a) $\qquad U[C(t)] = [C(t)]^n \qquad \text{where} \qquad 0 \le n \le 1$

(b) $\qquad\qquad C(t) \quad = \quad G[K(t)] \quad - \quad I(t)$
$$\qquad\qquad\qquad \downarrow \qquad\qquad \downarrow \qquad\qquad \downarrow$$

$$\text{Flow of consumption} = \text{production} - \text{investment}$$

where $G[K(t)] = aK(t)$, a linear production function with $a > 0$ and

(c) $\qquad\qquad I(t) = \dot{K}(t) + B + bK(t) \qquad 0 \le b \le 1 \qquad B > 0$

derived from

$$\dot{K}(t) \quad = \quad I(t) \quad - \quad [B + bK(t)]$$
$$\downarrow \qquad\qquad \downarrow \qquad\qquad \downarrow$$

$$\Delta \text{ in } K \text{ stock} = \text{investment} - \text{linear depreciation}$$

Substituting in (20.37), we wish to maximize

$$\int_0^T e^{-it}[aK(t) - \dot{K}(t) - B - bK(t)]^n\,dt$$

which by rearranging and omitting the arguments for simplicity becomes

$$\int_0^T e^{-it}(mK - \dot{K} - B)^n \, dt$$

where $m = a - b$. Letting $F = e^{-it}(mK - \dot{K} - B)^n$,

$$F_K = mne^{-it}(mK - \dot{K} - B)^{n-1} \qquad F_{\dot{K}} = -ne^{-it}(mK - \dot{K} - B)^{n-1}$$

Using Euler's equation,

$$mne^{-it}(mK - \dot{K} - B)^{n-1} = \frac{d}{dt}[-ne^{-it}(mK - \dot{K} - B)^{n-1}]$$

Using the product rule and the chain rule on the right,

$$mne^{-it}(mK - \dot{K} - B)^{n-1} = -ne^{-it}(n-1)(mK - \dot{K} - B)^{n-2}(m\dot{K} - \ddot{K}) + (mK - \dot{K} - B)^{n-1}(ine^{-it})$$

Dividing both sides by $ne^{-it}(mK - \dot{K} - B)^{n-1}$,

$$m = -(n-1)(mK - \dot{K} - B)^{-1}(m\dot{K} - \ddot{K}) + i$$

$$m - i = \frac{(1-n)(m\dot{K} - \ddot{K})}{mK - \dot{K} - B}$$

Cross-multiplying and simplifying,

$$(1-n)\ddot{K} + (i + mn - 2m)\dot{K} + (m^2 - im)K = (m-i)B$$

$$\ddot{K} + \frac{i + mn - 2m}{1-n}\dot{K} + \frac{m^2 - im}{1-n}K = \frac{m-i}{1-n}B \qquad (20.38)$$

Letting

$$Z_1 = \frac{i + mn - 2m}{1-n} \qquad Z_2 = \frac{m^2 - im}{1-n} \qquad Z_3 = \frac{m-i}{1-n}B$$

$$K_p = \frac{Z_3}{Z_2} = \frac{B(m-i)/(1-n)}{(m^2 - im)/(1-n)} = \frac{(m-i)B}{m(m-i)} = \frac{1}{m}B$$

where $m = a - b$, a is the marginal product of capital $(dG/dK = a)$, and b is the constant rate of depreciation.

$$K_c = A_1 e^{r_1 t} + A_2 e^{r_2 t}$$

where

$$r_1, r_2 = \frac{-Z_1 \pm \sqrt{Z_1^2 - 4Z_2}}{2}$$

and A_1 and A_2 can be computed from the boundary conditions.

20.23. Maximize

$$\int_0^T e^{-it} U[C(t)] \, dt$$

given the discount rate $i = 0.12$, the endpoints $K(0) = 320$ and $K(5) = 480$, and the utility function $U[C(t)] = [C(t)]^{0.5}$, where

$$C(t) = G[K(t)] - I(t)$$

$$G[K(t)] = 0.25K(t) \qquad I(t) = \dot{K} + 60 + 0.05K(t)$$

Substituting in the given functional, we seek to maximize

$$\int_0^5 e^{-0.12t}[0.25K(t) - \dot{K}(t) - 60 - 0.05K(t)]^{0.5} \, dt$$

Rearranging and omitting the arguments for simplicity,

$$\int_0^5 e^{-0.12t}(0.2K - \dot{K} - 60)^{0.5}\, dt$$

Letting

$$F = e^{-0.12t}(0.2K - \dot{K} - 60)^{0.5}$$

and using the chain rule or the generalized power function rule,

$$F_K = 0.1e^{-0.12t}(0.2K - \dot{K} - 60)^{-0.5} \qquad F_{\dot{K}} = -0.5e^{-0.12t}(0.2K - \dot{K} - 60)^{-0.5}$$

Substituting in Euler's equation, then using the product rule and the chain rule,

$$0.1e^{-0.12t}(0.2K - \dot{K} - 60)^{-0.5} = \frac{d}{dt}[-0.5e^{-0.12t}(0.2K - \dot{K} - 60)^{-0.5}]$$

$$= -0.5e^{-0.12t}[-0.5(0.2K - \dot{K} - 60)^{-1.5}(0.2\dot{K} - \ddot{K})]$$

$$+ (0.2K - \dot{K} - 60)^{-0.5}(0.06e^{-0.12t})$$

Dividing both sides by $0.5e^{-0.12t}(0.2K - \dot{K} - 60)^{-0.5}$ and rearranging,

$$0.2 = \frac{0.5(0.2\dot{K} - \ddot{K})}{0.2K - \dot{K} - 60} + 0.12$$

$$0.08(0.2K - \dot{K} - 60) = 0.1\dot{K} - 0.5\ddot{K}$$

$$\ddot{K} - 0.36\dot{K} + 0.032K = 9.6 \qquad\qquad (20.39)$$

Using (18.2) and (18.4)

$$K_p = \frac{9.6}{0.032} = 300$$

$$r_1, r_2 = \frac{-(-0.36) \pm \sqrt{(-0.36)^2 - 4(0.032)}}{2} = \frac{0.36 \pm \sqrt{0.0016}}{2}$$

$$r_1 = 0.2 \qquad r_2 = 0.16$$

Thus,

$$K(t) = A_1 e^{0.2t} + A_2 e^{0.16t} + 300$$

Applying the endpoint conditions,

$$K(0) = A_1 + A_2 + 300 = 320 \qquad A_2 = 20 - A_1$$
$$K(5) = A_1 e^{0.2(5)} + (20 - A_1)e^{0.16(5)} + 300 = 480$$

$$A_1(2.71828) + (20 - A_1)(2.22554) = 180$$

$$A_1 = 274.97 \approx 275 \qquad A_2 = 20 - 275 = -255$$

Substituting,

$$K(t) = 275e^{0.2t} - 255e^{0.16t} + 300$$

20.24. Since Problem 20.23 is a specific application of Problem 20.22, check the accuracy of the answer in Problem 20.23 by substituting the given values of $a = 0.25$, $b = 0.05$, $B = 60$, $i = 0.12$, $m = 0.2$, and $n = 0.5$ in Equation (20.38) to make sure it yields the same answer as Equation (20.39).

Substituting the specific values in (20.38),

$$\ddot{K} + \left[\frac{0.12 + (0.2)(0.5) - 2(0.2)}{1 - 0.5}\right]\dot{K} + \left[\frac{(0.2)^2 - (0.12)(0.2)}{1 - 0.5}\right]K = \left(\frac{0.2 - 0.12}{1 - 0.5}\right)60$$

$$\ddot{K} - 0.36\dot{K} + 0.032K = 9.6$$

On your own, check the values of r_1, r_2, K_c, and $K(t)$ by substituting the specific values in the equations immediately following (20.38) in Problem 20.19 and comparing them with the solutions found in Problem 20.20.

CONSTRAINED OPTIMIZATION

20.25. Minimize

$$\int_{t_0}^{t_1} e^{-it}(a\dot{x}^2 + bx)\, dt$$

subject to

$$\int_{t_0}^{t_1} \dot{x}(t)\, dt = N$$

where $x(t_0) = 0$ and $x(t_1) = N$

Setting up the Lagrangian function, as in Section 20.6,

$$\int_{t_0}^{t_1} [e^{-it}(a\dot{x}^2 + bx) + \lambda\dot{x}]\, dt$$

Letting H equal the integrand, the Euler equation is

$$\frac{\partial H}{\partial x} = \frac{d}{dt}\left(\frac{\partial H}{\partial \dot{x}}\right)$$

Taking the needed partial derivatives,

$$H_x = be^{-it} \qquad H_{\dot{x}} = 2a\dot{x}e^{-it} + \lambda$$

Substituting in Euler's equation,

$$be^{-it} = \frac{d}{dt}(2a\dot{x}e^{-it} + \lambda)$$

Using the product and generalized power function rules,

$$be^{-it} = -2ai\dot{x}e^{-it} + 2a\ddot{x}e^{-it}$$

Canceling the e^{-it} terms and rearranging,

$$\ddot{x} - i\dot{x} = \frac{b}{2a}$$

which is identical to what we found in (20.22) without using constrained dynamic optimization. This is another example of an isoperimetric problem, and it can be solved as it was in Section 20.8.

PROOFS AND DEMONSTRATIONS

20.26. In seeking an extremal for

$$\int_{t_0}^{t_2} F(t, x, \dot{x})\, dt$$

show that Euler's equation can also be expressed as

$$\frac{d}{dt}\left(F - \dot{x}\frac{\partial F}{\partial \dot{x}}\right) - \frac{\partial F}{\partial t} = 0 \qquad\qquad (20.40)$$

Taking the derivative with respect to t of each term within the parentheses and using the chain rule,

$$\frac{dF}{dt} = \frac{\partial F}{\partial t} + \frac{\partial F}{\partial x}\frac{dx}{dt} + \frac{\partial F}{\partial \dot{x}}\frac{d\dot{x}}{dt} = \frac{\partial F}{\partial t} + \frac{\partial F}{\partial x}\dot{x} + \frac{\partial F}{\partial \dot{x}}\ddot{x}$$

$$\frac{d}{dt}\left(\dot{x}\frac{\partial F}{\partial \dot{x}}\right) = \dot{x}\frac{d}{dt}\left(\frac{\partial F}{\partial \dot{x}}\right) + \frac{\partial F}{\partial \dot{x}}\ddot{x}$$

and substituting in (20.40),

$$\frac{\partial F}{\partial t} + \frac{\partial F}{\partial x}\dot{x} + \frac{\partial F}{\partial \dot{x}}\ddot{x} - \left[\dot{x}\frac{d}{dt}\left(\frac{\partial F}{\partial \dot{x}}\right) + \frac{\partial F}{\partial \dot{x}}\ddot{x}\right] - \frac{\partial F}{\partial t} = 0$$

$$\dot{x}\left[\frac{\partial F}{\partial x} - \frac{d}{dt}\left(\frac{\partial F}{\partial \dot{x}}\right)\right] = 0$$

$$\frac{\partial F}{\partial x} = \frac{d}{dt}\left(\frac{\partial F}{\partial \dot{x}}\right) \qquad \text{or} \qquad F_x = \frac{d}{dt}(F_{\dot{x}})$$

20.27. Show that if F is not an explicit function of t, the Euler equation can be expressed as

$$F - \dot{x}\frac{\partial F}{\partial \dot{x}} = c \qquad \text{a constant}$$

If t is not an explicit argument of F, $\partial F/\partial t = 0$ and Equation (20.40) reduces to

$$\frac{d}{dt}\left(F - \dot{x}\frac{\partial F}{\partial \dot{x}}\right) = 0$$

Integrating both sides with respect to t,

$$F - \dot{x}\frac{\partial F}{\partial \dot{x}} = c$$

20.28. (a) Show that if $F = F(t, \dot{x})$, with x not one of the arguments, the Euler equation reduces to

$$F_{\dot{x}} = c \qquad \text{a constant}$$

(b) Explain the significance.

a)   With $F = F(t, \dot{x})$

$$F_x = 0 \qquad F_{\dot{x}} = F_{\dot{x}}$$

Substituting in Euler's equation,

$$0 = \frac{d}{dt}(F_{\dot{x}})$$

Integrating both sides with respect to t,

$$F_{\dot{x}} = c \tag{20.41}$$

b)   Equation (20.41) is a first-order differential equation with arguments of t and $\dot{x}$ alone which, when solved, provides the desired extremal. See Problems 20.7 and 20.8.

20.29. (a) Show that if $F = F(\dot{x})$, that is, a function of $\dot{x}$ alone, the Euler equation reduces to

$$F_{\dot{x}\dot{x}}\ddot{x} = 0$$

and (b) explain the significance.

a)   Given

$$F = F(\dot{x})$$
$$F_x = 0 \qquad F_{\dot{x}} = F_{\dot{x}}$$

Using Euler's equation,

$$0 = \frac{d}{dt}[F_{\dot{x}}(\dot{x})] = F_{\dot{x}\dot{x}}\ddot{x} \tag{20.42}$$

b) From (20.42), either $\ddot{x} = 0$ or $F_{\dot{x}\dot{x}} = 0$. If $\ddot{x} = 0$, integrating twice yields $x(t) = c_1 t + c_2$, which is linear. If $F_{\dot{x}\dot{x}} = 0$, $F_{\dot{x}} = c$, a constant, which means that F is linear in $\dot{x}$. If F is linear in $\dot{x}$, the solution is trivial. See Problems 20.30 and 20.31.

20.30. Find extremals for

$$\int_{t_0}^{t_1} e^{-3\dot{x}^2}\, dt$$

subject to

$$x(t_0) = x_0 \qquad x(t_1) = x_1$$

$$F_x = 0 \qquad F_{\dot{x}} = -6\dot{x}e^{-3\dot{x}^2}$$

$$0 = \frac{d}{dt}(-6\dot{x}e^{-3\dot{x}^2})$$

By the product rule,

$$0 = -6\dot{x}(-6\dot{x}e^{-3\dot{x}^2}\ddot{x}) + e^{-3\dot{x}^2}(-6\ddot{x})$$

$$6\ddot{x}e^{-3\dot{x}^2}(6\dot{x}^2 - 1) = 0 \qquad\qquad (20.43)$$

which is a nonlinear second-order differential equation, not easily solved. However, since $\ddot{x}$ in (20.43) must equal zero for the equation to equal zero, from Problem 20.29 we know that the solution must be linear. Thus,

$$x(t) = c_1 t + c_2$$

20.31. Find extremals for

$$\int_{t_0}^{t_1} (27 - 5\dot{x})\, dt$$

subject to

$$x(t_0) = x_0 \qquad x(t_1) = x_1$$

$$F_x = 0 \qquad F_{\dot{x}} = -5$$

$$0 = \frac{d}{dt}(-5) = 0$$

The Euler equation is an identity which any admissible value of x satisfies trivially, as was indicated in Problem 20.29. This becomes clear upon direct integration of the extremal in this problem:

$$\int_{t_0}^{t_1} (27 - 5\dot{x})\, dt = 27(t_1 - t_0) - 5[x(t_1) - x(t_0)]$$

and any x satisfying the endpoint conditions yields the same value for this integrand.

20.32. (a) Show that if $F = F(t, x)$, with $\dot{x}$ not one of the arguments, the Euler equation reduces to

$$F_x = 0$$

(b) Explain the significance.

a) With $F = F(t, x)$,

$$F_x = F_x \qquad F_{\dot{x}} = 0$$

Substituting in Euler's equation,

$$F_x = \frac{d}{dt}(0) = 0$$

b) When there is no $\dot{x}$ term in F, the optimization problem is static and not dynamic. The condition for optimization, therefore, is the same as that in static optimization, namely,

$$F_x = 0$$

20.33. Show that if application of Euler's equation results in a second-order differential equation with no x or t terms, the second-order differential equation can be converted to a first-order differential equation and a solution found by means of equation (16.1). Demonstrate in terms of Equation (20.22) from Section 20.8.

From Equation (20.22),

$$\ddot{x} = i\dot{x} + \frac{b}{2a}$$

Since there are no x or t terms in (20.22), it can be converted to a first-order linear differential equation by letting

$$u = \dot{x} \qquad \text{and} \qquad \dot{u} = \ddot{x}$$

Substituting in (20.22) above and rearranging,

$$\dot{u} - iu = \frac{b}{2a}$$

which can be solved by means of the formula in (16.1). Letting $v = -i$ and $z = b/2a$.

$$u = e^{-\int(-i)\,dt}\left(A + \int \frac{b}{2a} e^{\int(-i)\,dt}\,dt\right)$$

$$= e^{it}\left(A + \int \frac{b}{2a} e^{-it}\,dt\right)$$

Taking the remaining integral,

$$u = e^{it}\left[A + \left(-\frac{1}{i}\right)\frac{b}{2a} e^{-it}\right]$$

$$u = Ae^{it} - \frac{b}{2ai}$$

But $u = \dot{x}(t)$ by definition, so we must integrate once again to find $x(t)$. Replacing A with c_1 for notational consistency with ordinary integration,

$$x(t) = \frac{c_1}{i} e^{it} - \frac{b}{2ai} t + c_2 \qquad\qquad (20.44)$$

Letting $t_0 = 0$ and $t_1 = T$, from the boundary conditions we have

$$x(0) = \frac{c_1}{i} + c_2 = 0 \qquad\qquad c_2 = -\frac{c_1}{i}$$

$$x(T) = \frac{c_1}{i} e^{iT} - \frac{b}{2ai} T + c_2 = N$$

Substituting $c_2 = -c_1/i$ in $x(T)$ and solving for c_1,

$$\frac{c_1}{i} e^{iT} - \frac{b}{2ai} T - \frac{c_1}{i} = N$$

$$\frac{c_1}{i} (e^{iT} - 1) = N + \frac{b}{2ai} T$$

$$c_1 = \frac{i\{N + [b/(2ai)]\,T\}}{e^{iT} - 1} \qquad\qquad (20.45)$$

Finally, substituting in (20.44) and noting that the i in (20.45) cancels out the i in the denominator of c_1/i, we have as a candidate for an extremal:

$$x(t) = \frac{N + [b/(2ai)]\,T}{e^{iT} - 1}\,e^{it} - \frac{b}{2ai}\,t - \frac{N + [b/(2ai)]\,T}{e^{iT} - 1}$$

$$= \left(N + \frac{b}{2ai}\,T\right)\left(\frac{e^{it} - 1}{e^{iT} - 1}\right) - \frac{b}{2ai}\,t \qquad 0 \le t \le T$$

Compare the work done here with the work done in Section 20.8, and note that conversion of a second-order differential equation to a first-order differential equation before integrating does not necessarily reduce the work involved in finding a solution.

VARIATIONAL NOTATION

20.34. Show that the operators δ and d/dt are commutative, i.e., show that

$$\delta\left(\frac{dx}{dt}\right) = \frac{d}{dt}(\delta x)$$

From (20.19) and (20.20),

$$\delta x = mh \qquad \text{and} \qquad \delta \dot{x} = m\dot{h}$$

where m is an arbitrary constant and h is an arbitrary function $h = h(t)$. Substituting dx/dt for $\dot{x}$ above on the right,

$$\delta\left(\frac{dx}{dt}\right) = m\dot{h}$$

Expressing $\dot{h}$ as dh/dt and recalling that m is a constant,

$$\delta\left(\frac{dx}{dt}\right) = \frac{d}{dt}(mh)$$

Then substituting from (20.19),

$$\delta\left(\frac{dx}{dt}\right) = \frac{d}{dt}(\delta x)$$

20.35. Given

$$\int_{t_0}^{t_1} F[t, x(t), \dot{x}(t)]\, dt$$

show that in terms of variational notation a necessary condition for an extremal is

$$\delta \int_{t_0}^{t_1} F[t, x(t), \dot{x}(t)]\, dt = 0$$

Moving δ within the integral sign,

$$\int_{t_0}^{t_1} \delta F[t, x(t), \dot{x}(t)]\, dt = 0$$

From (20.21),

$$\int_{t_0}^{t_1} \left(\frac{\partial F}{\partial x}\,\delta x + \frac{\partial F}{\partial \dot{x}}\,\delta \dot{x}\right) dt = 0$$

Substituting from (20.19) and (20.20) where

$$\delta x = mh \qquad \delta \dot{x} = m\dot{h}$$

$$\int_{t_0}^{t_1} \left(\frac{\partial F}{\partial x} mh + \frac{\partial F}{\partial \dot{x}} m\dot{h} \right) dt = 0$$

Dividing by m, an arbitrary constant,

$$\int_{t_0}^{t_1} \left(\frac{\partial F}{\partial x} h + \frac{\partial F}{\partial \dot{x}} \dot{h} \right) dt = 0 \qquad (20.46)$$

Equation (20.46) is identical to Equation (20.18) in the Euler equation proof from Example 2 and can be concluded in the same manner.

CHAPTER 21

Optimal Control Theory

21.1 TERMINOLOGY

Optimal control theory is a mid-twentieth-century advance in the field of dynamic optimization that can handle any problem the calculus of variations was designed for. More importantly, optimal control theory is more powerful than the calculus of variations because it can manage some problems the calculus of variations cannot, such as corner solutions, and other problems the calculus of variations cannot handle readily, such as constraints on the derivatives of the functions being sought.

In optimal control theory, the aim is to find the optimal time path for a *control variable*, which we shall denote as y. The variable for which we previously sought an optimal time path in the calculus of variations, known as a *state variable*, we shall continue to designate as x. State variables always have *equations of motion* or *transition* set equal to $\dot{x}$. The goal of optimal control theory is to select a stream of values over time for the control variable that will optimize the functional subject to the constraint set on the state variable.

Optimal control theory problems involving continuous time, a finite time horizon, and fixed endpoints are generally written:

Maximize

$$J = \int_0^T f[x(t), y(t), t]\, dt$$

subject to

$$\dot{x} = g[x(t), y(t), t]$$
$$x(0) = x_0 \qquad x(T) = x_T \qquad (21.1)$$

where J = the value of the functional to be optimized; $y(t)$ = the control variable, so called because its value is selected or controlled to optimize J; $x(t)$ = the state variable, which changes over time according to the differential equation set equal to $\dot{x}$ in the constraint, and whose value is indirectly determined by the control variable in the constraint; and t = time. The solution to an optimal control problem demarcates the optimal dynamic time path for the control variable $y(t)$.

21.2 THE HAMILTONIAN AND THE NECESSARY CONDITIONS FOR MAXIMIZATION IN OPTIMAL CONTROL THEORY

Dynamic optimization of a functional subject to a constraint on the state variable in optimal control involves a *Hamiltonian function H* similar to the Lagrangian function in concave programming. In terms of (*21.1*), the Hamiltonian is defined as

$$H[x(t), y(t), \lambda(t), t] = f[x(t), y(t), t] + \lambda(t)g[x(t), y(t), t] \qquad (21.2)$$

where $\lambda(t)$ is called the *costate* variable. Similar to the Lagrangian multiplier, the costate variable $\lambda(t)$ estimates the marginal value or shadow price of the associated state variable $x(t)$. Working from (*21.2*), formation of the Hamiltonian is easy. Simply take the integrand under the integral sign and add to it the product of the costate variable $\lambda(t)$ times the constraint.

Assuming the Hamiltonian is differentiable in y and strictly concave so there is an interior solution and not an endpoint solution, the necessary conditions for maximization are

1.
$$\frac{\partial H}{\partial y} = 0$$

2.   *a*) $\dfrac{\partial \lambda}{\partial t} = \dot{\lambda} = -\dfrac{\partial H}{\partial x}$ *b*) $\dfrac{\partial x}{\partial t} = \dot{x} = \dfrac{\partial H}{\partial \lambda}$

3.   *a*) $x(0) = x_0$ *b*) $x(T) = x_T$

The first two conditions are known as the *maximum principle* and the third is called the *boundary condition*. The two equations of motion in the second condition are generally referred to as the *Hamiltonian system* or the *canonical system*. For minimization, the objective functional can simply be multiplied by −1, as in concave programming. If the solution involves an end point, $\partial H/\partial y$ need not equal zero in the first condition, but H must still be maximized with respect to y. See Chapter 13, Example 9, and Fig. 13-1(*b*)–(*c*), for clarification. We shall generally assume interior solutions.

EXAMPLE 1. The conditions in Section 21.2 are used below to solve the following optimal control problem:

Maximize
$$\int_0^3 (4x - 5y^2)\, dt$$

subject to
$$\dot{x} = 8y$$
$$x(0) = 2 \qquad x(3) = 117.2$$

A. From (*21.1*), set up the Hamiltonian.

$$H = 4x - 5y^2 + \lambda(8y)$$

B. Assuming an interior solution, apply the maximum principle.

1.
$$\frac{\partial H}{\partial y} = 0$$

$$\frac{\partial H}{\partial y} = -10y + 8\lambda = 0$$

$$y = 0.8\lambda \qquad (21.3)$$

2. *a*)
$$\dot{\lambda} = -\frac{\partial H}{\partial x}$$

$$\dot{\lambda} = -4 \qquad (21.4)$$

b)
$$\dot{x} = \frac{\partial H}{\partial \lambda}$$

$$\dot{x} = 8y$$

But from (21.3), $y = 0.8\lambda$. So,

$$\dot{x} = 8(0.8\lambda) = 6.4\lambda \tag{21.5}$$

Having employed the maximum principle, we are left with two differential equations, which we now solve for the state variables $x(t)$ and the costate variable $\lambda(t)$.

By integrating (21.4) we find the costate variable.

$$\lambda(t) = \int \dot{\lambda}\, dt = \int -4\, dt = -4t + c_1 \tag{21.6}$$

Substituting (21.6) in (21.5),

$$\dot{x} = 6.4(-4t + c_1) = -25.6t + 6.4c_1$$

Integrating,

$$x(t) = \int (-25.6t + 6.4c_1)\, dt$$

$$x(t) = -12.8t^2 + 6.4c_1 t + c_2 \tag{21.7}$$

C. The boundary conditions can now be used to solve for the constants of integration. Applying $x(0) = 2$, $x(3) = 117.2$ successively to (21.7),

$$x(0) = -12.8(0)^2 + 6.4c_1(0) + c_2 = 2 \qquad c_2 = 2$$
$$x(3) = -12.8(3)^2 + 6.4c_1(3) + 2 = 117.2 \qquad c_1 = 12$$

Then by substituting $c_1 = 12$ and $c_2 = 2$ in (21.7) and (21.6), we have,

$$x(t) = -12.8t^2 + 76.8t + 2 \qquad \text{state variable} \tag{21.8}$$

$$\lambda(t) = -4t + 12 \qquad \text{costate variable}$$

D. Lastly, we can find the final solution for the control variable $y(t)$ in either of two ways.

1. From (21.3), $y(t) = 0.8\lambda$, so

$$y(t) = 0.8(-4t + 12) = -3.2t + 9.6 \qquad \text{control variable}$$

2. Or taking the derivative of (21.8),

$$\dot{x} = -25.6t + 76.8$$

we substitute for $\dot{x}$ in the equation of motion in the constraint,

$$\dot{x} = 8y$$
$$-25.6t + 76.8 = 8y$$
$$y(t) = -3.2t + 9.6 \qquad \text{control variable}$$

Evaluated at the endpoints,

$$y(0) = -3.2(0) + 9.6 = 9.6$$
$$y(3) = -3.2(3) + 9.6 = 0$$

The optimal path of the control variable is linear starting at $(0, 9.6)$ and ending at $(3, 0)$, with a slope of -3.2. For similar problems involving fixed endpoints, see also Problems 21.1 to 21.3.

21.3 SUFFICIENCY CONDITIONS FOR MAXIMIZATION IN OPTIMAL CONTROL

Assuming the maximum principle representing the necessary conditions for maximization in optimal control is satisfied, the sufficiency conditions will be fulfilled if:

1. Both the objective functional $f[x(t), y(t), t]$ and the constraint $g[x(t), y(t), t]$ are differentiable and jointly concave in x and y, and

2. $\lambda(t) \geq 0$, if the constraint is nonlinear in x or y. If the constraint is linear, λ may assume any sign.

Linear functions are always both concave and convex, but neither strictly concave or strictly convex. For nonlinear functions, an easy test for joint concavity is the simple discriminant test. Given the discriminant of the second-order derivatives of a function,

$$\mathbf{D} = \begin{vmatrix} f_{xx} & f_{xy} \\ f_{yx} & f_{yy} \end{vmatrix}$$

a function will be strictly concave if the discriminant is negative definite,

$$|\mathbf{D}_1| = f_{xx} < 0 \quad\quad \text{and} \quad\quad |\mathbf{D}_2| = |\mathbf{D}| > 0$$

and simply concave if the discriminant is negative semidefinite,

$$|\mathbf{D}_1| = f_{xx} \leq 0 \quad\quad \text{and} \quad\quad |\mathbf{D}_2| = |\mathbf{D}| \geq 0$$

A negative definite discriminant indicates a global maximum and is, therefore, always sufficient for a maximum. A negative semidefinite discriminant is indicative of a relative maximum and is sufficient for a maximum if the test is conducted for every possible ordering of the variables with similar results.

EXAMPLE 2.   The sufficiency conditions for the problem in Example 1 are demonstrated below. Starting with the objective functional which is nonlinear, we take the second derivatives and apply the discriminant test.

$$\mathbf{D} = \begin{vmatrix} f_{xx} & f_{xy} \\ f_{yx} & f_{yy} \end{vmatrix} = \begin{vmatrix} 0 & 0 \\ 0 & -10 \end{vmatrix} \quad\quad \text{where } |\mathbf{D}_1| = 0 \quad \text{and} \quad |\mathbf{D}_2| = |\mathbf{D}| = 0$$

$\mathbf{D}$ fails the strict negative-definite criteria but proves to be negative semidefinite with $|\mathbf{D}_1| \leq 0$ and $|\mathbf{D}_2| = |\mathbf{D}| \geq 0$. However, for the semidefinite test we must also test the variables in reverse order.

$$\mathbf{D} = \begin{vmatrix} f_{yy} & f_{yx} \\ f_{xy} & f_{xx} \end{vmatrix} = \begin{vmatrix} -10 & 0 \\ 0 & 0 \end{vmatrix} \quad\quad \text{where } |\mathbf{D}_1| = -10 \quad \text{and} \quad |\mathbf{D}_2| = |\mathbf{D}| = 0$$

With both discriminant tests negative semidefinite, the objective functional f is jointly concave in x and y. Since the constraint is linear, it is also jointly concave and does not need testing. We can conclude, therefore, that the functional is indeed maximized.

21.4   OPTIMAL CONTROL THEORY WITH A FREE ENDPOINT

The general format for an optimal control problem involving continuous time with a finite time horizon and a free endpoint is

Maximize
$$J = \int_0^T f[x(t), y(t), t]\, dt$$

subject to
$$\dot{x} = g[x(t), y(t), t]$$
$$x(0) = x_0 \quad\quad x(T) \quad\quad \text{free}$$

(21.9)

where the upper limit of integration $x(T)$ is free and unrestricted. Assuming an interior solution, the first two conditions for maximization, comprising the maximum condition, remain the same but the third or boundary condition changes:

1.
$$\frac{\partial H}{\partial y} = 0$$

2.   a) $\dfrac{\partial \lambda}{\partial t} = \dot{\lambda} = -\dfrac{\partial H}{\partial x}$ b) $\dfrac{\partial x}{\partial t} = \dot{x} = \dfrac{\partial H}{\partial \lambda}$

3.   *a)* $x(0) = x_0$ *b)* $\lambda(T) = 0$

where the very last condition is called the *transversality condition* for a free endpoint. The rationale for the transversality condition follows straightforward from what we learned in concave programming. If the value of x at T is free to vary, the constraint must be nonbinding and the shadow price λ evaluated at T must equal 0, i.e., $\lambda(T) = 0$. For problems involving free endpoints, see Examples 3 to 4 and Problems 21.4 to 21.6.

EXAMPLE 3.   The conditions in Section 21.4 are used below to solve the following optimal control problem with a free endpoint:

Maximize
$$\int_0^2 (3x - 2y^2)\, dt$$

subject to
$$\dot{x} = 8y$$
$$x(0) = 5 \qquad x(2) \qquad \text{free}$$

A.   From (*21.1*),

$$H = 3x - 2y^2 + \lambda(8y)$$

B.   Assuming an interior solution and applying the maximum principle.

1.
$$\frac{\partial H}{\partial y} = 0$$

$$\frac{\partial H}{\partial y} = -4y + 8\lambda = 0$$

$$y = 2\lambda \tag{21.10}$$

2.   *a)*
$$\dot{\lambda} = -\frac{\partial H}{\partial x}$$

$$\dot{\lambda} = -3 \tag{21.11}$$

b)
$$\dot{x} = \frac{\partial H}{\partial \lambda}$$

$$\dot{x} = 8y$$

But from (*21.10*), $y = 2\lambda$. So,

$$\dot{x} = 8(2\lambda) = 16\lambda \tag{21.12}$$

From the maximum principle, two differential equations emerge, which can now be solved for the state variable $x(t)$ and the costate variable $\lambda(t)$.
Integrating (*21.11*),

$$\lambda(t) = \int \dot{\lambda}\, dt = \int -3\, dt = -3t + c_1 \tag{21.13}$$

Substituting (*21.13*) in (*21.12*),

$$\dot{x} = 16(-3t + c_1) = -48t + 16c_1$$

Integrating,

$$x(t) = -24t^2 + 16c_1 t + c_2 \tag{21.14}$$

C.   We now use the boundary conditions to specify the constants of integration.

1.   Start with the transversality condition $\lambda(T) = 0$ for a free endpoint. Here

$$\lambda(2) = 0$$

Substituting in (21.13),

$$\lambda(2) = -3(2) + c_1 = 0$$
$$c_1 = 6$$

Therefore, $\lambda(t) = -3t + 6$ costate variable   (21.15)

2. Now substitute $c_1 = 6$ in (21.14),

$$x(t) = -24t^2 + 16(6)t + c_2$$
$$x(t) = -24t^2 + 96t + c_2$$

and apply the initial boundary condition, $x(0) = 5$.

$$x(0) = -24(0)^2 + 96(0) + c_2 = 5 \qquad c_2 = 5$$

So, $x(t) = -24t^2 + 96t + 5$ state variable   (21.16)

D. The control variable $y(t)$ can then be found in either of two ways.

1. From (21.10), $y(t) = 2\lambda$. Substituting from (21.15) for the final solution,

$$y(t) = 2(-3t + 6) = -6t + 12 \qquad \text{control variable} \qquad (21.17)$$

2. Or take the derivative of (21.16),

$$\dot{x} = -48t + 96$$

and substitute in the equation of transition in the constraint,

$$\dot{x} = 8y$$
$$-48t + 96 = 8y$$
$$y(t) = -6t + 12 \qquad \text{control variable}$$

Evaluated at the endpoints,

$$y(0) = -6(0) + 12 = 12$$
$$y(2) = -6(2) + 12 = 0$$

The optimal path of the control variable is linear starting at $(0, 12)$ and ending at $(2, 0)$, with a slope of -6.

EXAMPLE 4.   The sufficiency conditions for Example 3 are found in the same way as in Example 2. Taking the second derivatives of the objective functional and applying the discriminant test,

$$\mathbf{D} = \begin{vmatrix} f_{xx} & f_{xy} \\ f_{yx} & f_{yy} \end{vmatrix} = \begin{vmatrix} 0 & 0 \\ 0 & -4 \end{vmatrix} \qquad \text{where } |\mathbf{D}_1| = 0 \qquad \text{and} \qquad |\mathbf{D}_2| = |\mathbf{D}| = 0$$

$\mathbf{D}$ is not negative-definite but it is negative semidefinite with $|\mathbf{D}_1| \leq 0$ and $|\mathbf{D}_2| = |\mathbf{D}| \geq 0$. For the semidefinite test, however, we must test the variables in reverse order.

$$\mathbf{D} = \begin{vmatrix} f_{yy} & f_{yx} \\ f_{xy} & f_{xx} \end{vmatrix} = \begin{vmatrix} -4 & 0 \\ 0 & 0 \end{vmatrix} \qquad \text{where } |\mathbf{D}_1| = -4 \qquad \text{and} \qquad |\mathbf{D}_2| = |\mathbf{D}| = 0$$

With both discriminants negative semidefinite, the objective functional f is jointly concave in x and y. The constraint is linear and so needs no testing. The functional is maximized.

21.5   INEQUALITY CONSTRAINTS IN THE ENDPOINTS

If the terminal value of the state variable is subject to an inequality constraint, $x(T) \geq x_{\min}$, the optimal value $x^*(T)$ may be chosen freely as long as it does not violate the value set by the constraint $x_{\min}$. If $x^*(T) > x_{\min}$, the constraint is nonbinding and the problem reduces to a free endpoint problem. So

$$\lambda(T) = 0 \qquad \text{when } x^*(T) > x_{\min}$$

If $x^*(T) < x_{min}$, the constraint is binding and the optimal solution will involve setting $x(T) = x_{min}$, which is equivalent to a fixed-end problem with

$$\lambda(T) \geq 0 \qquad \text{when } x^*(T) = x_{min}$$

For conciseness, the endpoint conditions are sometimes reduced to a single statement analogous to the Kuhn-Tucker condition,

$$\lambda(T) \geq 0 \qquad x(T) \geq x_{min} \qquad [x(T) - x_{min}]\lambda(T) = 0$$

In practice, solving problems with inequality constraints on the endpoints is straightforward. First solve the problem as if it were a free endpoint problem. If the optimal value of the state variable is greater than the minimum required by the endpoint condition, i.e., if $x^*(T) \geq x_{min}$, the correct solution has been found. If $x^*(T) < x_{min}$, set the terminal endpoint equal to the value of the constraint, $x(T) = x_{min}$, and solve as a fixed endpoint problem. The method is illustrated in Examples 5 and 6 and further explained and developed in Example 7 and Problems 21.7 to 21.10.

EXAMPLE 5.

Maximize

$$\int_0^2 (3x - 2y^2) \, dt$$

subject to

$$\dot{x} = 8y$$
$$x(0) = 5 \qquad x(2) \geq 95$$

To solve an optimal control problem involving an inequality constraint, solve it first as an unconstrained problem with a free endpoint. This we did in Example 3 where we found the state variable in (21.16) to be

$$x(t) = -24t^2 + 96t + 5$$

Evaluating (21.16) at $x = 2$, the terminal endpoint, we have

$$x(2) = 101 > 95$$

Since the free endpoint solution satisfies the terminal endpoint constraint $x(T) \geq 95$, the constraint is not binding and we have indeed found the proper solution. From (21.17) in Example 3,

$$y(t) = -6t + 12$$

EXAMPLE 6. Redo the same problem in Example 5 with the new boundary conditions,

$$x(0) = 5 \qquad x(2) \geq 133$$

A. From Example 5 we know that the value for the state variable when optimized under free endpoint conditions is

$$x(2) = 101 < 133$$

which fails to meet the new endpoint constraints. This means the constraint is binding and we have now to optimize the functional as a fixed endpoint problem with the value of the constraint as the terminal endpoint,

$$x(0) = 5 \qquad x(2) = 133$$

B. The first two steps remain the same as when we solved the problem as a free endpoint in Example 3. Employing the maximum principle, we found:

in (21.10), $y = 2\lambda$
in (21.11), $\dot{\lambda} = -3$
in (21.12), $\dot{x} = 16\lambda$
in (21.13), $\lambda(t) = -3t + c_1$
in (21.14), $x(t) = -24t^2 + 16c_1 t + c_2$

Now we continue on with the new boundary conditions for a fixed endpoint.

C. Applying $x(0) = 5$ and $x(2) = 133$ successively in (21.14), we have

$$x(0) = -24(0)^2 + 16c_1(0) + c_2 = 5 \qquad c_2 = 5$$
$$x(2) = -24(2)^2 + 16c_1(2) + 5 = 133 \qquad c_1 = 7$$

Then, substituting $c_1 = 7$, $c_2 = 5$ in (21.13) and (21.14), we derive

$$\lambda(t) = -3t + 7 \qquad \text{costate variable}$$
$$x(t) = -24t^2 + 112t + 5 \qquad \text{state variable}$$

D. The control variable can be found in either of the two familiar ways. We opt once again for the first. From (21.10),

$$y(t) = 2\lambda = 2(-3t + 7) = -6t + 14 \qquad \text{control variable}$$

EXAMPLE 7.   With an inequality constraint as a terminal endpoint, in accord with the rules of Section 21.5, we first optimize the Hamiltonian subject to a free endpoint. With a free endpoint, we set $\lambda(T) = 0$, allowing the marginal value of the state variable to be taken down to zero. This, in effect, means that as long as the minimum value set by the constraint is met, the state variable is no longer of any value to us. Our interest in the state variable does not extend beyond time T.

Most variables have value, however, and our interest generally extends beyond some narrowly limited time horizon. In such cases we will not treat the state variable as a free good by permitting its marginal value to be reduced to zero. We will rather require some minimum value of the state variable to be preserved for use beyond time T. This means maximizing the Hamiltonian subject to a fixed endpoint determined by the minimum value of the constraint. In such cases, $\lambda(T) > 0$, the constraint is binding, and we will not use as much of the state variable as we would if it were a free good.

21.6   THE CURRENT-VALUED HAMILTONIAN

Optimal control problems frequently involve discounting, such as

Maximize
$$J = \int_0^T e^{-\rho t} f[x(t), y(t), t] \, dt$$

subject to
$$\dot{x} = g[x(t), y(t), t]$$
$$x(0) = x_0 \qquad x(T) \qquad \text{free}$$

The Hamiltonian for the discounted or present value follows the familiar format

$$H = e^{-\rho t} f[x(t), y(t), t] + \lambda(t) g[x(t), y(t), t]$$

but the presence of the discount factor $e^{-\rho t}$ complicates the derivatives in the necessary conditions. If we let $\mu(t) = \lambda(t) e^{\rho t}$, however, we can form a new, "current-valued" Hamiltonian

$$H_c = H e^{\rho t} = f[x(t), y(t), t] + \mu(t) g[x(t), y(t), t] \qquad (21.18)$$

which is generally easier to solve and requires only two adjustments to the previous set of necessary conditions. Converting condition $2(a)$ from Section 21.2 to correspond to the current-valued Hamiltonian, we have

$$\dot{\lambda} = -\frac{\partial H}{\partial x} = -\frac{\partial H_c}{\partial x} e^{-\rho t}$$

Taking the derivative of $\lambda(t) = \mu(t) e^{-\rho t}$, we have

$$\dot{\lambda} = \dot{\mu} e^{-\rho t} - \rho \mu e^{-\rho t}$$

Equating the λ's, canceling the common $e^{-\rho t}$ terms, and rearranging, we derive the adjusted condition for 2(a):

$$\dot{\mu} = \rho\mu - \frac{\partial H_c}{\partial x}$$

The second adjustment involves substituting $\lambda(t) = \mu(t)e^{-\rho t}$ in the boundary conditions. The transversality condition for a free endpoint then changes from $\lambda(T) = 0$ to the equivalent $\mu(T)e^{-\rho t} = 0$.

In short, given the current-valued Hamiltonian in (21.18) and assuming an interior solution, the necessary conditions for optimization are

1. $$\frac{\partial H_c}{\partial y} = 0$$

2. $a)$ $\dfrac{\partial \mu}{\partial t} = \dot{\mu} = \rho\mu - \dfrac{\partial H_c}{\partial x}$ $b)$ $\dfrac{\partial x}{\partial t} = \dot{x} = \dfrac{\partial H_c}{\partial \mu}$

3. $a)$ $x(0) = x_0$ $b)$ $\mu(T)e^{-\rho T} = 0$

If the solution does not involve an end point, $\partial H_c/\partial y$ need not equal zero in the first condition, but H_c must still be maximized with respect to y. With $H_c = He^{\rho t}$, the value of y that will maximize H_c will also maximize H since $e^{\rho t}$ is treated as a constant when maximizing with respect to y. The sufficiency conditions of Section 21.3 remain the same, as shown in Example 9. Maximization of a current-valued Hamiltonian is demonstrated in Examples 8 to 9 and followed up in Problems 21.11 to 21.12.

EXAMPLE 8.

Maximize

$$\int_0^2 e^{-0.02t}(x - 3x^2 - 2y^2)\, dt$$

subject to

$$\dot{x} = y - 0.5x$$
$$x(0) = 93.91 \qquad x(2) \qquad \text{free}$$

A. Set up the current-valued Hamiltonian.

$$H_c = x - 3x^2 - 2y^2 + \mu(y - 0.5x)$$

B. Assuming an interior solution, apply the modified maximum principle.

1. $$\frac{\partial H_c}{\partial y} = 0$$

$$\frac{\partial H_c}{\partial y} = -4y + \mu = 0$$

$$y = 0.25\mu \qquad\qquad (21.19)$$

2. $a)$

$$\dot{\mu} = 0.02\mu - (1 - 6x - 0.5\mu)$$
$$\dot{\mu} = 0.52\mu + 6x - 1 \qquad\qquad (21.20)$$

$b)$

$$\dot{x} = \frac{\partial H_c}{\partial \mu} = y - 0.5x$$

Substituting from (21.19),

$$\dot{x} = 0.25\mu - 0.5x \qquad\qquad (21.21)$$

Arranging the two simultaneous first-order differential equations from (21.20) and (21.21) in matrix form and solving with the techniques from Section 19.3,

$$\begin{bmatrix} \dot{\mu} \\ \dot{x} \end{bmatrix} = \begin{bmatrix} 0.52 & 6 \\ 0.25 & -0.5 \end{bmatrix} \begin{bmatrix} \mu \\ x \end{bmatrix} + \begin{bmatrix} -1 \\ 0 \end{bmatrix}$$

or $\dot{\mathbf{Y}} = \mathbf{AY} + \mathbf{B}$

The characteristic equation is

$$|\mathbf{A} - r\mathbf{I}| = \begin{vmatrix} 0.52 - r & 6 \\ 0.25 & -0.5 - r \end{vmatrix} = 0$$

From (19.3), the characteristic roots are,

$$r_1, r_2 = \frac{0.02 \pm \sqrt{(0.02)^2 - 4(-1.76)}}{2}$$

$$r_1 = 1.3367 \qquad r_2 = -1.3167$$

For $r_1 = 1.3367$, the eigenvector is

$$\begin{bmatrix} 0.52 - 1.3367 & 6 \\ 0.25 & -0.5 - 1.3367 \end{bmatrix} \begin{bmatrix} c_1 \\ c_2 \end{bmatrix} = \begin{bmatrix} -0.8167 & 6 \\ 0.25 & -1.8367 \end{bmatrix} \begin{bmatrix} c_1 \\ c_2 \end{bmatrix} = 0$$

$$-0.8167c_1 + 6c_2 = 0 \qquad c_1 = 7.3466c_2$$

$$y_c^1 = \begin{bmatrix} 7.3466 \\ 1 \end{bmatrix} k_1 e^{1.3367t}$$

For $r_2 = -1.3167$, the eigenvector is

$$\begin{bmatrix} 0.52 + 1.3167 & 6 \\ 0.25 & -0.5 + 1.3167 \end{bmatrix} \begin{bmatrix} c_1 \\ c_2 \end{bmatrix} = \begin{bmatrix} 1.8367 & 6 \\ 0.25 & 0.8167 \end{bmatrix} \begin{bmatrix} c_1 \\ c_2 \end{bmatrix} = 0$$

$$1.8367c_1 + 6c_2 = 0 \qquad c_1 = -3.2667c_2$$

$$y_c^2 = \begin{bmatrix} -3.2667 \\ 1 \end{bmatrix} k_2 e^{-1.3167t}$$

From (19.5), the particular solution is

$$\bar{\mathbf{Y}} = -\mathbf{A}^{-1}\mathbf{B}$$

$$\bar{\mathbf{Y}} = \begin{bmatrix} \bar{\mu} \\ \bar{x} \end{bmatrix} = -\left(\frac{1}{-1.76}\right) \begin{bmatrix} -0.5 & -6 \\ -0.25 & 0.52 \end{bmatrix} \begin{bmatrix} -1 \\ 0 \end{bmatrix} = \begin{bmatrix} 0.28 \\ 0.14 \end{bmatrix}$$

Adding the complementary and particular solutions, we have

$$\mu(t) = 7.3466k_1 e^{1.3667t} - 3.2667k_2 e^{-1.3167t} + 0.28 \tag{21.22}$$

$$x(t) = k_1 e^{1.3667t} + k_2 e^{-1.3167t} + 0.14 \tag{21.23}$$

C. Next we apply the boundary conditions.

   1. From the transversality condition for the free endpoint, $\mu(T)e^{-\rho t} = 0$, we have at $T = 2$,

$$\mu(2)e^{-0.02(2)} = 0$$

   Substituting for $\mu(2)$,

$$(7.3466k_1 e^{1.3667(2)} - 3.2667k_2 e^{-1.3167(2)} + 0.28)e^{-0.04} = 0$$

$$113.0282k_1 - 0.2350k_2 + 0.2690 = 0 \tag{21.24}$$

   2. Evaluating $x(t)$ at $x(0) = 93.91$,

$$k_1 + k_2 + 0.14 = 93.91 \tag{21.25}$$

   Solving (21.24) and (21.25) simultaneously,

$$k_1 = 0.2 \qquad k_2 = 93.57$$

Then substituting $k_1 = 0.2$, $k_2 = 93.57$ in (21.22) and (21.23), we get

$$\mu(t) = 1.4693e^{1.3667t} - 305.6651e^{-1.3167t} + 0.28 \qquad \text{costate variable}$$
$$x(t) = 0.2e^{1.3667t} + 93.57e^{-1.3167t} + 0.14 \qquad \text{state variable}$$

D. The solution for the control variable can now be found in either of the two usual ways. We choose the easier. From (21.19), $y(t) = 0.25\mu$. Substituting from the costate variable above,

$$y(t) = 0.3673e^{1.3667t} - 76.4163e^{-1.3167t} + 0.07 \qquad \text{control variable}$$

EXAMPLE 9. The sufficiency conditions follow the usual rules.

$$\mathbf{D} = \begin{vmatrix} f_{xx} & f_{xy} \\ f_{yx} & f_{yy} \end{vmatrix} = \begin{vmatrix} -6 & 0 \\ 0 & -4 \end{vmatrix}$$

With $|\mathbf{D}_1| = -6 < 0$, and $|\mathbf{D}_2| = 24 > 0$, $\mathbf{D}$ is negative definite, making f strictly concave in both x and y. With g linear in x and y, the sufficiency condition for a global maximum is fulfilled.

Solved Problems

FIXED ENDPOINTS

21.1. Maximize

$$\int_0^2 (6x - 4y^2)\, dt$$

subject to

$$\dot{x} = 16y$$
$$x(0) = 24 \qquad x(2) = 408$$

A. The Hamiltonian is

$$H = 6x - 4y^2 + \lambda(16y)$$

B. The necessary conditions from the maximum principle are

   1.

$$\frac{\partial H}{\partial y} = -8y + 16\lambda = 0$$

$$y = 2\lambda \tag{21.26}$$

   2. *a)*

$$\dot{\lambda} = -\frac{\partial H}{\partial x} = -6 \tag{21.27}$$

      b)

$$\dot{x} = \frac{\partial H}{\partial \lambda} = 16y \tag{}$$

From (21.26),

$$\dot{x} = 16(2\lambda) = 32\lambda \tag{21.28}$$

Integrating (21.27),

$$\lambda(t) = -6t + c_1 \tag{21.29}$$

Substituting in (21.28) and then integrating,

$$\dot{x} = 32(-6t + c_1) = -192t + 32c_1$$

$$x(t) = -96t^2 + 32c_1 t + c_2 \tag{21.30}$$

C. Applying the boundary conditions, $x(0) = 24$, $x(2) = 408$,

$$x(0) = c_2 = 24 \qquad\qquad c_2 = 24$$
$$x(2) = -96(2)^2 + 32c_1(2) + 24 = 408 \qquad c_1 = 12$$

Substituting $c_1 = 12$ and $c_2 = 24$ in (21.29) and (21.30),

$$\lambda(t) = -6t + 12 \qquad \text{costate variable} \qquad (21.31)$$

$$x(t) = -96t^2 + 384t + 24 \qquad \text{state variable} \qquad (21.32)$$

D.   For the control variable solution, we then either substitute (21.31) into (21.26),

$$y(t) = 2\lambda = 2(-6t + 12) = -12t + 24 \qquad \text{control variable}$$

or take the derivative of (21.32) and substitute it in the constraint $\dot{x} = 16y$.

$$\dot{x} = -192t + 384 = 16y$$
$$y(t) = -12t + 24 \qquad \text{control variable}$$

E.   The sufficiency conditions are met in analogous fashion to Example 2.

21.2.   Maximize

$$\int_0^1 (5x + 3y - 2y^2)\, dt$$

subject to

$$\dot{x} = 6y$$
$$x(0) = 7 \qquad x(1) = 70$$

A.   The Hamiltonian is

$$H = 5x + 3y - 2y^2 + \lambda(6y)$$

B.   The necessary conditions from the maximum principle are

1.
$$\frac{\partial H}{\partial y} = 3 - 4y + 6\lambda = 0$$

$$y = 0.75 + 1.5\lambda \qquad (21.33)$$

2. *a)*
$$\dot{\lambda} = -\frac{\partial H}{\partial x} = -5 \qquad (21.34)$$

b)
$$\dot{x} = \frac{\partial H}{\partial \lambda} = 6y$$

From (21.33), $\dot{x} = 6(0.75 + 1.5\lambda) = 4.5 + 9\lambda \qquad (21.35)$

Integrating (21.34), $\lambda(t) = -5t + c_1 \qquad (21.36)$

Substituting in (21.35) and then integrating,

$$\dot{x} = 4.5 + 9(-5t + c_1) = 4.5 - 45t + 9c_1$$

$$x(t) = 4.5t - 22.5t^2 + 9c_1 t + c_2 \qquad (21.37)$$

C.   Applying the boundary conditions, $x(0) = 7$, $x(1) = 70$,

$$x(0) = c_2 = 7 \qquad\qquad\qquad c_2 = 7$$
$$x(1) = 4.5 - 22.5 + 9c_1 + 7 = 70 \qquad c_1 = 9$$

Substituting $c_1 = 9$ and $c_2 = 7$ in (21.36) and (21.37),

$$\lambda(t) = -5t + 9 \qquad \text{costate variable} \qquad (21.38)$$

$$x(t) = -22.5t^2 + 85.5t + 7 \qquad \text{state variable} \qquad (21.39)$$

D.   For the final solution, we then simply substitute (21.38) into (21.33),

$$y(t) = 0.75 + 1.5(-5t + 9) = -7.5t + 14.25 \qquad \text{control variable}$$

E.   The sufficiency conditions are once again similar to Example 2.

21.3. Maximize

$$\int_0^5 (-8x - y^2)\, dt$$

subject to

$$\dot{x} = 0.2y$$
$$x(0) = 3 \qquad x(5) = 5.6$$

A.

$$H = -8x - y^2 + \lambda(0.2y)$$

B. 1.

$$\frac{\partial H}{\partial y} = -2y + 0.2\lambda$$

$$y = 0.1\lambda \tag{21.40}$$

2. a)

$$\dot{\lambda} = -\frac{\partial H}{\partial x} = -(-8) = 8 \tag{21.41}$$

b)

$$\dot{x} = \frac{\partial H}{\partial \lambda} = 0.2y$$

From (21.40),

$$\dot{x} = 0.2(0.1\lambda) = 0.02\lambda \tag{21.42}$$

Integrating (21.41),

$$\lambda(t) = 8t + c_1 \tag{21.43}$$

Substituting in (21.42) and then integrating,

$$\dot{x} = 0.02(8t + c_1) = 0.16t + 0.02c_1$$

$$x(t) = 0.08t^2 + 0.02c_1 t + c_2 \tag{21.44}$$

C. Applying the boundary conditions, $x(0) = 3$, $x(5) = 5.6$,

$$x(0) = c_2 = 3 \qquad\qquad\qquad c_2 = 3$$
$$x(5) = 0.08(5)^2 + 0.02c_1(5) + 3 = 5.6 \qquad c_1 = 6$$

Substituting $c_1 = 6$ and $c_2 = 3$ in (21.43) and (21.44),

$$\lambda(t) = 8t + 6 \qquad \text{costate variable} \tag{21.45}$$

$$x(t) = 0.08t^2 + 0.12t + 3 \qquad \text{state variable} \tag{21.46}$$

D. Substituting (21.45) into (21.40) for the final solution,

$$y(t) = 0.1\lambda = 0.8t + 0.6 \qquad \text{control variable}$$

or taking the derivative of (21.46) and substituting it in the constraint $\dot{x} = 0.2y$,

$$\dot{x} = 0.16t + 0.12 = 0.2y$$
$$y(t) = 0.8t + 0.6 \qquad \text{control variable}$$

FREE ENDPOINTS

21.4. Maximize

$$\int_0^4 (8x - 10y^2)\, dt$$

subject to

$$\dot{x} = 24y$$
$$x(0) = 7 \qquad x(4) \qquad \text{free}$$

A.

$$H = 8x - 10y^2 + \lambda(24y)$$

B. 1.

$$\frac{\partial H}{\partial y} = -20y + 24\lambda = 0$$

$$y = 1.2\lambda \tag{21.47}$$

2. a)

$$\dot{\lambda} = -\frac{\partial H}{\partial x} = -(8) = -8 \tag{21.48}$$

b) $$\dot{x} = \frac{\partial H}{\partial \lambda} = 24y$$

From (21.47), $$\dot{x} = 24(1.2\lambda) = 28.8\lambda$$ (21.49)

Integrating (21.48), $$\lambda(t) = -8t + c_1$$

Substituting in (21.49), $$\dot{x} = 28.8(-8t + c_1) = -230.4t + 28.8c_1$$

and then integrating, $$x(t) = -115.2t^2 + 28.8c_1t + c_2$$ (21.50)

C. We now apply the transversality condition $\lambda(4) = 0$.

$$\lambda(4) = -8(4) + c_1 = 0 \qquad c_1 = 32$$

Therefore, $$\lambda(t) = -8t + 32 \quad \text{costate variable}$$ (21.51)

Next we substitute $c_1 = 32$ in (21.50) and apply the initial condition $x(0) = 7$.

$$x(t) = -115.2t^2 + 921.6t + c_2$$
$$x(0) = 0 + 0 + c_2 = 7 \qquad c_2 = 7$$

So, $$x(t) = -115.2t^2 + 921.6t + 7 \quad \text{state variable}$$

D. Then substituting (21.51) in (21.47), we have the final solution,

$$y(t) = 1.2(-8t + 32) = -9.6t + 38.4 \quad \text{control variable}$$

E. The sufficiency conditions are easily confirmed as in Example 2.

21.5. Maximize $$\int_0^3 (2x + 18y - 3y^2)\, dt$$

subject to $$\dot{x} = 12y + 7$$
$$x(0) = 5 \qquad x(3) \quad \text{free}$$

A. $$H = 2x + 18y - 3y^2 + \lambda(12y + 7)$$

B. 1. $$\frac{\partial H}{\partial y} = 18 - 6y + 12\lambda = 0$$

$$y = 3 + 2\lambda$$ (21.52)

2. a) $$\dot{\lambda} = -\frac{\partial H}{\partial x} = -(2) = -2$$ (21.53)

b) $$\dot{x} = \frac{\partial H}{\partial \lambda} = 12y + 7$$

From (21.52), $$\dot{x} = 12(3 + 2\lambda) + 7 = 43 + 24\lambda$$ (21.54)

Integrating (21.53), $$\lambda(t) = -2t + c_1$$

Substituting in (21.54), $$\dot{x} = 43 + 24(-2t + c_1) = 43 - 48t + 24c_1$$

and then integrating, $$x(t) = 43t - 24t^2 + 24c_1t + c_2$$ (21.55)

C. We now resort to the transversality condition $\lambda(3) = 0$.

$$\lambda(3) = -2(3) + c_1 = 0 \qquad c_1 = 6$$

Therefore, $$\lambda(t) = -2t + 6 \quad \text{costate variable}$$ (21.56)

Next we substitute $c_1 = 6$ in (21.55) and apply the initial condition $x(0) = 5$.

$$x(t) = -24t^2 + 187t + c_2$$
$$x(0) = 0 + 0 + c_2 = 5 \qquad c_2 = 5$$

So, $x(t) = -24t^2 + 187t + 5$ state variable

D.   Then substituting (21.56) in (21.52), we have the final solution,

$$y(t) = 3 + 2(-2t + 6) = -4t + 15 \qquad \text{control variable}$$

E.   The sufficiency conditions once again follow Example 2.

21.6.   Maximize $$\int_0^1 (4y - y^2 - x - 2x^2) \, dt$$

subject to $$\dot{x} = x + y$$
$$x(0) = 6.15 \qquad x(1) \qquad \text{free}$$

A. $$H = 4y - y^2 - x - 2x^2 + \lambda(x + y)$$

B.   1. $$\frac{\partial H}{\partial y} = 4 - 2y + \lambda = 0$$

$$y = 2 + 0.5\lambda \tag{21.57}$$

2. $a)$ $$\dot{\lambda} = -\frac{\partial H}{\partial x} = -(-1 - 4x + \lambda) = 1 + 4x - \lambda$$

$b)$ $$\dot{x} = \frac{\partial H}{\partial \lambda} = x + y$$

From y in (21.57), $$\dot{x} = x + 2 + 0.5\lambda$$

In matrix form,

$$\begin{bmatrix} \dot{\lambda} \\ \dot{x} \end{bmatrix} = \begin{bmatrix} -1 & 4 \\ 0.5 & 1 \end{bmatrix} \begin{bmatrix} \lambda \\ x \end{bmatrix} + \begin{bmatrix} 1 \\ 2 \end{bmatrix}$$

$$\mathbf{Y} = \mathbf{AX} + \mathbf{B}$$

Using (19.3) for the characteristic roots,

$$r_1, r_2 = \frac{0 \pm \sqrt{0 - 4(-3)}}{2} = \frac{\pm 3.464}{2} = \pm 1.732$$

The eigenvector corresponding to $r_1 = 1.732$ is

$$|\mathbf{A} - r\mathbf{I}| = \begin{bmatrix} -1 - 1.732 & 4 \\ 0.5 & 1 - 1.732 \end{bmatrix} \begin{bmatrix} c_1 \\ c_2 \end{bmatrix} = \begin{bmatrix} -2.732 & 4 \\ 0.5 & -0.732 \end{bmatrix} \begin{bmatrix} c_1 \\ c_2 \end{bmatrix} = 0$$

$$-2.732c_1 + 4c_2 = 0 \qquad c_1 = 1.464c_2$$

$$y_c^1 = \begin{bmatrix} 1.464 \\ 1 \end{bmatrix} k_1 e^{1.732t}$$

The eigenvector corresponding to $r_2 = -1.732$ is

$$|\mathbf{A} - r\mathbf{I}| = \begin{bmatrix} -1 + 1.732 & 4 \\ 0.5 & 1 + 1.732 \end{bmatrix} \begin{bmatrix} c_1 \\ c_2 \end{bmatrix} = \begin{bmatrix} 0.732 & 4 \\ 0.5 & 2.732 \end{bmatrix} \begin{bmatrix} c_1 \\ c_2 \end{bmatrix} = 0$$

$$0.732c_1 + 4c_2 = 0 \qquad c_1 = -5.464c_2$$

$$y_c^2 = \begin{bmatrix} -5.464 \\ 1 \end{bmatrix} k_2 e^{-1.732t}$$

For the particular solution, $\bar{\mathbf{Y}} = -\mathbf{A}^{-1}\mathbf{B}$.

$$\bar{\mathbf{Y}} = \begin{bmatrix} \bar{\lambda} \\ \bar{x} \end{bmatrix} = -\left(\frac{1}{-3}\right)\begin{bmatrix} 1 & -4 \\ -0.5 & -1 \end{bmatrix}\begin{bmatrix} 1 \\ 2 \end{bmatrix} = \begin{bmatrix} -2.33 \\ -0.83 \end{bmatrix}$$

Combining the complementary and particular solutions,

$$\lambda(t) = 1.464k_1 e^{1.732t} - 5.464k_2 e^{-1.732t} - 2.33 \tag{21.58}$$

$$x(t) = k_1 e^{1.732t} + k_2 e^{-1.732t} - 0.83 \tag{21.59}$$

C. Applying the transversality condition for a free endpoint $\lambda(1) = 0$,

$$\lambda(1) = 8.2744k_1 - 0.9667k_2 - 2.33 = 0$$

From the initial condition $x(0) = 6.15$,

$$x(0) = k_1 + k_2 - 0.83 = 6.15$$

Solved simultaneously, $k_1 = 0.98,$ $k_2 = 5.95$

Substituting in (21.58) and (21.59),

$$\lambda(t) = 1.4347e^{1.732t} - 32.511e^{-1.732t} - 2.33 \qquad \text{costate variable} \tag{21.60}$$

$$x(t) = 0.98e^{1.732t} + 5.95e^{-1.732t} - 0.83 \qquad \text{state variable} \tag{21.61}$$

D. Finally, substituting (21.60) in (21.57), we find the solution.

$$y(t) = 0.7174e^{1.732t} - 16.256e^{-1.732t} + 0.835$$

E. For the sufficiency conditions, with $f = 4y - y^2 - x - 2x^2$, $f_x = -1 - 4x$, and $f_y = 4 - 2y$, we have

$$\mathbf{D} = \begin{vmatrix} f_{xx} & f_{xy} \\ f_{yx} & f_{yy} \end{vmatrix} = \begin{vmatrix} -4 & 0 \\ 0 & -2 \end{vmatrix}$$

$$|\mathbf{D}_1| = -4 < 0 \qquad |\mathbf{D}_2| = 8 > 0$$

Therefore, $\mathbf{D}$ is negative-definite and f is strictly concave. With the constraint $(x + y)$ linear and hence also concave, the sufficiency conditions for a global maximization in optimal control theory are satisfied.

INEQUALITY CONSTRAINTS

21.7. Maximize $\displaystyle\int_0^4 (8x - 10y^2)\, dt$

subject to $\dot{x} = 24y$

$$x(0) = 7 \qquad x(4) \geq 2000$$

1. For inequality constraints, we always start with a free endpoint. This problem was previously solved as a free endpoint in Problem 21.4 where we found

$$\lambda(t) = -8t + 32 \qquad\qquad \text{costate variable}$$
$$x(t) = -115.2t^2 + 921.6t + 7 \qquad \text{state variable}$$
$$y(t) = 1.2(-8t + 32) = -9.6t + 38.4 \qquad \text{control variable}$$

Evaluating the state variable at $t = 4$, we have

$$x(4) = 1850.2 < 2000 \qquad \text{a constraint violation}$$

2. Faced with a constraint violation, we have to redo the problem with a new fixed terminal endpoint:

$$x(0) = 7 \qquad x(4) = 2000$$

Solution of this new problem follows along exactly the same as in Problem 21.4 until the end of part B where we found

$$\lambda(t) = -8t + c_1$$
$$x(t) = -115.2t^2 + 28.8c_1 t + c_2$$

Now instead of the transversality condition, we apply each of the boundary conditions.

$$x(0) = -115.2t^2 + 28.8c_1 t + c_2 = 7 \qquad c_2 = 7$$

Substituting $c_2 = 7$ in the terminal boundary and solving,

$$x(4) = -115.2t^2 + 28.8c_1 t + 7 = 2000 \qquad c_1 = 33.3$$

This gives us

$$\lambda(t) = -8t + 33.3 \qquad\qquad \text{costate variable}$$
$$x(t) = -115.2t^2 + 959.04t + 7 \qquad \text{state variable}$$

Then for the final solution, from (21.46) we have

$$y(t) = 1.2(-8t + 33.3) = -9.6t + 39.96 \qquad \text{control variable}$$

21.8. Maximize

$$\int_0^1 (5x + 3y - 2y^2)\, dt$$

subject to

$$\dot{x} = 6y$$
$$x(0) = 7 \qquad x(1) \geq 70$$

1. For an inequality constraint, we always start with a free terminal endpoint. Here we can build on the work already done in solving this problem under fixed endpoint conditions in Problem 21.2. There we found in (21.36) and (21.37),

$$\lambda(t) = -5t + c_1$$
$$x(t) = 4.5t - 22.5t^2 + 9c_1 t + c_2$$

Now applying the transversality condition $\lambda(T) = 0$,

$$\lambda(1) = -5 + c_1 = 0 \qquad c_1 = 5$$

and then substituting $c_1 = 5$ and solving for the initial condition,

$$x(0) = 4.5t - 22.5t^2 + 45t + c_2 = 7 \qquad c_2 = 7$$

This leads to, $$\lambda(t) = -5t + 5 \qquad\qquad \text{costate variable}$$
$$x(t) = -22.5t^2 + 49.5t + 7 \qquad \text{state variable}$$

Then from $y = 0.75 + 1.5\lambda$ in (21.33), the solution is,

$$y(t) = -7.5t + 8.25 \qquad \text{control variable}$$

To see if the solution is acceptable, we evaluate the state variable at $t = 1$.

$$x(1) = -22.5t^2 + 49.5t + 7 = 34 < 70 \qquad \text{constraint violation}$$

2. With the terminal constraint violated, we must now rework the problem with a fixed endpoint $x(1) = 70$. We did this earlier in Problem 21.2 where we found

$$\lambda(t) = -5t + 9 \qquad\qquad \text{costate variable}$$
$$x(t) = -22.5t^2 + 85.5t + 7 \qquad \text{state variable}$$
$$y(t) = -7.5t + 14.25 \qquad\qquad \text{control variable}$$

To determine if the control variable is an acceptable solution, we evaluate the state variable at the terminal endpoint and see that it fulfills the constraint.

$$x(1) = -22.5t^2 + 85.5t + 7 = 70$$

21.9. Maximize
$$\int_0^1 (4y - y^2 - x - 2x^2)\,dt$$

subject to
$$\dot{x} = x + y$$
$$x(0) = 6.15 \qquad x(1) \geq 5$$

We have already optimized this function subject to free endpoint conditions in Problem 21.6. There we found,

$$\lambda(t) = 1.4347e^{1.732t} - 32.784e^{-1.732t} - 2.33 \qquad \text{costate variable}$$
$$x(t) = 0.98e^{1.732t} + 6e^{-1.732t} - 0.83 \qquad \text{state variable}$$
$$y(t) = 0.7174e^{1.732t} - 16.392e^{-1.732t} + 0.835 \qquad \text{control variable}$$

Evaluating the state variable at the terminal endpoint, we have

$$x(1) = 0.98e^{1.732t} + 6e^{-1.732t} - 0.83 = 5.7705 > 5$$

Since the endpoint constraint is satisfied, we have found the solution to the problem.

21.10. Redo Problem 21.9 with a new set of endpoint constraints,

$$x(0) = 6.15 \qquad x(1) \geq 8$$

From Problem 21.9 we know the free endpoint solution fails to meet the new terminal endpoint constraint $x(1) \geq 8$. We must optimize under fixed endpoint conditions, therefore, by setting $x(1) = 8$.
Starting from (21.55) and (21.56) where we found

$$\lambda(t) = 1.464k_1 e^{1.732t} - 5.464k_2 e^{-1.732t} - 2.33$$
$$x(t) = k_1 e^{1.732t} + k_2 e^{-1.732t} - 0.83$$

we apply the endpoint conditions where we have

$$\text{at } x(0) = 6.15, \qquad k_1 \ + \ k_2 \ - 0.83 = 6.15$$
$$\text{at } x(1) = 8, \qquad 5.6519k_1 + 0.1769k_2 - 0.83 = 8$$

When solved simultaneously,

$$k_1 = 1.3873 \qquad k_2 = 5.5927$$

Substituting in (21.55) and (21.56) repeated immediately above,

$$\lambda(t) = 2.0310e^{1.732t} - 30.5585e^{-1.732t} - 2.33 \qquad \text{costate variable}$$
$$x(t) = 1.3873e^{1.732t} + 5.5927e^{-1.732t} - 0.83 \qquad \text{state variable}$$

Then from $y = 2 + 0.5\lambda$ in (21.54), we derive the solution,

$$y(t) = 1.0155e^{1.732t} - 15.2793e^{-1.732t} - 0.835 \qquad \text{control variable}$$

Finally, to be sure the solution is acceptable, we evaluate the state variable at $t = 1$,

$$x(1) = 1.3873e^{1.732t} + 5.5927e^{-1.732t} - 0.83 = 8$$

and see that it satisfies the terminal endpoint constraint.

CURRENT-VALUED HAMILTONIANS

21.11. Maximize
$$\int_0^3 e^{-0.05t}(xy - x^2 - y^2)\,dt$$

subject to
$$\dot{x} = x + y$$
$$x(0) = 134.35 \qquad x(3) \qquad \text{free}$$

A. Setting up the current-valued Hamiltonian,

$$H_c = xy - x^2 - y^2 + \mu(x + y)$$

B.   Assuming an interior solution, we first apply the modified maximum principle.

1.
$$\frac{\partial H_c}{\partial y} = x - 2y + \mu = 0$$

$$y = 0.5(x + \mu) \tag{21.59}$$

2. *a)*
$$\dot{\mu} = \rho\mu - \frac{\partial H_c}{\partial x}$$

$$\dot{\mu} = 0.05\mu - (y - 2x + \mu)$$
$$\dot{\mu} = -0.95\mu + 2x - y$$

Using (21.59),
$$\dot{\mu} = -0.95\mu + 2x - 0.5(x + \mu)$$
$$\dot{\mu} = -1.45\mu + 1.5x$$

b)
$$\dot{x} = \frac{\partial H_c}{\partial \mu} = x + y$$

From (21.59),
$$\dot{x} = 1.5x + 0.5\mu$$

In matrix form,

$$\begin{bmatrix} \dot{\mu} \\ \dot{x} \end{bmatrix} = \begin{bmatrix} -1.45 & 1.5 \\ 0.5 & 1.5 \end{bmatrix} \begin{bmatrix} \mu \\ x \end{bmatrix} + \begin{bmatrix} 0 \\ 0 \end{bmatrix}$$

The characteristic equation is

$$|\mathbf{A} - r\mathbf{I}| = \begin{vmatrix} -1.45 - r & 1.5 \\ 0.5 & 1.5 - r \end{vmatrix} = 0$$

and from (19.3), the characteristic roots are

$$r_1, r_2 = \frac{0.05 \pm \sqrt{(0.05)^2 - 4(-2.925)}}{2}$$

$$r_1 = 1.7354 \qquad r_2 = -1.6855$$

For $r_1 = 1.7354$, the eigenvector is

$$\begin{bmatrix} -1.45 - 1.7354 & 1.5 \\ 0.5 & 1.5 - 1.7354 \end{bmatrix} \begin{bmatrix} c_1 \\ c_2 \end{bmatrix} = \begin{bmatrix} -3.1854 & 1.5 \\ 0.5 & -0.2354 \end{bmatrix} \begin{bmatrix} c_1 \\ c_2 \end{bmatrix} = 0$$

$$-3.1854c_1 + 1.5c_2 = 0 \qquad c_2 = 2.1236c_1$$

$$y_c^1 = \begin{bmatrix} 1 \\ 2.1326 \end{bmatrix} k_1 e^{1.7354t}$$

For $r_2 = -1.6855$, the eigenvector is

$$\begin{bmatrix} -1.45 + 1.6855 & 1.5 \\ 0.5 & 1.5 + 1.6855 \end{bmatrix} \begin{bmatrix} c_1 \\ c_2 \end{bmatrix} = \begin{bmatrix} 0.2355 & 1.5 \\ 0.5 & 3.1855 \end{bmatrix} \begin{bmatrix} c_1 \\ c_2 \end{bmatrix} = 0$$

$$0.2355c_1 + 1.5c_2 = 0 \qquad c_1 = -6.3694c_2$$

$$y_c^2 = \begin{bmatrix} -6.3694 \\ 1 \end{bmatrix} k_2 e^{-1.6855t}$$

With $\mathbf{B} = 0$ in $\bar{\mathbf{Y}} = -\mathbf{A}^{-1}\mathbf{B}$,
$$\bar{\mathbf{Y}} = \begin{bmatrix} \bar{\mu} \\ \bar{x} \end{bmatrix} = \begin{bmatrix} 0 \\ 0 \end{bmatrix}$$

Adding the complementary functions for the general solution, we have

$$\mu(t) = k_1 e^{1.7354t} - 6.3694k_2 e^{-1.6855t} \tag{21.60}$$

$$x(t) = 2.1236k_1 e^{1.7354t} + k_2 e^{-1.6855t} \tag{21.61}$$

C. Next we apply the transversality condition for the free endpoint, $\mu(T)e^{-\rho t} = 0$,

$$\mu(3)e^{-0.05(3)} = (k_1 e^{1.7354(3)} - 6.3694 k_2 e^{-1.6855(3)})e^{-0.15} = 0$$

$$156.9928 k_1 - 0.0349 k_2 = 0 \qquad\qquad (21.62)$$

and evaluate $x(t)$ at $x(0) = 134.25$,

$$x(0) = 2.1236 k_1 + k_2 = 134.25 \qquad\qquad (21.63)$$

Solving (21.62) and (21.63) simultaneously,

$$k_1 = 0.03 \qquad k_2 = 134.1819$$

Then substituting $k_1 = 0.03$, $k_2 = 134.1819$ in (21.60) and (21.61), we find

$$\mu(t) = 0.03 e^{1.7354t} - 854.6582 e^{-1.6855t} \qquad \text{costate variable}$$
$$x(t) = 0.0637 e^{1.7354t} + 134.1819 e^{-1.6855t} \qquad \text{state variable}$$

D. From (21.59), $y(t) = 0.5(x + \mu)$. Substituting from above,

$$y(t) = 0.04685 e^{1.7354t} - 360.2381 e^{-1.6855t} \qquad \text{control variable}$$

E. For the sufficiency conditions,

$$\mathbf{D} = \begin{vmatrix} f_{xx} & f_{xy} \\ f_{yx} & f_{yy} \end{vmatrix} = \begin{vmatrix} -2 & 1 \\ 1 & -2 \end{vmatrix}$$

With $|\mathbf{D}_1| = -2 < 0$, and $|\mathbf{D}_2| = 3 > 0$, $\mathbf{D}$ is negative-definite, making f strictly concave in both x and y. With g linear in x and y, the sufficiency condition for a global maximum is fulfilled.

21.12. Maximize

$$\int_0^1 e^{-0.08t}(10x + 4y + xy - 2x^2 - 0.5y^2)\, dt$$

subject to

$$\dot{x} = x + 2y$$
$$x(0) = 88.52 \qquad x(1) \qquad \text{free}$$

A. $$H_c = 10x + 4y + xy - 2x^2 - 0.5y^2 + \mu(x + 2y)$$

B. Assuming an interior solution,

1. $$\frac{\partial H_c}{\partial y} = 4 + x - y + 2\mu = 0$$
    $$y = x + 2\mu + 4 \qquad\qquad (21.64)$$

2. a) $$\dot{\mu} = \rho\mu - \frac{\partial H_c}{\partial x}$$
    $$\dot{\mu} = 0.08\mu - (y - 4x + \mu + 10)$$
    $$\dot{\mu} = -0.92\mu + 4x - y - 10$$

Substituting for y from (21.64)

$$\dot{\mu} = -2.92\mu + 3x - 14$$

b) $$\dot{x} = \frac{\partial H_c}{\partial \mu} = x + 2y$$

From (21.64),

$$\dot{x} = 4\mu + 3x + 8$$

In matrix form,

$$\begin{bmatrix} \dot{\mu} \\ \dot{x} \end{bmatrix} = \begin{bmatrix} -2.92 & 3 \\ 4 & 3 \end{bmatrix}\begin{bmatrix} \mu \\ x \end{bmatrix} + \begin{bmatrix} -14 \\ 8 \end{bmatrix}$$

where the characteristic equation is

$$|\mathbf{A} - r\mathbf{I}| = \begin{vmatrix} -2.92 - r & 3 \\ 4 & 3 - r \end{vmatrix} = 0$$

and the characteristic roots are

$$r_1, r_2 = \frac{0.08 \pm \sqrt{(0.08)^2 - 4(-20.76)}}{2}$$

$$r_1 = 4.5965 \qquad r_2 = -4.5165$$

The eigenvector for $r_1 = 4.5965$ is

$$\begin{bmatrix} -2.92 - 4.5965 & 3 \\ 4 & 3 - 4.5965 \end{bmatrix} \begin{bmatrix} c_1 \\ c_2 \end{bmatrix} = \begin{bmatrix} -7.5165 & 3 \\ 4 & -1.5965 \end{bmatrix} \begin{bmatrix} c_1 \\ c_2 \end{bmatrix} = 0$$

$$-7.5165c_1 + 3c_2 = 0 \qquad c_2 = 2.5055c_1$$

$$y_c^1 = \begin{bmatrix} 1 \\ 2.5055 \end{bmatrix} k_1 e^{4.5965t}$$

The eigenvector for $r_2 = -4.5165$ is

$$\begin{bmatrix} -2.92 + 4.5165 & 3 \\ 4 & 3 + 4.5165 \end{bmatrix} \begin{bmatrix} c_1 \\ c_2 \end{bmatrix} = \begin{bmatrix} 1.5965 & 3 \\ 4 & 7.5165 \end{bmatrix} \begin{bmatrix} c_1 \\ c_2 \end{bmatrix} = 0$$

$$1.5965c_1 + 3c_2 = 0 \qquad c_1 = -1.8791c_2$$

$$y_c^2 = \begin{bmatrix} -1.8791 \\ 1 \end{bmatrix} k_2 e^{-4.5165t}$$

For the particular solutions, $\bar{\mathbf{Y}} = -\mathbf{A}^{-1}\mathbf{B}$.

$$\bar{\mathbf{Y}} = \begin{bmatrix} \bar{\mu} \\ \bar{x} \end{bmatrix} = -\left(\frac{1}{-20.76}\right) \begin{bmatrix} 3 & -3 \\ -4 & -2.92 \end{bmatrix} \begin{bmatrix} -14 \\ 8 \end{bmatrix} = \begin{bmatrix} -3.1792 \\ 1.5723 \end{bmatrix}$$

Combining the complementary functions and particular solutions for the general solution,

$$\mu(t) = k_1 e^{4.5965t} - 1.8791k_2 e^{-4.5165t} - 3.1792 \qquad (21.65)$$

$$x(t) = 2.5055k_1 e^{4.5965t} + k_2 e^{-4.5165t} + 1.5723 \qquad (21.66)$$

C. Applying the transversality condition for the free endpoint, $\mu(T)e^{-\rho t} = 0$,

$$\mu(1)e^{-0.08(1)} = (k_1 e^{4.5965(1)} - 1.8791k_2 e^{-4.5165(1)} - 3.1792)e^{-0.08} = 0$$

$$91.5147k_1 - 0.0189k_2 - 2.9348 = 0 \qquad (21.67)$$

and evaluating $x(t)$ at $x(0) = 88.52$,

$$x(0) = 2.5055k_1 + k_2 + 1.5723 = 88.52 \qquad (21.68)$$

Solving (21.67) and (21.68) simultaneously,

$$k_1 = 0.05 \qquad k_2 = 86.8230$$

Then substituting these values in (21.65) and (21.66), we find

$$\mu(t) = 0.05e^{4.5965t} - 163.1491e^{-4.5165t} - 3.1792 \qquad \text{costate variable}$$
$$x(t) = 0.1253e^{4.5965t} + 86.8230e^{-4.5165t} + 1.5723 \qquad \text{state variable}$$

D. From (21.64), $y(t) = x + 2\mu + 4$. Substituting for x and μ from above,

$$y(t) = 0.2253e^{4.5965t} - 239.4752^{-4.5165t} - 0.7861 \qquad \text{control variable}$$

E. For the sufficiency conditons,

$$\mathbf{D} = \begin{vmatrix} f_{xx} & f_{xy} \\ f_{yx} & f_{yy} \end{vmatrix} = \begin{vmatrix} -4 & 1 \\ 1 & -1 \end{vmatrix}$$

$|\mathbf{D}_1| = -4 < 0$, and $|\mathbf{D}_2| = 3 > 0$. $\mathbf{D}$ is negative-definite and f is strictly concave in both x and y. Since the constraint is linear in x and y, the sufficiency condition for a global maximum is satisfied.

CHAPTER 22

Series in Economics: Descriptive Statistics and Linear Regression

22.1 SERIES

A series s_n is a summation of n elements. For example, the set $\{a_n\}$ has as elements 2, 3, 7, 13, 15. The series of the first four terms can be represented as s_4

$$s_4 = a_1 + a_2 + a_3 + a_4$$

$$s_4 = 2 + 3 + 7 + 13 = 25$$

Series s_n are usually expressed using the **summation operator** (Σ, which is capital letter sigma). Thus, s_n is also called **partial sum**. The use of this notation permits to represent a large number of elements. The series s_n within a set $\{x_n\}$ can be represented as

$$\sum_{i=1}^{n} x_i = x_1 + x_2 + x_3 + \cdots + x_n = s_n$$

Examples 1 and 2 provide examples on how to compute series.

EXAMPLE 1. Find the series s_4 given the set $\{6, 5, 4, 3, 2, 1\}$

s_4 is the summation of the first four elements.

$$s_4 = \sum_{i=1}^{4} a_i = a_1 + a_2 + a_3 + a_4 = 6 + 5 + 4 + 3$$

$$s_4 = \sum_{i=1}^{4} a_i = 18$$

EXAMPLE 2. Find the partial sum s_3 given the set $\{-3, 7, -4, 1, 4, -8\}$

$$s_3 = \sum_{i=1}^{3} a_i = a_1 + a_2 + a_3 = -3 + 7 - 4$$

$$s_3 = \sum_{i=1}^{3} a_i = 0$$

In many cases, the series is defined by a specific rule. Examples 3 and 4 show this notion.

EXAMPLE 3. Given the set $\{-1, 4, -2 \text{ and } 8\}$. Find

$$\sum_{i=1}^{3}(1 - x_i^2) = (1 - x_1^2) + (1 - x_2^2) + (1 - x_3^2)$$

$$\sum_{i=1}^{3}(1 - x_i^2) = (1 - (-1)^2) + (1 - (4)^2) + (1 - (-2)^2)$$

$$\sum_{i=1}^{3}(1 - x_i^2) = 2 - 15 - 3 = -16$$

EXAMPLE 4. Compute the series.

$$\sum_{i=1}^{5}(-1)^i \, i^2 = -1(1^2) + 1(2^2) - 1(3^2) + 1(4^2) - 1(5^2)$$

$$\sum_{i=1}^{5}(-1)^i i^2 = -1 + 4 - 9 + 16 - 25 = -15$$

22.2 PROPERTIES OF SUMMATIONS

Assuming a, b and c are constants, n is the number of observations within a dataset, x_i and y_i are elements within a defined set of samples, the *properties of summations* are outlined below and illustrated in Examples 5 to 7.

1. $\displaystyle\sum_{i=1}^{n} cx_i = c \sum_{i=1}^{n} x_i$

2. $\displaystyle\sum_{i=1}^{n} c = cn$

3. $\displaystyle\sum_{i=1}^{n}(ax_i \pm by_i) = a \sum_{i=1}^{n} x_i + b \sum_{i=1}^{n} y_i$

EXAMPLE 5. Compute the operation.

$$\sum_{i=1}^{4} 5i = 5 \sum_{i=1}^{4} i = 5(1 + 2 + 3 + 4) = 5(10) = 50$$

EXAMPLE 6.   Compute the summation.

$$\sum_{i=1}^{6} 8 = (8+8+8+8+8+8) = 6(8) = 48$$

EXAMPLE 7.   Compute the summation operations.

$$\sum_{i=1}^{3} (5x_i + 4y_i) = 5\sum_{i=1}^{3} x_i + 4\sum_{i=1}^{3} y_i$$

$$\sum_{i=1}^{3} (2i - (-1)^2) = 2\sum_{i=1}^{3} i - \sum_{i=1}^{3} (-1)^2 = 2(1+2+3) - (-1+1-1) = 12+1 = 13$$

22.3   DESCRIPTIVE STATISTICS: FREQUENCY, MEAN, AND VARIANCE

Economics relies on the use of statistical tools to present the analysis of studies. In particular, descriptive statistics are useful to summarize major characteristics of the data. Four measures are commonly discussed to described sample data: the frequency, the range, the mean, and the variance.

22.3.1   Frequency

The **frequency** (denoted as f) is the number of times that an observation appears in a dataset. The frequency can be arranged on a table, known as *frequency table*, that lists the repetition of each observation.

EXAMPLE 8.   Construct the frequency table for the dataset below:

2	2	3	4	7
1	2	1	3	6
4	3	4	2	6
1	4	1	5	1

$x_1 = 1$ appears 5 times, thus the frequency of 1 is 5 ($f_1 = 1$). The operation continues for all seven elements.

x_i	f_i
1	5
2	4
3	3
4	4
5	1
6	2
7	1

The frequency table can be visualized in a *bar chart*, whose height depends on the frequency of each element.

EXAMPLE 9.   The bar chart of the frequency table in example 5 is given in Fig. 22-1.

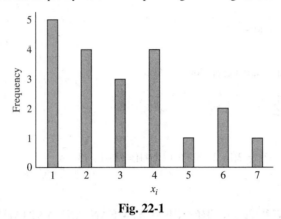

Fig. 22-1

Property: The sum of the frequencies of all elements must be equal to the number of elements.

EXAMPLE 10.   Verify that the frequency table is consistent.

$$\sum_{i=1}^{7} f_i = f_1 + f_2 + f_3 + f_4 + f_5 + f_6 + f_7 = 5 + 4 + 3 + 4 + 1 + 2 + 1$$

$$\sum_{i=1}^{7} f_i = 20 = n$$

22.3.2   Range

The **range** (symbolized as d) is the difference between the maximum and minimum value within the dataset.

EXAMPLE 11.   Find the range for the dataset given below:

21	19	23	40	17
11	5	3	23	26
27	33	34	14	15
19	4	12	12	9

The maximum value in the dataset is 40, whereas the minimum value is 3. Thus, the range is the difference of these values, $d = 40 - 3 = 37$. The range is 37.

22.3.3   Mean

The **mean** (denoted as $\overline{X}$) is a measure of tendency; it is defined as the average of all observations within the dataset.

$$\overline{X} = \frac{1}{n} \sum_{i=1}^{n} x_i$$

EXAMPLE 12. Compute the mean for the data below:

21	17
11	−27
−19	9

Here, the number of observations $n = 6$, the mean is

$$\bar{X} = \frac{1}{6}(21 + 17 + 11 - 26 - 19 + 9) = \frac{12}{6} = 2$$

22.3.4 Dispersion: Variance and Standard Deviation

The **variance** (denoted as S_x^2) is the sum of the squared difference of all observations from the mean $\bar{X}$,

$$S_x^2 = \frac{1}{n-1}\sum_i (x_i - \bar{X})^2$$

The **standard deviation** (represented as S_x) or *dispersion* is the squared root of the variance.

$$S_x = \sqrt{S_x^2} = \sqrt{\frac{1}{n-1}\sum_i (x_i - \bar{X})^2}$$

EXAMPLE 13. Compute the variance and standard deviation of the data below

21	−17
−19	11

First, compute the mean.

$$\bar{X} = \frac{1}{4}(21 - 17 - 19 + 11) = \frac{-4}{4} = -1$$

Thus, the variance is

$$S_x^2 = \frac{1}{4-1}\sum_i (x_i - (-1))^2 = \frac{1}{3}\sum_i (x_i + 1)^2$$

$$S_x^2 = \frac{(21+1)^2 + (-19+1)^2 + (-17+1)^2 + (11+1)^2}{3} = \frac{22^2 + 18^2 + 16^2 + 12^2}{3}$$

$$S_x^2 = \frac{1208}{3} = 402.6\bar{6}$$

The standard deviation is

$$S_x^2 = \sqrt{S_x^2} = \sqrt{402.6\bar{6}} = 20.067$$

22.4 PROBABILITY: THE DISCRETE CASE

The probability $\mathbf{P}(x_i = x)$ that an event x_i will occur is the frequency of the event divided by the total number of observations,

$$\mathbf{P}(x_i = x) = \frac{f_i}{n}$$

The profile of the probabilities of all events portray a probability density function (**pdf**), which is usually graphed with a bar chart.

EXAMPLE 14.   Given the frequency table below

x_i	f_i
1	5
2	4
3	3
4	4
5	1
6	2
7	1

Find the probability density function.

The total number of observations is $n = 20$. Thus, each frequency is divided by 20.

x_i	f_i	$P(x_i)$
1	5	0.25
2	4	0.20
3	3	0.15
4	4	0.20
5	1	0.05
6	2	0.10
7	1	0.05
Total	20	

EXAMPLE 15.   Graph the frequency table in Example 11; graph the probability density function.

See Fig. 22-2, which depicts the pdf in the last column of the table in Example 11.

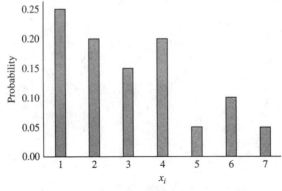

Fig. 22-2

This graph follows a similar proportion as the bar chart of the frequency.

A probability density function has three important properties:

1. $P(x_i) \geq 0$. Probability cannot be negative
2. $P(x_i) \leq 1$. A probability cannot be greater than 1.
3. $\sum_i P(x_i) = 1$. The sum of the probabilities of all events must be equal to 1.

EXAMPLE 16.   The sum of all probabilities in Example 11:

$$0.25 + 0.20 + 0.15 + 0.20 + 0.05 + 0.10 + 0.05 = 1$$

is equal to 1.

22.5   THE WEIGHTED AVERAGE

The weighted average ($\overline{W}$) is a type of average that takes into consideration the frequency of the observation within the dataset. It multiplies each observation by its frequency (or *weight*) and divides the sum by the number of observations,

$$\overline{W} = \frac{1}{n} \sum_{i=1}^{n} x_i f_i = \frac{1}{n}(x_1 f_1 + x_2 f_2 + \cdots + x_n f_n)$$

In a simple average $\overline{X}$, the weight is one (1), as each observation has the same importance.

EXAMPLE 17.   Find the weighted average of the following dataset.

x_i	f_i
10	4
20	5
30	1
40	2
50	3

The number of observations is $n = 4 + 5 + 1 + 2 + 3 = 15$. The weighted average is

$$\overline{W} = \frac{1}{15} \sum_{i=1}^{5} x_i f_i = \frac{1}{15}(10 \times 4 + 20 \times 5 + 30 \times 1 + 40 \times 2 + 50 \times 3)$$

$$\overline{W} = \frac{1}{15}(40 + 100 + 30 + 80 + 150) = \frac{400}{15}$$

$$\overline{W} = 26.\overline{66}$$

22.6   THE EXPECTED VALUE

The expected value (**EV**) is a weighted average in which the weight is the probability of the event. It is simply the sum of the multiplication of all terms with their respective probability,

$$EV(x) = \sum_{i=1}^{n} x_i p_i = x_1 p_1 + x_2 p_2 + \cdots + x_n p_n$$

EXAMPLE 18.   Find the expected value of the following dataset.

x_i	$P(x_i)$
1	0.25
2	0.20
3	0.15
4	0.05
5	0.35

There are five different events,

$$EV(x) = \sum_{i=1}^{5} x_i p_i = 1 \times 0.25 + 2 \times 0.20 + 3 \times 0.15 + 4 \times 0.05 + 5 \times 0.35$$

$$EV(x) = 0.25 + 0.40 + 0.45 + 0.20 + 1.75$$

$$EV(x) = 3.05$$

22.7   ECONOMIC APPLICATIONS: REGRESSION ANALYSIS

Regression modeling is used to study the relationship between variables. A simple linear regression model assumes that a variable y is linearly dependent on x, and it is represented by a linear equation.

$$y = \hat{a} + \hat{\beta}x + \varepsilon \tag{22.1}$$

y is referred to as the dependent variable and x is the independent variable. The goal is to estimate the intercept $\hat{a}$ and the slope $\hat{\beta}$ using data on (x, y), which can represent closely the database; in other words with the lowest amount of error ε, as shown in Fig. 22-3.

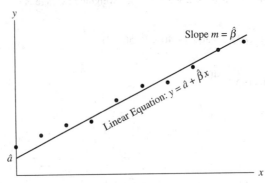

Fig. 22-3

Ordinary least square (**OLS**) is a method in econometrics commonly used to solve simple linear equations. The objective is to minimize the sum square of the errors by choosing appropriate estimates $\hat{a}$ and $\hat{\beta}$.

$$\min \sum_i \varepsilon_i^2 = \min \sum_i (y_i - \hat{a} - \hat{\beta} x_i)^2 \tag{22.2}$$

Apply the first-order condition on $\hat{a}$ and use the generalized power rule,

$$\frac{\partial}{\partial \hat{a}}\left[\sum_i (y_i - \hat{a} - \hat{\beta} x_i)^2 \right] = \sum_i 2(-1)(y_i - \hat{a} - \hat{\beta} x_i) = -2\sum_i (y_i - \hat{a} - \hat{\beta} x_i) = 0$$

Dividing by -2 and distributing the summation to each term,

$$= \sum_i y_i - \sum_i \hat{a} - \hat{\beta} \sum_i x_i = 0$$

Moving the summation term with $\hat{a}$ to the right-hand side.

$$= \sum_i y_i - \hat{\beta} \sum_i x_i = \sum_i \hat{a}$$

$\sum_i \hat{a} = \hat{a}n$ as $\hat{a}$ is a constant (not indexed by i). Dividing each by n,

$$\frac{\sum_i y_i}{n} - \frac{\hat{\beta} \sum_i x_i}{n} = \frac{\hat{a}n}{n}$$

$$\hat{a} = \bar{y} - \hat{\beta}\bar{x} \tag{22.3}$$

Apply the first-order condition on $\hat{\beta}$, and use the generalized power rule.

$$\frac{\partial}{\partial \hat{\beta}}\left[\sum_i (y_i - \hat{a} - \hat{\beta} x_i)^2 \right] = \sum_i 2(-x_i)(y_i - \hat{a} - \hat{\beta} x_i) = -2\sum_i (y_i x_i - \hat{a} x_i - \hat{\beta} x_i^2) = 0$$

Dividing by -2.

$$\sum_i (y_i x_i - \hat{a} x_i - \hat{\beta} x_i^2) = 0 \tag{22.4}$$

Substituting (*22.3*) in (*22.4*).

$$\sum_i (y_i x_i - (\bar{y} - \hat{\beta}\bar{x}) x_i - \hat{\beta} x_i^2) = 0$$

$$\sum_i (y_i x_i - \bar{y} x_i + \hat{\beta}\bar{x} x_i - \hat{\beta} x_i^2) = 0$$

$$\sum_i (y_i - \bar{y} + \hat{\beta}\bar{x} - \hat{\beta} x_i) x_i = 0$$

Split elements into two groups.

$$\sum_i (y_i - \bar{y} + \hat{\beta}(\bar{x} - x_i))x_i = 0$$

Separating terms with $\hat{\beta}$

$$\sum_i (y_i - \bar{y})x_i + \hat{\beta}\sum_i (\bar{x} - x_i)x_i = 0$$

Moving terms with $\hat{\beta}$ to the right-hand side (note the change in sign).

$$\sum_i (y_i - \bar{y})x_i = \hat{\beta}\sum_i (x_i - \bar{x})x_i$$

$$\hat{\beta} = \frac{\sum_i (y_i - \bar{y})x_i}{\sum_i (x_i - \bar{x})x_i} \tag{22.5}$$

The expression in (22.5) can be also represented as

$$\hat{\beta} = \frac{\sum_i^n (y_i - \bar{y})(x_i - \bar{x})}{\sum_i (x_i - \bar{x})^2} \tag{22.6}$$

In summary, equations (22.3) and (22.6) are used to find $\hat{a}$ and $\hat{\beta}$ that minimize the sum squared of the errors.

$$\hat{a} = \bar{y} - \hat{\beta}\bar{x}$$

$$\hat{\beta} = \frac{\sum_{i=1}^n (y_i - \bar{y})(x_i - \bar{x})}{\sum_{i=1}^n (x_i - \bar{x})^2}$$

EXAMPLE 19.   The following dataset of five observations is provided. Find the linear regression.

x	y
1	8
2	12
3	15
4	16
5	24

There are 5 observations; $n = 5$. The steps to solve this problem are shown below.

1) Find the mean of x and y

$$\bar{x} = \frac{1}{5}(1 + 2 + 3 + 4 + 5) = \frac{1}{5}(15) = 3$$

$$\bar{y} = \frac{1}{5}(8 + 12 + 15 + 16 + 24) = \frac{1}{5}(75) = 15$$

2) Find the numerator of $\hat{\beta}$, which is $\sum_{i}^{n}(y_i - \bar{y})(x_i - \bar{x})$

$$\sum_{i=1}^{5}(y_i - \bar{y})(x_i - \bar{x}) = \underbrace{(8-15)(1-3)}_{i=1} + \underbrace{(12-15)(2-3)}_{i=2} + \underbrace{(15-15)(3-3)}_{i=3}$$

$$+ \underbrace{(16-15)(4-3)}_{i=4} + \underbrace{(24-15)(5-3)}_{i=5}$$

$$\sum_{i=1}^{5}(y_i - \bar{y})(x_i - \bar{x}) = -7 \cdot (-2) - 3 \cdot (-1) + 0 \cdot 0 + 1 \cdot 1 + 9 \cdot 2$$

$$\sum_{i=1}^{5}(y_i - \bar{y})(x_i - \bar{x}) = 36$$

3) Find the denominator of $\hat{\beta}$, which is $\sum_{i=1}^{n}(x_i - \bar{x})^2$

$$\sum_{i=1}^{5}(x_i - \bar{x})^2 = (1-3)^2 + (2-3)^2 + (3-3)^2 + (4-3)^2 + (5-3)^2$$

$$\sum_{i=1}^{5}(x_i - \bar{x})^2 = 4 + 1 + 0 + 1 + 4$$

$$\sum_{i=1}^{5}(x_i - \bar{x})^2 = 10$$

4) Calculate $\hat{\beta}$ by using equation (22.6), which is the ratio of the result from part (2) and part (3).

$$\hat{\beta} = \frac{36}{10} = 3.6$$

5) Find $\hat{a}$ by using equation (22.3).

$$\hat{a} = \bar{y} - \hat{\beta}\bar{x} = 15 - (3.6) \cdot 3 = 15 - 10.8$$

$$\hat{a} = 4.2$$

The linear equation is $\hat{y} = 4.2 + 3.6x$

Linear regressions are often used for forecasting the value of y, referred to as $\hat{y}$, under the range of x

$$\hat{y} = \hat{a} + \hat{\beta}x$$

EXAMPLE 20.   From Example 18, find the predicted value for y when $x = 2.5$.

Using the linear equation found in Example 18.

$$\hat{y} = 4.2 + 3.6 \cdot (2.5)$$

$$\hat{y} = 13.2$$

Solved Problems

SERIES AND PROPERTIES OF SUMMATION

22.1.   Given the set $\{1, 64, -1, 0, -125, -8\}$, compute:

$$\sum_{i=1}^{6} (\sqrt[3]{x_i} - 1)$$

The summation has five elements ($i = 1$ to 6). Expanding the operation.

$$\sum_{i=1}^{6} (\sqrt[3]{x_i} - 1) = (\sqrt[3]{1} - 1) + (\sqrt[3]{64} - 1) + (\sqrt[3]{-1} - 1) + (\sqrt[3]{0} - 1) + (\sqrt[3]{-125} - 1) + (\sqrt[3]{-8} - 1)$$

$$\sum_{i=1}^{6} (\sqrt[3]{x_i - 1}) = 0 + 3 - 2 - 1 + 4 - 3 = 1$$

22.2.   Calculate the partial sum:

$$\sum_{k=1}^{4} \frac{2^k}{k+1}$$

The index is located in two components, 2^k and $k + 1$,

$$\sum_{k=1}^{4} \frac{2^k}{k+1} = \frac{2^1}{1+1} + \frac{2^2}{2+1} + \frac{2^3}{3+1} + \frac{2^4}{4+1}$$

$$\sum_{k=1}^{4} \frac{2^k}{k+1} = \frac{2}{2} + \frac{4}{3} + \frac{8}{4} + \frac{16}{5} = 7.533$$

22.3. Calculate the partial sum:

$$\sum_{i=1}^{3} \frac{(-1)^i}{i^2}$$

The index is located in two components.

$$\sum_{i=1}^{3} \frac{(-1)^i}{i^2} = \frac{-1}{1} + \frac{1}{2^2} + \frac{-1}{3^2} = -1 + \frac{1}{4} - \frac{1}{9}$$

$$\sum_{i=1}^{3} \frac{1}{i^{i+1} - 2i} = -0.8611$$

22.4. Compute the partial sum:

$$\sum_{k=1}^{4} (10k - 2)$$

Splitting the summation,

$$\sum_{k=1}^{4} (10k - 2) = 10 \sum_{k=1}^{4} k - \sum_{k=1}^{4} 2$$

Expanding the first term and using the summation property for the second term,

$$\sum_{k=1}^{4} (10k - 2) = 10(1 + 2 + 3 + 4) - (2 \cdot 4)$$

$$= 10(10) - 8$$

$$\sum_{k=1}^{4} (10k - 2) = 92$$

22.5. Compute the summation,

$$\sum_{k=1}^{3} (5k - k^2)$$

Splitting the summation,

$$\sum_{k=1}^{3} (5k - k^2) = 5 \sum_{k=1}^{3} k - \sum_{k=1}^{3} k^2$$

Expanding both terms,

$$\sum_{k=1}^{3} (5k - k^2) = 5(1 + 2 + 3) - (1^2 + 2^2 + 3^2)$$

$$= 5(6) - (1 + 4 + 9) = 30 - 14$$

$$\sum_{k=1}^{3} (5k - k^2) = 16$$

FREQUENCY, MEAN, AND STANDARD DEVIATION

22.6. Given the data below,

8	1	2	7
1	1	3	1
8	7	3	1
1	2	5	1

(*a*) Construct the frequency table and (*b*) the bar chart for the data below.

a)   The frequency table is constructed using the 16 observations ($n = 16$).

x_i	f_i
1	5
2	4
3	3
5	4
7	1
8	1

b)   The bar chart is depicted in Fig. 22-4.

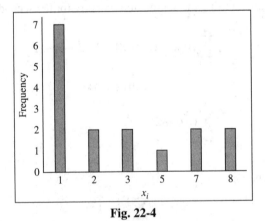

Fig. 22-4

22.7. Given the data below of scores from a test from 20 students,

A	B	C	C	D
B	B	F	A	B
A	C	A	C	B
A	D	B	B	A

(*a*) Construct the frequency table and (*b*) the bar chart for the data below.

a)   The frequency table is constructed using the 20 observations ($n = 20$).

x_i	f_i
A	6
B	7
C	4
D	2
F	1

b)   The bar chart is depicted in Fig. 22-5.

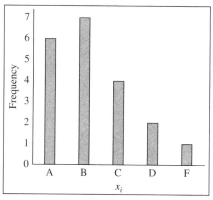

Fig. 22-5

22.8.   Compute the mean from the data given in Problem 22.6.

Computing the mean.

$$\overline{X} = \frac{1}{16} \sum_{i=1}^{16} x_i = \frac{8+1+2+7+1+1+3+1+8+7+3+1+1+2+5+1}{16}$$

$$\overline{X} = \frac{1}{16} \sum_{i=1}^{16} x_i = \frac{52}{16} = 3.25$$

22.9.   Find the range of the data given in Problem 22.6.

The maximum number is 8, the minimum is 1. The range is

$$d = 8 - 1 = 7$$

22.10. Compute the mean and standard deviation of the data below.

4	1
−3	0
1	3

Computing the mean, given 6 observations ($n = 6$)

$$\overline{X} = \frac{1}{6} \sum_{i=1}^{6} x_i = \frac{4 - 3 + 1 + 1 + 0 + 3}{6} = \frac{6}{6} = 1$$

To determine the standard deviation, first compute the variance.

$$S_x^2 = \frac{1}{6-1} \sum_{i=1}^{6} (x_i - \overline{X})^2$$

$$S_x^2 = \frac{(4-1)^2 + (-3-1)^2 + (1-1)^2 + (1-1)^2 + (0-1)^2 + (3-1)^2}{6-1}$$

$$= \frac{(3)^2 + (-4)^2 + (0)^2 + (0)^2 + (-1)^2 + (2)^2}{5} = \frac{9 + 16 + 0 + 0 + 1 + 4}{5}$$

$$S_x^2 = \frac{30}{5} = 6$$

The standard deviation is the square root of the variance.

$$S_x = \sqrt{S_x^2} = \sqrt{6} = 2.449$$

22.11. Compute the mean and standard deviation of the data below.

2	1
−3	0

Computing the mean, given 4 observations ($n = 4$)

$$\overline{X} = \frac{1}{4} \sum_{i=1}^{4} x_i = \frac{2 - 3 + 1 + 0}{4} = \frac{0}{6} = 0$$

To determine the standard deviation, first compute the variance.

$$S_x^2 = \frac{1}{4-1} \sum_{i=1}^{4} (x_i - \overline{X})^2$$

$$S_x^2 = \frac{(2-0)^2 + (-3-0)^2 + (1-0)^2 + (0-0)^2}{4-1}$$

$$= \frac{4 + 9 + 1 + 0}{3}$$

$$S_x^2 = \frac{14}{3} = 4.6\overline{6}$$

The standard deviation is the square root of the variance.

$$S_x = \sqrt{S_x^2} = \sqrt{4.6\overline{6}} = 2.160$$

THE WEIGHTED AVERAGE

22.12. Given the frequency table below,

x_i	f_i
2	1
4	1
5	5
7	7
8	4
10	2

(*a*) Construct the bar chart and (*b*) compute the weighted average.

a)   The bar chart is depicted in Fig. 22-6.

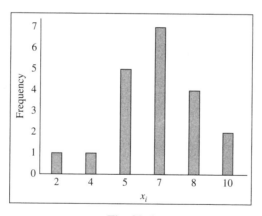

Fig. 22-6

b)   Compute the weighted average.

The number of observations is $n = 1 + 1 + 5 + 7 + 4 + 2 = 20$. The weighted average is

$$\overline{W} = \frac{1}{20} \sum_{i=1}^{6} x_i f_i = \frac{1}{20}(2 \times 1 + 4 \times 1 + 5 \times 5 + 7 \times 7 + 8 \times 4 + 10 \times 2)$$

$$\overline{W} = \frac{1}{20}(2 + 4 + 25 + 49 + 32 + 20) = \frac{132}{20}$$

$$\overline{W} = 6.6$$

22.13. A student wants to calculate his GPA given the grades below and the number of credits each class is worth,

Grades	Credits
A	4
B	3
C	1
A	3
B	2

Consider the grade points A = 4.0, B = 3.0, C = 2.0.

a) Construct the frequency table, in which x is the score and the frequency f is the total number of credits.

Grades	Frequency (total credits)
A	7
B	5
C	1

b) Compute the weighted average.

The number of observations is the total credits $n = 7 + 5 + 1 = 13$. Note that we will use the conversion to grade points. The weighted average is

$$\overline{W} = \frac{1}{13}\sum_{i=1}^{3}x_i f_i = \frac{1}{13}(4.0 \times 7 + 3.0 \times 5 + 2.0 \times 1)$$

$$\overline{W} = \frac{1}{13}(28 + 15 + 2) = \frac{45}{13}$$

$$\overline{W} = 3.4615$$

PROBABILITY AND EXPECTED VALUE

22.14. Given the frequency table,

x_i	f_i
2	1
4	1
5	5
7	7
8	4
10	2

(a) Construct the probability density function for x and *(b)* find the expected value of x.

1)   Build the probability density function by dividing each frequency by the total number of observations, which in this case is $n = 20$.

x_i	f_i	$P(x_i)$
2	1	0.05
4	1	0.05
5	5	0.25
7	7	0.35
8	4	0.20
10	2	0.10
Total	20	

2)   The expected value is

$$EV(x) = \sum_{i=1}^{6} x_i p_i = 2 \times 0.05 + 4 \times 0.05 + 5 \times 0.25 + 7 \times 0.35 + 8 \times 0.20 + 10 \times 0.10$$

$$EV(x) = 0.10 + 0.20 + 1.25 + 2.45 + 1.60 + 1.00$$

$$EV(x) = 6.60$$

22.15.   Given the incomplete probability density function below

x_i	p_i
1	0.28
2	
3	0.12
4	0.16
5	0.20

(*a*) Find the probability when $x = 2$ and (*b*) find the expected value of x

1)   The summation of all probabilities must be equal to 1; thus

$$0.28 + p_{x=2} + 0.12 + 0.16 + 0.20 = 1$$

$$p_{x=2} = 0.24$$

The probability of $x = 2$ is 0.24 or 24%.

2)   The expected value of x is

$$EV(x) = \sum_{i=1}^{5} x_i p_i = 1 \times 0.28 + 2 \times 0.24 + 3 \times 0.12 + 4 \times 0.16 + 5 \times 0.20$$

$$EV(x) = 0.28 + 0.48 + 0.36 + 0.64 + 1.00$$

$$EV(x) = 2.76$$

SIMPLE LINEAR REGRESSION

22.16.   Given the data table below

x	y
10	32
25	78
34	105
59	169

Fit the linear regression using the ordinary least square method.

There are 4 observations; $n = 4$.

(1) Find the mean of x and y

$$\bar{x} = \tfrac{1}{4}(10 + 25 + 34 + 59) = \tfrac{1}{4}(128) = 32$$

$$\bar{y} = \tfrac{1}{4}(32 + 78 + 105 + 169) = \tfrac{1}{4}(384) = 96$$

(2) Find the numerator of $\hat{\beta}$, which is $\displaystyle\sum_{i}^{n}(y_i - \bar{y})(x_i - \bar{x})$

$$\sum_{i=1}^{4}(y_i - \bar{y})(x_i - \bar{x}) = \underbrace{(32 - 96)(10 - 32)}_{i=1} + \underbrace{(78 - 96)(25 - 32)}_{i=2}$$

$$+ \underbrace{(105 - 96)(34 - 32)}_{i=4} + \underbrace{(169 - 96)(59 - 32)}_{i=5}$$

$$\sum_{i=1}^{4}(y_i - \bar{y})(x_i - \bar{x}) = -64 \cdot (-22) - 18 \cdot (-7) + 9 \cdot 2 + 73 \cdot 27$$

$$\sum_{i=1}^{4}(y_i - \bar{y})(x_i - \bar{x}) = 1408 + 126 + 18 + 1971 = 3523$$

(3) Find the denominator of $\hat{\beta}$, which is $\displaystyle\sum_{i=1}^{n}(x_i - \bar{x})^2$

$$\sum_{i=1}^{4}(x_i - \bar{x})^2 = (10 - 32)^2 + (25 - 32)^2 + (34 - 32)^2 + (59 - 32)^2$$

$$\sum_{i=1}^{4}(x_i - \bar{x})^2 = 484 + 49 + 4 + 729$$

$$\sum_{i=1}^{4}(x_i - \bar{x})^2 = 1266$$

(4) Calculate $\hat{\beta}$ by using equation (22.6), which is the ratio of the result from part (2) and part (3).

$$\hat{\beta} = \frac{3523}{1266} = 2.7828$$

(5) Find $\hat{a}$ by using equation (22.3).

$$\hat{a} = \bar{y} - \hat{\beta}\bar{x} = 96 - (2.7828) \cdot 32$$

$$\hat{a} = 6.9510$$

The linear equation is $\hat{y} = 6.9510 + 2.7828x$

22.17. Use the regression calculated in Problem 22.16 and predict the value of y if $x = 30$.

Use the linear equation found in Problem 22.16.

$$\hat{y} = 6.9510 + 2.7828(30)$$
$$\hat{y} = 90.4344$$

22.18. Given the data table below

x	y
2	0.3
3	0.6
5	0.9
7	1.4
9	1.6
10	2.4

Fit the linear regression using the ordinary least square method.

There are 6 observations; $n = 6$.

(1) Find the mean of x and y.

$$\bar{x} = \tfrac{1}{6}(2 + 3 + 5 + 7 + 9 + 10) = \tfrac{1}{6}(36) = 6$$

$$\bar{y} = \tfrac{1}{6}(0.3 + 0.6 + 0.9 + 1.4 + 1.6 + 2.4) = \tfrac{1}{6}(7.2) = 1.2$$

(2) Find the numerator of $\hat{\beta}$, which is $\displaystyle\sum_{i}^{n}(y_i - \bar{y})(x_i - \bar{x})$

$$\sum_{i=1}^{6}(y_i - \bar{y})(x_i - \bar{x}) = (0.3 - 1.2)(2 - 6) + (0.6 - 1.2)(3 - 6) + (0.9 - 1.2)(5 - 6)$$
$$+ (1.4 - 1.2)(7 - 6) + (1.6 - 1.2)(9 - 6) + (2.4 - 1.2)(10 - 6)$$

$$\sum_{i=1}^{6}(y_i - \bar{y})(x_i - \bar{x}) = -0.9 \cdot (-4) - 0.6 \cdot (-3) - 0.3 \cdot (-1) + 0.2 \cdot 1 + 0.4 \cdot 3 + 1.2 \cdot 4$$

$$\sum_{i=1}^{6}(y_i - \bar{y})(x_i - \bar{x}) = 3.6 + 1.8 + 0.3 + 0.2 + 1.2 + 4.8$$

$$\sum_{i=1}^{6}(y_i - \bar{y})(x_i - \bar{x}) = 11.9$$

(3) Find the denominator of $\hat{\beta}$, which is $\displaystyle\sum_{i=1}^{n}(x_i - \bar{x})^2$

$$\sum_{i=1}^{6}(x_i - \bar{x})^2 = (2-6)^2 + (3-6)^2 + (5-6)^2 + (7-6)^2 + (9-6)^2 + (10-6)^2$$

$$\sum_{i=1}^{6}(x_i - \bar{x})^2 = 4^2 + 3^2 + 1^2 + 1^2 + 3^2 + 4^2 = 16 + 9 + 1 + 1 + 9 + 16$$

$$\sum_{i=1}^{6}(x_i - \bar{x})^2 = 52$$

(4) Calculate $\hat{\beta}$ by using equation (22.6), which is the ratio of the result from part (2) and part (3).

$$\hat{\beta} = \frac{11.9}{52} = 0.2288$$

(5) Find $\hat{a}$ by using equation (22.3).

$$\hat{a} = \bar{y} - \hat{\beta}\bar{x} = 7.2 - (0.2288)\cdot 36$$

$$\hat{a} = -0.1731$$

The linear equation is $\hat{y} = -0.1731 + 0.2288x$.

22.19. Use the regression calculated in Problem 22.18 and predict the value of y if $x = 6$

Use the linear equation found in Problem 22.18.

$$\hat{y} = -0.1731 + 0.2288(6)$$

$$\hat{y} = 1.2$$

Excel Practice

Microsoft Excel™ Practice A: Input-Output Tables

Input-output tables summarize the production and demand for goods within an economy. Chapter 12.6 provides the foundation on input-output analysis, representing the total demand in matrix form is

$$X = AX + B,$$

where

$$X = \begin{bmatrix} x_1 \\ x_2 \\ \vdots \\ x_n \end{bmatrix} \quad A = \begin{bmatrix} a_{11} & a_{12} & \cdots & a_{1n} \\ a_{21} & a_{22} & & a_{2n} \\ & \vdots & \ddots & \vdots \\ a_{n1} & a_{n2} & \cdots & a_{nn} \end{bmatrix} \quad B = \begin{bmatrix} b_1 \\ b_2 \\ \vdots \\ b_n \end{bmatrix}$$

X is the total demand for each good i, A is the matrix of technical coefficients, and B represents the final demand of goods (i.e., goods consumed by households).

In economics, matrices A (expressed in total value or as input-output shares), B, and X are provided together arranged in an *input-output table*

$$\begin{bmatrix} a_{11} & a_{12} & \cdots & a_{1n} & b_1 & \bigm| & x_1 \\ a_{21} & a_{22} & & a_{2n} & b_2 & \bigm| & x_2 \\ & \vdots & \ddots & \vdots & \vdots & \bigm| & \vdots \\ a_{n1} & a_{n2} & \cdots & a_{nn} & b_n & \bigm| & x_n \end{bmatrix}$$

where

$$\text{for good 1: } x_1 = a_{11} + a_{12} + \cdots + a_{1n} + b_1$$

$$\text{for good } n\text{: } x_n = a_{n1} + a_{n2} + \cdots + a_{nn} + b_n$$

In output tables, the rows of A are called **sectors of origin** and its columns are the **sectors of destination**. Element a_{ij} is a technical coefficient that expresses the value of input i used to produce one dollar's worth of output j, as explained in section 12.6. However, a typical input-output table provides the dollar value of each input-output combination rather than the ratios, as represented below.

Sectors of origin	Sources of destination				Final demand	Total demand
	1	2	...	n		
1	m_{11}	m_{11}		m_{1n}	b_1	x_1
2	m_{21}	m_{22}		m_{2n}	b_2	x_2
$\vdots$	$\vdots$	$\vdots$		$\vdots$	$\vdots$	$\vdots$
n	m_{n1}	m_{n2}		m_{nn}	b_n	x_n
Value added						
Gross production	x_1	x_2		x_n		

For those cases, first it is necessary to convert the dollar values m_{ij} into technical coefficients a_{ij} by dividing each dollar value by the gross production of the sector of destination j,

$$a_{ij} = m_{ij}/x_j$$

An important identity of input output tables is that the gross production of each destination sector must be equal the total demand of that sector of origin.

To find the level of total output $\mathbf{X}$ required to satisfy final demand (i.e., sum of intermediate and final products) can be found by equation (*12.3*), which is

$$\mathbf{X} = (\mathbf{I} - \mathbf{A})^{-1}\mathbf{B}$$

in which $\mathbf{I} - \mathbf{A}$ was declared as the *Leontief matrix.*

This problem is challenging to solve manually when there are more than three industries, as the inverse of the Leontief matrix becomes much more complex. An alternative is the use of Microsoft Excel™, which provides useful functions to work with matrices, with a standard procedure that makes the computation straightforward. We will make use of three Excel™ functions:

- **MUNIT**: creates identity matrices

- **MINVERSE**: compute inverse of matrices

- **MMULT**: multiplication of two matrices

Examples A.1 to A.4 presented below show step-by-step how to solve input-output tables using Microsoft Excel™. Specifically, we will solve three problems that were presented in Chapter 12 (for comparison) and expand on a problem with four goods to show the usefulness of this method.

INPUT-OUTPUT TABLE WITH KNOWN MATRIX OF TECHNICAL COEFFICIENTS

A.1.   Determine the total demand for industries 1, 2, and 3, given the matrix of technical coefficients $\mathbf{A}$ and the final demand vector $\mathbf{B}$ below.

Output industry

$$\mathbf{A} = \begin{bmatrix} 0.2 & 0.3 & 0.2 \\ 0.4 & 0.1 & 0.3 \\ 0.3 & 0.5 & 0.2 \end{bmatrix} \quad \text{Input industry} \quad \mathbf{B} = \begin{bmatrix} 150 \\ 200 \\ 210 \end{bmatrix}$$

Given $\mathbf{X} = (\mathbf{I} - \mathbf{A})^{-1}\mathbf{B}$:

(*1*)   In this problem, originally presented in Problem 12.29, we assume that we already have the matrix of technical coefficients $\mathbf{A}$. Write $\mathbf{A}$ and $\mathbf{B}$ in the table in Excel™ in the positions described in the template of Fig. A-1.

	A	B	C	D	E	F	G	H
1								
2					**Matrix A**			**Vector B**
3								
4					Output industry			Final
5				1	2	3		demand
6		Input	1	0.2	0.3	0.2		150
7		industry	2	0.4	0.1	0.3		200
8			3	0.3	0.5	0.2		210

Fig. A-1

(2) To solve $\mathbf{X} = (\mathbf{I} - \mathbf{A})^{-1}\mathbf{B}$, the following order is required:

(a) Construction of the identity matrix $\mathbf{I}$

(b) Calculation of the Leontief matrix $\mathbf{I} - \mathbf{A}$

(c) Computation of the inverse $(\mathbf{I} - \mathbf{A})^{-1}$

(d) Matrix multiplication $(\mathbf{I} - \mathbf{A})^{-1}\mathbf{B}$

(a) *Construction of the identity matrix*: create a blank table with 3×3 dimensions for the identity matrix as in Fig. A-2.

Fig. A-2

Make the use of the **MUNIT()** function. Select cells K6 to M8. Then, write in cell K6

=MUNIT(3)

The number in parenthesis declares the dimension of the row of the identity matrix; in this case, the identity matrix has 3 columns. This is illustrated in Fig. A-3.

Fig. A-3

Once you finish typing the formula, press "ALT + SHIFT + ENTER" to proceed with the operation. Note that this is a standard method used in Excel™ for matrices, although recent versions only require pressing the ENTER button. The result is in Fig. A-4.

	J	K	L	M
2		**Identity Matrix I**		
3				
4			Output industry	
5		1	2	3
6	1	1	0	0
7	2	0	1	0
8	3	0	0	1

Fig. A-4

(b) *Calculation of the Leontief matrix*: Create a blank table with dimension 3×3 in the locations shown in Fig. A-5.

	O	P	Q	R
2		**Leontief Matrix I-A**		
3				
4			Output industry	
5		1	2	3
6	1			
7	2			
8	3			

Fig. A-5

In each cell, subtract the element from the corresponding identity matrix **I** minus the technical coefficient **A**. For example, for origin industry 1 – destination industry 1 in cell P6, write

$$= K6 - D6$$

This procedure is repeated for all nine elements, as shown in Fig. A-6.

	O	P	Q	R
1				
2			**Leontief Matrix I-A**	
3				
4			Output industry	
5		1	2	3
6	1	=K6-D6	=L6-E6	=M6-F6
7	2	=K7-D7	=L7-E7	=M7-F7
8	3	=K8-D8	=L8-E8	=M8-F8

Fig. A-6

The resultant matrix is shown in Fig. A-7.

	O	P	Q	R
1				
2		**Leontief Matrix I-A**		
3				
4			Output industry	
5		1	2	3
6	1	0.8	-0.3	-0.2
7	2	-0.4	0.9	-0.3
8	3	-0.3	-0.5	0.8

Fig. A-7

(c) *Inverse of the Leontief matrix*: Create a blank table with dimension 3×3 in the locations shown in Fig. A-8.

	T	U	V	W
1				
2		**Inverse Matrix (I-A)^-1**		
3				
4			Output industry	
5		1	2	3
6	1			
7	2			
8	3			

Fig. A-8

Make the use of the **MINVERSE()** function. Select cells U6 to W8. Then, write in cell U6

=MINVERSE(

In parenthesis, select all cell elements of the Leontief matrix or write P6:R8, and close parenthesis; in other words

=MINVERSE(P6:R8)

Once you finish typing the formula, press "ALT + SHIFT + ENTER" to proceed with the operation. The result is in Fig. A-9.

	T	U	V	W
1				
2		**Inverse Matrix (I-A)^-1**		
3				
4			Output industry	
5		1	2	3
6	1	2.38494	1.42259	1.12971
7	2	1.71548	2.42678	1.33891
8	3	1.96653	2.05021	2.51046

Fig. A-9

(d) *Matrix multiplication* $(\mathbf{I} - \mathbf{A})^{-1}\mathbf{B}$: Lastly, make use of the **MMULT()** function in order to multiply the inverse of the Leontief matrix $(\mathbf{I} - \mathbf{A})^{-1}$ and the vector of final demand $\mathbf{B}$. Because there are 3 industries, the total demand will be a column vector with dimension 3. Thus, create a blank table with dimension 3×1 in the locations shown in Fig. A-10.

	Y	Z
1		
2		**Vector X**
3		
4		Total
5		demand
6	1	
7	2	
8	3	

Fig. A-10

Make the use of the **MMULT()** function. Select cells Z6 to Z8. Then, write in cell Z6

=MMULT(

In parenthesis, select first elements of the Leontief matrix or write U6:W8, then write ",", which is the separation symbol comma, then select the elements of the vector of final demand or write H6:H8, and close parenthesis; in other words

=MMULT(U6:W8,H6:H8)

Once you finish typing the formula, press "ALT + SHIFT + ENTER" to proceed with the operation. The result is in Fig. A-11.

	Y	Z
1		
2		**Vector X**
3		
4		Total
5		demand
6	1	879.50
7	2	1023.85
8	3	1232.22

Fig. A-11

The answer presented in Fig. A-11 is exactly the same as the solution to Problem 12.29 when computed manually.

$$\mathbf{X} = \begin{bmatrix} 879.50 \\ 1023.85 \\ 1232.22 \end{bmatrix}$$

Note that all values were rounded up to only two decimals. You can do that by pressing the "**Decrease Decimals**" symbol $\overset{.00}{\to}.0$.

A.2. Determine the total demand for industries 1, 2, and 3, given the matrix of technical coefficients **A** and the final demand vector **B** below.

$$\begin{array}{c} \text{Output industry} \\ \begin{array}{ccc} 1 & 2 & 3 \end{array} \end{array}$$

$$\mathbf{A} = \begin{bmatrix} 0.4 & 0.3 & 0.1 \\ 0.2 & 0.2 & 0.3 \\ 0.2 & 0.4 & 0.2 \end{bmatrix} \text{ Input industry} \qquad \mathbf{B} = \begin{bmatrix} 140 \\ 220 \\ 180 \end{bmatrix}$$

Given $\mathbf{X} = (\mathbf{I} - \mathbf{A})^{-1}\mathbf{B}$:

(*1*) This problem is originally presented as Problem 12.31, assuming we already have the matrix of technical coefficients **A**. Write **A** and **B** in the table in Excel™ in the positions described in the template of Fig. A-12.

	A	B	C	D	E	F	G	H
1								
2					**Matrix A**			**Vector B**
3								
4					Output industry			Final
5				1	2	3		demand
6			1	0.4	0.3	0.1		140
7		Input industry	2	0.2	0.2	0.3		220
8			3	0.2	0.4	0.2		180

Fig. A-12

(2) To solve $\mathbf{X} = (\mathbf{I} - \mathbf{A})^{-1}\mathbf{B}$, the following order is required:

(*a*) *Construction of the identity matrix*: Create a blank table with dimension 3 × 3 in cells K6 to M8, similar to Fig. A-2. Then select those cells (K6 to M8) and write in cell K6

=MUNIT(3)

The number in parenthesis is the row dimension of the identity matrix, which is 3, similar to Fig. A-3. Then, press "ALT + SHIFT + ENTER" to proceed with the operation. The result is in Fig. A-13 (which mimics Fig. A-4), as both have the same dimensions.

	J	K	L	M
2			**Identity Matrix I**	
3				
4			Output industry	
5		1	2	3
6	1	1	0	0
7	2	0	1	0
8	3	0	0	1

Fig. A-13

(b) *Calculation of the Leontief matrix*: Create a blank table with dimension 3×3 (using cells P6 to R8), in similar positions as shown in Fig. A-5.

In each cell, subtract the element from the corresponding identity matrix **I** minus the technical coefficient **A**. For example, in cell P6, write

$$= K6 - D6$$

his procedure is repeated for all nine elements, as replicated in Fig. A-6.

The resultant matrix is shown in Fig. A-14.

	O	P	Q	R
1				
2		**Leontief Matrix I-A**		
3				
4		Output industry		
5		1	2	3
6	1	0.6	-0.3	-0.1
7	2	-0.2	0.8	-0.3
8	3	-0.2	-0.4	0.8

Fig. A-14

(c) *Inverse of the Leontief matrix*: Create a blank table with dimension 3×3 in the shown in Fig. A-8. Then, select cells U6 to W8, and write in cell U6

=MINVERSE(

In parenthesis, select all cell elements of the Leontief matrix or write P6:R8, and close parenthesis; in other words

=MINVERSE(P6:R8)

Once you finish typing the formula, press "ALT + SHIFT + ENTER" to proceed with the operation. The result is in Fig. A-15.

	T	U	V	W
1				
2		**Inverse Matrix (I-A)^-1**		
3				
4		Output industry		
5		1	2	3
6	1	2.342342	1.261261	0.765766
7	2	0.990991	2.072072	0.900901
8	3	1.081081	1.351351	1.891892

Fig. A-15

(d) *Matrix multiplication* $(\mathbf{I} - \mathbf{A})^{-1}\mathbf{B}$: the total demand is a column vector of dimension 3, as there are 3 industries. Create a blank table with dimension 1×3 in the locations shown in Fig. A-10. Select cells Z6 to Z8. Then, write in cell Z6

=MMULT(

In parenthesis, select first elements of the Leontief matrix or write U6:W8, then write "," which is the separation symbol comma, then select the elements of the vector of final demand or write H6:H8, and close parenthesis; in other words

=MMULT(U6:W8,H6:H8)

Once you finish typing the formula, press "ALT + SHIFT + ENTER" to proceed with the operation. The result is in Fig. A-16.

	Y	Z
1		
2		**Vector X**
3		
4		Total
5		demand
6	1	743.24
7	2	756.76
8	3	789.19

Fig. A-16

The answer presented in Fig. A-16 is exactly the same as the solution to Problem 12.31 when computed manually and rounded up to two decimals.

$$\mathbf{X} = \begin{bmatrix} 743.24 \\ 756.76 \\ 789.19 \end{bmatrix}$$

INPUT-OUTPUT TABLES THAT REQUIRE THE COMPUTATION OF TECHNICAL COEFFICIENTS

A.3. Given the interindustry transaction demand table below, find the matrix of technical coefficients and check your answer.

Sectors of Origin	Sectors of destination			Final demand	Total demand
	1	2	3		
1	20	60	10	50	**140**
2	50	10	80	10	**150**
3	40	30	20	40	**130**
Value added	30	50	20		
Gross production	**140**	**150**	**130**		

(*0*) Write the table in Excel™ as shown in Fig. A-17. Note that the first 9 elements are the input-output value in dollars (which will be used to compute **A**), followed by columns of final demand (vector **B**) and total demand (vector **X**).

	A	B	C	D	E	F	G
1							
2				**Matrix M**		**Vector B**	**Vector X**
3							
4		Sectors		Sectors of destination		Final	Total
5		of Origin	1	2	3	demand	demand
6		1	20	60	10	50	**140**
7		2	50	10	80	10	**150**
8		3	40	30	20	40	**130**
9		Value added	30	50	20		
10		Gross production	**140**	**150**	**130**		

Fig. A-17

(*1*) In this problem, to calculate the matrix of technical coefficients **A**, create a blank 3 × 3 table which will contain matrix **A**, as shown in Fig. A-18.

	I	J	K	L
1				
2			**Matrix A**	
3				
4			Sectors of destination	
5			1 2 3	
6	1			
7	2			
8	3			

Fig. A-18

Computation of matrix **A**: the computation of the matrix of technical coefficients requires dividing each cell from matrix M (input-output values) by the total gross production of the goods, which is the formula

$$a_{ij} = m_{ij}/x_j$$

Thus, write in cell J6

= C6/C\$10

Note the dollar sign ($) which is used to fix the row 10 that represents the location of the total (gross production of the goods). This technique permits to scroll to the right and down without the need of writing the formula eight more times. The matrix formulation should be similar to Fig. A-19.

	I	J	K	L
1				
2			**Matrix A**	
3				
4			Sectors of destination	
5		1	2	3
6	1	=C6/C$10	=D6/D$10	=E6/E$10
7	2	=C7/C$10	=D7/D$10	=E7/E$10
8	3	=C8/C$10	=D8/D$10	=E8/E$10

Fig. A-19

The resultant matrix of technical coefficients A is given in Fig. A-20 (rounded up to three decimals), which is the same result as part (*a*) of Problem 12.35.

	I	J	K	L
1				
2			**Matrix A**	
3				
4			Sectors of destination	
5		1	2	3
6	1	0.143	0.400	0.077
7	2	0.357	0.067	0.615
8	3	0.286	0.200	0.154

Fig. A-20

(*2*) To check the answer, the following property must hold, given

$$\mathbf{X} = \mathbf{AX} + \mathbf{B}$$

Then,

$$\mathbf{X} - \mathbf{B} = \mathbf{AX}$$

(*a*) First, create a column vector $\mathbf{X} - \mathbf{B}$ as in Fig. A-21.

	N	O
1		
2		**X-B**
3		
4		Output
5		industry
6	1	
7	2	
8	3	

Fig. A-21

Then, subtract each element of the final demand B from the total demand X. Write in cell O6 (that represents the first industry):

= G6-F6

Repeat the same operation for the next two industries. The resultant vector is in Fig. A-22.

	N	O
1		
2		**X-B**
3		
4		Output
5		industry
6	1	90
7	2	140
8	3	90

Fig. A-22

(*b*) Construct the matrix multiplication **AX**. The dimension is a vector column of 3 elements (**A** is 3×3 and **X** is 3×1). Thus, write the matrix as shown below in Fig. A-23

	Q	R
1		
2		**AX**
3		
4		Output
5		industry
6	1	
7	2	
8	3	

Fig. A-23

Select cells R6 to R8. Then, write in cell R6

=MMULT(

In parenthesis, select first elements of **A** or write J6:L8, then write ",", which is the separation symbol comma, then select the elements of the vector of final demand **B** or write G6:G8, and close parenthesis; in other words

=MMULT(J6:L8,G6:G8)

Once you finish typing the formula, press "ALT + SHIFT + ENTER" to proceed with the operation. The result is in Fig. A-24.

	Q	R
1		
2		**AX**
3		
4		Output
5		industry
6	1	90
7	2	140
8	3	90

Fig. A-24

Both vectors **AX** and **XB** result in the same answer.

A.4. Given the interindustry transaction demand (in millions of dollars) below, find the matrix of technical coefficients and verify the identity $\mathbf{X} = (\mathbf{I} - \mathbf{A})^{-1}\,\mathbf{B}$.

Sectors of Origin	Sectors of destination Steel	Coal	Iron	Auto	Final demand	Total demand
Steel	80	20	110	230	160	**600**
Coal	200	50	90	120	140	**600**
Iron	220	110	30	40	0	**400**
Auto	60	140	160	240	400	**1000**
Value added	40	280	10	370		
Gross production	**600**	**600**	**400**	**1000**		

Write the table in Excel™ as shown in Fig. A-25. Note that the first 16 elements are the input-output value in dollars (which will be used to compute **A**), followed by columns of final demand (vector **B**) and total demand (vector **X**).

	A	B	C	D	E	F	G	H
1								
2				**Matrix M**			**Vector B**	**Vector X**
3								
4		Sectors		Sectors of destination			Final	Total
5		of Origin	Steel	Coal	Iron	Auto	demand	demand
6		Steel	80	20	110	230	160	**600**
7		Coal	200	50	90	120	140	**600**
8		Iron	220	110	30	40	0	**400**
9		Auto	60	140	160	240	400	**1000**
10		Value added	40	280	10	370		
11		Gross production	**600**	**600**	**400**	**1000**		

Fig. A-25

(*1*) To calculate the matrix of technical coefficients **A**, create a blank 4×4 table as in Fig. A-26.

	J	K	L	M	N
1					
2			**Matrix A**		
3					
4			Sectors of destination		
5		Steel	Coal	Iron	Auto
6	Steel				
7	Coal				
8	Iron				
9	Auto				

Fig. A-26

Computation of matrix **A**: divide each cell from matrix M (input-output values) by the total gross production of the goods (row vector X). Thus, write in cell J6

= C6/C$111

Note the dollar sign ($) to fix the row. The matrix formulation should be similar to Fig. A-27.

	J	K	L	M	N
1					
2			**Matrix A**		
3					
4			Sectors of destination		
5		Steel	Coal	Iron	Auto
6	Steel	=C6/C$11	=D6/D$11	=E6/E$11	=F6/F$11
7	Coal	=C7/C$11	=D7/D$11	=E7/E$11	=F7/F$11
8	Iron	=C8/C$11	=D8/D$11	=E8/E$11	=F8/F$11
9	Auto	=C9/C$11	=D9/D$11	=E9/E$11	=F9/F$11

Fig. A-27

The matrix of technical coefficients A is in Fig. A-28 (rounded up to three decimals), which is the same solution as Problem 12.33.

	J	K	L	M	N
1					
2			**Matrix A**		
3					
4			Sectors of destination		
5		Steel	Coal	Iron	Auto
6	Steel	0.133	0.033	0.275	0.230
7	Coal	0.333	0.083	0.225	0.120
8	Iron	0.367	0.183	0.075	0.040
9	Auto	0.100	0.233	0.400	0.240

Fig. A-28

(2) To check the property $\mathbf{X} = (\mathbf{I} - \mathbf{A})^{-1} \mathbf{B}$, we follow steps similar to Problems A.1 and A.2.

(a) *Construction of the identity matrix*: create a blank 4×4 table as in Fig. A-29.

	P	Q	R	S	T
1					
2			**Identity Matrix I**		
3					
4			Sectors of destination		
5		Steel	Coal	Iron	Auto
6	Steel				
7	Coal				
8	Iron				
9	Auto				

Fig. A-29

Select cells Q6 to T9. Then, write in cell Q6

=MUNIT(4)

Here, 4 is the number of rows of the 4×4 identity. Then press "ALT + SHIFT + ENTER" to proceed with the operation. The result is in Fig. A-30.

	P	Q	R	S	T
1					
2		**Identity Matrix I**			
3					
4		Sectors of destination			
5		Steel	Coal	Iron	Auto
6	Steel	1	0	0	0
7	Coal	0	1	0	0
8	Iron	0	0	1	0
9	Auto	0	0	0	1

Fig. A-30

(*b*) *Calculation of the Leontief matrix*: Create a blank table with dimension 4×4 in the locations shown in Fig. A-31.

	V	W	X	Y	Z
1					
2		**Leontief Matrix I-A**			
3					
4		Sectors of destination			
5		Steel	Coal	Iron	Auto
6	Steel				
7	Coal				
8	Iron				
9	Auto				

Fig. A-31

In each cell, subtract the element of the identity matrix **I** from the technical coefficient **A**. For example, in cell W6, write

=Q6-K6

This procedure is repeated for all 16 elements. Thus, scroll right and down to copy this formula to all the remaining 15 cells. The resultant matrix is shown in Fig. A-32.

	V	W	X	Y	Z
1					
2			**Leontief Matrix I-A**		
3					
4			Sectors of destination		
5		Steel	Coal	Iron	Auto
6	Steel	0.867	-0.033	-0.275	-0.230
7	Coal	-0.333	0.917	-0.225	-0.120
8	Iron	-0.367	-0.183	0.925	-0.040
9	Auto	-0.100	-0.233	-0.400	0.760

Fig. A-32

(c) *Inverse of the Leontief matrix*: Create a blank 4×4 table with dimension in the locations shown in Fig. A-33.

	AB	AC	AD	AE	AF
1					
2			**Inverse Matrix (I-A)^-1**		
3					
4			Sectors of destination		
5		Steel	Coal	Iron	Auto
6	Steel				
7	Coal				
8	Iron				
9	Auto				

Fig. A-33

Select cells AC6 to AF9. Then, write in cell AC6

=MINVERSE(

In parenthesis, select all cell elements of the Leontief matrix or write W6:Z9, and close parenthesis; in other words

=MINVERSE(W6:Z8)

Once you finish typing the formula, press "ALT + SHIFT + ENTER" to proceed with the operation. The result is in Fig. A-34.

	Steel	Coal	Iron	Auto
Inverse Matrix (I-A)^-1				
Sectors of destination				
Steel	1.770087	0.410662	0.90645	0.648234
Coal	1.012694	1.458236	0.908551	0.58454
Iron	0.947451	0.484529	1.675829	0.451435
Auto	1.042479	0.756754	1.280226	1.818144

Fig. A-34

(d) *Matrix multiplication* $(\mathbf{I} - \mathbf{A})^{-1}\mathbf{B}$: Create a blank table with dimension 4×1 in the locations shown in Fig. A-35.

	Vector X
	Total demand
Steel	
Coal	
Iron	
Auto	

Fig. A-35

Select cells AI6 to AI9. Then, write in cell AI6

=MMULT(

In parenthesis, select first elements of the Leontief matrix or write AC6:AF9, then write ",", which is the separation symbol comma, then select the elements of the vector of final demand or write G6:G9, and close parenthesis; in other words

=MMULT(AC6:AF9,G6:G9)

Once you finish typing the formula, press "ALT + SHIFT + ENTER" to proceed with the operation. The result is in Fig. A-36.

	Vector X
	Total demand
Steel	600
Coal	600
Iron	400
Auto	1000

Fig. A-36

The answer presented in Fig. A-36 is exactly the same as the total demand.

$$\mathbf{X} = \begin{bmatrix} 600 \\ 600 \\ 400 \\ 1000 \end{bmatrix}$$

Therefore, the input-output table is verified.

Microsoft Excel™ Practice B: Simple Linear Regression

Simple linear regression models are often used in economics and finance to analyze the potential linear relationship between two variables x and y. This information permits you to understand policy consequences, the effect of economic variables, and to make predictions on future values of y. As seen in Chapter 22, a linear regression can be represented as

$$y = \hat{\alpha} + \hat{\beta}x + \varepsilon \tag{B.1}$$

where y is the dependent variable, x is the independent variable and ε is the error term.

Ordinary least square (OLS) is used to obtain the estimates for the intercept $\hat{\alpha}$ and the slope $\hat{\beta}$ using the formulas below, labeled as (B.2).

$$\hat{\alpha} = \bar{y} - \hat{\beta}\bar{x} \tag{B.2}$$

$$\hat{\beta} = \frac{\sum_{i=1}^{n}(y_i - \bar{y})(x_i - \bar{x})}{\sum_{i=1}^{n}(x_i - \bar{x})^2}$$

This method relies heavily on the use of summation operations. The computation depends on the number of observations of the data sample. As explained in Chapter 22, the procedure becomes cumbersome as the sample size increases, which is particularly true in economics, as we collect observational data from multiple sources. These limitations can be overcome by using MS Excel™ functions. Examples B.1 and B.2 explain how to use MS Excel™ formulations to solve simple linear regressions using formulas in (B.2). Example B.3 illustrates how to use a scatterplot to find the linear regression. Example B.4 solves the same problem by using preestablished MS Excel™ functions.

SIMPLE LINEAR REGRESSION: USING FORMULATIONS

B.1. The following dataset of five observations is provided. Find the linear regression.

x	y
1	8
2	12
3	15
4	16
5	24

This problem was presented in Example 19 of Chapter 22. To solve it using MS Excel™, follow the steps below.

(*0*)   Write the dataset in an Excel™ spreadsheet in the locations shown in Fig. B-1. Add two more rows at the end of the table: one for totals and the second for the averages.

	A	B	C	D
1				
2			*x*	*y*
3			1	8
4			2	12
5			3	15
6			4	16
7			5	24
8		Total		
9		Average		

Fig. B-1

(*1*)   Sum all elements of *x* in cell C8 by either selecting the AUTOSUM symbol $\sum$ AutoSum ∨ or writing in cell C8

$$=\text{SUM(C3:C7)}$$

In a similar manner, in cell D8, add all elements of *y* by writing

$$=\text{SUM(D3:D7)}$$

The totals for each variable must coincide with the ones presented in Fig. B-2.

	A	B	C	D
1				
2			*x*	*y*
3			1	8
4			2	12
5			3	15
6			4	16
7			5	24
8		Total	15	75
9		Average		

Fig. B-2

(2) To find the average of x, make use of the function **AVERAGE( )**. Thus, write in cell C9

$$=\text{AVERAGE(C3:C7)}$$

In a similar manner, in cell D9, take the average of all elements of y by writing

$$=\text{AVERAGE(D3:D7)}$$

The results must coincide with Fig. B-3.

	A	B	C	D
1				
2			*x*	*y*
3			1	8
4			2	12
5			3	15
6			4	16
7			5	24
8		Total	15	75
9		Average	3	15

Fig. B-3

Note that $\bar{x} = 3$ and $\bar{y} = 15$, similar to the solution in Example 19, part (1).

(3) Create two columns with labels **Dx** and **Dy**, which will represent the deviations of x and y from their respective mean, as in Fig. B-4.

	A	B	C	D	E	F
1						
2			*x*	*y*	*Dx*	*Dy*
3			1	8		
4			2	12		
5			3	15		
6			4	16		
7			5	24		
8		Total	15	75		
9		Average	3	15		

Fig. B-4

The column **Dx** represents the deviation of each x observation from the mean $\bar{x}$. There is a similar case for y. Mathematically,

$$Dx_i = x_i - \bar{x} \qquad\qquad Dy_i = y_i - \bar{y} \tag{B.3}$$

Note that in order to take the difference, we make use of the "Average" row 9 (which will be held constant). For the first observation of x, write in cell E3

=C3-C\$9

where C3 is the observation x_1 and C9 locates $\bar{x}$. We write \$ to lock the AVERAGE row and be able to scroll the formula down. The formulas in each cell should be similar to Fig. B-5.

	B	C	D	E
1				
2		*x*	*y*	***Dx***
3		1	8	=C3-C\$9
4		2	12	=C4-C\$9
5		3	15	=C5-C\$9
6		4	16	=C6-C\$9
7		5	24	=C7-C\$9
8	Total	=SUM(C3:C7)	=SUM(D3:D7)	
9	Average	=AVERAGE(C3:C7)	=AVERAGE(D3:D7)	

Fig. B-5

The results are shown in Fig. B-6.

	A	B	C	D	E	F
1						
2			*x*	*y*	***Dx***	***Dy***
3			1	8	-2	
4			2	12	-1	
5			3	15	0	
6			4	16	1	
7			5	24	2	
8		Total	15	75		
9		Average	3	15		

Fig. B-6

A similar method is used for y. Write in cell F3

=D3-D\$9

Then, scroll down the cell. The results are shown in Fig. B-7.

	A	B	C	D	E	F
1						
2			x	y	Dx	Dy
3			1	8	-2	-7
4			2	12	-1	-3
5			3	15	0	0
6			4	16	1	1
7			5	24	2	9
8		Total	15	75		
9		Average	3	15		

Fig. B-7

(3)   *Calculating* $\hat{\beta}$: this requires the use of the second formula in (*B.2*).

   (*a*)   Create two columns, including an additional total row, as in Fig. B-8. The first column contains the elements of numerator $(y_i - \bar{y})(x_i - \bar{x})$, here labeled as ***Dxy***, and the second column contains the elements of denominator $(x_i - \bar{x})^2$, labeled as ***Dxx***.

	A	B	C	D	E	F	G	H
1								
2			x	y	Dx	Dy	Dxy	Dxx
3			1	8	-2	-7		
4			2	12	-1	-3		
5			3	15	0	0		
6			4	16	1	1		
7			5	24	2	9		
8		Total	15	75				
9		Average	3	15			$SSxy$	$SSxx$

Fig. B-8

Note that each total is labeled ***SSxy*** and ***SSxx***, as they will represent $\sum_{i=1}^{n} (y_i - \bar{y})(x_i - \bar{x})$ and $\sum_{i=1}^{n} (x_i - \bar{x})^2$, respectively.

As mentioned earlier, cell G3 is the product of deviations $(y_1 - \bar{y})(x_1 - \bar{x}) = Dx \cdot Dy$. Thus, write in cell G3

$$=E3*F3$$

Do a similar procedure for cells G4 to G7.

Cell H3 is the square of the deviation of the first observation from x $(x_1 - \bar{x})^2 = (Dx)^2$. Thus, write in cell H3

=E3^2

Repeat this procedure for cells H4 to H7. The formulation is shown in Fig. B-9.

	G	H
1		
2	**Dxy**	**Dxx**
3	=E3*F3	=E3^2
4	=E4*F4	=E4^2
5	=E5*F5	=E5^2
6	=E6*F6	=E6^2
7	=E7*F7	=E7^2
8		
9	**SSxy**	**SSxx**

Fig. B-9

The results are shown in Fig. B-10.

	A	B	C	D	E	F	G	H
1								
2			**x**	**y**	**Dx**	**Dy**	**Dxy**	**Dxx**
3			1	8	-2	-7	14	4
4			2	12	-1	-3	3	1
5			3	15	0	0	0	0
6			4	16	1	1	1	1
7			5	24	2	9	18	4
8		Total	15	75				
9		Average	3	15			**SSxy**	**SSxx**

Fig. B-10

(*b*) Cell G8 represents **SSxy**, the total sum of **Dxy** or $\sum_{i=1}^{n}(y_i - \bar{y})(x_i - \bar{x})$. Thus, write in cell G8:

=SUM(G3:G7)

In a similar manner, SSxx is $\sum_{i=1}^{n}(x_i - \bar{x})^2$, the sum of the squared deviations of x. Thus, write in cell H8

=SUM(H3:H7)

The results are shown in Fig. B-11.

	A	B	C	D	E	F	G	H
1								
2			*x*	*y*	*Dx*	*Dy*	*Dxy*	*Dxx*
3			1	8	-2	-7	14	4
4			2	12	-1	-3	3	1
5			3	15	0	0	0	0
6			4	16	1	1	1	1
7			5	24	2	9	18	4
8		Total	15	75			36	10
9		Average	3	15			*SSxy*	*SSxx*
10								

Fig. B-11

(c) Add two more rows. The first row (row 11) will locate the answer for the slope $\hat{\beta}$; the second row (row 13) will contain the answer for the intercept $\hat{\alpha}$. This is displayed in Fig. B-12.

	G	H
10		
11	**Beta**	
12		
13	**Alpha**	

Fig. B-12

Recall the formula for $\hat{\beta}$, and replace with the labels in the Excel™ spreadsheet:

$$\hat{\beta} = \frac{\sum_{i=1}^{n}(y_i - \bar{y})(x_i - \bar{x})}{\sum_{i=1}^{n}(x_i - \bar{x})^2} = \frac{SSxy}{SSxx}$$

Thus, $\hat{\beta}$ is the ratio between $SSxy$ and $SSxx$. Therefore, write in cell H11

=G8/H8

The result appears in Fig. B-13, which is $36/10 = 3.6$.

	G	H
8	36	10
9	*SSxy*	*SSxx*
10		
11	**Beta**	3.6

Fig. B-13

(4)   *Calculation of $\hat{\alpha}$:* use the first formula in (B.2).

$$\hat{\alpha} = \bar{y} - \hat{\beta}\bar{x}$$

Recall that $\bar{y}$ is in D9, $\bar{x}$ is in C9, and $\hat{\beta}$ in H11. Thus, write in cell H13 (which represents $\hat{\alpha}$)

=D9-H11*C9

The result is $15 - 3.6(3) = 4.2$. The complete table appears in Fig. B-14.

	A	B	C	D	E	F	G	H
1								
2			*x*	*y*	*Dx*	*Dy*	*Dxy*	*Dxx*
3			1	8	-2	-7	14	4
4			2	12	-1	-3	3	1
5			3	15	0	0	0	0
6			4	16	1	1	1	1
7			5	24	2	9	18	4
8		Total	15	75			36	10
9		Average	3	15			*SSxy*	*SSxx*
10								
11							*Beta*	3.6
12								
13							*Alpha*	4.2

Fig. B-14

Thus, the linear equation is

$$\hat{y} = 4.2 + 3.6x$$

B.2.   Given the data table below

x	*y*
2	0.3
3	0.6
5	0.9
7	1.4
9	1.6
10	2.4

fit the linear regression using the ordinary least square method.

This problem was presented in Problem 22.18. To solve it using MS Excel™, follow the steps below.

(*1*) Create the template shown in Fig. B-15.

	A	B	C	D	E	F	G	H
1								
2			*x*	*y*	*Dx*	*Dy*	*Dxy*	*Dxx*
3			2	0.3				
4			3	0.6				
5			5	0.9				
6			7	1.4				
7			9	1.6				
8			10	2.4				
9		Total						
10		Average					*SSxy*	*SSxx*
11								
12							*Beta*	
13								
14							*Alpha*	

Fig. B-15

(*2*) *Calculate averages.*

(*a*) Compute the sum of all *x* elements in cell **C9** by writing **=SUM(C3:C8)**

(*b*) Find the sum of all *y* elements in cell **D9** by writing **=SUM(D3:D8)**

(*c*) Calculate the average of *x* or $\bar{x}$ in cell **C10** by writing **=AVERAGE(C3:C8)**

(*d*) Find the average of *y* elements or $\bar{y}$ in cell **D9** by writing **=AVERAGE (D3:D8)**

The results are displayed in Fig. B-16.

	A	B	C	D
1				
2			*x*	*y*
3			2	0.3
4			3	0.6
5			5	0.9
6			7	1.4
7			9	1.6
8			10	2.4
9		Total	36	7.2
10		Average	6	1.2

Fig. B-16

(*3*) *Calculating $\hat{\beta}$:* this requires the use of the formula in (*B.2*).

$$\hat{\beta} = \frac{\sum_{i=1}^{n}(y_i - \bar{y})(x_i - \bar{x})}{\sum_{i=1}^{n}(x_i - \bar{x})^2} = \frac{SSxy}{SSxx}$$

SSxy is the sum of the deviation of *y* times the deviation of *x* from their respective means.

SSxx is the sum squared deviations of *x* from the mean.

To calculate both, we need the deviation of *x* and *y* from their means. Call the deviation of each *x* observation from the mean **Dx** and the deviation of *y* label it as **Dy**.

(*a*) **Dx** is $x_i - \bar{x}$; in other words, the observation x_i minus the average of *x*. For the first observation (in E3) this is

=C3-C$10

Note again the dollar sign ($) used to lock the row column of averages. The result is $2 - 6 = -4$. A similar procedure is done for all elements in **Dx**. The result is shown in Fig. B-17.

(*b*) **Dy** is $y_i - \bar{y}$, the observation y_i minus the average of *y*. For the third observation (in F5) this is

=D5-D$10

The result is $0.9 - 1.2 = -0.3$. A similar procedure is done for all elements in **Dy**. The formulas are shown in Fig. B-17 and the results are in Fig. B-18.

	E	F
1		
2	***Dx***	***Dy***
3	=C3-C$10	=D3-D$10
4	=C4-C$10	=D4-D$10
5	=C5-C$10	=D5-D$10
6	=C6-C$10	=D6-D$10
7	=C7-C$10	=D7-D$10
8	=C8-C$10	=D8-D$10

Fig. B-17

	B	C	D	E	F
1					
2		*x*	*y*	***Dx***	***Dy***
3		2	0.3	-4	-0.9
4		3	0.6	-3	-0.6
5		5	0.9	-1	-0.3
6		7	1.4	1	0.2
7		9	1.6	3	0.4
8		10	2.4	4	1.2
9	Total	36	7.2		
10	Average	6	1.2		

Fig. B-18

(*c*) **Dxy** is the product $(x_i - \bar{x})(y_i - \bar{y}) = DxDy$. In other words, for the first observation (in G3)

=E3*F3

For G3, the value is $(-4)(-0.9) = -3.6$. A similar procedure is done for all elements in **Dxy**. The result is shown in Fig. B-19.

(d)   **Dxx** is the square of Dx $(x_i - \bar{x})^2 = (Dx)^2$. In other words, for the first observation (in H3)

$$=E3\text{^}2$$

For H3, the value is $(-4)^2 = 16$. The same is done for all elements in **Dxx**. The result is also displayed in Fig. B-19.

	B	C	D	E	F	G	H
1							
2		*x*	*y*	*Dx*	*Dy*	*Dxy*	*Dxx*
3		2	0.3	-4	-0.9	3.6	16
4		3	0.6	-3	-0.6	1.8	9
5		5	0.9	-1	-0.3	0.3	1
6		7	1.4	1	0.2	0.2	1
7		9	1.6	3	0.4	1.2	9
8		10	2.4	4	1.2	4.8	16
9	Total	36	7.2				
10	Average	6	1.2			*SSxy*	*SSxx*

Fig. B-19

(e)   **Sxy** is the sum of all elements of **Dxy** $= \sum_{i=1}^{n} (y_i - \bar{y})(x_i - \bar{x})$, which is located in cell G9. Thus, write

$$=SUM(G3:G8)$$

which is 11.9. The same is done for all elements in **Sxx**, which is the sum of all elements in

Dxx $\sum_{i=1}^{n} (x_i - \bar{x})^2$. Thus, write in cell H9

$$=SUM(H3:H8)$$

which is equal to 52. The results are displayed in Fig. B-20.

	A	B	C	D	E	F	G	H
1								
2			*x*	*y*	*Dx*	*Dy*	*Dxy*	*Dxx*
3			2	0.3	-4	-0.9	3.6	16
4			3	0.6	-3	-0.6	1.8	9
5			5	0.9	-1	-0.3	0.3	1
6			7	1.4	1	0.2	0.2	1
7			9	1.6	3	0.4	1.2	9
8			10	2.4	4	1.2	4.8	16
9		Total	36	7.2			11.9	52
10		Average	6	1.2			*SSxy*	*SSxx*

Fig. B-20

Finally, $\hat{\beta}$ is calculated as

$$\hat{\beta} = \frac{SSxy}{SSxx}.$$

Therefore, write in cell H12

=G9/H9

The result appears in Fig. B-21, which is $11.9/52 = 0.2288$.

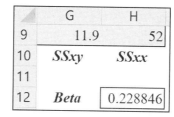

	G	H
9	11.9	52
10	*SSxy*	*SSxx*
11		
12	*Beta*	0.228846

Fig. B-21

(4) *Calculation of $\hat{\alpha}$:* use the first formula in (B.2).

$$\hat{\alpha} = \bar{y} - \hat{\beta}\bar{x}$$

Here, $\bar{y}$ is in D10, $\bar{x}$ is in C10, and $\hat{\beta}$ in H12. Thus, write in cell H14 (which represents $\hat{\alpha}$)

=D10-H12*C10

The result is $1.2 - 0.2288(6) = -0.1731$. The complete table appears in Fig. B-22.

	A	B	C	D	E	F	G	H
1								
2			*x*	*y*	*Dx*	*Dy*	*Dxy*	*Dxx*
3			2	0.3	-4	-0.9	3.6	16
4			3	0.6	-3	-0.6	1.8	9
5			5	0.9	-1	-0.3	0.3	1
6			7	1.4	1	0.2	0.2	1
7			9	1.6	3	0.4	1.2	9
8			10	2.4	4	1.2	4.8	16
9		Total	36	7.2			11.9	52
10		Average	6	1.2			*SSxy*	*SSxx*
11								
12							*Beta*	0.228846
13								
14							*Alpha*	-0.17308
15								

Fig. B-22

Thus, the linear equation is

$$\hat{y} = -0.1731 + 0.2288x$$

SIMPLE LINEAR REGRESSION: USING SCATTERPLOTS

B.3.   Solve Problem B.1 by using a scatterplot.

(*1*)   Write the dataset in an Excel™ spreadsheet in the locations shown in Fig. B-23. This is similar to Fig. B-1, except that there are no added rows or columns.

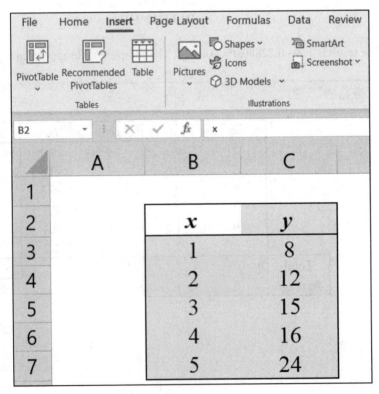

Fig. B-23

(*2*)   Select both columns that contain *x* and *y*, including the labels in row **2**.

(*3*)   Click on the Insert tab, as displayed in Fig. B-24.

Fig. B-24

(*4*) Then, click on the scatterplot icon 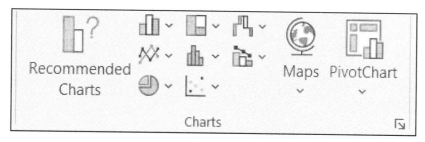, which can be found in the menu "Charts" shown in Fig. B-25.

Fig. B-25

(*5*) The graph in Fig. B-26 will appear, which represents the data points provided.

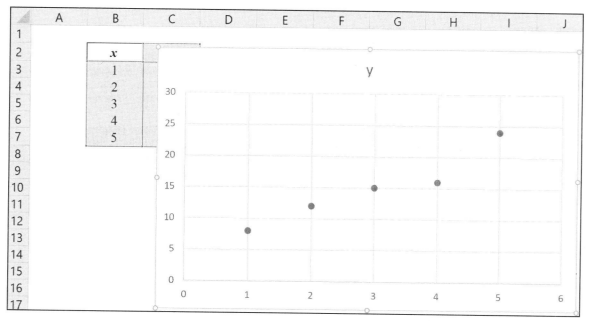

Fig. B-26

(*6*) Right-click on one of the data points; a menu will appear, select "Add Trendline." The menu "Format Trendline" will be displayed on the right of the window, as shown in Fig. B-27.

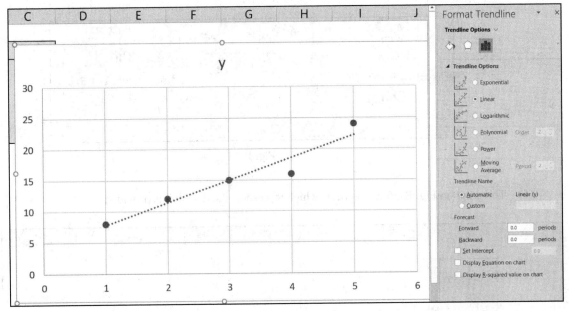

Fig. B-27

(*7*) In "**Trendline Options**," select **Linear**, and click on the box that says "**Display Equation on Chart**." Automatically, a linear equation will be displayed, as shown in Fig. B-28.

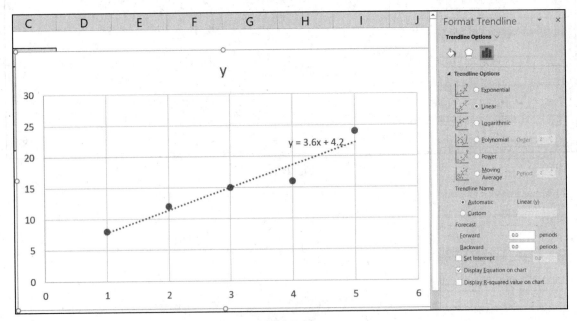

Fig. B-28

This equation is the estimation of the linear regression,

$$\hat{y} = 3.6x + 4.2$$

which is the same found in Problem B.1. Likewise, the line that is drawn in the scatterplot is the representation of the linear regression that was solved using OLS.

SIMPLE LINEAR REGRESSION: USING EXCEL™ FUNCTIONS

B.4.   Solve Problem B.1 by using an Excel™ spreadsheet.

(*1*)   Write the dataset in an Excel™ spreadsheet as in Fig. B-29, similar to Problem B.3, except with the addition of two new cells that will represent the locations of the estimates $\hat{\alpha}$ and $\hat{\beta}$.

	A	B	C	D	E	F
1						
2		*x*	*y*			
3		1	8		**Beta**	
4		2	12			
5		3	15		**Alpha**	
6		4	16			
7		5	24			
8						

Fig. B-29

(*2*)   Excel™ has a function called **SLOPE( )** that computes the slope of the linear regression, which is equivalent to $\hat{\beta}$. Thus, write in cell F3

=SLOPE(

Then, select the values of *y*, or write **C3:C7**. Posteriorly, write a comma "," and select the values of *x* (or write **B3:B7**), and close parenthesis. The formula in F3 should be **=SLOPE(C3:C7,B3:B7)**

The result of this function is 3.6, which is the result found in Exercises B.1 and B.3.

(*3*)   Excel™ also has a function called **INTERCEPT( )** that computes $\hat{\alpha}$, which is the intercept of the linear regression. Thus, write in cell F5

=INTERCEPT(C3:C7,B3:B7)

The outcome is 4.2, as displayed in Fig. B-30.

	A	B	C	D	E	F
1						
2		*x*	*y*			
3		1	8		**Beta**	3.6
4		2	12			
5		3	15		**Alpha**	4.2
6		4	16			
7		5	24			
8						

Fig. B-30

Additional Practice Problems

Chapter 1

THEORETICAL CONCEPTS

S1.1. The multiplication of two exponential terms with the same base x can be combined into the same base by adding the exponents of each term.

_____ True   _____ False

S1.2. Equations must be _____ (inconsistent/consistent) and _____ (independent/dependent), and there must _____ (the same number of/more/fewer) equations compared to variables.

S1.3. The graph of a quadratic equation is a:

a. parabola

b. upward line

c. downward line

d. ellipse

S1.4. The _____ of a function $f(x)$ is a rule that assigns a specific value to each x in the domain.

S1.5. _____ (One/Two/Three) points are required to determine the graph of a linear equation.

EXPONENTIAL TERMS

S1.6. Use the rules of exponents to simplify these fractions.

$$\frac{x^2}{\sqrt[6]{x^5} \cdot (x^{1/12})^2} = \frac{x^{\boxed{}}}{x^{\boxed{}}} = x^{\boxed{}} = x^{\boxed{}}$$

POLYNOMIALS

S1.7. Multiply these polynomials.

$$(3x + y)(x^2 - 5xy - 2)$$

EQUATIONS

S1.8. Solve the following linear equations by using properties of equality.

$$(1/2)x + 10 = 3(2x - 1) - 5/2$$

[   ] $x =$ [   ]

$x =$ [   ]

GRAPHS, SLOPES, AND INTERCEPTS

S1.9. Given the linear equation

$$4x + 6y = 48$$

determine the slope and intercepts, then graph the equation.

(1) Find the slope.

(2) Find the x- and y-intercepts

(3) Graph the equation.

S1.10. Graph the quadratic function:

$$y = (x - 1)^2$$

(*1*)   Plug values of x in the function and fill out the table.

x	$f(x)$
−3	
−2	
−1	
0	
1	
2	
3	
4	

(*2*)   Graph the function.

Chapter 2

THEORETICAL CONCEPTS

S2.1.   The isocost line is used in production economics to represent the combination of two inputs; typically these are _____ and _____.

S2.2.   In supply and demand analysis, the market equilibrium is found by setting the quantity supplied equal to the quantity demanded.

_____ True   _____ False

S2.3.   Which of the following factors is not explicitly included in the income determination model?

a.   consumption   c.   government expenditure

b.   foreign investment   d.   exports

S2.4. An increase in budget makes a _____ (parallel/rotational) shift toward the _____ (right/left) of the original line.

S2.5. The _____ analysis is used to interpret the relationship between income levels and interest rates.

ISOCOST LINES

S2.6. A company has a daily budget of $5,000 for capital and labor expenses. Cost of capital is $50 per day, and cost of labor is $100 per day (or $12.5 per hour).

    I.   Consider L as the workers and K as machinery rented per day. Write the company's cost budget equation in standard form.

    II.   Express capital K as a function of labor L.

    III.   Graph the isocost line, indicating what happens to the initial isocost if the budget increases.

SUPPLY AND DEMAND ANALYSIS

S2.7. Find the equilibrium price and quantity for the market.

$$\text{Supply: } Q_S = -16 + 4P \qquad\qquad \text{Demand: } Q_D = 64 - 2P$$

INCOME DETERMINATION MODEL

S2.8. Find the equilibrium level of income Y_e, given

$$Y = C + I + (X - Z) \qquad C = C_0 + bY \qquad Z = Z_0 + zY$$

$$I = I_0 \qquad\qquad\qquad X = X_0$$

where $C_0 = 80$, $I_0 = 120$, $X_0 = 100$, $Z_0 = 50$, $b = 0.9$, $z = 0.25$

   a. Find the reduced form of Y_e

   b. Determine the numerical value of Y_e

IS-LM EQUATIONS

S2.9. Given the commodity market for a two-sector economy

$$Y = C + I, \text{ where } C = 120 + 0.1Y \text{ and } I = 30 - 200i$$

and a money market system

$$M_s = 200, \ M_t = 0.2Y \text{ and } M_z = 175 - 100i$$

(*1*) Find the equilibrium income level and interest rate.

(*2*) Calculate the levels of C, I, M_t, and M_Z in equilibrium.

Chapter 3

THEORETICAL CONCEPTS

S3.1. A function is _____ (concave/differentiable/noncontinuous) at a point $x = a$ if the derivative exists at that particular point.

S3.2. For a function to be continuous at a point $x = a$, the limit must exist.

                                                      ____ True   ____ False

S3.3. The technique used to derive a multiplication of polynomials is called the _____ rule.

    a.   quotient   c.   chain

    b.   product   d.   power

S3.4. In an implicit function, the dependent variable y typically appears on the _____ (same/other) side of the equation with respect to the independent variable x. Thus, if the terms that include x are on the right side, then any term that includes y is on the _____ (right/left) side.

S3.5. The generalized power function rule and the chain rule provide the same derivative of the primitive function.

                                                        ____ True   ____ False

LIMITS AND DIFFERENTIATION

S3.6. Find $\lim\limits_{x \to -3} f(x)$, if it exists, given

$$f(x) = \frac{3x + 9}{x^2 - 9}$$

DERIVATIVE RULES

Find the first derivative of the following functions.

S3.7. Use the product rule to differentiate $f(x) = (4x + 1)(5x^{-1/3} + 7x^2)$

$$f'(x) = \underbrace{\boxed{}}_{g(x)} \cdot \underbrace{\boxed{}}_{h'(x)} + \underbrace{\boxed{}}_{g'(x)} \cdot \underbrace{\boxed{}}_{h(x)}$$

Simplify algebraically.

$f'(x) = $ [_____]

S3.8. Use the generalized power rule to differentiate $f(x) = \sqrt[4]{2x^3 - 11}$

Define the function in exponential terms.

$f(x) = (2x^3 - 11)^{\square}$

Takes the derivative with respect to x:

$$f'(x) = \underbrace{[\quad]}_{n} \cdot \underbrace{[\qquad\qquad]}_{[g(x)]^{n-1}} \cdot \underbrace{[\qquad\qquad]}_{g'(x)}$$

Simplify algebraically.

$f'(x) = $ [_____]

S3.9. Use the chain rule to differentiate $f(x) = (2x^3 + x^{1/2})^6$

Let $\qquad y = $ [_____] and $\qquad u = $ [_____]

Then $\qquad dy/du = $ [_____] and $du/dx = $ [_____]

Substitute in the formula.

$$\frac{dy}{dx} = [\qquad] \cdot [\qquad]$$

Replace u in terms of x.

$$\frac{dy}{dx} = [\qquad\qquad]$$

HIGHER-ORDER DERIVATIVES

S3.10. Given the function $f(x) = 24x^{1/8} - 20x^{-1/2} + 5x$, find the second-order derivative.

(*1*) Take the first derivative of the function.

[   ]

(*2*) Take the derivative of the first-order derivative to obtain the second-order derivative.

[   ]

Chapter 4

THEORETICAL CONCEPTS

S4.1. To maximize profits, marginal revenue must be _____ (greater than/lower than/equal to) marginal cost.

S4.2. The inflection point can be found by taking the first-order derivative of the function and setting it equal to zero.

                                                 ____ True   ____ False

S4.3. A differentiable function $f(x)$ whose first-order derivative evaluated at $x = a$ is positive can be classified as a(n) _____ function.

    a. convex   c. increasing

    b. concave   d. decreasing

S4.4. In an implicit function, the dependent variable y typically appears on the _____ (same/other) side of the equation with respect to the independent variable x. Thus, if the terms that include x are on the right side, then any term that includes y is on the _____ (right/left) side.

S4.5. The derivative of the total product with respect to labor is the average product of labor.

                                                   ____ True   ____ False

RELATIVE EXTREMA

S4.6. Find the relative extrema of the function $f(x) = x^3 - 9x^2 + 24x - 12$ and classify the critical points as relative minimum or maximum.

    (1) Take the first-order condition to find the critical values.

    (2) Evaluate each critical point in the second-order derivative and classify the points.

MARGINAL, AVERAGE, AND TOTAL COST

S4.7. Given the cost function $TC = Q^3 - 24Q^2 + 360Q$

    (1) Find the inflection point of the total cost function.

    (2) Find the average cost functions and the relative extrema.

    (3) Find the marginal cost functions and the relative extrema.

    (4) Verify the point of intersection between the average and the marginal cost functions (note $Q > 0$).

OPTIMIZATION IN BUSINESS AND ECONOMICS

S4.8. Maximize the profit function $\pi = -Q^3 - 15Q^2 + 1125Q - 6000$ by finding the critical values.

    (1) Find the critical value for $Q*$ (note $Q > 0$).

(2)   Verify that the critical value $Q*$ is a relative maximum point.

(3)   Compute the maximum profit level π^* by establishing $\pi^* = \pi(Q^*)$

MARGINAL RATE OF TECHNICAL SUBSTITUTION

S4.9.   Given labor L and capital K, the equation for the production isoquant is

$$L^{2/7} K^{5/7} = 21000$$

(1)   Find the marginal rate of technical substitution MRTS $= dK/dL$

(2)   Compute the MRTS at $K_0 = 300$ and $L_0 = 400$.

This means that if L is increased by one relatively small unit, K must decrease by ☐ units in order to remain on the isoquant production where the production level is constant.

Chapter 5

THEORETICAL CONCEPTS

S5.1. _____ is the method used to maximize constrained optimization problems.

S5.2. A bivariate function $z = f(x, y)$ has four second-order partial derivatives.

_____ True   _____ False

S5.3. Given a differentiable function $z = f(x, y)$, consider $f_{xx} \cdot f_{yy} < (f_{xy})^2$, $f_{xx} > 0$, and $f_{yy} < 0$ at the critical value (x^*, y^*). This point can be classified as the _____ point.

a. inflection   c. relative minimum

b. relative maximum   d. saddle

S5.4. Given $y = f(x)$, the _____ of y (dy) evaluates the change in y as a result of a small change in x; in other words, $dy = f'(x)dx$.

S5.5. To find the critical values of the differentiable function, $z = f(x, y)$ can be found by taking the first-order derivative with respect to x and y simultaneously and setting them to equal to zero.

_____ True   _____ False

PARTIAL DERIVATIVES

Find the first-order partial derivative with respect to x.

S5.6. Use the product rule to differentiate $f(x, y) = (2xy^2 + 4y)(10x + 2x^3 y)$

S5.7. Use the generalized power rule to differentiate $f(x, y) = \left(3x^2 - 5y + 8x^3 - 9x^{\frac{1}{3}} y \right)^4$

TOTAL DERIVATIVES

S5.8. Find the total derivative dz/dx of the bivariate function

$$z = f(x, y) = 48x^{1/2}y^3 + 32y - 12x^2$$

where $y = x^2 + 2$.

(*1*) The total derivative dz/dx is given by the formula

$$\frac{dz}{dx} = \boxed{} + \boxed{}\,\frac{dy}{dx}$$

(*2*) Find $\dfrac{dy}{dx}$, z_x, and z_y

IMPLICIT DIFFERENTIATION

S5.9. Find the derivative dy/dx of the implicit function

$$24x^{1/3}y^2 - 17 + 6x^2 + 48y^{1/2} = 0$$

CONSTRAINED OPTIMIZATION DIFFERENTIATION

S5.10. Given the function

$$z = x^{4/5} y^{1/5}$$

The constraint is

$$2x + y = 45$$

Find the optimal combination (x^*, y^*) that optimizes the Lagrange function.

(*1*) Establish the Lagrange function.

$$\mathcal{L} = \boxed{}$$

(*2*) Take the first-order partial derivatives for optimization to find (x^*, y^*).

$$\mathcal{L}_x = \boxed{}$$

$$\mathcal{L}_y = \boxed{}$$

$$\mathcal{L}_\lambda = \boxed{}$$

$$(x^*, y^*) = \boxed{}$$

(*3*) Find λ^* and interpret it.

With $\lambda^* = $ _____, a one-unit increase in the constant of the constraint will lead to a(n) _____ (decrease/increase) of approximately _____ in the value of the objective function.

Chapter 6

THEORETICAL CONCEPTS

S6.1. Given labor L and capital K needed to produce q, _____ is the production function with the form $q = AK^\alpha L^\beta$.

S6.2. The differential measures the effect of a change in a dependent variable when there is a variation in the independent variable.

_____ True   _____ False

S6.3. When all inputs are increased by the same proportion as the output, this is called _____ returns to scale.

a. constant   c. decreasing

b. increasing   d. economy

S6.4. _____ (Price/Income) elasticity of demand measures the change in the demand for a good due to a small change in the person's budget.

S6.5. In order to compute the marginal productivity of labor, all other factors may remain variable.

_____ True   _____ False

MARGINAL PRODUCTIVITY

S6.6. Given the production function

$$Q = 48KL - 4K^2 - 2L^2$$

find the marginal productivity of labor (L) and capital (K).

INCOME DETERMINATION MULTIPLIER AND COMPARATIVE STATISTICS

S6.7. Given the following economy

$$Y = C + I + (X - Z) \qquad C = C_0 + bY \qquad Z = Z_0 + zY$$

$$I = I_0 \qquad\qquad X = X_0$$

where $C_0, I_0, X_0, Z, b, z > 0$.

find how income Y changes if the propensity to import z changes by 0.1.

(*1*) Find the equilibrium for income Y^*.

(2)   Calculate $\partial Y^* / \partial z$

(3)   Consider $C_0 = 80$, $I_0 = 120$, $X_0 = 100$, $Z_0 = 50$, $b = 0.9$, $z = 0.25$. Compute Y^* from the reduced form found in (1).

(4)   Calculate the change in income Y^* (ΔY^*) as a result of result of $\Delta z = 0.1$

PARTIAL ELASTICITIES

S6.8.   Given the demand function

$$Q = 300 - 1.5P + 0.14Y$$

where $P = 30$ and $Y = 3000$,

find the income elasticity of demand.

OPTIMIZATION OF ECONOMIC FUNCTIONS

S6.9. Given the profit function for a firm producing two goods x and y

$$\pi = 160x - x^2 - xy - 3y^2 + 300y - 3000$$

find the optimal combination (x^*, y^*) that optimizes profits.

(*1*) Take the first-order condition.

$$\pi_x = \boxed{}$$

$$\pi_y = \boxed{}$$

$$(x^*, y^*) = \boxed{}$$

(*2*) Take the second-order condition to verify its concavity.

$$\pi_{xx} = \boxed{}$$

$$\pi_{yy} = \boxed{}$$

$$\pi_{xy} = \boxed{}$$

Using this information, compute the rule to determine its concavity.

$$\boxed{}$$

(*4*) Compute the maximum profit level π^* by establishing $\pi^* = \pi(x^*, y^*)$.

$$\boxed{}$$

HOMOGENEITY AND RETURNS TO SCALE

S6.10. Determine the level of homogeneity and returns to scale for the production function below:

$$Q = 2x^2 + 4xy - \frac{8x^3}{y^2}$$

CONSTRAINED OPTIMIZATION IN ECONOMICS

S6.11. Given the profit function

$$\pi = 40x - x^2 - 2xy - 2y^2 + 50y$$

The maximum output capacity (constraint) is

$$x + y = 25$$

Find the optimal combination (x^*, y^*) that maximizes the profit function using the Lagrange multiplier.

(1) Establish the Lagrange function.

$$\mathcal{L} = \boxed{}$$

(2) Take the first-order partial derivatives for optimization to find (x^*, y^*).

$$\mathcal{L}_x = \boxed{}$$

$$\mathcal{L}_y = \boxed{}$$

$$\mathcal{L}_\lambda = \boxed{}$$

$$(x^*, y^*) = \boxed{}$$

(3) Find λ^* and interpret it.

With $\lambda^* =$ _____, a one-unit increase in the constant of the output capacity will lead to a(n) _____ (decrease/increase) in profit of approximately _____.

Chapter 7

THEORETICAL CONCEPTS

S7.1. Exponential (power) functions cannot be negative.

_____ True   _____ False

S7.2. The inverse function of an exponential function is a logarithmic function.

_____ True   _____ False

S7.3. A multiplication of two terms in a logarithmic function can be split in the subtraction of two logarithmic functions.

_____ True   _____ False

S7.4. The base of the natural logarithmic function is e.

_____ True   _____ False

S7.5. Cobb-Douglas functions can be linearized using logarithms.

_____ True   _____ False

EXPONENTIAL-LOGARITHMIC CONVERSIONS

Change each of the following functions to their corresponding inverse function.

S7.6. Convert $y = e^{2x-5}$ into a logarithmic function by solving for x.

S7.7. Convert $y = 2\ln(x + 10)$ into an exponential function by solving for x.

PROPERTIES OF LOGARITHMIC AND EXPONENTIAL FUNCTIONS

Convert the expressions to sums, differences, or products.

S7.8. Convert $y = \ln(2x + 1) - \ln(4x + 7)$ into a ratio.

S7.9. Convert $z = \ln\left(\dfrac{2x^2}{\sqrt[4]{y^3}}\right)$ into additions and subtractions.

SOLVING LOGARITHMIC AND EXPONENTIAL FUNCTIONS

S7.10. Solve for x.

$$\ln(4x+7) = 3$$

S7.11. Find the two values for x.

$$\left(e^x\right)^{2x-3} = 1$$

Chapter 8

THEORETICAL CONCEPTS

S8.1. For a depreciation problem, the interest rate r is _____ (positive/negative).

S8.2. The effective annual rate is the rate used in finance to compare different sources of debt.

_____ True   _____ False

S8.3. _____ is the process of calculating the present value of a future cash flow.

    a. Compounding   c. Depreciating

    b. Discounting   d. Investing

S8.4. _____ (Discrete/Continuous) growth is the type of growth that takes place at the end of a determined period of the year, such as a quarter or at the end of the year.

S8.5. Positive consistent growth is usually represented by the exponential functions.

                                                                       ____ True   ____ False

COMPOUNDING INTEREST

S8.6. Given a principal P of $2,000 at 5 percent interest i for 7 years, find the future value S when the principal is compounded continually.

INTEREST RATE AND TIMING

S8.7. At what interest rate will $1,000 become 4 times if compounded continuously for 5 years?

DISCOUNTING

S8.8. Find the present value of $3,000 in 4 years at 12 percent when the interest rate is compounded.

EXPONENTIAL GROWTH FUNCTIONS

S8.9. Profits of a corporation are projected to increase by 3 percent every year over the next 5 years. With current profits of $100,000, what will the level of profits be at the end of the lustrum?

CONVERTING EXPONENTIAL FUNCTIONS

S8.10. Find the equivalent form for $S = Pe^{0.12t}$ under semiannual discrete compounding.

EXPONENTIAL GROWTH IN ECONOMICS

S8.11. Government spending G increased from 4 trillion in 2002 to 6 trillion in 2021. Find the annual growth rate.

(*1*) Express the final expenditure in terms of an ordinary annual exponential function.

(*2*) Find the annual rate of growth.

Chapter 9

THEORETICAL CONCEPTS

S9.1. The derivative of a natural exponential function is equal to the _____ times the derivative of the _____.

S9.2. To take the derivative of a general exponential function, the procedure follows the rule of taking the derivative of a natural exponential function but multiplied by the natural log of the base.

_____ True _____ False

S9.3. _____ is used to ease the process of differentiating products and quotients.

    a. Logarithmic differentiation   c. Chain rule

    b. Exponential differentiation   d. Power rule

S9.4. _____ (Exponential/Discrete) functions are used to determine the optimal timing for an investment.

S9.5. The partial derivative of a logarithmic function is found by differentiating the function with respect to that specific variable while holding all other independent variables constant.

_____ True _____ False

DERIVATIVES OF EXPONENTIAL FUNCTIONS

S9.6. Calculate the derivative of

$$y = 4e^{(2x-x^2)}$$

S9.7. Calculate the derivative of

$$y = 8^{(3x+1)}$$

DERIVATIVES OF NATURAL LOGARITHMIC FUNCTIONS

S9.8. Calculate the derivative of

$$y = \ln(x+3)^6$$

S9.9. Calculate the derivative of

$$y = \log_7(x - x^2)$$

SECOND DERIVATIVES

S9.10. Find the first and second derivatives of

$$y = xe^{2x}$$

PARTIAL DERIVATIVES

S9.11. Find the first and second partial derivatives of

$$z = e^{y - x^2}$$

$z_x = \boxed{}$ $\qquad$ $z_y = \boxed{}$

$z_{xx} = \boxed{}$ $\qquad$ $z_{yy} = \boxed{}$

$z_{xy} = \boxed{}$ $\qquad$ $z_{yx} = \boxed{}$

S9.12. Find the first and second partial derivatives of

$$z = \ln\left(\frac{x+y}{x-y}\right)$$

$z_x = \boxed{}$ $z_y = \boxed{}$

$z_{xx} = \boxed{}$ $z_{yy} = \boxed{}$

$z_{xy} = \boxed{}$ $z_{yx} = \boxed{}$

OPTIMIZATION OF EXPONENTIAL AND LOGARITHMIC FUNCTIONS

S9.13. Given $y = 2xe^{(4x+1)}$, find the critical value and classify it.

    (1)   Find the critical value using the product rule.

$$\boxed{}$$

    (2)   Classify the critical value by second-order condition.

$$\boxed{}$$

S9.14. Given $z = \ln(x^2 - 2x + 8y - y^2)$, find the critical value and classify it.

    (1)   Find the critical value (x^*, y^*).

$$\boxed{}$$

(2) Classify the critical value by second-order condition.

OPTIMAL TIMING

S9.15. An investment is increasing in value according to the formula

$$V = 100,000e^{\sqrt{t}}$$

The discount rate under continuous compounding is 0.05. How long should the investment be held to maximize the present value?

(1) Establish the present value formula given continuous discounting.

(2) Convert to natural logs and take the derivative to find the optimal timing.

(3) Take the second-order condition to verify that the critical point is a relative maximum point.

Chapter 10

THEORETICAL CONCEPTS

S10.1. In the multiplication of two matrices with the same dimensions A and B, the order the matrices does not affect the resultant matrix.

               ____ True   ____ False

S10.2. An identity matrix is symmetric and idempotent.

               ____ True   ____ False

S10.3. _____ is a matrix in which all elements are zero.

     a. Null matrix   c. Symmetric matrix

     b. Square matrix   d. Identity matrix

S10.4. The multiplication of two matrices is composed by _____ (inner/outer) products of row-column combinations.

S10.5. In order to subtract two matrices, both must have the same dimensions.

               ____ True   ____ False

MATRIX ADDITION AND SUBTRACTION

S10.6. Given the matrices below, find $A + B - C$.

$$A = \begin{bmatrix} 2 & 8 & 2 \\ 1 & 7 & 5 \end{bmatrix} \quad B = \begin{bmatrix} 2 & -4 & 5 \\ -1 & 15 & 2 \end{bmatrix} \quad C = \begin{bmatrix} 0 & 0 & 2 \\ -2 & 4 & 4 \end{bmatrix}$$

CONFORMABILITY

S10.7. Given the matrices below

$$A = \begin{bmatrix} 2 & 8 & 2 \\ 1 & 7 & 5 \end{bmatrix} \quad B = \begin{bmatrix} 2 & -4 & 5 \\ -1 & 15 & 2 \end{bmatrix} \quad C = \begin{bmatrix} 2 & 5 \\ 3 & 9 \end{bmatrix} \quad D = \begin{bmatrix} 2 & 0 \\ 1 & -1 \end{bmatrix} \quad E = \begin{bmatrix} 2 \\ 3 \\ 1 \end{bmatrix}$$

determine whether **AB**, **EC**, **AE**, **DC**, and **DE**, are conformable. If conformable, provide their dimensions.

(*a*) Matrix **AB** _____ (is/is not) conformable. The dimensions of **AB** are ____ × _____.

(*b*) Matrix **EC** _____ (is/is not) conformable. The dimensions of **EC** are ____ × _____.

(*c*) Matrix **AE** _____ (is/is not) conformable. The dimensions of **AE** are ____ × _____.

(*d*) Matrix **DC** _____ (is/is not) conformable. The dimensions of **DC** are ____ × _____.

(*e*) Matrix **DE** _____ (is/is not) conformable. The dimensions of **DE** are ____ × _____.

SCALAR MULTIPLICATION

S10.8. Determine **A**k, given

$$A = \begin{bmatrix} 2 & 8 & 2 \\ 1 & 7 & 5 \end{bmatrix} \qquad k = 1/4$$

VECTOR MULTIPLICATION

S10.9. Find **FE**, given

$$F = \begin{bmatrix} -1 & 0 & 5 \end{bmatrix} \qquad E = \begin{bmatrix} 2 \\ 3 \\ 1 \end{bmatrix}$$

MATRIX MULTIPLICATION

S10.10. Find **AB**, given

$$A = \begin{bmatrix} 2 & 8 & 2 \\ 1 & 7 & 5 \end{bmatrix} \qquad\qquad B = \begin{bmatrix} 2 & -1 \\ -4 & 15 \\ 5 & 2 \end{bmatrix}$$

Chapter 11

THEORETICAL CONCEPTS

S11.1. All the rows and columns of a singular matrix are linearly independent.

_____ True   _____ False

S11.2. A third-order determinant has as dimensions 2 × 3.

_____ True   _____ False

S11.3. The _____ of a matrix indicates the maximum number of linearly independent rows or columns.

a. rank   c. transpose

b. identity   d. singularity

S11.4. _____ (Laplace/Lagrange) expansion is used to calculate the determinant of a matrix based on the cofactors.

S11.5. A minor is a cofactor with a prescribed sign.

_____ True   _____ False

DETERMINANTS

S11.6. Find the determinant $|\mathbf{A}|$

$$\mathbf{A} = \begin{bmatrix} 20 & 0 \\ 14 & -1 \end{bmatrix}$$

S11.7. Find the determinant $|\mathbf{D}|$

$$\mathbf{D} = \begin{bmatrix} 1 & 4 & 3 \\ 0 & 5 & 0 \\ 2 & 8 & 6 \end{bmatrix}$$

PROPERTIES OF DETERMINANTS

S11.8. Given the lower-triangular matrix $\mathbf{B}$

$$\mathbf{B} = \begin{bmatrix} 2 & -4 & -3 \\ 0 & 5 & -2 \\ 0 & 0 & 6 \end{bmatrix}$$

find the determinant $|\mathbf{B}|$

MINORS AND COFACTORS

S11.9.   Given the matrix

$$\mathbf{D} = \begin{bmatrix} 1 & 4 & 3 \\ 0 & 5 & 0 \\ 2 & 8 & 6 \end{bmatrix}$$

find the minors and cofactors of the first column.

$\left|M_{11}\right| = \underline{\hspace{1cm}}$

$\left|M_{21}\right| = \underline{\hspace{1cm}}$

$\left|M_{31}\right| = \underline{\hspace{1cm}}$

$\left|C_{11}\right| = \underline{\hspace{1cm}}$

$\left|C_{21}\right| = \underline{\hspace{1cm}}$

$\left|C_{31}\right| = \underline{\hspace{1cm}}$

LAPLACE EXPANSION

S11.10.   Use Laplace expansion to find the determinant of matrix $\mathbf{F}$ using the second column.

$$\mathbf{F} = \begin{bmatrix} 1 & 4 & 3 \\ 2 & 5 & 1 \\ 2 & 8 & 6 \end{bmatrix}$$

INVERTING A MATRIX

S11.11.   Find the inverse $\mathbf{A}^{-1}$ for

$$\mathbf{A} = \begin{bmatrix} 20 & 0 \\ 14 & -1 \end{bmatrix}$$

S11.12. Find the inverse $\mathbf{A}^{-1}$ for the matrix using cofactors

$$\mathbf{A} = \begin{bmatrix} 1 & 4 & 3 \\ 0 & 5 & 2 \\ 2 & 1 & 1 \end{bmatrix}$$

The cofactor matrix $\mathbf{C}$ is

$$\mathbf{C} = \begin{bmatrix} & & \\ & & \end{bmatrix}$$

The adjoint matrix $\mathbf{A}$ is

$$\text{Adj } \mathbf{A} = \begin{bmatrix} & & \\ & & \end{bmatrix}$$

The determinant $|\mathbf{A}|$ is _____.

Then,

$$\mathbf{A}^{-1} = \begin{bmatrix} & & \\ & & \end{bmatrix}$$

MATRIX INVERSIONS IN EQUATION SOLUTIONS

S11.13. Use matrix inversions to solve the following system of linear equations:

$$4x + y = 21$$
$$3x - 2y = 13$$

(1) Set up the system into matrix form.

(2) Find the determinant of $\mathbf{A}$.

(3)   The cofactor matrix **C** is

$$\mathbf{C} = \begin{bmatrix} & \\ & \end{bmatrix}$$

(4)   The adjoint matrix **A** is

$$\text{Adj } \mathbf{A} = \begin{bmatrix} & \\ & \end{bmatrix}$$

(5)   The inverse matrix is

$$\mathbf{A}^{-1} = \begin{bmatrix} & \\ & \end{bmatrix}$$

(6)   Then substitute in $\mathbf{X} = \mathbf{A}^{-1}\mathbf{B}$.

$$X = \begin{bmatrix} & \\ & \end{bmatrix}\begin{bmatrix} \\ \end{bmatrix}$$

(7)   Multiply matrices; the solution for **X** is

$$X = \begin{bmatrix} \\ \end{bmatrix}$$

CRAMER'S RULE

S11.14. The market of a product is given by these linear equations:

$$\text{Supply: } -4P + Q = -16 \qquad \text{Demand: } 2P + Q = 64$$

where P is price and Q is quantity of the product. Find the equilibrium price (P^*) and quantity (Q^*) sold in this market by using Cramer's rule.

(1)   Express the equations in matrix form.

(2)   Find the determinant of **A**.

(*3*) To solve for *P*, replace column 1 of **A**, which contains the coefficients of *P*, with the column vector of constants **B**, forming a new matrix A_1.

Find the determinant of A_1.

Use Cramer's rule and find *P**.

(*4*) To solve for *Q*, replace column 2 of **A**, which contains the coefficients of *Q*, with the column vector of constants **B**, forming a new matrix A_2.

Find the determinant of A_2.

Use Cramer's rule and find Q^*.

(5)   The equilibrium price P^* is [   ] and quantity Q^* is [   ].

Chapter 12

THEORETICAL CONCEPTS

S12.1.   If the Jacobian determinant is not zero, then the equations are functionally independent.

            _____ True   _____ False

S12.2.   The Hessian matrix is used to test first-order conditions.

            _____ True   _____ False

S12.3.   For a function $y = f(x_1, x_2, x_3, x_4)$ with a nonsingular matrix, the number of principal minors to determine concavity is _____.

     a.   four   c.   one

     b.   three   d.   two

S12.4.   In an input-output table, matrix **A** is known as the _____.

S12.5.   Sign of definiteness of a matrix can be tested by using the characteristic roots or eigenvalues of the matrix.

            _____ True   _____ False

THE JACOBIAN

S12.6.   Test for functional dependence by means of the Jacobian.

$$y_1 = 4x_1 + 2x_2$$

$$y_2 = 16x_1^2 - 8x_1 x_2 - 4x_2^2$$

DISCRIMINANTS

S12.7. Use discriminants to determine if the quadratic function is positive or negative definite.

$$y = 16x_1^2 - 8x_1x_2 - 4x_2^2 + 12x_1x_3 + 8x_3^2 - 2x_2x_3$$

THE HESSIAN IN OPTIMIZATION PROBLEMS

S12.8. Maximize profits for a firm producing two goods x and y.

$$\pi = 160x - x^2 - xy - 3y^2 + 300y - 3000$$

    (1) Use Cramer's rule to find the critical points.

    (2) Use the Hessian to determine that the critical point is a relative maximum.

THE BORDERED HESSIAN IN CONSTRAINED OPTIMIZATION

S12.9. Given the profit function

$$\pi = 40x - x^2 - 2xy - 2y^2 + 50y$$

The maximum output capacity (constraint) is

$$x + y = 25$$

Find the optimal combination (x^*, y^*) that maximizes the profit function using the Lagrange multiplier.

(1) Establish the Lagrange function.

(2) Use Cramer's rule to find the critical points.

(3) Use the bordered Hessian matrix to test the second-order condition.

INPUT-OUTPUT ANALYSIS

S12.10. Determine the total demand **X** for industries 1, 2, and 3, given the matrix of technical coefficients **A** and the final demand vector **B** below.

Output industry

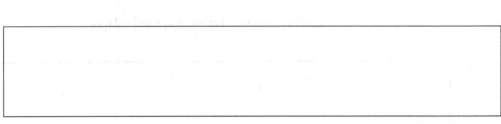

$$A = \begin{bmatrix} 0.00 & 0.20 & 0.00 \\ 0.35 & 0.00 & 0.25 \\ 0.10 & 0.10 & 0.00 \end{bmatrix} \quad \text{Input industry} \quad B = \begin{bmatrix} 600 \\ 1500 \\ 300 \end{bmatrix}$$

Given $X = (I - A)^{-1} B$

(1) The matrix $I - A$ is

$$I - A = $$

(2) Calculate the inverse matrix $(\mathbf{I} - \mathbf{A})^{-1}$

The cofactor matrix **C** is

$$\mathbf{C} = \boxed{}$$

The adjoint matrix Adj $(\mathbf{I} - \mathbf{A})$ is

$$\boxed{}$$

$$\text{Adj } (\mathbf{I} - \mathbf{A}) =$$

The determinant $|\mathbf{I} - \mathbf{A}| = \underline{\hspace{1cm}}$

With all this information, the inverse matrix $(\mathbf{I} - \mathbf{A})^{-1}$ is

$$(\mathbf{I} - \mathbf{A})^{-1} = \boxed{}$$

(3) Thus, **X** is

$$\mathbf{X} = \boxed{}$$

EIGENVECTORS

S12.11. Use eigenvalues (also known as characteristics roots) to determine the sign definiteness for

$$\mathbf{A} = \begin{bmatrix} 3 & 1 & 1 \\ 0 & 2 & 5 \\ 0 & 1 & 4 \end{bmatrix}$$

(1) Construct $|\mathbf{A} - c\mathbf{I}|$

(2) Expand along the first column to use the determinant.

(3) Solve for c and provide the characteristic roots.

Chapter 13

THEORETICAL CONCEPTS

S13.1. In comparative statics, the equilibrium is tested under changes in the endogenous variables.

____ True ____ False

S13.2. The envelope theorem allows us to evaluate the effect of a change in any exogenous variable on the optimal value of the objective function.

____ True ____ False

S13.3. The _____ is used for comparative statics with more than one endogenous variable.

a. Hessian   c. bordered Hessian

b. Jacobian   d. discriminant

S13.4. In a constrained maximization with two decision variables, if the second-order sufficient condition is met, then $|\bar{H}|$ __ 0 for maximization.

S13.5. Convex functions are excluded from concave programming.

____ True ____ False

COMPARATIVE STATICS WITH ONE ENDOGENOUS VARIABLE

S13.6. Given the following economy

$$Y = C + I + (X - Z) \qquad C = C_0 + bY \qquad Z = Z_0 + zY$$

$$I = I_0 \qquad\qquad X = X_0$$

where C_0, I_0, X_0, $Z > 0$ and $0 < b$, $z < 1$

Substituting for different variables, it can be found that

$$Y = C_0 + bY + I_0 + X_0 - Z_0 - zY$$

(*1*) Find the implicit function for the equilibrium condition.

(*2*) Find the effect of changing the propensity to import z in income $\bar{Y}$ (i.e., $\partial \bar{Y} / \partial z$).

COMPARATIVE STATISTICS FOR OPTIMIZATION PROBLEMS

S13.7. A firm seeks to optimize the present value of its profit from selling a good Z. The production of the good uses inputs X and Y and is represented by a Cobb-Douglas production function

$$\pi = P_0 (AX^\alpha Y^\beta) e^{rt} - P_x X - P_Y Y$$

where $0 < \alpha + \beta < 1$, P_0 is the price of good Z, whereas P_x and P_y are the prices of inputs X and Y.

Evaluate $\partial \bar{X} / \partial r$ and $\partial \bar{Y} / \partial r$.

(*1*) Find the first-order conditions and express them in implicit function form.

(*2*) Express the total derivatives of the function with respect to r in matrix form, recalling that P_0 and t are constants.

(*3*) Interpret the Jacobian with its the correct sign.

(*4*) Evaluate $\partial \bar{X} / \partial r$.

(*5*) Evaluate $\partial \bar{Y} / \partial r$.

COMPARATIVE STATISTICS USED IN CONSTRAINED OPTIMIZATION

S13.8. A consumer wants to maximize its Cobb-Douglas utility $u(x,y) = x^{1/2}y^{1/2}$ subject to the constant constraint $p_x x + p_y y = B$. Given the Lagrange function

$$U = x^{1/2}y^{1/2} + \lambda(B - p_x x - p_y y)$$

and assuming that the second-order sufficient condition is met:

(1) Express the total derivatives of the function with respect to p_y in matrix form.

(2) Find $\dfrac{\partial \bar{y}}{\partial p_y}$.

CONCAVE PROGRAMMING AND INEQUALITY CONSTRAINTS

S13.9.   Given the profit function

$$\pi = 40x - x^2 + 60y - 2y^2 - 120$$

The maximum output capacity (constraint) is

$$x + y \leq 37$$

Find the optimal combination (x^*, y^*) that maximizes the profit function using the Lagrange multiplier.

(*1*)   Establish the Lagrange function.

(*2*)   Set up the Kuhn-Tucker conditions.

(3) Check the possibility that $\bar{\lambda} = 0$ or $\bar{\lambda} > 0$.

(4) Check to see if either of the choice variables $\bar{x}$ or $\bar{y}$ can equal zero.

(5)   Using all the information collected, find the optimal solution.

Chapter 14

THEORETICAL CONCEPTS

S14.1.   Integration is the reverse operation of _____ (differentiation/antiderivatives).

S14.2.   The capital stock can be found by integrating the rate of net investment.

_____ True   _____ False

S14.3.   Integration by substitution is the inverse rule of the _____ rule.

    a.   chain   c.   quotient

    b.   product   d.   addition

S14.4. The integral of the marginal cost is the _____ (variable/total) cost.

S14.5. A constant can be taken out of the integral without changing the result.

_____ True   _____ False

INDEFINITE INTEGRALS

S14.6. Determine the integration of

$$\int \left(\frac{1}{2} + 4x^{-1/2} - \frac{2}{x} \right) dx$$

INTEGRATION BY SUBSTITUTION

S14.7. Use integration by substitution to determine the following indefinite integral

$$\int 72x^2(4x^3 - 4)^5 dx$$

(1) Select a suitable *u* function.

$$u(x) = \boxed{}$$

(2) Compute *du* and find *dx* in terms of *du*.

$$du = \boxed{} \, dx$$

$$dx = \boxed{} \, du$$

(3) Substitute the relationship in the integration.

(4) Integrate.

$$\boxed{\phantom{\rule{0pt}{6cm}}}$$

(5) Substitute back to have the integration in terms of x.

$$\boxed{\phantom{\rule{0pt}{5cm}}}$$

INTEGRATION BY PARTS

S14.8. Use integration by parts to determine the following definite integral:

$$\int 48(x+2)^{1/2}\,dx$$

(1) Pick suitable $f(x)$ and $g(x)$ functions.

$f(x) = \boxed{}$ $g'(x) = \boxed{}$

(2) Compute $f'(x)$ and find $g(x)$ in terms of dx.

Differentiate. $f'(x) = \boxed{}$

Integrate. $g(x) = \boxed{}$

(3) Substitute the relationship in the integration and create an indefinite integral.

$$\boxed{\phantom{\rule{0pt}{4cm}}}$$

(4)   Integrate the indefinite integral.

ECONOMIC APPLICATIONS

S14.9.   The rate of net investment is $I = 28t^{3/4}$ and a capital stock at $t = 1$ is 24.

Find the capital function K.

S14.10.   Marginal cost is given by $MC = 3Q^2 - 48Q + 360$. Fixed cost is 200. Find the variable and total costs.

Chapter 15

THEORETICAL CONCEPTS

S15.1.   The _____ sum is an approximation used to calculate areas under a curve, and one of the bases for integration.

S15.2.   The fundamental theorem of calculus permits to evaluate integrals within a range of values.

_____ True   _____ False

S15.3. _____ is the area that represents total benefit to consumers.

    a. Demand   c. Supply

    b. Consumer's surplus   d. Producer's surplus

S15.4. L'Hôpital's rule can be used when the function within the limit is indetermined (such as the form 0/0).

                                                            _____ True   _____ False

S15.5. An integral without limit on the upper or lower bound is called an improper integral.

                                                              _____ True   _____ False

DEFINITE INTEGRALS

S15.6. Evaluate the definite integral.

$$\int_{0}^{1}(e^{-3x}-x^3+2)\,dx$$

AREA BETWEEN CURVES

S15.7. Find the area between $y_1 = 12x - 3x^2$ and $y_2 = 3x$ from $x = 2$ to $x = 3$.

    (1) Graph both functions to know the order of the functions in the integration.

    (2) Evaluate the definite integral.

L'HÔPITAL'S RULE

S15.8. Use L'Hôpital's rule to evaluate the following limit.

$$\lim_{x \to \infty} \frac{3x^2 - 4x + 8}{4x^2 + 5x - 2}$$

CONSUMER AND PRODUCER'S SURPLUS

S15.9. Given the demand function $P = 40 - 2Q$, find the consumer's surplus CS when $P_0 = 30$ and $Q_0 = 5$.

S15.10. Given the supply function $P = 1 + 3Q^2$, find the producer's surplus PS when $P_0 = 28$ and $Q_0 = 3$.

Chapter 16

THEORETICAL CONCEPTS

S16.1. The differential equation

$$\left(\frac{dy}{dt}\right)^7 - 4t^2 + \left(\frac{d^2y}{dt^2}\right)^3 = 0$$

is of _____ order and _____ degree.

S16.2. When the constants can be specified in a differential equation, it is said that it has a definite solution.

_____ True   _____ False

S16.3. A(n) _____ equilibrium occurs when the function converges toward a specific value.

a. unstable   c. dimensional
b. irregular   d. intertemporal

S16.4. A first-order linear differential equation must have the derivative and the dependent variable on the

_____ (first/second) degree and _____ (no/some) product $y\left(\dfrac{dy}{dt}\right)$ may occur.

S16.5. Phase diagrams are used to provide qualitative information regarding the stability of equations.

_____ True   _____ False

LINEAR DIFFERENTIAL EQUATIONS

S16.6. Use the general solution to solve the following equation.

$$\frac{dy}{dt} + 6ty = 24t \qquad\qquad y(0) = 5$$

(*1*)   Select $v =$ ⬚ , $z =$ ⬚ , and $\int v\,dt =$ ⬚

(*2*)   Substitute in the formula.

(*3*)   Find the value of A and write the particular solution to this differential equation.

(*4*)   Verify your answer by taking the respective derivatives.

EXACT DIFFERENTIAL EQUATIONS AND PARTIAL INTEGRATION

S16.7. Solve the exact differential equation.

$$(2y - 12t^2)dy + (8 - 24yt)dt = 0$$

   (1) Check to see if it is an exact differential equation.

   (2) Integrate M partially with respect to y and add $Z(t)$ to get $F(y,t)$.

   (3) Differentiate $F(y,t)$ partially with respect to t and equate with N above to obtain $Z'(t)$.

   (4) Integrate $Z'(t)$ with respect to t to get $Z(t)$.

   (5) Use all the information collected and add the constant of integration to get $F(y,t)$.

SEPARATION OF VARIABLES

S16.8. Solve the following differential equation using the procedure for separating variables.

$$t\,dy - y^2\,dt = 0$$

USE OF DIFFERENTIAL EQUATIONS IN ECONOMICS

S16.9. Find the demand function $Q = f(P)$, if $\in = -(4P + 2P^2)/Q$ and $Q = 115$ when $P = 15$.

(1) Set up the elasticity function $\in$ to solve for dQ/dP.

(2) Use separation of variables to find the formula for quantity demanded Q.

(3) Use the initial condition to find the value of the constant of integration, then provide the particular solution for $Q = f(P)$.

PHASE DIAGRAMS OF VARIABLES

S16.10. Construct a phase diagram for $\dot{y} = y^2 - 5y + 4$, and test the dynamic.

    (1)   Find the intertemporal solution and test whether it represents a maximum or minimum.

    (2)   Construct a phase diagram, indicating the arrows of motion.

    (3)   Take the derivative of the equation to evaluate the stability of the critical points.

Chapter 17

THEORETICAL CONCEPTS

S17.1. Given the general form

$$y_t = Ab^t + c$$

assuming $A = 1$ and $c = 0$, if $b = 0.5$, then the time path _____ (converges/diverges) and _____ (is nonoscillating/oscillates).

S17.2. A first-order difference equation is linear when all variables are raised only to the first power, and there are no cross products between them.

_____ True   _____ False

S17.3. The _____ model is often used to forecast production of agricultural commodities based on last year's prices. When the slope of supply is greater than the absolute value of the demand and their ratio is greater than 1, then the time path _____.

a. Cobweb, converges   c. Cobweb, diverges

b. Harrod, converges   d. Harrod, diverges

S17.4. Δy_t is defined as a second difference of y or $\Delta y_t = y_t - y_{t-2}$.

_____ True   _____ False

S17.5. In a phase diagram, a 45°-line is drawn from the origin to capture all possible steady state equilibriums.

_____ True   _____ False

GENERAL FORMULA FOR FIRST-ORDER DIFFERENCE EQUATIONS

S17.6. Solve the difference equation given below.

$$y_{t+3} + \frac{1}{2} y_{t+2} + 6 = 0 \qquad\qquad y_0 = 10$$

(1) Move the time period back by _____ periods, and rearrange terms.

(2) Substitute in the formula and solve it. Determine the stability of y_t.

(3) Verify the answer.

LAGGED INCOME DETERMINATION MODELS

S17.7. Given the data below

$$C_t = 100 + 0.5Y_{t-1} \qquad I_t = 20 + 0.1Y_{t-1} \qquad \text{and} \qquad Y_0 = 1500$$

(1) Write the national income equation as a first-order difference equation.

(2) Find the national income formula Y_t.

(3) Verify your answer.

(4) Determine the stability of Y_t.

THE COBWEB MODEL

S17.8.   Given the data below

$$Q_{dt} = 120 - 0.5P_t \qquad Q_{st} = -20 + 0.2P_{t-1} \qquad \text{and} \qquad P_0 = 320$$

(*1*)   Equate demand and supply to obtain the first-order difference equation for price.

(*2*)   Find the price formula P_t.

(*3*)   Verify your answer at market equilibrium.

(*4*)   Determine the stability of Y_t.

PHASE DIAGRAMS FOR DIFFERENCE EQUATIONS

S17.9.   Given the nonlinear difference equation

$$y_t = y_{t-1}^{0.2}$$

(*1*)   Find the steady-state solutions.

(2) Take the first derivative to see if the slope is positive or negative.

(3) Take the second derivative to see if the phase line is concave or convex.

(4) Draw a rough sketch of the graph.

(5) Confirm the results of the phase diagram using a simple calculus test.

Chapter 18

THEORETICAL CONCEPTS

S18.1. The Euler equation transforms complex roots into trigonometric functions.

_____ True   _____ False

S18.2. A second-order differential equation is composed of two separate solutions, which are _____ and _____.

S18.3. For a second-order differential equation, if $b_1^2 < 4b_2$, then the solutions are _____.

a. distinct real roots   c. negative roots

b. repeated real roots   d. complex roots

S18.4. For second-order linear difference equations with distinct or repeated real roots, the root with the smallest absolute value is the dominant root.

_____ True   _____ False

S18.5. The derivative of a sine function is the negative cosine of the trigonometric function.

_____ True   _____ False

SECOND-ORDER LINEAR DIFFERENTIAL EQUATIONS

S18.6. Find the general solution of the following equation:

$$y''(t) + 4y'(t) + 3y(t) = 9$$

(1) Find the particular integral y_p

(2) Compute the complementary function y_c

(3) Write the general solution.

SECOND-ORDER LINEAR DIFFERENCE EQUATIONS

S18.7. Find the general solution of the following equation:

$$y_t - 4y_{t-1} + 4y_{t-2} = 9$$

(1) Find the particular integral y_p

(2) Compute the complementary function y_c

(3) Write the general solution.

DERIVATIVES OF TRIGONOMETRIC FUNCTIONS

S18.8. Compute the derivative of

$$y = \cos(x^2 - 2x + 1)$$

S18.9.   Compute the derivative of

$$y = \sin(x)\tan(12x)$$

COMPLEX ROOTS IN SECOND-ORDER DIFFERENTIAL EQUATIONS

S18.10.   Find the general solution of the following equation:

$$y''(t) + 2y'(t) + 8y(t) = 88$$

(1)   Find the particular integral y_p.

(2)   Compute the complementary function y_c.

(3)   Write the general solution.

Chapter 19

THEORETICAL CONCEPTS

S19.1. The coefficients of an autonomous linear differential equation are all constants.

_____ True   _____ False

S19.2. To determine the two characteristic roots, we need to obtain the _____ and _____ of matrix **A**.

S19.3. If r_1 and r_2 are characteristic roots, and $r_1 > 1$ but $r_2 < 0$, then _____.

    a. r_1 is locally stable but r_2 is not   c.   the system is stable

    b. r_2 is locally stable but r_1 is not   d.   the system is unstable

S19.4. Eigenvectors are used to solve systems of linear difference equations.

_____ True   _____ False

S19.5. The trace of the matrix of coefficient **A** is the multiplication of diagonal elements of **A**.

_____ True   _____ False

SYSTEMS OF LINEAR DIFFERENTIAL EQUATIONS

S19.6. Solve the system of first-order, autonomous, linear differential equations.

$$\dot{y}_1 = 5y_1 + 3y_2 - 8$$

$$\dot{y}_2 = \ y_1 + 3y_2 - 4$$

    _(1)_   Set up the system in matrix form.

    _(2)_   Find the characteristic roots.

(*3*) Determine the eigenvectors for each root and the general complementary function.

(*4*) Find the steady-state solution y_p.

(*5*) State the general solution.

S19.7. Find the definite solution for S19.6 using the initial conditions

$$y_1(0) = 12 \qquad\qquad y_2(0) = 10$$

Also state if the system is stable or unstable.

SYSTEMS OF LINEAR DIFFERENCE EQUATIONS

S19.8. Solve the system of first-order, autonomous, linear differential equations.

$$x_t = 0.7x_{t-1} - 0.4y_{t-1} + 4$$

$$y_t = 0.2x_{t-1} + 0.1y_{t-1} + 2$$

(1) Set up the system in matrix form.

(2) Find the characteristic roots.

(3)   Determine the eigenvectors for each root and the general complementary function.

(4)   Find the steady-state solution y_p.

(5)   State the general solution.

```




```

S19.9.   Find the definite solution for S19.8 using the initial conditions below:

$$x_0 = 18 \qquad\qquad y_0 = 9$$

```




```

Chapter 20

THEORETICAL CONCEPTS

S20.1.   In calculus of variation, the symbol _____ is used to express properties similar to the differential term *d* in standard calculus.

S20.2.   _____ optimization is used to solve isoperimetric problems.

S20.3.   Given the discriminant **D** for a functional $F[t, x(t), \dot{x}(t)]$ in a dynamic optimization, the condition for a global maximum is that $|\mathbf{D_1}|$ _____ 0, $|\mathbf{D}|$ _____ 0.

a.   >,≥   c.   ≥,≥

b.   >,>   d.   <,>

S20.4. The distance between two points on a plane has a similar formula as the arc length calculation in static standard calculus.

_____ True   _____ False

S20.5. The _____ is a necessary condition for dynamic optimization.

DISTANCE BETWEEN TWO POINTS ON A PLANE

S20.6. Minimize

$$\int_0^3 \sqrt{1+\dot{x}^2}\,dt$$

subject to $x(0) = 5$ $x(3) = 11$

(1) Write the formula of the candidate for the extremal.

(2) Apply the boundary conditions to find the constants.

(3) The solution is

S20.7. Estimate the optimal distance between the points (t_0, x_0) and (t_1, x_1) in Problem S20.6.

DYNAMIC OPTIMIZATION: FINDING CANDIDATES FOR EXTREMALS

S20.8. Optimize

$$\int_0^2 (36xt + 12t + \dot{x}^2)\,dt$$

subject to $x(0) = 4$ $x(2) = 10$

(1) Establish the function F.

(2) Compute Euler's equation.

(3) Integrate both sides of Euler's equation and find the functional form $x(t)$.

DYNAMIC OPTIMIZATION: THE SUFFICIENCY CONDITION

S20.9. Test the sufficient condition of Problem S20.8.

(*1*) Compute the discriminant.

(2) Evaluate $\left|\mathbf{D}^1\right|$ and $\left|\mathbf{D}^2\right|$

Chapter 21

THEORETICAL CONCEPTS

S21.1. In optimal control theory, one of the endpoints may be free.

_____ True   _____ False

S21.2. In an optimal control model, λ is the _____ variable.

S21.3. The _____ function in dynamic optimization is similar to the Lagrangian function in concave programming

S21.4. In optimal control theory, the objective is to find the optimal time path of the control variable subject to the constraint set on the state variable.

_____ True   _____ False

S21.5. Discounting in optimal control theory is solved using the _____.

FIXED ENDPOINTS

S21.6. Minimize

$$\int_0^2 (4x + 2y - y^2)\, dt$$

subject to $\dot{x} = 6y$

$x(0) = 3$ $x(2) = 9$

(*1*)   Write the Hamiltonian formula.

(*2*)   Assume an interior solution and apply the maximum principle.

(3) Solve the differential equations to find the general formulas for λ and $x(t)$.

(4) Apply the boundary conditions to find the formulas for λ and $x(t)$.

(5) Substitute your answers in *(4)* into *(2)* to find the control variable $y(t)$.

S21.7. Test the sufficient condition for Problem S21.6 using the discriminant test.

FREE ENDPOINTS

S21.8. Minimize

$$\int_0^3 (6x - 2y^2)\, dt$$

subject to $\qquad\qquad\qquad \dot{x} = 12y$

$\qquad x(0) = 9 \qquad\qquad\qquad\qquad x(3) \text{ free}$

(1) Write the Hamiltonian formula.

(2) Assume an interior solution and apply the maximum principle.

(*3*)   Solve the differential equations to find the general formulas for λ and $x(t)$

(*4*)   Apply the transversatility and boundary conditions to find the formulas for λ and $x(t)$

(*5*)   Substitute your answers in (*4*) into (*2*) to find the control variable $y(t)$

S21.9.   Test the sufficient condition for Problem S21.8 using the discriminant test.

Chapter 22

THEORETICAL CONCEPTS

S22.1. The sum of all probabilities of an event must be equal to _____ (0/2/1).

S22.2. The weighted average is a type of average that takes into account the frequency of an event.

_____ True _____ False

S22.3. Which of the following numbers cannot be a probability?

a. 0.25

c. −0.12

b. 1

d. 0

S22.4. For the expected value, the weights are the _____ (probabilities/frequencies) of each event.

S22.5. _____ is the method used to solve simple linear regressions.

SERIES AND SUMMATION

S22.6. Calculate the partial sum below:

$$\sum_{k=1}^{4}[(k-1)^2 + k]$$

Expand the polynomial in the summation and grouping terms.

$$\sum_{k=1}^{4}[(k-1)^2 + k] = \boxed{} + \boxed{} + \boxed{}$$

Use summation properties and computing operations.

$$\sum_{k=1}^{4}[(k-1)^2 + k] = \boxed{}$$

EXPECTED VALUE

S22.7. Find the expected value of the table below:

x_i	p_i
1	0.25
3	0.20
7	0.15
12	0.30
17	?

(*1*) Find the probability that $x = 17$.

(*2*) Compute the expected value.

MEAN AND STANDARD DEVIATION

S22.8. Find the mean and standard deviation of the table below:

x_i
1
3
7
12
17

(1) Compute the mean.

$$\bar{X} = \frac{1}{\Box}\sum_{i=1}^{5}x_i = \frac{\Box+\Box+\Box+\Box+\Box}{\Box} = \Box$$

(2) Find the variance.

$$S_x^2 = \frac{1}{\Box}\sum_{i=1}^{5}(x_i - \bar{X})^2 = \frac{\boxed{}+\boxed{}+\boxed{}+\boxed{}+\boxed{}}{\boxed{}}$$

$$S_x^2 = \boxed{}$$

(3) Compute the standard deviation.

$$S_x = \sqrt{S_x^2} = \boxed{}$$

SIMPLE LINEAR REGRESSION

S22.9. Given the data table below

x	y
1	10
3	25
5	40
7	60
9	95
11	100

fit the linear regression using the Ordinary Least Square method.

(0) There are _____ observations; $n =$ ____.

(1) Find the mean of x and y.

$$\bar{x} = \boxed{}$$

$$\bar{y} = \boxed{}$$

(2)   Find the numerator of $\hat{\beta}$, which is $\sum_i^n (y_i - \bar{y})(x_i - \bar{x})$

$$\sum_{i=1}^{6} (y_i - \bar{y})(x_i - \bar{x}) = \boxed{}$$

(3)   Find the denominator of $\hat{\beta}$, which is $\sum_{i=1}^{n} (x_i - \bar{x})^2$

$$\sum_{i=1}^{6} (x_i - \bar{x})^2 = \boxed{}$$

(4)   Calculate $\hat{\beta}$,

$$\hat{\beta} = \boxed{}$$

(5)   Find $\hat{a}$,

$$\hat{a} = \boxed{}$$

The linear equation is $\hat{y} = \boxed{} + \boxed{} \; x.$

S22.10. Use the regression calculated in Problem S22.9 and predict the value of y if $x = 4$.

$$\hat{y} = \boxed{} + \boxed{} \left(\boxed{} \right)$$

$$\hat{y} = \boxed{}$$

Additional Practice Problems: Solutions

Chapter 1

THEORETICAL CONCEPTS

S1.1. True.

S1.2. Equations must be *consistent* and *independent*, and there must *the same number of* equations compared to variables.

S1.3. The graph of a quadratic equation is a *parabola*.

S1.4. The *argument* of a function $f(x)$ is a rule that assigns a specific value to each x in the domain.

S1.5. *Two* points are required to determine the graph of a linear equation.

EXPONENTIAL TERMS

S1.6. Use the rules of exponents to simplify these fractions:

$$\frac{x^2}{\sqrt[6]{x^5 \cdot (x^{1/12})^2}} = \frac{x^2}{x^{5/6+1/6}} = x^{2-1} = x$$

POLYNOMIALS

S1.7. Multiply these polynomials:

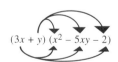

$$\boxed{\begin{aligned} &= 3x^3 - \underline{15x^2 y} - 6x + \underline{x^2 y} - 5xy^2 - 2y \\ &= 3x^3 - 14x^2 y - 6x - 5xy^2 - 2y \end{aligned}}$$

EQUATIONS

S1.8. Solve the following linear equations by using properties of equality:

$$(1/2)x + 10 = 3(2x - 1) - 5/2$$
$$(1/2)x + 10 = 6x - 11/2$$
$$(11/2)x = 31/2$$
$$x = \boxed{31/11}$$

GRAPHS, SLOPES, AND INTERCEPTS

S1.9. Given the linear equation

$$4x + 6y = 48$$

determine the slope and intercepts, then graph the equation.

(1) Find the slope.

> The slope can be found by converting the equation into slope-intercept form.
>
> $$6y = 48 - 4x \qquad \text{(Moving terms)}$$
> $$y = 8 - (2/3)x \qquad \text{(Division)}$$
>
> Slope: $m = -2/3$

(2) Find the x- and y- intercepts.

> The x-intercept occurs where the line crosses the x-axis, which is the point where $y = 0$. Setting $y = 0$ and solving for y, we have
>
> $$4x + 6(0) = 48 \qquad x = 12$$
>
> The x-intercept is $(12, 0)$.
> To find the y-intercept, we set $x = 0$ and solve for y.
>
> $$4(0) + 6y = 48 \qquad y = 8$$
>
> The y-intercept is $(0, 8)$.

(3) The graph of the linear equation is

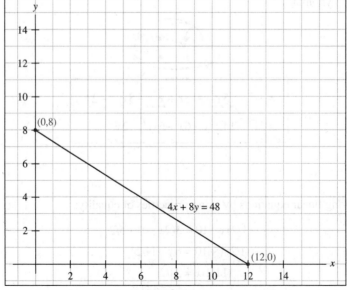

Fig. 1-5

S1.10. Graph the quadratic function below:

$$y = (x-1)^2$$

(1) Plug values of x in the function and fill out the table.

x	$f(x)$
−3	16
−2	9
−1	4
0	1
1	0
2	1
3	4
4	9

(2) The graph of the quadratic function is

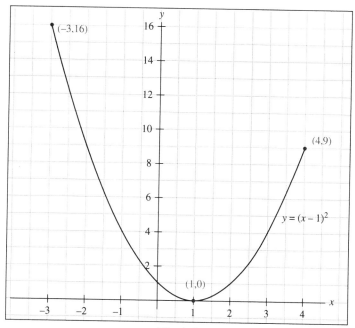

Fig. 1-6

Chapter 2

THEORETICAL CONCEPTS

S2.1. The isocost line is used in production economics to represent the combination of two inputs; typically these are *labor* and *capital*.

S2.2. True.

S2.3. Foreign investment.

S2.4. An increase in budget makes a *parallel* shift toward the *right* of the original line.

S2.5. The *IS-LM* analysis is used to interpret the relationship between income levels and interest rates.

ISOCOST LINES

S2.6. A company has a daily budget of $5,000 for capital and labor expenses. Cost of capital is $50 per day, and cost of labor is $100 per day (or $12.5 per hour).

    I. Consider L as the workers and K as machinery rented per day. Write the company's cost budget equation in standard form.

$$\boxed{50L + 100K = 5,000}$$

    II. Express capital K as a function of labor L.

> Convert into slope-intercept form.
>
> $$100K = 5,000 - 50L$$
> $$K = 50 - 0.5L$$

    III. An increase in budget will provoke a parallel shift of the isocost line toward the right.

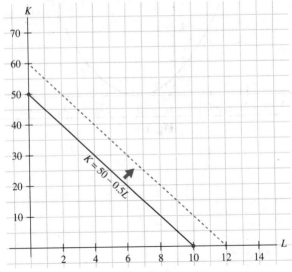

Fig. 2-11

SUPPLY AND DEMAND ANALYSIS

S2.7. Find the equilibrium price and quantity for the market.

$$\text{Supply: } Q_S = -16 + 4P \qquad \text{Demand: } Q_D = 64 - 2P$$

(1) Set quantity supplied Q_S equal to quantity demanded Q_D; solve for P.

$$-16 + 4P = 64 - 2P$$

$$6P = 80 \qquad P^* = 40/3$$

(2) Substitute P^* in either supply or demand equation.

$$Q = -16 + 4(40/3)$$

$$Q^* = 112/3$$

The market equilibrium is $P^* = 40/3$, $Q^* = 112/3$.

INCOME DETERMINATION MODEL

S2.8. Find the equilibrium level of income Y_e, given

$$Y = C + I + (X - Z) \qquad C = C_0 + bY \qquad Z = Z_0 + zY$$
$$I = I_0 \qquad X = X_0$$

where $C_0 = 80$, $I_0 = 120$, $X_0 = 100$, $Z_0 = 50$, $b = 0.9$, $z = 0.25$

(a) Find the reduced form of Y_e.

$$Y = C + I + X - Z = C_0 + bY + I_0 + X_0 - Z_0 - zY$$
$$Y - bY + zY = C_0 + I_0 + X_0 - Z_0$$
$$Y(1 - b + z) = C_0 + I_0 + X_0 - Z_0$$
$$Y_e = \frac{1}{1 - b + z}(C_0 + I_0 + X_0 - Z_0)$$

(b) Determine the numerical value of Y_e.

Substitute the numerical values of the constants in the reduced form of Y_e.

$$Y_e = \frac{1}{1 - 0.9 + 0.25}(80 + 120 + 100 - 50)$$

$$Y_e = \frac{1}{0.35}(250) = 714.29$$

IS-LM EQUATIONS

S2.9. Given the commodity market for a two-sector economy

$$Y = C + I, \text{ where } C = 120 + 0.1Y \text{ and } I = 30 - 200i$$

and a money market system

$$M_s = 200, \ M_t = 0.2Y \text{ and } M_z = 175 - 100i$$

(*1*) Find the equilibrium income level and interest rate.

For *IS*:

$$Y = 120 + 0.1Y + 30 - 200i$$

$$0.9Y + 200i = 150 \tag{2.23}$$

For *LM*:

$$M_S = M_t + M_Z$$

$$200 = 0.2Y + 175 - 100i$$

$$0.2Y - 100i = 25 \tag{2.24}$$

In equilibrium,

$$0.9Y + 200i = 150 \tag{2.23}$$

$$0.2Y - 100i = 25 \tag{2.24}$$

Multiply (*2.24*) by 2, and add this result to (*2.23*) to eliminate *i*.

$$0.9Y + 200i = 150$$

$$\underline{0.4Y - 200i = 50}$$

$$1.3Y \qquad\quad = 200$$

$$Y = 153.85$$

Substitute $Y = 153.85$ in (*2.23*) or (*2.24*).

$$0.9(153.85) + 200i = 150$$

$$i = 0.0577$$

---

(2) Calculate the levels of C, I, M_t, and M_Z in equilibrium.

At $Y = 153.84$ and $i = 0.0577$,

$$C = 120 + 0.1Y = 120 + 0.1(153.84) = 135.38$$

$$I = 30 - 200i = 30 - 200(0.0577) = 18.46$$

$$M_t = 0.2Y = 0.2(153.84) = 30.77$$

$$M_z = 175 - 100i = 169.23$$

and

$$C + I = Y \qquad\qquad M_t + M_z = M_s$$

$$135.38 + 18.46 = 153.84 \qquad 30.77 + 169.23 = 200$$

Chapter 3

THEORETICAL CONCEPTS

S3.1. A function is *differentiable* at a point $x = a$ if the derivative exists at that particular point.

S3.2. True.

S3.3. The technique used to derive a multiplication of polynomials is called the *Product* Rule.

S3.4. In an implicit function, the dependent variable y typically appears on the *same* side of the equation with respect to the independent variable x. Thus, if the terms that include x are on the right side, then any term that includes y is on the *right* side.

S3.5. True.

LIMITS AND DIFFERENTIATION

S3.6. Find $\lim\limits_{x \to -3} f(x)$, if it exists, given

$$f(x) = \frac{3x + 9}{x^2 - 9}$$

$$\lim_{x \to -3} \frac{3x + 9}{x^2 - 9} = \lim_{x \to -3} \frac{3(x + 3)}{(x + 3)(x - 3)} = \lim_{x \to -3} \frac{3}{(x - 3)} = \frac{3}{-6} = -\frac{1}{2}$$

DERIVATIVE RULES

Find the first derivative of the following functions.

S3.7. Use the product rule to differentiate $f(x) = (4x+1)(5x^{-1/3} + 7x^2)$.

$$g(x) \cdot h'(x) + g'(x) \cdot h(x)$$

$$f'(x) = (4x+1) \cdot (-(5/3)x^{-4/3} + 14x) + (4) \cdot (5x^{-1/3} + 7x^2)$$

$$f'(x) = -(20/3)x^{-1/3} + 56x^2 - (5/3)x^{-4/3} + 14x + 20x^{-1/3} + 28x^2$$

Simplify algebraically.

$$f'(x) = -(40/3)x^{-1/3} + 84x^2 - (5/3)x^{-4/3} + 14x$$

S3.8. Use the generalized power rule to differentiate $f(x) = \sqrt[4]{2x^3 - 11}$

Define the function in exponential terms.

$$f(x) = (2x^3 - 11)^{1/4}$$

Take the derivative with respect to x.

$$f'(x) = \boxed{(1/4)} \cdot \boxed{(2x^3 - 11)^{-3/4}} \cdot \boxed{(6x^2)}$$
$$\quad\quad\quad\; n \quad\quad\quad [g(x)]^{n-1} \quad\quad g'(x)$$

Simplify algebraically.

$$f'(x) = \boxed{\left(\frac{3}{2}x^2\right)(2x^3 - 11)^{-3/4}}$$

S3.9. Use the chain rule to differentiate $f(x) = (2x^3 + x^{1/2})^6$.

Let $\quad y = u^6 \quad$ and $\quad u = 2x^3 + x^{1/2}$.

Then $dy/du = \boxed{6u^5}$ and $du/dx = \boxed{6x^2 + 0.5x^{-1/2}}$

Substitute in the formula.

$$\frac{dy}{dx} = \boxed{6u^5} \cdot \boxed{(6x^2 + 0.5x^{-1/2})}$$

Replace u in terms of x.

$$\frac{dy}{dx} = \boxed{6(2x^3 + x^{1/2})^5(6x^2 + 0.5x^{-1/2})}$$

HIGHER-ORDER DERIVATIVES

S3.10. Given the function $f(x) = 24x^{1/8} - 20x^{-1/2} + 5x$, find the second-order derivative.

 (*1*) Take the first derivative of the function.

$$f'(x) = 24(1/8)x^{-7/8} - 20(-1/2)x^{-3/2} + 5$$

$$\boxed{f'(x) = 3x^{-7/8} + 10x^{-3/2} + 5}$$

 (*2*) Take the derivate of the first-order derivative to obtain the second-order derivative.

$$f''(x) = 3(-7/8)x^{-15/8} + 10(-3/2)x^{-5/2} + 0$$

$$\boxed{f''(x) = -\left(\frac{21}{8}\right)x^{-15/8} - 15x^{-5/2}}$$

Chapter 4

THEORETICAL CONCEPTS

S4.1. To maximize profits, marginal revenue must be *equal to* marginal cost.

S4.2. False; the inflection point is found by taking the second-order derivative of the function and setting it equal to zero.

S4.3. A differentiable function $f(x)$ whose first-order derivative evaluated at $x = a$ is positive can be classified as a(n) *increasing* function.

S4.4. In an implicit function, the dependent variable y typically appears on the *same* side of the equation with respect to the independent variable x. Thus, if the terms that include x are on the right side, then any term that includes y is on the *right* side.

S4.5. False; the derivative of the total product with respect to labor is the marginal product of labor.

RELATIVE EXTREMA

S4.6. Find the relative extrema of the function $f(x) = x^3 - 9x^2 + 24x - 12$ and classify the critical points as relative minimum or maximum.

 (*1*) Take the first-order derivative and set it equal to zero.

$$f'(x) = 3x^2 - 18x + 24 = 0$$

$$3(x^2 - 6x + 8) = 0$$

Factor:

$$3(x - 2)(x - 4) = 0$$

$$x = 2 \qquad x = 4 \qquad \text{critical values}$$

(2) Take the second derivative to evaluate the concavity of each critical value.

$$y'' = 6x - 18$$

$$y''(2) = 6(2) - 18 = -6 < 0 \qquad \text{concave, relative maximum}$$

$$y''(4) = 6(4) - 18 = \ 6 > 0 \qquad \text{convex, relative minimum}$$

$x = 2$ is a relative maximum point
$x = 4$ is a relative minimum point

MARGINAL, AVERAGE, AND TOTAL COST

S4.7. Given the cost function $TC = Q^3 - 24Q^2 + 360Q$

(1) Find the inflection point of the total cost function.

(a) Take the first-order derivative.

$$TC' = 3Q^2 - 48Q + 360$$

(b) Take the second-order derivative and set it equal to zero.

$$TC'' = 6Q - 48 = 0$$

$$Q = 8 \qquad \text{inflection point}$$

(2) Find the average cost functions and the relative extrema.

$$ATC = \frac{TC}{Q} = Q^2 - 24Q + 360$$

$$ATC' = 2Q - 24 = 0$$

$$Q = 12 \qquad \text{critical value}$$

$$ATC'' = 2 > 0 \qquad \text{convex, relative minimum}$$

(3) Find the marginal cost functions and the relative extrema.

$$MC = TC' = 3Q^2 - 48Q + 360$$

$$MC' = 6Q - 48 = 0$$

$$Q = 8 \qquad \text{critical value}$$

$$MC'' = 6 > 0 \qquad \text{convex, relative minimum}$$

(4) Verify the point of intersection between the average and the marginal cost functions (note $Q > 0$).

Set $AC = MC$: $\qquad Q^2 - 24Q + 360 = 3Q^2 - 48Q + 360$

$$2Q^2 - 24Q = 0$$

Factor: $\qquad 2Q(Q - 12) = 0$

Given $Q > 0$ $\qquad Q - 12 = 0$

$$Q = 12 \qquad \text{point of intersection}$$

OPTIMIZATION IN BUSINESS AND ECONOMICS

S4.8. Maximize the profit function $\pi = -Q^3 - 15Q^2 + 1125Q - 6000$ by finding the critical values.

(1) Find the critical values for Q^* by taking the first-order condition (note $Q > 0$).

$$\pi' = -3Q^2 - 30Q + 1125 = 0$$
$$= -3(Q^2 + 10Q - 1125) = 0$$
$$-3(Q + 25)(Q - 15) = 0$$
$$Q = -25 \qquad Q = 15 \qquad \text{critical values}$$

Because $Q > 0$, the only critical value for the profit function is $Q^* = 15$.

(2) Verify concavity of the critical value Q^*.

$$\pi'' = -6Q - 30$$
$$\pi''(15) = -6(15) - 30 = -120 < 0 \qquad \text{concave, relative maximum}$$

(3) Compute the maximum profit level π^* by establishing $\pi^* = \pi(Q^*)$.

$$\pi(Q^*) = -(15)^3 - 15(15)^2 + 1125(15) - 6000$$
$$\pi(15) = 4125$$

MARGINAL RATE OF TECHNICAL SUBSTITUTION

S4.9. Given labor L and capital K, the equation for the production isoquant is

$$L^{2/7} K^{5/7} = 21000$$

(1) Find the marginal rate of technical substitution MRTS = dK/dL

(a) Take the derivative of each term with respect to L and treat K as a function of L.

$$\frac{d}{dL}(L^{2/7} K^{5/7}) = \frac{d}{dL}(21000)$$

Use the product rule since K is being treated as a function of L.

$$K^{5/7} \cdot \frac{d}{dL}(L^{2/7}) + L^{2/7} \frac{d}{dL}(K^{5/7}) = \frac{d}{dL}(21000)$$

$$\frac{2}{7} L^{-5/7} K^{5/7} + \frac{5}{7} L^{2/7} K^{-2/7} \cdot \frac{dK}{dL} = 0$$

Solve algebraically for dK/dL

$$\frac{dK}{dL} = -\frac{(2/7)L^{-5/7} K^{5/7}}{(5/7)L^{2/7} K^{-2/7}} = \frac{-2K}{5L}$$

(2) Compute the MRTS at $K_0 = 300$ and $L_0 = 400$.

$$MRTS = \frac{dK}{dL} = \frac{-2(300)}{5(400)} = -0.3$$

This means that if L is increased by one relatively small unit, K must decrease by $\boxed{-0.3}$ units in order to remain on the isoquant production where the production level is constant.

Chapter 5

THEORETICAL CONCEPTS

S5.1. *Lagrange multiplier* is the method used to maximize constrained optimization problems.

S5.2. True.

S5.3. Given a differentiable function $z = f(x, y)$, consider $f_{xx} \cdot f_{yy} < (f_{xy})^2$, $f_{xx} > 0$, and $f_{yy} < 0$ at the critical value (x^*, y^*). This point can be classified as the *saddle* point.

S5.4. Given $y = f(x)$, the *differential* of y (dy) evaluates the change in y as a result of a small change in x; in other words $dy = f'(x)dx$.

S5.5. True.

PARTIAL DERIVATIVES

S5.6. Using the product rule to differentiate $f(x, y) = (2xy^2 + 4y)(10x + 2x^3 y)$ with respect to x.

$$f_x = \left[\frac{d}{dx}(2xy^2 + 4y)\right](10x + 2x^3 y) + (2xy^2 + 4y)\left[\frac{d}{dx}(10x + 2x^3 y)\right]$$

$$f_x = \quad (2[x]' y^2 + 0)(10x + 2x^3 y) + (2xy^2 + 4y)(10[x]' + 2[x^3]' y)$$

$$f_x = \qquad (2y^2)(10x + 2x^3 y) + (2xy^2 + 4y)(10 + 6x^2 y)$$

Simplify.

$$f_x = \quad 20xy^2 + 4x^3 y^3 + 20xy^2 + 40y + 12x^3 y^3 + 24x^2 y^2$$

$$f_x = \quad 40xy^2 + 16x^3 y^3 + 40y + 24x^2 y^2$$

S5.7. Use the generalized power rule to differentiate $f(x, y) = (3x^2 - 5y + 8x^3 - 9x^{1/3} y)^4$ with respect to x.

$$f_x = (4)(3x^2 - 5y + 8x^3 - 9x^{1/3} y)^3 \cdot (6x + 24x^2 - 3x^{-2/3} y)$$

Factor constants and simplify.

$$f_x = \quad 12(2x + 8x^2 - x^{-2/3} y)(3x^2 - 5y + 8x^3 - 9x^{1/3} y)^3$$

TOTAL DERIVATIVES

S5.8. Find the total derivative dz/dx of the bivariate function below:

$$z = f(x, y) = 48x^{1/2} y^3 + 32y - 12x^2$$

where $y = x^2 + 2$

(*1*) The total derivative dz/dx is given by the formula

$$\frac{dz}{dx} = \boxed{z_x} + \boxed{z_y}\,\frac{dy}{dx}$$

(*2*) Find $\dfrac{dy}{dx}$, z_x, and z_y

$$\frac{dy}{dx} = 2x$$

$$z_x = 24x^{-1/2}y^3 - 24x$$

$$z_y = 144x^{1/2}y^2 + 32$$

(*3*) Substitute in the formula.

$$\frac{dz}{dx} = (24x^{-1/2}y^3 - 24x) + (144x^{1/2}y^2 + 32)\,(2x)$$

$$\frac{dz}{dx} = 24x^{-1/2}y^3 - 24x + 288x^{5/2}y^2 + 64x$$

IMPLICIT DIFFERENTIATION

S5.9. Find the derivative dy/dx of the implicit function

$$24x^{1/3}y^2 - 17 + 6x^2 + 48y^{1/2} = 0$$

$$\frac{dy}{dx} = -\frac{f_x}{f_y} = \frac{8x^{-2/3}y^2 + 12x}{48x^{1/3}y + 24y^{-1/2}} = \frac{2x^{-2/3}y^2 + 3x}{12x^{1/3}y + 6y^{-1/2}}$$

CONSTRAINED OPTIMIZATION DIFFERENTIATION

S5.10. Given the function

$$z = x^{4/5}y^{1/5}$$

The constraint is

$$2x + y = 45$$

Find the optimal combination (x^*, y^*) that optimizes the Lagrange multiplier

(*1*) Establish the Lagrange function.

$$\mathcal{L} = x^{4/5}y^{1/5} + \lambda(45 - 2x - y)$$

$$\mathcal{L} = \boxed{x^{4/5}y^{1/5} + 45\lambda - 2x\lambda - y\lambda}$$

(2) Take the first-order partial derivatives for optimization to find (x^*, y^*)

$$\mathcal{L}_x = \boxed{(4/5)x^{-1/5}y^{1/5} - 2\lambda = 0} \qquad \lambda = (2/5)(x^{-1/5}y^{1/5}) \qquad (5.45)$$

$$\mathcal{L}_y = \boxed{(1/5)x^{4/5}y^{-4/5} - \lambda = 0} \qquad \lambda = (1/5)(x^{4/5}y^{-4/5}) \qquad (5.46)$$

$$\mathcal{L}_\lambda = \boxed{45 - 2x - y = 0} \qquad 2x + y = 45 \qquad (5.47)$$

Set equation (5.45) equal to (5.46).

$$(2/5)(x^{-1/5}y^{1/5}) = (1/5)(x^{4/5}y^{-4/5})$$

$$2y = x$$

Substitute in (5.47).

$$2(2y) + y = 45$$

$$y = 9$$

$$x = 2(9) = 18$$

$$(x^*, y^*) = \boxed{(18,9)}$$

(3) Find λ^* and interpret it.

$$\lambda = \frac{2}{5}\left((18)^{-1/5}(9)^{1/5}\right) = 0.3482$$

With $\lambda^* = 0.3482$, a one-unit increase in the constant of the constraint will lead to an increase of approximately 0.3482 in the value of the objective function.

Chapter 6

THEORETICAL CONCEPTS

S6.1. Given labor L and capital K needed to produce q, *Cobb-Douglas* is the production function with the form $q = AK^\alpha L^\beta$.

S6.2. True.

S6.3. When all inputs are increased by the same proportion as the output, this is called *constant* returns to scale.

S6.4. *Income* elasticity of demand measures the change in the demand for a good due to a small change in the person's budget.

S6.5. False. To compute the marginal productivity of labor, all other factors must remain fixed.

MARGINAL PRODUCTIVITY

S6.6. Given the production function

$$Q = 48KL - 4K^2 - 2L^2$$

find the marginal productivity of labor (L) and capital (K).

$$MP_L = \frac{\partial Q}{\partial L} = 48K - 4L$$

$$MP_K = \frac{\partial Q}{\partial K} = 48L - 8K$$

INCOME DETERMINATION MULTIPLIER AND COMPARATIVE STATISTICS

S6.7. Given the following economy

$$Y = C + I + (X - Z) \qquad C = C_0 + bY \qquad Z = Z_0 + zY$$
$$I = I_0 \qquad X = X_0$$

where $C_0, I_0, X_0, Z, b, z > 0$

find how income Y changes if the propensity to import z changes by 0.1.

(1) Find the equilibrium for income Y^*

$$Y = C + I + X - Z = C_0 + bY + I_0 + X_0 - Z_0 - zY$$
$$Y - bY + zY = C_0 + I_0 + X_0 - Z_0$$
$$Y(1 - b + z) = C_0 + I_0 + X_0 - Z_0$$
$$Y^* = \frac{1}{1 - b + z}(C_0 + I_0 + X_0 - Z_0)$$

(2) Calculate $\partial Y^* / \partial z$.

> Use the generalized power rule on the first term (which is the only term that contains z).
>
> $$\frac{\partial Y^*}{\partial z} = -\frac{1}{(1-b+z)^2}(C_0 + I_0 + X_0 - Z_0)$$
>
> This can alternatively be expressed as
>
> $$\frac{\partial Y^*}{\partial z} = -\frac{1}{(1-b+z)} \frac{(C_0 + I_0 + X_0 - Z_0)}{(1-b+z)} = -\frac{1}{(1-b+z)} \cdot Y^*$$
>
> $$\frac{\partial Y^*}{\partial z} = -\frac{Y^*}{(1-b+z)}$$

(3) Consider $C_0 = 80$, $I_0 = 120$, $X_0 = 100$, $Z_0 = 50$, $b = 0.9$, $z = 0.25$. Compute Y^* from the reduced form found in (1).

> Substituting the numerical values of the constants in the reduced form to find Y^*
>
> $$Y^* = \frac{1}{1-0.9+0.25}(80+120+100-50) = 714.29$$

(4) Calculate the change in income Y^* (ΔY^*) as a result of result of $\Delta z = 0.1$.

> The partial differential with respect to z is found by moving dz to the right-hand side.
>
> $$dY^* = -\frac{Y^*}{(1-b+z)}dz$$
>
> Hence
>
> $$\Delta Y^* = -\frac{Y^*}{(1-b+z)}\Delta z$$
>
> If the propensity of import increases by 0.1, substitute $Y^* = 714.29$, $b = 0.9$, and $z = 0.25$,
>
> $$\Delta Y^* = -\frac{714.29}{(1-0.9+0.25)}(0.1)$$
>
> $$\Delta Y^* = -204.08$$

PARTIAL ELASTICITIES

S6.8. Given the demand function

$$Q = 300 - 1.5P + 0.14Y$$

where $P = 30$ and $Y = 3000$

find the income elasticity of demand.

> Here, $\dfrac{\partial Q}{\partial P} = -1.5$, $\dfrac{\partial Q}{\partial Y} = 0.14$, and $Q = 300 - 1.5(30) + 0.14(3000) = 675$
>
> Substitute values in the income elasticity formula.
>
> $$\epsilon_Y = \frac{\partial Q}{\partial Y}\left(\frac{Y}{Q}\right) = (0.14)\left(\frac{3000}{675}\right) = 0.622$$

OPTIMIZATION OF ECONOMIC FUNCTIONS

S6.9. Given the profit function for a firm producing two goods x and y

$$\pi = 160x - x^2 - xy - 3y^2 + 300y - 3000$$

find the optimal combination (x^*, y^*) that optimizes profits.

(1) Take the first-order condition.

$$\pi_x = \boxed{160 - 2x - y = 0}$$

$$\pi_y = \boxed{300 - x - 6y = 0}$$

$$(x^*, y^*) = \boxed{(60, 40)}$$

(2) Take the second-order condition to verify its concavity.

$$\pi_{xx} = \boxed{-2 < 0}$$

$$\pi_{yy} = \boxed{-6 < 0}$$

$$\pi_{xy} = \boxed{-1}$$

Using this information, compute the rule to determine its concavity.

> With both second-order partial derivatives π_{xx} and π_{yy} negative, and $\pi_{xx}\pi_{yy} = 12 > 1 = \pi_{xy}^2$, π is maximized.

(3) Compute the maximum profit level π^* by establishing $\pi^* = \pi(x^*, y^*)$.

> $$\pi^* = 160(60) - 60^2 - (60)(40) - 3(40)^2 + 300(40) - 3000$$
>
> $$\pi^* = 7800$$

HOMOGENEITY AND RETURNS TO SCALE

S6.10. Determine the level of homogeneity and returns to scale for the production function below:

$$Q = 2x^2 + 4xy - \frac{8x^3}{y^2}$$

Here Q is homogeneous of degree 2, and returns to scare are increasing because

$$f(kx, ky) = 2(kx)^2 + 4(kx)(ky) - \frac{8(kx)^3}{(ky)^2} = 2k^2x^2 + 4k^2xy - 8k^2\frac{x^3}{y^2}$$

$$f(kx, ky) = k^2\left(2x^2 + 4xy - \frac{8x^3}{y^2}\right)$$

CONSTRAINED OPTIMIZATION IN ECONOMICS

S6.11. Given the profit function

$$\pi = 40x - x^2 - 2xy - 2y^2 + 50y$$

The maximum output capacity (constraint) is

$$x + y = 25$$

Find the optimal combination (x^*, y^*) that maximizes the profit function using the Lagrange multiplier.

(1) Establish the Lagrange function.

$$\mathcal{L} = 40x - x^2 - 2xy - 2y^2 + 50y + \lambda(25 - x - y)$$
$$\mathcal{L} = \boxed{40x - x^2 - 2xy - 2y^2 + 50y + 25\lambda - x\lambda - y\lambda}$$

(2) Take the first-order partial derivatives for optimization to find (x^*, y^*)

$\mathcal{L}_x = \boxed{-2x - 2y + 40 - \lambda = 0}$	$-\lambda = 2x + 2y - 40$	*(6.91)*
$\mathcal{L}_y = \boxed{-2x - 4y + 50 - \lambda = 0}$	$-\lambda = 2x + 4y - 50$	*(6.92)*
$\mathcal{L}_\lambda = \boxed{25 - x - y = 0}$	$x + y = 25$	*(6.93)*

Set equation *(6.91)* equal to *(6.92)*.

$$2x + 2y - 40 = 2x + 4y - 50$$
$$y = 5$$

Substitute in *(6.93)*.

$$x + 5 = 25$$
$$x = 20$$
$$(x^*, y^*) = \boxed{(20, 5)}$$

(3) Find λ^* and interpret it.

Substitute the values of x^* and y^* in either (6.91) or (6.92).

$$\lambda = -[2(20) + 2(5) - 40] = -10$$

With $\lambda^* = -10$, a one-unit increase in the constant of the output capacity will lead to an *decrease* in profit of approximately $\boxed{10}$.

Chapter 7

THEORETICAL CONCEPTS

S7.1. True.

S7.2. True.

S7.3. False. A multiplication of two terms in a logarithmic function can be split in the addition of two logarithmic functions.

S7.4. True.

S7.5. True.

EXPONENTIAL-LOGARITHMIC CONVERSIONS

Change each of the following functions to their corresponding inverse function.

S7.6. Convert $y = e^{2x-5}$ into a logarithmic function by solving for x.

Take the natural logarithm to both sides.

$$\ln(y) = \ln(e^{2x-5})$$

Simplify: $\ln(y) = 2x - 5$

Solve for x: $x = \left(\dfrac{1}{2}\right)\ln(y) + \dfrac{5}{2}$

S7.7. Convert $y = 2\ln(x+10)$ into an exponential function by solving for x.

Move the coefficient inside the logarithmic term.

$$y = \ln(x+10)^2$$

Raise both sides to natural exponentials.

$$e^y = e^{\ln(x+10)^2}$$

Simplify: $e^y = (x+10)^2$

Solve for x: $e^{y/2} = x + 10$

$$x = e^{y/2} - 10$$

PROPERTIES OF LOGARITHMIC AND EXPONENTIAL FUNCTIONS

Convert the expressions to sums, differences, or products.

S7.8. Convert $y = \ln(2x+1) - \ln(4x+7)$ into a ratio.

The addition of two terms can be converted into a ratio:

$$y = \ln\left(\frac{2x+1}{4x+7}\right)$$

S7.9. Convert $z = \ln\left(\dfrac{2x^2}{\sqrt[4]{y^3}}\right)$ into additions and subtractions.

Separate ratios as subtraction between terms and products as additions.

$$z = \ln\left(\frac{2x^2}{\sqrt[4]{y^3}}\right) = \ln(2) + \ln(x^2) - \ln\left(\sqrt[4]{y^3}\right)$$

Simplify terms.

$$z = \ln(2) + 2\ln(x) - \frac{3}{4}\ln(y)$$

SOLVING LOGARITHMIC AND EXPONENTIAL FUNCTIONS

S7.10. Solve for x.

$$\ln(4x+7) = 3$$

Raise both sides to natural exponentials.

$$e^{\ln(4x+7)} = e^3$$

Simplify: $4x + 7 = e^3$

Solve for x: $4x = e^3 - 7$

$$x = \frac{1}{4}e^3 - \frac{7}{4}$$

$$x = 3.2714$$

S7.11. Find the two values for x.

$$(e^x)^{2x-3} = 1$$

Multiply the terms in the exponential function.

$$e^{2x^2-3x} = 1$$

Take the natural logarithm to both sides.

$$\ln(e^{2x^2-3x}) = \ln(1)$$

Simplify:	$2x^2 - 3x = 0$
Factor:	$x(2x - 3) = 0$
	$x = 0 \qquad 2x - 3 = 0$
	$x = 1.5$

The values of x are 0 and 1.5.

Chapter 8

THEORETICAL CONCEPTS

S8.1. For a depreciation problem, the interest rate r is *negative*.

S8.2. True.

S8.3. *Discounting* is the process of calculating the present value of a future cash flow.

S8.4. *Discrete* growth is the type of growth that takes place at the end of a determined period of the year, such as a quarter or at the end of the year.

S8.5. True.

COMPOUNDING INTEREST

S8.6. Given a principal P of $2000 at 5 percent interest i for 7 years, find the future value S when the principal is compounded continually.

$$S = 2000e^{0.05(7)} = 2838.14$$

INTEREST RATE AND TIMING

S8.7. At what interest rate will $1,000 become 4 times that amount if it is compounded continuously for 5 years?

$$S = Pe^{rt}$$
$$4000 = 1000e^{r(5)}$$
$$4 = e^{r(5)}$$
$$\ln 4 = 5r$$
$$r = \frac{\ln 4}{5} = 0.2773$$
$$r = 27.73\%$$

DISCOUNTING

S8.8. Find the present value of $3,000 in 4 years at 12 percent when interest rate is compounded.

$$P = Se^{-rt} = 3000e^{-0.12(4)}$$
$$P = 1856.35$$

EXPONENTIAL GROWTH FUNCTIONS

S8.9. Profits of a corporation are projected to increase by 3 percent every year over the next 5 years. With current profits of $100,000, what will the level of profits be at the end of the lustrum?

$$S = P(1+r)^t = 100000(1+0.03)^5$$
$$S = 100000(1.03)^5$$
$$S = 115927.41$$

CONVERTING EXPONENTIAL FUNCTIONS

S8.10. Find the equivalent form for $S = Pe^{0.12t}$ under semiannual discrete compounding.

$$r = m \times \ln(1+i/m)$$
$$0.12 = 4 \times \ln(1+i/4)$$
$$0.12 = 4 \times \ln(1+i/4)$$

Raise both sides to natural exponentials.

$$e^{0.03} = e^{\ln\left(1+\frac{i}{4}\right)}$$
$$1.03045 = 1+i/4$$
$$0.03045 = i/4$$
$$i = 0.1218 = 12.18\%$$

EXPONENTIAL GROWTH IN ECONOMICS

S8.11. Government spending G increased from 4 trillion in 2002 to 6 trillion in 2021. Find the annual growth rate.

(*1*) Express the final expenditure in terms of an ordinary annual exponential function.

Consider G_0 the initial expense, and G the final government expenditure after the 19 years.

$$G = G_0 \times (1+i)^t$$
$$6 = 4 \times (1+i)^{19} \qquad (8.18)$$

(2) Find the annual rate of growth.

Simplify terms in (8.18) by dividing 4 into both terms.

$$1.5 = (1+i)^{19}$$

Take the common log of both sides.

$$\ln(1.5) = 19\ln(1+i)$$

$$\ln(1+i) = \frac{\ln(1.5)}{19} = 0.02134$$

Raise both sides to natural exponentials.

$$e^{\ln(1+i)} = e^{0.02134}$$

$$1+i = 1.02157$$

$$i = 0.02157$$

$$i = 2.16\%$$

Chapter 9

THEORETICAL CONCEPTS

S9.1. The derivative of a natural exponential function is equal to the *original exponential function* times the derivative of the *exponential function*.

S9.2. True.

S9.3. *Logarithmic differentiation* is used to ease the process of differentiating products and quotients.

S9.4. *Exponential* functions are used to determine the optimal timing for an investment.

S9.5. True.

DERIVATIVES OF EXPONENTIAL FUNCTIONS

S9.6. Calculate the derivative of

$$y = 4e^{(2x-x^2)}$$

$$y' = 4(2x - x^2)'e^{(2x-x^2)}$$

$$y' = 4(2 - 2x)e^{(2x-x^2)}$$

S9.7.   Calculate the derivative of

$$y = 8^{(3x+1)}$$

$$y' = (3x+1)'e^{(2x-x^2)}\ln 8$$
$$y' = 3\ln 8 e^{(2x-x^2)}$$

DERIVATIVES OF NATURAL LOGARITHMIC FUNCTIONS

S9.8.   Calculate the derivative of

$$y = \ln(x+3)^6$$

Move the exponent in the logarithmic function as a coefficient.

$$y = 6\ln(x+3)$$

Take the derivative of the function.

$$y' = \frac{6}{x+3}$$

S9.9.   Calculate the derivative of

$$y = \log_7(x - x^2)$$

$$y' = (x-x^2)'\frac{1}{(x-x^2)}\ln 7$$
$$y' = (1-2x)\frac{1}{(x-x^2)}\ln 7$$
$$y' = \ln 7 \times \left(\frac{1-2x}{x-x^2}\right)$$

SECOND DERIVATIVES

S9.10. Find the first and second derivatives of

$$y = xe^{2x}$$

Use the product rule to take the first-order derivative.

$$y' = [x]'(e^{2x}) + x(e^{2x})'$$
$$y' = e^{2x} + 2xe^{2x}$$

Use the exponential and product rules to take the second-order derivative.

$$y'' = [e^{2x}]' + [2xe^{2x}]'$$
$$y'' = 2e^{2x} + [2x]'e^{2x} + 2x[e^{2x}]'$$
$$y'' = 2e^{2x} + 2e^{2x} + 2x(2e^{2x})$$
$$y'' = 4e^{2x} + 4xe^{2x}$$

PARTIAL DERIVATIVES

S9.11. Find the first and second partial derivatives of

$$z = e^{y-x^2}$$

Calculate the first partial derivatives.

$$z_x = \left[\frac{\partial(y-x^2)}{\partial x}\right]e^{y-x^2} \qquad z_y = \left[\frac{\partial(y-x^2)}{\partial y}\right]e^{y-x^2}$$

$$z_x = \boxed{-2xe^{y-x^2}} \qquad z_y = \boxed{e^{y-x^2}}$$

Take the second partial derivatives by product rule for x and by exponential rule for y.

$$z_{xx} = [-2x]'e^{y-x^2} - 2x[e^{y-x^2}]' \qquad z_{yy} = (1)e^{y-x^2}$$

$$z_{xx} = -2e^{y-x^2} - 2x(-2x)(e^{y-x^2}) \qquad z_{yy} = e^{y-x^2}$$

$$z_{xx} = \boxed{-2e^{y-x^2} + 4x^2e^{y-x^2}} \qquad z_{yy} = \boxed{e^{y-x^2}}$$

Take the cross partial derivatives using the exponential rule for x and y.

$$z_{xy} = (1)\left[-2xe^{y-x^2}\right] \qquad z_{yx} = (-2x)e^{y-x^2}$$

$$z_{xy} = -2xe^{y-x^2} \qquad z_{yx} = -2xe^{y-x^2}$$

S9.12. Find the first and second partial derivatives of

$$z = \ln\left(\frac{x+y}{x-y}\right)$$

Split the logarithmic term into two terms.

$$z = \ln(x+y) - \ln(x-y)$$

Calculate the first partial derivatives.

$$z_x = (1)\frac{1}{x+y} - (1)\frac{1}{x-y}$$

$$\boxed{z_x = \frac{1}{x+y} - \frac{1}{x-y}}$$

$$z_y = (1)\frac{1}{x+y} - (-1)\frac{1}{x-y}$$

$$\boxed{z_y = \frac{1}{x+y} + \frac{1}{x-y}}$$

Calculate the second partial derivatives.

$$z_{xx} = (-1)\frac{1}{(x+y)^2} - (-1)\frac{1}{(x-y)^2}$$

$$\boxed{z_{xx} = -\frac{1}{(x+y)^2} + \frac{1}{(x-y)^2}}$$

$$z_{yy} = (-1)\frac{1}{(x+y)^2} + (+1)\frac{1}{(x-y)^2}$$

$$\boxed{z_{yy} = -\frac{1}{(x+y)^2} + \frac{1}{(x-y)^2}}$$

Calculate the cross partial derivatives.

$$z_{xy} = (-1)\frac{1}{(x+y)^2} - (+1)\frac{1}{(x-y)^2}$$

$$\boxed{z_{xy} = -\frac{1}{(x+y)^2} - \frac{1}{(x-y)^2}}$$

$$z_{yx} = (-1)\frac{1}{(x+y)^2} + (-1)\frac{1}{(x-y)^2}$$

$$\boxed{z_{yx} = -\frac{1}{(x+y)^2} - \frac{1}{(x-y)^2}}$$

OPTIMIZATION OF EXPONENTIAL AND LOGARITHMIC FUNCTIONS

S9.13. Given $y = 2xe^{(4x+1)}$, find the critical value and classify it.

(1) Find the critical value using the product rule.

$$y' = 2e^{4x+1} + (4)2xe^{4x+1} = 0$$

$$2e^{4x+1} + 8xe^{4x+1} = 0$$

$$e^{4x+1}(2+8x) = 0$$

Since there is no value of x for which $e^{4x+1} = 0$,

$$2+8x = 0 \qquad x = -\frac{1}{4}$$

(2) Classify the critical value by second-order condition.

$$y' = 2e^{4x+1} + 8xe^{4x+1}$$

By exponential and product rule

$$y'' = 2(4)e^{4x+1} + (8)e^{4x+1} + (4)8xe^{4x+1}$$

$$y'' = 16e^{4x+1} + 32xe^{4x+1}$$

At $x^* = \frac{1}{4}$

$$y'' = 16e^{4(1/4)+1} + 32(1/4)e^{4(1/4)+1}$$

$$y'' = 16e^2 + 8e^2 = 24e^2 > 0. \text{ The function is minimized.}$$

S9.14. Given $z = \ln(x^2 - 2x + 8y - y^2)$, find the critical value and classify it.

(1) Find the critical value (x^*, y^*).

First-order conditions.

$$z_x = \left[\frac{\partial(x^2 - 2x + 8y - y^2)}{\partial x}\right]\left(\frac{1}{x^2 - 2x + 8y - y^2}\right) = (2x - 2)\left(\frac{1}{x^2 - 2x + 8y - y^2}\right)$$

$$z_x = \frac{2x - 2}{x^2 - 2x + 8y - y^2} = 0 \quad \text{Here, } 2x - 2 = 0 \text{ or } x = 1$$

$$z_y = \left[\frac{\partial(x^2 - 2x + 8y - y^2)}{\partial y}\right]\left(\frac{1}{x^2 - 2x + 8y - y^2}\right) = (8 - 2y)\left(\frac{1}{x^2 - 2x + 8y - y^2}\right)$$

$$z_y = \frac{8 - 2y}{x^2 - 2x + 8y - y^2} = 0 \quad \text{Here, } 8 - 2y = 0 \text{ or } y = 4.$$

Thus, $x^* = 1$ and $y^* = 4$.

(2) Classify the critical value by second-order condition.

Using quotient rules.

$$z_{xx} = \frac{(2)(x^2 - 2x + 8y - y^2) - (2x - 2)(2x - 2)}{(x^2 - 2x + 8y - y^2)^2}$$

$$z_{yy} = \frac{(8)(x^2 - 2x + 8y - y^2) - (8 - 2y)(8 - 2y)}{(x^2 - 2x + 8y - y^2)^2}$$

$$z_{xy} = \frac{(0)(x^2 - 2x + 8y - y^2) - (2x - 2)(8 - 2y)}{(x^2 - 2x + 8y - y^2)^2} = \frac{-(2x - 2)(8 - 2y)}{(x^2 - 2x + 8y - y^2)^2}$$

Evaluated at $x^* = 1$, $y^* = 4$.

$$z_{xx} = \frac{(2)(1-2+32-16)-(2-2)(2-2)}{(1-2+32-16)^2} = \frac{(2)(15)-0}{15^2} = \frac{2}{15}$$

$$z_{yy} = \frac{(8)(1-2+32-16)-(8-8)(8-8)}{(1-2+32-16)} = \frac{(8)(15)-0}{15^2} = \frac{8}{15}$$

$$z_{xy} = \frac{-(2-2)(8-8)}{(1-2+32-16)^2} = 0$$

With $z_{xx} > 0$, $z_{xx} > 0$, and $z_{xy} = 0$, $z_{xx}z_{yy} > z_{xy}$; the function is at a minimum.

OPTIMAL TIMING

S9.15. An investment is increasing in value according to the formula

$$V = 100,000e^{\sqrt{t}}$$

The discount rate under continuous compounding is 0.05. How long should the investment be held to maximize the present value?

(*1*) Establish the present value formula given continuous discounting.

$$P = Ve^{-rt} = 100000e^{\sqrt{t}}e^{-0.05t}$$
$$P = 100000e^{(t^{1/2}-0.05t)}$$

(*2*) Convert to natural logs and take the derivative to find the optimal timing.

$$\frac{dP}{dt} = 100000(t^{1/2}-0.05t)'e^{(t^{1/2}-0.05t)} = 0$$
$$\frac{dP}{dt} = 100000(0.5t^{-1/2}-0.05)e^{(t^{1/2}-0.05t)} = 0$$

Since $P \neq 0$,

$$0.5t^{-1/2}-0.05 = 0$$
$$t^{-1/2} = 0.1$$
$$t = (0.1)^{-2}$$
$$t^* = 100$$

(3) Take the second-order condition to verify the critical point is a relative maximum point.

Using the product rule,

$$\frac{d^2P}{dt^2} = 100000(-0.25t^{-\frac{3}{2}})e^{(t^{1/2}-0.05t)} + 100000(0.5t^{-1/2}-0.05)^2 e^{(t^{1/2}-0.05t)}$$

$$\frac{d^2P}{dt^2} = 100000e^{(t^{1/2}-0.05t)}\left[(-0.25t^{-3/2}) + (0.5t^{-1/2}-0.05)^2\right]$$

Evaluate at $t^* = 100$.

$$\frac{d^2P}{dt^2} = 100000e^{(100^{1/2}-0.5)}\left[(-0.25(100)^{-3/2}) + (0.5(100)^{-1/2}-0.05)^2\right]$$

$$P'' = -237.5 < 0. \text{ The function is maximized.}$$

Chapter 10

THEORETICAL CONCEPTS

S10.1. False. The order does matter.

S10.2. True.

S10.3. *Null matrix* is a matrix in which all elements are zero.

S10.4. The multiplication of two matrices is composed of *inner* products of row-column combinations.

S10.5. True.

MATRIX ADDITION AND SUBTRACTION

S10.6. Given the matrices below, find $A+B-C$

$$A = \begin{bmatrix} 2 & 8 & 2 \\ 1 & 7 & 5 \end{bmatrix} \qquad B = \begin{bmatrix} 2 & -4 & 5 \\ -1 & 15 & 2 \end{bmatrix} \qquad C = \begin{bmatrix} 0 & 0 & 2 \\ -2 & 4 & 4 \end{bmatrix}$$

$$A+B-C = \begin{bmatrix} 2+2-0 & 8-4+0 & 2+5-2 \\ 1-1+2 & 7+15-4 & 5+2-4 \end{bmatrix}$$

$$A+B-C = \begin{bmatrix} 4 & 4 & 5 \\ 2 & 18 & 3 \end{bmatrix}$$

CONFORMABILITY

S10.7. Given the matrices below

$$A = \begin{bmatrix} 2 & 8 & 2 \\ 1 & 7 & 5 \end{bmatrix} \quad B = \begin{bmatrix} 2 & -4 & 5 \\ -1 & 15 & 2 \end{bmatrix} \quad C = \begin{bmatrix} 2 & 5 \\ 3 & 9 \end{bmatrix} \quad D = \begin{bmatrix} 2 & 0 \\ 1 & -1 \end{bmatrix} \quad E = \begin{bmatrix} 2 \\ 3 \\ 1 \end{bmatrix}$$

determine whether **AB**, **EC**, **AE**, **DC**, and **DE** are conformable. If conformable, provide their dimensions.

a) Matrix **AB** *is not* conformable. **AB** is given by $2 \times 3 \neq 2 \times 3$. The dimensions of **AB** cannot be determined.

b) Matrix **EC** *is not* conformable. **EC** is given by $3 \times 1 \neq 2 \times 2$. The dimensions of **EC** cannot be determined.

c) Matrix **AE** *is* conformable. **AE** is given by $2 \times 3 = 3 \times 1$. The dimensions of **AE** are 2×1.

d) Matrix **DC** *is* conformable. **DC** is given by $2 \times 2 = 2 \times 2$. The dimensions of **AE** are 2×2.

e) Matrix **DE** *is not* conformable. **DE** is given by $2 \times 2 \neq 3 \times 1$. The dimensions of **DE** cannot be determined.

SCALAR MULTIPLICATION

S10.8. Determine Ak, given

$$A = \begin{bmatrix} 2 & 8 & 2 \\ 1 & 7 & 5 \end{bmatrix} \quad k = 1/4$$

$$Ak = \begin{bmatrix} 2\left(\dfrac{1}{4}\right) & 8\left(\dfrac{1}{4}\right) & 2\left(\dfrac{1}{4}\right) \\ 1\left(\dfrac{1}{4}\right) & 7\left(\dfrac{1}{4}\right) & 5\left(\dfrac{1}{4}\right) \end{bmatrix}$$

$$Ak = \begin{bmatrix} 1/2 & 2 & 1/2 \\ 1/4 & 7/4 & 5/4 \end{bmatrix}$$

VECTOR MULTIPLICATION

S10.9. Find **FE**, given

$$F = \begin{bmatrix} -1 & 0 & 5 \end{bmatrix} \quad E = \begin{bmatrix} 2 \\ 3 \\ 1 \end{bmatrix}$$

Matrix **FE** *is* defined: $1 \times 3 = 3 \times 1$. The dimensions of **FE** are 1×1; in other words, the product will be a scalar, derived from multiplying each element of the row by its corresponding element in the column vector and then summing the products.

$$FE = -1 \times 2 + 0 \times 3 + 5 \times 1$$
$$FE = -2 + 0 + 5 = 3$$

MATRIX MULTIPLICATION

S10.10. Find **AB,** given

$$A = \begin{bmatrix} 2 & 8 & 2 \\ 1 & 7 & 5 \end{bmatrix} \qquad B = \begin{bmatrix} 2 & -1 \\ -4 & 15 \\ 5 & 2 \end{bmatrix}$$

Matrix **AB** is defined: $2 \times 3 = 3 \times 2$. The dimensions of **AB** are 2×2.

$$AB = \begin{bmatrix} R_1C_1 & R_1C_2 \\ R_2C_1 & R_2C_2 \end{bmatrix} = \begin{bmatrix} 2(2)+8(-4)+2(5) & 2(-1)+8(15)+2(2) \\ 1(2)+7(-4)+5(5) & 1(-1)+7(15)+5(2) \end{bmatrix}$$

$$AB = \begin{bmatrix} -18 & 122 \\ -1 & 114 \end{bmatrix}$$

Chapter 11

THEORETICAL CONCEPTS

S11.1. False. A singular matrix has at least one pair of rows or columns that are linearly dependent.

S11.2. False. A third-order determinant has 3×3 dimensions.

S11.3. The *rank* of a matrix indicates the maximum number of linearly independent rows or columns.

S11.4. *Laplace* expansion is used to calculate the determinant of a matrix based on the cofactors.

S11.5. False. A cofactor is a minor with a prescribed sign.

DETERMINANTS

S11.6. Find the determinant $|\mathbf{A}|$

$$A = \begin{bmatrix} 20 & 0 \\ 14 & -1 \end{bmatrix}$$

$$|\mathbf{A}| = 20(-1) - 14(0) = -20$$

S11.7. Find the determinant $|\mathbf{D}|$

$$\mathbf{D} = \begin{bmatrix} 1 & 4 & 3 \\ 0 & 5 & 0 \\ 2 & 8 & 6 \end{bmatrix}$$

$$|\mathbf{D}| = 1 \begin{vmatrix} 5 & 0 \\ 8 & 6 \end{vmatrix} + 4 \begin{vmatrix} 0 & 0 \\ 2 & 6 \end{vmatrix} + 3 \begin{vmatrix} 0 & 5 \\ 2 & 8 \end{vmatrix}$$

$$|\mathbf{D}| = 1[5(6) - 8(0)] + 4[0(6) - 2(0)] + 3[0(8) - 2(5)]$$

$$|\mathbf{D}| = 1(30) + 4(0) + 3(-10) = 30 - 30$$

$$|\mathbf{D}| = 0$$

PROPERTIES OF DETERMINANTS

S11.8. Given the lower-triangular matrix $\mathbf{B}$

$$\mathbf{B} = \begin{bmatrix} 2 & -4 & -3 \\ 0 & 5 & -2 \\ 0 & 0 & 6 \end{bmatrix}$$

find the determinant $|\mathbf{B}|$

The determinant of a triangular matrix can be found by multiplying the elements along the principal.

MINORS AND COFACTORS

S11.9. Given the matrix

$$\mathbf{D} = \begin{bmatrix} 1 & 4 & 3 \\ 0 & 5 & 0 \\ 2 & 8 & 6 \end{bmatrix}$$

find the minors and cofactors of the first column.

Delete row 1 and column 1.

$$|M_{11}| = \begin{vmatrix} 5 & 0 \\ 8 & 6 \end{vmatrix} = 30$$

Similarly,

$$|M_{21}| = \begin{vmatrix} 4 & 3 \\ 8 & 6 \end{vmatrix} = 0$$

$$|M_{31}| = \begin{vmatrix} 4 & 3 \\ 5 & 0 \end{vmatrix} = -15$$

$$|C_{11}| = (-1)^{1+1}|M_{11}| = (1)(30) = 30$$

$$|C_{21}| = (-1)^{2+1}|M_{21}| = (-1)(0) = 0$$

$$|C_{31}| = (-1)^{3+1}|M_{31}| = (1)(-15) = -15$$

LAPLACE EXPANSION

S11.10. Use Laplace expansion to find the determinant of matrix **F** using the second column.

$$\mathbf{F} = \begin{bmatrix} 1 & 4 & 3 \\ 2 & 5 & 1 \\ 2 & 8 & 6 \end{bmatrix}$$

Find the minors by deleting the second column.

$$|M_{12}| = \begin{vmatrix} 2 & 1 \\ 2 & 6 \end{vmatrix} = 10$$

$$|M_{22}| = \begin{vmatrix} 1 & 3 \\ 2 & 6 \end{vmatrix} = 0$$

$$|M_{32}| = \begin{vmatrix} 1 & 3 \\ 2 & 1 \end{vmatrix} = -5$$

$$|C_{12}| = (-1)^{1+2}|M_{12}| = (-1)(10) = -10$$

$$|C_{22}| = (-1)^{2+2}|M_{22}| = (1)(0) = 0$$

$$|C_{32}| = (-1)^{3+2}|M_{32}| = (-1)(-5) = 5$$

Expand along the second column.

$$|\mathbf{F}| = a_{12}|C_{12}| + a_{22}|C_{22}| + a_{32}|C_{22}| = 4(-10) + 5(0) + 8(5)$$
$$|\mathbf{F}| = -40 + 40 = 0$$

INVERTING A MATRIX

S11.11. Find the inverse $\mathbf{A}^{-1}$ for

$$\mathbf{A} = \begin{bmatrix} 20 & 0 \\ 14 & -1 \end{bmatrix}$$

given the formula

$$\mathbf{A}^{-1} = \frac{1}{|\mathbf{A}|} Adj\ \mathbf{A}$$

Evaluate the determinant: $|\mathbf{A}| = 20(-1) - 14(0) = -20$

Then, find the cofactor matrix to get the adjoint.

$$\mathbf{C} = \begin{bmatrix} -1 & -14 \\ 0 & 20 \end{bmatrix}$$

and $Adj\ \mathbf{A} = \mathbf{C}'$.

$$Adj\ \mathbf{A} = \begin{bmatrix} -1 & 0 \\ -14 & 20 \end{bmatrix}$$

Thus,

$$\mathbf{A}^{-1} = \frac{1}{-20} \begin{bmatrix} -1 & 0 \\ -14 & 20 \end{bmatrix}$$

$$\mathbf{A}^{-1} = \begin{bmatrix} -\dfrac{1}{20} & 0 \\ -\dfrac{14}{20} & 1 \end{bmatrix}$$

S11.12. Find the inverse $\mathbf{A}^{-1}$ for the matrix using cofactors.

$$\mathbf{A} = \begin{bmatrix} 1 & 4 & 3 \\ 0 & 5 & 2 \\ 2 & 1 & 1 \end{bmatrix}$$

The cofactor matrix $\mathbf{C}$ is

$$\mathbf{C} = \begin{bmatrix} \begin{vmatrix} 5 & 2 \\ 1 & 1 \end{vmatrix} & -\begin{vmatrix} 0 & 2 \\ 2 & 1 \end{vmatrix} & \begin{vmatrix} 0 & 5 \\ 2 & 1 \end{vmatrix} \\ -\begin{vmatrix} 4 & 3 \\ 1 & 1 \end{vmatrix} & \begin{vmatrix} 1 & 3 \\ 2 & 1 \end{vmatrix} & -\begin{vmatrix} 1 & 4 \\ 2 & 1 \end{vmatrix} \\ \begin{vmatrix} 4 & 3 \\ 5 & 2 \end{vmatrix} & -\begin{vmatrix} 1 & 3 \\ 0 & 2 \end{vmatrix} & \begin{vmatrix} 1 & 4 \\ 0 & 5 \end{vmatrix} \end{bmatrix} = \begin{bmatrix} 3 & 4 & -10 \\ -1 & -5 & 7 \\ -7 & -2 & 5 \end{bmatrix}$$

The adjoint matrix $Adj\ \mathbf{A}$ is

$$Adj\ \mathbf{A} = \begin{bmatrix} 3 & -1 & -7 \\ 4 & -5 & -2 \\ -10 & 7 & 5 \end{bmatrix}$$

To find the determinant $|\ \mathbf{A}\ |$, use the Laplace expansion along the first row.

$$|\mathbf{A}| = 1\begin{vmatrix} 5 & 2 \\ 1 & 1 \end{vmatrix} - 4\begin{vmatrix} 0 & 2 \\ 2 & 1 \end{vmatrix} + 3\begin{vmatrix} 0 & 5 \\ 2 & 1 \end{vmatrix} = 1(3) - 4(-4) + 3(-10)$$

$$|\mathbf{A}| = 3 + 16 - 30 = -11$$

Then,

$$\mathbf{A}^{-1} = -\frac{1}{11}\begin{bmatrix} 3 & -1 & -7 \\ 4 & -5 & -2 \\ -10 & 7 & 5 \end{bmatrix}$$

$$\mathbf{A}^{-1} = \begin{bmatrix} -\dfrac{3}{11} & \dfrac{1}{11} & \dfrac{7}{11} \\ -\dfrac{4}{11} & \dfrac{5}{11} & \dfrac{2}{11} \\ \dfrac{10}{11} & -\dfrac{7}{11} & -\dfrac{5}{11} \end{bmatrix}$$

MATRIX INVERSIONS IN EQUATION SOLUTIONS

S11.13. Use matrix inversions to solve the following system of linear equations:

$$4x + y = 21$$
$$3x - 2y = 13$$

(*1*) Set up the system into matrix form.

$$A \cdot X = B$$

$$\begin{bmatrix} 4 & 1 \\ 3 & -2 \end{bmatrix}\begin{bmatrix} x \\ y \end{bmatrix} = \begin{bmatrix} 21 \\ 13 \end{bmatrix}$$

(*2*) Find the determinant of **A**.

$$|A| = 4(-2) - 3(1) = -11$$

(*3*) The cofactor matrix **C** is

$$C = \begin{bmatrix} -2 & -3 \\ -1 & 4 \end{bmatrix}$$

(*4*) The adjoint matrix **A** is

$$Adj\ \mathbf{A} = \begin{bmatrix} -2 & -1 \\ -3 & 4 \end{bmatrix}$$

(*5*) The inverse matrix is

$$\mathbf{A}^{-1} = \frac{1}{-11}\begin{bmatrix} -2 & -1 \\ -3 & 4 \end{bmatrix} = \begin{bmatrix} \dfrac{2}{11} & \dfrac{1}{11} \\ \dfrac{3}{11} & -\dfrac{4}{11} \end{bmatrix}$$

(*6*) Then, substitute in $\mathbf{X} = \mathbf{A}^{-1}\mathbf{B}$

$$X = \begin{bmatrix} \dfrac{2}{11} & \dfrac{1}{11} \\ \dfrac{3}{11} & -\dfrac{4}{11} \end{bmatrix}\begin{bmatrix} 21 \\ 13 \end{bmatrix}$$

(*7*) Multiply matrices; the solution for **X** is

$$X = \begin{bmatrix} \dfrac{2}{11}(21) + \dfrac{1}{22}(13) \\ \dfrac{3}{11}(21) - \dfrac{4}{11}(13) \end{bmatrix} = \begin{bmatrix} \dfrac{42+13}{11} \\ \dfrac{63-52}{11} \end{bmatrix}$$

$$X = \begin{bmatrix} 5 \\ 1 \end{bmatrix}$$

Thus $\bar{x} = 5$, $\bar{y} = 1$.

CRAMER'S RULE

S11.14. The market of a product is given by these linear equations:

Supply: $-4P + Q = -16$ Demand: $2P + Q = 64$

where P is price and Q is quantity of the product. Find the equilibrium price (P^*) and quantity (Q^*) sold in this market by using Cramer's rule.

(*1*) Express the equations in matrix form.

$$\begin{bmatrix} -4 & 1 \\ 2 & 1 \end{bmatrix}\begin{bmatrix} P \\ Q \end{bmatrix} = \begin{bmatrix} -16 \\ 64 \end{bmatrix}$$

(*2*) Find the determinant of **A**

$$|\mathbf{A}| = -4(1) - 2(1) = -6$$

(3) Replacing the first column of **A**, which contains the coefficients of P, with the column vector of constants **B**,

$$A_1 = \begin{bmatrix} -16 & 1 \\ 64 & 1 \end{bmatrix}$$

The determinant of A_1 is

$$|A_1| = -16(1) - 64(1) = -80$$

Use Cramer's rule and find $P*$

$$P^* = \frac{|A_1|}{|\mathbf{A}|} = \frac{-80}{-6} = \frac{40}{3}$$

(4) Replace the second column of **A,** which contains the coefficients of Q, with the column vector of constants **B**.

$$A_2 = \begin{bmatrix} -4 & -16 \\ 2 & 64 \end{bmatrix}$$

The determinant of A_2 is

$$|A_2| = -4(64) - 2(-16) = -256 + 32 = -224$$

Use Cramer's rule and find $Q*$

$$Q^* = \frac{|A_1|}{|\mathbf{A}|} = \frac{-224}{-6} = \frac{112}{3}$$

(5) The equilibrium price $P*$ is 40/3 and quantity $Q*$ is 112/3.

Chapter 12

THEORETICAL CONCEPTS

S12.1. True.

S12.2. False. The Hessian matrix is used to test second-order condition.

S12.3. For a function $y = f(x_1, x_2, x_3, x_4)$ with a nonsingular matrix, the number of principal minors to determine concavity is *four*.

S12.4. In an input-output table, matrix **A** is known as the *matrix of technical coefficients.*

S12.5. True.

THE JACOBIAN

S12.6. Test for functional dependence by means of the Jacobian.

$$y_1 = 4x_1 + 2x_2$$
$$y_2 = 16x_1^2 - 8x_1 x_2 - 4x_2^2$$

Take the first-order partials to set up the Jacobian $|\mathbf{J}|$

$$\frac{\partial y_1}{\partial x_1} = 4 \qquad \frac{\partial y_1}{\partial x_2} = 2 \qquad \frac{\partial y_2}{\partial x_1} = 32x_1 - 8x_2 \qquad \frac{\partial y_2}{\partial x_2} = -8x_1 - 8x_2$$

$$|\mathbf{J}| = \begin{bmatrix} 4 & 2 \\ 32x_1 - 8x_2 & -8x_1 - 8x_2 \end{bmatrix}$$

$$|\mathbf{J}| = 4(-8x_1 - 8x_2) - 2(32x_1 - 8x_2) = -32x_1 - 32x_2 - 64x_1 + 16x_2$$

$$|\mathbf{J}| = -96x_1 - 16x_2 \neq 0$$

The equations are functionally independent.

DISCRIMINANTS

S12.7. Use discriminants to determine if the quadratic function is positive or negative definite.

$$y = 16x_1^2 - 8x_1 x_2 - 4x_2^2 + 12x_1 x_3 + 8x_3^2 - 2x_2 x_3$$

$$|\mathbf{D}| = \begin{bmatrix} 16 & -4 & 6 \\ -4 & -4 & -1 \\ 6 & -1 & 8 \end{bmatrix}$$

where $\qquad |D_1| = 16 > 0 \qquad |D_2| = \begin{bmatrix} 16 & -4 \\ -4 & -4 \end{bmatrix} = -96 < 0$

and $\qquad |D_3| = |\mathbf{D}| = -464 < 0$

Thus, y is not sign definite and y may assume both positive and negative values.

THE HESSIAN IN OPTIMIZATION PROBLEMS

S12.8. Maximize profits for a firm producing two goods x and y.

$$\pi = 160x - x^2 - xy - 3y^2 + 300y - 3000$$

(1) Use Cramer's rule to find the critical points.

The first-order conditions are

$$\pi_x = 160 - 2x - y = 0 \qquad 2x + y = 160$$
$$\pi_y = 300 - x - 6y = 0 \qquad x + 6y = 300$$

The matrix form is $\begin{bmatrix} 2 & 1 \\ 1 & 6 \end{bmatrix} \begin{bmatrix} x \\ y \end{bmatrix} = \begin{bmatrix} 160 \\ 300 \end{bmatrix}$

The determinant $|\mathbf{A}| = 2(6) - 1(1) = 11$

$$A_1 = \begin{bmatrix} 160 & 1 \\ 300 & 6 \end{bmatrix} \text{ and } |A_1| = 160(6) - 300(1) = 660$$

$$A_2 = \begin{bmatrix} 2 & 160 \\ 1 & 300 \end{bmatrix} \text{ and } |A_2| = 2(300) - 1(160) = 440$$

Thus, $\quad x^* = \dfrac{|A_1|}{|\mathbf{A}|} = \dfrac{660}{11} = 60 \qquad y^* = \dfrac{|A_2|}{|\mathbf{A}|} = \dfrac{440}{11} = 40$

(2) Test the second-order condition by forming the Hessian.

$$\pi_{xx} = -2 \qquad\qquad \pi_{yy} = -6$$
$$\pi_{xy} = -1 \qquad\qquad \pi_{yx} = -1$$

Thus,

$$|\mathbf{H}| = \begin{bmatrix} -2 & -1 \\ -1 & -6 \end{bmatrix}$$

where $|H_1| = -2 < 0$

and $\quad |H_2| = |\mathbf{H}| = |\mathbf{A}| = 11 > 0$

With $|\mathbf{H}|$ negative definite, π is maximized.

THE BORDERED HESSIAN IN CONSTRAINED OPTIMIZATION

S12.9. Given the profit function

$$\pi = 40x - x^2 - 2xy - 2y^2 + 50y$$

The maximum output capacity (constraint) is

$$x + y = 25$$

Find the optimal combination (x^*, y^*) that maximizes the profit function using the Lagrange multiplier.

(*1*) Establish the Lagrange function.

$$\mathcal{L} = 40x - x^2 - 2xy - 2y^2 + 50y + \lambda(25 - x - y)$$

$$\mathcal{L} = \boxed{40x - x^2 - 2xy - 2y^2 + 50y + 25\lambda - x\lambda - y\lambda}$$

(*2*) Use Cramer's rule to find the critical points.

$\mathcal{L}_x = \boxed{-2x - 2y + 40 - \lambda = 0}$ $2x + 2y + \lambda = 40$ (*12.23*)

$\mathcal{L}_y = \boxed{-2x - 4y + 50 - \lambda = 0}$ $2x + 4y + \lambda = 50$ (*12.24*)

$\mathcal{L}_\lambda = \boxed{25 - x - y = 0}$ $x + y = 25$ (*12.25*)

The matrix form is $\begin{bmatrix} 2 & 2 & 1 \\ 2 & 4 & 1 \\ 1 & 1 & 0 \end{bmatrix} \begin{bmatrix} x \\ y \\ \lambda \end{bmatrix} = \begin{bmatrix} 40 \\ 50 \\ 25 \end{bmatrix}$

The determinant $|\mathbf{A}| = 2 \begin{vmatrix} 4 & 1 \\ 1 & 0 \end{vmatrix} - 2 \begin{vmatrix} 2 & 1 \\ 1 & 0 \end{vmatrix} + 1 \begin{vmatrix} 2 & 4 \\ 1 & 1 \end{vmatrix} = 2(-1) - 2(-1) + 1(-2) = -2$

$A_1 = \begin{bmatrix} 40 & 2 & 1 \\ 50 & 4 & 1 \\ 25 & 1 & 0 \end{bmatrix}$ and $|A_1| = 40 \begin{vmatrix} 4 & 1 \\ 1 & 0 \end{vmatrix} - 2 \begin{vmatrix} 50 & 1 \\ 25 & 0 \end{vmatrix} + 1 \begin{vmatrix} 50 & 4 \\ 25 & 1 \end{vmatrix} = -40 + 50 - 50 = -40$

$A_2 = \begin{bmatrix} 2 & 40 & 1 \\ 2 & 50 & 1 \\ 1 & 25 & 0 \end{bmatrix}$ and $|A_2| = 2 \begin{vmatrix} 50 & 1 \\ 25 & 0 \end{vmatrix} - 40 \begin{vmatrix} 2 & 1 \\ 1 & 0 \end{vmatrix} + 1 \begin{vmatrix} 2 & 50 \\ 1 & 25 \end{vmatrix} = -50 + 40 + 0 = -10$

$A_3 = \begin{bmatrix} 2 & 2 & 40 \\ 2 & 4 & 50 \\ 1 & 1 & 25 \end{bmatrix}$ and $|A_3| = 2 \begin{vmatrix} 4 & 50 \\ 1 & 25 \end{vmatrix} - 2 \begin{vmatrix} 2 & 50 \\ 1 & 25 \end{vmatrix} + 40 \begin{vmatrix} 2 & 4 \\ 1 & 1 \end{vmatrix} = 100 + 0 - 80 = 20$

Thus, $x^* = \dfrac{|A_1|}{|\mathbf{A}|} = \dfrac{-40}{-2} = 20$ $y^* = \dfrac{|A_2|}{|\mathbf{A}|} = \dfrac{-10}{-2} = 5$ $\lambda^* = \dfrac{|A_2|}{|\mathbf{A}|} = \dfrac{-20}{-2} = -10$

(*3*) Use the bordered Hessian matrix to test the second-order condition.

Since $\mathcal{L}_{xx} = -2$, $\mathcal{L}_{yy} = -4$, $\mathcal{L}_{xy} = -2$, $c_x = 1$, $c_y = 1$

$$|\overline{\mathbf{H}}| = \begin{bmatrix} -2 & -2 & 1 \\ -2 & -4 & 1 \\ 1 & 1 & 0 \end{bmatrix}$$

$$|\overline{H}_2| = |\overline{\mathbf{H}}| = -2 \begin{vmatrix} -4 & 1 \\ 1 & 0 \end{vmatrix} + 2 \begin{vmatrix} -2 & 1 \\ 1 & 0 \end{vmatrix} + 1 \begin{vmatrix} -2 & -4 \\ 1 & 1 \end{vmatrix} = 2 - 2 + 2 = 2$$

With $|\overline{H}_2| > 0$, $|\overline{\mathbf{H}}|$ is negative definite, π is maximized.

INPUT-OUTPUT ANALYSIS

S12.10. Determine the total demand **X** for industries 1, 2, and 3, given the matrix of technical coefficients **A** and the final demand vector **B** below.

Output industry

$$\begin{array}{ccc} 1 & 2 & 3 \end{array}$$

$$\mathbf{A} = \begin{bmatrix} 0.00 & 0.20 & 0.00 \\ 0.35 & 0.00 & 0.25 \\ 0.10 & 0.10 & 0.00 \end{bmatrix} \qquad \text{Input industry} \qquad \mathbf{B} = \begin{bmatrix} 600 \\ 1500 \\ 300 \end{bmatrix}$$

Given $\mathbf{X} = (\mathbf{I} - \mathbf{A})^{-1}\mathbf{B}$

(*1*) The matrix $\mathbf{I} - \mathbf{A}$ is

$$\mathbf{I} - \mathbf{A} = \begin{bmatrix} 1-0 & 0-0.2 & 0 \\ 0-0.35 & 1-0 & 0-0.25 \\ 0-0.10 & 0-0.10 & 1-0 \end{bmatrix} = \begin{bmatrix} 1 & -0.2 & 0 \\ -0.35 & 1 & -0.25 \\ -0.10 & -0.10 & 1 \end{bmatrix}$$

(*2*) And the inverse matrix $(\mathbf{I} - \mathbf{A})^{-1}$ is

The cofactor matrix **C** is

$$\mathbf{C} = \begin{bmatrix} \begin{vmatrix} 0 & -0.25 \\ -0.1 & 1 \end{vmatrix} & -\begin{vmatrix} -0.35 & -0.25 \\ -0.1 & 1 \end{vmatrix} & \begin{vmatrix} -0.35 & 0 \\ -0.1 & -0.1 \end{vmatrix} \\[4mm] -\begin{vmatrix} -0.2 & 0 \\ -0.1 & 1 \end{vmatrix} & \begin{vmatrix} 1 & 0 \\ -0.1 & 1 \end{vmatrix} & -\begin{vmatrix} 1 & -0.2 \\ -0.1 & -0.1 \end{vmatrix} \\[4mm] \begin{vmatrix} -0.2 & 0 \\ -0.1 & -0.25 \end{vmatrix} & -\begin{vmatrix} 1 & 0 \\ -0.35 & -0.25 \end{vmatrix} & \begin{vmatrix} 1 & -0.2 \\ -0.35 & 1 \end{vmatrix} \end{bmatrix} = \begin{bmatrix} 0.975 & 0.375 & 0.135 \\ 0.2 & 1 & 0.12 \\ 0.05 & 0.25 & 0.93 \end{bmatrix}$$

The adjoint matrix Adj $(\mathbf{I} - \mathbf{A})$ is

$$\text{Adj } (\mathbf{I} - \mathbf{A}) = \begin{bmatrix} 0.975 & 0.2 & 0.05 \\ 0.375 & 1 & 0.25 \\ 0.135 & 0.12 & 0.93 \end{bmatrix}$$

The determinant $|\mathbf{I} - \mathbf{A}|$ is

$$|\mathbf{I} - \mathbf{A}| = 0.975 \begin{vmatrix} 1 & 0.25 \\ 0.12 & 0.93 \end{vmatrix} - 0.2 \begin{vmatrix} 0.375 & 0.25 \\ 0.135 & 0.93 \end{vmatrix} + 0.05 \begin{vmatrix} 0.375 & 1 \\ 0.135 & 0.12 \end{vmatrix}$$

$$|\mathbf{I} - \mathbf{A}| = 0.9$$

The inverse matrix $(\mathbf{I} - \mathbf{A})^{-1}$ is

$$(\mathbf{I} - \mathbf{A})^{-1} = \frac{1}{0.9} \begin{bmatrix} 0.975 & 0.2 & 0.05 \\ 0.375 & 1 & 0.25 \\ 0.135 & 0.12 & 0.93 \end{bmatrix}$$

(*3*) Thus **X** is

$$\mathbf{X} = \frac{1}{0.9} \begin{bmatrix} 0.975 & 0.2 & 0.05 \\ 0.375 & 1 & 0.25 \\ 0.135 & 0.12 & 0.93 \end{bmatrix} \begin{bmatrix} 600 \\ 1500 \\ 300 \end{bmatrix} = \frac{1}{0.9} \begin{bmatrix} 900 \\ 1800 \\ 540 \end{bmatrix}$$

$$\mathbf{X} = \begin{bmatrix} 1000 \\ 2000 \\ 600 \end{bmatrix}$$

EIGENVECTORS

S12.11. Use eigenvalues (also known as characteristics roots) to determine the sign definiteness for

$$\mathbf{A} = \begin{bmatrix} 3 & 1 & 1 \\ 0 & 2 & 5 \\ 1 & 0 & 4 \end{bmatrix}$$

(*1*) Construct $|\mathbf{A} - c\mathbf{I}|$

$$|\mathbf{A} - c\mathbf{I}| = \begin{bmatrix} 3-c & 1 & 1 \\ 0 & 2-c & 5 \\ 0 & 1 & 4-c \end{bmatrix} = 0$$

(*2*) Expand along the first column to use the determinant.

$$|\mathbf{A} - c\mathbf{I}| = (3-c) \begin{vmatrix} 2-c & 5 \\ 1 & 4-c \end{vmatrix} - 0 \begin{vmatrix} 0 & 1 \\ 1 & 4-c \end{vmatrix} + 0 \begin{vmatrix} 1 & 1 \\ 2-c & 5 \end{vmatrix} = 0$$

$$|\mathbf{A} - c\mathbf{I}| = (3-c)[(2-c)(4-c)-5] = 0$$

$$|\mathbf{A} - c\mathbf{I}| = (3-c)[8-6c+c^2-5] = 0$$

$$|\mathbf{A} - c\mathbf{I}| = (3-c)[3-6c+c^2] = 0 \qquad (12.26)$$

(*3*) Solve for c in (*12.26*).

$$3-c = 0 \quad \text{or} \quad 3-6c+c^2 = 0$$

$$c = \frac{-(-6) \pm \sqrt{(-6)^2 - 4(1)(3)}}{2(1)} = \frac{6 \pm \sqrt{36-12}}{2} = \frac{6 \pm 2\sqrt{6}}{2} = 3 \pm \sqrt{6}$$

Thus, $c_1 = 3$ $c_2 = 5.44$ $c_3 = 0.55$

With all eigenvalues positive, **A** is positive definite.

Chapter 13

THEORETICAL CONCEPTS

S13.1. False. In comparative statics, the equilibrium is tested under changes in the exogenous variables.

S13.2. True.

S13.3. The *Jacobian* is used for comparative statics with more than one endogenous variable.

S13.4. In a constrained maximization with two decision variables, if the second-order sufficient condition is met, then $|\overline{\mathbf{H}}| > 0$ for maximization.

S13.5. False. Convex functions can be solved using concave programming.

COMPARATIVE STATICS WITH ONE ENDOGENOUS VARIABLE

S13.6. Given the following economy

$$Y = C + I + (X - Z) \quad C = C_0 + bY \quad Z = Z_0 + zY$$
$$I = I_0 \qquad\qquad X = X_0$$

where $C_0, I_0, X_0, Z > 0$ and $0 < b, z < 1$

Substituting for different variables, it can be found that

$$Y = C_0 + bY + I_0 + X_0 - Z_0 - zY$$

(1) Find the implicit function **F** for the equilibrium condition, by moving all terms to the left.

$$F = Y - bY + zY - C_0 - I_0 - X_0 + Z_0 = 0$$

(2) Under the optimal solution for income $\overline{Y}$, find the effect of changing the propensity to import z in income $\overline{Y}$ (i.e., $\partial \overline{Y}/\partial z$).

From the Implicit Function Rule:

$$\frac{\partial \overline{Y}}{\partial z} = -\frac{F_z}{F_{\overline{Y}}} = -\frac{\overline{Y}}{1-b+z} = -\frac{(+)}{(+)} < 0$$

An increase in the marginal propensity to import z will decrease income $\overline{Y}$.

COMPARATIVE STATISTICS FOR OPTIMIZATION PROBLEMS

S13.7. A firm seeks to optimize the present value of its profit from selling a good Z. The production of the good uses inputs X and Y and is represented by a Cobb-Douglas production function

$$\pi = P_0(AX^\alpha Y^\beta)e^{rt} - P_x X - P_Y Y$$

where $0 < \alpha + \beta < 1$, P_0 is the price of good Z, whereas P_x and P_y are the prices of inputs X and Y.

Evaluate $\partial \overline{X} / \partial r$ and $\partial \overline{Y} / \partial r$.

(*1*) Given that $Q = X^{\alpha} Y^{\beta}$, take the first-order conditions, which is expressed as implicit function.

$$F^1(X,Y;P_0,P_x,P_y,A,r,t) = \pi_x = \alpha P_0 (AX^{\alpha-1}Y^{\beta})e^{-rt} - P_x = 0$$

$$F^2(X,Y;P_0,P_x,P_y,A,r,t) = \pi_y = \beta P_0 (AX^{\alpha}Y^{\beta-1})e^{-rt} - P_y = 0$$

(2) Express the total derivatives of the function with respect to r in matrix form, recalling that P_0 and t are constants.

The total derivative expression is

$$|\mathbf{H}| \begin{bmatrix} \dfrac{\partial \overline{X}}{\partial r} \\[2mm] \dfrac{\partial \overline{Y}}{\partial r} \end{bmatrix} = \begin{bmatrix} -\dfrac{\partial F^1}{\partial r} \\[2mm] -\dfrac{\partial F^2}{\partial r} \end{bmatrix}$$

The Jacobian for the optimization problem is the Hessian

$$|\mathbf{H}| = \begin{bmatrix} F_x^1 & F_y^1 \\ F_x^2 & F_2^2 \end{bmatrix} = \begin{bmatrix} \pi_{xx} & \pi_{xy} \\ \pi_{yx} & \pi_{yy} \end{bmatrix} = \begin{bmatrix} \alpha(\alpha-1)P_0(AX^{\alpha-2}Y^{\beta})e^{-rt} & \alpha\beta P_0(AX^{\alpha-1}Y^{\beta-1})e^{-rt} \\ \alpha\beta P_0(AX^{\alpha-1}Y^{\beta-1})e^{-rt} & \beta(\beta-1)P_0(AX^{\alpha}Y^{\beta-2})e^{-rt} \end{bmatrix}$$

The right-hand side is $\begin{bmatrix} -\dfrac{\partial F^1}{\partial r} \\[2mm] \dfrac{\partial F^2}{\partial r} \end{bmatrix} = \begin{bmatrix} \alpha t P_0(AX^{\alpha-1}Y^{\beta})e^{-rt} \\[2mm] \beta t P_0(AX^{\alpha}Y^{\beta-1})e^{-rt} \end{bmatrix}$

Thus, the representation is

$$\begin{bmatrix} \alpha(\alpha-1)P_0(AX^{\alpha-2}Y^{\beta})e^{-rt} & \alpha\beta P_0(AX^{\alpha-1}Y^{\beta-1})e^{-rt} \\ \alpha\beta P_0(AX^{\alpha-1}Y^{\beta-1})e^{-rt} & \beta(\beta-1)P_0(AX^{\alpha}Y^{\beta-2})e^{-rt} \end{bmatrix} \begin{bmatrix} \dfrac{\partial \overline{X}}{\partial r} \\[2mm] \dfrac{\partial \overline{Y}}{\partial r} \end{bmatrix} = \begin{bmatrix} \alpha t P_0(AX^{\alpha-1}Y^{\beta})e^{-rt} \\[2mm] \beta t P_0(AX^{\alpha}Y^{\beta-1})e^{-rt} \end{bmatrix}$$

(3) Here the Jacobian is the Hessian matrix

$$|\mathbf{H}| = \begin{bmatrix} \alpha(\alpha-1)P_0(AX^{\alpha-2}Y^{\beta})e^{-rt} & \alpha\beta P_0(AX^{\alpha-1}Y^{\beta-1})e^{-rt} \\ \alpha\beta P_0(AX^{\alpha-1}Y^{\beta-1})e^{-rt} & \beta(\beta-1)P_0(AX^{\alpha}Y^{\beta-2})e^{-rt} \end{bmatrix}$$

The determinant is

$$|\mathbf{H}| = \alpha(\alpha-1)P_0(AX^{\alpha-2}Y^{\beta})e^{-rt}\beta(\beta-1)P_0(AX^{\alpha}Y^{\beta-2})e^{-rt} - [\alpha\beta P_0(AX^{\alpha-1}Y^{\beta-1})e^{-rt}]^2$$

$$|\mathbf{H}| = \alpha(\alpha-1)\beta(\beta-1)P_0^2 A^2 (X^{2\alpha-2}Y^{\beta-2})e^{-2rt} - \alpha^2\beta^2 P_0^2 (A^2X^{2\alpha-2}Y^{2\beta-2})e^{-2rt}$$

$$|\mathbf{H}| = \alpha\beta P_0^2 A^2 (X^{2\alpha-2}Y^{\beta-2})e^{-2rt}[(\alpha-1)(\beta-1)-\alpha\beta]$$

$$\mathbf{H}| = \alpha\beta P_0^2 A^2 (X^{2\alpha-2}Y^{\beta-2})e^{-2rt}[\alpha\beta-\alpha-\beta+1-\alpha\beta]$$

$$|\mathbf{H}| = \alpha\beta P_0^2 A^2 (X^{2\alpha-2}Y^{\beta-2})e^{-2rt}[1-\alpha-\beta]$$

Because all terms are positive and $\alpha+\beta < 1$ (which means $1-\alpha-\beta > 0$), then $|\mathbf{H}| > 0$.

(4) Evaluate $\partial\overline{X}/\partial r$:

Use Cramer's rule.

$$|H_1| = \begin{bmatrix} \alpha t P_0(AX^{\alpha-1}Y^\beta)e^{-rt} & \alpha\beta P_0(AX^{\alpha-1}Y^{\beta-1})e^{-rt} \\ \beta t P_0(AX^\alpha Y^{\beta-1})e^{-rt} & \beta(\beta-1)P_0(AX^\alpha Y^{\beta-2})e^{-rt} \end{bmatrix}$$

The determinant $|H_1|$:

$$|H_1| = \alpha\beta(\beta-1)tP_0^2(A^2X^{2\alpha-1}Y^{2\beta-2})e^{-2rt} - \alpha\beta^2 tP_0^2(A^2X^{2\alpha-1}Y^{2\beta-2})e^{-2rt}$$

$$|H_1| = \alpha\beta P_0^2 A^2(X^{2\alpha-1}Y^{2\beta-2})te^{-2rt}[(\beta-1)-\beta]$$

$$|H_1| = \alpha\beta P_0^2 A^2(X^{2\alpha-1}Y^{2\beta-2})te^{-2rt}[1] > 0$$

The sign is positive since all terms are positive.

$$\frac{\partial\overline{X}}{\partial r} = \frac{|H_1|}{|\mathbf{H}|} = \frac{(+)}{(+)} < 0$$

An increase in interest rate r will increase the demand for good X

(5) Evaluate $\partial\overline{Y}/\partial r$

Use Cramer's rule.

$$|H_2| = \begin{bmatrix} \alpha(\alpha-1)P_0(AX^{\alpha-2}Y^\beta)e^{-rt} & \alpha t P_0(AX^{\alpha-1}Y^\beta)e^{-rt} \\ \alpha\beta P_0(AX^{\alpha-1}Y^{\beta-1})e^{-rt} & \beta t P_0(AX^\alpha Y^{\beta-1})e^{-rt} \end{bmatrix}$$

The determinant $|H_2|$:

$$|H_2| = \alpha\beta(\alpha-1)tP_0^2(A^2X^{2\alpha-2}Y^{2\beta-1})e^{-2rt} - \alpha^2\beta tP_0^2(A^2X^{2\alpha-2}Y^{2\beta-1})e^{-2rt}$$

$$|H_2| = \alpha\beta P_0^2 A^2(X^{2\alpha-2}Y^{2\beta-1})te^{-2rt}[(\alpha-1)-\alpha]$$

$$|H_2| = \alpha\beta P_0^2 A^2(X^{2\alpha-2}Y^{2\beta-1})te^{-2rt}[1] > 0$$

The sign is negative since all terms are positive.

$$\frac{\partial\overline{Y}}{\partial r} = \frac{|H_2|}{|\mathbf{H}|} = \frac{(+)}{(+)} > 0$$

An increase in interest rate r will increase the demand for good Y.

COMPARATIVE STATISTICS USED IN CONSTRAINED OPTIMIZATION

S13.8. A consumer wans to maximize its Cobb-Douglas utility $u(x, y) = x^{1/2} y^{1/2}$ subject to the constant constraint $p_x x + p_y y - B$. Given the Lagrange function

$$U = x^{1/2} y^{1/2} + \lambda(B - p_x x - p_y y)$$

Assume that the second-order sufficient condition is met.

(*1*) Take the first-order condition to the Lagrange function.

$$F^1 = 0.5x^{-1/2} y^{1/2} - \lambda p_x = 0$$
$$F^2 = 0.5x^{1/2} y^{-1/2} - \lambda p_y = 0$$
$$F^3 = B - p_x x - p_y y = 0$$

Set the Jacobian, which in this case is the bordered Hessian matrix

$$|\overline{\mathbf{H}}| = \begin{bmatrix} -0.25x^{-3/2} y^{1/2} & 0.25x^{-1/2} y^{-1/2} & -p_x \\ 0.25x^{-1/2} y^{-1/2} & -0.25x^{1/2} y^{-3/2} & -p_y \\ -p_x & -p_y & 0 \end{bmatrix}$$

The determinant is

$$|\overline{\mathbf{H}}| = -0.25 p_x \left(-p_y x^{-1/2} y^{-1/2} + p_x x^{1/2} y^{-3/2} \right) + 0.25 p_y (-p_y x^{-3/2} y^{1/2} + p_x x^{-1/2} y^{-1/2})$$
$$|\overline{\mathbf{H}}| = 0.25[\left(-p_x p_y x^{-1/2} y^{-1/2} + p_x^2 x^{1/2} y^{-3/2} \right) + (p_y^2 x^{-3/2} y^{1/2} + p_x p_y x^{-1/2} y^{-1/2})]$$

Simplify terms.

$$|\overline{\mathbf{H}}| = 0.25\left[p_x^2 x^{1/2} y^{-3/2} + p_y^2 x^{-3/2} y^{1/2} \right]$$
$$|\overline{\mathbf{H}}| = 0.25 x^{-3/2} y^{-3/2} \left[p_x^2 x^2 + p_y^2 y^2 \right] > 0$$

Because all terms are greater than zero, $|\overline{\mathbf{H}}| > 0$, which suffices the conditions for maximization.

The total derivative expression is

$$|\overline{\mathbf{H}}| \begin{bmatrix} \dfrac{\partial \overline{x}}{\partial p_y} \\[2ex] \dfrac{\partial \overline{y}}{\partial p_y} \\[2ex] \dfrac{\partial \overline{\lambda}}{\partial p_y} \end{bmatrix} = \begin{bmatrix} -\dfrac{\partial F^1}{\partial p_y} \\[2ex] -\dfrac{\partial F^2}{\partial p_y} \\[2ex] -\dfrac{\partial F^3}{\partial p_y} \end{bmatrix}$$

The right-hand side is $\begin{bmatrix} -\dfrac{\partial F^1}{\partial p_y} \\[2mm] -\dfrac{\partial F^2}{\partial p_y} \\[2mm] -\dfrac{\partial F^3}{\partial p_y} \end{bmatrix} = \begin{bmatrix} 0 \\[1mm] \bar{\lambda} \\[1mm] \bar{y} \end{bmatrix}$

Thus, the expression is

$$\begin{bmatrix} -\dfrac{1}{4}x^{-3/2}y^{1/2} & \dfrac{1}{4}x^{-1/2}y^{-1/2} & -p_x \\[2mm] \dfrac{1}{4}x^{-1/2}y^{-1/2} & -\dfrac{1}{4}x^{1/2}y^{-3/2} & -p_y \\[2mm] -p_x & -p_y & 0 \end{bmatrix} \begin{bmatrix} \dfrac{\partial \bar{x}}{\partial p_y} \\[2mm] \dfrac{\partial \bar{y}}{\partial p_y} \\[2mm] \dfrac{\partial \bar{\lambda}}{\partial p_y} \end{bmatrix} = \begin{bmatrix} 0 \\[1mm] \bar{\lambda} \\[1mm] \bar{y} \end{bmatrix}$$

(2) To calculate $\partial \bar{y}/\partial p_y$:

Use Cramer's rule for $\bar{y}$ by replacing the second column of the Jacobian (bordered Hessian).

$$|H_2| = \begin{vmatrix} -\dfrac{1}{4}x^{-3/2}y^{1/2} & 0 & -p_x \\[2mm] \dfrac{1}{4}x^{-1/2}y^{-1/2} & \lambda & -p_y \\[2mm] -p_x & y & 0 \end{vmatrix}$$

The determinant $|H_2|$ along the third row:

$$|H_2| = -p_x(+p_x\lambda) - y\left(\dfrac{1}{4}x^{-3/2}y^{1/2}p_y + \dfrac{1}{4}x^{-1/2}y^{-1/2}p_x\right) + 0$$

$$|H_2| = -p_x^2\lambda - 0.25(x^{-3/2}y^{3/2}p_y + x^{-1/2}y^{1/2}p_x)$$

In this case, all terms are negative; thus, $|H_2| < 0$.

$$\dfrac{\partial \bar{y}}{\partial p_y} = \dfrac{|H_2|}{|\mathbf{H}|} = \dfrac{(-)}{(+)} < 0$$

An increase in the price of y will decrease the demand for good y.

CONCAVE PROGRAMMING AND INEQUALITY CONSTRAINTS

S13.9. Given the profit function

$$\pi = 40x - x^2 + 60y - 2y^2 - 120$$

The maximum output capacity (constraint) is

$$x + y \leq 37$$

Find the optimal combination (x^*, y^*) that maximizes the profit function using the Lagrange multiplier.

(*1*) Establish the Lagrange function.

$$\Pi = 40x - x^2 + 60y - 2y^2 - 120 + \lambda(50 - x - y)$$

(*2*) Set up Kuhn-Tucker conditions.

I. *a)* $\Pi_x = 40 - 2\bar{x} - \bar{\lambda} \leq 0$ $\Pi_y = 60 - 4\bar{y} - \bar{\lambda} \leq 0$

   b) $\bar{x} \geq 0$ $\bar{y} \geq 0$

   c) $\bar{x}(40 - 2\bar{x} - \bar{\lambda}) = 0$ $\bar{y}(60 - 4\bar{y} - \bar{\lambda}) = 0$

II. *a)* $\Pi_\lambda = 37 - \bar{x} - \bar{y} \geq 0$

   b) $\bar{\lambda} \geq 0$

   c) $\bar{\lambda}(37 - \bar{x} - \bar{y}) = 0$

(*3*) Check the possibility that $\bar{\lambda} = 0$ or $\bar{\lambda} > 0$.

Test Kuhn-Tucker conditions methodically.

I. Check $\bar{\lambda} = 0$ from I(*a*).

$$40 - 2\bar{x} \leq 0 \qquad 60 - 4\bar{y} \leq 0$$

Therefore, I(*b*) holds.

$$\bar{x} \geq 20 \qquad \bar{y} \geq 15$$

Considering the complementary slackness in I(c), $\bar{x} = 20$ and $\bar{y} = 15$.

Since $\bar{x} + \bar{y} = 35 < 37$, no condition is violated. Therefore, it is possible for $\bar{\lambda} = 0$.

II. Check $\bar{\lambda} > 0$ from 2(*c*); the constraint holds as an equality.

$$37 - \bar{x} - \bar{y} = 0$$

(*4*) Check to see if either of the choice variables $\bar{x}$ or $\bar{y}$ can equal zero.

I. If $\bar{x} = 0$, $\bar{y} = 37$, and the second condition in I(*c*) is violated.

$$\bar{y}(60 - 4\bar{y} - \bar{\lambda}) = 37[60 - 4(37) - (\bar{\lambda} \geq 0)] \neq 0$$

II. If $\bar{y} = 0$, $\bar{x} = 37$, and the first condition in I(c) is violated.

$$\bar{x}(40 - 2\bar{x} - \bar{\lambda}) = 37[40 - 2(37) - (\bar{\lambda} \geq 0)] \neq 0$$

Therefore, neither choice variable can equal zero, and from I(b)

$$\bar{x} > 0 \text{ and } \bar{y} > 0$$

(5) Use all the information collected to find the optimal solution.

I. Check the solution if $\bar{\lambda} > 0$, $\bar{x}, \bar{y} > 0$, then from the Kuhn-Tucker conditions listed under (c):

$$40 - 2\bar{x} - \bar{\lambda} = 0$$
$$60 - 4\bar{y} - \bar{\lambda} = 0$$
$$37 - \bar{x} - \bar{y} = 0$$

In matrix form:

$$\begin{bmatrix} -2 & 0 & -1 \\ 0 & -4 & -1 \\ -1 & -1 & 0 \end{bmatrix} \begin{bmatrix} \bar{x} \\ \bar{y} \\ \bar{\lambda} \end{bmatrix} = \begin{bmatrix} -40 \\ -60 \\ -37 \end{bmatrix}$$

Use Cramer's rule where $|\mathbf{A}| = 6$, $|\mathbf{A}_1| = 128$, $|\mathbf{A}_2| = 94$, $|\mathbf{A}_3| = -16$

This is $\bar{x} = 21.33$, $\bar{y} = 15.67$, $\bar{\lambda} = -2.67$

This cannot optimal because $\bar{\lambda} < 0$ violates rule 2(b) of the Kuhn-Tucker conditions. This would make the constraint a strict equality and decreasing the level of output would increase the level of profit.

II. Check the solution if $\bar{\lambda} = 0$, $\bar{x}, \bar{y} > 0$, then from I(c):

$$40 - 2\bar{x} = 0, \bar{x} = 20$$
$$60 - 4\bar{y} = 0, \bar{y} = 15$$

This gives the optimal solution $\bar{x} = 20$, $\bar{y} = 15$, $\bar{\lambda} = 0$, which we know is optimal because it violates none of the Kuhn-Tucker conditions. With $\bar{\lambda} = 0$, the constraint is nonbinding as we see from the optimal solution $\bar{x} + \bar{y} = 35 < 37$.

Chapter 14

THEORETICAL CONCEPTS

S14.1. Integration is the reverse operation of *differentiation*.

S14.2. True.

S14.3. Integration by substitution is the inverse rule of the *chain* rule.

S14.4. The integral of the marginal cost is the *variable* cost.

S14.5. True.

INDEFINITE INTEGRALS

S14.6. Determine the integration of

$$\int \left(\frac{1}{2} + 4x^{-1/2} - \frac{2}{x} \right) dx$$

$$\int \left(\frac{1}{2} + 4x^{-1/2} - \frac{2}{x} \right) dx = \left(\frac{1}{2} \right) x - 4(2)x^{1/2} - 2\ln|x| + c$$
$$= 0.5x - 8x^{1/2} - 2\ln|x| + c$$

INTEGRATION BY SUBSTITUTION

S14.7. Use integration by substitution to determine the following indefinite integral

$$\int 72x^2 (4x^3 - 4)^5 dx$$

(*1*) Select a suitable *u* function.

$$u(x) = \boxed{4x^3 - 4}$$

(*2*) Compute *du* and find *dx* in terms of *du*

$$du = \boxed{12x^2} \; dx$$

$$dx = \boxed{1/(12x^2)} \; du$$

(*3*) Substitute the relationship in the integration.

$$\int 72x^2 (4x^3 - 4)^5 dx = \int 72x^2 \cdot u^5 \cdot \frac{du}{12x^2} = \int 6u^5 \cdot du = 6\int u^5 du$$

(*4*) Integrate.

$$6\int u^5 \, du = u^6 + c$$

(*5*) Substitute back to have the integration in terms of *x*.

$$\int 72x^2 (4x^3 - 4)^5 dx = u^6 + c = (4x^3 - 4)^6 + c$$
$$\int 72x^2 (4x^3 - 4)^5 dx = (4x^3 - 4)^6 + c$$

INTEGRATION BY PARTS

S14.8. Use integration by parts to determine the following definite integral:

$$\int 48x(x+2)^{1/2} dx$$

(*1*) Pick suitable $f(x)$ and $g(x)$ functions.

$$f(x) = \boxed{x} \qquad g'(x) = \boxed{(x+2)^{1/2}}$$

(*2*) Compute $f'(x)$ and find $g(x)$ in terms of dx

Differentiate: $f'(x) = \boxed{1}$

Integrate: $g(x) = \boxed{\dfrac{2}{3}(x+2)^{3/2}}$

(*3*) Substitute the relationship in the integration to create the indefinite integral.

First, the indefinite integration is required

$$\int [48x(x+2)^{1/2}]\, dx = 48(x)\left(\frac{2}{3}\right)(x+2)^{3/2} - 48\int \left[(1) \cdot \frac{2}{3}(x+2)^{3/2} \right] dx$$

$$\int [48x(x+2)^{1/2}]\, dx = 32x(x+2)^{3/2} - 32\int [(x+2)^{3/2}]\, dx$$

(*4*) Integrate the indefinite integral.

$$\int \left[48x(x+2)^{1/2} \right] dx = 32x(x+2)^{3/2} - 32\left(\frac{2}{5}\right)(x+2)^{5/2} + c$$

$$\int \left[48x(x+2)^{1/2} \right] dx = 32x(x+2)^{3/2} - \frac{64}{5}(x+2)^{5/2} + c$$

ECONOMIC APPLICATIONS

S14.9. The rate of net investment is $I = 28t^{3/4}$, and a capital stock at $t = 1$ is 24.

Find the capital function K.

$$K = \int I\, dt = \int 28t^{3/4} dt = 28\left(\frac{4}{7}\right)t^{7/4} + c$$

$$K = 16t^{7/4} + c$$

Substitute $t = 1$ and $K = 24$.

$$24 = 16(1)^{7/4} + c$$

$$c = 8$$

Thus,

$$K = 16t^{7/4} + 8$$

S14.10. Marginal cost is given by $MC = 3Q^2 - 48Q + 360$. Fixed cost is 200. Find the variable and total costs.

$$VC = \int MCdQ = \int (3Q^2 - 48Q + 360)dQ = Q^3 - 24Q^2 + 360Q + c$$

For variable cost, there is no constant; thus, $c = 0$.

$$VC = Q^3 - 24Q^2 + 360Q$$

Total cost is the addition of variable and fixed costs.

$$TC = VC + FC = Q^3 - 24Q^2 + 360Q + 200$$

Thus,

$$TC = Q^3 - 24Q^2 + 360Q + 200$$

Chapter 15

THEORETICAL CONCEPTS

S15.1. The *Riemann* sum is an approximation used to calculate areas under a curve, and one of the bases for integration.

S15.2. True.

S15.3. Consumer's surplus is the area that represents total benefit to consumers.

S15.4. True.

S15.5. True.

DEFINITE INTEGRALS

S15.6. Evaluate the definite integral.

$$\int_0^1 (e^{-3x} - x^3 + 2)dx$$

$$\int_0^1 (e^{-3x} - x^3 + 2)dx = \left(-\frac{1}{3}e^{-3x} - \frac{1}{4}x^4 + 2x \right)\Bigg|_0^1$$

$$= \left(-\frac{1}{3}e^{-3(1)} - \frac{1}{4}(1)^4 + 2(1) \right) - \left(-\frac{1}{3}e^{-3(0)} - \frac{1}{4}(0)^4 + 2(0) \right)$$

$$= \left(-\frac{1}{3}e^{-3} - \frac{1}{4} + 2 \right) - \left(-\frac{1}{3}(1) - 0 + 0 \right)$$

$$= -\frac{1}{3}e^{-3} + \frac{7}{4} + \frac{1}{3} =$$

$$= \frac{25}{12} - \frac{1}{3}e^{-3}$$

AREA BETWEEN CURVES

S15.7. Find the area between $y_1 = 12x - 3x^2$ and $y_2 = 3x$ from $x = 2$ to $x = 3$.

(1) Graph both functions to know the order of the functions in the integration.

   See Fig. 15-10.

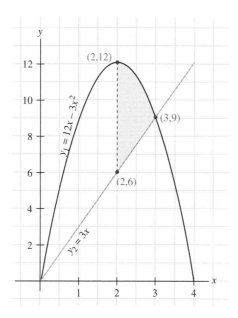

Fig. 15-10

(2) Evaluate the definite integral.

$$A = \int_2^3 [(12x - 3x^2) - (3x)]dx = \int_2^3 [9x - 3x^2]dx = \left(\frac{9}{2}x^2 - x^3 \right) \Big|_2^3$$

$$A = \left(\frac{9}{2}(3)^2 - (3)^3 \right) - \left(\frac{9}{2}(2)^2 - (2)^3 \right)$$

$$A = \frac{27}{2} - 10 = \frac{7}{2}$$

L'HÔPITAL'S RULE

S15.8. Use L'Hôpital's rule to evaluate the limit.

$$\lim_{x \to \infty} \frac{3x^2 - 4x + 8}{4x^2 + 5x - 2}$$

$$\lim_{x \to \infty} \frac{3x^2 - 4x + 8}{4x^2 + 5x - 2} = \lim_{x \to \infty} \frac{6x - 4}{8x + 5} = \lim_{x \to \infty} \frac{6}{8}$$

$$\lim_{x \to \infty} \frac{3x^2 - 4x + 8}{4x^2 + 5x - 2} = \frac{3}{4}$$

CONSUMER AND PRODUCER'S SURPLUS

S15.9. Given the demand function $P = 40 - 2Q$, find the consumer's surplus CS when $P_0 = 30$ and $Q_0 = 5$.

$$CS = \int_0^{Q^*} D(Q)\,dQ - P^*Q^* = \int_0^5 (40 - 2Q)\,dQ - (30)(5)$$

$$= (40Q - Q^2)\Big|_0^5 - 150 = (40(5) - 5^2) - (0) - 150$$

$$CS = 175 - 150 = 25$$

S15.10. Given the supply function $P = 1 + 3Q^2$, find the producer's surplus PS when $P_0 = 28$ and $Q_0 = 3$.

$$PS = P^*Q^* - \int_0^{Q^*} S(Q)\,dQ = (28)(3) - \int_0^3 (1 + 3Q^2)\,dQ$$

$$= 84 - (Q + Q^3)\Big|_0^3 = 84 - (3 + (3)^3) + (0)$$

$$PS = 84 - 30 = 54$$

Chapter 16

THEORETICAL CONCEPTS

S16.1. The differential equation

$$\left(\frac{dy}{dt}\right)^7 - 4t^2 + \left(\frac{d^2y}{dt^2}\right)^3 = 0$$

is of *second* order and *third* degree.

S16.2. True.

S16.3. An *intertemporal* equilibrium occurs when the function converges toward a specific value.

S16.4. A first-order linear differential equation must have the derivative and the dependent variable on the *first* degree and *no* product $y\left(\dfrac{dy}{dt}\right)$ may occur.

S16.5. True.

LINEAR DIFFERENTIAL EQUATIONS

S16.6. Use the general solution to solve the following equation.

$$\frac{dy}{dt} + 6ty = 24t \qquad\qquad y(0) = 5 \qquad\qquad (16.78)$$

(1) Select $v = 6t$, $z = 24t$, and $\displaystyle\int v\,dt = \int 6t\,dt = 3t^2$

(2) Substitute in the formula.

$$y(t) = e^{-\int v dt}\left(A + \int z e^{\int v dt} dt\right) = e^{-3t^2}\left(A + \int 24t e^{3t^2} dt\right)$$

Integrate the remaining integral $24\int t e^{3t^2} dt$ by the substitution method. Let $u = 3t^2$, $du = 6t dt$, and $dt = du/6t$

$$24\int t e^{-3t^2} dt = 24\int \frac{t e^u du}{6t} = 4\int e^u du = 4e^u = 4e^{3t^2}$$

Finally, substitute back.

$$y(t) = e^{-3t^2}(A + 4e^{3t^2}) = Ae^{-3t^2} + 4$$

(3) Find the value of A and write the particular solution to this differential equation.

As $t \to \infty$, the complementary function $y_c = Ae^{-3t^2} \to 0$, and $y(t) \to 4$. The equilibrium is stable.

Substitute; $t = 0$, $y = 5$:

$$y(0) = Ae^{-3(0)^2} + 4 = A + 12 = 5$$

$$A = 1$$

Thus,

$$y(t) = e^{-3t^2} + 4 \tag{16.79}$$

(4) Verify your answer.

Take the respective derivatives of (16.79), $\dfrac{dy}{dt} = -6t e^{-3t^2}$

From (16.78), $\dfrac{dy}{dt} = 24t - 6ty$. Substitute y from (16.79).

$$\frac{dy}{dt} = 24t - 6ty = 24t - 6t(e^{-3t^2} + 4) = 24t - 6t e^{-3t^2} - 24$$

$$\frac{dy}{dt} = 24t - 6ty = -6t e^{-3t^2}$$

EXACT DIFFERENTIAL EQUATIONS AND PARTIAL INTEGRATION

S16.7. Solve the exact differential equation.

$$(2y - 12t^2)dy + (8 - 24yt)dt = 0$$

(1) Check to see if it is an exact differential equation.

Let $\qquad M = 2y - 12t^2 \qquad$ and $\qquad N = 8 - 24yt$.

$$\frac{\partial M}{\partial t} = -24t \qquad\qquad \frac{\partial N}{\partial y} = -24t$$

It is a differential equation, as $\dfrac{\partial M}{\partial t} = \dfrac{\partial N}{\partial y}$

(2) Integrate M partially with respect to y and add $Z(t)$ to get $F(y,t)$

$$F(y,t) = \int (2y - 12t^2)dy + Z'(t)$$

$$F(y,t) = y^2 - 12t^2 y + Z(t) \tag{16.80}$$

(3) Differentiate $F(y,t)$ partially with respect to t.

$$\frac{\partial F}{\partial t} = N = -24ty + Z(t)$$

Equate with N to obtain $Z'(t)$.

$$-24ty + Z'(t) = 8 - 24ty$$

$$Z'(t) = 8$$

(4) Integrate $Z'(t)$ with respect to t to get $Z(t)$

$$Z(t) = \int Z'(t)dt = \int 8dt$$

$$Z(t) = 8t \tag{16.81}$$

(5) Use all the information collected and add the constant of integration to get $F(y,t)$

Substitute (16.80) in (16.80) and add the constant of integration.

$$F(y,t) = y^2 - 12t^2 y + 8t + c$$

SEPARATION OF VARIABLES

S16.8. Solve the following differential equation using the procedure for separating variables.

$$tdy - y^2 dt = 0$$

Divide both terms by ty^2 so each term is composed of only one variable.

$$\frac{1}{y^2}dy - \frac{dt}{t} = 0$$

Integrate each term separately.

$$\int \frac{1}{y^2}dy - \int \frac{dt}{t} = 0$$

$$-\frac{1}{y} - \ln(t) = c$$

Move terms.

$$\frac{-1}{y} = c + \ln(t)$$

$$y = -\frac{1}{\ln(t) + c}$$

USE OF DIFFERENTIAL EQUATIONS IN ECONOMICS

S16.9. Find the demand function $Q = f(P)$, if $\in = -(4P + 2P^2)/Q$ and $Q = 115$ when $P = 15$.

(1) Set up the elasticity function $\in$ to solve for dQ/dP,

$$\in = \frac{dQ}{dP}\frac{P}{Q} = -\frac{4P + 2P^2}{Q}$$

$$\frac{dQ}{dP} = -\frac{4P + 2P^2}{Q}\left(\frac{Q}{P}\right)$$

$$\frac{dQ}{dP} = -(4 + 2P)$$

(2) Use separation of variables to find the formula for quantity demanded Q.

$$dQ = -(4 + 2P)dP$$

$$dQ + (4 + 2P)dP = 0$$

Integrate.

$$Q + 4P + P^2 = c$$

$$Q = c - 4P - P^2$$

(3) Use the initial condition, where $P = 15$, $Q = 115$

$$115 = c - 4(15) - (15^2) = c - 285$$

$$c = 400$$

Thus, the demand function is

$$Q = 400 - 4P - P^2$$

PHASE DIAGRAMS OF VARIABLES

S16.10. Construct a phase diagram for $\dot{y} = y^2 - 5y + 4$, and test the dynamic.

(1) Find the intertemporal solutions by setting $\dot{y} = 0$

$$y^2 - 5y + 4 = 0$$

$$(y - 1)(y - 4) = 0$$

$$\bar{y}_1 = 1 \qquad \bar{y}_2 = 4$$

Test whether the critical values represent a maximum or minimum.

$$\frac{d\dot{y}}{dy} = 2y - 5 = 0 \qquad y = 2.5 \qquad \text{critical value}$$

$$\frac{d^2\dot{y}}{dy^2} = 2 > 0 \qquad\qquad \text{relative minimum}$$

(*2*)   Construct a phase diagram, in Fig. 16-6, indicating the arrows.

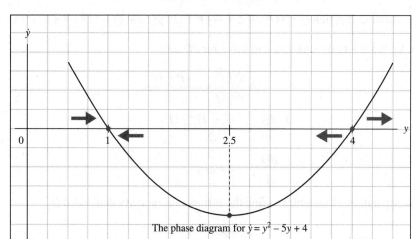

The phase diagram for $\dot{y} = y^2 - 5y + 4$

Fig. 16-6

As explained before, when $\dot{y} > 0$, the arrow points out to the right; when $\dot{y} < 0$, the arrow points out to the left.

(*3*)   Take the derivative of $\dot{y}$ with respect to y

$$\frac{d\dot{y}}{dy} = 2y - 5$$

Insert each solution.

$$\frac{d\dot{y}}{dy}(1) = 2(1) - 5 = -3 < 0 \qquad \bar{y}_1 = 1 \text{ is stable}$$

$$\frac{d\dot{y}}{dy}(4) = 2(4) - 5 = 3 > 0 \qquad \bar{y}_1 = 1 \text{ is unstable}$$

Chapter 17

THEORETICAL CONCEPTS

S17.1. Given the general form

$$y_t = Ab^t + c$$

assuming $A = 1$ and $c = 0$, if $b = 0.5$, then the time path *converges* and *is nonoscillating*.

S17.2. True.

S17.3. The *Cobweb* model is often used to forecast production of agricultural commodities based on last year's prices. When the slope of supply is greater than the absolute value of the demand and their ratio is greater than 1, then the time path *explodes*.

S17.4. False. Δy_t is a first difference of y where $\Delta y_t = y_t - y_{t-1}$.

S17.5. True.

GENERAL FORMULA FOR FIRST-ORDER DIFFERENCE EQUATIONS

S17.6. Solve the difference equation given below.

$$y_{t+3} + \frac{1}{2}y_{t+2} + 6 = 0 \qquad y_0 = 10$$

(*1*) Move the time period back by 3 periods, and rearrange terms.

$$y_t = -\frac{1}{2}y_{t-1} - 6$$

where $b = -1/2$ and $a = -6$

(*2*) Substitute in the formula and solve it. Determine the stability of y_t.

$$y_t = \left(y_0 - \frac{a}{1-b}\right)b^t + \frac{a}{1-b}$$

$$y_t = \left(10 - \frac{-6}{1-(-1/2)}\right)\left(-\frac{1}{2}\right)^t + \frac{-6}{1-(-1/2)} = \left(10 + \frac{6}{3/2}\right)\left(-\frac{1}{2}\right)^t + \frac{-6}{3/2}$$

$$y_t = (14)\left(-\frac{1}{2}\right)^t - 4$$

With $b = -1/2$, $b < 0$ and $|b| < 1$, so y_t is oscillating and convergent.

(*3*) Verify the answer.

At $t = 0, y_0 = 10$. At $t = 1, y_1 = -11$. Substitute y_1 for y_{t+3} and y_0 for y_{t+2} in the original equation, so

$$y_{t+3} + \frac{1}{2}y_{t+2} + 6 = -11 + \frac{1}{2}(10) + 6 = -11 + 5 + 6 = 0$$

LAGGED INCOME DETERMINATION MODELS

S17.7. Given the data below

$$C_t = 100 + 0.5Y_{t-1} \qquad I_t = 20 + 0.1Y_{t-1} \qquad \text{and} \qquad Y_0 = 1500$$

(*1*) Write the national income equation as a first-order difference equation
In equilibrium $Y_t = C_t + I_t$. Thus,

$$Y_t = 100 + 0.5Y_{t-1} + 20 + 0.1Y_{t-1}$$
$$Y_t = 120 + 0.6Y_{t-1}$$

where $b = 0.6$ and $a = 120$

(*2*) Find the national income formula Y_t.

$$Y_t = \left(Y_0 - \frac{a}{1-b}\right)b^t + \frac{a}{1-b} = \left(1500 - \frac{120}{1-0.6}\right)(0.6)^t + \frac{120}{1-0.6}$$

$$Y_t = (1200)(0.6)^t + 300$$

(3) Verify your answer.

Use the national income formula.

$$Y_0 = 1200(1) + 300 = 1500$$

$$Y_1 = 1200(0.6) + 300 = 1020$$

(4) Determine the stability of Y_t.

With $b = 0.6$, $b > 0$ and $|b| < 1$. So, the time path Y_t is nonoscillatory and convergent.

THE COBWEB MODEL

S17.8. Given the data below

$$Q_{dt} = 120 - 0.5P_t \qquad Q_{st} = -20 + 0.2P_{t-1} \qquad \text{and} \qquad P_0 = 320$$

(1) Equate $Q_{dt} = Q_{st}$ to obtain the first-order difference equation for price.

$$120 - 0.5P_t = -20 + 0.2P_{t-1}$$

$$-0.5P_t = -140 + 0.2P_{t-1}$$

$$P_t = -0.4P_{t-1} + 280$$

(2) Find the price formula P_t.

$$P_t = \left(P_0 - \frac{a}{1-b} \right) b^t + \frac{a}{1-b} = \left(320 - \frac{280}{1-(-0.4)} \right)(-0.4)^t + \frac{280}{1-(-0.4)}$$

$$P_t = \left(320 - \frac{280}{1.4} \right)(-0.4)^t + \frac{280}{1.4}$$

$$P_t = 120(-0.4)^t + 200$$

(3) Verify the answer at market equilibrium $Q_{dt} = Q_{st}$ with intertemporal price P_e where $P_t = P_{t-1}$

$$120 - 0.5P_e = -20 + 0.2P_e$$

$$140 = 0.7P_e$$

$$P_e = 200$$

This is the second term on the right-hand side of the general formula for price.

(4) Determine the stability of Y_t.

With $b = -0.4$, $b < 0$ and $|b| < 1$. So the time path Y_t is oscillatory and convergent.

PHASE DIAGRAMS FOR DIFFERENCE EQUATIONS

S17.9. Given the nonlinear difference equation

$$y_t = y_{t-1}^{0.2}$$

(1) Set $y_t = y_{t-1} = \bar{y}$ for the intertemporal equilibrium solutions.

$$\bar{y} = \bar{y}^{-0.2}$$

$$\bar{y}^{-0.2} - \bar{y} = 0$$

$$\bar{y}(\bar{y}^{-1.2} - 1) = 0$$

$$\bar{y}_1 = 0 \qquad \bar{y}_2 = 1 \qquad \text{steady-state solutions}$$

The phase diagram must intersect the 45°-line at $\bar{y}_1 = 0$ and $\bar{y}_2 = 1$.

(2) Take the first derivative to see if the slope is positive or negative.

$$\frac{dy_t}{dy_{t-1}} = 0.2y_{t-1}^{-0.8} = \frac{0.2}{\sqrt[5]{y_{t-1}^4}} > 0$$

Assume y_t, $y_{t-1} > 0$, the phase diagram slope is positive.

(3) Take the second derivative to see if the phase line is concave or convex.

$$\frac{d^2y_t}{dy_{t-1}^2} = -0.16y_{t-1}^{-1.8} = \frac{-0.16}{\sqrt[5]{y_{t-1}^9}} < 0 \qquad \text{concave}$$

(4) Draw a rough sketch of the graph.

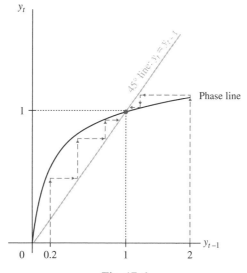

Fig. 17-6

Starting from a value of $y_t = 0.2$ and following the series of moves indicated by the arrows (where arrows from the left of the 45°-line go up and to the right), we see that the function converges at $\bar{y}_2 = 1$. It also converges to $\bar{y}_2 = 1$ when $y_t = 0.2$. Thus $\bar{y}_2 = 1$ equilibrium is locally stable. On the other hand, $\bar{y}_1 = 0$ is locally unstable.

(5) Confirm the results using the simple calculus test.

Take the derivative $\dfrac{dy_t}{dy_{t-1}}$

$$\frac{dy_t}{dy_{t-1}} = 0.2y_{t-1}^{-0.8}$$

Evaluate the absolute value at $\bar{y}_2 = 1$.

$$\left| \frac{dy_t}{dy_{t-1}}(1) \right| = \left| 0.2(1)^{-0.8} \right| = 0.2 < 1 \qquad \text{locally stable}$$

Evaluate simply as $\bar{y}_2 = 1$.

$$\frac{dy_t}{dy_{t-1}}(1) = 0.2(1)^{-0.8} = 0.2 < 1 \qquad \text{no oscillation}$$

Chapter 18

THEORETICAL CONCEPTS

S18.1.   True.

S18.2.   A second-order differential equation is composed of two separate solutions: *a particular integral* and *a complementary function*.

S18.3.   For a second-order differential equation, if $b_1^2 < 4b_2$, then the solutions are *complex roots*.

S18.4.   False. For a second-order linear difference equation with distinct or repeated real roots, the root with the largest absolute value is the dominant root.

S18.5.   False. The derivative of a sine function is the cosine of the trigonometric function.

SECOND-ORDER LINEAR DIFFERENTIAL EQUATIONS

S18.6.   Find the general solution of the following equation:

$$y''(t) + 4y'(t) + 3y(t) = 9$$

(*1*)   Find the particular integral y_p.

Since $b_1 = 4$, $b_2 = 3 \neq 0$, and $a = 9$, then

$$y_p = \frac{a}{b_2} = \frac{9}{3} = 3$$

(*2*)   Compute the complementary function y_c.

$$r_1, r_2 = \frac{-4 \pm \sqrt{16 - 4(3)}}{2} = \frac{-4 \pm \sqrt{4}}{2} = -2 \pm 1$$

$$r_1 = -3, \ r_2 = -1$$

$$y_1 = A_1 e^{-3t}$$

$$y_2 = A_2 e^{-1t}$$

$$y_c = y_1 + y_2 = A_1 e^{-3t} + A_1 e^{-t}$$

(*3*)   Write the general solution.

$$y(t) = y_c + y_p = A_1 e^{-3t} + A_1 e^{-t} + 3$$

SECOND-ORDER LINEAR DIFFERENCE EQUATIONS

S18.7. Find the general solution of the following equation:

$$y_t - 4y_{t-1} + 4y_{t-2} = 9$$

(*1*) Find the particular integral y_p.

Since $b_1 = -4$, $b_2 = 4$, and $a = 9$, then

$$y_p = \frac{a}{1 + b_1 + b_2} = \frac{9}{1 - 4 + 4} = 9$$

(*2*) Compute the complementary function y_c.

$$r_1, r_2 = \frac{+4 \pm \sqrt{16 - 4(4)}}{2} = \frac{+4 \pm \sqrt{0}}{2} = \pm 2$$

$$r_1 = r_2 = 2$$

$$y_c = A_1(r)^t + A_2 t(r)^t = A_1(2)^t + A_2 t(2^t)$$

(*3*) Write the general solution.

$$y(t) = y_c + y_p = A_1(2)^t + A_2 t(2^t) + 9$$

DERIVATIVES OF TRIGONOMETRIC FUNCTIONS

S18.8. Compute the derivative of

$$y = \cos(x^2 - 2x + 1)$$

By generalized power rule,

$$y' = -(2x - 2)\sin(x^2 - 2x + 1)$$

S18.9. Compute the derivative of

$$y = \sin(x)\tan(12x)$$

By product rule,

$$y' = \frac{d[\sin(x)]}{dx}\tan(12x) + \sin(x)\frac{d[\tan(12x)]}{dx} = \cos(x)\tan(12x) + \sin(x)(12\sec^2(12x))$$

$$y' = \cos(x)\tan(12x) + 12\sin(x)\sec^2(12x)$$

COMPLEX ROOTS IN SECOND-ORDER DIFFERENTIAL EQUATIONS

S18.10. Find the general solution of the following equation:

$$y''(t) + 2y'(t) + 8y(t) = 88$$

(*1*) Find the particular integral y_p:

Since $b_1 = 2$, $b_2 = 8 \neq 0$, and $a = 88$, then

$$y_p = \frac{a}{b_2} = \frac{88}{8} = 11$$

(*2*) Compute the complementary function y_c:

Since $b_1^2 = 4 \leq 32 = 4b_2$, then

$$g = -\frac{1}{2}b_1 = -\frac{1}{2}(2) = -1$$

$$h = \frac{1}{2}\sqrt{4b_2 - b_1^2} = \frac{1}{2}\sqrt{4(8) - (2)^2} = \frac{1}{2}\sqrt{32 - 4} = \frac{1}{2}\sqrt{28} = \sqrt{7}$$

Thus,

$$r_1 = -1 + \sqrt{7} \qquad\qquad r_2 = -1 - \sqrt{7}$$

Substitute in the general formula.

$$y_c = e^{-1}(B_1\cos\sqrt{7}t + B_2\sin\sqrt{7}t)$$

(*3*) Write the general solution.

$$y(t) = y_c + y_p = e^{-1}(B_1\cos\sqrt{7}t + B_2\sin\sqrt{7}t) + 11$$

Chapter 19

THEORETICAL CONCEPTS

S19.1. True.

S19.2. To determine the two characteristic roots, we need to obtain the *trace* tr(**A**) and the *determinant* |A| of matrix **A**.

S19.3. If r_1 and r_2 are characteristic roots, and $r_1 > 1$ but $r_2 < 0$, then the *system is unstable*.

S19.4. True.

S19.5. False. The trace of the matrix of coefficients **A** is the sum of the diagonal elements of **A**.

SYSTEMS OF LINEAR DIFFERENTIAL EQUATIONS

S19.6. Solve the system of first-order, autonomous, linear differential equations:

$$\dot{y}_1 = 5y_1 + 3y_2 - 8$$
$$\dot{y}_2 = y_1 + 3y_2 - 4$$

(*1*) Set up the system in matrix form.

$$\begin{bmatrix} \dot{y}_1 \\ \dot{y}_2 \end{bmatrix} = \begin{bmatrix} 5 & 3 \\ 1 & 3 \end{bmatrix}\begin{bmatrix} y_1 \\ y_2 \end{bmatrix} + \begin{bmatrix} -8 \\ -4 \end{bmatrix}$$

$$\dot{\mathbf{Y}} = \qquad \mathbf{A} \quad \mathbf{Y} \quad + \quad \mathbf{B}$$

(2) Find the characteristic roots, assuming distinct real roots.

The trace is $Tr(\mathbf{A}) = 5 + 3 = 8$; the determinant of A or $|\mathbf{A}| = 5(3)-1(3) = 12$.

The characteristic roots are found by solving this equation:

$$r_1, r_2 = \frac{Tr(\mathbf{A}) \pm \sqrt{[Tr(\mathbf{A})]^2 - 4|\mathbf{A}|}}{2} = \frac{8 \pm \sqrt{8^2 - 4(12)}}{2} = \frac{8 \pm \sqrt{16}}{2} = 4 \pm 2$$

$$r_1 = 2 \qquad r_2 = 6 \qquad \text{eigenvalues}$$

(3) Determine the eigenvectors for each root, given that

$$(\mathbf{A} - r_i\mathbf{I})\mathbf{C}_i = 0$$

Substitute first for $r_1 = 2$.

$$\begin{bmatrix} 5-2 & 3 \\ 1 & 3-2 \end{bmatrix} \begin{bmatrix} c_1 \\ c_2 \end{bmatrix} = \begin{bmatrix} 3 & 3 \\ 1 & 1 \end{bmatrix} \begin{bmatrix} c_1 \\ c_2 \end{bmatrix} = 0$$

Then, by simple multiplication of row and columns, we have

$$3c_1 + 3c_2 = 0 \qquad c_2 = -c_1$$
$$c_1 + c_2 = 0 \qquad c_2 = -c_1$$

If $c_1 = 1$, then $c_2 = -1$

Thus, the eigenvector $\mathbf{C}_1$ corresponding to $r_1 = 2$ is

$$\mathbf{C}_1 = \begin{bmatrix} c_1 \\ c_2 \end{bmatrix} = \begin{bmatrix} 1 \\ -1 \end{bmatrix}$$

The first elements of the complementary function of the general solution are

$$y_c^1 = k_1 \mathbf{C}_1 e^{r_1 t} = k_1 \begin{bmatrix} 1 \\ -1 \end{bmatrix} e^{2t} = \begin{bmatrix} k_1 e^{2t} \\ -k_1 e^{2t} \end{bmatrix}$$

Substitute next for $r_1 = 6$

$$\begin{bmatrix} 5-6 & 3 \\ 1 & 3-6 \end{bmatrix} \begin{bmatrix} c_1 \\ c_2 \end{bmatrix} = \begin{bmatrix} -1 & 3 \\ 1 & -3 \end{bmatrix} \begin{bmatrix} c_1 \\ c_2 \end{bmatrix} = 0$$

Then, by simple multiplication of row and columns, we have

$$-c_1 + 3c_2 = 0 \qquad c_2 = \frac{1}{3}c_1$$

$$c_1 - 3c_2 = 0 \qquad c_2 = \frac{1}{3}c_1$$

If $c_1 = 3$, then $c_2 = 1$

Thus, the eigenvector $\mathbf{C}_2$ corresponding to $r_2 = 6$ is

$$\mathbf{C}_2 = \begin{bmatrix} c_1 \\ c_2 \end{bmatrix} = \begin{bmatrix} 3 \\ 1 \end{bmatrix}$$

The second elements of the complementary function of the general solution are

$$y_c^2 = k_2 \mathbf{C}_2 e^{r_2 t} = k_2 \begin{bmatrix} 3 \\ 1 \end{bmatrix} e^{6t} = \begin{bmatrix} 3k_2 e^{6t} \\ k_2 e^{6t} \end{bmatrix}$$

Put them together for the complete complementary solution for the system.

$$y_c = y_c^1 + y_c^2 = \begin{bmatrix} k_1 e^{2t} \\ -k_1 e^{2t} \end{bmatrix} + \begin{bmatrix} 3k_2 e^{6t} \\ k_2 e^{6t} \end{bmatrix} = \begin{bmatrix} k_1 e^{2t} + 3k_2 e^{6t} \\ -k_1 e^{2t} + k_2 e^{6t} \end{bmatrix}$$

$$y_1(t) = k_1 e^{2t} + 3k_2 e^{6t}$$

$$y_2(t) = -k_1 e^{2t} + k_2 e^{6t}$$

(*4*) Find the steady-state solution y_p

$$y_p = \overline{\mathbf{Y}} = -\mathbf{A}^{-1} \mathbf{B} \qquad \text{where } \mathbf{B} = \begin{bmatrix} -8 \\ -4 \end{bmatrix} \text{ and } \mathbf{A} = \begin{bmatrix} 5 & 3 \\ 1 & 3 \end{bmatrix}, |\mathbf{A}| = 12$$

The cofactor matrix is $\mathbf{C} = \begin{bmatrix} 3 & -1 \\ -3 & 5 \end{bmatrix}$, the adjoint matrix is $\text{Adj } \mathbf{A} = \mathbf{C}' = \begin{bmatrix} 3 & -3 \\ -1 & 5 \end{bmatrix}$

and the inverse is $\mathbf{A}^{-1} = \dfrac{1}{|\mathbf{A}|} \cdot \text{Adj } \mathbf{A} = \dfrac{1}{12} \begin{bmatrix} 3 & -3 \\ -1 & 5 \end{bmatrix}$

Substitute in the formula for the particular solution.

$$\overline{\mathbf{Y}} = -\frac{1}{12} \begin{bmatrix} 3 & -3 \\ -1 & 5 \end{bmatrix} \begin{bmatrix} -8 \\ -4 \end{bmatrix} = -\frac{1}{12} \begin{bmatrix} -24+12 \\ 8-20 \end{bmatrix} = \frac{1}{12} \begin{bmatrix} -12 \\ -12 \end{bmatrix}$$

$$\overline{\mathbf{Y}} = \begin{bmatrix} 1 \\ 1 \end{bmatrix}$$

(*5*) The general solution, $y(t) = y_c + y_p$, is

$$y_1(t) = k_1 e^{2t} + 3k_2 e^{6t} + 1 \qquad\qquad (19.51)$$

$$y_2(t) = -k_1 e^{2t} + k_2 e^{6t} + 1$$

S19.7. Find the definite solution for S19.6 using the initial conditions

$$y_1(0) = 12 \qquad y_2(0) = 10$$

$$y_1(0) = k_1(1) + 3k_2(1) + 1 = 12 \qquad k_1 + 3k_2 = 11$$

$$y_2(0) = -k_1(1) + k_2(1) + 1 = 10 \qquad -k_1 + k_2 = 9$$

Solved simultaneously. $k_1 = -4$, $k_2 = 5$

Substitute in (*19.51*); we have the solution

$$y_1(t) = -4e^{2t} + 5e^{6t} + 1$$

$$y_2(t) = 4e^{2t} + 5e^{6t} + 1$$

which remains dynamically unstable because of the positive roots.

SYSTEMS OF LINEAR DIFFERENCE EQUATIONS

S19.8. Solve the system of first-order, autonomous, linear differential equations.

$$x_t = 0.7x_{t-1} - 0.4y_{t-1} + 4$$

$$y_t = 0.2x_{t-1} + 0.1y_{t-1} + 2$$

(1) Setting up the system in matrix form.

$$\begin{bmatrix} x_t \\ y_t \end{bmatrix} = \begin{bmatrix} 0.7 & -0.4 \\ 0.2 & 0.1 \end{bmatrix} \begin{bmatrix} x_{t-1} \\ y_{t-1} \end{bmatrix} + \begin{bmatrix} 4 \\ 2 \end{bmatrix}$$

$$\mathbf{Y}_t = \mathbf{A}\,\mathbf{Y}_{t-1} + \mathbf{B}$$

(2) Use the matrix on the characteristic equation $|\mathbf{A} - r_i\mathbf{I}| = 0$ to find the characteristic roots.

The trace is $\text{Tr}(\mathbf{A}) = 0.7 + 0.1 = 0.8$; the determinant of A or $|\mathbf{A}| = 0.7(0.1) - 0.2(-0.4) = 0.15$

$$r_1, r_2 = \frac{\text{Tr}(\mathbf{A}) \pm \sqrt{[\text{Tr}(\mathbf{A})]^2 - 4|\mathbf{A}|}}{2} = \frac{0.8 \pm \sqrt{0.8^2 - 4(0.15)}}{2} = \frac{0.8 \pm \sqrt{0.04}}{2} = 0.4 \pm 0.1$$

$$r_1 = 0.3 \qquad\qquad r_2 = 0.5 \qquad\qquad \text{eigenvalues}$$

(3) Determine the eigenvectors for each root

Substitute first for $r_1 = 0.3$

$$\begin{bmatrix} 0.7 - 0.3 & -0.4 \\ 0.2 & 0.1 - 0.3 \end{bmatrix} \begin{bmatrix} c_1 \\ c_2 \end{bmatrix} = \begin{bmatrix} 0.4 & -0.4 \\ 0.2 & -0.2 \end{bmatrix} \begin{bmatrix} c_1 \\ c_2 \end{bmatrix} = 0$$

Then, by simple multiplication of row and columns, we have

$$0.4c_1 - 0.4c_2 = 0 \qquad c_2 = c_1$$

$$0.2c_1 - 0.2c_2 = 0 \qquad c_2 = c_1$$

If $c_1 = 1$, then $c_2 = 11$.

Thus, the eigenvector $\mathbf{C}_1$ corresponding to $r_1 = 0.3$ is

$$\mathbf{C}_1 = \begin{bmatrix} c_1 \\ c_2 \end{bmatrix} = \begin{bmatrix} 1 \\ 1 \end{bmatrix}$$

The first elements of the complementary function of the general solution are

$$y_c^1 = k_1\mathbf{C}_1(r_1)^t = k_1 \begin{bmatrix} 1 \\ 1 \end{bmatrix} (0.3)^t = \begin{bmatrix} k_1(0.3)^t \\ k_1(0.3)^t \end{bmatrix}$$

Substitute next for $r_1 = 0.5$.

$$\begin{bmatrix} 0.7 - 0.5 & -0.4 \\ 0.2 & 0.1 - 0.5 \end{bmatrix} \begin{bmatrix} c_1 \\ c_2 \end{bmatrix} = \begin{bmatrix} 0.2 & -0.4 \\ 0.2 & -0.4 \end{bmatrix} \begin{bmatrix} c_1 \\ c_2 \end{bmatrix} = 0$$

Then, by simple multiplication of row and columns, we have

$$0.2c_1 - 0.4c_2 = 0 \qquad c_1 = 2c_2$$

$$0.2c_1 - 0.4c_2 = 0 \qquad c_1 = 2c_2$$

If $c_2 = 1$, then $c_1 = 2$.

Thus, the eigenvector $\mathbf{C}_2$ corresponding to $r_2 = 0.5$ is

$$\mathbf{C}_2 = \begin{bmatrix} c_1 \\ c_2 \end{bmatrix} = \begin{bmatrix} 2 \\ 1 \end{bmatrix}$$

The second elements of the complementary function of the general solution are

$$y_c^2 = k_2 \mathbf{C}_2 (r_2)^t = k_2 \begin{bmatrix} 2 \\ 1 \end{bmatrix} (0.5)^t = \begin{bmatrix} 2k_2(0.5)^t \\ k_2(0.5)^t \end{bmatrix}$$

Put them together for the complete complementary solution for the system.

$$y_c = y_c^1 + y_c^2 = \begin{bmatrix} k_1(0.3)^t \\ k_1(0.3)^t \end{bmatrix} + \begin{bmatrix} 2k_2(0.5)^t \\ k_2(0.5)^t \end{bmatrix} = \begin{bmatrix} k_1(0.3)^t + 2k_2(0.5)^t \\ k_1(0.3)^t + k_2(0.5)^t \end{bmatrix}$$

$$x_c = k_1(0.3)^t + 2k_2(0.5)^t$$

$$y_c = k_1(0.3)^t + k_2(0.5)^t$$

(4) Find the steady-state solution y_p

$$y_p = (\mathbf{I} - \mathbf{A})^{-1} \mathbf{B} \qquad \text{where } \mathbf{B} = \begin{bmatrix} 4 \\ 2 \end{bmatrix}$$

where $(\mathbf{I} - \mathbf{A}) = \begin{bmatrix} 1 & 0 \\ 0 & 1 \end{bmatrix} - \begin{bmatrix} 0.7 & -0.4 \\ 0.2 & 0.1 \end{bmatrix} = \begin{bmatrix} 0.3 & 0.4 \\ -0.2 & 0.9 \end{bmatrix}$

$$(\mathbf{I} - \mathbf{A})^{-1} = \frac{1}{0.27 + 0.08} \begin{bmatrix} 0.9 & -0.4 \\ 0.2 & 0.3 \end{bmatrix} = \frac{1}{0.35} \begin{bmatrix} 0.9 & -0.4 \\ 0.2 & 0.3 \end{bmatrix}$$

Substitute in the formula for the particular solution.

$$y_p = \frac{1}{0.35} \begin{bmatrix} 0.9 & -0.4 \\ 0.2 & 0.3 \end{bmatrix} \begin{bmatrix} 4 \\ 2 \end{bmatrix} = \frac{1}{0.35} \begin{bmatrix} 3.6 - 0.8 \\ 0.8 + 0.6 \end{bmatrix} = \frac{1}{0.35} \begin{bmatrix} 2.8 \\ 1.4 \end{bmatrix}$$

$$y_p = \begin{bmatrix} 8 \\ 4 \end{bmatrix}$$

(5) This makes the complete general solution $y(t) = y_c + y_p$.

$$x_c = k_1(0.3)^t + 2k_2(0.5)^t + 8$$

$$y_c = k_1(0.3)^t + k_2(0.5)^t + 1$$

(19.52)

S19.9. Find the definite solution for S19.8 using the initial conditions below:

$$x_0 = 18 \qquad y_0 = 9$$

$$x_c(0) = k_1(0.3)^0 + 2k_2(0.5)^0 + 8 = 18 \qquad k_1 + 2k_2 = 10$$

$$y_c(0) = k_1(0.3)^0 + k_2(0.5)^0 + 1 = 9 \qquad k_1 + k_2 = 8$$

Solved simultaneously, $k_1 = 6$, $k_2 = 2$

Substitute in (19.52); we have the solution

$$x_c = 6(0.3)^t + 4(0.5)^t + 8$$

$$y_c = 6(0.3)^t + 2(0.5)^t + 1$$

with $|0.3| < 1$ and $|0.5| < 1$, the time path is convergent. With both roots positive, there will be oscillation.

Chapter 20

THEORETICAL CONCEPTS

S20.1. In calculus of variation, the symbol δ is used to express properties similar to the differential term d in standard calculus.

S20.2. *Constrained* optimization is used to solve isoperimetric problems.

S20.3. Given the discriminant **D** for a functional $F[t, x(t), \dot{x}(t)]$ in a dynamic optimization, the condition for a global maximum is that $|\mathbf{D}_1| < 0$, $|\mathbf{D}| > 0$.

S20.4. True.

S20.5. *Euler's equation* is a necessary condition for dynamic optimization.

DISTANCE BETWEEN TWO POINTS ON A PLANE

S20.6. Minimize

$$\int_0^3 \sqrt{1 + \dot{x}^2}\, dt$$

subject to $\qquad x(0) = 5 \qquad x(3) = 11$

(*1*) Write the formula of the candidate for the extremal.

$$\boxed{x(t) = k_1 t + k_2}$$

(*2*) Apply the boundary conditions to find the constants.

$$x(0) = k_1(0) + k_2 = 5$$
$$x(3) = k_1(3) + k_2 = 9 \qquad 3k_1 + 5 = 11 \qquad \boxed{\begin{array}{l} k_2 = 5 \\ k_1 = 2 \end{array}}$$

(*3*) The solution is

$$\boxed{x(t) = 2t + 5}$$

S20.7. Estimate the optimal distance between the points (t_0, x_0) and (t_1, x_1) in Problem S20.6.

Given

$$\int_0^3 \sqrt{1+\dot{x}^2}\, dt \qquad \text{and} \qquad x(t) = 2t+5$$

by taking the derivative $\dot{x}(t) = 2$ and substituting, we have

$$\int_0^3 \sqrt{1+2^2}\, dt = \int_0^3 \sqrt{5}\, dt = \sqrt{5}\, t\big|_0^3 = \sqrt{5}\,(3-0) = \sqrt{5}\,(3)$$

$$\int_0^3 \sqrt{1+2^2}\, dt = 6.708$$

DYNAMIC OPTIMIZATION: FINDING CANDIDATES FOR EXTREMALS

S20.8. Optimize

$$\int_0^2 (36xt + 12t + \dot{x}^2)\, dt$$

subject to $\qquad\qquad\qquad x(0) = 4 \qquad\qquad\qquad\qquad x(2) = 10$

(*1*) Establish the function F.

$$F = 36xt + 12t + \dot{x}^2$$

(*2*) Compute Euler's equation.

Find the derivatives of F.

$$F_x = 36t \qquad F_{\dot{x}} = 2\dot{x}$$

Substitute in Euler's equation.

$$F_x = \frac{d}{dt}(F_{\dot{x}})$$
$$36t = 2\ddot{x}$$
$$\ddot{x} = 18t$$

(*3*) Integrate both sides of Euler's equation and find the functional form $x(t)$.

$$\dot{x} = 9t^2 + c_1$$

Integrate again to solve for x.

$$x(t) = 3t^3 + c_1 t + c_2$$

Apply the initial conditions.

$$x(0) = 3(0) + c_1(0) + c_2 = 4 \qquad\qquad c_2 = 4$$

$$x(2) = 3(8) + c_1(2) + c_2 = 10 \qquad 2c_1 + 4 = -14 \qquad c_1 = -9$$

Then substitute above.

$$\boxed{x(t) = 3t^3 - 9t + 4}$$

DYNAMIC OPTIMIZATION: THE SUFFICIENCY CONDITION

S20.9. Test the sufficient condition for Problem S20.8.

(*1*) Compute the discriminant.

$$|\mathbf{D}^1| = \begin{vmatrix} F_{xx} & F_{x\dot{x}} \\ F_{\dot{x}x} & F_{\dot{x}\dot{x}} \end{vmatrix} = \begin{vmatrix} 0 & 0 \\ 0 & 2 \end{vmatrix}$$

$$|\mathbf{D}^2| = \begin{vmatrix} F_{\dot{x}\dot{x}} & F_{\dot{x}x} \\ F_{x\dot{x}} & F_{xx} \end{vmatrix} = \begin{vmatrix} 2 & 0 \\ 0 & 0 \end{vmatrix}$$

(*2*) Evaluate $|\mathbf{D}^1|$ and $|\mathbf{D}^2|$

$$|\mathbf{D}_1^1| = 0 \qquad\qquad |\mathbf{D}_2^1| = 0(2) - 0 = 0$$

$$|\mathbf{D}_1^2| = 2 > 0 \qquad\qquad |\mathbf{D}_2^2| = 2(0) - 0 = 0$$

When tested for both orderings of the variables, the discriminant is negative semidefinite ($|\mathbf{D}_1^2| > 0$, $|\mathbf{D}_2^2| = 0$). This fulfills the sufficiency conditions for a relative maximum.

Chapter 21

THEORETICAL CONCEPTS

S21.1. True.

S21.2. In an optimal control model, λ is the *costate* variable.

S21.3. The *Hamiltonian* function in dynamic optimization is similar to the Lagrangian function in concave programming

S21.4. True.

S21.5. Discounting in optimal control theory is solved using the *current-valued Hamiltonian*.

FIXED ENDPOINTS

S21.6. Minimize

$$\int_0^2 (4x+2y-y^2)\,dt$$

subject to $\dot{x}=6y$

$$x(0)=3 \qquad x(2)=9$$

(*1*) Write the Hamiltonian formula.

$$H = 4x+2y-y^2+\lambda(6y)$$

(*2*) Assume an interior solution and apply the maximum principle.
    (*a*) Condition on the control variable:

$$H_y = 2-2y+6\lambda = 0$$

$$y = 3\lambda+1 \tag{21.69}$$

    (*b*) Condition on the state and costate variables:

$$\dot{\lambda} = -H_x = -4 \tag{21.70}$$

$$\dot{x} = -H_\lambda = 6y \tag{21.71}$$

(*3*) Solve the differential equations to find $\lambda(t)$ and $x(t)$

From (*21.69*), substitute in (*21.71*).

$$\dot{x} = 6(3\lambda+1) = 18\lambda+3 \tag{21.72}$$

Integrate (*21.70*).

$$\lambda(t) = -4t+c_1 \tag{21.73}$$

Substitute in (*21.72*) and then integrate.

$$\dot{x} = 18(-4t+c_1)+3$$

$$\dot{x} = -72t+18c_1+3$$

$$x(t) = -36t^2+3t+18tc_1+c_2 \tag{21.74}$$

(*4*) Apply the boundary conditions to find the formulas for λ and $x(t)$

$$x(0) = -36(0)+3(0)+18(0)c_1+c_2 = 3 \qquad c_2 = 3$$

$$x(2) = -36(4)+3(2)+18(2)c_1+c_2 = 3$$

$$-144+6+36c_1+3 = 9 \qquad c_1 = 4$$

Substitute the constants.

$$\lambda(t) = -4t + 4 \qquad \text{costate variable}$$

$$x(t) = -36t^2 + 3t + 72t + 3 \qquad \text{state variable}$$

(5) Substitute your answers in (4) into (2) to find the control variable $y(t)$

$$y(t) = 3\lambda + 1 = 3(-4t + 4) + 1$$

$$y(t) = -12t + 13 \qquad \text{control variable}$$

S21.7. Test the sufficient condition for Problem S21.6 using the discriminant test.

$$\mathbf{D}| = \begin{vmatrix} f_{yy} & f_{yx} \\ f_{xy} & f_{xx} \end{vmatrix} = \begin{vmatrix} -2 & 0 \\ 0 & 0 \end{vmatrix}$$

$$\mathbf{D}^1| = -2 < 0 \qquad \mathbf{D}^2| = 0$$

D is negative semidefinite, and the objective functional f is jointly concave in x and y. Since the constraint is linear, the functional is indeed maximized.

FREE ENDPOINTS

S21.8. Minimize

$$\int_0^3 (6x - 2y^2)\,dt$$

subject to
$$\dot{x} = 12y$$

$$x(0) = 9 \qquad\qquad x(3) \text{ free}$$

(1) Write the Hamiltonian formula.

$$H = 6x - 2y^2 + \lambda(12y)$$

(2) Assuming an interior solution and apply the maximum principle.

   (a) Condition on the control variable

$$H_y = -4y + 12\lambda = 0$$

$$y = 3\lambda \qquad\qquad (21.75)$$

   (b) Condition on the state and costate variables

$$\dot{\lambda} = -H_x = -6 \qquad\qquad (21.76)$$

$$\dot{x} = -H_\lambda = 12y \qquad\qquad (21.77)$$

(*3*) Solve the differential equations to find the general formulas for λ and $x(t)$

From (*21.75*), substitute in (*21.77*).

$$\dot{x} = 12(3\lambda) = 36\lambda \tag{21.78}$$

Integrate (*21.76*).

$$\lambda(t) = -6t + c_1 \tag{21.79}$$

Substitute in (*21.78*) and then integrate.

$$\dot{x} = 36(-6t + c_1)$$
$$\dot{x} = -216t + 36c_1$$
$$x(t) = -108t^2 + 36tc_1 + c_2 \tag{21.80}$$

(*4*) Apply the transversatility and boundary conditions to find the formulas for λ and $x(t)$

Start with the transversatility condition for a free point.

$$\lambda(3) = 0 = -6(3) + c_1$$
$$c_1 = 18$$

Therefore,

$$\lambda(t) = -6t + 18 \qquad \text{costate variable} \tag{21.81}$$

Substitute in (*21.80*).

$$x(t) = -108t^2 + 648t + c_2 \tag{21.82}$$

Apply the initial boundary condition in (*21.82*).

$$x(0) = -108(0)^2 + 648(0) + c_2 = 9$$
$$c_2 = 9$$

So, $$x(t) = -108t^2 + 648t + 9 \qquad \text{state variable} \tag{21.83}$$

(*5*) Substitute your answers in (*4*) into (*2*) to find the control variable $y(t)$

Substitute (*21.81*) in (*21.75*).

$$y = 3\lambda = 3(-6t + 18)$$
$$y(t) = -18t + 54 \qquad \text{control variable} \tag{21.84}$$

Evaluate at the endpoints.

$$y(0) = -18(0) + 54 = 54$$
$$y(3) = -18(3) + 54 = 0$$

The optimal path of the control variable is linear starting at (0, 54) and ending at (54, 0), with a slope of -18.

S21.9. Test the sufficient condition for Problem S21.8 using the discriminant test.

$$|\mathbf{D}| = \begin{vmatrix} f_{yy} & f_{yx} \\ f_{xy} & f_{xx} \end{vmatrix} = \begin{vmatrix} -4 & 0 \\ 0 & 0 \end{vmatrix}$$

$$|\mathbf{D}^1| = -4 < 0 \qquad |\mathbf{D}^2| = 0$$

$\mathbf{D}$ is negative semidefinite, and the objective functional f is jointly concave in x and y. Since the constraint is linear, the functional is indeed maximized.

Chapter 22

THEORETICAL CONCEPTS

S22.1. The sum of all probabilities of an event must be equal to **1**.

S22.2. True.

S22.3. -0.12 is not a feasible probability because it is less than 0.

S22.4. For the expected value, the weights are the *probabilities* of each event.

S22.5. *Ordinary Least Square* is the method used to solve simple linear regressions.

SERIES AND SUMMATION

S22.6. Calculate the partial sum.

$$\sum_{k=1}^{4} [(k-1)^2 + k \,]$$

Expand the polynomial in the summation and grouping terms.

$$\sum_{k=1}^{4} [(k-1)^2 + k] = \sum_{k=1}^{4} [k^2 - 2k + 1 + k] = \sum_{k=1}^{4} [k^2 - k + 1]$$

$$\sum_{k=1}^{4} [(k-1)^2 + k] = \boxed{\sum_{k=1}^{4} k^2} - \boxed{\sum_{k=1}^{4} k} + \boxed{\sum_{k=1}^{4} 1}$$

Use summation properties and computing operations.

$$\sum_{k=1}^{4} [(k-1)^2 + k] = (1^2 + 2^2 + 3^2 + 4^2) - (1+2+3+4) + 8 \cdot (1)$$

$$\sum_{k=1}^{4} [(k-1)^2 + k] = 30 - 10 + 8$$

$$\sum_{k=1}^{4} [(k-1)^2 + k] = \boxed{28}$$

Note that if you solve it directly, the solution must give the same answer.

$$\sum_{k=1}^{4}[(k-1)^2 + k] = (0^2 + 1) + (1^2 + 2) + (2^2 + 3) + (3^2 + 4) = 1 + 3 + 7 + 16 = \boxed{28}$$

EXPECTED VALUE

S22.7. Find the expected value of the table below:

x_i	p_i
1	0.25
3	0.20
7	0.15
12	0.30
17	?

(*1*) Find the probability that $x = 17$.

> The summation of all probabilities must be equal to 1; thus
>
> $$0.25 + 0.20 + 0.15 + 0.30 + p_{x=17} = 1$$
>
> $$p_{x=17} = 0.10$$
>
> The probability of $x = 17$ is 0.10, or 10%.

(*2*) Compute the expected value.

> $$EV(x) = \sum_{i=1}^{5} x_i p_i = 1 \times 0.25 + 3 \times 0.20 + 7 \times 0.15 + 12 \times 0.30 + 17 \times 0.10$$
>
> $$EV(x) = 0.25 + 0.60 + 1.05 + 3.60 + 1.70$$
>
> $$EV(x) = 7.20$$

MEAN AND STANDARD DEVIATION

S22.8. Find the mean and standard deviation of the table below:

x_i
1
3
7
12
17

(*1*) Computing the mean

$$\overline{X} = \frac{1}{5}\sum_{i=1}^{5} x_i = \frac{1 + 3 + 7 + 12 + 17}{5} = \frac{40}{5} = \boxed{8}$$

(2) Finding the variance

$$S_x^2 = \frac{1}{5-1} \sum_{i=1}^{5} (x_i - \overline{X})^2$$

$$S_x^2 = \frac{(1-8)^2 + (3-8)^2 + (7-8)^2 + (12-8)^2 + (17-8)^2}{4}$$

$$S_x^2 = \frac{49 + 25 + 1 + 16 + 81}{4}$$

$$S_x^2 = \frac{172}{4} = \boxed{43}$$

(3) Computing the standard deviation

$$S_x = \sqrt{S_x^2} = \sqrt{43} = \boxed{6.557}$$

SIMPLE LINEAR REGRESSION

S22.9.   Given the data table below,

x	y
1	10
3	25
5	40
7	60
9	95
11	100

Fit the linear regression using the Ordinary Least Square method

(0) There are 6 observations; $n = 6$.

(1) Find the mean of x and y.

$$\overline{x} = \frac{1}{6}(1 + 3 + 5 + 7 + 9 + 11) = \frac{1}{6}(36) = \boxed{6}$$

$$\overline{y} = \frac{1}{6}(10 + 25 + 40 + 60 + 95 + 100) = \frac{1}{6}(330) = \boxed{55}$$

(2) Find the numerator of $\hat{\beta}$, which is $\sum_{i}^{n} (y_i - \overline{y})(x_i - \overline{x})$

$$\sum_{i=1}^{6} (y_i - \overline{y})(x_i - \overline{x}) = (10 - 55)(1 - 6) + (25 - 55)(3 - 6) + (40 - 55)(5 - 6)$$

$$+ (60 - 55)(7 - 6) + (95 - 55)(9 - 6) + (100 - 55)(11 - 6)$$

$$\sum_{i=1}^{6} (y_i - \overline{y})(x_i - \overline{x}) = -45 \cdot (-5) - 30 \cdot (-3) - 15 \cdot (-1) + 5 \cdot 1 + 40 \cdot 3 + 45 \cdot 5$$

$$\sum_{i=1}^{6}(y_i - \bar{y})(x_i - \bar{x}) = 225 + 90 + 15 + 5 + 120 + 225$$

$$\sum_{i=1}^{6}(y_i - \bar{y})(x_i - \bar{x}) = \boxed{680}$$

(3) Find the denominator of $\hat{\beta}$, which is $\sum_{i=1}^{n}(x_i - \bar{x})^2$

$$\sum_{i=1}^{6}(x_i - \bar{x})^2 = (1-6)^2 + (3-6)^2 + (5-6)^2 + (7-6)^2 + (9-6)^2 + (11-6)^2$$

$$\sum_{i=1}^{6}(x_i - \bar{x})^2 = 5^2 + 3^2 + 1^2 + 1^2 + 3^2 + 5^2 = 25 + 9 + 1 + 1 + 9 + 25$$

$$\sum_{i=1}^{6}(x_i - \bar{x})^2 = \boxed{70}$$

(4) Calculate $\hat{\beta}$,

$$\hat{\beta} = \frac{\sum_{i=1}^{6}(y_i - \bar{y})(x_i - \bar{x})}{\sum_{i=1}^{6}(x_i - \bar{x})^2} = \frac{680}{70} = \boxed{9.7143}$$

(5) Find $\hat{a}$.

$$\hat{a} = \bar{y} - \hat{\beta}\bar{x} = 55 - 9.7143 \times 6$$

$$\boxed{\hat{a} = -3.2857}$$

The linear equation is $\boxed{\hat{y} = -3.2857 + 9.7143x}$.

S22.10. Use the regression calculated in Problem S22.9 and predict the value of y if $x = 4$

$$\hat{y} = -3.2857 + 9.7143 \times (4)$$

$$\boxed{\hat{y} = 35.5714}$$

INDEX

Notes

Notes